The Rademacher System in Function Spaces

Sergey V. Astashkin

The Rademacher System in Function Spaces

Birkhäuser

Sergey V. Astashkin
Samara National Research University
Samara, Russia

ISBN 978-3-030-47889-6 ISBN 978-3-030-47890-2 (eBook)
https://doi.org/10.1007/978-3-030-47890-2

Mathematics Subject Classification: 46B70, 46M35, 46E30, 47B10, 46B42, 46B15, 446B20

Original Russian edition published by Fizmatlit (Физматлит), Moscow, Russia, 2017.

This book is published under the imprint Birkhäuser, www.birkhauser-science.com, by the registered company Springer Nature Switzerland AG.
The registered company address is: Gewerbestrasse 11, 6330 Cham, Switzerland

This book is dedicated to
ASYA
for her understanding,
patience, and long-lasting support

Preface

... "and what is the use of a book," thought Alice, "without pictures or conversations?"

— Lewis Carroll. Alice's adventures in Wonderland.

The increasing trend of diversity in mathematics topics has led to the appearance of a large number of articles and monographs dedicated to very specialized issues and consequently results with a rather limited scope of applications. For this reason, general concepts, methods, and ideas that allow to link the different fields of mathematics, thereby ensuring their unity, have become increasingly important. One of those unifying concepts in mathematics is the Rademacher system or, alternatively, the Bernoulli sequence of independent identically and symmetrically distributed random variables, taking the values ± 1. Appearing for the first time in 1922 in a work of Hans Rademacher, it became a classical object of probability theory and the theory of orthogonal series with numerous applications within these fields and then in many others, first of all, in Banach space theory, operator theory, harmonic analysis, mathematical statistics, and number theory.

Almost immediately after its discovery, the Rademacher system began to be applied to the study of various function spaces. By 1923, Khintchine had obtained his famous inequality, showing that this system generates a Hilbert subspace in all L_p-spaces with $p < \infty$. This key fact was the starting point of further fruitful research, which eventually led to the introduction of the Banach space theory notions of Rademacher type and cotype, thus reflecting the interplay between geometry and probability in this field. These ideas, which appeared implicitly as early as in the 1930s in the works of Orlicz, were formalized much later by Hoffmann-Jørgensen, Maurey, and Pisier. Rademacher functions play an important role in the study of lattice and symmetric structures in Banach spaces, which as was demonstrated in full through the profound results obtained by Johnson, Maurey, Schechtman, and Tzafriri in their memoir [142].

The Rademacher functions and their related inequalities appear as useful tools in many monographs devoted to functional analysis, function theory, and probability theory. In particular, Rademacher functions play a prominent role in the books by Albiac and Kalton [1], Carothers [95], Diestel [109], Golubov, Efimov, and Skvortsov [126], Kaczmarz and Steinhaus [146], Kashin and Saakyan [126], Wojtaszczyk [288], Zygmund [293]. The monograph by Blei [76], where in as much as the Rademacher system also plays a central role in the study of many issues, related to harmonic analysis, functional analysis, and probability theory, on the base of the concept of combinatorial dimension, seems to be the closest to the point of view of the present book. Other highly relevant publications include the surveys by Gaposhkin [122] and Peskir and Shiryaev [233]. Of course, the above list is far from being complete.

All this said, we should note that the above-mentioned sources deal only with a narrow set of properties of the Rademacher system (mostly, connected with classical Khintchine's inequality and its applications). Therefore, this book seems to be the first attempt to present a systematic treatment of the Rademacher system from the point of view of Banach space theory. In a nutshell, the main idea of this project is to show the relationship between properties of this system and the geometry of function spaces. To achieve our goal, we included a number of fairly recent beautiful and important results related to the Rademacher functions that are currently scattered in journal articles and have not appeared in book form before.

The book consists of three parts, in which roughly speaking the Rademacher system is considered, respectively, in L_p-spaces, in general symmetric spaces and in non-symmetric spaces of certain classes (BMO, Paley, Cesáro, Morrey). Although this is the first book completely devoted to the Rademacher system, our main goal was not to give a complete treatment of the field (which is really impossible) but to arouse interest in the subject. So, we strive to make the presentation as clear and transparent as possible, providing the main results with detailed proofs and, if possible, giving substantial examples illustrating the theory.

Even though this book is not an "encyclopaedia about Rademacher system", its properties in various function spaces are represented quite fully. To the greatest extent, this relates to symmetric (or rearrangement invariant) spaces. An important step in the study of the behaviour of the Rademacher functions in symmetric spaces was taken in 1975 paper by Rodin and Semenov. On the one hand, these authors solved some of the main problems concerning this topic, while on the other, they outlined directions for further research. Recall that during the 1950s and 1960s in the pioneering works by Lorentz and Semenov investigations of general symmetric spaces were started. Some results of these studies were summed up in the books by Krein–Petunin–Semenov [168], Lindenstrauss–Tzafriri [182], and Bennett–Sharpley [70]. More recently, the book by Rubshtein, Grabarnik, Muratov, and Pashkova [253] was published, where some of the contents of the above-mentioned monographs are given in an exposition suitable for those who are familiar with basic measure theory and functional analysis as presented in the standard university courses. A few words should be said to support the choice of symmetric spaces as the most natural "arena" for the Rademacher system and to explain why an interesting and substantial theory arose in this context. The point is that the Rademacher functions generate in this case a symmetric sequence space, which is just ℓ^2 for all symmetric spaces located "sufficiently far" from L_∞. Furthermore, using this system, one can also construct other useful subspaces whose properties are closely related to the geometric structure of the spaces themselves.

Interest in the Rademacher system received a fairly powerful new impulse in the 1990s, due to the work of Montgomery-Smith and Hitczenko, who managed to refine and strengthen Khintchine's inequality. It is worth to emphasize the central role played in these investigations, explicitly or implicitly, by Peetre's $\mathcal{K}$-functional, one of the main tools in the modern theory of interpolation of operators. Later on, in a series of papers by the author of this book it was shown that the real $\mathcal{K}$-method of interpolation allows us effectively to describe the subspaces generated by the Rademacher system in symmetric spaces as well as to compare it with general systems of random variables. Namely, the relationship between properties of this system and interpolation constructions was the main topic of the author's survey [27]. On the other hand, many important topics such as estimates for Rademacher vector-valued sums, optimal constants in the Khintchine and Kahane–Khintchine inequalities, extremal properties of the Rademacher system, versions of Khintchine's inequality among many other related results were not been included in [27]. In particular, we note the remarkable achievement by the Polish mathematicians Bednorz and Latała, who gave in 2013 a positive answer to the so-called Bernoulli conjecture that had been formulated by Talagrand and had remained an open problem for more than 25 years. Sharp upper estimates for the supremum of a Bernoulli process that were obtained by these authors have led to the discovery of new deep properties of the Rademacher system. Moreover, in the last decades, some publications, dealing with the behaviour of Rademacher functions in non-symmetric function spaces, have been appeared. In this case, the subspaces, generated by the Rademacher system, may not have a symmetric basis and can exhibit sometimes a rather nontrivial structure. However, applying the techniques of Banach space theory and, in particular, methods of the theory of bases and basic sequences, it is possible to find, in many cases, a description of the geometrical structure of Rademacher subspaces similar to the classical Kadec–Pełczyński alternative for L_p-spaces.

Let us describe the contents of the book in more detail. Chapter 1 is intended, mostly, to introduce the reader to basic classical facts concerning the Rademacher system; they are mainly related to the convergence of series and estimates for the L_p-norms of Rademacher sums. The focus here is the famous Khintchine inequality and some of its immediate consequences. In the last section of the chapter, Sect. 1.6, one can find more recent results, which refine and strengthen Khintchine's inequality and have not been appeared earlier in the book form. It is at this point that an intensive use of the methods of interpolation theory starts.

The next two chapters are devoted to a detailed study of some properties of the subspace of a symmetric space, generated by the Rademacher system. As it turns out, they heavily depend on how "close" this space is located to the end points of the scale of symmetric spaces, namely to the spaces L_∞ and L_1. From this point of view, the symmetric space G, being the closure of L_∞ in the Orlicz space L_{N_2}, where $N_2(u) = e^{u^2} - 1$, is of particular importance. In Sects. 3.1 and 3.2, applying the interpolation theory allows us to establish a one-to-one correspondence between interpolation symmetric spaces with respect to the couple (L_∞, G) and their Rademacher subspaces as well to find a way to identify a space containing those and only those Rademacher sums whose sequences of coefficients belong to a given coordinate space. In addition, in Sect. 3.5 we present another possible description of Rademacher subspaces as some cones of step-functions.

In Chap. 4, we study properties of Rademacher sums with vector coefficients; much of the material in the beginning of the chapter is very classical indeed. So, in Sect. 4.1 we present the Kahane–Khintchine inequality and its consequences. In the remaining sections, we include somewhat less known results by Dilworth and Montgomery-Smith, Hitczenko, Oleszkiewicz, and Talagrand. In the last section of the chapter, we study geometrical properties of the subspace of a symmetric space X on the square $[0, 1] \times [0, 1]$, consisting of sums of Rademacher series with coefficients from X.

Chapter 5 deals with the problem of finding optimal constants in the Khintchine and Kahane–Khintchine inequalities. In particular, in Sect. 5.2, we consider in detail a spectacular result by Latała and Oleszkiewicz on the optimal constant in the Kahane–Khintchine L_1-inequality (using an approach proposed by Reinov and Podkorytov).

Chapter 6 concentrates on studying the Rademacher chaos consisting of the products of distinct Rademacher functions; the main result gives necessary and sufficient conditions for a symmetric space X, under which the Rademacher chaos is an unconditional system in X. Chapters 7 and 8 are somewhat less connected with the theory of function spaces, but interpolation methods still play an important role. Here, we consider the problems related to comparing arbitrary systems of random variables with the Rademacher system and to the selection of lacunary subsystems equivalent in distribution to this "model" system. In Chap. 9, we discuss some extreme properties of the Rademacher system.

Chapter 10 covers some issues related to the theory of Bernoulli processes; in particular, we include a proof of the minorant Sudakov type estimate for them. In Sect. 10.3, we touch upon the well-known Bernoulli conjecture, as well as its recent affirmative solution. In the last section of the chapter, we obtain, as consequence of the above results, the equivalence of scalar and vector domination of systems of random variables by the Rademacher system.

In Chap. 11, we introduce and study the so-called Rademacher multiplicator spaces and some others associated with them. We are primarily interested in finding the conditions, which assure that these spaces are symmetric.

Chapters 12 and 13 contain some versions of Khintchine's inequality. In the first of them, in particular, necessary and sufficient conditions are found, under which in a symmetric space a local version of this estimate holds (the proof of this statement relies heavily on results of the previous chapter); in the second, Khintchine's inequality extends to martingale transforms of the Rademacher sequence.

Finally, the last three chapters of the book deal with the behaviour of Rademacher functions in various *non*-symmetric spaces. We concentrate mostly on considering the following two problems: a description of the subspace, generated by the Rademacher system in such a space, and the study of geometrical properties of that subspace, in particular, its complementability.

Observe that numerous and diverse applications of the Rademacher functions in the Banach space theory, approximation theory, function theory, and other fields of mathematics are beyond the scope of this book (we only occasionally touch or mention them). In our opinion, they might become a sufficient basis for writing a separate book.

One of the features of the book is the diversity of methods used, which reflects the "presence" of the Rademacher functions in a number of areas of mathematics. We apply operator methods (in particular, interpolation and extrapolation tools), methods of Banach space theory, combinatorial and probability methods. In this regard, reading of the book requires a certain familiarity with some basics of Banach space theory (in particular, with methods of bases and basic sequences), the theory of symmetric spaces, the interpolation theory of operators, and probability theory. The reader will find all the prerequisites we assume (basically, without proofs) in Appendices A—D.

Unless otherwise stated, we consider real Banach spaces. In practice, almost all the results in the book are equally valid for real and complex scalars. We leave to the reader the extension to the complex case when needed.

Each chapter has its own numbering and ends with literary and historical comments. Their role is not so much in listing primary sources, but to indicate the relationship of issues, covered in this book, with related problems. Of course, we could not list all works worthy of attention. We strived to mention mainly for key publications, which to some extent compensates this incompleteness.

The material of the first few chapters of the book served as a basis for special courses given by the author at the Samara State University for several years. We believe that this book will be suitable for graduate students and researchers interested in functional analysis, function theory, and Banach space theory.

The author wants to express his deep gratitude to his teacher Evgeniy Mikhailovich Semenov, to whom he completely owes his interest in the subject of this book, for many years of unconditional support. The author is also very grateful to his co-authors Guillermo Curbera, Konstantin Lykov, and Lech Maligranda, whose joint results are reflected in this book.

Samara, Russia Sergey V. Astashkin

Contents

List of Symbols and Conventions

The symbols $\mathbb{N}$, $\mathbb{Z}$, $\mathbb{R}$, $\mathbb{R}^n$, ℓ_p, c_0, L_p are used in the conventional sense.

We apply the following abbreviations: a.e.—almost everywhere, r.v.—random variable, s.s.—symmetric space.

To indicate that an equality is a definition, instead of the equality sign, we use the sign $:=$.

$f \asymp g$, $t \in T$ $(f(t) \geq 0,\ g(t) \geq 0)$ means that $C^{-1}f(t) \leq g(t) \leq Cf(t)$ for some constant $C > 0$ and all $t \in T$.

$f \sim g$ as $t \to a$ $(a \in \mathbb{R})$ means that $\lim_{t\to a} f(t)/g(t) = 1$.

$\chi_B(t)$—the characteristic function of a set B.

card E—the number of elements of a finite set E.

$[a]$—the integer part of a number $a \in \mathbb{R}$.

$\operatorname{supp} f := \{t : f(t) \neq 0\}$—the support of a function f.

$m(B)$—the Lebesgue measure of a measurable set $B \subset \mathbb{R}^n$.

$\operatorname{sign} t = 1$ if $t \geq 0$, and $\operatorname{sign} t = -1$ if $t < 0$.

$\langle t\rangle^p := |t|^p \operatorname{sign} t$ $(t \in \mathbb{R},\ p > 0)$.

$C_n^k := \frac{n!}{k!(n-k)!}$ $(n \in \mathbb{N}, k = 0, 1, \ldots, n)$.

$\Gamma(t)$—the Γ-function.

$a^* = (a_k^*)_{k=1}^\infty$—the decreasing permutation of a sequence $(a_i)_{i=1}^\infty$, $a_i \geq 0$.

$a^\sharp = (a_i^\sharp)_{i=1}^n$—the decreasing permutation of a vector $a = (a_i)_{i=1}^n \in \mathbb{R}^n$.

$\Delta_n^k := ((k-1)2^{-n}, k2^{-n})$, $n = 0, 1, 2, \ldots$, $k = 1, 2, \ldots, 2^n$.

$\mathfrak{D} := \{\Delta_n^k,\ \ n = 0, 1, 2, \ldots,\ \ k = 1, 2, \ldots, 2^n\}$.

$\mathcal{D}_n := \{\sum_{k=1}^{2^n} a_k \chi_{\Delta_n^k},\ \ a_k \in \mathbb{R}\}$ $(n = 0, 1, 2, \ldots)$.

$\mathcal{D} := \bigcup_{n=0}^\infty \mathcal{D}_n$.

$r_n(t)$ $(n = 1, 2, \ldots)$—the Rademacher functions.

$\mathcal{R} := \{\sum_{n=1}^\infty a_n r_n,\ a_n \in \mathbb{R}\ \ \text{(a.e. convergence on } [0,1])\}$.

Let Y be a normed linear space:

$[x_n]_Y$—the closed linear span of a sequence $\{x_n\}_{n=1}^\infty$ in Y;

Y^*—the (Banach) dual space;

$x_n \xrightarrow{w} x$ $(x_n, x \in Y)$—x_n converges to x weakly in Y;

$x_n^* \xrightarrow{w^*} x^*$ $(x_n^*, x^* \in Y^*)$—x_n^* converges to x^* weakly* in Y^*;
$\ell_p^w(Y) := \{\{x_n\}_{n=1}^\infty \subset Y : \|\{x_n\}\|_{\ell_p^w(Y)} := \sup_{\|x^*\|_{Y^*} \le 1} \|(x^*(x_n))\|_{\ell_p} < \infty\}$, $1 \le p \le \infty$;
A *subspace* of Y means a *closed linear subspace* of Y.

$\mathbb{E}\xi$—the expectation of an r.v. ξ.
$\mathbb{E}[\xi \,|\, \mathfrak{R}]$—the conditional expectation of an r.v. ξ with respect to a σ-algebra $\mathfrak{R}$.
$\mathbb{E}[\xi \,|\, \eta_1, \ldots, \eta_n] = \mathbb{E}_{\eta_1,\ldots,\eta_n}\xi$—the conditional expectation of an r.v. ξ with respect to the σ-algebra, generated by r.v.'s $\eta_1, \ldots, \eta_n$.
$\mathfrak{g}$—a standard Gaussian r.v.

$S(T, \Sigma, \mu)$ (resp. $S(\Omega, \Sigma, \mathbb{P})$)—the linear topological space of all a.e. finite measurable real-valued functions (equivalence classes) on a σ-finite measure space (T, Σ, μ) (resp. on a probability space $(\Omega, \Sigma, \mathbb{P})$) with natural algebraic operations and the convergence in measure μ on sets of finite measure (resp. in probability $\mathbb{P}$).
E' (resp. E'')—the Köthe dual (resp. second Köthe dual) space for a normed lattice E.
$n_x(\tau) := m\{t \in [0, l] : x(t) > \tau\}$ $(\tau > 0)$—the distribution function of $x \in S([0, l], m)$, $x \ge 0$ $(0 < l \le \infty)$.
$x^*(t) := \inf\{\tau \ge 0 : n_{|x|}(\tau) < t\}$ $(0 < t \le l)$—the decreasing left-continuous rearrangement of $x \in S([0, l], m)$.
$y \simeq x$—x and y are equimeasurable functions on $[0, l]$, i.e., $n_{|x|}(\tau) = n_{|y|}(\tau)$ for all $\tau > 0$.
$\tilde{\varphi}(t) := \varphi(t)/t$.
$\hat{\varphi}(t) := t^{-1/2}\varphi(2^{1-1/t})$.
$\bar{\varphi}(t)$—the least concave majorant of a function $\varphi(t)$.
$\mathcal{M}_f$—the dilation function of a function $f(t) \ge 0$, $t \in (0, l)$ ($l = 1$ or $l = \infty$).
γ_f and δ_f—the lower and upper dilation exponents of a function f.
σ_τ $(\tau > 0)$—the dilation operator.
$\|x\|_p$—the norm of $x \in L_p = L_p[0, l]$, $0 < l \le \infty$, $1 \le p \le \infty$.
$L_{p,q}$ $(1 < p < \infty,\ 1 \le q \le \infty)$—the s.s. on [0, 1] with the norm

$$\|x\|_{p,q} := \begin{cases} \left[\frac{q}{p}\int_0^1 \left(t^{1/p}x^*(t)\right)^q \frac{dt}{t}\right]^{1/q}, & 1 \le q < \infty, \\ \operatorname{ess\,sup}_{t\in[0,1]} \left(t^{1/p}x^*(t)\right), & q = \infty. \end{cases}$$

$\ell_{p,q}$ $(1 < p < \infty,\ 1 \le q \le \infty)$—the sequence s.s. with the norm

$$\|a\|_{\ell_{p,q}} := \begin{cases} \left\{\sum_{k=1}^\infty (a_k^*)^q k^{q/p-1}\right\}^{1/q}, & 1 \le q < \infty, \\ \sup_{k=1,2,\ldots} a_k^* k^{1/p}, & q = \infty. \end{cases}$$

$\Lambda_p(\varphi)$ (φ is a quasiconcave function, $1 \le p < \infty$)—the Lorentz space with the norm

$$\|x\|_{\varphi,p} := \left(\int_0^1 (x^*(t))^p \, d\bar{\varphi}(t)\right)^{1/p}.$$

$\Lambda(\varphi) := \Lambda_1(\varphi)$.

$M(\varphi)$ (φ is a quasiconcave function)—the Marcinkiewicz space with the norm

$$\|x\|_{M(\varphi)} := \sup_{0<t\le 1} \frac{1}{\varphi(t)} \int_0^t x^*(s)\, ds.$$

L_M (M is an Orlicz function on $[0,\infty)$)—the Orlicz space with the norm

$$\|x\|_{L_M} := \inf\left\{\lambda > 0 : \int_0^1 M\left(\frac{|x(t)|}{\lambda}\right) dt \le 1\right\}.$$

L_{N_α} ($\alpha > 0$)—the Orlicz space, where $N_\alpha(t) \asymp \exp(t^\alpha) - 1$ for all sufficiently large $t > 0$.

$\operatorname{Exp} L^F$—the Orlicz space L_M, where $M(t)$ is a convex function such that $M(t) \asymp e^{F(t)} - 1$ for all sufficiently large $t > 0$.

Let X be a s.s. on $[0, 1]$*:*

X°—the separable part of X (the closure of L_∞ in X);

$G := L^\circ_{N_2}$; $H := L^\circ_{N_1}$;

$\phi_X(t) := \|\chi_{[0,t]}\|_X$ ($0 \le t \le 1$)—the fundamental function of X;

$X_{\ln^{1/2}} := \{x \in S([0,1],m) : \|x\|_{\ln^{1/2}} := \|x^*(t)\ln^{1/2}(e/t)\|_X < \infty\}$;

$X_{\ln^{-1/2}} := \{x \in S([0,1],m) : \|x\|_{\ln^{-1/2}} := \|x^*(t)\ln^{-1/2}(e/t)\|_X < \infty\}$.

α_X and β_X—the lower and upper Boyd indices of X;

$\mathcal{R}(X) := \mathcal{R} \cap X$;

$E_X := \{(a_k)_{k=1}^\infty \subset \mathbb{R} : \sum_{k=1}^\infty a_k r_k \in X\}$;

$H_X := \{f \in X : f = \sum_{k=1}^\infty b_k \chi_{(0,2^{-k+1}]}, b_1 \ge b_2 \ge \cdots \ge 0\}$;

$\operatorname{Rad} X := \{\sum_{i=1}^\infty x_i(s) r_i(t),\ \ x_i \in X$ (a.e. convergence on $[0,1]\times[0,1]$)$\}$;

$\mathcal{R}_{\mathrm{ch}}(X) := \{f \in X : f = \sum_{1\le i<j<\infty} a_{i,j} r_i r_j, (a_{i,j})_{1\le i<j<\infty} \in \ell_2\}$;

$\overline{X(\ell_2)} := \{\{x_n\}_{n=1}^\infty \subset X : \|\{x_n\}\|_{\overline{X(\ell_2)}} := \|(\sum_{n=1}^\infty x_n^2)^{1/2}\|_X < \infty\}$;

$X(\ell_2)$—the closure of the set of all finitely nonzero sequences in $\overline{X(\ell_2)}$;

$\mathcal{M}(X) = \mathcal{M}(\mathcal{R}, X)$—the Rademacher multiplicator space of X;

$\operatorname{Sym}(X) = \operatorname{Sym}(\mathcal{R}, X)$—the symmetric kernel of $\mathcal{M}(X)$;

$\mathcal{M}_T(X)$—the tail Rademacher multiplicator space of X;

$X \in \mathbb{K}$—X has the Kruglov property.

$L(\vec{X}, \vec{Y})$—the space of all linear operators bounded from a Banach couple $\vec{X} = (X_0, X_1)$ in a Banach couple $\vec{Y} = (Y_0, Y_1)$.

Let (X_0, X_1) be a Banach couple:

$\mathcal{I}(X_0, X_1)$—the set of all interpolation Banach spaces with respect to (X_0, X_1);

$\mathcal{K}(t, x; X_0, X_1)$—Peetre's $\mathcal{K}$-functional for (X_0, X_1);

$(X_0, X_1)_F^{\mathcal{K}} := \{x \in X_0 + X_1 : \|x\| := \|(\mathcal{K}(2^k, x; X_0, X_1))_{k=-\infty}^{\infty}\|_F < \infty\}$ (F is a parameter of the real $\mathcal{K}$-method);

$\widetilde{X} := (L_\infty, G)_F^{\mathcal{K}}$, where X is a s.s. and the parameter F is defined by $E_X = (\ell_1, \ell_2)_F^{\mathcal{K}}$;

$(X_0, X_1)_{\theta,p}$ $(0 < \theta < 1, 1 \le p \le \infty)$—the Lions–Peetre interpolation space with the norm

$$\|x\|_{(X_0,X_1)_{\theta,p}} := \begin{cases} \left(\sum_{k=-\infty}^{\infty}\left(\mathcal{K}(2^k, x; X_0, X_1)2^{-k\theta}\right)^p\right)^{1/p}, & 1 \le p < \infty, \\ \sup_{k=0,\pm1,\pm2,\dots}\left(2^{-k\theta}\mathcal{K}(2^k, x; X_0, X_1)\right), & p = \infty. \end{cases}$$

$$\mathcal{K}\{X_0, X_1\} := \{f(t) = \mathcal{K}(t, x; X_0, X_1), t > 0, \text{ where } x \in X_0 + X_1\}.$$

$\kappa(t, a) := \mathcal{K}(t, a; \ell_1, \ell_2)$ $(a = (a_k)_{k=1}^\infty \in \ell_2, t > 0)$.

$\kappa_w(z, \{x_n\}) := \sup_{\|x^*\|_{Y^*} \le 1} \kappa(z, (x^*(x_n)))$ ($\{x_n\} \in \ell_2^w(Y)$, Y is a Banach space).

$\|a\|_{Q(m)} := \sup\{\sum_{j=1}^m (\sum_{n\in A_j} a_n^2)^{1/2}\}$ $(a = (a_n)_{n=1}^\infty \in \ell_2, m = 1, 2, \dots)$, where the supremum is taken over all partitions $\{A_j\}_{j=1}^m$ of $\mathbb{N}$.

$\mathcal{P}_0$—the cone of all quasiconcave functions f on $(0, \infty)$ such that $\lim_{t\to0+} f(t) = \lim_{t\to\infty} f(t)/t = 0$.

$\mathcal{P}_1 := \{f \in \mathcal{P}_0 : f(t) = f(1)t, \ 0 < t \le 1\}$.

$\mathcal{P}_2$—the cone of all quasiconcave functions f on $(0, \infty)$ such that $f(t) = f(1)t$ for $0 < t \le 1$.

$\{f_i\} \overset{\mathbb{P}}{\prec} \{g_i\}$ (resp. $\{f_i\} \overset{\vec{\mathbb{P}}}{\prec} \{g_i\}$) means that a system of r.v.'s $\{f_i\}_{i=1}^\infty$ is dominated in distribution (resp. is dominated in distribution in the vector sense) by a system of r.v.'s $\{g_i\}_{i=1}^\infty$.

$\{f_i\} \overset{\mathbb{P}}{\sim} \{g_i\}$ (resp. $\{f_i\} \overset{\vec{\mathbb{P}}}{\sim} \{g_i\}$) means that a system of r.v.'s $\{f_i\}_{i=1}^\infty$ is equivalent in distribution (resp. is equivalent in distribution in the vector sense) to a system of r.v.'s $\{g_i\}_{i=1}^\infty$..

A sequence $(a_k)_{k=1}^\infty$ is increasing (strictly increasing) if $a_{k_1} \le a_{k_2}$ $(a_{k_1} < a_{k_2})$ for $k_1, k_2 \in \mathbb{N}$, $k_1 < k_2$.

A function $f : T \to \mathbb{R}$, $T \subset \mathcal{R}$, is increasing (strictly increasing) if $f(t_1) \le f(t_2)$ $(f(t_1) < f(t_2))$ for $t_1, t_2 \in T$, $t_1 < t_2$.

In a similar way, a decreasing (strictly decreasing) sequence (function) is defined.

If X and Y are linear normed spaces, then $X \overset{C}{\subset} Y$ means that $X \subset Y$ as sets and $\|x\|_Y \le C\|x\|_X$, $x \in X$; $X \subset Y$ means that $X \overset{C}{\subset} Y$ for some $C > 0$.

Sometimes the value of a constant C (or some other letter) may change from case to case.

Chapter 1
Rademacher Functions in L_p-Spaces

1.1 Definition and Basic Properties

In what follows, Δ_n^k are the dyadic subintervals of the interval [0, 1], i.e.,

$$\Delta_n^k = \left(\frac{k-1}{2^n}, \frac{k}{2^n}\right), \quad n = 1, 2, \ldots, \quad k = 1, 2, \ldots, 2^n.$$

Definition 1.1 The Rademacher functions $r_n(t)$, $t \in [0, 1]$, $n = 1, 2, \ldots$, are defined as follows:

$$r_n(t) = (-1)^{k-1} \text{ if } t \in \Delta_n^k \ (k = 1, 2, \ldots, 2^n), \quad \text{and } r_n(k2^{-n}) = 0 \ (k = 0, 1, 2, \ldots, 2^n).$$

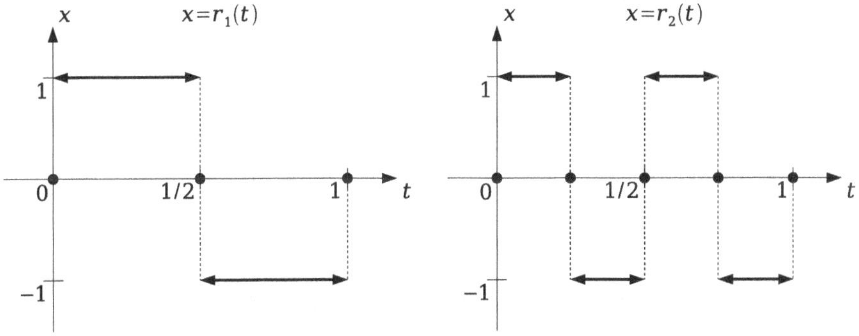

More briefly (and equivalently),

$$r_n(t) = \text{sign } \sin 2^n \pi t, \quad 0 \le t \le 1.$$

S. V. Astashkin, *The Rademacher System in Function Spaces*,
https://doi.org/10.1007/978-3-030-47890-2_1

First of all, we observe a connection between the Rademacher functions and the binary expansion of real numbers. Let $t \in [0, 1]$ be a binary-irrational number (i.e., t cannot be written as $p/2^n$, $p, n = 0, 1, 2, \ldots$). Then,

$$t = \sum_{i=1}^{\infty} \theta_i 2^{-i}, \quad \text{where } \theta_i = \theta_i(t) = 0 \text{ or } 1.$$

Clearly, for each $n = 1, 2, \ldots$ we have

$$\sum_{i=1}^{n} \theta_i 2^{-i} < t < \sum_{i=1}^{n} \theta_i 2^{-i} + \sum_{i=n+1}^{\infty} 2^{-i} = \sum_{i=1}^{n} \theta_i 2^{-i} + 2^{-n},$$

and so there is a positive integer m such that

$$\frac{\theta_n + 2m}{2^n} < t < \frac{\theta_n + 2m + 1}{2^n}.$$

This inequality shows that $\theta_n(t) = 0$ (resp. $\theta_n(t) = 1$) if and only if $t \in \Delta_n^{2m+1}$ (resp. $t \in \Delta_n^{2m+2}$). Therefore, from the definition of r_n, $n = 1, 2, \ldots$, it follows that $\theta_n(t) = 0$ if and only if $r_n(t) = 1$. This yields

$$r_n(t) = 1 - 2\theta_n(t) = (-1)^{\theta_n(t)} \text{ if } t \in (0, 1),\ t \neq \frac{p}{2^n}, n \in \mathbb{N}, p = 0, 1, \ldots, 2^n. \tag{1.1}$$

Many important properties of the Rademacher functions follow from their (probabilistic) independence (see Definition A.1).

Proposition 1.1 *The functions* r_k, $k = 1, 2, \ldots$, *are independent on* $[0, 1]$. *Moreover, for any* $n \in \mathbb{N}$ *and arbitrary functions* $\varphi_k(x)$, $k = 1, 2, \ldots, n$, *with domains containing the set* $\{\pm 1\}$, *it holds*

$$\int_0^1 \varphi_1(r_1(t))\varphi_2(r_2(t)) \ldots \varphi_n(r_n(t))\, dt = \prod_{k=1}^{n} \int_0^1 \varphi_k(r_1(t))\, dt. \tag{1.2}$$

Proof Let $\{I_k\}_{k=1}^n$, $n \in \mathbb{N}$, be an arbitrary finite set of intervals. Then, we claim

$$m\{t \in [0, 1] : r_k(t) \in I_k,\ k = 1, 2, \ldots, n\} = \prod_{k=1}^{n} m\{t \in [0, 1] : r_k(t) \in I_k\}. \tag{1.3}$$

Indeed, if there exists $1 \le k \le n$ such that $\{+1, -1\} \cap I_k = \varnothing$, then both parts of (1.3) are equal to 0. Otherwise, (1.3) can be rewritten as follows:

$$m\{t \in [0,1] : r_{k_i}(t) = \theta_i,\ i = 1, 2, \dots, l\} = \prod_{i=1}^{l} m\{t \in [0,1] : r_{k_i}(t) = \theta_i\},$$

where $1 \le k_1 < k_2 < \dots < k_l \le n$ and $\theta_i = \pm 1$, $i = 1, 2, \dots, l$. Clearly, from the definition of r_n, $n = 1, 2, \dots$, it follows that both parts of the last relation are equal to 2^{-l}. Thus, (1.3) is proved. This means that the Rademacher functions are independent.

One may easily check that (1.2) is an immediate consequence of the independence of r_n, $n = 1, 2, \dots$, and the fact that these functions are identically distributed. At the same time, this can be proved also by an easy calculation. Indeed, we have

$$\begin{aligned}\prod_{k=1}^{n} \int_0^1 \varphi_k(r_1(t))\,dt &= \prod_{k=1}^{n} \frac{1}{2}(\varphi_k(1) + \varphi_k(-1)) \\ &= \frac{1}{2^n} \sum_{\theta_k = \pm 1, k=1,\dots,n} \varphi_1(\theta_1)\varphi_2(\theta_2)\dots\varphi_n(\theta_n) \\ &= \int_0^1 \varphi_1(r_1(t))\varphi_2(r_2(t))\dots\varphi_n(r_n(t))\,dt.\end{aligned}$$ □

Corollary 1.1 *Let $\{A_j\}_{j=1}^m$ be an arbitrary set of disjoint finite subsets of $\mathbb{N}$, and let $\varphi_j(x_1^j, \dots, x_{k_j}^j)$ be a function of real variables, $k_j = \operatorname{card} A_j$, $j = 1, 2, \dots, m$. Then, the functions $f_j(t) := \varphi_j(r_i(t), i \in A_j)$ form a system of independent functions.*

Corollary 1.2 *For any $k_1, k_2, .., k_s \in \mathbb{Z}$ and pairwise different $i_1, i_2, .., i_s \in \mathbb{N}$ the integral*

$$\int_0^1 r_{i_1}^{k_1}(t) r_{i_2}^{k_2}(t) \dots r_{i_s}^{k_s}(t)\,dt$$

is equal to 1 if all numbers $k_1, k_2, .., k_s$ are even, and to 0 if at least one of them is odd. In particular, $\{r_k\}_{k=1}^\infty$ is a strongly multiplicative system of functions on $[0, 1]$ (see Definition B.7).

Corollary 1.3 *The Rademacher functions form an orthonormal system on $[0, 1]$.*

Remark 1.1 For every $1 \le p \le \infty$, the Rademacher system is not complete in the space $L_p[0, 1]$. Indeed, according to Corollary 1.2,

$$\int_0^1 r_1(t) r_2(t) r_k(t)\,dt = 0, \quad k = 1, 2, \dots$$

Therefore, the product $r_1 r_2$ is orthogonal to all Rademacher functions and thus it does not belong to the closed linear span $[r_k]$ in $L_p[0, 1]$ for each $1 \le p \le \infty$.

1.2 A.e. Convergence of Rademacher Series

Here, we prove two classical results, which give rather full information related to convergence of Rademacher series a.e. on $[0, 1]$.

Theorem 1.1 *From the condition*

$$\sum_{k=1}^{\infty} a_k^2 < \infty \tag{1.4}$$

it follows that the series

$$\sum_{k=1}^{\infty} a_k r_k(t) \tag{1.5}$$

converges a.e. on $[0, 1]$.

Let us begin with the proof of the following auxiliary assertion.

Lemma 1.1 *Let a function* $F : [0, 1] \to \mathbb{R}$ *be differentiable at a point* t_0. *If* $0 \le a_k < t_0 < b_k \le 1$ *and* $b_k - a_k \to 0$ *as* $k \to \infty$, *then we have*

$$\lim_{k\to\infty} \frac{F(b_k) - F(a_k)}{b_k - a_k} = F'(t_0).$$

Proof Let $\varepsilon > 0$. By the hypothesis, denoting $\mu_k := (b_k - t_0)/(b_k - a_k)$ and $\nu_k := (t_0 - a_k)/(b_k - a_k)$, $k = 1, 2, \ldots$, we obtain

$$\left|\frac{F(b_k) - F(a_k)}{b_k - a_k} - F'(t_0)\right| = \left|\frac{F(b_k) - F(t_0)}{b_k - t_0}\mu_k + \frac{F(t_0) - F(a_k)}{t_0 - a_k}\nu_k - F'(t_0)\right| \le$$

$$\le \left|\frac{F(b_k) - F(t_0)}{b_k - t_0} - F'(t_0)\right|\mu_k + \left|\frac{F(t_0) - F(a_k)}{t_0 - a_k} - F'(t_0)\right|\nu_k \le \varepsilon(\mu_k + \nu_k) = \varepsilon$$

whenever $k \in \mathbb{N}$ is large enough. Since $\varepsilon > 0$ is arbitrary, the proof is completed. □

Proof of Theorem 1.1 Let $S_n = S_n(t)$ be the n-th partial sum of series (1.5), that is, $S_n = \sum_{k=1}^{n} a_k r_k$. Since $\{r_k\}_{k=1}^{\infty}$ is an orthonormal system on $[0, 1]$ by Corollary 1.3, for arbitrary $1 \le n < m$ it follows

$$\|S_m - S_n\|_2 = \Big(\sum_{k=n+1}^{m} a_k^2\Big)^{1/2}.$$

By (1.4), the right-hand side tends to zero as $n \to \infty$ uniformly with respect to m. Therefore, $\{S_n\}_{n=1}^{\infty}$ is a Cauchy sequence and, due to the completeness of $L_2[0, 1]$, there exists a function $f \in L_2[0, 1]$ such that $\|S_n - f\|_2 \to 0$. Hence, for arbitrary $0 \le a < b \le 1$, we have

$$\lim_{n\to\infty} \int_a^b (f(t) - S_n(t))^2\, dt = 0.$$

Combining this with the Cauchy–Schwarz-Bunyakovskii inequality, we get

$$\lim_{n\to\infty} \int_a^b (f(t) - S_n(t))\, dt = 0.$$

Moreover, let $k = 1, 2, \dots$ be fixed and let I_k be an interval of constancy of the function r_k (and, hence, also of the partial sum S_k). One can see that

$$\int_{I_k} (S_n(t) - S_k(t))\, dt = 0$$

for any $n > k$. Hence, according to the preceding equation, we have

$$\int_{I_k} f(t)\, dt = \int_{I_k} S_k(t)\, dt. \tag{1.6}$$

Putting $F(t) := \int_0^t f(u)\, du$, $0 \le t \le 1$, denote by E the set of all binary-irrational points $t \in [0, 1]$ such that the derivative $F'(t)$ exists and $F'(t) = f(t)$. Since $f \in L_1[0, 1]$, it follows that $m(E) = 1$ (see e.g. [215, Theorem IX.4.2]). Given $t_0 \in E$ there exists a sequence of intervals $\{I_k\}_{k=1}^{\infty}$ such that $t_0 \in I_k$ for all $k = 1, 2, \dots$ and the function $S_k(t)$ is constant on every interval I_k. Therefore, if $I_k = (a_k, b_k)$, then from (1.6) it follows that

$$S_k(t_0) = \frac{1}{b_k - a_k} \int_{I_k} S_k(t)\, dt = \frac{1}{b_k - a_k} \int_{I_k} f(t)\, dt.$$

Since $b_k - a_k \to 0$ as $k \to \infty$, then applying Lemma 1.1 yields

$$\lim_{k\to\infty} S_k(t_0) = F'(t_0) = f(t_0).$$

Thus, series (1.5) converges at any point $t_0 \in E$, which concludes the proof. □

Theorem 1.2 *If condition* (1.4) *does not hold, i.e.,* $\sum_{k=1}^{\infty} a_k^2 = \infty$, *then series* (1.5) *diverges a.e. on* $[0, 1]$.

Proof On the contrary, assume that series (1.5) converges on a set $B \subset [0, 1]$ of positive measure. As above, denote by S_n the n-th partial sum of series (1.5). Moreover, let B_k, $k = 1, 2, \ldots$, be the set of all $t \in B$ such that

$$|S_{n+p}(t) - S_n(t)| \le 1 \tag{1.7}$$

for all $n \ge k$ and $p \in \mathbb{N}$. Then $B = \bigcup_{k=1}^{\infty} B_k$, and so $m(B_{k_0}) > 0$ for some k_0. Squaring and integrating (1.7) for $k = k_0$ over the set B_{k_0}, we obtain the inequality

$$\int_{B_{k_0}} (S_{n+p}(t) - S_n(t))^2 \, dt \le m(B_{k_0}), \quad n \ge k_0,$$

whence

$$m(B_{k_0}) \ge \int_{B_{k_0}} \Big(\sum_{i=n+1}^{n+p} a_i r_i(t) \Big)^2 dt = m(B_{k_0}) \sum_{i=n+1}^{n+p} a_i^2 +$$

$$+ 2 \sum_{n+1 \le i < j \le n+p} a_i a_j \int_{B_{k_0}} r_i(t) r_j(t) \, dt, \quad n \ge k_0, \, p \in \mathbb{N}. \tag{1.8}$$

According to Corollary 1.2, the functions $r_i(t) r_j(t)$, $1 \le i < j < \infty$, form an orthonormal system on $[0, 1]$. Therefore, by the Bessel inequality, if $k_1 \ge k_0$ is large enough, for all $n \ge k_1$ we have

$$\sum_{n+1 \le i < j \le n+p} \Big(\int_{B_{k_0}} r_i(t) r_j(t) \, dt \Big)^2 \le \frac{(m(B_{k_0}))^2}{9}$$

(because the left-hand side of this inequality is the sum of squares of Fourier coefficients of the function $\chi_{B_{k_0}}$ with respect to the system $\{r_i(t) r_j(t)\}_{1 \le i < j < \infty}$). Hence, by the Cauchy–Schwarz-Bunyakovskii inequality, we obtain

$$\Big| \sum_{n+1 \le i < j \le n+p} a_i a_j \int_{B_{k_0}} r_i(t) r_j(t) \, dt \Big| \le \frac{m(B_{k_0})}{3} \Big(\sum_{n+1 \le i < j \le n+p} (a_i a_j)^2 \Big)^{1/2}$$

$$\le \frac{m(B_{k_0})}{3} \sum_{i=n+1}^{n+p} a_i^2, \quad n \ge k_1, \, p \in \mathbb{N}.$$

Therefore, from (1.8) it follows

$$m(B_{k_0}) \geq \frac{m(B_{k_0})}{3} \sum_{i=n+1}^{n+p} a_i^2,$$

i.e., $\sum_{i=n+1}^{n+p} a_i^2 \leq 3$ for $n \geq k_1$ and arbitrary $p \in \mathbb{N}$. Thus, (1.4) holds, which contradicts the assumption of the theorem. □

1.3 Khintchine's Inequality and Related Estimates

The following classical result, being the source of numerous investigations and applications, will play further a very important role in this book.

Theorem 1.3 *For arbitrary $p > 0$, $n \in \mathbb{N}$, and $a_i \in \mathbb{R}$, $i = 1, 2, \ldots, n$, the following inequality holds*

$$\int_0^1 \Big|\sum_{i=1}^n a_i r_i(t)\Big|^p dt \leq \Big(\frac{p}{2}+1\Big)^{p/2} \Big(\sum_{i=1}^n a_i^2\Big)^{p/2}. \tag{1.9}$$

Proof First, we assume that $p = 2m$, $m \in \mathbb{N}$. Due to the polynomial theorem and Corollary 1.2, we have

$$\int_0^1 \Big(\sum_{i=1}^n a_i r_i(t)\Big)^{2m} dt = \sum_{k_1+\cdots+k_n=m, k_i\geq 0} \frac{(2m)!}{(2k_1)!\ldots(2k_n)!} |a_1|^{2k_1} \ldots |a_n|^{2k_n},$$

where summation is extended to all integers $k_i \geq 0$ such that $k_1 + \cdots + k_n = m$. Since $2^{k_i} k_i! \leq (2k_i)!$, $i = 1, 2, \ldots, n$, we have $2^m k_1! \ldots k_n! \leq (2k_1)! \ldots (2k_n)!$ Therefore, again by the polynomial theorem, we get the inequality

$$\begin{aligned}\int_0^1 \Big(\sum_{i=1}^n a_i r_i(t)\Big)^{2m} dt &\leq \frac{(2m)!}{2^m m!} \sum_{k_1+\cdots+k_n=m, k_i\geq 0} \frac{m!}{k_1!\ldots k_n!} |a_1|^{2k_1} \ldots |a_n|^{2k_n} \\ &= \frac{(2m)!}{2^m m!} \Big(\sum_{i=1}^n a_i^2\Big)^m. \end{aligned} \tag{1.10}$$

Thus, (1.9) is proved with the constant $C_{2m} := (2m)!/(2^m m!)$. Moreover, since

$$(m+1)(m+2)\ldots 2m \leq (2m)^m,$$

we have $C_{2m} \leq m^m$.

Now, we proceed with the case of any $p > 0$. Let us find $m \in \mathbb{N}$ such that $2m - 2 < p \le 2m$. Since the norm $\|f\|_p$ in the space $L_p[0,1]$ increases if p increases, then setting $S_n := \sum_{i=1}^n a_i r_i$, $n = 1, 2, \dots$, and using inequality (1.10) together with the above estimate for C_{2m}, we get

$$\|S_n\|_p = \Big(\int_0^1 |S_n(t)|^p\,dt\Big)^{1/p} \le \Big(\int_0^1 |S_n(t)|^{2m}\,dt\Big)^{1/(2m)}$$
$$\le m^{1/2}\Big(\sum_{i=1}^n a_i^2\Big)^{1/2} \le \Big(\frac{p}{2}+1\Big)^{1/2}\Big(\sum_{i=1}^n a_i^2\Big)^{1/2}. \qquad \square$$

Remark 1.2 Let $\mathfrak{g}$ be a standard Gaussian r.v., i.e.,

$$\mathbb{P}\{\mathfrak{g} < x\} = \frac{1}{\sqrt{2\pi}}\int_{-\infty}^x e^{-t^2/2}\,dt, \quad x \in \mathbb{R}.$$

Then, for every $m \in \mathbb{N}$, the moment $\mathbb{E}\mathfrak{g}^{2m}$ coincides with the constant C_{2m} from the last proof (see Appendix A). Moreover, arguing precisely in the same way as in the proof of Theorem 1.3, we see that the right-hand side of inequality (1.10) is equal to the expectation

$$\mathbb{E}\Big(\sum_{k=1}^n a_k \mathfrak{g}_k\Big)^{2m},$$

where $\mathfrak{g}_1, \dots, \mathfrak{g}_n$ are independent standard Gaussian r.v.'s. Indeed, due to the above property of a standard Gaussian r.v. and the independence of $\mathfrak{g}_k$, $k = 1, 2, \dots, n$, we have

$$\mathbb{E}\Big(\sum_{k=1}^n a_k \mathfrak{g}_k\Big)^{2m} = \sum_{k_1+\dots+k_n=m,\,k_i\ge 0} \frac{(2m)!}{(2k_1)!\dots(2k_n)!}|a_1|^{2k_1}\dots|a_n|^{2k_n}\mathbb{E}\mathfrak{g}_1^{2k_1}\dots\mathbb{E}\mathfrak{g}_n^{2k_n}$$
$$= \sum_{k_1+\dots+k_n=m,\,k_i\ge 0} \frac{(2m)!}{(2k_1)!\dots(2k_n)!}\cdot\frac{(2k_1)!\dots(2k_n)!}{2^{k_1}\dots 2^{k_n}(k_1)!\dots(k_n)!}|a_1|^{2k_1}\dots|a_n|^{2k_n}$$
$$= \frac{(2m)!}{2^m m!}\sum_{k_1+\dots+k_n=m,\,k_i\ge 0}\frac{m!}{k_1!\dots k_n!}|a_1|^{2k_1}\dots|a_n|^{2k_n} = \frac{(2m)!}{2^m m!}\Big(\sum_{i=1}^n a_i^2\Big)^m.$$

Thus, (1.10) can be rewritten as

$$\mathbb{E}\Big(\sum_{k=1}^n a_k r_k\Big)^{2m} \le \mathbb{E}\Big(\sum_{k=1}^n a_k \mathfrak{g}_k\Big)^{2m}, \quad m = 1, 2, \dots \tag{1.11}$$

Corollary 1.4 *If $\sum_{k=1}^\infty a_k^2 < \infty$, then the series $\sum_{k=1}^\infty a_k r_k$ converges in the space $L_p[0,1]$ for every $p > 0$.*

Proof As above, $S_n := \sum_{k=1}^{n} a_k r_k$, $n = 1, 2, \ldots$. Applying inequality (1.9) to the sums $S_m - S_n = \sum_{k=n+1}^{m} a_k r_k$, with $m > n \geq 1$, and taking into account the assumption, we see that $\{S_n\}$ is a Cauchy sequence in $L_p[0, 1]$. Since the latter space is complete, everything is done. □

Let us show that the moment inequalities for Rademacher sums, proved in Theorem 1.3, imply also rather sharp distribution estimates.

Corollary 1.5 *For arbitrary* $n \in \mathbb{N}$ *and* $a_i \in \mathbb{R}$, $i = 1, 2, \ldots, n$, *it holds*

$$m\Big\{t \in [0, 1] : \Big|\sum_{i=1}^{n} a_i r_i(t)\Big| > u\Big\} \leq e\sqrt{2}\exp\left(-\frac{u^2}{2\sum_{k=1}^{n} a_k^2}\right), \quad u > 0. \tag{1.12}$$

Proof Applying the Stirling formula

$$n! = \sqrt{2\pi n} n^n e^{-n} e^{\theta_n/(12n)}, \quad \text{where } 0 < \theta_n < 1, \tag{1.13}$$

we can estimate the constant C_{2m} from the proof of Theorem 1.3 as follows:

$$C_{2m} \leq \sqrt{2} \cdot 2^m e^{-m} m^m.$$

Combining this with Chebyshev's inequality and (1.10) for $m = \left[\frac{u^2}{2\sum_{k=1}^{n} a_k^2}\right]$, we infer

$$m\Big\{t \in [0, 1] : \Big|\sum_{i=1}^{n} a_i r_i(t)\Big| > u\Big\} \leq u^{-2m}\mathbb{E}\Big|\sum_{i=1}^{n} a_i r_i\Big|^{2m} \leq C_{2m} u^{-2m}\Big(\sum_{k=1}^{n} a_k^2\Big)^m \leq$$

$$\leq \sqrt{2}\left(\frac{2m\sum_{k=1}^{n} a_k^2}{eu^2}\right)^m \leq \sqrt{2}e^{-m} \leq e\sqrt{2}\exp\left(-\frac{u^2}{2\sum_{k=1}^{n} a_k^2}\right). \qquad \square$$

Remark 1.3 Conversely, exponential estimate (1.12) implies inequality (1.9). To see this, it suffices to use the following well-known formula for the L_p-norm (see e.g. [153, Appendix 1, § 1, Proposition 1]):

$$\|f\|_p^p = p\int_0^\infty u^{p-1} m\{t \in [0, 1] : |f(t)| > u\}\, du.$$

Using another argument based on the use of some properties of the function e^t, we can get a slightly more precise version of the estimate, proved in Corollary 1.5.

Proposition 1.2 *For any $n \in \mathbb{N}$ and $a_i \in \mathbb{R}$, $i = 1, 2, \ldots, n$,*

$$m\Big\{t \in [0,1] : \sum_{i=1}^{n} a_i r_i(t) > u\Big\} \le \exp\left(-\frac{u^2}{2\sum_{i=1}^{n} a_i^2}\right), \quad u > 0. \tag{1.14}$$

Proof Denoting $S_n := \sum_{i=1}^{n} a_i r_i$, by the independence of the Rademacher functions, for each $\lambda > 0$ we have

$$\mathbb{E}\exp(\lambda S_n) = \prod_{i=1}^{n} \mathbb{E}\exp(\lambda a_i r_i) = \prod_{i=1}^{n} \frac{\exp(\lambda a_i) + \exp(-\lambda a_i)}{2}.$$

Hence, comparing the Taylor coefficients of the functions $(e^x + e^{-x})/2$ and $e^{x^2/2}$, we conclude that

$$\mathbb{E}\exp(\lambda S_n) \le \prod_{i=1}^{n} \exp\Big(\frac{\lambda^2}{2} a_i^2\Big) = \exp\Big(\frac{\lambda^2}{2} \sum_{i=1}^{n} a_i^2\Big).$$

Therefore, Chebyshev's inequality immediately gives

$$\exp\Big(\frac{\lambda^2}{2} \sum_{i=1}^{n} a_i^2\Big) \ge m\{t \in [0,1] : S_n(t) > u\} \exp(\lambda u),$$

or

$$m\{t \in [0,1] : S_n(t) > u\} \le \exp\Big(\frac{\lambda^2}{2} \sum_{i=1}^{n} a_i^2 - \lambda u\Big)$$

for each $\lambda > 0$. Finally, minimizing the right-hand side of this inequality over λ, we come to (1.14). □

One more consequence of Theorem 1.3 implies the fundamental fact of the equivalence of the L_p-norms of Rademacher sums for all finite p. This important result will be repeatedly used in the future.

Theorem 1.4 (Khinthine's Inequality) *For arbitrary finite $p > 0$, $n \in \mathbb{N}$, and $a_i \in \mathbb{R}$, $i = 1, 2, \ldots, n$, we have*

$$\min(1, 3^{\frac{p-2}{2p}}) \Big(\sum_{i=1}^{n} a_i^2\Big)^{1/2} \le \Big\|\sum_{i=1}^{n} a_i r_i\Big\|_p \le \max(1, \sqrt{p}) \Big(\sum_{i=1}^{n} a_i^2\Big)^{1/2}. \tag{1.15}$$

Proof In fact, the right-hand side of (1.15) is already proved in Theorem 1.3. Indeed, in the case when $p \ge 2$, it is sufficient to observe that $(p/2+1)^{1/2} \le \sqrt{p}$. If $1 \le p < 2$, the desired inequality follows easily from the inequality $\|f\|_p \le \|f\|_2$.

For the same reason, the left-hand side of (1.15) holds for $p \geq 2$. Thus, it remains to prove only the latter inequality in the case when $0 < p < 2$.

Let $S_n := \sum_{i=1}^n a_i r_i$. Since the Rademacher system is orthonormal, due to Hölder's inequality with parameters $q = (4-p)/2$ and $q' = (4-p)/(2-p)$, we have

$$\sum_{i=1}^n a_i^2 = \mathbb{E}S_n^2 = \Big(\int_0^1 |S_n(t)|^{\frac{2p}{4-p}} \cdot |S_n(t)|^{\frac{4(2-p)}{4-p}}\,dt\Big)$$

$$\leq \Big(\int_0^1 |S_n(t)|^p\,dt\Big)^{\frac{2}{4-p}} \Big(\int_0^1 (S_n(t))^4\,dt\Big)^{\frac{2-p}{4-p}}.$$

Then, applying (1.10) for $m = 2$, we obtain the inequality

$$\sum_{i=1}^n a_i^2 \leq \|S_n\|_p^{\frac{2p}{4-p}} \cdot 3^{\frac{2-p}{4-p}} \Big(\sum_{i=1}^n a_i^2\Big)^{\frac{2(2-p)}{4-p}},$$

whence

$$\Big(\sum_{i=1}^n a_i^2\Big)^{1/2} \leq 3^{\frac{2-p}{2p}} \|S_n\|_p.$$

Thus, the proof is completed. □

Corollary 1.6 *Let $0 < p < \infty$. For an arbitrary set of functions $\{f_i\}_{i=1}^n \subset L_p(T, \Sigma, \mu)$, where (T, Σ, μ) is a σ-finite measure space, the following inequality holds:*

$$c_p \Big\|\Big(\sum_{i=1}^n f_i^2\Big)^{1/2}\Big\|_p \leq \Big\{\int_0^1 \Big\|\sum_{i=1}^n f_i r_i(s)\Big\|_p^p\,ds\Big\}^{1/p} \leq \max(1, \sqrt{p})\Big\|\Big(\sum_{i=1}^n f_i^2\Big)^{1/2}\Big\|_p, \tag{1.16}$$

where $c_p := \min\Big(1, 3^{\frac{p-2}{2p}}\Big)$.

Proof By Theorem 1.4, for each fixed $t \in T$, we have

$$c_p^p\Big(\sum_{i=1}^n f_i^2(t)\Big)^{p/2} \leq \int_0^1 \Big|\sum_{i=1}^n f_i(t) r_i(s)\Big|^p\,ds \leq \max(1, p^{p/2})\Big(\sum_{i=1}^n f_i^2(t)\Big)^{p/2}.$$

Since due to Fubini's theorem

$$\int_T \int_0^1 \Big|\sum_{i=1}^n f_i(t) r_i(s)\Big|^p\,ds\,d\mu = \int_0^1 \int_T \Big|\sum_{i=1}^n f_i(t) r_i(s)\Big|^p\,d\mu\,ds,$$

integration of the preceding inequality over $t \in T$ gives us (1.16). □

Remark 1.4 Clearly, from inequality (1.15) it follows that

$$\min(1, 3^{\frac{p-2}{2p}})\|a\|_{\ell_2} \le \Big\| \sum_{i=1}^{\infty} a_i r_i \Big\|_p \le \max(1, \sqrt{p})\|a\|_{\ell_2}$$

for every finite $p > 0$ and all $a = (a_i)_{i=1}^{\infty} \in \ell_2$. This means that the Rademacher system is equivalent in L_p to the unit vector basis of the sequence space ℓ_2 whenever $p > 0$ is finite. It is not the case for $p = \infty$. Indeed, in view of the definition of the Rademacher functions for every $n \in \mathbb{N}$ and arbitrary $\theta_i = \pm 1$, $i = 1, 2, \dots, n$, there exists $k = 1, 2, .., 2^n$ such that $\theta_i = r_i(t)$, $i = 1, 2, \dots, n$, for all $t \in \Delta_n^k$. Therefore, for arbitrary $n \in \mathbb{N}$ and real numbers $a_1, \dots, a_n$ we have

$$\Big\| \sum_{k=1}^{n} a_k r_k \Big\|_\infty = \sum_{k=1}^{n} |a_k|. \tag{1.17}$$

Thus, the Rademacher sequence is equivalent in L_∞ to the unit vector basis of the sequence space ℓ_1. This implies also that $\{r_k\}_{k=1}^{\infty}$ is a Sidon system on $[0, 1]$ (see Definition B.8). Indeed, if a series $\sum_{k=1}^{\infty} a_k r_k$ converges in L_2 and its sum belongs to $L_\infty[0, 1]$, then $\lim_{k\to\infty} a_k = 0$. Consequently, since $\{r_k\}_{k=1}^{\infty}$ is uniformly bounded, then from (1.17) it follows easily absolute convergence of the series of coefficients. The next result shows that a similar assertion is true if we replace $[0, 1]$ with each its subinterval.

Proposition 1.3 *Suppose that there is an interval $\Delta \subset [0, 1]$ of positive measure such that $\sum_{k=1}^{\infty} a_k r_k \in L_\infty(\Delta)$. Then, we have $\sum_{k=1}^{\infty} |a_k| < \infty$.*

Proof Note first that, by Theorems 1.1 and 1.2, the given series $\sum_{k=1}^{\infty} a_k r_k$ converges a.e. on $[0, 1]$. Let us prove that $\sum_{k=1}^{\infty} a_k r_k \in L_\infty[0, 1]$.

Indeed, without loss of generality, we may assume that Δ is a dyadic interval Δ_m^j for some $m = 1, 2, \dots$ and $j = 1, 2, \dots, 2^m$. Then, since the functions

$$\sum_{k=m+1}^{\infty} a_k r_k \cdot \chi_{\Delta_m^i}, \quad i = 1, 2, \dots, 2^m,$$

are identically distributed, we have for all $t \in [0, 1]$

$$\Big| \sum_{k=1}^{\infty} a_k r_k(t) \Big| \le \sum_{k=1}^{m} |a_k| + \Big| \sum_{k=m+1}^{\infty} a_k r_k(t) \Big| \le C,$$

where

$$C := 2\sum_{k=1}^{m} |a_k| + \Big\| \sum_{k=1}^{\infty} a_k r_k \Big\|_{L_\infty(\Delta)}.$$

Next, fixing an arbitrary positive integer n, we consider a dyadic interval Δ_n^l, $l = 1, 2, \ldots, 2^n$, such that $r_k(t) = \operatorname{sign} a_k$ for all $k = 1, 2, \ldots, n$ and $t \in \Delta_n^l$. Since $\int_{\Delta_n^l} r_k(t)\,dt = 0$ if $k > n$, then from the preceding inequality it follows

$$\sum_{k=1}^{n} |a_k| 2^{-n} = \sum_{k=1}^{n} a_k \int_{\Delta_n^l} r_k(t)\,dt = \int_{\Delta_n^l} \sum_{k=1}^{\infty} a_k r_k(t)\,dt \le \int_{\Delta_n^l} \Big|\sum_{k=1}^{\infty} a_k r_k(t)\Big|\,dt \le C 2^{-n}.$$

As a result, $\sum_{k=1}^{n} |a_k| \le C$ for any $n \in \mathbb{N}$, and so the proof is completed. □

1.4 Paley–Zygmund Inequalities

An important role in the study of systems of independent functions (including the Rademacher system) is played by the Paley–Zygmund type inequalities. Unlike Khintchine's inequality that gives upper distribution estimates for Rademacher sums (see Corollary 1.5 and Proposition 1.2), these inequalities make it possible to establish lower distribution estimates for them.

Proposition 1.4 (Paley–Zygmund Inequalities) *Let* $(\Omega, \Sigma, \mathbb{P})$ *be a probability space,* $f \in L_2(\Omega, \Sigma, \mathbb{P})$, $f \ge 0$. *Then, for any* $0 < \lambda < 1$

$$\mathbb{P}\{f \ge \lambda \mathbb{E} f\} \ge (1-\lambda)^2 \frac{(\mathbb{E} f)^2}{\mathbb{E}(f^2)}. \tag{1.18}$$

In particular, for every sequence $(a_k)_{k=1}^{\infty} \in \ell_2$ *we have*

$$m\Big\{t \in [0,1] : \Big|\sum_{k=1}^{\infty} a_k r_k(t)\Big|^2 \ge \lambda \sum_{k=1}^{\infty} a_k^2\Big\} \ge \frac{1}{9}(1-\lambda)^2. \tag{1.19}$$

Proof Define the r.v. g as follows:

$$g(\omega) = f(\omega) \text{ if } f(\omega) \ge \lambda \mathbb{E} f, \quad \text{and} \quad g(\omega) = 0 \text{ if } f(\omega) < \lambda \mathbb{E} f.$$

Due to the Cauchy–Schwarz-Bunyakovskii inequality, we have

$$(\mathbb{E} g)^2 \le \mathbb{E}(g^2)\mathbb{P}\{g \ne 0\} \le \mathbb{E}(f^2)\mathbb{P}\{f \ge \lambda \mathbb{E} f\}.$$

Moreover, $\mathbb{E} f \le \mathbb{E} g + \lambda \mathbb{E} f$. Therefore, from the preceding inequality it follows that

$$(1-\lambda)^2 (\mathbb{E} f)^2 \le \mathbb{E}(f^2)\mathbb{P}\{f \ge \lambda \mathbb{E} f\},$$

and inequality (1.18) is proved.

In order to obtain (1.19), we set

$$f(t) := \Big|\sum_{i=1}^{n} a_i r_i(t)\Big|^2.$$

Then, clearly $\mathbb{E}f = \sum_{k=1}^{n} a_k^2$. In addition, by inequality (1.9) for $p = 4$, we have

$$(\mathbb{E}f)^2 \geq \frac{1}{9}\mathbb{E}(f^2).$$

Therefore, (1.19) is a partial case of inequality (1.18). □

With inequality (1.19) in hand, we are able to show that the lower estimate in (1.15) is preserved even when one replaces the interval $[0, 1]$ with each its subset E whenever the measure of E is large enough.

Corollary 1.7 *Suppose $E \subset [0, 1]$ and $m(E) > 31/32$. Then, for every sequence $a = (a_k)_{k=1}^{\infty} \in \ell_2$ we have*

$$\int_E \Big|\sum_{k=1}^{\infty} a_k r_k(t)\Big|\, dt \geq \frac{1}{64}\Big(\sum_{k=1}^{\infty} a_k^2\Big)^{1/2}.$$

Proof First, from inequality (1.19) with $\lambda = 1/4$ it follows that

$$m\Big\{t : \Big|\sum_{k=1}^{\infty} a_k r_k(t)\Big| \geq \frac{1}{2}\Big(\sum_{i=1}^{\infty} a_k^2\Big)^{1/2}\Big\} \geq \frac{1}{16}.$$

Then, setting

$$F := \Big\{t : \Big|\sum_{k=1}^{\infty} a_k r_k(t)\Big| \geq \frac{1}{2}\Big(\sum_{i=1}^{\infty} a_k^2\Big)^{1/2}\Big\},$$

we obtain $m(E \cap F) \geq 1/32$ provided if $m(E) > 31/32$. Therefore, from the definition of F it follows

$$\int_E \Big|\sum_{k=1}^{\infty} a_k r_k(t)\Big|\, dt \geq \int_{E\cap F} \Big|\sum_{k=1}^{\infty} a_k r_k(t)\Big|\, dt \geq \frac{1}{64}\Big(\sum_{k=1}^{\infty} a_k^2\Big)^{1/2},$$

as we wished. □

The next consequence of Proposition 1.4, showing the equivalence of convergence of Rademacher series in measure and a.e., complements Theorems 1.1 and 1.2.

Corollary 1.8 *A series* $\sum_{k=1}^{\infty} a_k r_k$ *converges in measure on* $[0, 1]$ *if and only if* $\sum_{k=1}^{\infty} a_k^2 < \infty$.

Proof Let a given series $\sum_{k=1}^{\infty} a_k r_k$ converge in measure on $[0, 1]$. Assuming that $\sum_{k=1}^{\infty} a_k^2 = \infty$, we find an increasing sequence of positive integers $\{n_i\}_{i=1}^{\infty}$ such that

$$\sum_{k=n_i+1}^{n_{i+1}} a_k^2 > 1, \quad i = 1, 2, \dots$$

Then, by inequality (1.19) for $\lambda = 1/4$, we get

$$m\Big\{\Big|\sum_{k=n_i+1}^{n_{i+1}} a_k r_k(t)\Big| \geq \frac{1}{2}\Big(\sum_{k=n_i+1}^{n_{i+1}} a_k^2\Big)^{1/2}\Big\} \geq m\Big\{\Big|\sum_{k=n_i+1}^{n_{i+1}} a_k r_k(t)\Big| \geq \frac{1}{2}\Big\} \geq \frac{1}{16}$$

for all $i = 1, 2, \dots$. Thus, the series $\sum_{k=1}^{\infty} a_k r_k$ fails to converge in measure, which contradicts the hypothesis.

The converse implication is a consequence of Theorem 1.1 because the convergence in measure follows from the convergence a.e. □

Our next aim is to get more precise estimates for the L_p-norms of Rademacher sums than those obtained by Khintchine's inequality. Their proof will be based on using connections between these norms and Peetre's $\mathcal{K}$-functional for the couple (ℓ_1, ℓ_2). That is why we start with the definition and some properties of this functional.

1.5 Peetre's $\mathcal{K}$-Functional for the Couple (ℓ_1, ℓ_2)

Let us recall the definition of Peetre's $\mathcal{K}$-functional (see also Appendix D). For an arbitrary Banach couple (X_0, X_1), any $x \in X_0 + X_1$ and $t > 0$ we set

$$\mathcal{K}(t, x; X_0, X_1) := \inf\{\|x_0\|_{X_0} + t\|x_1\|_{X_1} : x = x_0 + x_1, x_0 \in X_0, x_1 \in X_1\}.$$

Further, we shall be interested mainly in the $\mathcal{K}$-functional $\mathcal{K}(t, a; \ell_1, \ell_2)$, corresponding to the couple of sequence spaces (ℓ_1, ℓ_2), which will be denoted for brevity by $\kappa(t, a)$.

In this section, we shall look for quantities equivalent to the functional $\mathcal{K}(t, a; \ell_1, \ell_2)$. To this end, we need clearly to find for a given $t > 0$ an "almost optimal" expansion $a = a_0(t) + a_1(t)$, where $a_0(t) \in \ell_1$, $a_1(t) \in \ell_2$. We consider first this task for an analogue of $\kappa(t, a)$, the $\mathcal{K}$-functional $\mathcal{K}(t, f; L_1, L_2)$, corresponding to the L_p-spaces of functions defined on the semi-axis $(0, \infty)$.

Proposition 1.5 *For an arbitrary function* $f \in L_1(0,\infty) + L_2(0,\infty)$ *and any* $t > 0$ *we have*

$$\frac{1}{4}\Big\{\int_0^{t^2} f^*(s)\,ds + t\Big(\int_{t^2}^{\infty} f^*(s)^2\,ds\Big)^{1/2}\Big\} \le \mathcal{K}(t, f; L_1, L_2)$$
$$\le \int_0^{t^2} f^*(s)\,ds + t\Big(\int_{t^2}^{\infty} f^*(s)^2\,ds\Big)^{1/2}, \qquad (1.20)$$

where f^* *is the decreasing left-continuous rearrangement of* f *(see Appendix C).*

Proof Let $t > 0$ be fixed. Denote $E := \{s > 0 : |f(s)| \ge f^*(t^2)\}$, $f_0(s) := f(s)\chi_E(s)$, $f_1(s) := f(s) - f_0(s)$. Without loss of generality, we can assume that $m\{s > 0 : f(s) = \tau\} = 0$ for any $\tau \in \mathbb{R}$. Then f_0 (resp. f_1) is equimeasurable with $f^*\chi_{(0,t^2)}$ (resp. $f^*\chi_{(t^2,\infty)}$). Therefore, we have

$$\|f_0\|_1 + t\|f_1\|_2 = \int_0^{t^2} f^*(s)\,ds + t\Big(\int_{t^2}^{\infty} f^*(s)^2\,ds\Big)^{1/2},$$

whence the right-hand side inequality in (1.20) follows immediately from the definition of Peetre's $\mathcal{K}$-functional.

To prove the opposite inequality, consider an arbitrary representation $f = g + h$, where $g \in L_1$, $h \in L_2$. Let $u > 0$ be arbitrary. Given $\varepsilon > 0$, by the definition of the decreasing rearrangement of a measurable function, we have

$$m\{s > 0 : |g(s)| > g^*(u/2 - \varepsilon)\} \le u/2 - \varepsilon$$

and

$$m\{s > 0 : |h(s)| > h^*(u/2 - \varepsilon)\} \le u/2 - \varepsilon.$$

Therefore, it holds

$$m\{s > 0 : |f(s)| > g^*(u/2 - \varepsilon) + h^*(u/2 - \varepsilon)\} \le u - 2\varepsilon < u,$$

whence

$$f^*(u) \le g^*(u/2 - \varepsilon) + h^*(u/2 - \varepsilon).$$

Since the function f^* is left-continuous and $\varepsilon > 0$ is arbitrary, we conclude that

$$f^*(u) \le g^*(u/2) + h^*(u/2), \quad u > 0. \qquad (1.21)$$

Therefore, applying the Cauchy–Schwarz-Bunyakovskii inequality, we obtain

$$\|g\|_1 + t\|h\|_2 \geq \frac{1}{2}\Big(\int_0^{t^2} g^*(u/2)\,du + t\Big(\int_0^{t^2} h^*(u/2)^2\,du\Big)^{1/2}\Big) \geq$$

$$\geq \frac{1}{2}\Big(\int_0^{t^2} g^*(u/2)\,du + \int_0^{t^2} h^*(u/2)\,du\Big) \geq \frac{1}{2}\int_0^{t^2} f^*(u)\,du. \tag{1.22}$$

Moreover, since $g^*(u)$ is decreasing, we have

$$t^2\int_{t^2}^{\infty} \big(g^*(u/2)\big)^2\,du \leq \int_0^{t^2} g^*(u/2)\,du \int_{t^2}^{\infty} g^*(u/2)\,du$$
$$\leq \Big(\int_0^{\infty} g^*(u/2)\,du\Big)^2 = 4\|g\|_1^2.$$

This inequality, (1.21), and the triangle inequality for the L_2-norm yield

$$\|g\|_1 + t\|h\|_2 \geq \frac{t}{2}\Big(\int_{t^2}^{\infty} \big(g^*(u/2)\big)^2\,du\Big)^{1/2}$$
$$+ \frac{t}{2}\Big(\int_{t^2}^{\infty} \big(h^*(u/2)\big)^2\,du\Big)^{1/2}$$
$$\geq \frac{t}{2}\Big(\int_{t^2}^{\infty} f^*(u)^2\,du\Big)^{1/2}.$$

Combining the latter inequality together with (1.22) and taking the infimum over all representations $f = g + h$, $g \in L_1$, $h \in L_2$, we obtain the left-hand side inequality in (1.20) by the definition of Peetre's $\mathcal{K}$-functional. □

In order to obtain a similar result for the functional $\kappa(t, a)$, we need the following auxiliary assertion, where for every sequence $a = (a_k)_{k=1}^{\infty} \in \ell_2$ we set

$$f_a(t) := \sum_{k=1}^{\infty} a_k\,\chi_{(k-1,k]}(t), \quad t > 0. \tag{1.23}$$

Lemma 1.2 *For any $a = (a_k)_{k=1}^{\infty} \in \ell_2$ and $t > 0$, we have*

$$\mathcal{K}(t, f_a; L_1, L_2) = \kappa(t, a). \tag{1.24}$$

Proof First of all, it may be easily checked that

$$\|f_a\|_1 = \|a\|_{\ell_1} \text{ and } \|f_a\|_2 = \|a\|_{\ell_2}. \tag{1.25}$$

Therefore, if $a = b+c$, where $b \in \ell_1$, $c \in \ell_2$, then denoting $f_b = \sum_{k=1}^{\infty} b_k \chi_{(k-1,k]}$ and $f_c = \sum_{k=1}^{\infty} c_k \chi_{(k-1,k]}$, we get

$$\|b\|_{\ell_1} + t\|c\|_{\ell_2} = \|f_b\|_1 + t\|f_c\|_2.$$

Since $f_b + f_c = f_a$, it follows that

$$\mathcal{K}(t, f_a; L_1, L_2) \le \kappa(t, a), \quad t > 0.$$

Conversely, let $f_a = g + h$, where $g \in L_1$, $h \in L_2$. Immediate estimates show that the average operator

$$Qx(t) = \sum_{k=1}^{\infty} a_k(x)\, \chi_{(k-1,k]}(t), \quad t > 0, \quad \text{where } a_k(x) = \int_{k-1}^{k} x(s)\, ds,$$

is a projection in L_1 and in L_2 with norm 1. Therefore, due to (1.25), we have

$$\|g\|_1 + t\|h\|_2 \ge \|Qg\|_1 + t\|Qh\|_2 = \|a(g)\|_{\ell_1} + t\|a(h)\|_{\ell_2},$$

where $a(x) = (a_k(x))_{k=1}^{\infty}$. Moreover, since $f_a = Qf_a = Qg + Qh$, it follows that $a = a(g) + a(h)$. Hence, the latter inequality and the definition of the $\mathcal{K}$-functional imply

$$\kappa(t, a) \le \mathcal{K}(t, f_a; L_1, L_2), \quad t > 0,$$

and so (1.24) is proved. □

Theorem 1.5 (Holmstedt Inequality) *For any $a = (a_k)_{k=1}^{\infty} \in \ell_2$ and $t > 0$, we have*

$$\frac{1}{4}\Big\{\sum_{i=1}^{[t^2]} a_i^* + t\Big(\sum_{i=[t^2]+1}^{\infty} (a_i^*)^2\Big)^{1/2}\Big\} \le \kappa(t, a)$$

$$\le \Big\{\sum_{i=1}^{[t^2]} a_i^* + t\Big(\sum_{i=[t^2]+1}^{\infty} (a_i^*)^2\Big)^{1/2}\Big\}, \quad (1.26)$$

where $(a_i^*)_{i=1}^{\infty}$ is the decreasing permutation of the sequence $(|a_k|)_{k=1}^{\infty}$.

Proof If $0 < t \le 1$, then the inequality $\|a\|_{\ell_2} \le \|a\|_{\ell_1}$ implies $\kappa(t, a) = t\|a\|_{\ell_2}$. Hence, we clearly get (1.26).

Now, suppose $t \ge 1$. Observe that ℓ_1 and ℓ_2 are symmetric sequence spaces (see Appendix C). Consequently, it follows from the definition of the $\mathcal{K}$-functional that, for any sequence $a = (a_k)_{k=1}^{\infty}$ and every permutation π of $\mathbb{N}$, we have $\kappa(t, |a_\pi|) = \kappa(t, a)$, where $a_\pi = (a_{\pi(k)})_{k=1}^{\infty}$. Hence, we may (and shall) assume that $a_k = a_k^*$, $k = 1, 2, \ldots$.

Further, for a fixed $t \geq 1$, we define the sequence $b^t = (b_k^t)$, where $b_k^t := a_k$ if $k \leq [t^2]$, and $b_k^t := 0$ if $k > [t^2]$. Also, let $c^t := a - b^t$. Then, the right-hand side inequality in (1.26) is an immediate consequence of the definition of $\kappa(t, a)$. As to the left-hand side, then this inequality follows from Proposition 1.5 and Lemma 1.2. Indeed, if f_a is still defined by (1.23), then we have

$$\begin{aligned}\kappa(t, a) \geq \kappa(\sqrt{[t^2]}, a) &= \mathcal{K}(\sqrt{[t^2]}, f_a; L_1, L_2)\\ &\geq \frac{1}{4}\Big\{\int_0^{[t^2]} f_a^*(s)\,ds + t\Big(\int_{[t^2]}^{\infty} f_a^*(s)^2\,ds\Big)^{1/2}\Big\}\\ &= \frac{1}{4}\Big\{\sum_{i=1}^{[t^2]} a_i^* + t\Big(\sum_{i=[t^2]+1}^{\infty} (a_i^*)^2\Big)^{1/2}\Big\}.\end{aligned}$$ □

We shall use also one more equivalent expression for the functional $\kappa(t, a)$. For any $a = (a_k)_{k=1}^{\infty} \in \ell_2$ and $p \in \mathbb{N}$ we set

$$\|a\|_{Q(p)} = \sup\Big\{\sum_{j=1}^{p}\Big(\sum_{n\in A_j} a_n^2\Big)^{1/2}\Big\}, \tag{1.27}$$

where the supremum is taken over all partitions $\{A_j\}_{j=1}^{p}$ of the set of positive integers.

Lemma 1.3 *If $a = (a_k)_{k=1}^{\infty} \in \ell_2$ and $t^2 \in \mathbb{N}$, then we have*

$$\|a\|_{Q(t^2)} \leq \kappa(t, a) \leq \sqrt{2}\|a\|_{Q(t^2)}. \tag{1.28}$$

Proof First of all, it follows from (1.27) that $\|a\|_{Q(t^2)} \leq \|a\|_{\ell_1}$ and $\|a\|_{Q(t^2)} \leq t\|a\|_{\ell_2}$. Indeed, the first inequality is obvious. For proving the second one it is sufficient to apply the Cauchy–Schwarz-Bunyakovskii inequality to get that for an arbitrary partition $\{A_j\}_{j=1}^{t^2}$ of $\mathbb{N}$

$$\sum_{j=1}^{t^2}\Big(\sum_{n\in A_j} a_n^2\Big)^{1/2} \leq t\Big(\sum_{j=1}^{t^2}\sum_{n\in A_j} a_n^2\Big)^{1/2} = t\|a\|_{\ell_2}.$$

Now taking the supremum over all such partitions gives us the desired result.

From the above inequalities we obtain

$$\begin{aligned}\kappa(t,a) &= \inf\{\|b\|_{\ell_1} + t\|c\|_{\ell_2} : b+c=a, b\in\ell_1, c\in\ell_2\}\\ &\ge \inf\{\|b\|_{Q(t^2)} + \|c\|_{Q(t^2)} : b+c=a, b\in\ell_1, c\in\ell_2\} \ge \|a\|_{Q(t^2)},\end{aligned}$$

and so the left-hand side inequality in (1.28) is proved.

To prove the right-hand side inequality in (1.28), we shall show that

$$\kappa(t,a) = \sup\Big\{\sum_{k=1}^{\infty} a_k b_k : b=(b_k)_{k=1}^{\infty}\in\ell_2\,,\ \mathcal{J}_{\infty,2}(1/t,b)\le 1\Big\}, \tag{1.29}$$

where $\mathcal{J}_{\infty,2}(s,b) := \max\{\|b\|_{\ell_\infty}, s\|b\|_{\ell_2}\}$, $s>0$. Denote by ℓ_2^t the space ℓ_2 with the norm $\kappa(t,a)$. Then, by the Hahn–Banach theorem, we have

$$\kappa(t,a) = \sup\{|f(a)| :\ f\in(\ell_2^t)^*\,,\ \|f\|_{(\ell_2^t)^*}\le 1\}.$$

Therefore, it suffices to prove that the dual space $(\ell_2^t)^*$ can be described as the set of all functionals of the form

$$f(a) = \sum_{k=1}^{\infty} a_k b_k,\quad a=(a_k)\in\ell_2, \tag{1.30}$$

and $\|f\|_{(\ell_2^t)^*} = \mathcal{J}_{\infty,2}(1/t,b)$.

To prove this claim, we suppose first that $f\in(\ell_2^t)^*$. Since $\ell_1 \overset{1}{\subset} \ell_2^t$, then $f\in\ell_1^*$. Therefore, the functional f is representable in the form (1.30), where the sequence $b=(b_k)_{k=1}^{\infty}$ satisfies the inequality $\|b\|_\infty = \|f\|_{\ell_1^*} \le \|f\|_{(\ell_2^t)^*}$. Similarly, we have $t\cdot\ell_2 \overset{1}{\subset} \ell_2^t$ (we set $\|a\|_{t\cdot\ell_2} := t\|a\|_{\ell_2}$), and hence $\|b\|_{\ell_2} = \|f\|_{\ell_2^*} \le t\|f\|_{(\ell_2^t)^*}$. Summing up, we get $\mathcal{J}_{\infty,2}(1/t,b)\le\|f\|_{(\ell_2^t)^*}$.

Conversely, let $a\in\ell_2$. Then, if $a=c+d$ and $c\in\ell_1$, $d\in\ell_2$, we have

$$\begin{aligned}|f(a)| &= \Big|\sum_{k=1}^{\infty} a_k b_k\Big| \le \Big|\sum_{k=1}^{\infty} c_k b_k\Big| + \Big|\sum_{k=1}^{\infty} d_k b_k\Big| \le \sup_{k=1,2,\dots}|b_k|\cdot\sum_{k=1}^{\infty}|c_k|\\ &\quad + \frac{1}{t}\Big(\sum_{k=1}^{\infty} b_k^2\Big)^{1/2}\cdot t\Big(\sum_{k=1}^{\infty} d_k^2\Big)^{1/2} \le \mathcal{J}_{\infty,2}(1/t,b)(\|c\|_{\ell_1} + t\|d\|_2).\end{aligned}$$

Taking the infimum over all representations $a=c+d$, $c\in\ell_1$, $d\in\ell_2$, by the definition of ℓ_2^t, we obtain the estimate

$$\|f\|_{(\ell_2^t)^*} \le \mathcal{J}_{\infty,2}(1/t,b).$$

So, (1.29) is proved.

Now, according to (1.29), for any $\varepsilon > 0$ there exists a sequence $b \in \ell_2$ such that

$$(1-\varepsilon)\kappa(t,a) \le \sum_{k=1}^{\infty} a_k b_k \text{ and } \mathcal{J}_{\infty,2}(t^{-1}, b) \le 1.$$

Choose $n_0, n_1, n_2, \ldots, n_{t^2} \in \{0, 1, \ldots, \infty\}$ by induction as follows: if $0 = n_0 < n_1 < \ldots < n_m$ have been already found, then

$$n_{m+1} := 1 + \sup\left\{k : \sum_{n=n_m+1}^{k} b_n^2 \le 1\right\}.$$

Since $\|b\|_{\ell_\infty} \le 1$, we have $\sum_{n=n_m+1}^{n_{m+1}} b_n^2 \le 2$. Moreover, from the inequality $\|b\|_{\ell_2} \le t$ it follows that $n_{t^2} = \infty$. Thus, by definition, we obtain

$$(1-\varepsilon)\kappa(t,a) \le \sum_{k=1}^{\infty} a_k b_k \le \sum_{m=1}^{t^2} \Big(\sum_{n=n_{m-1}+1}^{n_m} b_n^2\Big)^{1/2} \Big(\sum_{n=n_{m-1}+1}^{n_m} a_n^2\Big)^{1/2} \le \sqrt{2}\|a\|_{Q(t^2)}.$$

Since $\varepsilon > 0$ is arbitrary, the proof is completed. □

1.6 Hitczenko's and Montgomery-Smith's Inequalities

Now, we are ready to state one of the most important results of this chapter.

Theorem 1.6 *There exists a universal constant $\gamma > 0$ such that for all $p \ge 1$ and $a = (a_k)_{k=1}^{\infty} \in \ell_2$ the following inequality holds:*

$$\gamma \kappa_a(\sqrt{p}) \le \Big\|\sum_{k=1}^{\infty} a_k r_k\Big\|_p \le \kappa_a(\sqrt{p}). \tag{1.31}$$

Before proving this statement, we shall make sure that the combination of inequalities (1.26) and (1.31) gives much more precise information related to dependence of the norms $\left\|\sum_{k=1}^{\infty} a_k r_k\right\|_p$ on p, than Khintchine's inequality. Let us consider the following three sequences of coefficients: $a)$ $a_k = 1/k$, $k = 1, 2, \ldots$; $b)$ $a_k = 2^{-k}$, $k = 1, 2, \ldots$; $c)$ $a_k = 1/\sqrt{n}$ if $k = 1, 2, \ldots, n$, and $a_k = 0$ if $k > n$. In all these cases from inequality (1.15) it follows only that, for $p \ge 1$,

$$\frac{1}{\sqrt{3}}\|(a_k)\|_{\ell_2} \le \Big\|\sum_{k=1}^{\infty} a_k r_k\Big\|_p \le C\sqrt{p}\|(a_k)\|_{\ell_2}.$$

However, the behaviour of the norms $\left\|\sum_{k=1}^{\infty} a_k r_k\right\|_p$ in the cases $a)$, $b)$, and $c)$ is quite different. Namely, inequalities (1.26) and (1.31) combined with easy calculations show that, with constants independent of $p \geq 1$ and $n \in \mathbb{N}$,

$$a)\left\|\sum_{k=1}^{\infty} a_k r_k\right\|_p \asymp \ln(1+p), \quad b)\left\|\sum_{k=1}^{\infty} a_k r_k\right\|_p \asymp 1,$$

and

$$c)\left\|\sum_{k=1}^{\infty} a_k r_k\right\|_p \asymp \sqrt{\min(n, p)}.$$

Proof of Theorem 1.6 The right-hand side inequality in (1.31) is an immediate consequence of the definition of Peetre's $\mathcal{K}$-functional and Khintchine's inequality. Indeed, for arbitrary $\varepsilon > 0$ we find $b = (b_k) \in \ell_1$ and $c = (c_k) \in \ell_2$ satisfying

$$\|b\|_{\ell_1} + \sqrt{p}\|c\|_{\ell_2} \leq \kappa_a(\sqrt{p}) + \varepsilon.$$

Then, applying inequality (1.15) gives us

$$\left\|\sum_{k=1}^{\infty} a_k r_k\right\|_p \leq \left\|\sum_{k=1}^{\infty} b_k r_k\right\|_p + \left\|\sum_{k=1}^{\infty} c_k r_k\right\|_p \leq \|b\|_{\ell_1} + \sqrt{p}\|c\|_{\ell_2} \leq \kappa_a(\sqrt{p}) + \varepsilon.$$

As $\varepsilon > 0$ is arbitrary, the desired inequality follows.

Next, we prove the left-hand side inequality in the case when $p \in \mathbb{N}$. By the definition of the $Q(p)$-norm, we can choose pairwise disjoint subsets $A_1, \ldots, A_p$ of $\mathbb{N}$ so that

$$\|a\|_{Q(p)} \leq \frac{3}{2} \cdot \sum_{j=1}^{p} \Big(\sum_{k \in A_j} a_k^2\Big)^{1/2}. \tag{1.32}$$

Then, using Chebyshev's inequality and the independence of the Rademacher functions (see Proposition 1.1), we obtain

$$\left\|\sum_{k=1}^{\infty} a_k r_k\right\|_p^p = \left\|\sum_{j=1}^{p} \sum_{k \in A_j} a_k r_k\right\|_p^p \geq$$

$$\geq 2^{-p}\Big\{\sum_{j=1}^{p}\Big(\sum_{k \in A_j} a_k^2\Big)^{1/2}\Big\}^p \cdot m\Big(\bigcap_{j=1}^{p}\Big\{\sum_{k \in A_j} a_k r_k \geq \frac{1}{2}\Big(\sum_{k \in A_j} a_k^2\Big)^{1/2}\Big\}\Big) =$$

$$= 2^{-p}\Big\{\sum_{j=1}^{p}\Big(\sum_{k \in A_j} a_k^2\Big)^{1/2}\Big\}^p \cdot \prod_{j=1}^{p} m\Big\{\sum_{k \in A_j} a_k r_k \geq \frac{1}{2}\Big(\sum_{k \in A_j} a_k^2\Big)^{1/2}\Big\}.$$

Moreover, since r_k, $k = 1, 2, \ldots$, are symmetrically distributed, from inequality (1.19) it follows

$$m\Big\{\sum_{k\in A_j} a_k r_k \ge \frac{1}{2}\Big(\sum_{k\in A_j} a_k^2\Big)^{1/2}\Big\} = \frac{1}{2}m\Big\{\Big|\sum_{k\in A_j} a_k r_k\Big|^2 \ge \frac{1}{4}\sum_{k\in A_j} a_k^2\Big\} \ge \frac{1}{32}.$$

Hence, inequality (1.32) and Lemma 1.3 imply that

$$\Big\|\sum_{k=1}^{\infty} a_k r_k\Big\|_p \ge \frac{1}{64}\cdot\sum_{j=1}^{p}\Big(\sum_{k\in A_j} a_k^2\Big)^{1/2} \ge \frac{1}{96\sqrt{2}}\kappa_a(\sqrt{p}), \quad p \in \mathbb{N}.$$

Finally, for arbitrary $p \ge 1$, taking into account the concavity of the function $t \mapsto \kappa(t, a)$ and the elementary inequality $\sqrt{p} \le 2\sqrt{[p]}$, we get

$$\frac{1}{192\sqrt{2}}\kappa_a(\sqrt{p}) \le \frac{1}{96\sqrt{2}}\kappa_a(\sqrt{[p]}) \le \Big\|\sum_{k=1}^{\infty} a_k r_k\Big\|_{[p]} \le \Big\|\sum_{k=1}^{\infty} a_k r_k\Big\|_p. \qquad \square$$

The next result will make possible in subsequent chapters to characterize the subspaces of symmetric spaces, generated by the Rademacher system.

Theorem 1.7 *There exists a universal constant* $\delta \in (0, 1)$ *such that for all* $a = (a_k)_{k=1}^{\infty} \in \ell_2$ *and* $t > 0$ *we have*

$$m\Big\{s \in [0, 1] : \sum_{k=1}^{\infty} c_k r_k(s) > \kappa(t, a)\Big\} \le e^{-t^2/2} \tag{1.33}$$

and

$$m\Big\{s \in [0, 1] : \sum_{k=1}^{\infty} c_k r_k(s) \ge \delta\kappa(t, a)\Big\} \ge \delta e^{-t^2/\delta}. \tag{1.34}$$

Proof First, we prove inequality (1.33). For arbitrary $\varepsilon > 0$ we can find a representation $a = b + c$, with $b \in \ell_1$, $c \in \ell_2$, such that

$$(1 + \varepsilon)\kappa(t, a) > \|b\|_{\ell_1} + t\|c\|_{\ell_2}. \tag{1.35}$$

Then, from Proposition 1.2 and the fact that the series $\sum_{k=1}^{\infty} c_k r_k$ converges in measure on [0, 1] it follows

$$m\Big\{s : \sum_{k=1}^{\infty} c_k r_k(s) > t\|c\|_{\ell_2}\Big\} \le e^{-t^2/2}.$$

Moreover (see Remark 1.4),

$$m\Big\{s: \sum_{k=1}^{\infty} b_k r_k(s) > \|b\|_{\ell_1}\Big\} = 0.$$

Therefore, according to (1.35), we have

$$m\Big\{s: \sum_{k=1}^{\infty} a_k r_k(s) > (1+\varepsilon)\kappa(t,a)\Big\} \le m\Big\{s: \sum_{k=1}^{\infty} b_k r_k(s) > \|b\|_{\ell_1}\Big\}$$
$$+\, m\Big\{s: \sum_{k=1}^{\infty} c_k r_k(s) > t\|c\|_{\ell_2}\Big\} \le e^{-t^2/2}.$$

Passing to the limit as $\varepsilon \to 0$, we come to (1.33).

Now, proceed with the proof of inequality (1.34), assuming firstly that $t \in \mathbb{N}$. For an arbitrary $\varepsilon > 0$ we find pairwise disjoint sets $A_1, A_2, \dots, A_{t^2}$ such that

$$\bigcup_{j=1}^{t^2} A_j = \mathbb{N} \quad \text{and} \quad \|a\|_{Q(t^2)} \le (1+\varepsilon)\sum_{j=1}^{t^2}\Big(\sum_{k\in A_j} a_k^2\Big)^{1/2}.$$

Then, from Lemma 1.3 and Proposition 1.1 it follows that

$$m\Big\{s: \sum_{k=1}^{\infty} a_k r_k(s) \ge \frac{1}{2}\kappa(t,a)\Big\} \ge m\Big\{s: \sum_{k=1}^{\infty} a_k r_k(s) \ge \frac{1}{\sqrt{2}}\|a\|_{Q(t^2)}\Big\} \ge$$
$$\ge m\Big\{s: \sum_{j=1}^{t^2}\sum_{k\in A_j} a_k r_k(s) \ge \frac{1}{\sqrt{2}}(1+\varepsilon)\sum_{j=1}^{t^2}\Big(\sum_{k\in A_j} a_k^2\Big)^{1/2}\Big\} \ge$$
$$\ge \prod_{j=1}^{t^2} m\Big\{s: \sum_{k\in A_j} a_k r_k(s) \ge \frac{1}{\sqrt{2}}(1+\varepsilon)\Big(\sum_{k\in A_j} a_k^2\Big)^{1/2}\Big\}.$$

As the functions $\sum_{k\in A_j} a_k r_k$, $j = 1, 2, \dots, t^2$, are symmetrically distributed, then this estimate and the Paley–Zygmund inequality (1.19) yield

$$m\Big\{s: \sum_{k=1}^{\infty} a_k r_k(s) \ge \frac{1}{2}\kappa(t,a)\Big\} \ge \left(\frac{1}{18}\Big(1-\frac{1}{2}(1+\varepsilon)^2\Big)^2\right)^{t^2}.$$

Hence, passing to the limit as $\varepsilon \to 0$, we get

$$m\Big\{s : \sum_{k=1}^{\infty} a_k r_k(s) > \frac{1}{2}\kappa(t,a)\Big\} \geq \exp\Big(-(\ln 72)t^2\Big).$$

Thus, (1.34) is proved for $t \in \mathbb{N}$.

In the case when $t \geq 1$, writing (1.34) for $[t]+1 \in \mathbb{N}$ and then applying elementary inequalities

$$\kappa(t,a) \leq \kappa([t]+1,a) \ \text{ and } \ ([t]+1)^2 \leq 4t^2,$$

we obtain (1.34) with $\delta < 1/(4\ln 72)$. Finally, if $0 < t < 1$, we have $\kappa(t,a) = t\|a\|_{\ell_2}$. Therefore, using the Paley–Zygmund inequality once more, we obtain

$$m\left\{s : \sum_{k=1}^{\infty} a_k r_k(s) \geq \frac{1}{2}\kappa(t,a)\right\} = m\left\{s : \sum_{k=1}^{\infty} a_k r_k(s) \geq \frac{t}{2}\|a\|_{\ell_2}\right\} \geq \frac{1}{18}\left(1-\frac{t^2}{4}\right)^2 \geq \frac{1}{32}.$$

As a result, inequality (1.34) is valid for all $t > 0$ with the constant $\delta = 1/32$. □

The above tail distribution estimates for Rademacher sums may be equivalently restated by using the notion of decreasing rearrangement of a measurable function.

Corollary 1.9 *There exists a universal constant $\beta \in (0,1)$ such that for every Rademacher sum $f = \sum_{k=1}^{\infty} a_k r_k$, where $a = (a_k)_{k=1}^{\infty} \in \ell_2$, and each $0 < u \leq 1$ we have*

$$f^*(u) \leq \beta^{-1}\kappa_a(\ln^{1/2}(e/u)) \tag{1.36}$$

and

$$f^*(\beta u) \geq \beta\kappa_a(\ln^{1/2}(e/u)). \tag{1.37}$$

Proof First, due to the symmetry of distribution of Rademacher sums and inequality (1.33),

$$m\{s \in [0,1] : |f(s)| > \kappa(t,a)\} = 2m\{s \in [0,1] : f(s) > \kappa(t,a)\} < e^{-t^2/2+1}.$$

Hence, by the definition of the decreasing rearrangement and the concavity of the function $t \mapsto \kappa(t,a)$, by changing variables to $u = e^{-t^2/2+1}$, we obtain

$$f^*(u) \leq \kappa_a(\sqrt{2}\ln^{1/2}(e/u)) \leq \sqrt{2}\kappa(\ln^{1/2}(e/u),a).$$

Thus, inequality (1.36) holds with $\beta \leq 1/\sqrt{2}$.

Let us prove (1.37). As in the first part of the proof, using inequality (1.34), we come to the inequality

$$f^*\left(\delta e^{-t^2/\delta}\right) \geq \frac{\delta}{2}\kappa(t, a), \quad t > 0.$$

Again due to the concavity of the function $t \mapsto \kappa(t, a)$, after the appropriate change of variables this yields

$$f^*(\delta u/e) \geq \frac{\delta^{3/2}}{2}\kappa(\ln^{1/2}(e/u), a), \quad 0 < u \leq 1.$$

Taking into account that $0 < \delta < 1$, we conclude that inequalities (1.36) and (1.37) are valid with the constant $\beta = \min(\delta/e, \delta^{3/2}/2)$. □

Comments and References For the first time, the functions r_k, $k = 1, 2, \ldots$, appeared explicitly in 1922 in the work [248] by Rademacher (that is why they call "Rademacher functions"). He used these functions to identify the conditions, under which a series $\sum_{k=1}^{\infty} \pm a_k$ converges for a random choice of signs. As is proved in [248], this holds if $\sum_{n=1}^{\infty} a_n^2 < \infty$ (see Theorem 1.1). In the definition of r_k, $k = 1, 2, \ldots$, Rademacher exploited the binary expansion of real numbers (cf. (1.1)). However, in fact, long time before the publication of the paper [248], the Rademacher system was well known in probability theory as a sequence of independent, identically and symmetrically distributed random variables taking on values ± 1. In particular, a special attention had been paid to studying the behaviour of the sums $\sum_{k=1}^{n} r_k(t)$ as $n \to \infty$ in connection with the strong Borel law of large numbers and its refinements (cf. [233, § 1]).

There is also another customary definition of the Rademacher functions as the characters defined on the Cantor group $\mathcal{K} = \{-1, 1\}^{\mathbb{N}}$ consisting of all functions mapping $\mathbb{N}$ to the two-point set $\{-1, 1\}$ (see, for instance, [126, § 1.2], [115, § 14.1], [76]). Here, the group $\mathcal{K}$ is equipped with the natural multiplication operation and the discrete topology; the n-th Rademacher character r_n is defined by $r_n(\omega) := \omega(n)$, where $\omega = (\omega(k))_{k\in\mathbb{N}} \in \mathcal{K}$.

Moreover, the Rademacher system is a special example of functional sequences from the following class. Let a function $\psi \in L_2(-\infty, \infty)$ be 1-periodic and satisfy the condition $\psi(t) + \psi(t + 1/2) = 0$ for all $t \in \mathbb{R}$. We set $\psi_n(t) := \psi(2^{n-1}t)$, $0 \leq t \leq 1$, $n = 1, 2, \ldots$ One can easily check that $\{\psi_n\}_{n=1}^{\infty}$ is an orthogonal system of functions on $[0, 1]$ [146, § II.2, p. 43-44]. Clearly, choosing $\psi(t) = \operatorname{sign}\sin 2\pi t$, we get the Rademacher system.

Section 1.1 contains basically standard and well-known facts related to the Rademacher functions. The system $\{r_n\}_{n=1}^{\infty}$ is a simple but however very important example of a sequence of independent functions. The notion of independence is central in probability theory and as well it is very useful in the study of the geometry of Banach spaces. In the 1930's, an explicit and transparent method of the construction of sequences of independent functions was developed by Polish mathematicians Banach, Marcinkiewicz, Steinhaus, Zygmund and others; its detailed account is given in § II.1 and § II.2 of the monograph [153]. Observe that equation (1.2) holds for an arbitrary set of independent functions (see (A.1) or [153, Theorem II.4]).

As was already said, Theorem 1.1 has been obtained by Rademacher in [248]. The converse result, Theorem 1.2, is due to Kolmogorov [155] (in fact, he has proved there a similar result for general systems of independent functions satisfying certain conditions). The proofs of the above theorems are borrowed from [146, § 4.5].

Classical inequality (1.9) was actually proved (although it had not been highlighted there as a separate statement) by Khintchine in [154], when he tried to get a "good" estimate for the following distribution function

$$m\Big\{t \in [0,1] : \Big|\sum_{k=1}^{n} a_k r_k(t)\Big| \geq \tau\Big\}, \quad \tau > 0.$$

Later on, this direction of research led to the famous "law of the iterated logarithm", which summarized successive refinements of the strong Borel law of large numbers [233, § 1]. For the first time (in 1930), as a separate result, the right-hand side of Khintchine's inequality (1.15) with a detailed proof appeared either in the work [184] by Littlewood, or in the work [227] by Paley and Zygmund, devoted to the study of random series (these authors, apparently, were not aware about a much earlier paper due to Khintchine). Note that Paley–Zygmund's proof is usually repeated when discussing the above upper estimate. As for the left-hand side of inequality (1.15), then in the case $p = 1$ it had been obtained by Littlewood in [183]. Observe that the idea of its proof used here (see Theorem 1.4) is taken also from the paper [183].

Theorems 1.3 and 1.4 are already a long time an important source of numerous further investigations. In 1937, Marcinkiewicz and Zygmund [193] generalized Khintchine's inequality to the case of uniformly bounded independent r.v.'s (see also [153, Theorems II.2.5 and II.2.6] and [192, Theorem 5 on p. 176; see further references therein]): Let $\{f_k\}_{k=1}^{\infty}$ be a sequence of independent r.v.'s on $[0, 1]$ satisfying the conditions:

$$\int_0^1 f_k(t)\,dt = 0, \quad \|f_k\|_2 = \Big(\int_0^1 f_k(t)^2\,dt\Big)^{1/2} = 1, \quad \|f_k\|_\infty \leq M, \quad k = 1, 2, \ldots$$

Then, for every $1 \leq p < \infty$ there exists $C = C(M, p) > 0$ such that for all $n \in \mathbb{N}$, $a_k \in \mathbb{R}$, $k = 1, 2, \ldots, n$

$$\frac{1}{C}\Big(\sum_{k=1}^{n} a_k^2\Big)^{1/2} \leq \Big\|\sum_{k=1}^{n} a_k f_k(t)\Big\|_p \leq C\Big(\sum_{k=1}^{n} a_k^2\Big)^{1/2}.$$

Note that already in 1929 distribution estimates for sequences of uniformly bounded independent r.v.'s similar to those for Rademacher functions proved in Corollary 1.5

and Proposition 1.2 were established by Kolmogorov [161, p. 127]. In [193, Theorem 13], Marcinkiewicz and Zygmund have obtained also a version of Khintchine's inequality for sequences of independent and not necessarily uniformly bounded r.v.'s. See also the paper by Burkholder and Gandy [93], where rather general modular inequalities similar to the above Marcinkiewicz–Zygmund theorem have been proved, and the survey [233], which contains extensions of the Khintchine type inequalities to the class of martingale differences and much other interesting related information.

Let us mention briefly non-commutative generalizations of the classical Khintchine inequality to the case of matrix-valued coefficients. They were first proved by Lust-Piquard [187] in the case $1 < p < \infty$, and by Pisier and Lust-Piquard [188] for $p = 1$. Later on, in [128], Haagerup and Musat proved similar estimates with better (and in some cases with the best) constants. For further results and references we refer to [129, 144, 241, 242, 245].

Khintchine's inequality, its versions and refinements have found a lot of applications in various fields of analysis and will play a crucial role also in the present book. The vector-valued version of Khintchine's inequality, proved in Corollary 1.6, is widely used in studying spaces of measurable functions (see e.g. the monograph [182]). Proposition 1.3, showing that the Rademacher functions form a strong Sidon–Zygmund system, has been proved in [145]. As it is well known (see e.g. the monographs [148] and [87]), the Paley–Zygmund inequalities (Proposition 1.4) are extremely important in the study of systems of independent functions, and they will be further repeatedly used.

The main results of Sects. 1.5 and 1.6, which deal with the behaviour of Rademacher sums in the L_p-spaces as well as with their tail distribution estimates, were obtained much later, in the 1990's. They substantially refine Khintchine's inequality, making possible to estimate the asymptotic behaviour of the L_p-norms of Rademacher sums as $p \to \infty$ (see the examples presented after the statement of Theorem 1.6).

Starting with this chapter, we widely use the interpolation theory of operators and, in particular, Peetre's $\mathcal{K}$-functional introduced in 1963 [228]. As we shall see, in matters, related to the Rademacher system, a key role is played by the $\mathcal{K}$-functional for the Banach couple (ℓ_1, ℓ_2). Proposition 1.5 and Theorem 1.5, which will be applied repeatedly further, are special cases of the well-known result proved by Holmstedt in 1970 [136]. Theorem 1.6 is due to Hitczenko [135]. This interesting result allows to find the order of the norms $\left\|\sum_{k=1}^{\infty} a_k r_k\right\|_p$ as $p \to \infty$ for any fixed sequence of coefficients $(a_k)_{k=1}^{\infty} \in \ell_2$. Its proof is based on using an equivalent expression for the $\mathcal{K}$-functional of the couple (ℓ_1, ℓ_2) from Lemma 1.3, which has been obtained by Montgomery-Smith in [209]. In the latter paper, it has been proved also important Theorem 1.7, which allows to identify (up to equivalence) the distribution of Rademacher sums. This result and Corollary 1.9 will be frequently used in this book.

Chapter 2
Rademacher System in Symmetric Spaces Located "far" from L_∞

In the next two chapters we shall examine the Rademacher functions in general symmetric spaces (s.s.'s) on $[0, 1]$ (see Appendix C). As we shall see, their behaviour in such a space X depends largely on how "near" X is located to the space L_∞. The concept of "nearness" in the question is determined by the position of X with respect to the special s.s. G, which can be defined by different ways. On the one hand, G is the separable part of the Orlicz space L_{N_2}, generated by the function $N_2(u) = e^{u^2} - 1$, i.e., the closure of L_∞ in L_{N_2}. On the other hand, one can check (see also Corollary 2.1) that G consists of all measurable functions $x(t)$ satisfying the condition

$$\lim_{t\to 0} \frac{x^*(t)}{\ln^{1/2}(e/t)} = 0.$$

In particular, the latter description implies that $G \subset L_p$ for each $p < \infty$. In this chapter, we shall show that in the case when $X \supset G$ (i.e., X is located "far away" from L_∞) the behaviour of the Rademacher system does not depend so much on X. The most important issue in this relation that $\{r_k\}_{k=1}^\infty$ is equivalent in such a space to the unit vector basis of ℓ_2 just like in L_p with a finite p. Quite different properties this system has in s.s.'s X provided that $X \not\supset G$ (i.e., when X is located "near" to L_∞).

We begin with studying general properties of the sequence $\{r_k\}_{k=1}^\infty$ and some related subspaces of s.s.'s.

2.1 Rademacher Subspaces of Symmetric Spaces

One of the main objects of this book is the subspace $\mathcal{R}(X)$ of a s.s. X consisting of all measurable functions $f : [0, 1] \to \mathbb{R}$ from X, representable in the form $f(t) = \sum_{k=1}^\infty a_k r_k(t)$ a.e. on $[0, 1]$ (we shall refer $\mathcal{R}(X)$ as to the *Rademacher* subspace

S. V. Astashkin, *The Rademacher System in Function Spaces*,
https://doi.org/10.1007/978-3-030-47890-2_2

of X). Simultaneously $\mathcal{R}(X)$ can be naturally described as a space of sequences of real numbers. Specifically, let the set E_X consist of all sequences $a = (a_k)_{k=1}^\infty$, $a_k \in \mathbb{R}$, such that $\sum_{k=1}^\infty a_k r_k \in X$. It is easy to check that E_X is a linear space and the functional

$$\|a\|_{E_X} := \Big\| \sum_{k=1}^{\infty} a_k r_k \Big\|_X \tag{2.1}$$

is a norm on E_X. In particular, if $\|a\|_{E_X} = 0$, then the fact that $a = 0$ is a consequence of orthogonality of the Rademacher functions. Since the spaces $\mathcal{R}(X)$ (with the X-norm) and E_X are isometric, then in order to prove that $\mathcal{R}(X)$ is closed in X it suffices to show that E_X is complete.

Let a sequence $\{a^n\}_{n=1}^\infty$, $a^n = (a_k^n)_{k=1}^\infty$, be a Cauchy sequence in E_X. This means that the functions $x_a^n := \sum_{k=1}^\infty a_k^n r_k$, $n = 1, 2, \ldots$, form a Cauchy sequence in X. Hence, there exists a function $x \in X$ such that $x_a^n \to x$ in X as $n \to \infty$.

On the other hand, since every s.s. on $[0, 1]$ is continuously embedded into the space $L_1 = L_1[0, 1]$ (see Appendix C), then due to Khintchine's inequality from Theorem 1.4 it follows that

$$\frac{1}{\sqrt{3}}\Big(\sum_{k=1}^{\infty}(a_k^m - a_k^n)^2\Big)^{1/2} \le \Big\|\sum_{k=1}^{\infty}(a_k^m - a_k^n)\, r_k\Big\|_1 \le C\Big\|\sum_{k=1}^{\infty}(a_k^m - a_k^n)\, r_k\Big\|_X$$

for all positive integers $m > n$. Thus, the sequence $\{a^n\}$ is a Cauchy sequence in ℓ_2; denote its limit by $a = \{a_k\}_{k=1}^\infty$. Then, for the function $x_a := \sum_{k=1}^\infty a_k r_k$ we have

$$\|x_a^n - x_a\|_1 \le \|x_a^n - x_a\|_2 = \|a^n - a\|_{\ell_2},$$

whence $x_a^n \to x_a$ in L_1. Since X is continuously embedded into L_1, then we have also that $x_a^n \to x$ in L_1, and so $x = x_a$. As a result, $x_a^n \to x_a$ in X, or equivalently $a^n \to a$ in E_X. Thus, the space E_X is complete and hence $\mathcal{R}(X)$ is a closed subspace of X.

From Theorem 1.4 and Remark 1.4 it follows that the Rademacher system is a symmetric basic sequence in the space $L_p = L_p[0, 1]$ for any $1 \le p \le \infty$. We prove that, in fact, $\{r_k\}_{k=1}^\infty$ possesses this property in every s.s.

Proposition 2.1 *For arbitrary positive integers $1 \le n < l$, $a_k \in \mathbb{R}$, $k = 1, 2, \ldots, l$, and $z > 0$ it holds:*

$$m\Big\{t \in [0, 1] : \Big|\sum_{k=1}^{n} a_k r_k(t)\Big| > z\Big\} \le 2m\Big\{t \in [0, 1] : \Big|\sum_{k=1}^{l} a_k r_k(t)\Big| > z\Big\}.$$

Moreover, $\{r_k\}_{k=1}^{\infty}$ is a monotone basic sequence in every s.s. X, i.e.,

$$\Big\| \sum_{k=1}^{n} a_k r_k \Big\|_X \le \Big\| \sum_{k=1}^{l} a_k r_k \Big\|_X$$

for all positive integers $1 \le n < l$ and $a_k \in \mathbb{R}$, $k = 1, 2, \dots, l$.

Proof Let $z > 0$. Since the sum $\sum_{k=1}^{n} a_k r_k$ is constant for any $a_k \in \mathbb{R}$, $k = 1, 2, \dots, n$, on each of the dyadic intervals Δ_n^j, $j = 1, 2, \dots, 2^n$, we have

$$m\Big\{t \in [0,1] : \Big|\sum_{k=1}^{n} a_k r_k(t)\Big| > z\Big\} = 2^{-n} \cdot \operatorname{card} J, \tag{2.2}$$

where J is some subset of $\{1, 2, \dots, 2^n\}$. Assuming that $l > n$, we write

$$\sum_{k=1}^{l} a_k r_k = \sum_{k=1}^{n} a_k r_k + \sum_{k=n+1}^{l} a_k r_k.$$

Observe that the second sum from the right-hand side of this equation is symmetrically distributed on each of the intervals Δ_n^j, $j = 1, 2, \dots, 2^n$. Therefore, given $j \in J$ there exists a set $E_j \subset \Delta_n^j$ such that $m(E_j) \ge 2^{-n-1}$ and for all $t \in E_j$

$$\Big|\sum_{k=1}^{l} a_k r_k(t)\Big| \ge \Big|\sum_{k=1}^{n} a_k r_k(t)\Big| > z.$$

Consequently,

$$m\Big\{t \in [0,1] : \Big|\sum_{k=1}^{l} a_k r_k(t)\Big| > z\Big\} \ge 2^{-n-1} \cdot \operatorname{card} J,$$

and the first desired inequality follows from (2.2).

Clearly, the second inequality will be proved once we show that for every $n \in \mathbb{N}$ and all $a_k \in \mathbb{R}$, $k = 1, 2, \dots, n+1$, it holds

$$\Big\| \sum_{k=1}^{n} a_k r_k \Big\|_X \le \Big\| \sum_{k=1}^{n+1} a_k r_k \Big\|_X. \tag{2.3}$$

Indeed, consider the transformation ω, mapping the interval $[0, 1]$ onto itself and permuting the neighboring dyadic intervals Δ_{n+1}^{2j-1} and Δ_{n+1}^{2j}, $j = 1, 2, \dots, 2^n$, where the function r_{n+1} takes values $+1$ and -1, respectively. Obviously, ω preserves measure and, therefore, denoting $x := \sum_{k=1}^{n+1} a_k r_k$, by the definition of s.s.'s, we have $\|x(\omega)\|_X = \|x\|_X$. Finally, applying the triangle inequality to the right-hand side of the equation

$$\sum_{k=1}^{n} a_k r_k = \frac{1}{2}(x + x(\omega)),$$

we get (2.3). □

Proposition 2.2 *If $(a_k)_{k=1}^{\infty} \in \ell_2$ and π is an arbitrary permutation of the set of positive integers, then the functions $\sum_{k=1}^{\infty} a_k r_k$ and $\sum_{k=1}^{\infty} a_k r_{\pi(k)}$ are identically distributed on* $[0, 1]$ *and therefore the Rademacher system is a symmetric basic sequence with constant* 1 *in every s.s. on* $[0, 1]$ *(see Definitions B.1 and B.6).*

Moreover, if $(a_k^*)_{k=1}^{\infty}$ is the decreasing permutation of the sequence $(|a_k|)_{k=1}^{\infty}$, where $(a_k)_{k=1}^{\infty} \in \ell_2$, then the functions $\sum_{k=1}^{\infty} a_k r_k$ and $\sum_{k=1}^{\infty} a_k^* r_k$ are identically distributed on $[0, 1]$. *In particular, the Rademacher system is unconditional with constant* 1 *in every s.s. on* $[0, 1]$ *(see Definition B.4).*

Proof First of all, since $(a_k)_{k=1}^{\infty} \in \ell_2$, by Corollary 1.4, the series $\sum_{k=1}^{\infty} a_k r_k$ converges in measure on $[0, 1]$ to an integrable function ξ. Next, denote by $\theta_{\xi_{n,m}}(t)$ and $\theta_{\eta_{n,m}}(t)$ the characteristic functions of the r.v.'s $\xi_{n,m} := \sum_{k=n}^{m} a_k r_k$ and $\eta_{n,m} := \sum_{k=n}^{m} a_k r_{\pi(k)}$, $1 \le n \le m$ (see Appendix A). Then, for any $t \in \mathbb{R}$, it follows from Proposition 1.1

$$\theta_{\eta_{n,m}}(t) = \theta_{\xi_{n,m}}(t) = \prod_{k=n}^{m} \int_0^1 \exp(ita_k r_1(u))\,du = \prod_{k=n}^{m} \cos(t|a_k|). \tag{2.4}$$

Observe that, according to the uniqueness theorem (see Appendix A), the characteristic function of an r.v. identifies uniquely its distribution function. Consequently, the latter equation implies that the r.v.'s $\xi_{n,m}$ and $\eta_{n,m}$ are identically distributed for arbitrary $1 \le n \le m$. Hence, the series $\sum_{k=1}^{\infty} a_k r_{\pi(k)}$ also converges in measure on $[0, 1]$. Let η be its sum. Then, by the continuity theorem for the convergence of characteristic functions (see Appendix A), we have

$$\lim_{m\to\infty} \theta_{\xi_{1,m}}(t) = \theta_{\xi}(t) \quad \text{and} \quad \lim_{m\to\infty} \theta_{\eta_{1,m}}(t) = \theta_{\eta}(t), \quad t \in \mathbb{R}.$$

Therefore, taking into account equation (2.4), we deduce that $\theta_\eta(t) = \theta_\xi(t)$ for all $t \in \mathbb{R}$. Thus, from the uniqueness theorem it follows that the r.v.'s $\xi = \sum_{k=1}^{\infty} a_k r_k$ and $\eta = \sum_{k=1}^{\infty} a_k r_{\pi(k)}$ are identically distributed, and hence, by the definition of s.s's,

$$\Big\|\sum_{k=1}^{\infty} a_k r_{\pi(k)}\Big\|_X = \Big\|\sum_{k=1}^{\infty} a_k r_k\Big\|_X$$

for arbitrary X and π. As a result, the Rademacher functions form a symmetric basic sequence with constant 1 in every s.s.

Now, we prove the second assertion of the proposition. As above, putting $\xi = \sum_{k=1}^{\infty} a_k r_k$, by (2.4),

$$\theta_\xi(t) = \prod_{k=1}^{\infty} \cos(t|a_k|). \tag{2.5}$$

Additionally, since $(a_k)_{k=1}^{\infty} \in \ell_2$, we have $a_k^* = |a_{\sigma(k)}|$, $k = 1, 2, \dots$, for some permutation σ of the set of positive integers. Therefore, by Theorem 1.1, the series $\sum_{k=1}^{\infty} a_k^* r_k$ converges in measure on $[0, 1]$; denote by ζ its sum. Then, the same reasoning as in the case of ξ shows that

$$\theta_\zeta(t) = \prod_{k=1}^{\infty} \cos(t|a_{\sigma(k)}|).$$

Since $\cos z = 1 - 2\sin^2(z/2)$ and the series $\sum_{k=1}^{\infty} \sin^2(t|a_k|/2)$ converges (thanks to the hypothesis), then the product (2.5) converges absolutely for every $t \in \mathbb{R}$. Hence, $\theta_\zeta(t) = \theta_\xi(t)$ for all $t \in \mathbb{R}$, which implies, as above, that $\xi = \sum_{k=1}^{\infty} a_k r_k$ and $\zeta = \sum_{k=1}^{\infty} a_k^* r_k$ are identically distributed r.v.'s and so their norms in each s.s. are equal. This completes the proof. □

Summarizing all, we get the following result.

Theorem 2.1 *Let X be an arbitrary s.s. on* $[0, 1]$. *Then, the Rademacher subspace $\mathcal{R}(X)$ is closed in X, and the system $\{r_k\}_{k=1}^{\infty}$ is a symmetric basis in $\mathcal{R}(X)$ with constant* 1.

In the sequel, we shall show that properties of the subspace $\mathcal{R}(X)$ depend largely on how X is "near" to the space L_∞. As was said already in the beginning of the chapter, a natural "bound" here is the space G, the separable part of the Orlicz space L_{N_2}, generated by the function $N_2(t) = e^{t^2} - 1$.

2.2 Exponential Integrability of Rademacher Series and the Rodin–Semenov Theorem

By Theorem 1.4, the function

$$f := \sum_{k=1}^{\infty} a_k r_k \tag{2.6}$$

is well-defined and belongs to the space $L_p[0, 1]$ for every $1 \le p < \infty$ provided that a sequence of coefficients $a = (a_k)_{k=1}^{\infty} \in \ell_2$. We show that all functions of the form (2.6) are integrable in a stronger exponential sense, or more precisely, we prove that $\exp(f^2) \in L_1[0, 1]$.

Theorem 2.2 *Any function f of the form (2.6), with $a = (a_k)_{k=1}^\infty \in \ell_2$, belongs to the space G, and $\|f\|_{L_{N_2}} \le \sqrt{2e}\|a\|_{\ell_2}$.*

Proof Combining the Taylor expansion

$$\exp(u^2 z^2) = \sum_{k=0}^\infty \frac{u^{2k}}{k!} z^{2k}, \quad \text{where } u > 0,$$

inequality (1.10), and the estimate $C_{2k} \le k^k$ from the proof of Theorem 1.3, we obtain

$$\int_0^1 \left(\exp(u^2 f(t)^2) - 1\right) dt = \sum_{k=1}^\infty \frac{u^{2k}}{k!} \int_0^1 f(t)^{2k}\, dt \le \sum_{k=1}^\infty \frac{u^{2k}}{k!} k^k \|a\|_{\ell_2}^{2k}.$$

If $u \le (\sqrt{2e}\|a\|_{\ell_2})^{-1}$, then, due to the Stirling formula (see (1.13)), the sum in the right-hand side of this inequality does not exceed

$$(2\pi)^{-1/2} \sum_{k=1}^\infty 2^{-k} = (2\pi)^{-1/2} < 1,$$

and so, by the definition of the (Luxemburg) L_{N_2}-norm, we have

$$\|f\|_{L_{N_2}} \le \sqrt{2e}\|a\|_{\ell_2}.$$

Hence, due to the condition $a \in \ell_2$, a standard reasoning shows that the partial sums $S_n = \sum_{k=1}^n a_k r_k$, $n = 1, 2, \ldots$, of series (2.6), form a Cauchy sequence in L_{N_2}. Thus, $S_n \to f$ in L_{N_2} as $n \to \infty$. Since $S_n \in L_\infty$, $n = 1, 2, \ldots$, we have $f \in G$, and the proof is completed. □

Remark 2.1 It is clear that Theorem 2.2 still holds if we replace the space G with an arbitrary s.s. X such that $G \subset X$. More precisely, if $a = (a_k)_{k=1}^\infty \in \ell_2$, then series (2.6) converges in such a space X and with some constant C, depending only on X, we have

$$\left\| \sum_{k=1}^\infty a_k r_k \right\|_X \le C\|a\|_{\ell_2}. \tag{2.7}$$

The following simple observation shows that, if a s.s. X contains G, the subspace $\mathcal{R}(X)$ can be identified with the closed linear span of the Rademacher sequence in X.

Proposition 2.3 *If a s.s. $X \supset G$, then the subspace $\mathcal{R}(X)$ coincides with the closed linear span $[r_n]_X$.*

Proof The embedding $[r_n]_X \subset \mathcal{R}(X)$ is a direct consequence of the definitions and the fact that $\{r_n\}$ is a basic sequence in every s.s. (see Proposition 2.1).

Now, assume that $f = \sum_{k=1}^{\infty} a_k r_k \in \mathcal{R}(X)$. In particular, this series converges a.e. on $[0, 1]$, whence it follows from Theorem 1.2 that $a = (a_k)_{k=1}^{\infty} \in \ell_2$. Then, by hypothesis, we obtain (2.7). Therefore, the series $\sum_{k=1}^{\infty} a_k r_k$ converges in X, which implies that $f \in [r_n]_X$. □

Note that the opposite inequality to (2.7) holds in every s.s. X. Indeed, from the embedding $X \subset L_1$ (see Appendix C) and Theorem 1.4 it follows

$$\Big\|\sum_{k=1}^{\infty} a_k r_k\Big\|_X \geq \frac{1}{C}\Big\|\sum_{k=1}^{\infty} a_k r_k\Big\|_1 \geq \frac{1}{\sqrt{3}C}\|a\|_{\ell_2}.$$

Thus, the embedding $G \subset X$ assures that

$$\Big\|\sum_{k=1}^{\infty} a_k r_k\Big\|_X \asymp \|a\|_{\ell_2} \tag{2.8}$$

with a constant depending only on the space X. From the point of view of Banach space theory, equivalence (2.8) means that the Rademacher system is equivalent in X to the unit vector basis of the sequence space ℓ_2 or, in other words, $E_X = \ell_2$. The next important result shows that the converse assertion, which is not so obvious, holds as well.

Theorem 2.3 (Rodin–Semenov) *Equivalence* (2.8) *holds in a s.s. X on* $[0, 1]$ *if and only if $G \subset X$.*

We begin with several lemmas, which will be useful also in the future. Recall that $n_x(z) := m\{t \in [0, 1] : x(t) > z\}$, where $x(t)$ is a measurable function on $[0, 1]$ and $z > 0$.

Lemma 2.1 *Suppose $x_k(t)$, $k = 1, 2, \ldots$, and $x(t)$ are measurable functions on* $[0, 1]$ *satisfying at least one of the following conditions:*

(a) $x_k \to x$ *in measure on* $[0, 1]$;
(b) $\lim_{k\to\infty} n_{|x_k|}(z) = n_{|x|}(z)$ *for each* $z > 0$.

Then, $\lim_{k\to\infty} x_k^*(t_0) = x^*(t_0)$ *provided that t_0 is a point of continuity of the function $x^*(t)$. Therefore, $x_k^*(t)$ tends to $x^*(t)$ as $k \to \infty$ a.e. on* $[0, 1]$.

Proof We assume first that the condition (a) holds. Then, for arbitrary $\varepsilon > 0$ and $\delta > 0$ we can find a positive integer k_0 such that for all $k \geq k_0$

$$m\{s \in [0, 1] : |x_k(s) - x(s)| > \varepsilon\} < \delta.$$

Hence, from the definition of the decreasing rearrangement of a measurable function it follows that $(x - x_k)^*(\delta) \le \varepsilon$ whenever $k \ge k_0$. Moreover, in the same way as in the proof of Proposition 1.5 (see inequality (1.21)) one can check that for each $t \in [0, 1]$ we have

$$x^*(t + \delta) \le x_k^*(t) + (x - x_k)^*(\delta)$$

and

$$x_k^*(t) \le x^*(t - \delta) + (x - x_k)^*(\delta).$$

These inequalities and the preceding estimate imply that for arbitrarily small $\varepsilon > 0$ and $\delta > 0$ and for all $k \ge k_0$

$$x^*(t + \delta) - \varepsilon \le x_k^*(t) \le x^*(t - \delta) + \varepsilon.$$

Hence, if $x^*(t)$ is continuous at a point t_0, then $\lim_{k\to\infty} x_k^*(t_0) = x^*(t_0)$.

Proceed now with the case when the condition (b) is fulfilled. Let $x^*(t)$ be continuous at a point t_0 and $\varepsilon > 0$. Denoting $z_1 := x^*(t_0)+\varepsilon$, we have $n_{|x|}(z_1) < t_0$, and therefore $n_{|x_k|}(z_1) < t_0$ if $k \ge k_1$, where a positive integer k_1 is large enough. Then, again by the definition of the decreasing rearrangement, we have

$$x_k^*(t_0) \le z_1 = x^*(t_0) + \varepsilon \ \text{ if } \ k \ge k_1. \tag{2.9}$$

On the other hand, due to continuity of the function $x^*(t)$ at t_0, there exists $\delta > 0$ such that

$$x^*(t_0 + \delta) \ge x^*(t_0) - \varepsilon. \tag{2.10}$$

If $z_2 := x^*(t_0 + \delta) - \varepsilon$, then $n_{|x|}(z_2) \ge t_0 + \delta$, and so there is $k_2 \in \mathbb{N}$ such that $n_{|x_k|}(z_2) \ge t_0$ for $k \ge k_2$. Therefore, by (2.10),

$$x_k^*(t_0) \ge z_2 \ge x^*(t_0) - 2\varepsilon \ \text{ if } \ k \ge k_2.$$

Combining this estimate with inequality (2.9) for $k \ge \max(k_1, k_2)$ and taking into account that $x^*(t)$ decreases, we get the first desired result. Since each decreasing function on $[0, 1]$ is continuous a.e., the second assertion of this lemma follows as well. □

In the next proposition we modify the definition of Orlicz spaces that are located "near" to L_∞.

Let F be an increasing function on $(0, \infty)$ such that $F(0) = 0$. Assume that the function $e^{F(t)} - 1$ is equivalent to some convex function for large $t > 0$. Then F generates an Orlicz space on $[0, 1]$, which will be denoted by $\mathrm{Exp}\, L^F$.

Proposition 2.4 *The space* $\mathrm{Exp}\, L^F$ *coincides with the Marcinkiewicz space* $M(\psi_F)$, $\psi_F(t) := tF^{-1}(\ln(e/t))$ *(*F^{-1} *is the inverse function to* F*) and with* *a constant depending only on F we have*

$$\|x\|_{\operatorname{Exp} L^F} \asymp \sup_{0<s\le 1} \frac{x^*(s)}{F^{-1}(\ln(e/s))}.$$

In particular, for each $\alpha > 0$*, the Orlicz space* L_{N_α}*, where* $N_\alpha(t)$ *is equivalent to the function* $\exp(t^\alpha) - 1$ *for large* $t > 0$*, coincides with the Marcinkiewicz space* $M(\psi_\alpha)$*,* $\psi_\alpha(t) = t \ln^{1/\alpha}(e/t)$*, and with a constant depending only on* α

$$\|x\|_{L_{N_\alpha}} \asymp \sup_{0<s\le 1} \big(x^*(s) \ln^{-1/\alpha}(e/s)\big).$$

Proof Denote by $M(\psi_F)_w$ the space consisting of all measurable functions $x(s)$ on $[0, 1]$ such that the following quasi-norm

$$\|x\|_{M(\psi_F)_w} := \sup_{0<s\le 1} \frac{x^*(s)}{F^{-1}(\ln(e/s))}$$

is finite. By the inequality $x^*(t) \le t^{-1} \int_0^t x^*(s)\,ds$ and the definition of Marcinkiewicz spaces, we have

$$M(\psi_F) \subset M(\psi_F)_w \text{ and } \|x\|_{M(\psi_F)_w} \le \|x\|_{M(\psi_F)}, \quad x \in M(\psi_F). \tag{2.11}$$

Moreover, since the spaces $M(\psi_F)$ and $\operatorname{Exp} L^F$ have the same fundamental function, equal to $1/F^{-1}(\ln(e/t))$, and the Marcinkiewicz space is the largest one in the set of all s.s.'s with the same fundamental function (see Appendix C), then

$$\operatorname{Exp} L^F \subset M(\psi_F) \text{ and } \|x\|_{M(\psi_F)} \le \|x\|_{\operatorname{Exp} L^F}, \quad x \in \operatorname{Exp} L^F. \tag{2.12}$$

On the other hand, one can easily check that the function $F^{-1}(\ln(e/t))$ belongs to the Orlicz space $\operatorname{Exp} L^F$ and it has the largest (up to a constant) decreasing rearrangement in the space $M(\psi_F)_w$, or more precisely, for all $x \in M(\psi_F)_w$

$$x^*(t) \le \|x\|_{M(\psi_F)_w} \cdot F^{-1}(\ln(e/t)), \quad 0 < t \le 1.$$

This implies that

$$M(\psi_F)_w \subset \operatorname{Exp} L^F \text{ and } \|x\|_{\operatorname{Exp} L^F} \le C\|x\|_{M(\psi_F)_w}, \quad x \in M(\psi_F)_w,$$

where $C := \left\|F^{-1}(\ln(e/t))\right\|_{\operatorname{Exp} L^F}$. Combining these relations with (2.11) and (2.12), we obtain both statements of the proposition. □

According to Proposition 2.4, the space $(L_{N_\alpha})^\circ$ coincides with the separable part of the Marcinkiewicz space $M(\psi_\alpha)$, where $\psi_\alpha(t) = t \ln^{1/\alpha}(e/t)$ (equivalently, of the space $M(\psi_\alpha)_w$ from the proof of this proposition). Consequently (see Appendix C), we get

Corollary 2.1 *The space* $(L_{N_\alpha})^\circ$ *consists of all measurable functions* $x(t)$ *satisfying the condition*

$$\lim_{t\to 0} \frac{x^*(t)}{\ln^{1/\alpha}(e/t)} = 0.$$

In the last auxiliary assertion we consider the second Köthe dual space X''. Recall (for details, see Appendix C) that the Köthe dual space X' for a function Banach lattice X consists of all measurable functions $y = y(t)$ on $[0, 1]$ satisfying

$$\|y\|_{X'} := \sup\left\{\int_0^1 x(t)y(t)\,dt : x \in X, \|x\|_X \le 1\right\} < \infty.$$

Lemma 2.2 *For an arbitrary s.s.* X *on* $[0, 1]$ *we have* $(X'')^\circ = X^\circ$ *isometrically.*

Proof Since the statement is obvious for $X = L_\infty$, we can assume that

$$X \ne L_\infty. \tag{2.13}$$

Then, for the fundamental function ϕ_X of X we have

$$\lim_{s\to 0} \phi_X(s) = 0. \tag{2.14}$$

Indeed, assuming that $\phi_X(s) \ge c > 0$, $0 < s \le 1$, with some $c > 0$, we get for every $x \in X$

$$\|x\|_X = \|x^*\|_X \ge \|x^*\chi_{(0,s)}\|_X \ge x^*(s)\phi_X(s) \ge cx^*(s) \text{ for all } 0 < s \le 1,$$

whence $\|x\|_\infty \le c^{-1}\|x\|_X$ and so $X \subset L_\infty$. Since the opposite embedding holds for every s.s. (see Appendix C), we arrive at a contradiction with the assumption (2.13).

Next, we show that, under condition (2.13), the space X° is separable. Let $x \in X^\circ$ and let $\varepsilon > 0$ be arbitrary. First of all, by the definition of X°, there exists a function $y \in L_\infty$, for which $\|x - y\|_X < \varepsilon/3$. Assume that $\|y\|_\infty = M$, and, using (2.14), choose $\delta > 0$ so that $\phi_X(\delta) < \varepsilon/(6M)$. By the Lusin theorem [215, Theorem IV.4.4], there exists a continuous on $[0, 1]$ function z satisfying the conditions $\|z\|_\infty \le M$ and

$$m\{t \in [0, 1] : z(t) \ne y(t)\} < \delta.$$

Then, thanks to the choice of δ, we have

$$\|y - z\|_X = \|(y - z) \cdot \chi_{\{y \neq z\}}\|_X \le 2M \cdot \phi_X(\delta) < \frac{\varepsilon}{3}.$$

Hence, by the uniform continuity of z, we can find a step-function

$$u := \sum_{k=1}^{2^n} c_k \chi_{[(k-1)2^{-n}, k2^{-n}))}, \tag{2.15}$$

where $n \in \mathbb{N}$ and c_k are rational numbers, so that $\|z - u\|_X < \varepsilon/3$. From the above inequalities, clearly, it follows $\|x - u\|_X < \varepsilon$. Since the set of functions of the form (2.15) is countable, we conclude that X° is a separable space.

The separability of X° assures that the norm of this space is order semi-continuous (see Appendix C), and so X° is embedded into the space $(X^\circ)''$ isometrically [151, Theorem VI.1.6]. On the other hand, from the definition of X' and the properties of Lebesgue's integral, it follows that $(X^\circ)' = X'$, and therefore $(X^\circ)'' = X''$. Thus, X° is embedded isometrically into the second Köthe dual X''. Since $L_\infty \subset X^\circ$, this implies that the norms of X and X'' are equal on L_∞. As a result, by the definition of the separable part of a s.s., we get $(X'')^\circ = X^\circ$. □

Proof of Theorem 2.3 In view of Theorem 2.2 and the subsequent observation, we need to prove only the following: if a s.s. X satisfies the inequality

$$\Big\| \sum_{k=1}^{\infty} a_k r_k \Big\|_X \le C \|a\|_{\ell_2} \tag{2.16}$$

for some $C > 0$ and all $a = (a_k)_{k=1}^\infty$, then $G \subset X$.

The key idea consists in applying the central limit theorem (see Theorem A.1 or [81, Theorem 8.2.1]) to the sequence of functions

$$v_n := \frac{1}{\sqrt{n}} \sum_{k=1}^{n} r_k, \quad n = 1, 2, \ldots,$$

defined on the interval [0, 1], which is equipped, as usual, with the Lebesgue measure. According to the above theorem, taking into account that $v_n, n = 1, 2, \ldots,$ are symmetrically distributed, we have

$$\lim_{n\to\infty} n_{|v_n|}(\tau) = \lim_{n\to\infty} m\{t \in [0, 1] : |v_n(t)| > \tau\} = \Psi(\tau), \quad \tau > 0,$$

where

$$\Psi(\tau) := \frac{2}{\sqrt{2\pi}} \int_\tau^\infty e^{-t^2/2}\, dt.$$

Then, Lemma 2.1 yields

$$\lim_{n\to\infty} v_n^*(s) = \Psi^{-1}(s), \quad 0 < s \le 1$$

(here, Ψ^{-1} is the inverse function for Ψ). Moreover, condition (2.16) guarantees that $\|v_n\|_X \le C$ for all $n = 1, 2, \ldots$ Therefore, since X'' has the Fatou property, we deduce that $\Psi^{-1} \in X''$.

On the other hand, from the estimate

$$\int_\tau^\infty e^{-t^2/2}\,dt \ge \int_\tau^\infty te^{-t^2}\,dt = \frac{1}{2}e^{-\tau^2}, \tau \ge 0,$$

it follows that $\Psi(\tau) \ge (1/\sqrt{2\pi}) \cdot e^{-\tau^2}$, $\tau \ge 0$. Hence, passing to the inverse functions, we obtain

$$\Psi^{-1}(s) \ge \ln^{1/2}\left(\frac{1}{s\sqrt{2\pi}}\right), \quad 0 < s \le \frac{1}{\sqrt{2\pi}}.$$

Consequently, since $\Psi^{-1} \in X''$ and the space X'' is symmetric, the function $\ln^{1/2}(e/t)$ belongs to X''. As a result, by Proposition 2.4, $L_{N_2} \subset X''$, and thus (see Lemma 2.2) $G \subset (X'')^\circ = X^\circ \subset X$. □

Remark 2.2 An inspection of the proof of Theorem 2.3 shows that the embedding $G \subset X$ is equivalent to the condition

$$\sup_{n\in\mathbb{N}} \frac{1}{\sqrt{n}} \left\| \sum_{k=1}^n r_k \right\|_X < \infty,$$

which is formally weaker than inequality (2.16).

Denote by $\mathcal{R}$ the set of all functions f such that $f = \sum_{k=1}^\infty a_k r_k$, where $(a_k)_{k=1}^\infty \in \ell_2$ (equivalently, f can be represented as the sum of a Rademacher series, converging a.e. on $[0, 1]$; see Theorems 1.1 and 1.2). Then, Theorem 2.3 can be restated as follows:

Corollary 2.2 *If X is a s.s. on* $[0, 1]$, *then* $\mathcal{R}(X) = \mathcal{R}$ *if and only if* $G \subset X$.

2.3 Equivalence of the Rademacher System to the Unit Vector ℓ_1-Basis

Here, we focus on the problem of identification of s.s.’s X such that the Rademacher system is equivalent in X to the unit vector basis of ℓ_1 (equivalently, $E_X = \ell_1$). By Remark 1.4, L_∞ satisfies this condition. It turns out that this space is a unique (up to equivalence of norms) s.s. with the above property.

Theorem 2.4 *The equivalence*

$$\Big\|\sum_{k=1}^{\infty} a_k r_k\Big\|_X \asymp \|a\|_{\ell_1}$$

holds in a s.s. X on $[0, 1]$ if and only if $X = L_\infty$.

We begin with the following simple statement.

Lemma 2.3 *Let X be a s.s. on $[0, 1]$. Then, $X = L_\infty$ (with equivalence of norms) if and only if there exist a constant $c > 0$ and a sequence $\{x_n(t)\} \subset X$ such that $|x_n(t)| \le 1$, $x_n(t) \to 0$ a.e. on $[0, 1]$ and $\|x_n\|_X \ge c$.*

Proof First, if $X = L_\infty$, then the functions $x_n := \chi_{(0,1/n)}$, $n = 1, 2, \dots$, clearly form a sequence with the required properties.

Conversely, assume that X contains a sequence $\{x_n\}_{n=1}^\infty$ such that $|x_n(t)| \le 1$, $x_n(t) \to 0$ a.e. on $[0, 1]$ and $\|x_n\|_X \ge c$ for some $c > 0$. Since a.e. convergence implies convergence in measure, then for every $\varepsilon > 0$ and $n_0 \in \mathbb{N}$ large enough

$$m\left\{t \in [0, 1] : |x_{n_0}(t)| \ge \frac{c}{2}\right\} < \varepsilon.$$

Then, denoting

$$E := \left\{t \in [0, 1] : |x_{n_0}(t)| \ge \frac{c}{2\|\chi_{[0,1]}\|_X}\right\} \quad \text{and} \quad E^c := [0, 1] \setminus E,$$

we have

$$\|\chi_E\|_X \ge \|x_{n_0}\chi_E\|_X \ge \|x_{n_0}\|_X - \|x_{n_0}\chi_{E^c}\|_X \ge c - \frac{c\|\chi_{E^c}\|_X}{2\|\chi_{[0,1]}\|_X} \ge \frac{c}{2}.$$

Since $m(E) < \varepsilon$ and $\varepsilon > 0$ can be taken arbitrarily small, the latter inequality implies that $\phi_X(s) \ge c/2 > 0$ for all $s \in (0, 1]$. Hence, quite in the same way as in the proof of Lemma 2.2, one can deduce that $X \subset L_\infty$. Combining this with the fact that the opposite embedding holds as well, we obtain $X = L_\infty$, which concludes the proof. □

Proof of Theorem 2.4 It suffices to show that from the inequality

$$\Big\|\sum_{k=1}^{\infty} a_k r_k\Big\|_X \ge c\|a\|_{\ell_1}, \quad a = (a_k)_{k=1}^\infty \in \ell_1, \tag{2.17}$$

it follows $X = L_\infty$.

Let $x_n := n^{-1}\sum_{k=1}^n r_k$, $n = 1, 2, \dots$. Obviously, $|x_n(t)| \le 1$ and, by (2.17), $\|x_n\|_X \ge c$. Moreover, from the Strong Law of Large Numbers (see e.g. [81,

Theorem 5.1.2]) it follows that $x_n(t) \to 0$ a.e. on $[0, 1]$. Therefore, by Lemma 2.3, $X = L_\infty$ (with equivalence of norms). □

In the concluding part of this chapter we consider the complementability problem for the subspace $\mathcal{R}(X)$ in a s.s. X.

2.4 Complementability of Rademacher Subspaces in Symmetric Spaces

Recall that a linear operator Q acting in a linear space E is called a *projection* if $Q^2 = Q$. A closed linear subspace H of a linear normed space X is called *complemented* whenever there exists a projection Q bounded in X such that the image of Q coincides with H.

In the next theorem, along with already familiar space G, we shall consider also its Köthe dual G'. To characterize the latter space, recall that, by Corollary 2.1, $G = M(\psi_2)^\circ$, where $\psi_2(t) = t \ln^{1/2}(e/t)$. Therefore, G' coincides with the Lorentz space $\Lambda(\psi_2)$ (see Appendix C). A simple calculation shows that $\psi_2'(t) \asymp \ln^{1/2}(e/t)$, $0 < t \le 1$, and hence G' consists of all measurable functions $x(t)$ such that

$$\|x\|_{G'} \asymp \int_0^1 x^*(t) \ln^{1/2}(e/t)\, dt < \infty.$$

Theorem 2.5 *The subspace $\mathcal{R}(X)$ is complemented in a s.s. X if and only if*

$$G \subset X \subset G'. \tag{2.18}$$

Proof Assume that condition (2.18) holds. Then $G'' \subset X'$, whence $G \subset X'$. Therefore, by Theorem 2.3, the Rademacher system is equivalent both in X and X'' to the unit vector basis of ℓ_2. This means that there exists a constant $A > 0$ such that

$$\Big\| \sum_{k=1}^{\infty} a_k r_k \Big\|_X \le A \|a\|_{\ell_2} \quad \text{and} \quad \Big\| \sum_{k=1}^{\infty} a_k r_k \Big\|_{X'} \le A \|a\|_{\ell_2}. \tag{2.19}$$

Show that the orthogonal projection

$$Pf(t) := \sum_{k=1}^{\infty} \int_0^1 f(u) r_k(u)\, du \cdot r_k(t) \tag{2.20}$$

is bounded in X. Indeed, denote $c_k(f) := \int_0^1 f(u)r_k(u)\,du$, $k = 1, 2, \ldots$. Then, in view of (2.19) and Proposition 2.1, for any $n \in \mathbb{N}$ we have

$$\begin{aligned}\sum_{k=1}^n c_k(f)^2 &= \int_0^1 f(u)\sum_{k=1}^n c_k(f)r_k(u)\,du \\ &\le \|f\|_X \Big\|\sum_{k=1}^n c_k(f)r_k\Big\|_{X'} \\ &\le A\|f\|_X\Big(\sum_{k=1}^\infty c_k(f)^2\Big)^{1/2},\end{aligned}$$

whence

$$\Big(\sum_{k=1}^\infty c_k(f)^2\Big)^{1/2} \le A\|f\|_X, \quad f \in X.$$

Combining this with the first inequality in (2.19), we deduce that

$$\|Pf\|_X \le A\Big(\sum_{k=1}^\infty c_k(f)^2\Big)^{1/2} \le A^2\|f\|_X$$

for all $f \in X$. Thus, since the image of P, obviously, coincides with $\mathcal{R}(X)$, this subspace is complemented in X.

Now, proceed with the proof of the converse statement. We show firstly that it suffices to prove the boundedness in X of the orthogonal projection P defined by (2.20).

Since every s.s. on $[0, 1]$ is embedded into L_1, then by Khintchine's L_1-inequality (see Theorem 1.4) we have

$$\Big\|\sum_{k=1}^\infty a_k r_k\Big\|_X \ge \frac{1}{\sqrt{3}}\|a\|_{\ell_2} \quad\text{and}\quad \Big\|\sum_{k=1}^\infty a_k r_k\Big\|_{X'} \ge \frac{1}{\sqrt{3}}\|a\|_{\ell_2}.$$

Hence, if $\|Pf\|_X \le B\|f\|_X$, then, for arbitrary $f \in X$, $n \in \mathbb{N}$, and $a_k \in \mathbb{R}$, $k = 1, 2, \ldots, n$, we obtain

$$\begin{aligned}\int_0^1 f(t)\cdot\sum_{k=1}^n a_k r_k(t)\,dt &= \sum_{k=1}^n a_k c_k(f) \le \|a\|_{\ell_2}\Big(\sum_{k=1}^n c_k(f)^2\Big)^{1/2} \\ &\le \sqrt{3}\|a\|_{\ell_2}\cdot\|Pf\|_X \le \sqrt{3}B\|a\|_{\ell_2}\cdot\|f\|_X,\end{aligned}$$

whence

$$\Big\|\sum_{k=1}^{n} a_k r_k\Big\|_{X'} \le \sqrt{3}B\|a\|_{\ell_2}, \quad n \in \mathbb{N}.$$

By using this inequality for sums $\sum_{k=n}^{m} a_k r_k$, $1 \le n < m$, where $a = (a_k)_{k=1}^{\infty} \in \ell_2$, we see that the series $\sum_{k=1}^{\infty} a_k r_k$ converges in the space X' and moreover

$$\Big\|\sum_{k=1}^{\infty} a_k r_k\Big\|_{X'} \le \sqrt{3}B\|a\|_{\ell_2}. \tag{2.21}$$

Next, by Fubini's theorem, for arbitrary $f \in X$, $g \in X'$ and any $n \in \mathbb{N}$ we have

$$\int_0^1 f(u)\cdot\sum_{k=1}^{n} c_k(g) r_k(u)\,du = \int_0^1 g(t)\cdot\sum_{k=1}^{n} c_k(f) r_k(t)\,dt \le \|Pf\|_X\|g\|_{X'} \le B\|f\|_X\|g\|_{X'},$$

and therefore

$$\Big\|\sum_{k=1}^{n} c_k(g) r_k\Big\|_{X'} \le B\|g\|_{X'}, \quad n \in \mathbb{N}.$$

Since the space X' has the Fatou property, the last inequality implies that the projection P is bounded in X' with the norm not exceeding B. Thus, replacing X with X' and repeating the previous reasoning, we obtain

$$\Big\|\sum_{k=1}^{\infty} a_k r_k\Big\|_{X''} \le \sqrt{3}B\|a\|_{\ell_2}. \tag{2.22}$$

Now, applying Theorem 2.3 to the spaces X' and X'', we see that inequalities (2.21) and (2.22) imply the embeddings $G \subset X'$ and $G \subset X''$. Hence, we have $X \subset X'' \subset G'$ and $G \subset (X'')^\circ$. Since $(X'')^\circ = X^\circ$ by Lemma 2.2, from the latter embeddings it follows $G \subset X^\circ \subset X$. As a result, we come to (2.18). Thus, it remains to show that the boundedness of P in X is a consequence of the complementability of $\mathcal{R}(X)$ in this space.

Let $t = \sum\limits_{i=1}^{\infty} \alpha_i 2^{-i}$ and $u = \sum\limits_{i=1}^{\infty} \beta_i 2^{-i}$ $(\alpha_i, \beta_i = 0, 1)$ be the binary expansions of numbers $t, u \in [0, 1]$. Setting

$$t \oplus u := \sum_{i=1}^{\infty} 2^{-i}[(\alpha_i + \beta_i) mod 2],$$

one can easily verify that the interval $[0, 1]$ together with this operation is a compact Abelian group. Moreover, for every $u \in [0, 1]$ the transformation $w_u(s) = s \oplus u$ preserves the Lebesgue measure on $[0, 1]$, i.e., for any measurable $E \subset [0, 1]$, its inverse image $w_u^{-1}(E)$ is also measurable and $m(w_u^{-1}(E)) = m(E)$. Therefore, the operators $T_u f := f \circ w_u$, $0 \le u \le 1$, act isometrically in X. It follows from the definition of the Rademacher functions that the subspace $\mathcal{R}(X)$ is invariant with respect to these operators. Thus, by Rudin's theorem [255, Theorem 1.5.18], there exists a bounded linear projection Q acting from X onto $\mathcal{R}(X)$ and commuting with all operators T_u, $0 \le u \le 1$. Show that $Q = P$ (as above, P is the orthogonal projection defined in (2.20)).

First of all, represent the projection Q as follows:

$$Qf(t) = \sum_{i=1}^{\infty} Q_i(f) r_i(t), \tag{2.23}$$

observing that, by Proposition 2.1, the linear functionals Q_i, $i = 1, 2, \ldots$, are bounded on X. Since the image of Q is the subspace $\mathcal{R}(X)$, we have

$$Q_i(r_j) = \begin{cases} 1, i = j, \\ 0, i \neq j. \end{cases} \tag{2.24}$$

Consider the sets

$$E_i := \Big\{u \in [0, 1] : u = \sum_{j=1}^{\infty} \alpha_j 2^{-j}, \ \ \alpha_i = 0\Big\} \ \text{ and } \ E_i^c := [0, 1] \setminus E_i, \ \ i = 1, 2, \ldots$$

One can easily check that

$$r_i(t \oplus u) = \begin{cases} r_i(t), u \in E_i, \\ -r_i(t), u \in E_i^c. \end{cases}$$

Due to the equation $T_u Q = Q T_u$, $0 \le u \le 1$, this implies

$$Q_i(T_u f) = \begin{cases} Q_i(f), u \in E_i, \\ -Q_i(f), u \in E_i^c. \end{cases}$$

Taking into account that $m(E_i) = m(E_i^c) = 1/2$, we get

$$\int\limits_{E_i} Q_i(T_u f)\, du = \frac{1}{2} Q_i(f) \ \text{ and } \int\limits_{E_i^c} Q_i(T_u f)\, du = -\frac{1}{2} Q_i(f).$$

Further, thanks to the boundedness of Q_i on X, this functional can be moved outside the integral in the latter equations. Hence, we obtain

$$Q_i(f) = Q_i\Big(\int\limits_{E_i} T_u f\, du - \int\limits_{E_i^c} T_u f\, du\Big). \tag{2.25}$$

Since

$$\{s \in [0,1] : s = t \oplus u, \quad u \in E_i\} = \begin{cases} E_i, t \in E_i, \\ E_i^c, t \in E_i^c, \end{cases}$$

$$\{s \in [0,1] : s = t \oplus u, \quad u \in E_i^c\} = \begin{cases} E_i^c, t \in E_i, \\ E_i, t \in E_i^c, \end{cases}$$

and the transformation ω_u preserves the Lebesgue measure on $[0, 1]$, we have

$$\int\limits_{E_i} T_u f(t) du = \int\limits_{E_i} f(s)\, ds \cdot \chi_{E_i}(t) + \int\limits_{E_i^c} f(s)\, ds \cdot \chi_{E_i^c}(t)$$

and

$$\int\limits_{E_i^c} T_u f(t)\, du = \int\limits_{E_i^c} f(s)\, ds \cdot \chi_{E_i}(t) + \int\limits_{E_i} f(s)\, ds \cdot \chi_{E_i^c}(t).$$

It is easy to see that $r_i(t) = \chi_{E_i}(t) - \chi_{E_i^c}(t)$, $i = 1, 2, \ldots$ Therefore, we obtain

$$\int\limits_{E_i} T_u f(t)\, du - \int\limits_{E_i^c} T_u f(t)\, du = \int\limits_0^1 f(s) r_i(s)\, ds \cdot r_i(t).$$

Finally, this and equations (2.23)–(2.25) yield

$$Q_i(f) = \int\limits_0^1 f(s) r_i(s)\, ds \quad \text{for all} \quad i = 1, 2, \ldots,$$

i.e., $Q = P$, which concludes the proof. □

Comments and References In this chapter, we begin a fairly detailed study of properties of the Rademacher system in general s.s.'s. Propositions 2.1 and 2.2 show that the latter class of spaces is a highly natural "habitat" for this system. Theorem 2.2 was proved in 1930 by Paley and Zygmund [227] (see also [293,

Theorem V.8.7]). The main result of the chapter is Theorem 2.3, which was proved by Rodin and Semenov in 1975 [251]. In terms of embeddings, this theorem identifies s.s.'s X such that the Rademacher system spans in X the space ℓ_2. It proved to be extremely useful when studying various problems related to the geometrical structure of s.s.'s. For instance, in [134] (see also [43]), this property was used to prove that the canonical inclusion $X \subset L_1$ is not strictly singular if and only if a s.s. X contains the space G. The central limit theorem, which plays a crucial role in the proof of Theorem 2.3, will be used repeatedly in this book in similar situations. Later on, we shall also frequently exploit Proposition 2.4 about the coincidence of exponential Orlicz spaces with appropriate Marcinkiewicz spaces (a characterization of Orlicz spaces, satisfying this property, can be found in the papers [256] and [185]).

Theorem 2.4 has been obtained also in [251]. Theorem 2.5 on the complementability of the Rademacher subspaces has been proved by Rodin and Semenov [252] and independently by Lindenshtraus and Tzafriri [182, Theorem 2.b.4(ii)] (our proof follows the scheme given in the note [252]). Theorems 2.3 and 2.5 have been become a source of numerous further investigations. Let us mention only some of them (others, directly related to the Rademacher system, will be considered in detail in the subsequent chapters of the book).

In [85] (see also [87]), Braverman has shown that if a sequence $\{f_k\}_{k=1}^{\infty}$ of independent identically distributed r.v.'s spans a subspace isomorphic to ℓ_2 in a s.s. X, then $f_1 \in L_2$ and $G \subset X$. Moreover, we have $G \subset X \subset G'$ whenever this subspace is complemented in X. More general issues in connection with some Semenov's problems were considered by Raynaud in the paper [249]. Assume that a separable s.s. X does not contain ℓ_2 as a sublattice. Then the condition $G \subset X$ is necessary and sufficient for X to contain ℓ_2 as a subspace. Similarly, if a separable s.s. X does not contain ℓ_2 as a complemented sublattice, then the condition $G \subset X \subset G'$ is necessary and sufficient for X to contain ℓ_2 as a complemented subspace. Close but somewhat more restricted results were obtained previously by Tokarev in [279].

Another direction of investigations is connected with replacing scalar multiples of the Rademacher functions in the left-hand side of inequality (2.7) by some other sequences of functions. Let X be a s.s. As it has been proved in [143], the inequality

$$\Big\|\sum_{k=1}^{\infty} f_k\Big\|_X \le C\Big\|\Big(\sum_{k=1}^{\infty} f_k^2\Big)^{1/2}\Big\|_X, \tag{2.26}$$

holds for some $C > 0$ and all sequences of martingale differences $\{f_k\}_{k=1}^{\infty} \subset X$ if and only if the lower Boyd index α_X of the space X is positive. If instead of the class of martingale differences we restrict ourselves to considering sequences of mean zero (i.e., $\int_0^1 f_k(t)\,dt = 0$, $k = 1, 2, \dots$) independent functions $\{f_k\}_{k=1}^{\infty} \subset X$, then inequality (2.26) is valid if and only if the space X has the so-called Kruglov property [26] (more about the Kruglov property and its applications to the study of s.s.'s see Chap. 13 and [31, 59–62, 64, 87]).

Chapter 3
Rademacher System in Symmetric Spaces Located "close" to L_∞

In Chap. 2, we have proved that in the case when a s.s. X contains the separable part G of the Orlicz space L_{N_2}, $N_2(t) = e^{t^2} - 1$, the Rademacher system is equivalent in X to the unit vector basis in ℓ_2. It turns out that the behaviour of this system in spaces located "nearer" to L_∞ (in other words, when $X \not\supset G$) is more diverse and complicated. To clarify this, we formulate the main results of the previous chapter, using the notation introduced there: $E_X = \ell_2$ if and only if $X \supset G$ (Theorem 2.3); $E_X = \ell_1$ if and only if $X = L_\infty$ (Theorem 2.4). What sequence spaces else could be obtained in a similar way? In this chapter we give a fairly complete answer to this question: these are exactly the interpolation spaces between the spaces ℓ_1 and ℓ_2. Moreover, we establish a one-to-one correspondence between the above class of sequence spaces and the class of interpolation function spaces between L_∞ and G. The results obtained will allow us to identify those s.s.'s that contain Rademacher sums with sequences of coefficients belonging to a given sequence space. In the final part of the chapter we find necessary and sufficient conditions, under which the Rademacher subspace $\mathcal{R}(X)$ can be described as a certain cone of step-functions. A crucial role in the study of all these issues will be played by methods of the interpolation theory of operators.

3.1 A Description of Subspaces Generated by the Rademacher System

To state the main results of this section we need the notion of real $\mathcal{K}$-method of interpolation, by which one can describe all interpolation spaces with respect to every $\mathcal{K}$-monotone couple (see Definition D.1). Recall that $\mathcal{I}(X_0, X_1)$ is the set of all interpolation Banach spaces with respect to a Banach couple (X_0, X_1).

S. V. Astashkin, *The Rademacher System in Function Spaces*,
https://doi.org/10.1007/978-3-030-47890-2_3

The next result allows us to identify a s.s. $X \in \mathcal{I}(L_\infty, G)$ that contains exactly those Rademacher sums, whose sequences of coefficients belong to a given coordinate space $E \in \mathcal{I}(\ell_1, \ell_2)$.

Theorem 3.1 *Let $E \in \mathcal{I}(\ell_1, \ell_2)$. Then, there is a parameter F of the real $\mathcal{K}$-method such that*

$$E = (\ell_1, \ell_2)_F^{\mathcal{K}} \text{ and } \|a\|_E \asymp \|(\mathcal{K}(2^k, a; \ell_1, \ell_2))\|_F. \tag{3.1}$$

Moreover, if $X = (L_\infty, G)_F^{\mathcal{K}}$ with the norm $\|x\|_X = \|(\mathcal{K}(2^k, x; L_\infty, G))\|_F$, then we have

$$\Big\|\sum_{k=1}^{\infty} a_k r_k\Big\|_X \asymp \|a\|_E$$

with constants independent of a sequence $a = (a_k)_{k=1}^{\infty}$.

We shall obtain also the following characterization of the class $\mathcal{I}(\ell_1, \ell_2)$.

Theorem 3.2 *Let E be a Banach sequence space. Then $E \in \mathcal{I}(\ell_1, \ell_2)$ if and only if there exists a s.s. X on $[0, 1]$ such that $E = E_X$ (with equivalence of norms), where E_X consists of all sequences $a = (a_k)_{k=1}^{\infty}$ such that $\sum_{k=1}^{\infty} a_k r_k \in X$, and it is equipped with the norm*

$$\|(a_k)\|_{E_X} := \Big\|\sum_{k=1}^{\infty} a_k r_k\Big\|_X. \tag{3.2}$$

The proof of the above results will be based on the use of a special property of the certain subcouple of the Banach couple (L_∞, G) generated by the Rademacher system. We begin with the following definition.

Definition 3.1 Let (X_0, X_1) be a Banach couple, Y_0 (resp. Y_1) be a closed subspace of X_0 (resp. X_1). The Banach couple (Y_0, Y_1) is called a *$\mathcal{K}$-subcouple* (or *$\mathcal{K}$-closed subcouple*) of the couple (X_0, X_1) if

$$\mathcal{K}(t, y; Y_0, Y_1) \asymp \mathcal{K}(t, y; X_0, X_1)$$

with a constant independent of $y \in Y_0 + Y_1$ and $t > 0$.

Note that the inequality $\mathcal{K}(t, y; Y_0, Y_1) \geq \mathcal{K}(t, y; X_0, X_1)$ follows immediately from the definition of the $\mathcal{K}$-functional. Hence, (Y_0, Y_1) is a $\mathcal{K}$-subcouple of the couple (X_0, X_1) whenever there is $C > 0$ such that for all $y \in Y_0 + Y_1$ and $t > 0$ we have

$$\mathcal{K}(t, y; Y_0, Y_1) \leq C\mathcal{K}(t, y; X_0, X_1).$$

Define the operator R as follows: if a sequence $a = (a_k)_{k=1}^{\infty} \in \ell_2$, then

$$Ra(t) := \sum_{k=1}^{\infty} a_k r_k(t), \quad t \in [0, 1]. \tag{3.3}$$

Theorem 1.1 implies that, for a fixed $a \in \ell_2$, $Ra(t)$ is a measurable a.e. finite function. Moreover, by Theorem 2.2, R is a linear bounded operator from ℓ_2 into G satisfying

$$||Ra||_G \asymp \|a\|_{\ell_2}, \tag{3.4}$$

and, due to Remark 1.4, this operator acts from ℓ_1 into L_∞ isometrically, i.e.,

$$\|Ra\|_\infty = \|a\|_{\ell_1}. \tag{3.5}$$

Therefore, $R(\ell_2)$ (resp. $R(\ell_1)$) is a subspace of G (resp. L_∞), and so $(R(\ell_1), R(\ell_2))$ is a subcouple of the Banach couple (L_∞, G).

Taking into account the definition of the $\mathcal{K}$-functional, we conclude that from relations (3.4) and (3.5) it follows

$$\mathcal{K}(t, Ra; R(\ell_1), R(\ell_2)) \asymp \mathcal{K}(t, a; \ell_1, \ell_2) \tag{3.6}$$

with a constant independent of $a = (a_k)_{k=1}^{\infty} \in \ell_2$ and $t > 0$.

The proof of Theorems 3.1 and 3.2 will be based on using the following important result.

Theorem 3.3 *The Banach couple $(R(\ell_1), R(\ell_2))$ is a $\mathcal{K}$-subcouple of the Banach couple (L_∞, G), i.e., the equivalence*

$$\mathcal{K}(t, Ra; L_\infty, G) \asymp \mathcal{K}(t, a; \ell_1, \ell_2)$$

holds with a constant independent of $a = (a_k)_{k=1}^{\infty} \in \ell_2$ and $t > 0$.

Proof According to Proposition 2.4, the Orlicz space L_{N_2} coincides with the Marcinkiewicz space $M(\varphi_1)$, generated by the function $\varphi_1(u) = u\ln^{1/2}(e/u)$. Also, $L_\infty = M(\varphi_0)$, where $\varphi_0(u) = u$. Therefore, from (D.3) it follows for every $x \in G$

$$\begin{aligned} \mathcal{K}(t, x; L_\infty, G) &= \mathcal{K}(t, x; L_\infty, L_{N_2}) \\ &\asymp \sup_{0<u\leq 1} \left\{ \frac{1}{u} \int_0^u x^*(s)\, ds \cdot \min\left(1, t \cdot \ln^{-1/2}(e/u)\right) \right\}. \end{aligned} \tag{3.7}$$

Since $x^*(u) \le u^{-1} \int_0^u x^*(s)\,ds$, this equivalence implies

$$\mathcal{K}(t, x; L_\infty, G) \ge c \sup_{0<u\le 1} \left\{ x^*(u) \min\left(1, t \cdot \ln^{-1/2}(e/u)\right)\right\}. \tag{3.8}$$

Let $a = (a_k)_{k=1}^\infty \in \ell_2$. By Corollary 1.9 (see (1.37)), we have with a universal constant $\beta \in (0, 1)$

$$(Ra)^*(\beta u) \ge \beta \mathcal{K}(\ln^{1/2}(e/u), a; \ell_1, \ell_2), \quad 0 < u \le 1,$$

or

$$(Ra)^*(\beta e^{1-t^2}) \ge \beta \mathcal{K}(t, a; \ell_1, \ell_2), \quad t \ge 1. \tag{3.9}$$

On the other hand, by (3.8),

$$\mathcal{K}(t, Ra; L_\infty, G) \ge c(Ra)^*(\beta e^{1-t^2}) \min(1, t \cdot \ln^{-1/2}(\beta^{-1} e^{t^2})) \ge$$

$$\ge c(Ra)^*(\beta e^{1-t^2}) \min\left(1, \frac{t}{\sqrt{t^2 - \ln \beta}}\right).$$

Since the function $t/\sqrt{t^2 - \ln \beta}$ increases for $t > 0$, from this inequality it follows

$$\mathcal{K}(t, Ra; L_\infty, G) \ge c_1 (Ra)^*(\beta e^{1-t^2}), \quad t \ge 1,$$

where $c_1 := c/\sqrt{1 - \ln \beta}$. Therefore, due to inequality (3.9), we infer that

$$\mathcal{K}(t, Ra; L_\infty, G) \ge c_1 \beta \mathcal{K}(t, a; \ell_1, \ell_2), \quad t \ge 1.$$

If $0 < t \le 1$, then combining the inequalities $||a||_{\ell_2} \le ||a||_{\ell_1}$ and $||x||_G \le ||x||_\infty$ together with the definition of the $\mathcal{K}$-functional and equivalence (3.4), we have

$$\mathcal{K}(t, Ra; L_\infty, G) = t||Ra||_G \ge c_2 t||a||_{\ell_2} = c_2 \mathcal{K}(t, a; \ell_1, \ell_2).$$

Thus,

$$\mathcal{K}(t, Ra; L_\infty, G) \ge \min(c_2, c_1\beta)\mathcal{K}(t, a; \ell_1, \ell_2)$$

for all $t > 0$. The opposite inequality

$$\mathcal{K}(t, Ra; L_\infty, G) \le C\mathcal{K}(t, a; \ell_1, \ell_2), \quad t > 0,$$

follows from equivalence (3.6) and the definition of the $\mathcal{K}$-functional. □

Remark 3.1 The statement of Theorem 3.3 gets false if we replace the space G with L_q, $q < \infty$. Indeed, assuming that $(R(\ell_1), R(\ell_2))$ is a $\mathcal{K}$-subcouple of the couple (L_∞, L_q), by (3.6), we obtain

$$\mathcal{K}(t, a; \ell_1, \ell_2) \asymp \mathcal{K}(t, Ra; L_\infty, L_q),$$

where the equivalence constants do not depend on $a \in \ell_2$ and $t > 0$. Therefore, calculating the classical Lions–Peetre spaces for the couples (L_∞, L_q) and (ℓ_1, ℓ_2) with $\theta \in (0, 1)$ and $p = q/\theta$, by Theorem D.5, we obtain $(L_\infty, L_q)_{\theta,p} = L_p$ and $(\ell_1, \ell_2)_{\theta,p} = \ell_{r,p}$. Here, $r = 2/(2 - \theta)$ and $\ell_{r,p}$ is the sequence space equipped with the norm

$$\|a\|_{\ell_{r,p}} = \Big\{\sum_{k=1}^{\infty}(a_k^*)^p k^{p/r-1}\Big\}^{1/p}.$$

So, $\|Ra\|_p \asymp \|a\|_{\ell_{r,p}}$. On the other hand, from Theorem 1.4 it follows $\|Ra\|_p \asymp \|a\|_{\ell_2}$. One can easily check that $r < 2$ and hence the norms of the spaces ℓ_2 and $\ell_{r,p}$ fail to be equivalent to each other. This contradiction proves the claim.

Proof of Theorem 3.1 By the Sparr theorem [268], an arbitrary couple of weighted L_p-spaces over a σ-finite measure space is $\mathcal{K}$-monotone and so is the couple (ℓ_1, ℓ_2). Therefore, since $E \in \mathcal{I}(\ell_1, \ell_2))$, we have (3.1) for some parameter F of the real $\mathcal{K}$-method. It remains to apply Theorem 3.3 and the definition of the real $\mathcal{K}$-method of interpolation. □

Proof of Theorem 3.2 If $E \in \mathcal{I}(\ell_1, \ell_2)$, then we have (3.1) for some parameter of the real $\mathcal{K}$-method F. Hence, from Theorem 3.1 it follows $E = E_X$, where $X = (L_\infty, G)_F^{\mathcal{K}}$.

Conversely, suppose that $E = E_X$ for a s.s. X. Let T be a linear operator bounded in ℓ_1 and ℓ_2. We need to prove that T is bounded in E, i.e., for some constant $D > 0$ and all $a = (a_k)_{k=1}^{\infty} \in E$ we have

$$\Big\|\sum_{k=1}^{\infty}(Ta)_k r_k\Big\|_X \le D\|T\|\Big\|\sum_{k=1}^{\infty} a_k r_k\Big\|_X, \tag{3.10}$$

where $\|T\| := \max\big(\|T\|_{\ell_1\to\ell_1}, \|T\|_{\ell_2\to\ell_2}\big)$. Due to the positive homogeneity of inequality (3.10), we can assume that $\|T\| = 1$. Denote $x = \sum_{k=1}^{\infty} a_k r_k$ and $y = \sum_{k=1}^{\infty}(Ta)_k r_k$. From the definition of the $\mathcal{K}$-functional it follows that

$$\mathcal{K}(t, Ta; \ell_1, \ell_2) \le \mathcal{K}(t, a; \ell_1, \ell_2), \quad t > 0.$$

Indeed, if $a = b + c$, $b \in \ell_1$, $c \in \ell_2$, then for all $t > 0$

$$\mathcal{K}(t, Ta; \ell_1, \ell_2) \le \inf\{\|Tb\|_1 + t\|Tc\|_2 : a = b + c, b \in \ell_1, c \in \ell_2\} \le$$

$$\inf\{\|b\|_{\ell_1} + t\|c\|_{\ell_2} : a = b + c, b \in \ell_1, c \in \ell_2\} = \mathcal{K}(t, a; \ell_1, \ell_2).$$

Therefore, applying Corollary 1.9, we obtain the inequality

$$y^*(u) \le \beta^{-2} x^*(\beta u) = \beta^{-2} \sigma_{1/\beta} x^*(u), \quad 0 < u \le 1.$$

Since $\|\sigma_\tau\|_{X \to X} \le \max(1, \tau)$ (see Appendix C), we have $\|y\|_X \le \beta^{-3} \|x\|_X$. Thus, we get inequality (3.10) and so the proof is completed. □

3.2 Symmetric Spaces with Identical Rademacher Subspaces: The Interpolation Case

In this section, we examine properties of the mapping $X \mapsto E_X$ (see (3.2)) restricted to the class of all interpolation spaces X with respect to the couple (L_∞, G). We shall prove that, in contrast to the fact that the Rademacher system is incomplete in an arbitrary s.s, this mapping is one-to-one.

Theorem 3.4 *Let X_0 and X_1 be two interpolation s.s's between the spaces L_∞ and G. If*

$$\left\| \sum_{k=1}^{\infty} a_k r_k \right\|_{X_0} \asymp \left\| \sum_{k=1}^{\infty} a_k r_k \right\|_{X_1}, \tag{3.11}$$

then $X_0 = X_1$ (with equivalence of norms).

The proof of Theorem 3.4 will be based on a rather deep investigation of interpolation properties of the Banach couples (ℓ_1, ℓ_2) and (L_∞, G).

Denote by $\mathcal{K}\{X_0, X_1\}$ the set of all $\mathcal{K}$-functionals corresponding to a Banach couple (X_0, X_1). More explicitly, $\mathcal{K}\{X_0, X_1\}$ consists of all functions of the form $f(t) = \mathcal{K}(t, x; X_0, X_1)$, $x \in X_0 + X_1$. Our first goal is to prove the coincidence (with respect to equivalence) of the sets $\mathcal{K}\{\ell_1, \ell_2\}$ and $\mathcal{K}\{L_\infty, G\}$. To this end, let us introduce two cones of quasiconcave functions defined on the semi-axis $(0, \infty)$. Recall that a function $\varphi(t)$ defined for $t \ge 0$ is said to be *quasiconcave* whenever $\varphi(0) = 0$, $\varphi(t)$ increases, and $\varphi(t)/t$ decreases. In particular, one can check that any increasing concave function, which is equal to zero at zero (in particular, an arbitrary $\mathcal{K}$-functional $\mathcal{K}(t, x; X_0, X_1)$) is a quasiconcave function.

Let $\mathcal{P}_0$ be the set of all quasiconcave functions f such that

$$\lim_{t \to 0+} f(t) = \lim_{t \to \infty} \frac{f(t)}{t} = 0.$$

Also, denote by $\mathcal{P}_1$ the cone consisting of all $f \in \mathcal{P}_0$ such that $f(t) = f(1)t$, $0 < t \le 1$. The latter cone appears in a natural way when studying the set of $\mathcal{K}$-functionals generated by a Banach couple (X_0, X_1) satisfying the condition: $X_0 \overset{1}{\subset} X_1$.

Proposition 3.1 *Suppose X_0 and X_1 are two Banach spaces such that $X_0 \overset{1}{\subset} X_1$ and X_0 is dense in X_1. Then, $\mathcal{K}\{X_0, X_1\} \subset \mathcal{P}_1$.*

Proof Denote $f(t) := \mathcal{K}(t, x; X_0, X_1)$, where $x \in X_1 = X_0 + X_1$. Since X_0 is dense in X_1, for any $\varepsilon > 0$ there is a vector $x_0 \in X_0$ such that $x_1 := x - x_0$ satisfies the inequality: $\|x_1\|_{X_1} \le \varepsilon$. Choose t_0 so that $\|x_0\|_{X_0}/t \le \varepsilon$ for all $t \ge t_0$. Then we have $x = x_0 + x_1$ and

$$\frac{f(t)}{t} = \mathcal{K}\Big(\frac{1}{t}, x; X_1, X_0\Big) \le \|x_1\|_{X_1} + \frac{1}{t}\|x_0\|_{X_0} \le 2\varepsilon \ \text{ if } \ t \ge t_0.$$

Since $\varepsilon > 0$ is arbitrary, this implies $\lim_{t\to\infty} f(t)/t = 0$. Moreover, by the definition of the $\mathcal{K}$-functional, for each $x \in X_1$ we have

$$\mathcal{K}(t, x; X_0, X_1) = t\|x\|_{X_1} \ \text{ if } \ 0 < t \le 1.$$

Thus, $f \in \mathcal{P}_1$, and so everything is done. □

Since the couples (ℓ_1, ℓ_2) and (L_∞, G) satisfy the hypothesis of the latter proposition, we obtain

Corollary 3.1 *We have $\mathcal{K}\{\ell_1, \ell_2\} \subset \mathcal{P}_1$ and $\mathcal{K}\{L_\infty, G\} \subset \mathcal{P}_1$.*

Next, we aim to prove the inverse embeddings. Let us introduce the following useful notion.

Definition 3.2 Let $\mathcal{U}$ be a cone of quasiconcave functions. A Banach couple (X_0, X_1) is called *$\mathcal{U}$-abundant* if for any function $f \in \mathcal{U}$ there exists $x \in X_0 + X_1$ such that

$$\mathcal{K}(t, x; X_0, X_1) \asymp f(t), \quad t > 0.$$

In other words, the set $\mathcal{K}\{X_0, X_1\}$ of all $\mathcal{K}$-functionals of a $\mathcal{U}$-abundant couple (X_0, X_1) contains (up to equivalence) the cone $\mathcal{U}$.

To prove the $\mathcal{P}_1$-abundance of the couples (ℓ_1, ℓ_2) and (L_∞, G), we need to modify slightly the equivalent expression for the $\mathcal{K}$-functional $\mathcal{K}(t, f; L_1, L_2)$ from Proposition 1.5.

Lemma 3.1 *With a constant independent of $f \in L_1(0, \infty) + L_2(0, \infty)$ and $t > 0$ we have*

$$\mathcal{K}(t, f; L_1, L_2) \asymp t\Big(\int_{t^2}^{\infty} \Big(\frac{1}{s}\int_0^s f^*(u)\,du\Big)^2 ds\Big)^{1/2}. \tag{3.12}$$

In addition, for $t \geq 1$

$$\mathcal{K}(t, f; L_1, L_2) \asymp t\Big(\sum_{l=[t^2]}^{\infty} \frac{1}{l^2}\Big(\int_0^l f^*(u)\, du\Big)^2\Big)^{1/2}. \tag{3.13}$$

Proof Denote by $G(t, f)$ the right-hand side of equivalence (3.12) and by $F(t, f)$ the following expression equivalent to the $\mathcal{K}$-functional $\mathcal{K}(t, f; L_1, L_2)$:

$$F(t, f) := \int_0^{t^2} f^*(s)\, ds + t\Big(\int_{t^2}^{\infty} f^*(s)^2\, ds\Big)^{1/2}$$

(see inequality (1.20)). By Proposition 1.5, equivalence (3.12) will be proved once we show that $G(t, f) \asymp F(t, f)$, $t > 0$.

Since $f^*(s) \leq s^{-1} \int_0^s f^*(u)\, du$, we have

$$G(t, f) \geq t\Big(\int_{t^2}^{\infty} f^*(u)^2\, du\Big)^{1/2}.$$

Also,

$$G(t, f) \geq t\Big(\int_{t^2}^{\infty} \Big(\frac{1}{s} \int_0^{t^2} f^*(u)\, du\Big)^2 ds\Big)^{1/2} = \int_0^{t^2} f^*(u)\, du,$$

and therefore $2G(t, f) \geq F(t, f)$.

To get an inequality going in other way, we estimate

$$G(t, f) = t\Big(\int_{t^2}^{\infty} \Big(\frac{1}{s} \int_0^{t^2} f^*(u)\, du + \frac{1}{s} \int_{t^2}^{s} f^*(u)\, du\Big)^2 ds\Big)^{1/2} \leq$$

$$\leq t\Big(\int_{t^2}^{\infty} \Big(\frac{1}{s} \int_0^{t^2} f^*(u)\, du\Big)^2 ds\Big)^{1/2} + t\Big(\int_{t^2}^{\infty} \Big(\frac{1}{s} \int_{t^2}^{s} f^*(u)\, du\Big)^2 ds\Big)^{1/2} =$$

$$= \int_0^{t^2} f^*(u)\, du + t\Big(\int_{t^2}^{\infty} \Big(\frac{1}{s} \int_{t^2}^{s} f^*(u)\, du\Big)^2 ds\Big)^{1/2}.$$

Observe that

$$\Big(\int_{t^2}^{\infty} \Big(\frac{1}{s} \int_{t^2}^{s} f^*(u)\, du\Big)^2 ds\Big)^{1/2} = \Big(\int_0^{\infty} \Big(\frac{1}{s+t^2} \int_0^{s} f^*(u+t^2)\, du\Big)^2 ds\Big)^{1/2}$$

$$\leq \Big(\int_0^{\infty} \big(\mathcal{H} f^*(\cdot + t^2)(s)\big)^2 ds\Big)^{1/2},$$

where $\mathcal{H}$ is the Hardy–Littlewood operator defined by $\mathcal{H}f(s) = \frac{1}{s}\int_0^s f(u)\,du$. Hence, applying Theorem C.4, we conclude

$$G(t,f) \le \int_0^{t^2} f^*(u)\,du + 2t\Big(\int_0^{\infty} \big(f^*(s+t^2)\big)^2\,ds\Big)^{1/2}$$
$$= \int_0^{t^2} f^*(u)\,du + 2t\Big(\int_{t^2}^{\infty} (f^*(s))^2\,ds\Big)^{1/2} \le 2F(t,f),$$

and so (3.12) is proved.

The second equivalence (3.13) is an immediate consequence of (3.12). Indeed, we have

$$\mathcal{K}(m,f;L_1,L_2) \asymp m\Big(\sum_{l=m^2}^{\infty}\int_l^{l+1}\Big(\frac{1}{s}\int_0^s f^*(u)\,du\Big)^2\,ds\Big)^{1/2}, \quad m\in\mathbb{N},$$

and it is left only to use the concavity of the functions $\int_0^s f^*(u)\,du$ and $\mathcal{K}(s,f;L_1,L_2)$. □

Proposition 3.2 *The Banach couple $(L_1(0,\infty), L_2(0,\infty))$ is $\mathcal{P}_0$-abundant.*

Proof We shall make use of the following general result from the interpolation theory of operators (see [89, Theorem 4.5.7, p. 614]): a Banach couple (X_0,X_1) is $\mathcal{P}_0$-abundant whenever the set $\mathcal{K}\{X_0,X_1\}$ contains a function equivalent to the function t^θ for some $\theta\in(0,1)$. Meanwhile, by equivalence (3.12) and direct calculations, one can easily see that the function $f_0(u) := u^{-3/4}$ belongs to the space $L_1(0,\infty)+L_2(0,\infty)$ and $\mathcal{K}(t,f_0;L_1,L_2)\asymp t^{1/2}$, $t>0$. □

Corollary 3.2 *The Banach couples (ℓ_1,ℓ_2) and (L_∞,G) are $\mathcal{P}_1$-abundant.*

Proof We claim that the set $\mathcal{K}\{\ell_1,\ell_2\}$ contains (up to equivalence) the cone $\mathcal{P}_1$.

Let $f\in\mathcal{P}_1$. Since $\mathcal{P}_1\subset\mathcal{P}_0$, then by Proposition 3.2 there exists a function $y\in L_1(0,\infty)+L_2(0,\infty)$ such that

$$\mathcal{K}(t,y;L_1,L_2)\asymp f(t), \quad t>0. \tag{3.14}$$

Further, to each function $x\in L_1(0,\infty)+L_2(0,\infty)$ we associate the sequence

$$a(x) = (a_k(x))_{k=1}^{\infty}, \quad \text{where } a_k(x) := \int_{k-1}^{k} x(s)\,ds,$$

and denote by Q the averaging operator with respect to the system of intervals $\{(k-1,k]\}_{k=1}^{\infty}$, i.e.,

$$Qx(t) := \sum_{k=1}^{\infty} a_k(x)\,\chi_{(k-1,k]}(t), \quad t>0.$$

Since $\int_0^k Qx^*(u)\,du = \int_0^k x^*(u)\,du$ for each $k = 1.2, \ldots,$ then according to Lemma 3.1 (see (3.13)), we have

$$\mathcal{K}(t, Qx^*; L_1, L_2) \asymp \mathcal{K}(t, x^*; L_1, L_2) = \mathcal{K}(t, x; L_1, L_2), \quad t \geq 1.$$

Moreover, by Lemma 1.2,

$$\mathcal{K}(t, Qx^*; L_1, L_2) = \mathcal{K}(t, a(x^*); \ell_1, \ell_2), \quad t > 0.$$

Hence,

$$\mathcal{K}(t, x; L_1, L_2) \asymp \mathcal{K}(t, a(x^*); \ell_1, \ell_2), \quad t \geq 1.$$

In particular, the latter equivalence holds for a function y satisfying (3.14). Therefore, taking into account that $\|a(y)\|_{\ell_2} \leq \|y\|_2$, for every $f \in \mathcal{P}_1$ we can find a sequence $a \in \ell_2$ such that

$$\mathcal{K}(t, a; \ell_1, \ell_2) \asymp f(t), \quad t \geq 1.$$

This equivalence can be extended to all values $t > 0$, because in the case $0 < t \leq 1$ we have

$$\mathcal{K}(t, a; \ell_1, \ell_2) = t\|a\|_2 = t\mathcal{K}(1, a; \ell_1, \ell_2) \asymp tf(1) = f(t).$$

Thus, our claim is proved, and hence the couple (ℓ_1, ℓ_2) is $\mathcal{P}_1$-abundant. According to Theorem 3.3, the set $\mathcal{K}\{L_\infty, G\}$ contains (up to equivalence) the set $\mathcal{K}\{\ell_1, \ell_2\}$. Therefore, the same assertion holds also for the couple (L_∞, G). □

We now proceed with proving the $\mathcal{K}$-monotonicity of the couple (L_∞, G). One of the most common methods, allowing to reach that, is based on using the following notion.

Definition 3.3 A Banach couple $\vec{X} = (X_0, X_1)$ is called a *partial retract* of a couple $\vec{Y} = (Y_0, Y_1)$ if every vector $x \in X_0 + X_1$ is *orbitally equivalent* to some $y \in Y_0 + Y_1$. This means that there exist linear operators $U \in L(\vec{X}, \vec{Y})$ and $V \in L(\vec{Y}, \vec{X})$ such that $Ux = y$ and $Vy = x$.

Lemma 3.2 *If a Banach couple $\vec{X} = (X_0, X_1)$ is a partial retract of a $\mathcal{K}$-monotone Banach couple $\vec{Y} = (Y_0, Y_1)$, then the couple $\vec{X}$ is $\mathcal{K}$-monotone as well.*

Proof Suppose that $x_i \in X_0 + X_1$, $i = 1, 2$, and

$$\mathcal{K}(t, x_2; \vec{X}) \leq \mathcal{K}(t, x_1; \vec{X}), \quad t > 0. \tag{3.15}$$

By the hypothesis, there are vectors $y_i \in Y_0 + Y_1$, $i = 1, 2$, and linear operators $U_i \in L(\vec{X}, \vec{Y})$, $V_i \in L(\vec{Y}, \vec{X})$ satisfying $y_i = U_i x_i$ and $x_i = V_i y_i$, $i = 1, 2$. Then,

we have

$$\mathcal{K}(t, x_i; \vec{X}) = \mathcal{K}(t, V_i y_i; \vec{X}) \le \inf\{\|V_i z_0\|_{X_0} + t\|V_i z_1\|_{X_1} : y_i = z_0 + z_1\}$$
$$\le \|V_i\|_{L(\vec{Y},\vec{X})} \cdot \mathcal{K}(t, y_i; \vec{Y}), \quad i = 1, 2.$$

Similarly,

$$\mathcal{K}(t, y_i; \vec{Y}) \le \|U_i\|_{L(\vec{X},\vec{Y})} \cdot \mathcal{K}(t, x_i; \vec{X}), \quad i = 1, 2.$$

Therefore, from (3.15) it follows

$$\mathcal{K}(t, y_2; \vec{Y}) \le \|U_2\|\|V_1\|\mathcal{K}(t, y_1; \vec{Y}), \quad t > 0.$$

Since the couple $\vec{Y}$ is $\mathcal{K}$-monotone, there exists a linear operator $W \in L(\vec{Y}, \vec{Y})$ such that $y_2 = W y_1$ (see Appendix D). As a result, we obtain $x_2 = Sx_1$, where $S := V_2 W U_1 \in L(\vec{X}, \vec{X})$, and so the lemma is proved. □

Before stating the next auxiliary result, we introduce one more definition.

Definition 3.4 Let E and F be measurable subsets of the real line, $m(E) = m(F)$. We say that a mapping $\omega : E \to F$ is a *measure-preserving transformation* if for any measurable $C \subset F$ its inverse image $\omega^{-1}(C)$ is a measurable set such that $m(\omega^{-1}(C)) = m(C)$.

Lemma 3.3 *Let functions $x(t)$ and $y(t)$, $0 \le t \le 1$, be equimeasurable, i.e.,*

$$m\{t \in [0, 1] : |x(t)| > z\} = m\{t \in [0, 1] : |y(t)| > z\}, \quad z > 0.$$

Then, for arbitrary $\varepsilon > 0$, we can find a linear operator T such that $y = Tx$ and $\|T\|_{X \to X} \le 1 + \varepsilon$ for every s.s. X.

Proof First, we show that for any equimeasurable on $[0, 1]$ functions $u(t) \ge 0$, $v(t) \ge 0$ and each $\delta > 0$ there exists a measure-preserving transformation $\tau : [0, 1] \to [0, 1]$ such that

$$\|v - u_\tau\|_\infty \le \delta, \quad \text{where } u_\tau(t) = u(\tau(t)). \tag{3.16}$$

To this end, let us consider the sets

$$U_k := \{t \in [0, 1] : \delta k < u(t) \le \delta(k + 1)\}$$

and

$$V_k := \{t \in [0, 1] : \delta k < v(t) \le \delta(k + 1)\}, \quad k = 0, 1, \ldots$$

Since $u(t)$ and $v(t)$ are equimeasurable, then $m(U_k) = m(V_k)$ and hence there is a measure-preserving transformation $\tau_k : V_k \to U_k$ for each $k = 0, 1, \dots$ Also, denote

$$U_{-1} := [0, 1] \setminus \bigcup_{k=0}^{\infty} U_k \quad \text{and} \quad V_{-1} = [0, 1] \setminus \bigcup_{k=0}^{\infty} V_k.$$

Clearly, $u(t) = 0$ (resp. $v(t) = 0$) if $t \in U_{-1}$ (resp. $t \in V_{-1}$), and $m(U_{-1}) = m(V_{-1})$. We take for $\tau_{-1} : V_{-1} \to U_{-1}$ any measure-preserving transformation. Further, define $\tau : [0, 1] \to [0, 1]$ by "gluing" the mappings τ_k, $k = -1, 0, 1, \dots$, that is, by setting $\tau(t) = \tau_k(t)$ if $t \in V_k$, $k = -1, 0, 1, \dots$ One can easily check that τ is a measure-preserving transformation such that inequality (3.16) holds.

Furthermore, introduce the notation:

$$E_0 = \{t : |x(t)| > 1\} \quad \text{and} \quad E_n = \left\{t : \frac{1}{n+1} < |x(t)| \le \frac{1}{n}\right\}, \quad n \in \mathbb{N}.$$

Let us denote also by F_n, $n = 0, 1, \dots$, similar sets for the function y. Then, due to the assumption, we have $m(E_n) = m(F_n)$ for all n. Moreover, clearly the functions $x\chi_{E_n}$ and $y\chi_{F_n}$ are equimeasurable for any $n = 0, 1, \dots$. Therefore, as was proved above, for arbitrary $\varepsilon \in (0, 1)$ and $n = 0, 1 \dots$ there exists a measure-preserving transformation $\omega_n : F_n \to E_n$ such that

$$\| |y|\chi_{F_n} - (|x|\chi_{E_n})(\omega_n) \|_\infty \le \frac{\varepsilon}{n+1}. \tag{3.17}$$

Denote $z_n := |y|\chi_{F_n} - (|x|\chi_{E_n})(\omega_n)$, $\alpha_n(t) := z_n(t)/(|x|\chi_{E_n})(\omega_n(t))$ for $t \in F_n$ and $\alpha_n(t) = 0$ for $t \notin F_n$. From inequality (3.17) and the definition of the sets E_n it follows immediately that

$$\|\alpha_n\|_\infty \le \varepsilon, \quad n = 0, 1, 2, \dots \tag{3.18}$$

Moreover,

$$|y|\chi_{F_n} = (|x|\chi_{E_n})(\omega_n) + z_n = (1 + \alpha_n)(|x|\chi_{E_n})(\omega_n), \quad n = 0, 1, \dots \tag{3.19}$$

Further, setting

$$E_{-1} := [0, 1] \setminus \bigcup_{n=0}^{\infty} E_n \quad \text{and} \quad F_{-1} := [0, 1] \setminus \bigcup_{n=0}^{\infty} F_n,$$

we have $x(t) = 0$ (resp. $y(t) = 0$) if $t \in E_{-1}$ (resp. $t \in F_{-1}$). Since $m(E_{-1}) = m(F_{-1})$, then there is a measure-preserving transformation $\omega_{-1} : F_{-1} \to E_{-1}$.

Now, let $\omega : [0, 1] \to [0, 1]$ and $\alpha : [0, 1] \to \mathbb{R}$ be obtained again by "gluing" the mappings ω_n and α_n, respectively, i.e., $\omega(t) = \omega_n(t)$ and $\alpha(t) = \alpha_n(t)$

if $t \in F_n$, $n = -1, 0, 1, \ldots$ (α_{-1} is defined on the set F_{-1} in an arbitrary way). Then, by (3.19), we have

$$|y(t)| = (1 + \alpha(t))|x(\omega(t))|, \quad t \in [0, 1]. \tag{3.20}$$

One can easily check that ω is a measure-preserving transformation. Moreover, from inequality (3.18) it follows $\|\alpha\|_\infty \le \varepsilon$.

Show that the linear operator

$$Tf(t) = \operatorname{sign} x(\omega(t)) \operatorname{sign} y(t)(1 + \alpha(t)) f(\omega(t))$$

possesses all the required properties. First, by the definition of ω and α, we have

$$\begin{aligned} m\{t \in [0, 1] : |Tf(t)| > z\} &\le m\{t \in [0, 1] : (1 + \varepsilon)|f(\omega(t))| > z\} \\ &= m\{t \in [0, 1] : (1 + \varepsilon)|f(t)| > z\}, \quad z > 0. \end{aligned}$$

Therefore, $(Tf)^*(t) \le (1 + \varepsilon) f^*(t)$, $t \in [0, 1]$, whence we immediately deduce $\|Tf\|_X \le (1 + \varepsilon)\|f\|_X$ for every s.s. X. Finally, from (3.20) and the definition of T it follows $y = Tx$. □

Remark 3.2 Comparing to this lemma, the Calderón–Mityagin Theorem D.6 implies a somewhat weaker result: If functions x and y are equimeasurable on $[0, 1]$, then there exists a linear operator T, with $y = Tx$, that is bounded on every interpolation s.s. between L_1 and L_∞.

Remark 3.3 Lemma 3.3 cannot be strengthened as follows: for an arbitrary pair of equimeasurable on $[0, 1]$ functions $x(t)$ and $y(t)$ there exists a measure-preserving transformation $\omega : [0, 1] \to [0, 1]$ such that $|y(t)| = |x(\omega(t))|$. Indeed, let $z(t) = 1 - 2t \pmod 1$. Then, $z^*(t) = 1 - t$, $0 < t < 1$. Moreover, it can be easily checked that the transformation $\sigma(t) = 2t \pmod 1$ preserves measure and $z(t) = z^*(\sigma(t))$ a.e. on $[0, 1]$. At the same time, there is no measure-preserving transformation $\omega : [0, 1] \to [0, 1]$ such that $z^*(t) = z(\omega(t))$ a.e. [70, Example 2.7.7]. The same example shows that the statement from the beginning of this remark is still false even if we consider instead of measure-preserving transformations a wider class of mappings satisfying the inequality $m(\omega^{-1}(E)) \le m(E)$ for each measurable set $E \subset [0, 1]$.

The $\mathcal{K}$-monotonicity of the couple (L_∞, G) will be a consequence of a more general result. Recall (see Appendix C) that the dilation exponents of a function f on $[0, 1]$ are defined as follows:

$$\gamma_f = \lim_{t \to 0+} \frac{\ln \mathcal{M}_f(t)}{\ln t} \quad \text{and} \quad \delta_f = \lim_{t \to \infty} \frac{\ln \mathcal{M}_f(t)}{\ln t},$$

where

$$\mathcal{M}_f(t) = \sup\left\{\frac{f(st)}{f(s)} : 0 < s \le \min(1, t^{-1})\right\}, \quad t > 0.$$

We have always $0 \le \gamma_f \le \delta_f \le 1$.

Proposition 3.3 *If $M(\varphi)$ is the Marcinkiewicz space on $[0, 1]$ generated by a quasiconcave function φ, whose lower dilation exponent γ_φ is strictly positive, then the Banach couple $\vec{X} = (L_\infty, M(\varphi))$ is $\mathcal{K}$-monotone.*

Proof It is sufficient to show that $\vec{X}$ is a partial retract of the Banach couple $\vec{Y} = (L_\infty, L_\infty(\tilde{\varphi}))$, where $L_\infty(\tilde{\varphi})$ is the weighted L_∞-space, $\widetilde{\varphi}(t) := t/\varphi(t)$, equipped with the natural norm

$$\|x\|_{L_\infty(\tilde{\varphi})} = \sup_{0<t\le 1} \tilde{\varphi}(t)|x(t)|.$$

Indeed, taking into account the above-mentioned Sparr theorem [268], we see that the couple $\vec{Y}$ is $\mathcal{K}$-monotone, and therefore, applying Lemma 3.2, we get at once the desired result.

Let now $x \in M(\varphi)$ (since $L_\infty \subset M(\varphi)$, we have $L_\infty + M(\varphi) = M(\varphi)$). By Lemma 3.3, we can find a linear operator T_1, bounded in the couple $\vec{X}$, such that $x^* = T_1 x$. For $y \in M(\varphi)$ we set

$$U_1 y(t) := \sum_{k=1}^{\infty} 2^k \int_0^{2^{-k}} y(s)ds \cdot \chi_{(2^{-k}, 2^{-k+1}]}(t).$$

It follows easily from the definition of the decreasing rearrangement of a measurable function (see also [168, § II.2, Property 6°, p. 64]) that $\int_0^t |y(s)|\, ds \le \int_0^t y^*(s)\, ds$ for all $0 < t \le 1$. Therefore, due to the quasiconcavity of φ, we have

$$||U_1 y||_{L_\infty(\tilde{\varphi})} \le 2 \sup_{k=1,2,\ldots} (\varphi(2^{-k+1}))^{-1} \int_0^{2^{-k}} y^*(s)ds \le 2||y||_{M(\varphi)},$$

whence $U_1 \in L(\vec{X}, \vec{Y})$. Further, since $x^*(t)$ is decreasing, it follows $U_1 x^*(t) \ge x^*(t)$. Hence, the linear operator

$$U y(t) := \frac{x^*(t)}{U_1 x^*(t)} U_1 T_1 y(t)$$

is also bounded from the couple $\vec{X}$ into $\vec{Y}$ and $Ux(t) = x^*(t)$.

On the other hand, since $\gamma_\varphi > 0$, according to Theorem C.2, we get

$$||y||_{M(\varphi)} \le C \sup_{0<t\le 1} \tilde{\varphi}(t) y^*(t).$$

Therefore, as $\tilde{\varphi}$ increases and, vice versa, y^* decreases, we infer that

$$||y||_{M(\varphi)} \le C \sup_{0<t\le 1} \tilde{\varphi}(t)|y(t)| = C||y||_{L_\infty(\tilde{\varphi})}.$$

Thus, the identity operator I is bounded from the couple $\vec{Y}$ in $\vec{X}$. Applying Lemma 3.3 once more, we can find a linear operator T_2, bounded in the couple $\vec{X}$, such that $x = T_2 x^*$. Then, the operator $V := T_2 I$ is bounded from $\vec{Y}$ in $\vec{X}$ and $Vx^* = x$.

Summing up, we see that an arbitrary function $x \in M(\varphi)$ is orbitally equivalent in the couple $\vec{X}$ to its rearrangement x^* in $\vec{Y}$. Hence, the first couple is a partial retract of the second one. □

Corollary 3.3 *Any Banach couple of the form $(L_\infty, M(\varphi)^\circ)$ is $\mathcal{K}$-monotone whenever $\gamma_\varphi > 0$. In particular, the couple (L_∞, G) is $\mathcal{K}$-monotone.*

Proof First of all, we claim that the space $M(\varphi)^\circ$ has the interpolation property with respect to the couple $(L_\infty, M(\varphi))$. Indeed, let A be a linear operator, $A : L_\infty \to L_\infty$ and $A : M(\varphi) \to M(\varphi)$. For each $x \in M(\varphi)^\circ$ there is a sequence $\{x_n\} \subset L_\infty$ such that $x_n \to x$ in the space $M(\varphi)$. Clearly, $\{Ax_n\} \subset L_\infty$ and $Ax_n \to Ax$ in $M(\varphi)$. Therefore, $Ax \in M(\varphi)^\circ$, whence $A : M(\varphi)^\circ \to M(\varphi)^\circ$.

Assume that $x, y \in M(\varphi)^\circ$ and

$$\mathcal{K}(t, y; L_\infty, M(\varphi)^\circ) \le \mathcal{K}(t, x; L_\infty, M(\varphi)^\circ), \quad t > 0.$$

Combining this inequality with the fact that for any function $z \in M(\varphi)^\circ$

$$\mathcal{K}(t, z; L_\infty, M(\varphi)^\circ) = \mathcal{K}(t, z; L_\infty, M(\varphi)), \quad t > 0,$$

we get

$$\mathcal{K}(t, y; L_\infty, M(\varphi)) \le \mathcal{K}(t, x; L_\infty, M(\varphi)), \quad t > 0.$$

Therefore, by Proposition 3.3, there exists a linear operator S, bounded in the couple $(L_\infty, M(\varphi))$, such that $y = Sx$. Applying the claim from the beginning of the proof, we infer that S is bounded in the couple $(L_\infty, M(\varphi)^\circ)$, and hence the first assertion is proved.

Further, recall that the Orlicz space L_{N_2} coincides with the Marcinkiewicz space $M(\psi_2)$, where $\psi_2(u) = u \ln^{1/2}(e/u)$ (see Proposition 2.4). Since $\gamma_{\psi_2} = 1$, this observation completes the proof of the second assertion of the corollary. □

Now we are ready to prove the main result of this section.

Proof of Theorem 3.4 Since the Banach couple (L_∞, G) is $\mathcal{K}$-monotone, due to the Brudnyi–Kruglyak Theorem D.7, there exist parameters of the real $\mathcal{K}$-method F_0 and F_1 such that

$$X_0 = (L_\infty, G)^{\mathcal{K}}_{F_0} \text{ and } X_1 = (L_\infty, G)^{\mathcal{K}}_{F_1}. \tag{3.21}$$

Therefore, by Theorem 3.1, we have

$$\Big\| \sum_{k=1}^{\infty} a_k r_k \Big\|_{X_i} \asymp \|(a_k)\|_{E_i}, \quad \text{where } E_i = (\ell_1, \ell_2)^{\mathcal{K}}_{F_i}, \ i = 0, 1.$$

Combining these relations together with condition (3.11), we deduce that

$$(\ell_1, \ell_2)^{\mathcal{K}}_{F_0} = (\ell_1, \ell_2)^{\mathcal{K}}_{F_1}. \tag{3.22}$$

This means that the F_0- and F_1-norms are equivalent on the set $\mathcal{K}\{\ell_1, \ell_2\}$. Taking into account that

$$\mathcal{K}\{\ell_1, \ell_2\} = \mathcal{K}\{L_\infty, G\} = \mathcal{P}_1, \tag{3.23}$$

(see Corollaries 3.1 and 3.2), we conclude that these norms are equivalent also on the set $\mathcal{K}\{L_\infty, G\}$.

Let $x \in X_0$ be arbitrary. From (3.21) it follows $(\mathcal{K}(2^k, x; L_\infty, G))_k \in F_0$. Then, (3.23) assures the existence of a sequence $a \in \ell_2$ such that

$$\mathcal{K}(t, a; \ell_1, \ell_2) \asymp \mathcal{K}(t, x; L_\infty, G), \quad t > 0. \tag{3.24}$$

Since each parameter of the $\mathcal{K}$-method is a Banach lattice, this implies $(\mathcal{K}(2^k, a; \ell_1, \ell_2))_k \in F_0$, and hence, according to (3.22), we have $(\mathcal{K}(2^k, a; \ell_1, \ell_2))_k \in F_1$. Therefore (see (3.24)), $(\mathcal{K}(2^k, x; L_\infty, G))_k \in F_1$, or $x \in X_1$. As a result, we get $X_0 \subset X_1$. By changing places of X_0 and X_1, we obtain that $X_1 \subset X_0$, i.e., $X_0 = X_1$ as sets. Since every s.s. on [0, 1] is embedded continuously into $L_1[0, 1]$ (see Appendix C), then by the Closed Graph theorem the norms of Banach spaces X_0 and X_1 are equivalent, and so Theorem 3.4 is proved. □

The following simple example: $X_0 = G$ and $X_1 = L_{N_2}$, $N_2(t) = e^{t^2} - 1$, exhibits that the interpolation property of the spaces X_0 and X_1 with respect to the couple (L_∞, G) in Theorem 3.4 cannot be dropped. Indeed, as it is known, $E_G = E_{L_{N_2}} = \ell_2$, but $G \neq L_{N_2}$ (for instance, $\ln^{1/2}(e/t) \in L_{N_2} \setminus G$). In the next section, we show that this interpolation condition is exact in a much stronger sense.

3.3 Symmetric Spaces with Identical Rademacher Subspaces: The General Case

Let X be an arbitrary s.s. on $[0, 1]$. Then, according to Theorem 3.2, the sequence space E_X, with the norm defined by (3.2), has the interpolation property with respect to the couple (ℓ_1, ℓ_2). Therefore, from the Sparr theorem [268] it follows that $E_X = (\ell_1, \ell_2)_F^{\mathcal{K}}$ for some parameter of the $\mathcal{K}$-method F and so, by Theorem 3.1, we have

$$\Big\| \sum_{k=1}^{\infty} a_k r_k \Big\|_{\widetilde{X}} \asymp \|(a_k)\|_{E_X},$$

where $\widetilde{X} := (L_\infty, G)_F^{\mathcal{K}}$. Hence, $E_{\widetilde{X}} = E_X$. In particular, assuming that $X \notin \mathcal{I}(L_\infty, G)$, obviously, we conclude that $\widetilde{X} \neq X$. Thus, to every s.s. X can be assigned a unique (by Theorem 3.4) s.s. $\widetilde{X} \in \mathcal{I}(L_\infty, G)$ such that $E_{\widetilde{X}} = E_X$. Moreover, $X = \widetilde{X}$ if and only if $X \in \mathcal{I}(L_\infty, G)$.

Next, we intend to recover a more explicit connection between spaces X and $\widetilde{X}$, which will imply, in particular, the fact that $\widetilde{X} \subset X$. To this end, we introduce the operator S defined on the exponential Orlicz space L_{N_2} as follows:

$$Sx(t) := \ln^{1/2}(e/t) \sup_{0<u\le t} \big(x^*(u)\ln^{-1/2}(e/u)\big), \quad 0 < t \le 1. \tag{3.25}$$

Theorem 3.5 *For every s.s. X we have $\|x\|_{\widetilde{X}} \asymp \|Sx\|_X$.*

Firstly, we obtain an equivalent expression for the $\mathcal{K}$-functional of the couple (L_∞, L_{N_α}) (here, as above, $N_\alpha(t) = \exp(t^\alpha) - 1$), which will be used repeatedly further.

Lemma 3.4 *Let $\alpha \ge 1$. With a constant independent of $x \in L_{N_\alpha}$ and $t > 0$, we have*

$$\mathcal{K}(t, x; L_\infty, L_{N_\alpha}) \asymp t \sup_{0<u\le \min(1, e^{1-t^\alpha})} \big(x^*(u)\ln^{-1/\alpha}(e/u)\big). \tag{3.26}$$

Proof Recall that $L_{N_\alpha} = M(\psi_\alpha)$, where $\psi_\alpha(u) \asymp u\ln^{-1/\alpha}(e/u)$ (see Proposition 2.4). Therefore, applying once more the equivalent expression (D.3) for the $\mathcal{K}$-functional of a couple of Marcinkiewicz spaces, we get

$$\mathcal{K}(t, x; L_\infty, L_{N_\alpha}) \asymp \|x\|_{M(\psi_t)}, \quad \text{where } \psi_t(u) = u\max\{1, t^{-1}\ln^{1/\alpha}(e/u)\}. \tag{3.27}$$

To prove the lemma it suffices to show that $\sup_{t>0} \mathcal{M}_{\psi_t}(1/2) < 1$, where $\mathcal{M}_{\psi_t}(s)$ is the dilation function of ψ_t. Indeed, applying the second assertion of Theorem C.2, we obtain

$$\begin{aligned}\|x\|_{M(\psi_t)} &\asymp \sup_{0<u\le 1}\left\{x^*(u)\min\left(1, t\cdot \ln^{-1/\alpha}(e/u)\right)\right\}\\ &= t \sup_{0<u\le \min(1,e^{1-t^\alpha})}\left(x^*(u)\ln^{-1/\alpha}(e/u)\right),\end{aligned}$$

with a constant depending only on α. As a result, (3.26) follows from (3.27).

In order to estimate $\mathcal{M}_{\psi_t}(1/2)$, we write:

$$\mathcal{M}_{\psi_t}\left(\frac{1}{2}\right) = \frac{1}{2}\sup_{0<u\le 1} F_t(u), \quad \text{where } F_t(u) = \frac{\max(1, t^{-1}\ln^{1/\alpha}(2e/u))}{\max(1, t^{-1}\ln^{1/\alpha}(e/u))}.$$

Since $\alpha \ge 1$, then $\ln^{1/\alpha}(2e/u) \le (1+\ln 2)\ln^{1/\alpha}(e/u)$, $0 < u \le 1$. Therefore, $F_t(u) \le 1 + \ln 2$ for every $t > 0$, whence $\sup_{t>0}\mathcal{M}_{\psi_t}(1/2) \le (1+\ln 2)/2 < 1$.

□

Remark 3.4 Further, for technical reasons, we need the following simple modification of equivalence (3.26).

It is easy to see that for any $1 < a < b < \infty$ we have

$$\frac{\log_a(a/u)}{\log_a b} \le \log_b(b/u) \le \log_a(a/u), \quad 0 < u \le 1.$$

Therefore, from Proposition 2.4 it follows

$$\|x\|_{L_{N_\alpha}} \asymp \sup_{0<u\le 1}\left(x^*(u)\ln^{-1/\alpha}(e/u)\right) \asymp \sup_{0<u\le 1}\left(x^*(u)\log_b^{-1/\alpha}(b/u)\right).$$

Thus, in the same way, as in the proof of the last lemma, one can show that for each $b > 1$

$$\mathcal{K}(t, x; L_\infty, L_{N_\alpha}) \asymp t \sup_{0<u\le \min(1,b^{1-t^\alpha})}\left(x^*(u)\log_b^{-1/\alpha}(b/u)\right) \tag{3.28}$$

(with a constant depending on b and α).

Proof of Theorem 3.5 Recall that E_X is the space equipped with the norm $\|a\|_{E_X} = \|\sum_{k=1}^\infty a_k r_k\|_X$. Observe firstly that from the estimates, proved in Corollary 1.9, it follows

$$\|a\|_{E_X} \asymp \left\|\mathcal{K}\left(\ln^{1/2}(e/t), a, \ell_1, \ell_2\right)\right\|_X. \tag{3.29}$$

Let $E_X = (\ell_1, \ell_2)_E^{\mathcal{K}}$ and $\widetilde{X} = (L_\infty, G)_F^{\mathcal{K}}$ for some parameter of the real $\mathcal{K}$-method F. If $x \in \widetilde{X}$, then, by the definition of the real $\mathcal{K}$-method, we have

$\mathcal{K}(t, x, L_\infty, G) \in F$. Moreover, according to Corollaries 3.1 and 3.2, the sets of $\mathcal{K}$-functionals $\mathcal{K}\{L_\infty, G\}$ and $\mathcal{K}\{\ell_1, \ell_2\}$ coincide (up to equivalence). Therefore, there exists a sequence $a \in \ell_2$ such that

$$\mathcal{K}(t, x, L_\infty, G) \asymp \mathcal{K}(t, a, \ell_1, \ell_2), \quad t > 0. \tag{3.30}$$

Hence, $a \in E_X$ and so, by (3.29), $\mathcal{K}\big(\ln^{1/2}(e/t), a, \ell_1, \ell_2\big) \in X$. As a result, from (3.30)) it follows $\mathcal{K}\big(\ln^{1/2}(e/t), x, L_\infty, G\big) \in X$.

A quite similar reasoning shows that the converse is also holds, i.e., the assumption $\mathcal{K}\big(\ln^{1/2}(e/t), x, L_\infty, G\big) \in X$ implies $x \in \widetilde{X}$. Thus, the s.s. $\widetilde{X}$ and the (Banach) space with the norm $\big\|\mathcal{K}\big(\ln^{1/2}(e/t), x, L_\infty, G\big)\big\|_X$ coincide as the sets. Hence, by the Closed Graph theorem,

$$\|x\|_{\widetilde{X}} \asymp \big\|\mathcal{K}\big(\ln^{1/2}(e/t), x, L_\infty, G\big)\big\|_X.$$

Moreover, for each function $x \in G$ we have

$$\mathcal{K}\left(t, x, L_\infty, G\right) = \mathcal{K}\left(t, x, L_\infty, L_{N_2}\right), \quad t > 0, \tag{3.31}$$

and so, by Lemma 3.4,

$$\mathcal{K}\big(\ln^{1/2}(e/t), x, L_\infty, L_{N_2}\big) \asymp \ln^{1/2}(e/t) \sup_{0<u\le t} \big(x^*(u)\ln^{-1/2}(e/u)\big) = Sx(t).$$

Combining the last relations, we complete the proof. □

From Theorem 3.5 it follows immediately the monotonicity of the mapping $X \mapsto \widetilde{X}$.

Corollary 3.4 *For arbitrary s.s.'s X and Y, satisfying $X \subset Y$, we have $\widetilde{X} \subset \widetilde{Y}$.*

One can easily see that $Sx(t) \ge x^*(t)$. Therefore, we obtain

Corollary 3.5 *For every s.s. X we have $\widetilde{X} \subset X$ and $\|x\|_X \le C\|x\|_{\widetilde{X}}$ for some $C > 0$ and each $x \in X$. In particular, $\phi_X(t) \le C\phi_{\widetilde{X}}(t)$, $t \in [0, 1]$.*

These results indicate a special role of the class $\mathcal{I}(L_\infty, G)$ of interpolation s.s.'s with respect to the couple (L_∞, G) in studying the Rademacher system. In the next proposition, we give a characterization of Lorentz and Marcinkiewicz spaces that belong to this class.

Proposition 3.4 *Let ψ be an increasing convex function on $[0, 1]$, $\psi(0) = 0$, and let $\widetilde{\psi}(t) = t/\psi(t)$. Then*

(a) A Lorentz space $\Lambda(\psi) \in \mathcal{I}(L_\infty, G)$ if and only if $\phi_{\widetilde{\Lambda(\psi)}}(t) \le C\psi(t)$ for some $C > 0$ and all $0 \le t \le 1$.
(b) A Marcinkiewicz space $M(\psi) \in \mathcal{I}(L_\infty, L_{N_2})$ if and only if $\phi_{\widetilde{M(\psi)}}(t) \le C\widetilde{\psi}(t)$ for some $C > 0$ and all $0 \le t \le 1$.

Proof

(a) If $\Lambda(\psi) \in \mathcal{I}(L_\infty, G)$, we have $\widetilde{\Lambda(\psi)} = \Lambda(\psi)$. Since ψ is the fundamental function of $\Lambda(\psi)$, hence, it follows that the functions $\phi_{\widetilde{\Lambda(\psi)}}$ and ψ are equivalent.

Conversely, assume that $\phi_{\widetilde{\Lambda(\psi)}}(t) \leq C\psi(t)$. Then, by Theorem 3.5, the operator S, defined by (3.25), is bounded on the set of characteristic functions with respect to the norm of $\Lambda(\psi)$. Though S is not subadditive, in general (precisely as the decreasing rearrangement of a measurable function), however the proof of Theorem 3.5 shows that, for every s.s. X, the functional $\|Sx\|_X$ is equivalent to the $\widetilde{X}$-norm. Therefore, by Theorem C.1, S is bounded on the whole space $\Lambda(\psi)$, i.e., $\|Sx\|_{\Lambda(\psi)} \leq C\|x\|_{\Lambda(\psi)}$ for some $C > 0$. Finally, from Theorem 3.5 and Corollary 3.5 it follows that $\widetilde{\Lambda(\psi)} = \Lambda(\psi)$, whence $\Lambda(\psi) \in \mathcal{I}(L_\infty, G)$.

(b) Let now $M(\psi) \in \mathcal{I}(L_\infty, L_{N_2})$. We claim that $M(\psi) \cap G \in \mathcal{I}(L_\infty, G)$. Indeed, since the Banach couple (L_∞, L_{N_2}) is $\mathcal{K}$-monotone (see Proposition 3.3), we have $M(\psi) = (L_\infty, L_{N_2})_F^{\mathcal{K}}$ for some parameter F. This and equation (3.31) give us $M(\psi) \cap G = (L_\infty, G)_F^{\mathcal{K}}$, which proves the claim.

Since $M(\psi) \cap G \in \mathcal{I}(L_\infty, G)$, we have $M(\psi) \cap G = \widetilde{M(\psi) \cap G}$. Moreover, from the embedding $M(\psi) \subset L_{N_2}$ it follows that the fundamental functions of the spaces $M(\psi)$ and $M(\psi) \cap G$ are equivalent, and

$$\left\|\sum_{k=1}^{\infty} a_k r_k\right\|_{M(\psi)} \asymp \left\|\sum_{k=1}^{\infty} a_k r_k\right\|_{M(\psi)\cap G}.$$

Hence, by Theorem 3.4, $\widetilde{M(\psi)} = \widetilde{M(\psi) \cap G}$, and consequently

$$\phi_{\widetilde{M(\psi)}}(t) \asymp \phi_{\widetilde{M(\psi)\cap G}}(t) \asymp \phi_{M(\psi)\cap G}(t) \asymp \phi_{M(\psi)}(t) = \widetilde{\psi}(t), \quad 0 < t \leq 1.$$

Let us prove the converse. Suppose that $\phi_{\widetilde{M(\psi)}}(t) \leq C\widetilde{\psi}(t)$, $0 < t \leq 1$. Then, applying Theorem 3.5, we obtain for some $C > 0$

$$\|S\chi_{(0,\tau)}\|_{M(\psi)} \leq C\widetilde{\psi}(\tau), \quad 0 < \tau \leq 1. \tag{3.32}$$

Furthermore, an easy calculation gives

$$S\chi_{(0,\tau)}(t) = \begin{cases} 1, & 0 < t \leq \tau \\ \left(\dfrac{\ln(e/t)}{\ln(e/\tau)}\right)^{1/2}, & \tau \leq t \leq 1, \end{cases}$$

for each $\tau \in (0, 1]$. Therefore, the function $S\chi_{(0,\tau)}$ decreases and from the definition of a Marcinkiewicz norm it follows

$$\|S\chi_{(0,\tau)}\|_{M(\psi)} \geq \sup_{0<t\leq 1} \big(\widetilde{\psi}(t)S\chi_{(0,\tau)}\big) \geq \big(\ln(e/\tau)\big)^{-1/2} \sup_{\tau\leq t\leq 1} \big(\widetilde{\psi}(t)\ln^{1/2}(e/t)\big).$$

Combining this together with inequality (3.32), we get

$$\frac{\widetilde{\psi}(t)}{\widetilde{\psi}(\tau)} \leq C_2 \left(\frac{\ln(e/\tau)}{\ln(e/t)}\right)^{1/2}, \quad 0 < \tau \leq t \leq 1,$$

or

$$\frac{\psi(\tau)}{\psi(t)} \leq C_2 \max\left\{\frac{\tau}{t}, \frac{\tau\ln^{1/2}(e/\tau)}{t\ln^{1/2}(e/t)}\right\}, \quad 0 < \tau, t \leq 1.$$

Since $L_{N_2} = M(\varphi_2)$, where $\varphi_2(u) = u\cdot\ln^{1/2}(e/u)$ (see Proposition 2.4), and $L_\infty = M(\varphi_0)$, where $\varphi_0(u) = u$, then, combining the latter inequality with Theorem D.4, we conclude that $M(\psi) \in \mathcal{I}(L_\infty, L_{N_2})$. □

Corollary 3.6 *If a s.s.* $X \in \mathcal{I}(L_\infty, G)$, *then* $M(\widetilde{\phi_X}) \in \mathcal{I}(L_\infty, L_{N_2})$.

Proof Recalling that $X \subset M(\widetilde{\phi_X})$ (see Appendix C), we observe that the hypothesis, combined with Corollaries 3.4 and 3.5, implies

$$X = \widetilde{X} \subset \widetilde{M(\widetilde{\phi_X})} \subset M(\widetilde{\phi_X}).$$

Since X and $M(\widetilde{\phi_X})$ have the identical fundamental functions, from these embeddings it follows that the fundamental functions of the spaces $\widetilde{M(\widetilde{\phi_X})}$ and $M(\widetilde{\phi_X})$ are equivalent. Thus, applying Proposition 3.4, we obtain $M(\widetilde{\phi_X}) \in \mathcal{I}(L_\infty, L_{N_2})$. □

Corollary 3.7 *Let* $M(\varphi)$ *be any Marcinkiewicz space. Then, there is a Marcinkiewicz space* $M(\psi)$ *such that* $M(\psi) \in \mathcal{I}(L_\infty, L_{N_2})$ *and* $\widetilde{M(\varphi)} = M(\psi) \cap G$.

Proof Denote by ψ the function such that $\widetilde{\psi}$ is the fundamental function of the space $\widetilde{M(\varphi)}$. Since $\widetilde{M(\varphi)} \in \mathcal{I}(L_\infty, G)$, then, by Corollary 3.6, we have $M(\psi) \in \mathcal{I}(L_\infty, L_{N_2})$. Moreover, from the embedding $\widetilde{M(\varphi)} \subset M(\varphi)$ it follows the inequality $\psi(t) \leq C\varphi(t)$, $0 < t \leq 1$. Therefore, taking into account that $M(\psi)$ is the largest space among all s.s.'s, whose fundamental function is equivalent to $\widetilde{\psi}$, we have

$$\widetilde{M(\varphi)} \subset M(\psi) \cap G \subset M(\varphi).$$

Consequently, by the definition of $\widetilde{X}$, we get

$$\Big\|\sum_{k=1}^{\infty} a_k r_k\Big\|_{\widetilde{M(\varphi)}} \asymp \Big\|\sum_{k=1}^{\infty} a_k r_k\Big\|_{M(\psi)\cap G} \asymp \Big\|\sum_{k=1}^{\infty} a_k r_k\Big\|_{M(\varphi)}$$

for all sequences $a = (a_k)_{k=1}^{\infty} \in \ell_2$. Since $M(\psi) \cap G \in \mathcal{I}(L_\infty, G)$ (see the proof of Proposition 3.4), we finally have $\widetilde{M(\varphi)} = M(\psi) \cap G$. □

Now, we are in position to prove that for every Marcinkiewicz (resp. Lorentz) space, which is contained in G and fails to have the interpolation property with respect to the Banach couple (L_∞, G), there exists a different Marcinkiewicz (resp. Lorentz) space with the same Rademacher subspace.

Theorem 3.6 *If $M(\varphi) \subset G$, then $\widetilde{M(\varphi)} = M(\psi)$ for some quasiconcave function ψ, and, in consequence, $E_{M(\psi)} = E_{M(\varphi)}$. Moreover, $M(\psi) \neq M(\varphi)$ provided that $M(\varphi) \notin \mathcal{I}(L_\infty, G)$.*

Proof By Corollary 3.5, we have $\widetilde{M(\varphi)} \subset M(\varphi) \subset G$. Therefore, assuming that ψ is such that $\widetilde{\psi}$ is the fundamental function of the space $\widetilde{M(\varphi)}$, by Corollary 3.7, we infer $\widetilde{M(\varphi)} = M(\psi) \cap G = M(\psi)$. The remaining assertions are obvious. □

Theorem 3.7 *Let φ be an increasing concave function on $[0, 1]$ such that the Lorentz space $\Lambda(\varphi)$ is embedded into G. Then, if $\Lambda(\varphi) \notin \mathcal{I}(L_\infty, G)$, there exists a Lorentz space $\Lambda(\psi)$ such that $\Lambda(\psi) \overset{\neq}{\subset} \Lambda(\varphi)$ and $E_{\Lambda(\varphi)} = E_{\Lambda(\psi)}$.*

Proof Denoting $X := \widetilde{\Lambda(\varphi)}$, observe that $X \in \mathcal{I}(L_\infty, G)$ and, by condition, $\Lambda(\varphi) \notin \mathcal{I}(L_\infty, G)$. Hence, $X \overset{\neq}{\subset} \Lambda(\varphi)$. According to the definition of X, to prove the theorem, it is sufficient to find a Lorentz space $\Lambda(\psi)$ such that $X \subset \Lambda(\psi) \overset{\neq}{\subset} \Lambda(\varphi)$.

First of all, we know that $\Lambda(\varphi)' = M(\varphi)$ (see Appendix C), whence by duality we have $M(\varphi) \overset{\neq}{\subset} X'$. Therefore, we can find a decreasing positive function $a(t) \in X'$ satisfying the following conditions:

$$a(t) \geq \varphi'(t) \ \text{ for all } \ 0 < t \leq 1 \quad \text{and} \quad \limsup_{t \to +0} \frac{1}{\varphi(t)} \int_0^t a(s)\, ds = \infty. \tag{3.33}$$

Setting $\psi(t) := \int_0^t a(s)\, ds$, $0 \leq t \leq 1$, observe that, by the definition of a Köthe dual space, we have for every function $x \in X$

$$\int_0^1 x^*(t)\, d\psi(t) = \int_0^1 x^*(t) a(t)\, dt < \infty.$$

Therefore, $X \subset \Lambda(\psi)$ and, in addition, from (3.33) and the choice of ψ it follows that $\Lambda(\psi) \overset{\neq}{\subset} \Lambda(\varphi)$. As was mentioned, this implies that $E_{\Lambda(\varphi)} = E_{\Lambda(\psi)}$, which completes the proof. □

3.4 Examples of Rademacher Subspaces

As was already mentioned, the results, obtained in the previous sections, allow us to characterize the subspaces, spanning by the Rademacher system in function s.s.'s. Before considering the most important examples of Rademacher subspaces, we prove a useful statement on the comparison of Rademacher sums with the dyadic step-functions that will be needed not only here, but also and even more in the next section of this chapter.

Proposition 3.5 *Let $x := \sum_{k=1}^{\infty} a_k r_k$, where $a = (a_k)_{k=1}^{\infty} \in \ell_2$. Then,*

$$\sum_{k=1}^{\infty} a_k^* \chi_{(0,2^{-k-1})}(t) \le x^*(t), \quad 0 < t \le 1.$$

Here, as above, $(a_k^*)_{k=1}^{\infty}$ is the decreasing permutation of the sequence $(|a_k|)_{k=1}^{\infty}$.

Proof Let $y := \sum_{k=1}^{\infty} a_k^* r_k$. For each $l \in \mathbb{N}$, we split this function as follows:

$$y(t) = \sum_{k=1}^{l} a_k^* r_k(t) + \sum_{k=l+1}^{\infty} a_k^* r_k(t).$$

Since the second sum in the right-hand side of this equation is symmetrically distributed on the interval $[0, 2^{-l}]$, we can find a set $E_l \subset [0, 2^{-l}]$, $m(E_l) \ge 2^{-l-1}$, such that this sum is nonnegative on E_l. Therefore, $y(t) \ge a_1^* + a_2^* + \cdots + a_l^*$, $t \in E_l$, and hence for all $l \in \mathbb{N}$ we have

$$\sum_{k=1}^{l} a_k^* \chi_{(0,2^{-k-1})}(t) \le y^*(t), \quad 0 \le t \le 1.$$

Since $y^*(t) = x^*(t)$ (see Proposition 2.2), this implies that

$$\sum_{k=1}^{l} a_k^* \chi_{(0,2^{-k-1})}(t) \le x^*(t), \quad 0 \le t \le 1,$$

As a result, passing to the limit as $l \to \infty$, we come to the desired inequality. □

Corollary 3.8 *Let X be a s.s. on $[0,1]$, and let $a = (a_k)_{k=1}^\infty \in \ell_2$. Then, setting*

$$x_a := \sum_{k=1}^\infty a_k r_k \quad \text{and} \quad f_a := \sum_{k=1}^\infty a_k^* \chi_{(0,2^{-k+1})},$$

we have $\|f_a\|_X \le 4\|x_a\|_X$.

Proof It suffices only to observe that

$$f_a(t) = \sigma_4\Big(\sum_{k=1}^\infty a_k^* \chi_{(0,2^{-k-1})}(t)\Big),$$

where $\sigma_\tau x(t) := x(t/\tau)$, and then to use Proposition 3.5 and the fact that $\|\sigma_\tau\|_{X\to X} \le \max(1,\tau)$ (see Appendix C). □

Example 3.1 Recall that the Lorentz space $\Lambda_p(\varphi)$, where φ is an increasing concave function on $[0,1]$, $\varphi(0) = 0$, and $1 \le p < \infty$, consists of all measurable functions $x = x(s)$ such that the following norm

$$\|x\|_{\varphi,p} := \Big(\int_0^1 x^*(s)^p \, d\varphi(s)\Big)^{1/p}$$

is finite. We show that in the case, when $\varphi(s) = \ln^{1-p}(e/s)$ and $1 < p < 2$, the Rademacher system is equivalent in $\Lambda_p(\varphi)$ to the unit vector basis in ℓ_p, i.e.,

$$\Big\|\sum_{k=1}^\infty a_k r_k\Big\|_{\varphi,p} \asymp \|(a_k)\|_{\ell_p}. \tag{3.34}$$

First of all, we have $\ell_p = (\ell_1, \ell_2)_{\theta,p}$, where $\theta = 2(p-1)/p$ (see Theorem D.5). Moreover, by Theorem 3.1,

$$\Big\|\sum_{k=1}^\infty a_k r_k\Big\|_{(L_\infty,G)_{\theta,p}} \asymp \|(a_k)\|_{\ell_p}. \tag{3.35}$$

Further, it follows from Lemma 3.4 that for every $x \in G$

$$\begin{aligned} \mathcal{K}(\sqrt{k}, x; L_\infty, G) &= \mathcal{K}(\sqrt{k}, x; L_\infty, L_{N_2}) \\ &\asymp \sqrt{k} \cdot \sup_{0<u\le e^{1-k}} \big(x^*(u)\ln^{-1/2}(e/u)\big) \\ &\ge \frac{1}{2} x^*(e^{-k}), \quad k = 1, 2, \ldots \end{aligned}$$

Therefore, by using successively the definition of the real $\mathcal{K}$-method with a function parameter (see Appendix D), the concavity of the function $t \mapsto \mathcal{K}(t, x; L_\infty, G)$, and the equation $\theta p = 2(p-1)$, we get

$$\|x\|^p_{(L_\infty,G)_{\theta,p}} \geq c_p \int_1^\infty \left(t^{-\theta}\mathcal{K}(t,x;L_\infty,G)\right)^p \frac{dt}{t} = c_p \sum_{k=1}^\infty \int_{\sqrt{k}}^{\sqrt{k+1}} \left(t^{-\theta}\mathcal{K}(t,x;L_\infty,G)\right)^p \frac{dt}{t}$$

$$\geq \frac{c_p}{2^p} \sum_{k=1}^\infty x^*(e^{-k})^p \int_{\sqrt{k}}^{\sqrt{k+1}} t^{-\theta p} \frac{dt}{t} = \frac{c_p}{2^{p+1}(p-1)} \sum_{k=1}^\infty x^*(e^{-k})^p (k^{1-p} - (k+1)^{1-p}).$$

On the other hand,

$$\|x\|^p_{\varphi,p} = \int_0^1 x^*(s)^p \, d\varphi(s) \leq \sum_{k=1}^\infty x^*(e^{-k})^p (\varphi(e^{-k+1}) - \varphi(e^{-k}))$$

$$= \sum_{k=1}^\infty x^*(e^{-k})^p (k^{1-p} - (k+1)^{1-p}).$$

From these inequalities it follows that $(L_\infty, G)_{\theta,p} \subset \Lambda_p(\varphi)$, whence taking into account (3.35), we deduce

$$\left\| \sum_{k=1}^\infty a_k r_k \right\|_{\varphi,p} \leq C \|(a_k)\|_{\ell_p}. \tag{3.36}$$

Let us prove the opposite inequality. Denoting

$$x_a := \sum_{k=1}^\infty a_k r_k \quad \text{and} \quad f_a := \sum_{k=1}^\infty a_k^* \chi_{(0,2^{-k-1})},$$

according to Proposition 3.5, we have $f_a(t) \leq x_a^*(t)$, $0 < t \leq 1$. Clearly, this implies that

$$x_a^*(e^{-k-1}) \geq f_a(e^{-k-1}) \geq \sum_{i=1}^k a_i^*, \quad k = 1, 2, \ldots$$

Therefore, by using the elementary inequality $(1+x)^\beta \geq 1 + \frac{\beta}{2}x$, $0 \leq x, \beta \leq 1$, we get

$$\|x_a\|^p_{\varphi,p} \geq e^{-2p} \|\sigma_{e^2} x_a\|^p_{\varphi,p} = e^{-2p} \sum_{k=1}^\infty \int_{e^{-k}}^{e^{-k+1}} x_a^*(e^{-2}s)^p \, d\varphi(s)$$

$$\geq e^{-2p} \sum_{k=1}^\infty x_a^*(e^{-k-1})^p (k^{1-p} - (k+1)^{1-p})$$

$$\geq e^{-2p}2^{-p}(p-1)\sum_{k=1}^{\infty} k^{-p} x_a^*(e^{-k-1})^p$$

$$\geq c_p' \sum_{k=1}^{\infty}\Big(\frac{1}{k}\sum_{i=1}^{k} a_i^*\Big)^p \geq c_p' \|(a_k)\|_{\ell_p}^p.$$

Combining this together with (3.36), we come to equivalence (3.34).

Example 3.2 Consider the exponential Orlicz spaces L_{N_α}, $\alpha > 0$. In the case when $0 < \alpha \leq 2$ we have $L_{N_\alpha} \supset L_{N_2}$, and hence, by Theorem 2.2 (see also Remark 2.1),

$$\Big\|\sum_{k=1}^{\infty} a_k r_k\Big\|_{L_{N_\alpha}} \asymp \|(a_k)\|_{\ell_2}.$$

In contrast to that, for $\alpha > 2$ we have

$$\Big\|\sum_{k=1}^{\infty} a_k r_k\Big\|_{L_{N_\alpha}} \asymp \|(a_k)\|_{\ell_{\beta,\infty}}, \quad \frac{1}{\beta} + \frac{1}{\alpha} = 1, \tag{3.37}$$

where $\ell_{\beta,\infty}$ is the Marcinkiewicz sequence space equipped with the norm

$$\|(a_k)\|_{\ell_{\beta,\infty}} = \sup_{k=1,2,\ldots} k^{1/\beta-1} \cdot \sum_{i=1}^{k} a_i^*.$$

As above, a crucial role in proving (3.37) will be played by an interpolation description of the spaces $\ell_{\beta,\infty}$ and L_{N_α}.

First, from Theorem D.5 it follows that $\ell_{\beta,\infty} = (\ell_1, \ell_2)_{\theta,\infty}$, with $\theta = 2(\beta - 1)/\beta$. Hence, thanks to Theorem 3.1, it is sufficient to check that for the same θ we have

$$(L_\infty, G)_{\theta,\infty} = L_{N_\alpha}. \tag{3.38}$$

Indeed, according to the definition of the real $\mathcal{K}$-method and equivalence (3.28) (see also Remark 3.4),

$$\|x\|_{(L_\infty,G)_{\theta,\infty}} \asymp \sup_{k=0,1,2,\ldots} 2^{-k\theta}2^k \sup_{0<u\leq\min(1,2^{1-2^{2k}})} \big(x^*(u)\log_2^{-1/2}(2/u)\big).$$

Since $1/\beta + 1/\alpha = 1$, we have $\alpha = 2/\theta$. Therefore, reversing the order of the supremums and applying Proposition 2.4, we obtain

$$\begin{aligned}\|x\|_{(L_\infty,G)_{\theta,\infty}} &= \sup_{0<u\le 1} \sup_{0<k\le 2^{-1}\log_2\log_2(2/u)} \left(2^{k(1-\theta)}x^*(u)\log_2^{-1/2}(2/u)\right)\\ &\asymp \sup_{0<u\le 1}\left(x^*(u)\log_2^{-1/2}(2/u)\log_2^{(1-\theta)/2}(2/u)\right)\\ &= \sup_{0<u\le 1}\left(x^*(u)\log_2^{-1/\alpha}(2/u)\right) \asymp \|x\|_{L_{N_\alpha}},\end{aligned}$$

that is, (3.38) is proved. As was said above, this gives (3.37).

Example 3.3 Let $M(\psi)$ be the Marcinkiewicz space generated by the function $\psi(t) = t\log_2\log_2(4/t)$, $0 < t \le 1$. We prove that

$$\left\|\sum_{k=1}^{\infty} a_k r_k\right\|_{M(\psi)} \asymp \|(a_k)\|_{\ell_{\log,\infty}}, \tag{3.39}$$

where $\ell_{\log,\infty}$ is the Marcinkiewicz sequence space with the norm

$$\|(a_k)\|_{\ell_{\log,\infty}} := \sup_{k=1,2,\dots} \log_2^{-1}(2k)\sum_{i=1}^{k} a_i^*.$$

In order to establish this equivalence, again by Theorem 3.1, it is sufficient to find a parameter of the real $\mathcal{K}$-method F satisfying both

$$(\ell_1, \ell_2)_F^{\mathcal{K}} = \ell_{\log,\infty} \tag{3.40}$$

and

$$(L_\infty, G)_F^{\mathcal{K}} = M(\psi). \tag{3.41}$$

We show that equations (3.40) and (3.41) hold if F is the weighted space $\ell_\infty(u_k)$, where $u_k = 1/(k+1)$ for $k \ge 0$, and $u_k = 1$ for $k < 0$.

Let us begin with proving equation (3.40). Without loss of generality, we can assume that $a_i = a_i^*$, $i = 1, 2, \dots$ Then, by inequality (1.26) from Theorem 1.5, we get

$$\kappa(2^k, a) \le \sum_{i=1}^{2^{2k}} a_i + 2^k\Big(\sum_{i=2^{2k}+1}^{\infty} a_i^2\Big)^{1/2} \le 4\kappa(2^k, a), \tag{3.42}$$

where $\kappa(t, a) := \mathcal{K}(t, a; \ell_1, \ell_2)$.

Suppose first that $B := \|a\|_{\ell_{\log,\infty}} < \infty$. Let us estimate from above the expression from the middle part of inequality (3.42). Taking into account definition of the $\ell_{\log,\infty}$-norm, we have

$$\sum_{i=1}^{2^{2k}} a_i \le 2B(k+1), \tag{3.43}$$

whence $a_{2^{2k}} \le 2^{-2k+1}(k+1)B$, $k = 0, 1, \dots$ Therefore,

$$\begin{aligned}\sum_{i=2^{2k}+1}^{\infty} a_i^2 &= \sum_{j=k}^{\infty} \sum_{i=2^{2j}+1}^{2^{2(j+1)}} a_i^2 \le 3\sum_{j=k}^{\infty} 2^{2j} a_{2^{2j}}^2 \\ &\le 12B^2 \sum_{j=k}^{\infty} 2^{-2j}(j+1)^2 \\ &\le 192B^2 \int_k^{\infty} x^2 2^{-2x}\, dx.\end{aligned}$$

Then, integration by parts gives us

$$\sum_{i=2^{2k}+1}^{\infty} a_i^2 \le 24^2 B^2 (k+1)^2 2^{-2k},$$

and hence

$$2^k \Big(\sum_{i=2^{2k}+1}^{\infty} a_i^2 \Big)^{1/2} \le 24B(k+1).$$

Combining this inequality with estimate (3.43) and the left-hand side inequality from (3.42), we get

$$\|a\|_{(\ell_1,\ell_2)_F^{\mathcal{K}}} = \sup_{k=0,1,..} \frac{\kappa(2^k, a)}{k+1} \le 26\|a\|_{\ell_{\log,\infty}}.$$

Conversely, if $2^{2j} + 1 \le k \le 2^{2(j+1)}$, $j = 0, 1, \dots$, then the right-hand side inequality from (3.42) implies that

$$\sum_{i=1}^{k} a_i \le 4\kappa(2^{j+1}, a) \le 4(j+2)\|a\|_{(\ell_1,\ell_2)_F^{\mathcal{K}}} \le 4\log_2(2k)\|a\|_{(\ell_1,\ell_2)_F^{\mathcal{K}}}.$$

Thus, $\|a\|_{\ell_{\log,\infty}} \le 4\|a\|_{(\ell_1,\ell_2)_F^{\mathcal{K}}}$, and so (3.40) is proved.

Next, we prove (3.41). As above, by (3.28), we have

$$\|x\|_{(L_\infty,G)^{\mathcal{K}}_F} \asymp \sup_{k=0,1,2,\ldots} \frac{2^k}{k+1} \sup_{0<u\le\min(1,2^{1-2^{2k}})} \left(x^*(u)\log_2^{-1/2}(2/u)\right)$$

$$\asymp \sup_{0<u\le 1} x^*(u)\log_2^{-1/2}(2/u) \sup_{0\le k\le 2^{-1}\log_2\log_2(2/u)} \frac{2^k}{k+1}$$

$$\asymp \sup_{0<u\le 1} \frac{x^*(u)}{\log_2\log_2(4/u)} \asymp \|x\|_{M(\psi)},$$

where the last equivalence follows from Theorem C.2. Thus, (3.41) is also proved. Finally, as was mentioned, (3.39) is an immediate consequence of (3.40) and (3.41).

3.5 The Rademacher System and Cones of Step-Functions

The results obtained show that the Rademacher subspace of a s.s. X can be described by using the corresponding sequence space E_X. It turns out that in the case when X is located sufficiently "near" to L_∞ Rademacher subspaces can be characterized as well as some cones of dyadic step-functions.

For an arbitrary s.s. X on $[0, 1]$, denote by H_X the set of all sequences $a = (a_k)_{k=1}^\infty$ such that

$$\|a\|_{H_X} := \left\| \sum_{k=1}^\infty a_k^* \chi_{(0,2^{-k+1}]} \right\|_X < \infty,$$

where $(a_k^*)_{k=1}^\infty$ is still the decreasing permutation of the sequence $(|a_k|)_{k=1}^\infty$. Clearly, H_X can be identified also with the cone consisting of all dyadic step-functions $v \in X$ of the form $v = \sum_{k=1}^\infty b_k \chi_{(0,2^{-k+1}]}$, $b_1 \ge b_2 \ge \cdots \ge 0$.

By Corollary 3.8, for every s.s. X we have

$$\|a\|_{H_X} \le 4 \left\| \sum_{k=1}^\infty a_k r_k \right\|_X. \tag{3.44}$$

In this regard, the following natural question arises: When does the opposite inequality hold? Or equivalently: Under which conditions on X the Rademacher subspace of a s.s. X can be described as the cone H_X (this means that a function $\sum_{k=1}^\infty a_k r_k \in X$ if and only if the sequence of coefficients $(a_k)_{k=1}^\infty$ belongs to H_X with equivalence of respective norms)? A crucial role in addressing these issues will be played by the sublinear operator U defined on the set of measurable functions on $[0, 1]$ by

$$Ux(t) := \Big(\frac{1}{\varphi_1(t)}\int_0^t x(s)^2\,d\varphi_1(s)\Big)^{1/2},$$

where $\varphi_1(s) = \log_2^{-1}(2/s)$, $0 < s \le 1$ (for technical reasons, we prefer here to use logarithms with base 2). We state now the main result of this section.

Theorem 3.8 *Let X be an interpolation space with respect to the couple (L_∞, L_{N_1}). Then, the inequality*

$$\Big\|\sum_{k=1}^{\infty} a_k r_k\Big\|_X \le C\|a\|_{H_X}, \tag{3.45}$$

holds for some $C > 0$, depending only on X, if and only if the operator U is bounded in X.

Sufficiency In view of the definition of H_X and Proposition 2.2, without loss of generality, it can be assumed that $a_k = a_k^*$, $k = 1, 2, \dots$ Putting for such a sequence

$$f_a(s) = \sum_{k=1}^{\infty} a_k \chi_{(0,2^{-k+1}]}(s) \text{ and } x_a(s) = \sum_{k=1}^{\infty} a_k r_k(s), \tag{3.46}$$

let us prove that

$$\|Uf_a\|_X \asymp \|x_a\|_X, \tag{3.47}$$

with universal constants (not depending on a s.s. X).

We replace the functions Uf_a and x_a^* with their discrete substitutes

$$D_a(s) = \sum_{k=1}^{\infty}(Uf_a)(2^{-k})\chi_{(2^{-k},2^{-k+1}]}(s) \text{ and } d_a(s) = \sum_{k=1}^{\infty} x_a^*(2^{-k})\chi_{(2^{-k},2^{-k+1}]}(s), \quad 0 < s \le 1.$$

As φ_1 increases, the function

$$Ux^*(t) = \Big(\frac{1}{\varphi_1(t)}\int_0^{\varphi_1(t)} \big(x^*(\varphi_1^{-1}(s))\big)^2\,ds\Big)^{1/2}$$

decreases on $[0, 1]$. Therefore, Uf_a decreases together with f_a, and we infer that

$$\sigma_{1/2}(D_a)(s) \le (Uf_a)(s) \le D_a(s), \quad 0 < s \le 1,$$

and

$$\sigma_{1/2}(d_a)(s) \le x_a^*(s) \le d_a(s), \quad 0 < s \le 1,$$

where σ_τ is the dilation operator (Appendix C). Since X is a s.s. and $\|\sigma_2\|_{X\to X} \le 2$, relation (3.47) will be proved once we establish the equivalence of sequences $\{Uf_a(2^{-k})\}_{k=1}^\infty$ and $\{x_a^*(2^{-k})\}_{k=1}^\infty$.

To this end, rewriting the formula for f_a in the following way

$$f_a(s) = \sum_{k=1}^\infty \Big(\sum_{j=1}^k a_j\Big)\chi_{(2^{-k},2^{-k+1}]}(s), \tag{3.48}$$

we obtain

$$\begin{aligned} Uf_a(2^{-k}) &= \left(\frac{1}{\varphi_1(2^{-k})}\sum_{i=k+1}^\infty \Big(\sum_{j=1}^i a_j\Big)^2 (\varphi_1(2^{-i+1}) - \varphi_1(2^{-i}))\right)^{1/2} \\ &\asymp \sqrt{k}\Big(\sum_{i=k}^\infty \Big(\frac{1}{i}\sum_{j=1}^i a_j\Big)^2\Big)^{1/2}, \quad k = 1, 2, \ldots \end{aligned}$$

On the other hand, successively applying Corollary 1.9, Lemma 1.2, and Lemma 3.1, we have

$$\begin{aligned} x_a^*(2^{-k}) \asymp \mathcal{K}(\sqrt{k}, a; \ell_1, \ell_2) &= \mathcal{K}\Big(\sqrt{k}, \sum_{i=1}^\infty a_i\chi_{(i-1,i)}; L_1(0,\infty), L_2(0,\infty)\Big) \\ &\asymp \sqrt{k}\Big(\sum_{i=k}^\infty \Big(\frac{1}{i}\sum_{j=1}^i a_j\Big)^2\Big)^{1/2}, \quad k = 1, 2, \ldots \end{aligned}$$

Thus, the sequences $\{Uf_a(2^{-k})\}$ and $\{x_a^*(2^{-k})\}$ are equivalent (with universal constants), and so (3.47) is proved.

Now, if the operator U is bounded in a s.s. X, then inequality (3.45) is an immediate consequence of equivalence (3.47) and the notation. Therefore, the sufficiency is proved. □

The proof of the necessity is much more complicated. First of all, we note that from (3.47) and (3.45) it follows the boundedness of the restriction of U to the cone H_X. We show first that, under certain conditions, this implies the boundedness of the operator U on the whole space X. For this purpose, we define one more sublinear operator V by

$$Vx(t) := \frac{1}{\varphi_1(t)} \int_0^t x^*(s)\, d\varphi_1(s),$$

where still $\varphi_1(s) = \log_2^{-1}(2/s)$.

Proposition 3.6 *Suppose the operator V is bounded in a s.s. X. Then an operator (not necessarily linear) A is bounded in X whenever it possesses the following properties:*

(a) if $0 \le x(t) \le y(t)$, $0 < t \le 1$, then $0 \le Ax(t) \le Ay(t)$, $0 < t \le 1$;
(b) $|Ax(t)| \le Ax^*(t)$, $0 < t \le 1$;
(c) there is a constant $C > 0$ such that $\|Av\|_X \le C\|v\|_X$ for all $v \in H_X$.

Proof Let $x \in X$ be arbitrary. Since the function φ_1 increases and, by definition,

$$Vx(t) = \frac{1}{\varphi_1(t)} \int_0^{\varphi_1(t)} x^*\big(\varphi_1^{-1}(s)\big)\, ds,$$

then the function $Vx(t)$ decreases on $[0, 1]$. Moreover, one can easily check that the sequence $\{Vx(2^{-k})\}_{k=1}^\infty$ increases, while the sequence $\{\frac{1}{k} Vx(2^{-k})\}_{k=1}^\infty$ decreases. Similarly as a quasiconcave function defined on $[0, 1]$ or $[0, \infty)$ is equivalent to some concave function (see Appendix C), such a sequence is equivalent to some concave sequence. More precisely, there is an increasing sequence $(c_k)_{k=1}^\infty$ such that the differences $b_k := c_k - c_{k-1}$ ($c_0 = 0$) form a decreasing sequence and also

$$Vx(2^{-k}) \le c_k \le 2Vx(2^{-k}), \quad k = 1, 2, \ldots$$

Therefore, setting

$$z_1(t) := \sum_{k=1}^\infty c_k \chi_{(2^{-k}, 2^{-k+1}]}(t) = \sum_{k=1}^\infty b_k \chi_{(0, 2^{-k+1}]}(t), \quad 0 < t \le 1,$$

and

$$z_2(t) := \sum_{k=1}^\infty (Vx)(2^{-k}) \chi_{(2^{-k}, 2^{-k+1}]}(t), \quad 0 < t \le 1,$$

by the inequalities $z_2(t) \le (Vx)(t/2)$, $0 < t \le 1$, $\|\sigma_2\|_{X \to X} \le 2$, and the boundedness of the operator V in X, we obtain

$$\|z_1\|_X \le 2\|z_2\|_X \le 2\|\sigma_2 Vx\|_X \le 4\|V\| \|x\|_X.$$

Since the sequence $(b_k)_{k=1}^\infty$ decreases, this inequality implies that $z_1 \in H_X$. Moreover, it is clear that

$$x^*(t) \le Vx(t) \le z_2(t) \le z_1(t), \quad 0 < t \le 1.$$

Thus, from the hypotheses of the proposition and the preceding inequality it follows that

$$\|Ax\|_X \le \|Ax^*\|_X \le \|Az_1\|_X \le C\|z_1\|_X \le 4C\|V\|\|x\|_X. \qquad \square$$

Since the operator U satisfies all the conditions of Proposition 3.6, then the proof of Theorem 3.8 will be completed once we show that the operator V is bounded in the given interpolation space X with respect to the couple (L_∞, L_{N_1}), provided if hypothesis (3.45) holds. This will require the greatest technical effort. Before starting the detailed proof, we give its outline.

Firstly, we define a certain s.s. $Y \in \mathcal{I}(L_\infty, L_{N_1})$ (depending on the given space X) such that V is bounded in Y. Then, by using condition (3.45), we shall establish the coincidence of the cones H_Y and H_X. Finally, we prove that the mapping $X \mapsto H_X$ is one-to-one on the set $\mathcal{I}(L_\infty, L_{N_1})$. Since $X, Y \in \mathcal{I}(L_\infty, L_{N_1})$, summing up, we conclude that $X = Y$. Then, an application of Proposition 3.6 will finish the proof.

Let X be a s.s. on $[0, 1]$. As above, by E_X we denote the sequence space equipped with the norm $\|a\|_{E_X} := \|\sum_{k=1}^\infty a_k r_k\|_X$, $a = (a_k)_{k=1}^\infty$. By Theorem 3.2, we have $E_X \in \mathcal{I}(\ell_1, \ell_2)$. Hence, as above,

$$E_X = (\ell_1, \ell_2)_F^{\mathcal{K}} \tag{3.49}$$

for some parameter of the real $\mathcal{K}$-method F. Define now the s.s. Y by

$$Y := (L_\infty, \Lambda_2)_F^{\mathcal{K}}, \tag{3.50}$$

where Λ_2 denotes the Lorentz space $\Lambda_2(\varphi_1)$, that is,

$$\|x\|_{\Lambda_2} := \left(\int_0^1 x^*(t)^2\, d\varphi_1(t)\right)^{1/2}.$$

Let us prove that the operator V is bounded in Y. To this end, observe that the real $\mathcal{K}$-method "interpolates" not only linear, but also sublinear operators (see Appendix D or [88, § 11]). Therefore, since Y is a real $\mathcal{K}$-interpolation space with respect to the couple (L_∞, Λ_2), it suffices to establish the following result.

Lemma 3.5 *The operator V is bounded in the spaces L_∞ and Λ_2.*

Proof Since the boundedness of V in L_∞ is obvious, we need only to prove the same property of V in the space Λ_2.

Since $Vx(t)$ is a decreasing function, we have

$$\|Vx\|_{\Lambda_2} \le \Big\|\sum_{k=1}^{\infty}(Vx)(2^{-k})\chi_{(2^{-k},2^{-k+1}]}\Big\|_{\Lambda_2} \le \left(\sum_{k=1}^{\infty}\big((Vx)(2^{-k})\big)^2\frac{1}{k^2}\right)^{1/2}.$$

Further, as it is well known,

$$\|(a_k)\|_{\ell_2} = \sup\left\{\sum_{k=1}^{\infty} a_k b_k :\ \|(b_k)\|_{\ell_2} \le 1\right\}.$$

Hence, there is a sequence $(d_k)_{k=1}^{\infty}$ such that $d_k \ge 0$, $\|(d_k)\|_{\ell_2} \le 1$, and

$$\left(\sum_{k=1}^{\infty}\big((Vx)(2^{-k})\big)^2\frac{1}{k^2}\right)^{1/2} \le 2\sum_{k=1}^{\infty}\frac{d_k}{k}(Vx)(2^{-k}).$$

Combining the last inequalities with the definition of V, we get

$$\begin{aligned}
\|Vx\|_{\Lambda_2} &\le 4\sum_{k=1}^{\infty} d_k \int_0^{2^{-k}} x^*(s)\, d\varphi_1(s)\\
&= 4\Big(\sum_{j=1}^{k} d_j\Big)\sum_{k=1}^{\infty}\int_{2^{-k-1}}^{2^{-k}} x^*(s)\, d\varphi_1(s)\\
&\le 4\sum_{k=1}^{\infty}\Big(\frac{1}{k}\sum_{j=1}^{k} d_j\Big)x^*(2^{-k-1})\frac{1}{k}\\
&= 4\sum_{k=1}^{\infty}\Big(\tilde{\mathcal{H}}(d_j)\Big)_k x^*(2^{-k-1})\frac{1}{k}.
\end{aligned}$$

By Theorem C.4, the Hardy operator $\tilde{\mathcal{H}}$ is bounded in ℓ_2. Therefore, applying Cauchy–Schwarz-Bunyakovskii inequality combined with the fact that $\|\sigma_4\|_{X\to X} \le 4$, we have

$$\begin{aligned}
\|Vx\|_{\Lambda_2} &\le 4\|\tilde{\mathcal{H}}\|_{\ell_2\to\ell_2}\left(\sum_{k=1}^{\infty}\big(x^*(2^{-k-1})\big)^2\frac{1}{k^2}\right)^{1/2}\\
&\le 8\|\tilde{\mathcal{H}}\|\Big\|\sum_{k=1}^{\infty} x^*(2^{-k-1})\chi_{(2^{-k},2^{-k+1}]}\Big\|_{\Lambda_2}
\end{aligned}$$

$$= 8\|\tilde{\mathcal{H}}\| \Big\| \sum_{k=1}^{\infty} (\sigma_4 x^*)(2^{-k+1}) \chi_{(2^{-k}, 2^{-k+1}]} \Big\|_{\Lambda_2}$$

$$\leq 8\|\tilde{\mathcal{H}}\| \|\sigma_4 x^*\|_{\Lambda_2} \leq 32\|\tilde{\mathcal{H}}\| \|x\|_{\Lambda_2}. \qquad \square$$

Proceed now with the proof of the equation

$$H_Y = H_X. \tag{3.51}$$

First of all, from (3.44), (3.45) and the definition of E_X it follows that

$$H_X = E_X. \tag{3.52}$$

Hence, for the proof of (3.51) it is sufficient to show that

$$H_Y = E_X. \tag{3.53}$$

In the next lemma, all functions defined on the interval $[0, 1]$ are extended by zero to the semi-axis $[0, \infty)$.

Lemma 3.6 *With constants independent of $x \in \Lambda_2$ and $t > 0$ we have*

$$\mathcal{K}(t, x; L_\infty, \Lambda_2) \asymp t\Big(\int_0^{2^{1-t^2}} x^*(s)^2 \, d\varphi_1(s)\Big)^{1/2}.$$

Proof Let us define, for a fixed $t > 0$, the function x_1 by

$$x_1(s) := x(s) - x^*(2^{1-t^2})\frac{x(s)}{|x(s)|} \quad \text{if } |x(s)| \geq x^*(2^{1-t^2}),$$

and

$$x_1(s) := 0 \ \text{ if } |x(s)| < x^*(2^{1-t^2}).$$

Moreover, we set $x_0 := x - x_1$. Then, if $E = \{s \in [0, 1] : x_1(s) \neq 0\}$, by definition of the decreasing rearrangement, we have $x_1^*(s) = x^*(s) - x^*(2^{1-t^2})$, $0 < s < m(E)$, and $m(E) \leq 2^{1-t^2}$. Consequently,

$$\mathcal{K}(t, x; L_\infty, \Lambda_2) \leq t\|x_1\|_{\Lambda_2} + \|x_0\|_\infty \leq t\Big(\int_0^{2^{1-t^2}} (x^*(s) - x^*(2^{1-t^2}))^2 \, d\varphi_1(s)\Big)^{1/2}$$

$$+ x^*(2^{1-t^2}) = t\Big(\int_0^{2^{1-t^2}} (x^*(s) - x^*(2^{1-t^2}))^2 \, d\varphi_1(s)\Big)^{1/2}$$

$$+ t\Big(\int_0^{2^{1-t^2}} x^*(2^{1-t^2})^2 \, d\varphi_1(s)\Big)^{1/2} \leq 2t\Big(\int_0^{2^{1-t^2}} x^*(s)^2 \, d\varphi_1(s)\Big)^{1/2}.$$

To get an inequality going the other way, we consider any representation $x = y_0 + y_1$, where $y_0 \in L_\infty$, $y_1 \in \Lambda_2$. Precisely as in the proof of Proposition 1.5 (see (1.21)), we have $x^*(s) \le y_0^*(s/2) + y_1^*(s/2)$, $0 < s \le 1$. Therefore,

$$\Big(\int_0^{2^{1-t^2}} x^*(s)^2\,d\varphi_1(s)\Big)^{1/2} \le \Big(\int_0^{2^{1-t^2}} y_0^*(s/2)^2\,d\varphi_1(s)\Big)^{1/2} + \Big(\int_0^{2^{1-t^2}} y_1^*(s/2)^2\,d\varphi_1(s)\Big)^{1/2} \le \frac{\|y_0\|_\infty}{t} + \sqrt{2}\|y_1\|_{\Lambda_2},$$

whence from definition of the $\mathcal{K}$-functional it follows

$$t\Big(\int_0^{2^{1-t^2}} x^*(s)^2\,d\varphi_1(s)\Big)^{1/2} \le \sqrt{2}\mathcal{K}(t, x; L_\infty, \Lambda_2). \qquad \square$$

In the concluding part of the proof of Theorem 3.8 we shall need also an equivalent expression for the $\mathcal{K}$-functional of the couple $(L_\infty, \Lambda(\varphi_1))$, where $\Lambda(\varphi_1)$ is the Lorentz space generated by the function φ_1. Since the proof of this result is quite similar to that of Lemma 3.6, we omit it.

Lemma 3.7 *With constants independent of $x \in \Lambda(\varphi_1)$ and $t > 0$ we have*

$$\mathcal{K}(t, x; L_\infty, \Lambda(\varphi_1)) \asymp t\int_0^{2^{1-t}} x^*(s)\,d\varphi_1(s).$$

For proving (3.53) we need one more assertion.

Proposition 3.7 *With constants independent of $a = (a_k)_{k=1}^\infty \in \ell_2$ and $t > 0$ the equivalence*

$$\mathcal{K}(t, f_a; L_\infty, \Lambda_2) \asymp \kappa(t, a) \tag{3.54}$$

holds, where the function f_a is defined by the first of equations (3.46) *and* $\kappa(t, a) = \mathcal{K}(t, a; \ell_1, \ell_2)$.

Proof Without loss of generality, we can (and will) assume that $a_k = a_k^*$, $k = 1, 2, \ldots$ Firstly, taking into account (3.48), by Lemma 3.6, we have

$$\mathcal{K}(m, f_a; L_\infty, \Lambda_2) \asymp m\left(\sum_{k=m^2}^\infty \Big(\sum_{i=1}^k a_i\Big)^2 \int_{2^{-k}}^{2^{-k+1}} d\varphi_1(s)\right)^{1/2} \asymp m\left(\sum_{k=m^2}^\infty \Big(\frac{1}{k}\sum_{i=1}^k a_i\Big)^2\right)^{1/2}, \quad m = 1, 2, \ldots$$

On the other hand, precisely as in the proof of sufficiency, according to Lemmas 1.2 and 3.1, we obtain

$$\kappa(t,a) = \mathcal{K}\Big(t, \sum_{k=1}^{\infty} a_k \chi_{(k-1,k)}; L_1(0,\infty), L_2(0,\infty)\Big)$$
$$\asymp t\left(\sum_{k=[t^2]}^{\infty}\left(\frac{1}{k}\sum_{i=1}^{k} a_i\right)^2\right)^{1/2}, \quad t \geq 1.$$

Comparing these equivalences and using the concavity of the function $t \mapsto \kappa(t,a)$, we get (3.54) for $t \geq 1$.

To obtain (3.54) for the values $0 < t < 1$, we observe that

$$H_{\Lambda_2} = \ell_2 \text{ (with equivalence of norms).} \tag{3.55}$$

Indeed, applying (3.48) once more, we have

$$\|f_a\|_{\Lambda_2} = \left(\sum_{k=1}^{\infty}\Big(\sum_{i=1}^{k} a_i\Big)^2 \int_{2^{-k}}^{2^{-k+1}} d\varphi_1(s)\right)^{1/2} \asymp \left(\sum_{k=1}^{\infty}\Big(\frac{1}{k}\sum_{i=1}^{k} a_i\Big)^2\right)^{1/2}.$$

On the one hand, this relation and the inequality $a_k \leq k^{-1}\sum_{i=1}^{k} a_i$ imply $\|a\|_{\ell_2} \leq C_1\|f_a\|_{\Lambda_2}$. On the other hand, combining the latter equivalence with Theorem C.4, we get $\|f_a\|_{\Lambda_2} \leq C_2\|\tilde{\mathcal{H}}\|_{\ell_2\to\ell_2}\|a\|_{\ell_2}$, where $\tilde{\mathcal{H}}$ is the Hardy operator. Thus, (3.55) is proved. Finally, from (3.55) and the inequalities $\|a\|_{\ell_2} \leq \|a\|_{\ell_1}$, $\|x\|_{\Lambda_2} \leq \|x\|_{\infty}$ it follows that

$$\mathcal{K}(t, f_a; L_\infty, \Lambda_2) = t\|f_a\|_{\Lambda_2} \asymp t\|a\|_{\ell_2} = \kappa(t,a)$$

if $0 < t < 1$. So, the proof is completed. □

Corollary 3.9 *With constants independent of $a = (a_k)_{k=1}^{\infty} \in \ell_2$ and $t > 0$ we have*

$$\mathcal{K}(t, x_a; L_\infty, G) \asymp \mathcal{K}(t, f_a; L_\infty, \Lambda_2).$$

Here, the functions f_a and x_a are defined by (3.46), *and, as above, G is the separable part of the Orlicz space L_{N_2}, $N_2(u) = e^{u^2} - 1$.*

Moreover, if $X = (L_\infty, G)_F^{\mathcal{K}}$ for some parameter of the $\mathcal{K}$-method F, then

$$\Big\|\sum_{k=1}^{\infty} a_k r_k\Big\|_X \asymp \|(a_k)\|_{H_Y},$$

where $Y = (L_\infty, \Lambda_2)_F^{\mathcal{K}}$.

Proof The required equivalence of the $\mathcal{K}$-functionals follows at once from Proposition 3.7 and Theorem 3.3. As for the second of the statements of the corollary, it is a direct consequence of this equivalence and definition of the real $\mathcal{K}$-method. □

Now we are ready to prove (3.51). As in the preceding section, we set

$$\widetilde{X} = (L_\infty, G)_F^{\mathcal{K}},$$

where the Banach sequence lattice F is defined by (3.49). Then, by Corollaries 3.9 and 3.1, we have

$$\|a\|_{H_Y} \asymp \left\| \sum_{k=1}^\infty a_k r_k \right\|_{\widetilde{X}} \asymp \left\| \sum_{k=1}^\infty a_k r_k \right\|_X = \|a\|_{E_X},$$

whence we get (3.53). As was mentioned above, (3.51) follows immediately from (3.53) and (3.52).

Thus, it remains to show that (3.51) implies the coincidence of the spaces X and Y. To this end, we need to prove that H_Z is an interpolation space with respect to the couple (ℓ_1, ℓ_∞) provided that $Z \in \mathcal{I}(L_\infty, L_{N_1})$.

Proposition 3.8 *With a constant independent of* $a = (a_k)_{k=1}^\infty \in \ell_\infty$ *and* $t > 0$ *we have*

$$\mathcal{K}(t, f_a; L_\infty, L_{N_1}) \asymp \mathcal{K}(t, a, \ell_1, \ell_\infty).$$

Proof Once again, without loss of generality, we can assume that $a_k = a_k^*$, $k = 1, 2, \dots$

Firstly, let us check that

$$\|f_a\|_{L_{N_1}} \asymp \|a\|_{\ell_\infty}, \quad a \in \ell_\infty. \tag{3.56}$$

Since

$$f_a(s) \le \sum_{k=1}^\infty k\|a\|_{\ell_\infty} \chi_{(2^{-k}, 2^{-k+1}]}(s), \quad 0 < s \le 1,$$

we get

$$\int_0^1 \left\{ \exp\left(\frac{f_a(s)}{\|a\|_{\ell_\infty}} \ln(4/3) \right) - 1 \right\} ds \le \sum_{k=1}^\infty \int_{2^{-k}}^{2^{-k+1}} \{\exp(k \ln(4/3)) - 1\} \, ds = 1.$$

Therefore, $\|f_a\|_{L_{N_1}} \le \frac{1}{\ln(4/3)}\|a\|_{\ell_\infty}$. On the other hand, if for some $\lambda > 0$

$$\int_0^1 \left\{\exp\left(\frac{f_a(s)}{\lambda}\right) - 1\right\} ds \le 1,$$

then from the inequality $f_a(s) \ge \|a\|_{\ell_\infty}$, $0 < s \le 1$, it follows $\exp(\|a\|_{\ell_\infty}/\lambda) \le 2$, whence $\lambda \ge (\ln 2)^{-1}\|a\|_{\ell_\infty}$. Thus, by definition of the Luxemburg norm, we infer $\|f_a\|_{L_{N_1}} \ge (\ln 2)^{-1}\|a\|_{\ell_\infty}$, and so equivalence (3.56) is proved.

Further, (3.48) and Lemma 3.4 (see also Remark 3.4) imply that

$$\mathcal{K}(m, f_a; L_\infty, L_{N_1}) \asymp m \sup_{k \ge m} \left(\frac{1}{k}\sum_{i=1}^{k} a_i\right), \quad m = 1, 2, \ldots$$

Hence, as the sequence $\left(1/k \sum_{i=1}^{k} a_i\right)_{k=1}^{\infty}$ decreases, we get

$$\mathcal{K}(m, f_a; L_\infty, L_{N_1}) \asymp \sum_{i=1}^{m} a_i, \quad m = 1, 2, \ldots$$

Moreover, one can easily verify that

$$\mathcal{K}(t, a; \ell_1, \ell_\infty) = \sum_{k=1}^{[t]} a_k + (t - [t])a_{[t]+1}, \quad t > 0 \tag{3.57}$$

(see also equation (D.1) for sequence spaces in the case $p = 1$). Combining this with the preceding equivalence, we conclude that

$$\mathcal{K}(m, f_a; L_\infty, L_{N_1}) \asymp \mathcal{K}(m, a, \ell_1, \ell_\infty), \quad m = 1, 2, \ldots$$

Since the function $t \mapsto \mathcal{K}(t, a, \ell_1, \ell_\infty)$ is concave, a standard reasoning allows us to extend now the latter equivalence to the set $[1, \infty)$. Finally, if $0 < t < 1$, then from the inequalities $\|a\|_{\ell_\infty} \le \|a\|_{\ell_1}$, $\|x\|_{L_{N_1}} \le \|x\|_\infty$ and equivalence (3.56) it follows

$$\mathcal{K}(t, f_a; L_\infty, L_{N_1}) = t\|f_a\|_{L_{N_1}} \asymp t\|a\|_{\ell_\infty} = \mathcal{K}(t, a, \ell_1, \ell_\infty). \qquad \square$$

Taking into account definition of the real $\mathcal{K}$-method, from Proposition 3.8 we obtain

Corollary 3.10 *Suppose that* $Z = (L_\infty, L_{N_1})_F^{\mathcal{K}}$ *for some parameter of the real* $\mathcal{K}$*-method* F*. Then,* $H_Z = (\ell_1, \ell_\infty)_F^{\mathcal{K}}$ *(with equivalence of norms).*

The next statement shows that the mapping $Z \mapsto H_Z$, restricted to the set of all interpolation spaces with respect to the couple (L_∞, L_{N_1}), is one-to-one (it is instructive to compare this result with Theorem 3.4).

Proposition 3.9 *Let X_0 and X_1 be interpolation spaces with respect to the Banach couple (L_∞, L_{N_1}) and let $H_{X_0} = H_{X_1}$. Then, $X_0 = X_1$ (with equivalence of norms).*

Proof Denote by $\mathcal{P}_2$ the cone of all increasing concave functions $f(t)$, $0 < t < \infty$, such that $f(t) = f(1)t$, for $0 < t \le 1$. We show that

$$\mathcal{K}\{L_\infty, L_{N_1}\} = \mathcal{K}\{\ell_1, \ell_\infty\} = \mathcal{P}_2, \tag{3.58}$$

where $\mathcal{K}\{X_0, X_1\}$ is still the set of all functions of the form $f(t) = \mathcal{K}(t, x; X_0, X_1)$, $x \in X_0 + X_1$.

Suppose $f \in \mathcal{P}_2$. Then, $a_f := (f(k) - f(k-1))_{k=1}^\infty$ is a decreasing positive sequence and from (3.57) it follows $\mathcal{K}(k, a_f; \ell_1, \ell_\infty) = f(k)$, $k = 1, 2, \ldots$ Hence, taking into account the concavity of the functions in question and as well the equation $\mathcal{K}(t, a; \ell_1, \ell_\infty) = t\|a\|_{\ell_\infty}$, $0 < t \le 1$, we deduce that $f(t) \asymp \mathcal{K}(t, a_f; \ell_1, \ell_\infty)$, $t > 0$, with a constant independent of $f \in \mathcal{P}_2$. Therefore, $\mathcal{P}_2 \subset \mathcal{K}\{\ell_1, \ell_\infty\}$. On the other hand, by Proposition 3.8, we have $\mathcal{K}\{\ell_1, \ell_\infty\} \subset \mathcal{K}\{L_\infty, L_{N_1}\}$. Finally, the inequality $\|x\|_{L_{N_1}} \le \|x\|_\infty$, combined with definition of the $\mathcal{K}$-functional (see the proof of Proposition 3.1) implies that $\mathcal{K}\{L_\infty, L_{N_1}\} \subset \mathcal{P}_2$. Thus, (3.58) is proved.

Next, since the couple (L_∞, L_{N_1}) is $\mathcal{K}$-monotone (see Propositions 2.4 and 3.3), from Theorem D.7 it follows that

$$X_0 = (L_\infty, L_{N_1})^{\mathcal{K}}_{F_0} \text{ and } X_1 = (L_\infty, L_{N_1})^{\mathcal{K}}_{F_1} \tag{3.59}$$

for some parameters of the $\mathcal{K}$-method F_0 and F_1. Hence, applying Corollary 3.10, we obtain

$$H_{X_0} = (\ell_1, \ell_\infty)^{\mathcal{K}}_{F_0} \text{ and } H_{X_1} = (\ell_1, \ell_\infty)^{\mathcal{K}}_{F_1}. \tag{3.60}$$

Assume now that $x \in X_0$. Then, $(\mathcal{K}(2^k, x; L_\infty, L_{N_1}))_k \in F_0$ and according to (3.58) there is a sequence $a \in \ell_\infty$ satisfying

$$\mathcal{K}(2^k, a; \ell_1, \ell_\infty) \asymp \mathcal{K}(2^k, x; L_\infty, L_{N_1}), \quad k = 0, \pm 1, \pm 2, \ldots \tag{3.61}$$

Therefore, $(\mathcal{K}(2^k, a; \ell_1, \ell_\infty))_k \in F_0$, whence $a \in (\ell_1, \ell_\infty)^{\mathcal{K}}_{F_0}$. Thus, equations (3.60), combined together with the hypothesis of the proposition, give $a \in (\ell_1, \ell_\infty)^{\mathcal{K}}_{F_1}$. Hence, (3.59) and (3.61) imply $x \in X_1$. As a result, $X_0 \subset X_1$. Changing the places of these spaces, we can come, in the same way, to the opposite embedding. So, $X_0 = X_1$ as sets. The equivalence of the norms in X_0 and X_1 follows now from the Closed Graph theorem, because of every s.s. is continuously embedded into L_1. □

Remark 3.5 For an arbitrary s.s. Z we have $H_Z \in \mathcal{I}(\ell_1, \ell_\infty)$ (cf. Corollary 3.2). Indeed, if B is a linear operator such that $\|B\|_{\ell_1 \to \ell_1} \le 1$ and $\|B\|_{\ell_\infty \to \ell_\infty} \le 1$, then applying once more (3.48) and (3.57), we get

$$\|Ba\|_{H_Z} = \Big\|\sum_{k=1}^{\infty}(Ba)_k^* \chi_{(0,2^{-k+1}]}\Big\|_Z = \Big\|\sum_{k=1}^{\infty}\mathcal{K}(k, Ba; \ell_1, \ell_\infty)\chi_{(2^{-k},2^{-k+1}]}\Big\|_Z \le$$

$$\le \Big\|\sum_{k=1}^{\infty}\mathcal{K}(k, a; \ell_1, \ell_\infty)\chi_{(2^{-k},2^{-k+1}]}\Big\|_Z = \Big\|\sum_{k=1}^{\infty}\Big(\sum_{i=1}^{k} a_i^*\Big)\chi_{(2^{-k},2^{-k+1}]}\Big\|_Z = \|a\|_{H_Z}.$$

Hence, in particular, it follows that the interpolation property of X_0 and X_1 with respect to the couple (L_∞, L_{N_1}) in the previous corollary cannot be dropped. Indeed, let X_0 be a non-interpolation s.s. with respect to the last couple. Then, $H_{X_0} \in \mathcal{I}(\ell_1, \ell_\infty)$ and therefore $H_{X_0} = (\ell_1, \ell_\infty)_F^{\mathcal{K}}$ for some parameter F. Setting $X_1 := (L_\infty, L_{N_1})_F^{\mathcal{K}}$, observe that, by Corollary 3.10, we have $H_{X_1} = (\ell_1, \ell_\infty)_F^{\mathcal{K}} = H_{X_0}$. At the same time, $X_1 \ne X_0$, because $X_0 \notin \mathcal{I}(L_\infty, L_{N_1})$ while $X_1 \in \mathcal{I}(L_\infty, L_{N_1})$.

Finally, we are ready to finish the proof of Theorem 3.8.

Necessity In view of the above discussion, it remains to establish the interpolation property of the s.s. Y, defined by (3.50), with respect to the Banach couple (L_∞, L_{N_1}).

Let a linear operator A be bounded in the spaces L_∞ and L_{N_1}, $\|A\|_{L_\infty \to L_\infty} \le 1$ and $\|A\|_{L_{N_1} \to L_{N_1}} \le 1$. Then, by definition of the $\mathcal{K}$-functional, we have

$$\mathcal{K}(t, Ax; L_\infty, L_{N_1}) \le \mathcal{K}(t, x; L_\infty, L_{N_1}), \quad t > 0.$$

Moreover, applying Lemma 3.4 (see also Remark 3.4), we get the estimate

$$\mathcal{K}(t, Ax; L_\infty, L_{N_1}) \asymp t \sup_{0<u\le 2^{1-t}} \big((Ax)^*(u)\log_2^{-1}(2/u)\big) \ge (Ax)^*(2^{1-t}), \quad t \ge 1.$$

On the other hand, since φ_1 is equivalent to the fundamental function of the space L_{N_1}, we have $\Lambda(\varphi_1) \subset L_{N_1}$ (see Appendix C). Therefore, by definition of the operator V and Lemma 3.7, we get

$$\begin{aligned}\mathcal{K}(t, x; L_\infty, L_{N_1}) &\le C_1\mathcal{K}(t, x; L_\infty, \Lambda(\varphi_1))\\ &\le C_2 t\int_0^{2^{1-t}} x^*(s)\,d\varphi_1(s)\\ &\le C_2 Vx(2^{1-t}), \quad t \ge 1.\end{aligned}$$

From the last inequalities it follows that $(Ax)^*(s) \le CVx(s)$, $0 < s \le 1$, for all $x \in \Lambda_2$. Since the operator V is bounded in Y (see Lemma 3.5), so is A. As a result, we have $Y \in \mathcal{I}(L_\infty, L_{N_1})$, and the proof is completed. □

Comments and References A key role in this chapter is played by the methods of the interpolation theory of operators. Theorems 3.1, 3.3, and 3.4 were proved by Astashkin in 1997 (see [12, 16]). As it follows from the proofs given here, a clue, in this regard, is the notion of a $\mathcal{K}$-closed subcouple, which has been introduced by Peetre in [230] and which widely used in studying various problems of the theory of function spaces and operators (see, in particular, [3, 140, 157–160, 240, 257]). A description of sequence spaces generated by the Rademacher system in function s.s.'s, given in Theorem 3.2, has been appeared, for the first time, in the paper [23].

The proof of Theorem 3.4 has a rather long prehistory. As early as in 1973, under much stronger restrictions imposed on s.s.'s X_0 and X_1, Rodin proved a weaker version of this theorem, namely, he showed the equivalence of the fundamental functions of these spaces [250, Theorem 3.11]. Later on, the result was obtained in [251] (see Theorem 5), but still under the same restrictions to s.s.'s X_0 and X_1. Note that the proof of Theorem 3.4 in the paper [12] somewhat differs from the proof given here, although in both cases the main tool is the $\mathcal{K}$-monotonicity of the Banach couples in question as well as their abundance with respect to some special function cones.

Lemma 3.3 has been obtained in [10]. By using a somewhat more sophisticated reasoning, the statement of Proposition 3.3 can be proved even for any couple of the form $\vec{X} = (L_\infty, M(\varphi))$ [103, Theorem 1] (however, we preferred here to give another, shorter proof based on using the techniques of partial retracts). Results related to Rademacher subspaces of non-interpolation s.s.'s with respect to the couple (L_∞, G) (Theorems 3.5, 3.6, 3.7, and their consequences) have been proved in the paper [43]. The equivalent expression for the $\mathcal{K}$-functional $\mathcal{K}(t, x; L_\infty, L_{N_\alpha})$ from Lemma 3.4 has appeared, for the first time, apparently in [25].

Proposition 3.5, repeatedly used in this book, is borrowed from the paper [251]. The same work contains also equivalence (3.34) (earlier, a slightly weaker statement was obtained by Nikishin [219], who considered a more general case of sequences of independent functions). Note that the Lorentz space $\Lambda_p(\varphi)$, where $\varphi(s) = \ln^{1-p}(e/s)$ and $1 < p < 2$ (see Example 3.1), is an interpolation space with respect to the couple (L_∞, L_{N_2}) [13, Theorem 1]. Hence, in view of equivalences (3.34), (3.35), and Theorem 3.4, we have $\Lambda_p(\varphi) = (L_\infty, G)_{\theta,p}$, with $\theta = 2(p-1)/p$. By using an extrapolation description of the space $\Lambda_p(\varphi)$ [24, Remark 3], we obtain the following criterion for the sequence of coefficients of a series $S = \sum_{k=1}^\infty a_k r_k$ to belong to the space ℓ_p, $1 < p < 2$, in terms of the L_p-norms of S: $(a_k)_{k=1}^\infty \in \ell_p$ if and only if

$$\sum_{k=1}^{\infty} \left(\frac{\|S\|_k}{k} \right)^p < \infty.$$

Another proof of this statement, based on some results of the famous Rudin's work [254], has been given by Orlov in [224].

Equivalence (3.37) from Example 3.2 has been proved independently in the papers [251] and [194] and it is being the most common illustration of the behavior of Rademacher functions in s.s.'s located "near" to L_∞ (see, for instance, [233, § 9] and [177, p. 94]). Notice that the proof of this result in the book differs from those given in [251] and [194]. Arguing in the same way as in the end of the proof of Proposition 3.4 (see Lemma 2.4 and [7, Theorem 5]), one can prove that, for each $\alpha > 2$, L_{N_α} is an interpolation space between L_∞ and L_{N_2}. Then, from Proposition 3.3 and the coincidence of the $\mathcal{K}$-functionals of the couples (L_∞, G) and (L_∞, L_{N_2}) on G it follows that L_{N_α} is an interpolation space between L_∞ and G as well. Therefore, by Theorem 3.4, for every $\alpha > 2$, L_{N_α} is a unique interpolation space between L_∞ and G satisfying equivalence (3.37). Lastly, equivalence (3.39) from Example 3.3 has been proved in [12].

The idea to identify the Rademacher subspace of a s.s. as a cone of dyadic step-functions goes back to the work [251] by Rodin and Semenov. Inequality (3.44) and the assertion of Theorem 3.8, under some stronger conditions imposed on a s.s., were proved also in [251]. Observe that the latter result has been used in [251] to prove the uniqueness of a s.s. having the given Rademacher subspace. Here, in contrast to that, Theorems 3.4 and 3.8 are proved independently. A somewhat weaker version of Theorem 3.8 (namely, for interpolation spaces between L_∞ and G) was announced in [55]. Note that Lemma 3.7 and Proposition 3.7 may be deduced also from the formula for the $\mathcal{K}$-functional of the couple (L_∞, X), where X is a s.s. with a strictly increasing fundamental function ϕ_X, from the paper [189] (see also a remark from [190]). The comparison of Proposition 3.9 with Theorem 3.4 indicates an interesting analogy between properties of the Rademacher subspaces of s.s.'s and the corresponding cones of dyadic step-functions. In the proof of Proposition 3.6 we make use of that of Theorem 5 from the paper [251].

Chapter 4
Rademacher Sums with Vector Coefficients

In this chapter we deal with Rademacher sums, whose coefficients are elements of a Banach space. In contrast to the scalar case, we have no longer the following "natural" orthogonality equation

$$\int_0^1 \left|\sum_{i=1}^n a_i r_i(t)\right|^2 dt = \sum_{i=1}^n a_i^2 \text{ if } a_i \in \mathbb{R},$$

and so we cannot compare directly averages of Rademacher sums with the appropriate quantities determined by their coefficients. It is worth to note that the classification of Banach spaces with respect to a similar condition underlies of the concepts of Rademacher type and cotype, which play an important role in the study of their geometric properties. However, as it will be shown here, the analogues of almost all results obtained in previous chapters in the scalar case hold as well for Rademacher sums with vector coefficients.

4.1 The Kahane–Khintchine Inequality and Its Consequences

By Khintchine's inequality, for each $1 \le p < \infty$ the Rademacher system is equivalent in $L_p[0, 1]$ to the unit vector basis in ℓ_2. In consequence, the norms of scalar Rademacher sums in $L_p[0, 1]$ are equivalent for different (finite) p. The following classical result due to Kahane shows that this statement extends as well to the vector-valued case.

S. V. Astashkin, *The Rademacher System in Function Spaces*,
https://doi.org/10.1007/978-3-030-47890-2_4

Theorem 4.1 (Kahane–Khintchine Inequality) *For every $r > 1$ there exists a constant $K_r > 0$ such that for arbitrary Banach space F, all $n \in \mathbb{N}$ and $x_1, x_2, \dots, x_n \in F$ we have*

$$\int_0^1 \Big\| \sum_{i=1}^n r_i(t)x_i \Big\|_F dt \le \Big(\int_0^1 \Big\| \sum_{i=1}^n r_i(t)x_i \Big\|_F^r dt \Big)^{1/r} \le K_r \int_0^1 \Big\| \sum_{i=1}^n r_i(t)x_i \Big\|_F dt. \tag{4.1}$$

In addition, if a series $\sum_{i=1}^\infty r_i(t)x_i$ converges in F a.e. on $[0, 1]$, then the inequality

$$\int_0^1 \Big\| \sum_{i=1}^\infty r_i(t)x_i \Big\|_F dt \le \Big(\int_0^1 \Big\| \sum_{i=1}^\infty r_i(t)x_i \Big\|_F^r dt \Big)^{1/r} \le K_r \int_0^1 \Big\| \sum_{i=1}^\infty r_i(t)x_i \Big\|_F dt$$

holds.

To prove this, we need the following numerical inequality.

Lemma 4.1 *Let $1 < p \le q < \infty$ and $l := \sqrt{(p-1)/(q-1)}$. Then, for any $u \ge 0$ and $v \ge 0$ we have*

$$\left(\frac{|v + lu|^q + |v - lu|^q}{2} \right)^{1/q} \le \left(\frac{|v + u|^p + |v - u|^p}{2} \right)^{1/p}. \tag{4.2}$$

Proof By homogeneity, it suffices to consider only the case when $v = 1$ (the inequality is obvious for $v = 0$). First, assume that $1 < p \le q \le 2$ and $0 \le u \le 1$. For each $k = 1, 2, \dots$ we define the "binomial coefficients" by

$$C_{2k}^{(q)} := \frac{q(q-1)\dots(q-2k+1)}{(2k)!} \quad \text{and} \quad C_{2k}^{(p)} := \frac{p(p-1)\dots(p-2k+1)}{(2k)!}.$$

In view of the conditions imposed on p and q, they are nonnegative. Therefore, since $C_{2k}^{(2)} = 0$, $k = 2, 3, \dots$, a direct inspection shows that the inequality

$$C_{2k}^{(q)} l^{2k} \le \frac{q}{p} C_{2k}^{(p)}, \quad k = 1, 2, \dots \tag{4.3}$$

holds for $q = 2$. Furthermore, the last inequality is valid also for $k = 1$ and $1 < p \le q < 2$. For the remaining values of k we get, after cancellations, that (4.3) is equivalent to the inequality

$$\prod_{i=2}^{2k-1} \frac{q-i}{\sqrt{q-1}} \le \prod_{i=2}^{2k-1} \frac{p-i}{\sqrt{p-1}}. \tag{4.4}$$

Observe that

$$\frac{p-i}{\sqrt{p-1}} \le \frac{q-i}{\sqrt{q-1}} < 0, \quad i = 2, 3, \ldots$$

Therefore, since the number of factors in the both sides of inequality (4.4) is the same and even, it holds. Thus, (4.3) is proved for all $1 < p \le q \le 2$ and $k = 1, 2, \ldots$ Consequently, by using the Taylor expansion of the function $f(x) = (1+x)^r$, the elementary inequality $(1+x)^{p/q} \le 1 + px/q$ for $x \ge 0$, and (4.3), we obtain

$$\left(\frac{(1+lu)^q + (1-lu)^q}{2}\right)^{p/q} = \left(1 + \sum_{k=1}^{\infty} C_{2k}^{(q)} l^{2k} u^{2k}\right)^{p/q} \le$$

$$\le 1 + \sum_{k=1}^{\infty} C_{2k}^{(p)} u^{2k} = \frac{(1+u)^p + (1-u)^p}{2}.$$

Thus, the lemma is proved in the case $0 \le u \le 1$.

Let now $u > 1$. It can be easily checked that $|1 \pm lu| \le |u \pm l|$ because of $0 < l \le 1$. Therefore, from inequality (4.2), where u is replaced with $1/u (\le 1)$, it follows that

$$\left(\frac{|1+lu|^q + |1-lu|^q}{2}\right)^{1/q} \le \left(\frac{|u+l|^q + |u-l|^q}{2}\right)^{1/q} \le$$

$$\le u \left(\frac{|1+1/u|^p + |1-1/u|^p}{2}\right)^{1/p} = \left(\frac{|1+u|^p + |1-u|^p}{2}\right)^{1/p}.$$

Summing up, we conclude that (4.2) is proved if $1 < p \le q \le 2$.

Next, in order to get the desired result for the other values of p and q, we define the mapping T on $L_1[0,1]$ by

$$Tf(s) := \int_0^1 f(t)\,dt + l \int_0^1 f(t) r_1(t)\,dt \cdot r_1(s).$$

Then, inequality (4.2) holds if and only if the norm of T as an operator from $L_p[0,1]$ into $L_q[0,1]$ does not exceed 1. Indeed, firstly, one can check that this statement is true if we replace T with its restriction to the two-dimensional space $M := \{a + br_1,\ a, b \in \mathbb{R}\}$. Also, let us observe that $T = TR$, where R is the orthogonal projection in L_p with the image M, i.e.,

$$Rf(s) := \int_0^1 f(t)\,dt + \int_0^1 f(t) r_1(t)\,dt \cdot r_1(s).$$

Secondly, direct calculations show that the norm of R in L_p is equal to 1, and hence the norm of T from L_p into L_q is the same as that of the restriction of T to M. This completes the proof of our claim.

Further, assume that $2 \le p \le q < \infty$. It is easily to see that the operator T is self-adjoint, i.e., $T^* = T$, and $l = \sqrt{(q'-1)/(p'-1)}$ (here, $1/p + 1/p' = 1/q + 1/q' = 1$). Since $1 < q' \le p' \le 2$, then, as was proved, the norm of T from $L_{q'}$ into $L_{p'}$, and thereby from L_p into L_q does not exceed 1. Thus, (4.2) is proved.

The remaining case when $1 < p < 2 < q < \infty$ can be reduced to the two preceding ones. Indeed, since $l = l_1 l_2$, where $l_1 = \sqrt{p-1}$ and $l_2 = 1/\sqrt{q-1}$, we have

$$\left(\frac{|v+lu|^q + |v-lu|^q}{2}\right)^{1/q} \le \left(\frac{|v+l_1u|^2 + |v-l_1u|^2}{2}\right)^{1/2} \le \left(\frac{|v+u|^p + |v-u|^p}{2}\right)^{1/p}.$$

□

Corollary 4.1 *Let $1 < p \le q < \infty$ and $l = \sqrt{(p-1)/(q-1)}$. Then, for every Banach space F and all $y_1, y_2 \in F$ we have*

$$\left(\frac{\|y_1+ly_2\|^q + \|y_1-ly_2\|^q}{2}\right)^{1/q} \le \left(\frac{\|y_1+y_2\|^p + \|y_1-y_2\|^p}{2}\right)^{1/p}.$$

Proof Denote $z_1 = y_1 + y_2$, $z_2 = y_1 - y_2$, $u = (|\|z_1\| - \|z_2\||)/2$, and $v = (\|z_1\| + \|z_2\|)/2$. Since

$$y_1 + ly_2 = \frac{(1+l)z_1}{2} + \frac{(1-l)z_2}{2}, \quad y_1 - ly_2 = \frac{(1-l)z_1}{2} + \frac{(1+l)z_2}{2}$$

and $0 < l \le 1$, by Lemma 4.1, we get

$$\left(\frac{\|y_1+ly_2\|^q + \|y_1-ly_2\|^q}{2}\right)^{1/q}$$

$$\le \left(\frac{((1+l)\|z_1\|/2 + (1-l)\|z_2\|/2)^q + ((1-l)\|z_1\|/2 + (1+l)\|z_2\|/2)^q}{2}\right)^{1/q}$$

$$\le \left(\frac{|v+lu|^q + |v-lu|^q}{2}\right)^{1/q} \le \left(\frac{|v+u|^p + |v-u|^p}{2}\right)^{1/p} = \left(\frac{\|z_1\|^p + \|z_2\|^p}{2}\right)^{1/p}.$$

□

Proof of Theorem 4.1 As above, we set $l = \sqrt{(p-1)/(q-1)}$, $1 < p \le q < \infty$. It suffices to show that for every Banach space F and all $x_0, x_1, \dots, x_n \in F$ we have

$$\left(\int_0^1 \Big\|x_0 + l\sum_{i=1}^n r_i(t)x_i\Big\|^q dt\right)^{1/q} \le \left(\int_0^1 \Big\|x_0 + \sum_{i=1}^n r_i(t)x_i\Big\|^p dt\right)^{1/p}. \tag{4.5}$$

Indeed, putting $p = r$, $q = 2r$, where $r > 1$, and $x_0 = 0$, we deduce from (4.5) that

$$\left(\int_0^1 \Big\|\sum_{i=1}^n r_i(t)x_i\Big\|^{2r} dt\right)^{1/(2r)} \le \sqrt{\frac{2r-1}{r-1}}\left(\int_0^1 \Big\|\sum_{i=1}^n r_i(t)x_i\Big\|^r dt\right)^{1/r}.$$

Moreover, by Hölder's inequality,

$$\left(\int_0^1 \Big\|\sum_{i=1}^n r_i(t)x_i\Big\|^r dt\right)^{1/r} = \left(\int_0^1 \Big\|\sum_{i=1}^n r_i(t)x_i\Big\|^{r/(2r-1)} \Big\|\sum_{i=1}^n r_i(t)x_i\Big\|^{(2r^2-2r)/(2r-1)} dt\right)^{1/r} \le$$

$$\le \left(\int_0^1 \Big\|\sum_{i=1}^n r_i(t)x_i\Big\| dt\right)^{1/(2r-1)} \left(\int_0^1 \Big\|\sum_{i=1}^n r_i(t)x_i\Big\|^{2r} dt\right)^{(r-1)/(2r^2-r)}.$$

Combining this with the preceding inequality, after easy calculations we get

$$\left(\int_0^1 \Big\|\sum_{i=1}^n r_i(t)x_i\Big\|^r dt\right)^{1/r} \le \left(\frac{2r-1}{r-1}\right)^{r-1} \int_0^1 \Big\|\sum_{i=1}^n r_i(t)x_i\Big\| dt,$$

whence (4.1) follows with the constant $K_r = ((2r-1)/(r-1))^{r-1}$.

We prove now inequality (4.5) by induction. Clearly, for $n = 1$ it is sufficient to apply Corollary 4.1. Assuming that (4.5) is fulfilled for n, we prove it for $n+1$. Let $x_0, x_1, \dots, x_{n+1} \in F$. Since the Rademacher functions $r_i(t)$, $1 \le i \le n$, are constant on each of the dyadic intervals Δ_n^k, $k = 1, 2, \dots, 2^n$, then, by the inductive hypothesis, we have

$$\begin{aligned} U_{n+1} &:= \left(\int_0^1 \Big\|x_0 + l\sum_{i=1}^{n+1} r_i(t)x_i\Big\|^q dt\right)^{1/q} \\ &= \left(\frac{1}{2}\int_0^1 \left(\Big\|x_0 + lx_{n+1} + l\sum_{i=1}^n r_i(t)x_i\Big\|^q \right.\right. \\ &\quad \left.\left. + \Big\|x_0 - lx_{n+1} + l\sum_{i=1}^n r_i(t)x_i\Big\|^q\right) dt\right)^{1/q} \\ &\le 2^{-1/q}(\|\Phi\|_p^q + \|\Psi\|_p^q)^{1/q}, \end{aligned}$$

where

$$\Phi(t) := \Big\|x_0 + lx_{n+1} + \sum_{i=1}^{n} r_i(t)x_i\Big\| \quad \text{and} \quad \Psi(t) := \Big\|x_0 - lx_{n+1} + \sum_{i=1}^{n} r_i(t)x_i\Big\|.$$

Note that

$$(\|\Phi\|_p^q + \|\Psi\|_p^q)^{1/q} \le \|(\Phi^q + \Psi^q)^{1/q}\|_p.$$

Indeed, as it is easily seen, this inequality can be rewritten in the form

$$\|\Phi^q\|_\alpha + \|\Psi^q\|_\alpha \le \|\Phi^q + \Psi^q\|_\alpha, \quad \text{where } \alpha := p/q,$$

which is just (reverse) Minkowski's inequality for the L_α-norm with $0 < \alpha \le 1$. Therefore, from the preceding estimate for U_{n+1} and Corollary 4.1, for $y_1 = x_0 + \sum_{i=1}^{n} r_i(t)x_i$ (with a fixed $t \in [0, 1]$) and $y_2 = x_{n+1}$ it follows that

$$U_{n+1} \le 2^{-1/q}\|(\Phi^q + \Psi^q)^{1/q}\|_p \le \Big(\frac{1}{2}\int_0^1 \Big(\Big\|x_0 + x_{n+1} + \sum_{i=1}^{n} r_i(t)x_i\Big\|^p$$

$$+ \Big\|x_0 - x_{n+1} + \sum_{i=1}^{n} r_i(t)x_i\Big\|^p\Big)\,dt\Big)^{1/p} = \Big(\int_0^1 \Big\|x_0 + \sum_{i=1}^{n+1} r_i(t)x_i\Big\|^p\,dt\Big)^{1/p},$$

and (4.5) is proved.

The second assertion of the theorem is an immediate consequence of the first one and the fact that the a.e. convergence on $[0, 1]$ of a series $\sum_{i=1}^{\infty} r_i(t)x_i$ in a Banach space F implies its convergence in the space $L_1(F)$ (see e.g. [280, Theorem 5.3.1]). □

Setting $x_0 = 0$ in inequality (4.5), we obtain the following result on the comparison of the L_p- and L_q-norms of vector-valued Rademacher sums.

Corollary 4.2 *Let F be a Banach space, $x_i \in F$, $i = 1, 2, \ldots$ Suppose the series $\sum_{i=1}^{\infty} r_i(t)x_i$ converges in F a.e. on $[0, 1]$. Then, for arbitrary $1 < p \le q < \infty$ we have*

$$\Big(\int_0^1 \Big\|\sum_{i=1}^{\infty} r_i(t)x_i\Big\|^q\,dt\Big)^{1/q} \le \sqrt{\frac{q-1}{p-1}}\Big(\int_0^1 \Big\|\sum_{i=1}^{\infty} r_i(t)x_i\Big\|^p\,dt\Big)^{1/p}. \tag{4.6}$$

Furthermore, under the same conditions as in Corollary 4.2, from Theorem 4.1 it follows the exponential integrability of the mapping $t \in [0, 1] \longmapsto \|\sum_{i=1}^{\infty} r_i(t)x_i\|_F$.

Corollary 4.3 *If a series* $\sum_{i=1}^{\infty} r_i(t)x_i$, $x_i \in F$, $i = 1, 2, \ldots$, *converges in* F *a.e. on* $[0, 1]$, *then for every* $\alpha > 0$

$$\int_0^1 \exp\left(\alpha \Big\| \sum_{i=1}^{\infty} r_i(t)x_i \Big\|^2\right) dt < \infty.$$

Proof By inequality (4.6) for $p = 2$ and $q = 2k$, $k \in \mathbb{N}$, we have

$$\int_0^1 \exp\left(\alpha \Big\| \sum_{i=1}^{\infty} r_i x_i \Big\|^2\right) dt = \sum_{k=0}^{\infty} \frac{\alpha^k}{k!} \int_0^1 \Big\| \sum_{i=1}^{\infty} r_i x_i \Big\|^{2k} dt$$
$$\leq \sum_{k=0}^{\infty} \frac{\alpha^k (2k-1)^k}{k!} \left(\int_0^1 \Big\| \sum_{i=1}^{\infty} r_i x_i \Big\|^2 dt \right)^k. \quad (4.7)$$

Since the series $\sum_{i=1}^{\infty} r_i x_i$ converges in the space $L_2(F)$ (see Corollary 4.2), then the desired result follows whenever $\alpha > 0$ is sufficiently small.

Let now $\alpha > 0$ be arbitrary. Applying estimate (4.7) to the tail $\sum_{i=n+1}^{\infty} r_i x_i$ for any $n \in \mathbb{N}$ and taking 2α instead of α, we get

$$\int_0^1 \exp\left(2\alpha \Big\| \sum_{i=n+1}^{\infty} r_i x_i \Big\|^2\right) dt \leq \sum_{k=0}^{\infty} \frac{2^k \alpha^k (2k-1)^k}{k!} \left(\int_0^1 \Big\| \sum_{i=n+1}^{\infty} r_i x_i \Big\|^2 dt \right)^k. \quad (4.8)$$

As above, according to the hypothesis and Theorem 4.1, the integral

$$\int_0^1 \Big\| \sum_{i=n+1}^{\infty} r_i x_i \Big\|^2 dt$$

can be made arbitrarily small whenever n is sufficiently large. Consequently, from inequality (4.8) it follows that for every $\alpha > 0$ and n large enough

$$\int_0^1 \exp\left(2\alpha \Big\| \sum_{i=n+1}^{\infty} r_i x_i \Big\|^2\right) dt < \infty.$$

Furthermore, since the Rademacher functions are independent (see Corollary 1.1 and equation (A.1)), easy calculations show that

$$\int_0^1 \exp\left(\alpha \Big\| \sum_{i=1}^{\infty} r_i x_i \Big\|^2\right) dt \leq \int_0^1 \exp\left(2\alpha \Big\| \sum_{i=1}^{n} r_i x_i \Big\|^2\right) dt \cdot \int_0^1 \exp\left(2\alpha \Big\| \sum_{i=n+1}^{\infty} r_i x_i \Big\|^2\right) dt.$$

As a result, the integral from the left-hand side of this inequality is finite for each $\alpha > 0$, and this finishes the proof. □

4.2 A Distribution Deviation Inequality

In Chapter 1, we proved some distribution estimates for Rademacher sums with scalar coefficients. Let now F be a Banach space, $x_1, x_2, \dots, x_n \in F$. We shall get here a similar estimate for the deviation of the r.v. $t \in [0,1] \mapsto \|\sum_{i=1}^n x_i r_i(t)\|_F$ from its median.

Let $n \in \mathbb{N}$ be fixed. Rather than the Rademacher functions r_k, $k = 1, 2, \dots, n$, on the interval $[0,1]$ we shall consider the sign arrangements, i.e., the collections $\{\varepsilon_k\}_{k=1}^n$, $\varepsilon_k = \pm 1$, as elements of the probability space $(\{-1,1\}^n, \mathbb{P}_n)$, where $\mathbb{P}_n$ is the canonical probability measure that gives mass 2^{-n} to each point of the set $\{-1,1\}^n$. Ultimately, we are interested in estimation of distribution of the r.v. $\|\sum_{k=1}^n x_k \varepsilon_k\|_F$, where $x_1, \dots, x_n$ are arbitrary vectors from a Banach space F. However, the main result of the section concerns with a slightly different r.v.

Let us consider $\{-1,1\}^n$ as a subset of the n-dimensional Hilbert space $\mathbb{R}^n$ equipped with the usual Euclidean inner product. Moreover, for every nonempty set $A \subset \{-1,1\}^n$ by $\zeta_A(w)$ we denote the distance from a point $w \in \{-1,1\}^n$ to the convex hull of A, i.e.,

$$\zeta_A(w) := \inf\{\|w - y\|_e : y \in \operatorname{conv} A\},$$

where $\|z\|_e$ is the usual Euclidean norm of a vector $z \in \mathbb{R}^n$. Clearly, ζ_A is an r.v. on the probability space $(\{-1,1\}^n, \mathbb{P}_n)$. Finally, by $\mathbb{E}_n f$ we denote the expectation of an r.v. f defined on $(\{-1,1\}^n, \mathbb{P}_n)$.

Theorem 4.2 *For every $A \subset \{-1,1\}^n$ we have*

$$\mathbb{E}_n \exp\left(\frac{\zeta_A^2}{8}\right) \leqslant \frac{1}{\mathbb{P}_n(A)}.$$

Proof We assume first that card $A = 1$. If i is the number of coordinates where the elements w and y of $\{-1,1\}^n$ differ, then $\|w - y\|_e^2 = 4i$. Therefore,

$$\mathbb{E}_n \exp\left(\frac{\zeta_A^2}{8}\right) = 2^{-n} \sum_{0 \leqslant i \leqslant n} C_n^i e^{i/2} = \left(\frac{1 + e^{1/2}}{2}\right)^n \leqslant 2^n = \frac{1}{\mathbb{P}_n(A)}$$

(as usual, $C_n^i = n!/(i!(n-i)!)$). Thus, it is left to consider the case when card $A \geq 2$.

Next, we use induction over n. If $n = 1$, then from the preceding observation it follows that $A = \{-1,1\}$, and hence $\zeta_A(w) \equiv 0$ for each $w \in \{-1,1\}$, and we get the desired result. Assuming that it holds for n, we prove it for $n+1$. Observe that A contains at least 2 points. Therefore, identifying the set $\{-1,1\}^{n+1}$ with the product $\{-1,1\}^n \times \{-1,1\}$, we have $A = A_{-1} \times \{-1\} \cup A_1 \times \{1\}$, where $A_{-1}, A_1 \subset \{-1,1\}^n$ and $A_{-1} \neq \varnothing, A_1 \neq \varnothing$. For definiteness, we shall assume further that $\mathbb{P}_n(A_{-1}) \leqslant \mathbb{P}_n(A_1)$.

From the definition of ζ_A it follows that for each point $w \in \{-1, 1\}^n$ we have

$$\zeta_A((w, 1)) \leqslant \zeta_{A_1}(w). \tag{4.9}$$

Moreover, we claim that for any $w \in \{-1, 1\}^n$ and $0 \leqslant \alpha \leqslant 1$ it holds

$$\zeta_A^2((w, -1)) \leqslant 4\alpha^2 + \alpha\zeta_{A_1}^2(w) + (1-\alpha)\zeta_{A_{-1}}^2(w). \tag{4.10}$$

Indeed, for $i = -1, 1$ choose $z_i \in conv A_i$ so that $||w - z_i||_e = \zeta_{A_i}(w)$. Then, $(z_i, i) \in conv A$, and hence $z = (\alpha z_1 + (1-\alpha)z_{-1}, -1 + 2\alpha) \in conv A$. Further, by the triangle inequality and the convexity of the function $t \mapsto t^2$, we have

$$\begin{aligned} ||(w, -1) - z||_e^2 &= 4\alpha^2 + ||w - (\alpha z_1 + (1-\alpha)z_{-1})||_e^2 \\ &= 4\alpha^2 + ||\alpha(w - z_1) + (1-\alpha)(w - z_{-1})||_e^2 \\ &\leqslant 4\alpha^2 + \alpha||w - z_1||_e^2 + (1-\alpha)||w - z_{-1}||_e^2, \end{aligned}$$

and (4.10) is proved.

Now, denoting $u_i := \mathbb{E}_n \exp(\zeta_{A_i}^2/8)$ and $v_i := 1/\mathbb{P}_n(A_i)$, we see that $u_i \leqslant v_i$, $i = -1, 1$, by the induction hypothesis. Therefore, from (4.9), (4.10) and Hölder's inequality for all $0 \leqslant \alpha \leqslant 1$ we get

$$\begin{aligned} \mathbb{E}_{n+1} \exp\left(\frac{\zeta_A^2}{8}\right) &\leqslant \frac{1}{2}\mathbb{E}_n \exp(\zeta_{A_1}^2/8) + \frac{1}{2}\mathbb{E}_n \exp(\alpha^2/2 + \alpha\zeta_{A_1}^2/8 + (1-\alpha)\zeta_{A_{-1}}^2/8) \\ &\leqslant \frac{1}{2}u_1 + \frac{1}{2}e^{\alpha^2/2}\, \mathbb{E}_n \left\{(\exp(\zeta_{A_1}^2/8))^\alpha (\exp(\zeta_{A_{-1}}^2/8))^{1-\alpha}\right\} \\ &\leqslant \frac{1}{2}u_1 + \frac{1}{2}e^{\alpha^2/2}\, u_1^\alpha\, u_{-1}^{1-\alpha} \leqslant \frac{1}{2}v_1\left(1 + e^{\alpha^2/2}(v_1/v_{-1})^{\alpha-1}\right). \end{aligned}$$

Note that from the assumption $\mathbb{P}_n(A_{-1}) \leqslant \mathbb{P}_n(A_1)$ it follows that the number $\alpha = 1 - v_1/v_{-1}$ satisfies the inequality $0 \leqslant \alpha \leqslant 1$. Thus, the last estimate implies

$$\mathbb{E}_{n+1} \exp\left(\frac{\zeta_A^2}{8}\right) \leqslant \frac{1}{2}v_1\left(1 + e^{\alpha^2/2}(1-\alpha)^{\alpha-1}\right). \tag{4.11}$$

Furthermore, for all $0 \leqslant \alpha < 1$

$$1 + e^{\alpha^2/2}(1-\alpha)^{\alpha-1} \leqslant \frac{4}{2-\alpha}. \tag{4.12}$$

Indeed, this is equivalent to the inequality

$$e^{\alpha^2/2}(1-\alpha)^{\alpha-1} \leqslant \frac{2+\alpha}{2-\alpha}$$

or

$$\alpha^2/2 - (1-\alpha)\ln(1-\alpha) \leqslant \ln(1+\alpha/2) - \ln(1-\alpha/2).$$

We set

$$g(\alpha) := \ln(1+\alpha/2) - \ln(1-\alpha/2) - \alpha^2/2 + (1-\alpha)\ln(1-\alpha).$$

Then, $g(0) = 0$, and the desired inequality is an immediate consequence of the fact that the function g increases for $0 \leqslant \alpha < 1$, which can be easily verified by its differentiation. From (4.11) and (4.12) it follows

$$\begin{aligned}\mathbb{E}_{n+1} \exp\left(\frac{\zeta_A^2}{8}\right) &\leqslant \frac{2v_1}{2-\alpha} = \frac{2v_1}{1+v_1/v_{-1}} = \frac{2}{1/v_1 + 1/v_{-1}} \\ &= \frac{2}{\mathbb{P}_n(A_1) + \mathbb{P}_n(A_{-1})} = \frac{1}{\mathbb{P}_{n+1}(A)},\end{aligned}$$

and the proof is completed. □

By Chebyshev's inequality, Theorem 4.2 implies

Corollary 4.4 *For all $\tau \geqslant 0$ we have*

$$\mathbb{P}_n(\{\zeta_A \geqslant \tau\}) \leqslant \frac{1}{\mathbb{P}_n(A)} \cdot e^{-\tau^2/8}.$$

Recall that a function $f : T \to \mathbb{R}$ defined on a metric space (T, ρ) satisfies the *Lipschitz condition with constant σ* if for all $x, y \in T$

$$|f(x) - f(y)| \leq \sigma\rho(x, y).$$

Moreover, one says that a real number M is a *median* of an r.v. f if the following inequalities hold:

$$\mathbb{P}_n(\{f \geq M\}) \geq 1/2 \ \text{ and } \ \mathbb{P}_n(\{f \leq M\}) \geq 1/2$$

(see e.g. [118, § 2.7]).

Theorem 4.3 *Let a convex function f on $(\mathbb{R}^n, ||\cdot||_e)$ satisfy the Lipschitz condition with constant σ, and let M be a median of f (more precisely, of its restriction to the set $\{-1, 1\}^n$) with respect to $\mathbb{P}_n$. Then, for all $\tau \geqslant 0$ we have*

$$\mathbb{P}_n(\{|f - M| > \tau\}) \leqslant 4e^{-\tau^2/8\sigma^2}.$$

Proof Denote $A := \{y \in \{-1, 1\}^n : f(y) \leq M\}$. Since f is convex, then $f(y) \leq M$ for all $y \in \operatorname{conv} A$. Assume that $f(z) \geq M + \tau$, where $\tau > 0$. Then, choosing $y \in \operatorname{conv} A$ in such a way that $\zeta_A(z) = \|z - y\|_e$, by hypothesis, we get

$$|f(z) - f(y)| \leq \sigma \|z - y\|_e = \sigma \zeta_A(z),$$

whence

$$\zeta_A(z) \geq \frac{|f(z) - f(y)|}{\sigma} \geq \frac{\tau}{\sigma}.$$

Therefore, since $\mathbb{P}_n(A) \geq 1/2$, Corollary 4.4 implies that

$$\mathbb{P}_n(\{f \geq M + \tau\}) \leq \mathbb{P}_n(\{\zeta_A \geq \tau/\sigma\}) \leq 2e^{-\tau^2/8\sigma^2}. \tag{4.13}$$

Furthermore, consider the set $B := \{y \in \{-1, 1\}^n : f(y) \leq M - \tau\}$, $\tau > 0$. Let $u < \tau$. As above, one can deduce that from the inequality $f(z) \geq M - \tau + u$ it follows $\zeta_B(z) \geq u/\sigma$. Hence, again by Corollary 4.4, we have

$$\frac{1}{2} \leq \mathbb{P}_n(\{f \geq M - \tau + u\}) \leq \frac{1}{\mathbb{P}_n(B)} e^{-u^2/8\sigma^2},$$

whence $\mathbb{P}_n(B) \leq 2e^{-u^2/8\sigma^2}$. Tending now u to τ, we infer

$$\mathbb{P}_n(\{f \leq M - \tau\}) \leq 2e^{-\tau^2/8\sigma^2}.$$

Combining this inequality with (4.13), we complete the proof. □

Remark 4.1 The convexity condition imposed on f in Theorem 4.3 cannot be skipped. Indeed, let $A := \{y \in \{-1, 1\}^n : \sum_{i=1}^n y_i \leqslant 0\}$ and $f(w) := \inf\{\|w - y\|_e : y \in A\}$, $w \in \mathbb{R}^n$. To verify that f satisfies the Lipschitz condition with constant 1, take $w, z \in \mathbb{R}^n$ and for definiteness assume that $f(w) \geq f(z)$. Choosing $y \in A$ so that $f(z) = \|z - y\|_e$, we get

$$|f(w)-f(z)| = f(w)-f(z) \leq \|w-y\|_e-f(z) = \|w-y\|_e-\|z-y\|_e \leq \|w-z\|_e.$$

Also, observe that from $\mathbb{P}_n(A) = 1/2$ it follows $M = 0$.

On the other hand, for each $w \in \{-1, 1\}^n$, by definition,

$$f(w) = \sqrt{2} \cdot \left(\left(\sum_{i=1}^n w_i\right)_+\right)^{1/2},$$

where $a_+ := \max(a, 0)$. Since independent r.v.'s w_i, $1 \leq i \leq n$, are identically and symmetrically distributed on the probability space $(\{-1, 1\}^n, \mathbb{P}_n)$, by the central limit theorem (see Theorem A.1), we have

$$\mathbb{P}_n(\{f > cn^{1/4}\}) \geqslant \frac{1}{4}$$

with some $c > 0$ independent of n. Thus, the result of Theorem 4.3 fails for f.

Let F be an arbitrary Banach space, $x_1, x_2, \ldots, x_n \in F$. We define

$$\sigma^2 = \sigma^2\left(\{x_i\}_{i=1}^n\right) := \sup\left\{\sum_{i=1}^n x^*(x_i)^2 : x^* \in F^*, ||x^*|| \leqslant 1\right\}. \tag{4.14}$$

Corollary 4.5 *If M is a median of the r.v. $||\sum_{i=1}^n x_i r_i(s)||_F$, $0 \le s \le 1$, then for all $\tau \geqslant 0$ we have the estimate*

$$m\left(\left\{s \in [0,1] : \left|\left\|\sum_{i=1}^n x_i r_i(s)\right\|_F - M\right| \geqslant \tau\right\}\right) \leqslant 4e^{-\tau^2/8\sigma^2}.$$

Proof We apply Theorem 4.3 to the function

$$f : u = (u_i)_{i=1}^n \in \mathbb{R}^n \mapsto \left\|\sum_{i=1}^n u_i x_i\right\|_F.$$

Clearly, f is a convex function. Moreover, for any $u, v \in \mathbb{R}^n$ and $x^* \in F^*$, $\|x^*\| \le 1$, we have

$$\left|x^*\left(\sum_{i=1}^n u_i x_i\right) - x^*\left(\sum_{i=1}^n v_i x_i\right)\right| \leqslant \left|\sum_{i=1}^n (u_i - v_i) x^*(x_i)\right|$$
$$\leqslant \left(\sum_{i=1}^n x^*(x_i)^2\right)^{1/2} \cdot ||u - v||_e.$$

Combining this with the triangle inequality and (4.14), we get

$$|f(u) - f(v)| \leqslant \left\|\sum_{i=1}^n u_i x_i - \sum_{i=1}^n v_i x_i\right\| = \sup_{\|x^*\| \le 1}\left|x^*\left(\sum_{i=1}^n u_i x_i - \sum_{i=1}^n v_i x_i\right)\right| \le \sigma ||u - v||_e,$$

i.e., f satisfies the Lipschitz condition with constant σ. Thus, by Theorem 4.3, for all $\tau \geqslant 0$ it holds

$$\mathbb{P}\left(\left\{\left|\left\|\sum_{i=1}^n \varepsilon_i x_i\right\|_F - M\right| \geqslant \tau\right\}\right) \leqslant 4e^{-\tau^2/8\sigma^2},$$

where M is a median of the r.v. $||\sum_{i=1}^{n} \varepsilon_i x_i||_F$ defined on the probability space $(\{-1, 1\}^n, \mathbb{P}_n)$. Observe that for every Banach space F and any $x_1, x_2, \dots, x_n \in F$ the latter r.v. has the same distribution as the r.v. $||\sum_{i=1}^{n} x_i r_i(s)||_F$ on the probability space $([0, 1], m)$. Therefore, the proof is finished. □

Remark 4.2 A similar result holds also for a sequence $\{\mathfrak{g}_i\}_{i=1}^{n}$ of standard independent Gaussian r.v.'s. Namely, according to the Borell isoperimetric inequality [80], we have

$$\mathbb{P}\Big(\Big\{\Big|\Big\|\sum_{i=1}^{n} x_i \mathfrak{g}_i\Big\|_F - M\Big| \geqslant \tau\Big\}\Big) \leqslant \frac{2}{\sigma\sqrt{2\pi}} \int_{\tau}^{\infty} e^{-u^2/2\sigma^2}\, du \leqslant 2e^{-\tau^2/2\sigma^2},$$

where the second inequality holds for $\tau \geqslant \sigma(2\pi)^{1/2}$ (here, M is a median of the r.v. $||\sum_{i=1}^{n} x_i \mathfrak{g}_i||_F$ and σ is defined by (4.14)).

Using the same formula for the L_p-norm as in Remark 1.3, by Corollary 4.5, one can easily deduce the following variant of the Kahane–Khintchine inequality.

Corollary 4.6 *There exists a universal constant $K > 0$ such that for every Banach space F, any $n \in \mathbb{N}$, $x_1, x_2, \dots, x_n \in F$, and $p \geqslant 1$ we have*

$$\mathbb{E}\Big(\Big\|\sum_{i=1}^{n} x_i r_i\Big\|_F^p\Big)^{1/p} \leqslant \mathbb{E}\Big\|\sum_{i=1}^{n} x_i r_i\Big\|_F + K\sigma \cdot p^{1/2},$$

where σ is still defined by (4.14).

4.3 The Hitczenko and Montgomery-Smith Type Inequalities for Vector-Valued Rademacher Sums

Let $1 \leqslant p < \infty$, and let F be a Banach space. The set of all sequences $\{x_n\}_{n=1}^{\infty}$ from F such that for every $x^* \in F^*$ it holds

$$\|(x^*(x_n))\|_{\ell_p} := \Big(\sum_{n=1}^{\infty} |x^*(x_n)|^p\Big)^{1/p} < \infty$$

is a Banach space $\ell_p^w(F)$ equipped with the norm

$$\|\{x_n\}\|_{\ell_p^w(F)} := \sup_{\|x^*\| \leq 1} \|(x^*(x_n))\|_{\ell_p}.$$

The finiteness of this supremum for each $\{x_n\} \in \ell_p^w(F)$ is a consequence of the Closed Graph theorem. Indeed, for any such a sequence let us define the linear operator $T : F^* \to \ell_p$ by $Tx^* := (x^*(x_n))_{n=1}^{\infty}$. Suppose $x_k^* \to x^*$ in F^* and

$Tx_k^* \to u$ in ℓ_p. Then, if $u = (u_n)_{n=1}^\infty$, we have both $x_k^*(x_n) \to x^*(x_n)$ and $(Tx_k^*)_n = x_k^*(x_n) \to u_n$ as $k \to \infty$ for each $n \in \mathbb{N}$. Hence, $u_n = x^*(x_n)$ for all $n \in \mathbb{N}$, which implies $u = Tx^*$. Therefore, T has the closed graph, and so this operator is bounded. But the latter is equivalent to the fact that $\|\{x_n\}\|_{\ell_p^w(F)} < \infty$.

As above, by $\kappa(z, a)$ we denote the $\mathcal{K}$-functional $\mathcal{K}(z, a; \ell_1, \ell_2)$, where $a = (a_n) \in \ell_2$ and $z > 0$. Moreover, for every sequence $\{x_n\} \in \ell_2^w(F)$ we set

$$\kappa_w(z, \{x_n\}) := \sup_{\|x^*\| \leqslant 1} \kappa(z, (x^*(x_n))). \tag{4.15}$$

From the definition it follows that the mapping $z \mapsto \kappa_w(z, \{x_n\})$ is a continuous increasing function for $z > 0$.

The first result of this section gives the distribution estimates for norms of vector-valued Rademacher sums, which are similar to Montgomery-Smith's inequalities proved in Chap. 1 (see Theorem 1.7).

Theorem 4.4 *Let F be a Banach space. Suppose that a series $S := \sum_{n=1}^\infty x_n r_n$, where $x_n \in F$, $n = 1, 2, \ldots$, converges in F a.e. on* $[0, 1]$. *Then, for every $z > 0$ we have*

$$m\{t : \|S(t)\|_F > 2\mathbb{E}\|S\|_F + 6\kappa_w(z, \{x_n\})\} \leqslant 4e^{-z^2/8}, \tag{4.16}$$

and for a universal constant $c > 0$

$$m\left\{t : \|S(t)\|_F > \frac{1}{2}\mathbb{E}\|S\|_F + c\kappa_w(z, \{x_n\})\right\} \geqslant ce^{-z^2/c}. \tag{4.17}$$

We start with some auxiliary assertions.

Lemma 4.2 *Let $S_N = \sum_{n=1}^N x_n r_n$, $x_n \in F$, $N = 1, 2, \ldots$ For every $z > 0$ it holds*

$$m\{t : \|S_N(t)\|_F > 2\mathbb{E}\|S_N\|_F + 3\mathcal{K}(z, \{x_n\}_{n=1}^N; \ell_1^w(F), \ell_2^w(F))\} \leqslant 4e^{-z^2/8}.$$

Proof Suppose that $N \in \mathbb{N}$ and $y_1, \ldots, y_N \in F$ are fixed. Denote by M a median of the r.v. $S_{N,y}$ defined by

$$S_{N,y}(t) := \left\| \sum_{n=1}^N y_n r_n(t) \right\|_F, \quad t \in [0, 1].$$

By the definition of M (see e.g. [118, § II.7, p. 49]), we have

$$\mathbb{E}\|S_{N,y}\| \geq \int_{\{\|S_{N,y}(t)\| \geq M\}} \|S_{N,y}(t)\|\,dt \geq Mm\{t : \|S_{N,y}(t)\| \geq M\} \geq \frac{M}{2}.$$

Therefore, since the quantity $\sigma = \sigma(\{y_n\})$ (see (4.14)) coincides with $\|\{y_n\}\|_{\ell_2^w(F)}$, then from Corollary 4.5 it follows

$$m\left\{t: \|S_{N,y}(t)\| > 2\mathbb{E}\|S_{N,y}\| + z\|\{y_n\}\|_{\ell_2^w(F)}\right\} \leqslant 4e^{-z^2/8}, \quad z > 0. \tag{4.18}$$

On the other hand,

$$\max_{t\in[0,1]} \|S_{N,y}(t)\| = \max_{t\in[0,1]} \sup_{\|y^*\|\le 1} \left|\sum_{n=1}^{N} y^*(y_n) r_n(t)\right| = \sup_{\|y^*\|\le 1} \sum_{n=1}^{N} |y^*(y_n)| = \|\{y_n\}\|_{\ell_1^w(F)},$$

whence we get the following trivial estimate

$$m\left\{t: \|S_{N,y}(t)\| > \|\{y_n\}\|_{\ell_1^w(F)}\right\} = 0. \tag{4.19}$$

Further, by the definition of Peetre's $\mathcal{K}$-functional, for arbitrary $\varepsilon > 0$ we can find an expansion $x_n = x_n^{(1)} + x_n^{(2)}$, $1 \leqslant n \leqslant N$, such that

$$\|\{x_n^{(1)}\}\|_{\ell_1^w(F)} + z\|\{x_n^{(2)}\}\|_{\ell_2^w(F)} \le (1+\varepsilon)\mathcal{K}(z, \{x_n\}_{n=1}^N; \ell_1^w(F), \ell_2^w(F)). \tag{4.20}$$

Then, if $S_N^{(1)} = \sum_{n=1}^N x_n^{(1)} r_n$ and $S_N^{(2)} = \sum_{n=1}^N x_n^{(2)} r_n$, by (4.19), we have

$$\begin{aligned} \|\{x_n^{(1)}\}\|_{\ell_1^w(F)} &+ z\|\{x_n^{(2)}\}\|_{\ell_2^w(F)} + 2\mathbb{E}\|S_N^{(2)}\| \leqslant \|\{x_n^{(1)}\}\|_{\ell_1^w(F)} + z\|\{x_n^{(2)}\}\|_{\ell_2^w(F)} \\ &+ 2\mathbb{E}\|S_N^{(1)}\| + 2\mathbb{E}\|S_N\| \leqslant 3\|\{x_n^{(1)}\}\|_{\ell_1^w(F)} + z\|\{x_n^{(2)}\}\|_{\ell_2^w(F)} + 2\mathbb{E}\|S_N\| \\ &\leqslant 2\mathbb{E}\|S_N\| + 3\left(\|\{x_n^{(1)}\}\|_{\ell_1^w(F)} + z\|\{x_n^{(2)}\}\|_{\ell_2^w(F)}\right). \end{aligned}$$

Now, if

$$Q := 2\mathbb{E}\|S_N\| + 3\left(\|\{x_n^{(1)}\}\|_{\ell_1^w(F)} + z\|\{x_n^{(2)}\}\|_{\ell_2^w(F)}\right),$$

from the preceding inequality, (4.18) and (4.19) it follows that

$$\begin{aligned} m\{t: \|S_N(t)\| > Q\} &\leqslant m\{t: \|S_N^{(1)}(t)\| + \|S_N^{(2)}(t)\| \\ &> \|\{x_n^{(1)}\}\|_{\ell_1^w(F)} + z\|\{x_n^{(2)}\}\|_{\ell_2^w(F)} + 2\mathbb{E}\|S_N^{(2)}\|\} \\ &\leqslant m\{t: ||S_N^{(1)}(t)|| > \|\{x_n^{(1)}\}\|_{\ell_1^w(F)}\} \\ &+ m\{t: ||S_N^{(2)}(t)|| > 2\mathbb{E}\|S_N^{(2)}\| + z\|\{x_n^{(2)}\}\|_{\ell_2^w(F)}\} \le 4e^{-z^2/8}. \end{aligned}$$

Combining this with (4.20), we deduce that

$$m\{t : \|S_N(t)\|_F > 2\mathbb{E}\|S_N\|_F + 3(1+\varepsilon)\mathcal{K}(z, \{x_n\}_{n=1}^N; \ell_1^w(F), \ell_2^w(F))\} \leqslant 4e^{-z^2/8}.$$

Finally, taking into account that $\varepsilon > 0$ is arbitrary, we come to the desired inequality. □

Lemma 4.3 *Let $x_n = (x_{n,j})_{j=1}^\infty \in \ell_\infty$, $n = 1, 2, \ldots$ Then, for every $1 \le p < \infty$*

$$\|\{x_n\}\|_{\ell_p^w(\ell_\infty)} = \sup_{1 \leqslant j < \infty} \Big(\sum_{n=1}^\infty |x_{n,j}|^p\Big)^{1/p}.$$

Proof First, by definition, we have

$$\|\{x_n\}\|_{\ell_p^w(\ell_\infty)} = \sup_{\|x^*\| \le 1} \Big(\sum_{n=1}^\infty |x^*(x_n)|^p\Big)^{1/p}. \tag{4.21}$$

Therefore, substituting the vectors of the canonical basis of ℓ_1 for x^* in the right-hand side, we get

$$\|\{x_n\}\|_{\ell_p^w(\ell_\infty)} \ge \sup_{1 \leqslant j < \infty} \Big(\sum_{n=1}^\infty |x_{n,j}|^p\Big)^{1/p}.$$

Let us prove the opposite inequality. Assuming that $1/p + 1/q = 1$, by (4.21), we have

$$\begin{aligned}
\|\{x_n\}\|_{\ell_p^w(\ell_\infty)} &= \sup_{\|x^*\| \le 1} \sup_{\|(b_n)\|_{\ell_q} \le 1} \sum_{n=1}^\infty b_n x^*(x_n) = \sup_{\|(b_n)\|_{\ell_q} \le 1} \sup_{\|x^*\| \le 1} x^*\Big(\sum_{n=1}^\infty b_n x_n\Big) \\
&\le \sup_{\|(b_n)\|_{\ell_q} \le 1} \Big\|\sum_{n=1}^\infty b_n x_n\Big\|_{\ell_\infty} = \sup_{\|(b_n)\|_{\ell_q} \le 1} \sup_{1 \leqslant j < \infty} \Big|\sum_{n=1}^\infty b_n x_{n,j}\Big| \\
&= \sup_{1 \leqslant j < \infty} \sup_{\|(b_n)\|_{\ell_q} \le 1} \Big|\sum_{n=1}^\infty b_n x_{n,j}\Big| = \sup_{1 \leqslant j < \infty} \Big(\sum_{n=1}^\infty |x_{n,j}|^p\Big)^{1/p},
\end{aligned}$$

and everything is done. □

Lemma 4.4 *For arbitrary sequences $x_1, \ldots, x_N$ from ℓ_∞ and each $z > 0$ it holds*

$$\mathcal{K}(z, \{x_n\}_{n=1}^N; \ell_1^w(\ell_\infty), \ell_2^w(\ell_\infty)) \leqslant 2\kappa_w(z, \{x_n\}_{n=1}^N).$$

Proof For every $1 \le p < \infty$ denote by $\ell_\infty(\ell_p)$ the Banach space of all infinite matrices $(a_{n,j})_{n,j=1}^\infty$ of real numbers with the mixed norm

$$\|(a_{n,j})\|_{\ell_\infty(\ell_p)} := \sup_{1\leqslant j<\infty} \Big(\sum_{n=1}^{\infty} |a_{n,j}|^p\Big)^{1/p}.$$

Denote by ϕ the "transposition operator", i.e., the mapping, which associates an element $\{y_n\}_{n=1}^\infty = \{(y_{n,j})_j\}_{n=1}^\infty \in \ell_p^w(\ell_\infty)$ with the element $\phi(\{y_n\}) = \{(y_{n,j})_n\}_{j=1}^\infty \in \ell_\infty(\ell_p)$. By Lemma 4.3, ϕ maps $\ell_p^w(\ell_\infty)$ onto $\ell_\infty(\ell_p)$ isometrically. Hence, for every $\{y_n\}_{n=1}^\infty \in \ell_2^w(\ell_\infty)$ we have

$$\mathcal{K}(z, \{y_n\}; \ell_1^w(\ell_\infty), \ell_2^w(\ell_\infty)) = \mathcal{K}(z, \phi(\{y_n\}); \ell_\infty(\ell_1), \ell_\infty(\ell_2)). \tag{4.22}$$

Consequently, since $\{x_n\}_{n=1}^N = \{(x_{n,j})\}_{n=1}^N \in \ell_2^w(\ell_\infty)$, we see that $\phi(\{x_n\}) = \{(x_{n,j})\}_{j=1}^\infty \in \ell_\infty(\ell_2)$. Moreover, again by the definition of Peetre's $\mathcal{K}$-functional, for arbitrary $\varepsilon > 0$ and each $j \in \mathbb{N}$ there exists a representation $(x_{n,j})_n = (x_{n,j}^{(1)})_n + (x_{n,j}^{(2)})_n$, $n = 1, 2, \ldots, N$, such that

$$\begin{aligned}
\|\{(x_{n,j}^{(1)})_n\}_{j=1}^\infty\|_{\ell_\infty(\ell_1)} + z\|\{(x_{n,j}^{(2)})_n\}_{j=1}^\infty\|_{\ell_\infty(\ell_2)} &= \sup_{1\leqslant j<\infty} \|(x_{n,j}^{(1)})_{n=1}^N\|_{\ell_1} + z \sup_{1\leqslant j<\infty} \|(x_{n,j}^{(2)})_{n=1}^N\|_{\ell_2} \\
&\leqslant 2 \sup_{1\leqslant j<\infty} \kappa\left(z, (x_{n,j})_{n=1}^N\right) + \varepsilon \\
&\leqslant 2 \sup_{\|x^*\|\le 1} \kappa\left(z, (x^*(x_n))\right) + \varepsilon \\
&= 2\kappa_w(z, \{x_n\}_{n=1}^N) + \varepsilon.
\end{aligned}$$

Therefore, we get

$$\mathcal{K}(z, \phi(\{y_n\}); \ell_\infty(\ell_1), \ell_\infty(\ell_2)) \le 2\kappa_w(z, \{x_n\}_{n=1}^N) + \varepsilon.$$

Since $\varepsilon > 0$ is arbitrary, combining this inequality together with equation (4.22), we complete the proof. □

Proof of Theorem 4.4 Let us fix $x_n \in F$, $n = 1, 2, \ldots$. Rather the whole of F we shall consider the closed linear span in F of the set

$$\Big\{\sum_{n=1}^{N} c_n x_n, \quad c_n \text{ are reals}, \quad N = 1, 2, \ldots\Big\}.$$

The latter space is separable and hence it is isometrically embedded into ℓ_∞ [1, Theorem 2.5.7]. Therefore, we can assume that F is a closed linear subspace of ℓ_∞. Moreover, by the Hahn-Banach theorem, the supremum in the definition of the functional $\kappa_w(t, \{x_n\})$ (see (4.15)) is the same whether it is taken over elements of F^* or over elements of the whole ℓ_∞^*. Therefore, inequality (4.16) for every finite sum $\sum_{n=1}^N x_n r_n$, $N = 1, 2, \ldots$, is an immediate consequence of Lemmas 4.2 and 4.4. Analogous result for an infinite Rademacher series $S = \sum_{n=1}^\infty x_n r_n$ follows from the inequalities for its partial sums by taking the limit as $N \to \infty$.

In the proof of inequality (4.17) we make use of the following similar estimate for scalar Rademacher sums, proved in Theorem 1.7: there exists a universal constant $\delta > 0$ such that for every sequence $(a_n)_{n=1}^\infty \in \ell_2$

$$m\Big\{t : \sum_{n=1}^\infty a_n r_n(t) > \delta \cdot \kappa(z, (a_n))\Big\} \geqslant \delta e^{-z^2/\delta}, \quad z > 0.$$

According to this inequality, we have

$$\begin{aligned} m\Big\{t : \|S(t)\| > \frac{\delta}{2}\kappa_w(z, \{x_n\})\Big\} &\geqslant \inf_{\|x^*\|\leqslant 1} m\Big\{t : \Big\|\sum_{n=1}^\infty x_n r_n(t)\Big\| > \delta \cdot \kappa(z, (x^*(x_n)))\Big\} \\ &\geqslant \inf_{\|x^*\|\leqslant 1} m\Big\{\sum_{n=1}^\infty x^*(x_n) r_n(t) > \delta \cdot \kappa\left(z, (x^*(x_n))\right)\Big\} \\ &\geqslant \delta \cdot e^{-z^2/\delta}. \end{aligned}$$

Since

$$\begin{aligned} & m\left\{t : \|S(t)\| > \frac{1}{2}\mathbb{E}\|S\| + \frac{\delta}{6}\kappa_w(z, \{x_n\})\right\} \\ &\geq m\left(\left\{t : \|S(t)\| > \frac{3}{4}\mathbb{E}\|S\|\right\} \bigcap \left\{t : \|S(t)\| > \frac{\delta}{2}\kappa_w(z, \{x_n\})\right\}\right) \\ &= m\left\{t : \|S(t)\| > \max\left(\frac{3}{4}\mathbb{E}\|S\|, \frac{\delta}{2}\kappa_w(z, \{x_n\})\right)\right\} \\ &= \min\left(m\left\{t : \|S(t)\| > \frac{3}{4}E\|S\|\right\}, m\left\{t : \|S(t)\| > \frac{\delta}{2}\kappa_w(z, \{x_n\})\right\}\right), \end{aligned}$$

the preceding estimate, combined with the Paley–Zygmund and Kahane–Khintchine inequalities (see Proposition 1.4 and Theorem 4.1), yields

$$m\left\{t : \|S(t)\| > \frac{1}{2}\mathbb{E}\|S\| + \frac{\delta}{6}\kappa_w(z, \{x_n\})\right\} \geqslant \min\left(\frac{1}{16K_2^2}, \delta e^{-z^2/\delta}\right)$$

(K_2 is the constant from (4.1)), and so (4.17) is proved. □

Just like in the scalar case (see Corollary 1.9), the result of Theorem 4.4 can be restated by using the notion of decreasing rearrangement.

Corollary 4.7 *Let F be a Banach space, $x_n \in F$, $n = 1, 2, \ldots$, and let the series $S := \sum_{n=1}^{\infty} x_n r_n$ converge in F a.e. on $[0, 1]$. Then, with a universal constant, for all $0 < t \leqslant 1/(2K_2^2)$ we have*

$$\mathcal{S}^*(t) \asymp \mathbb{E}\|S\| + \kappa_w\left(\sqrt{\ln(e/t)}, \{x_n\}\right), \tag{4.23}$$

where K_2 is the constant from inequality (4.1) *and $\mathcal{S}(t) := \|S(t)\|_F$.*

Proof From Theorem 4.4 and definition of the decreasing rearrangement it follows that

$$\mathcal{S}^*(4e^{-z^2/8}) \leqslant 2\mathbb{E}\|S\| + 6\kappa_w(z, \{x_n\})$$

and

$$\mathcal{S}^*(ce^{-t^2/c}) \geqslant \frac{1}{2}\mathbb{E}\|S\| + c\kappa_w(z, \{x_n\}).$$

Hence, changing variables, we obtain (4.23) for all sufficiently small $t > 0$. It is left to show that the appropriate choice of a constant assures the lower estimate in equivalence (4.23) for all $0 < t \leqslant 1/(2K_2^2)$.

Similarly as in the proof of Theorem 4.4, we apply Proposition 1.4 and Theorem 4.1. Then, in view of inequality (1.18) with $\lambda = 1 - 1/\sqrt{2}$, we get

$$m\left\{t : \mathcal{S}(t) > \left(1 - \frac{1}{\sqrt{2}}\right)\mathbb{E}\|S\|\right\} \geqslant \frac{1}{2K_2^2},$$

and the proof is completed. □

Remark 4.3 In the next chapter (see Theorem 5.4), we shall show that for K_2 can be taken $\sqrt{2}$.

In conclusion of the section, we prove the vector-valued estimates similar to Hitczenko's inequality from Theorem 1.6.

Theorem 4.5 *With a universal constant, for arbitrary Banach space F, every Rademacher series $S := \sum_{n=1}^{\infty} x_n r_n$, $x_n \in F$, $n = 1, 2, \ldots$, converging in F a.e. on $[0, 1]$, and all $1 \leqslant p < \infty$ it holds*

$$(\mathbb{E}\|S\|^p)^{1/p} \asymp \mathbb{E}\|S\| + \kappa_w(\sqrt{p}, \{x_n\}). \tag{4.24}$$

Proof By Theorem 4.1, without loss of generality, we can assume that $p \geqslant 2$. A key step in the proof is the replacement of the norm $(\mathbb{E}\|S\|^p)^{1/p}$ with its "weak" counterpart.

Recall that the "weak" L_p-norm of a measurable function f on $[0, 1]$ is defined by

$$\|f\|_{p,\infty} := \sup_{0<t\leq 1} t^{1/p} f^*(t)$$

(from Theorem C.2(a) it follows that $\|f\|_{p,\infty}$ is equivalent to the norm of f in the Marcinkiewicz space $M(t^{1-1/p})$).

We show that for every $1 < p < \infty$ and each measurable function f on $[0, 1]$

$$\frac{1}{2}\|f\|_p \leqslant \|f\|_{2p,\infty} \leqslant \|f\|_{2p}. \tag{4.25}$$

Indeed, the right-hand side of (4.25) is a consequence of the estimate

$$\|f\|_{2p} \geq \left(\int_0^t f^*(s)^{2p}\,ds\right)^{1/(2p)} \geq f^*(t)t^{1/(2p)}, \quad 0 < t \leq 1.$$

Moreover, from the inequality $\|f\|_{2p,\infty} \leq 1$ it follows that $f^*(t) \leq t^{-1/(2p)}$ for all $0 < t \leq 1$. Therefore,

$$\|f\|_p^p = \int_0^1 f^*(t)^p\,dt \leq \int_0^1 t^{-1/2}\,dt = 2,$$

and, by homogeneity, the left-hand side of (4.25) is also proved.

Further, since $p \geqslant 2$, then according to Corollary 4.2 we have

$$(\mathbb{E}\|S\|^{2p})^{1/2p} \leqslant \sqrt{3}(\mathbb{E}\|S\|^p)^{1/p}.$$

Therefore, applying (4.25) in the case $f(t) = \|S(t)\|_F$ yields

$$\frac{1}{2}(\mathbb{E}\|S\|^p)^{1/p} \leqslant \|\|S(\cdot)\|_F\|_{2p,\infty} \leqslant \sqrt{3}(\mathbb{E}\|S\|^p)^{1/p}.$$

As a result, since the function $p \mapsto \kappa_w(\sqrt{p}, \{x_n\})$ is concave, equivalence (4.24) will be proved once we show that

$$\|\|S(\cdot)\|_F\|_{p,\infty} \asymp \mathbb{E}\|S\| + \kappa_w(\sqrt{p}, \{x_n\}). \tag{4.26}$$

To this end, by using Corollary 4.7 and the definition of the "weak" L_p-norm, we write

$$\begin{aligned}\|\|S(\cdot)\|_F\|_{p,\infty} &\asymp \mathbb{E}\|S\| + \sup_{0<z\leq 1} z^{1/p}\kappa_w(\sqrt{\ln(e/z)}, \{x_n\}) \\ &= \mathbb{E}\|S\| + \sup_{0<z\leq 1}\left\{z^{1/p} \sup_{\|x^*\|\leqslant 1} \kappa\left(\sqrt{\ln(e/z)}, (x^*(x_n))\right)\right\}\end{aligned}$$

$$= \mathbb{E}\|S\| + \sup_{\|x^*\|\leqslant 1}\left\{\sup_{0<z\leq 1} z^{1/p}\kappa\left(\sqrt{\ln(e/z)}, (x^*(x_n))\right)\right\}.$$

To estimate the inner supremum, we consider separately two cases: $z \in (0, e^{-p+1}]$ and $z \in (e^{-p+1}, 1]$.

From the definition of $\kappa(s, (a_n))$ it follows easily that

$$\kappa(s, (a_n)) \leqslant \max(1, s/t)\kappa(t, (a_n)), \quad s, t > 0.$$

Hence,

$$\sup_{0<z\leqslant e^{-p+1}} z^{1/p}\kappa\left(\sqrt{\ln(e/z)}, (x^*(x_n))\right) \leqslant \left(\sup_{0<z\leqslant e^{-p+1}} z^{1/p}\sqrt{\frac{\ln(e/z)}{p}}\right)\kappa(\sqrt{p}, (x^*(x_n))).$$

Since

$$\sup_{0<z\leqslant e^{-p+1}} z^{1/p}\sqrt{\frac{\ln(e/z)}{p}} = e^{1/p}\sup_{u\geq 1} e^{-u}\sqrt{u} = e^{1/p-1},$$

we get

$$\sup_{0<z\leqslant e^{-p+1}} z^{1/p}\kappa\left(\sqrt{\ln(e/z)}), (x^*(x_n))\right) \leqslant e^{1/p-1}\kappa(\sqrt{p}, (x^*(x_n))).$$

As for estimation of the supremum over the interval $(e^{-p+1}, 1]$, it is clear that

$$\sup_{e^{-p+1}<z\leq 1} z^{1/p}\kappa\left(\sqrt{\ln(e/z)}), (x^*(x_n))\right) \leqslant \kappa(\sqrt{p}, (x^*(x_n))).$$

In the opposite direction, we have

$$\sup_{0<z\leq 1} z^{1/p}\kappa\left(\sqrt{\ln(e/z)}, (x^*(x_n))\right) \geq \frac{1}{e}\kappa(\sqrt{p}, (x^*(x_n))).$$

Combining the above inequalities and taking into account definition of the functional $\kappa_w(z, \{x_n\})$, we obtain that

$$\begin{aligned}\frac{1}{e}\kappa_w(\sqrt{p}, \{x_n\}) &\leqslant \sup_{\|x^*\|\leqslant 1}\left\{\sup_{0<z\leq 1} z^{1/p}\kappa(\sqrt{\ln(e/z)}, (x^*(x_n)))\right\} \\ &\leqslant \kappa_w(\sqrt{p}, \{x_n\}),\end{aligned}$$

and so equivalence (4.26) is proved. □

4.4 Rademacher Sums with Vector Coefficients in Exponential Orlicz Spaces

In Chap. 3, we have got a description of the Rademacher subspaces for some classes of s.s.'s, including the Orlicz exponential spaces L_{N_α}, $\alpha > 0$ (see Example 3.2 in Sect. 3.4). Here, we consider the L_{N_α}-norms of Rademacher sums $S = \sum_{n=1}^{\infty} x_n r_n$, where x_n, $n = 1, 2, \ldots$, are vectors from a Banach space F.

Observe that in the case $0 < \alpha \leq 2$ we have $L_{N_2} \subset L_{N_\alpha}$, and so, by Corollary 4.3, $\| \|S(\cdot)\|_F \|_{L_{N_\alpha}} \asymp \mathbb{E}\|S\|$. Let us consider the more interesting situation when $\alpha > 2$.

Recall that $\ell_{\beta,\infty}$, $1 < \beta < \infty$, is the coordinate Marcinkiewicz space consisting of all sequences $a = (a_k)_{k=1}^{\infty}$ such that

$$\|(a_k)\|_{\ell_{\beta,\infty}} := \sup_{k=1,2,\ldots} k^{1/\beta-1} \sum_{i=1}^{k} a_i^* < \infty,$$

where $(a_i^*)_{i=1}^{\infty}$ is the decreasing permutation of the sequence $(|a_k|)_{k=1}^{\infty}$.

Theorem 4.6 *Let $\alpha > 2$, $1/\alpha + 1/\beta = 1$, and let F be a Banach space, $x_n \in F$, $n = 1, 2, \ldots$. If the series $S := \sum_{n=1}^{\infty} x_n r_n$ converge a.e. on $[0, 1]$, then, with some constant depending only on $\alpha > 0$, we have*

$$\| \|S(\cdot)\|_F \|_{L_{N_\alpha}} \asymp \mathbb{E}\|S\| + \sup_{\|x^*\| \leqslant 1} \|(x^*(x_n))\|_{\ell_{\beta,\infty}}.$$

Proof In view of Proposition 2.4, the L_{N_α}-norm satisfies the following equivalence:

$$\|f\|_{L_{N_\alpha}} \asymp \sup_{0<t\leq 1} (\ln(e/t))^{-1/\alpha} f^*(t).$$

Therefore, by Corollary 4.7, we get

$$\begin{aligned}
\| \|S(\cdot)\|_F \|_{L_{N_\alpha}} &\asymp \mathbb{E}\|S\| + \sup_{0<t\leq 1} (\ln(e/t))^{-1/\alpha} \kappa_w \left(\sqrt{\ln(e/t)}, \{x_n\}\right) \\
&\asymp \mathbb{E}\|S\| + \sup_{0<t\leq 1} \left\{ (\ln(e/t))^{-1/\alpha} \sup_{\|x^*\| \leqslant 1} \kappa \left(\sqrt{\ln(e/t)}, (x^*(x_n))\right) \right\} \\
&\asymp \mathbb{E}\|S\| + \sup_{\|x^*\| \leqslant 1} \left\{ \sup_{0<t\leq 1} (\ln(e/t))^{-1/\alpha} \kappa \left(\sqrt{\ln(e/t)}, (x^*(x_n))\right) \right\}.
\end{aligned}$$

Moreover, with a universal constant, from Corollary 1.9 it follows that

$$\left(\sum_{n=1}^{\infty} a_n r_n \right)^* (s) \asymp \kappa \left(\sqrt{\ln(e/s)}, (a_n) \right), \quad 0 < s \leq 1,$$

for all $(a_n) \in \ell_2$. Combining this with the preceding equivalence and using Example 3.2, we obtain

$$\begin{aligned} \| \|S(\cdot)\|_F \|_{L_{N_\alpha}} &\asymp \mathbb{E}\|S\| + \sup_{\|x^*\| \leqslant 1} \Big\| \sum_{n=1}^{\infty} x^*(x_n) r_n \Big\|_{L_{N_\alpha}} \\ &\asymp \mathbb{E}\|S\| + \sup_{\|x^*\| \leqslant 1} \|(x^*(x_n))\|_{\ell_{\beta,\infty}}, \end{aligned}$$

as we wished to show. □

4.5 The Comparison of Distributions of Vector-Valued Rademacher Sums and of Those of Their "weak" Counterparts

Some results of Sect. 4.3 (see e.g. Theorem 4.4) indicate that, to some extent, the distribution of norms of vector-valued Rademacher sums is determined by the certain "weak" characteristics of the sequences of their coefficients. Here, we show that there is a direct connection between the distributions of vector-valued and corresponding scalar Rademacher sums.

Throughout this section, F is a Banach space. For every Rademacher sum $S = \sum_{k=1}^n x_k r_k$, where $n \in \mathbb{N}$, $x_1, x_2, \ldots, x_n \in F$, and $z > 0$ we define the following distribution functions

$$T_S(z) := m\{t : \|S(t)\| > z\} \text{ and } \tau_S(z) := \sup_{x^* \in F^*: \|x^*\| \leq 1} m\{t : |x^*(S(t))| > z\}.$$

Obviously, $\tau_S(z) \leq T_S(z)$ for all $z > 0$. The following theorem shows that some opposite estimates hold as well.

Theorem 4.7 *For every $\delta \in (0, 1]$ and an arbitrary Rademacher sum $S = \sum_{k=1}^n x_k r_k$, where $n \in \mathbb{N}$, $x_1, x_2, \ldots, x_n \in F$, it holds*

$$T_S\left(16\delta^{-1}\mathbb{E}\|S\| + (1+\delta)z\right) \leq 13\tau_S(z), \quad z > 0.$$

Moreover, assuming that $\mathbb{E}\|S\| = 1$, we have

$$T_S(z + 10\sqrt{z}) \leq \tau_S(z), \quad z > 16,$$

and

$$T_S(z) \leq \tau_S(z - 10\sqrt{z}), \quad z > 100.$$

Further, we need several auxiliary assertions. We start with proving a lower bound for the distribution of a sum of mean zero independent r.v.'s, which is similar to the Paley–Zygmund inequality for the Rademacher sums from Sect. 1.4.

Lemma 4.5 *Let $h_1, h_2, \ldots, h_n$ be independent r.v.'s with finite fourth moments such that $\mathbb{E}h_k = 0$ for all $k = 1, 2, \ldots, n$. Suppose also that $\mathbb{E}h_k^4 \le 3(\mathbb{E}h_k^2)^2$, $k = 1, 2, \ldots, n$. Then, if $h = \sum_{k=1}^n h_k$, we have $\mathbb{P}\{h \ge 0\} > 1/8$.*

Proof First, the case $h = 0$ a.e. is trivial. Therefore, we can assume that $\mathbb{E}|h| > 0$. By induction on n, one can readily check that $\mathbb{E}h^4 \le 3(\mathbb{E}h^2)^2$. Moreover, since $\mathbb{E}h = 0$, from Hölder's inequality it follows

$$\mathbb{E}|h| = 2\mathbb{E}(h \cdot \chi_{\{h\ge 0\}}) \le 2(\mathbb{E}h^4)^{1/4}\mathbb{P}\{h \ge 0\}^{3/4}.$$

Applying Hölder's inequality once more, we get

$$\mathbb{E}h^2 = \mathbb{E}(|h|^{4/3}|h|^{2/3}) \le (\mathbb{E}h^4)^{1/3}(\mathbb{E}|h|)^{2/3}.$$

Putting these estimates together gives

$$16\mathbb{P}\{h \ge 0\}^3 \ge \frac{(\mathbb{E}|h|)^4}{\mathbb{E}h^4} \ge \left(\frac{(\mathbb{E}h^2)^2}{\mathbb{E}h^4}\right)^3 \ge \frac{1}{27} > \frac{1}{32},$$

which ends the proof. □

Lemma 4.6 *Let $\delta \in (0, 1]$. Suppose $f_1, f_2, \ldots, f_n$ be identically distributed independent r.v.'s such that $\mathbb{P}\{f_k = -1\} = 1/(2 + 2\delta)$ and $\mathbb{P}\{f_k = 1\} = (1 + 2\delta)/(2 + 2\delta)$, $k = 1, 2, \ldots, n$. Then, for any $x_1, \ldots, x_n \in F$ we have*

$$\mathbb{P}\left\{\left\|\sum_{k=1}^n x_k f_k\right\| \ge \frac{\delta}{1+\delta}\left\|\sum_{k=1}^n x_k\right\|\right\} > \frac{1}{8}.$$

Proof Clearly, there is a norm-one functional $x^* \in F^*$ satisfying

$$x^*\Big(\sum_{k=1}^n x_k\Big) = \Big\|\sum_{k=1}^n x_k\Big\|.$$

Then, since $\mathbb{E}f_k = \delta/(1 + \delta)$, we have

$$\mathbb{P}\left\{\left\|\sum_{k=1}^n x_k f_k\right\| \ge \frac{\delta}{1+\delta}\left\|\sum_{k=1}^n x_k\right\|\right\} \ge \mathbb{P}\left\{x^*\Big(\sum_{k=1}^n x_k f_k\Big) \ge \frac{\delta}{1+\delta}x^*\Big(\sum_{k=1}^n x_k\Big)\right\}$$
$$= \mathbb{P}\left\{\sum_{k=1}^n (f_k - \mathbb{E}f_k)x^*(x_k) \ge 0\right\}.$$

Furthermore, it can be immediately checked that

$$\mathbb{E}(f_k - \mathbb{E} f_k)^4 = (2\delta + (1+2\delta)^{-1}) \cdot (\mathbb{E}(f_k - \mathbb{E} f_k)^2)^2.$$

Therefore, $\mathbb{E}(f_k - \mathbb{E} f_k)^4 \le 3 \cdot (\mathbb{E}(f_k - \mathbb{E} f_k)^2)^2$, and applying Lemma 4.5 to the r.v.'s $h_k := (f_k - \mathbb{E} f_k) x^*(x_k)$, $k = 1, 2, \ldots, n$, finishes the proof. □

The following tail estimate is classical.

Lemma 4.7 *For every vector-valued Rademacher sum* $S = \sum_{j=1}^n x_j r_j$ *and any* $u, v > 0$ *it holds*

$$T_S(u+v) \le 4 T_S(u) T_S(v).$$

Proof For each $k = 1, 2, \ldots, n$ we set $S_k := \sum_{j=1}^k x_j r_j$ and $U_k := \sum_{j=k}^n x_j r_j$. Moreover, let $A_1 := \{\|S_1\| > u\}$ and

$$A_k := \{\|S_1\| \le u, \ldots, \|S_{k-1}\| \le u, \|S_k\| > u\}, \quad k = 2, \ldots, n,$$

where $u > 0$ is fixed. From definition of the Rademacher functions it follows that the r.v. $\|U_k\|$ does not depend on r_k and so on the σ-algebra $\sigma(r_1, \ldots, r_k)$, containing the set A_k. Furthermore, the sets A_k, $k = 1, 2, \ldots, n$, are pairwise disjoint. Hence, by using twice Levy's maximal inequality (see Theorem A.2), we get

$$\begin{aligned}
T_S(u+v) &= \sum_{k=1}^n m\left(A_k \cap \{\|S\| > u+v\}\right) \\
&\le \sum_{k=1}^n m\left(A_k \cap \{\|U_k\| > v\}\right) = \sum_{k=1}^n m(A_k) m\{\|U_k\| > v\} \\
&\le m\Big(\bigcup_{k=1}^n A_k\Big) \max_{1 \le k \le n} m\{\|U_k\| > v\} \\
&= m\{\max_{1 \le k \le n} \|S_k\| > u\} \cdot \max_{1 \le k \le n} m\{\|U_k\| > v\} \\
&\le 4 m\{\|S\| > u\} \cdot m\{\|S\| > u\} = 4 T_S(u) T_S(v).
\end{aligned}$$

□

Lemma 4.8 *For an arbitrary vector-valued Rademacher sum* $S = \sum_{j=1}^n x_j r_j$ *we have* $\mathbb{E}\|S\|^4 \le 12(\mathbb{E}\|S\|)^4$, $T_S(5\mathbb{E}\|S\|) \le 1/52$ *and for every* $u > 0$

$$T_S(5\mathbb{E}\|S\| + u) \le \frac{T_S(u)}{13}.$$

Proof In Chap. 5, we shall show that the optimal constants $K_{1,2}$ and $K_{2,4}$ in the Kahane–Khintchine inequality

$$\left(\mathbb{E}\Big\|\sum_{k=1}^{n} x_k r_k\Big\|_F^q\right)^{1/q} \le K_{p,q}\left(\mathbb{E}\Big\|\sum_{k=1}^{n} x_k r_k\Big\|_F^p\right)^{1/p}$$

are $\sqrt{2}$ and $\sqrt[4]{3}$, respectively (see Theorem 5.4 and Corollary 5.6). This implies at once that $\mathbb{E}\|S\|^4 \le 12(\mathbb{E}\|S\|)^4$. Further, applying this inequality together with Chebyshev's inequality, we obtain

$$T_S(5\mathbb{E}\|S\|) = m\{t : \|S(t)\|^4 > (5\mathbb{E}\|S\|)^4\} \le \frac{\mathbb{E}\|S\|^4}{(5\mathbb{E}\|S\|)^4} \le \frac{12}{625} < \frac{1}{52}$$

(we have omitted the trivial case when $S = 0$ a.e.). The last inequality of the lemma is an immediate consequence of the proved second one and Lemma 4.7. □

Lemma 4.9 *Let a function $\Phi : [-c, c] \to \mathbb{R}$ be convex and let a mean zero r.v. ξ be such that $|\xi| \le c$ a.e. Then, $\mathbb{E}\Phi(\xi) \le (\Phi(c) + \Phi(-c))/2$.*

Proof From the convexity of Φ and the fact that $\mathbb{E}\xi = 0$ it follows

$$\mathbb{E}\Phi(\xi) = \mathbb{E}\Phi\left(\frac{c+\xi}{2c}\cdot c + \frac{c-\xi}{2c}\cdot(-c)\right) \le \mathbb{E}\left(\frac{c+\xi}{2c}\cdot\Phi(c) + \frac{c-\xi}{2c}\cdot\Phi(-c)\right)$$
$$= \frac{\Phi(c)+\Phi(-c)}{2}.$$

□

Lemma 4.10 *Suppose that a function $\Phi : [-c, c]^n \to \mathbb{R}$ is coordinate-wise convex and $\xi_1, \xi_2, \dots, \xi_n$ are mean zero independent r.v.'s satisfying the inequality $|\xi_k| \le c$ a.e. for all $k = 1, 2, \dots, n$. Then, $\mathbb{E}\Phi(\xi_1, \xi_2, \dots, \xi_n) \le \mathbb{E}\Phi(c\eta_1, c\eta_2, \dots, c\eta_n)$, where each of the independent r.v.'s $\eta_1, \eta_2, \dots, \eta_n$ has the same distribution as the Rademacher function $r_1(t)$, $t \in [0, 1]$.*

Proof This follows from Lemma 4.9 by using an easy induction argument. □

Now, we introduce a system of random vectors, which will play a key role in the proof of Theorem 4.7.

Let $\delta \in (0, 1]$. By $(f_1, g_1), (f_2, g_2), \dots, (f_n, g_n)$ we denote $\mathbb{R}^2$-valued independent vectors satisfying the following conditions:

$$\mathbb{P}\{(f_k, g_k) = (1, 1-\delta)\} = \frac{1}{2},$$

$$\mathbb{P}\{(f_k, g_k) = (1, 1+\delta)\} = \frac{\delta}{2(1+\delta)},$$

and

$$\mathbb{P}\{(f_k, g_k) = (-1, -1 - \delta)\} = \frac{1}{2(1+\delta)}.$$

One can easily check that $\mathbb{E} f_k = \delta/(1+\delta)$, $\mathbb{E} g_k = 0$, $|g_k| \le 1 + \delta$ a.e. and that the following assertion holds.

Lemma 4.11 *Let each of independent r.v.'s $\epsilon_1, \epsilon_2, \dots, \epsilon_n$ have the same distribution as the function r_1 on $[0, 1]$. Then, if $r'_1, r'_2, \dots, r'_n$ are the Rademacher functions independent with respect to the functions $r_1, r_2, \dots, r_n$, then the vectors $(\epsilon_k f_k, \epsilon_k g_k)$ and $(r_k, r_k - \delta r'_k)$ are identically distributed for each $k = 1, 2, \dots, n$.*

Now everything is ready to prove Theorem 4.7.

Proof of Theorem 4.7 Let $\delta \in (0, 1]$, $x \in F$ and $S' = \sum_{k=1}^n x_k r'_k$, where $r'_1, r'_2, \dots, r'_n$ are the same as in Lemma 4.11. If $x^* \in F^*$ is a functional such that $\|x^*\| = 1$ and $x^*(x) = \|x\|$, then

$$|x^*(\delta S')| \ge |x^*(x)| - |x^*(x - \delta S')| \ge \|x\| - \|x - \delta S'\|.$$

Hence, as the sums $S = \sum_{k=1}^n x_k r_k$ and S' are identically distributed, we have

$$\tau_S(z) \ge m\{t \in [0, 1] : \|x\| - \|x - \delta S'(t)\| > \delta z\}$$

for all $x \in F$ and $z > 0$. Thus, in view of the independence of S and S', Fubini's theorem yields

$$\tau_S(z) \ge m\{t \in [0, 1] : \|S(t)\| - \|S(t) - \delta S'(t)\| > \delta z\}.$$

Assume that r.v.'s ϵ_k, $k = 1, 2, \dots, n$, are independent and each of them has the same distribution as the function r_1 on $[0, 1]$. Let $\{(f_k, g_k)\}_{k=1}^n$ be such a collection of $\mathbb{R}^2$-valued independent vectors as was described just before Lemma 4.11. Moreover, we may (and will) assume that the collections $(\epsilon_k)_{k=1}^n$ and $\{(f_k, g_k)\}_{k=1}^n$ are independent. Then, by Lemma 4.11, the vectors $(S, S - \delta S')$ and $\left(\sum_{k=1}^n \epsilon_k x_k f_k, \sum_{k=1}^n \epsilon_k x_k g_k\right)$ are identically distributed, and hence the last inequality implies

$$\tau_S(z) \ge \mathbb{P}\Big\{\Big\|\sum_{k=1}^n \epsilon_k x_k f_k\Big\| - \Big\|\sum_{k=1}^n \epsilon_k x_k g_k\Big\| > \delta z\Big\}.$$

Further, since the sum $\sum_{k=1}^n x_k \epsilon_k$ has the same distribution as S, then denoting by $\mathbb{E}_\epsilon$ the expectation with respect to the σ-algebra $\sigma(\epsilon_1, \epsilon_2, \dots, \epsilon_n)$ (the notations

$\mathbb{E}_{f,g}$, $\mathbb{P}_{f,g}$,.. will be understood similarly) and using Fubini's theorem and Chebyshev's inequality together with Lemma 4.6, we get

$$\tau_S(z) \geq \mathbb{E}_\epsilon \mathbb{P}_{f,g}\left\{\left\|\sum_{k=1}^n \epsilon_k x_k f_k\right\| > \delta z + \left\|\sum_{k=1}^n \epsilon_k x_k g_k\right\|\right\}$$

$$\geq \mathbb{E}_\epsilon \mathbb{P}_{f,g}\left\{\left\|\sum_{k=1}^n \epsilon_k x_k f_k\right\| \geq \frac{\delta}{1+\delta}\left\|\sum_{k=1}^n x_k \epsilon_k\right\|\right.$$

$$\left. > \delta z + 16(1+\delta)^{-1}\mathbb{E}\|S\|, \left\|\sum_{k=1}^n \epsilon_k x_k g_k\right\| \leq 8\mathbb{E}\|S\|\right\}$$

$$\geq \mathbb{E}_\epsilon \mathbb{P}_f\left\{\left\|\sum_{k=1}^n \epsilon_k x_k f_k\right\| \geq \frac{\delta}{1+\delta}\left\|\sum_{k=1}^n x_k \epsilon_k\right\|\right\}\chi_{\{\|\sum_{k=1}^n x_k\epsilon_k\| > (1+\delta)z + 16\delta^{-1}\mathbb{E}\|S\|\}}$$

$$- \mathbb{E}_\epsilon \mathbb{P}_g\left\{\left\|\sum_{k=1}^n \epsilon_k x_k g_k\right\| > 8\mathbb{E}\|S\|\right\}\chi_{\{\|\sum_{k=1}^n x_k\epsilon_k\| > (1+\delta)z + 16\delta^{-1}\mathbb{E}\|S\|\}}$$

$$\geq \mathbb{E}_\epsilon\left(\frac{1}{8} - \frac{\mathbb{E}_g(\|\sum_{k=1}^n x_k\epsilon_k g_k\|^4)}{2^{12}(\mathbb{E}\|S\|)^4}\right)\chi_{\{\|\sum_{k=1}^n x_k\epsilon_k\| > (1+\delta)z + 16\delta^{-1}\mathbb{E}\|S\|\}}.$$

Now, since $|g_k| \leq 1 + \delta$ a.e., by Lemma 4.10, applied to the function $\Phi(y_1, \ldots, y_n) = \|\sum_{k=1}^n y_k x_k \epsilon_k\|^4$, and Lemma 4.8, for arbitrary $\epsilon_k = \pm 1$, $k = 1, 2, \ldots, n$, we have

$$\mathbb{E}_g\left(\left\|\sum_{k=1}^n x_k\epsilon_k g_k\right\|^4\right) \leq (1+\delta)^4 \cdot \mathbb{E}_\eta\left(\left\|\sum_{k=1}^n x_k\epsilon_k \eta_k\right\|^4\right)$$

$$= (1+\delta)^4 \cdot (\mathbb{E}\|S\|)^4 \leq 3 \cdot 2^6 \cdot (\mathbb{E}\|S\|)^4$$

(here, each of the independent r.v.'s $\eta_1, \ldots, \eta_n$ has the same distribution as $r_1(t)$, $t \in [0, 1]$). Combining this together with the preceding inequality, we obtain

$$\tau_S(z) \geq \frac{5}{64}\mathbb{E}_\epsilon \chi_{\{\|\sum_{k=1}^n x_k\epsilon_k\| > (1+\delta)z + 16\delta^{-1}\mathbb{E}\|S\|\}}.$$

Therefore,

$$T_S\left(16\delta^{-1}\mathbb{E}\|S\| + (1+\delta)z\right) = \mathbb{E}_\epsilon \chi_{\{\|\sum_{k=1}^n x_k\epsilon_k\| > (1+\delta)z + 16\delta^{-1}\mathbb{E}\|S\|\}} \leq 13\tau_S(z),$$

and the proof of the first inequality of the theorem is completed.

To prove the second inequality, we set $\delta := 4(\mathbb{E}\|S\|/z)^{1/2}$. Then, from the inequality $z \geq 16\mathbb{E}\|S\|$ it follows $\delta \leq 1$. Also, since

$$16\delta^{-1}\mathbb{E}\|S\| + \delta z = 8\sqrt{z\mathbb{E}\|S\|},$$

then

$$T_S(z + 8\sqrt{z\mathbb{E}\|S\|}) \le 13\tau_S(z).$$

Hence, taking into account that $5\mathbb{E}\|S\| \le 2\sqrt{z\mathbb{E}\|S\|}$ provided if $z \ge 16\mathbb{E}\|S\|$ and applying Lemmas 4.7 and 4.8, we get

$$\begin{aligned} T_S(z + 10\sqrt{z\mathbb{E}\|S\|}) &\le T_S(z + 8\sqrt{z\mathbb{E}\|S\|} + 5\mathbb{E}\|S\|) \\ &\le 4T_S(z + 8\sqrt{z\mathbb{E}\|S\|})T_S(5\mathbb{E}\|S\|) \le \tau_S(z), \end{aligned}$$

which implies the second inequality. Since the third inequality of the theorem is an immediate consequence of the second one, the proof is finished. □

4.6 Subspaces of Symmetric Spaces on the Square Formed by Rademacher Sums with Vector Coefficients

In the final section of the chapter we consider Rademacher sums whose coefficients are functions from a s.s. on [0, 1].

Let X be an arbitrary s.s. on $I = [0, 1]$. Denote by $X(I \times I)$ the space of measurable on $I \times I$ functions $f(s, t)$ equipped with the natural norm

$$\|f\|_{X(I\times I)} := \|f^*\|_X,$$

where $f^*(u)$ is the decreasing rearrangement of the function $f(s, t)$, $(s, t) \in I \times I$ (see Appendix C). Denote by Rad X the set of all functions $f(s, t) \in X(I \times I)$ of the form

$$f(s,t) = \sum_{i=1}^{\infty} x_i(s) r_i(t), \quad \text{where } x_i \in X \text{ (the series converges a.e. on } I \times I). \tag{4.27}$$

We show that Rad X is a closed linear subspace of $X(I \times I)$, then find sharp conditions, under which Rad X can be identified with a certain space of sequences of functions from X, and finally we clarify when this subspace is complemented in $X(I \times I)$.

Let x_i, $i = 1, 2, \ldots$ be measurable functions on I. By Fubini's theorem, the series $\sum_{i=1}^{\infty} x_i(s) r_i(t)$ converges a.e. on $I \times I$ if and only if there is a set $A \subset I$, $m(A) = 1$, such that for each fixed $s \in A$ this series converges for almost all $t \in I$. Therefore, applying Theorems 1.1 and 1.2, we get the following simple result, which will be used repeatedly further.

Lemma 4.12 *The series* $\sum_{i=1}^{\infty} x_i(s)r_i(t)$, *where* x_i *are measurable functions on* I, *converges a.e. in* $I \times I$ *if and only if*

$$\sum_{i=1}^{\infty} x_i^2(s) < \infty \quad a.e. \ on \ I. \tag{4.28}$$

Proposition 4.1 Rad X *is a closed linear subspace of the space* $X(I \times I)$ *for every s.s.* X.

Proof Since the linearity of Rad X is obvious, it is sufficient to prove that this set is closed in $X(I \times I)$. Let $\{f_n\} \subset$ Rad X be such that $\|f_n - f\|_{X(I\times I)} \to 0$ as $n \to \infty$ for some $f \in X(I \times I)$. Recall (see Appendix C) that any s.s. on I is continuously embedded into $L_1(I)$. Hence, we have $\|f_n - f\|_{L_1(I\times I)} \to 0$. In consequence, if

$$f_n(s,t) = \sum_{i=1}^{\infty} x_{n,i}(s)r_i(t), \quad x_{n,i} \in X,$$

then for arbitrary $\varepsilon > 0$ there exists a positive integer N such that for all $m > n \geq N$ we have

$$\|f_m - f_n\|_{L_1(I\times I)} = \int_0^1 \int_0^1 \left|\sum_{i=1}^{\infty} \big(x_{m,i}(s) - x_{n,i}(s)\big)r_i(t)\right| ds\, dt \ \leq \ \varepsilon.$$

Next, by Fubini's theorem and Khintchine's inequality (see Theorem 1.4), the integral from the last inequality is not less than

$$\frac{1}{\sqrt{3}} \int_0^1 \Big(\sum_{i=1}^{\infty} \big(x_{m,i}(s) - x_{n,i}(s)\big)^2\Big)^{1/2} ds,$$

and so we get

$$\int_0^1 \Big(\sum_{i=1}^{\infty} \big(x_{m,i}(s) - x_{n,i}(s)\big)^2\Big)^{1/2} ds \leq \sqrt{3} \cdot \varepsilon \tag{4.29}$$

for all $m > n \geq N$. In particular, this implies that, for each $i \in \mathbb{N}$, $\{x_{m,i}\}_{m=1}^{\infty}$ is a Cauchy sequence in $L_1(I)$, and hence there is a function $x_i \in L_1(I)$ such that $\|x_{m,i} - x_i\|_{L_1(I)} \to 0$ as $m \to \infty$. Thus, fixing $n \geq N$ and passing to the limit as $m \to \infty$, according to the Fatou lemma, we get

$$\int_0^1 \Big(\sum_{i=1}^{\infty} \big(x_i(s) - x_{n,i}(s)\big)^2\Big)^{1/2} ds \leq \sqrt{3} \cdot \varepsilon \ \text{ for all } \ n \geq N, \tag{4.30}$$

whence

$$\sum_{i=1}^{\infty} \big(x_i(s) - x_{n,i}(s)\big)^2 < \infty \ \text{ a.e. on } I \ \text{ for all } n \geq N.$$

Since for each $n \in \mathbb{N}$, by Lemma 4.12,

$$\sum_{i=1}^{\infty} x_{n,i}^2(s) < \infty \ \text{ a.e. on } I,$$

then from the preceding inequality we deduce (4.28). Therefore, applying Lemma 4.12 once more, we get that the series

$$g(s,t) := \sum_{i=1}^{\infty} x_i(s) r_i(t)$$

converges a.e. on $I \times I$, which means that $g \in \text{Rad } X$.

Further, by using Parseval's Identity and the fact that the norm $\|x\|_{L_p(I)}$ increases with increasing p, we have

$$\int_0^1 \Big(\sum_{i=1}^{\infty} \big(x_i(s) - x_{n,i}(s)\big)^2\Big)^{1/2} ds = \int_0^1 \Big(\int_0^1 \Big(\sum_{i=1}^{\infty} \big(x_i(s) - x_{n,i}(s)\big) r_i(t)\Big)^2 dt\Big)^{1/2} ds$$
$$\geq \int_0^1 \int_0^1 \Big|\sum_{i=1}^{\infty} \big(x_i(s) - x_{n,i}(s)\big) r_i(t)\Big|\, dt\, ds.$$

Combining this with Fubini's theorem, we deduce that (4.30) gives $\|f_n - g\|_{L_1(I \times I)} \leq \sqrt{3}\varepsilon$ whenever $n \geq N$. Hence, we have $\|f_n - g\|_{L_1(I \times I)} \to 0$ as $n \to \infty$. Thus, $f = g \in \text{Rad } X$, and the proof is completed. □

Our next goal is to find conditions on a s.s. X, under which elements of Rad X can be identified in a certain way with some sequences of functions from X.

Denote by $X(\ell_2)$ the closure of the set of all finite sequences from X with respect to the norm

$$\|\{x_i\}\|_{X(\ell_2)} := \Big\|\Big(\sum_{i=1}^{\infty} x_i^2\Big)^{1/2}\Big\|_X.$$

Equivalently, $X(\ell_2)$ is the set of all sequences $x = \{x_i\}_{i=1}^{\infty}$, $x_i \in X$, such that

$$\lim_{n\to\infty} \sup_{m>n} \Big\|\Big(\sum_{i=n+1}^{m} x_i^2\Big)^{1/2}\Big\|_X = 0. \tag{4.31}$$

Recall that the norm of a s.s. X on I is said to be *order semi-continuous* if from $x_n \in X$, $n = 1, 2, \ldots$, $x \in X$, and $x_n \to x$ a.e. on I, it follows: $||x||_X \le \liminf_{n\to\infty} ||x_n||_X$ (see Appendix C).

Theorem 4.8 *Let X be a s.s. on $I = [0, 1]$ with an order semi-continuous norm. The following conditions are equivalent:*

1) there exists $M > 0$ such that for any sequence $x = \{x_k\}_{k=1}^{\infty} \in X(\ell_2)$

$$M^{-1}\|x\|_{X(\ell_2)} \le \Big\| \sum_{i=1}^{\infty} x_i(s) r_i(t) \Big\|_{X(I\times I)} \le M\|x\|_{X(\ell_2)}; \tag{4.32}$$

2) the lower Boyd index $\alpha_X > 0$.

In the proof of this theorem we shall make use of some properties of the tensor product in s.s.'s. Namely, we set $T_\varphi x(s,t) := x(s)\varphi(t)$, where φ is a fixed measurable function on I. Observe that functions $T_\varphi x$ and $T_\varphi x^*$ are equimeasurable for every measurable function $x(s)$ on I.

Lemma 4.13 *Let X be a s.s. on I.*

(a) If $\varphi \notin L_\infty(I)$ and the operator T_φ is bounded from X into $X(I \times I)$, then $\alpha_X > 0$;
(b) Let $\varphi(t) = \ln^a(e/t)$, where $a > 0$. The operator T_φ is bounded from X into $X(I \times I)$ if and only if $\alpha_X > 0$.

Proof

(*a*) On the contrary, assume that $\alpha_X = 0$. Then, $\|\sigma_\tau\|_{X\to X} = 1$ for all $0 < \tau \le 1$; therefore, for any $n = 1, 2, \ldots$ there is a function $x_n \in X$, $\|x_n\| = 1$, such that

$$\|\sigma_{1/n} x_n\| \ge \frac{1}{2}.$$

One can readily check that the functions $x_n(s)\chi_{(0,1/n)}(t)$ and $\sigma_{1/n}x_n(s)$ are identically distributed. Hence, as the operator T_φ is bounded from X into $X(I \times I)$, for all $n = 1, 2, \ldots$ we get

$$\begin{aligned}\|T_\varphi\| &\ge \|T_\varphi x_n\|_{X(I\times I)} = \|x_n(s)\varphi(t)\|_{X(I\times I)} = \|x_n(s)\varphi^*(t)\|_{X(I\times I)} \\ &\ge \|x_n(s)\varphi^*(1/n)\chi_{(0,1/n)}(t)\|_{X(I\times I)} = \varphi^*(1/n)\|\sigma_{1/n}x_n\|_X \ge \frac{1}{2}\varphi^*(1/n).\end{aligned}$$

This inequality, clearly, implies that the function $\varphi(t)$ is bounded, which contradicts the assumption. Thus, $\alpha_X > 0$.

(*b*) By definition of the Boyd indices (see Appendix C), there exists $C > 0$ such that for all $\tau \in (0, 1)$ the dilation operator σ_τ satisfies the condition

$$\|\sigma_\tau\|_{X\to X} \le C\tau^{\alpha_X/2}.$$

Therefore, if $\varphi(t) = \ln^a(e/t)$, $a > 0$, we have

$$\|T_\varphi x\|_{X(I\times I)} = \|T_\varphi x^*\|_{X(I\times I)} \le \Big\| x^*(s) \cdot \sum_{k=1}^{\infty} (k+1)^a \chi_{(e^{-k}, e^{-k+1}]}(t) \Big\|_{X(I\times I)}$$

$$\le \sum_{k=1}^{\infty} (k+1)^a \|x^*(s) \cdot \chi_{(e^{-k}, e^{-k+1}]}(t)\|_{X(I\times I)}$$

$$\le \sum_{k=1}^{\infty} (k+1)^a \|\sigma_{e^{-k+1}} x\|_X \le C \sum_{k=1}^{\infty} e^{(1-k)\alpha_X/2}(k+1)^a \|x\|_X = C(a)\|x\|_X.$$

Thus, $T_\varphi : X \to X(I\times I)$ is bounded. Since the converse is a consequence of the assertion (a), everything is done.

□

Proof of Theorem 4.8 First of all, show that the left-hand side of inequality (4.32) holds in every s.s. X with an order semi-continuous norm.

For every s.s. X on I and all $x_i \in X$, $i = 1, 2, \ldots$, the sequence $\{x_i(s) r_i(t)\}_{i=1}^{\infty}$ is unconditional in the space $X(I\times I)$ with constant 1, i.e.,

$$\Big\| \sum_{i=1}^{n} \theta_i x_i(s) r_i(t) \Big\|_{X(I\times I)} = \Big\| \sum_{i=1}^{n} x_i(s) r_i(t) \Big\|_{X(I\times I)} \tag{4.33}$$

for all $n \in \mathbb{N}$ and $\theta_i = \pm 1$, $i = 1, 2, \ldots, n$ (see Definition B.4). Indeed, by Fubini's theorem and Proposition 2.2, for each $\tau > 0$ we have

$$m\Big\{(s,t) \in I\times I : \Big|\sum_{i=1}^{\infty} \theta_i x_i(s) r_i(t)\Big| > \tau\Big\} = \int_0^1 m\Big\{t \in I : \Big|\sum_{i=1}^{\infty} \theta_i x_i(s) r_i(t)\Big| > \tau\Big\}\, ds$$

$$= \int_0^1 m\Big\{t \in I : \Big|\sum_{i=1}^{\infty} x_i(s) r_i(t)\Big| > \tau\Big\}\, ds$$

$$= m\Big\{(s,t) \in I\times I : \Big|\sum_{i=1}^{\infty} x_i(s) r_i(t)\Big| > \tau\Big\},$$

whence (4.33) follows immediately from the definition of s.s.'s.

Further, using subsequently equation (4.33), Minkowski's and Khintchine's inequalities (see Theorem 1.4), for all $n = 1, 2, \ldots$ we get

$$\Big\| \sum_{i=1}^{n} x_i(s) r_i(t) \Big\|_{X(I\times I)} = \int_0^1 \Big\| \sum_{i=1}^{n} x_i(s) r_i(t) r_i(u) \Big\|_{X(I\times I)} du$$

$$\ge \Big\| \int_0^1 \Big| \sum_{i=1}^{n} x_i(s) r_i(t) r_i(u) \Big| \, du \Big\|_{X(I\times I)}$$

$$\geq \frac{1}{\sqrt{3}} \Big\| \Big(\sum_{i=1}^{n} x_i(s)^2 \Big)^{1/2} \Big\|_{X(I\times I)}$$

$$= \frac{1}{\sqrt{3}} \Big\| \Big(\sum_{i=1}^{n} x_i^2 \Big)^{1/2} \Big\|_{X}. \tag{4.34}$$

Since $x = \{x_i\}_{i=1}^{\infty} \in X(\ell_2)$, we have $\sum_{i=1}^{\infty} x_i^2(s) < \infty$ a.e. on I. Hence, by Lemma 4.12, the series $\sum_{i=1}^{\infty} x_i(s) r_i(t)$ converges a.e. on the square $I \times I$. Assume that its sum belongs to the space $X(I \times I)$. According to (4.33), the sequence

$$\Big\{ \Big\| \sum_{i=1}^{n} x_i(s) r_i(t) \Big\|_{X(I\times I)} \Big\}_{n=1}^{\infty}$$

increases. Consequently, as the norm in X is order semi-continuous, we infer

$$\Big\| \sum_{i=1}^{\infty} x_i(s) r_i(t) \Big\|_{X(I\times I)} = \lim_{n\to\infty} \Big\| \sum_{i=1}^{n} x_i(s) r_i(t) \Big\|_{X(I\times I)}.$$

Therefore, passing to the limit in inequality (4.34) as $n \to \infty$, we obtain

$$\Big\| \sum_{i=1}^{\infty} x_i(s) r_i(t) \Big\|_{X(I\times I)} \geq \frac{1}{\sqrt{3}} \|\{x_i\}_{i=1}^{\infty}\|_{X(\ell_2)}. \tag{4.35}$$

Now, we proceed with the proof of the theorem. Firstly, we assume that $\alpha_X > 0$. For every $\{x_i\}_{i=1}^{\infty} \in X(\ell_2)$ we set $f := (\sum_{i=1}^{\infty} x_i^2)^{1/2}$. Clearly, $f \in X$ and $\|f\|_X = \|\{x_i\}\|_{X(\ell_2)}$. Applying the distribution exponential estimate from Proposition 1.2, for almost all $s \in I$ we have

$$m\Big\{t \in I : \Big| \sum_{i=1}^{n} x_i(s) r_i(t) \Big| > \tau \Big\} \leq 2\exp\Big(-\frac{\tau^2}{2f(s)^2} \Big) \leq 2m\{t \in I : 2f(s)\ln^{1/2}(e/t) > \tau\}.$$

Integration of this inequality over I and applying Fubini's theorem then imply that

$$m\Big\{(s,t) \in I \times I : \Big| \sum_{i=1}^{n} x_i(s) r_i(t) \Big| > \tau \Big\} \leq 2m\{(s,t) \in I \times I : 2f(s)\ln^{1/2}(e/t) > \tau\}.$$

Hence, by Theorem C.3, we get

$$\Big\| \sum_{i=1}^{n} x_i(s) r_i(t) \Big\|_{X(I\times I)} \leq 4\|T_\varphi f\|_{X(I\times I)},$$

where $\varphi(t) = \ln^{1/2}(e/t)$, and consequently from Lemma 4.13(b) it follows

$$\Big\|\sum_{i=1}^{n} x_i(s)r_i(t)\Big\|_{X(I\times I)} \le 4\|T_\varphi\|_{X(I\times I)\to X(I\times I)}\|f\|_X = M\|\{x_i\}\|_{X(\ell_2)}, \quad n=1,2,\ldots,$$

with $M = 4\|T_\varphi\|_{X(I\times I)\to X(I\times I)}$. Since $\{x_i\} \in X(\ell_2)$, we have (4.31). Therefore, from the last inequality it follows that the series $\sum_{i=1}^{\infty} x_i(s)r_i(t)$ converges in the space $X(I \times I)$ as well. So, passing to the limit in this inequality as $n \to \infty$, we get the right-hand side of (4.32). As a result, taking into account (4.35), we see that inequality (4.32) is proved.

Conversely, assume that (4.32) holds (in fact, only the right-hand side of this inequality will be used). Let $\psi(t)$ be the inverse to the function

$$\Psi(u) := \frac{2}{\sqrt{2\pi}} \int_u^\infty e^{-v^2/2}dv, \quad 0 \le u < \infty.$$

Clearly, $\psi(t)$ is defined on $(0, 1]$ and $\lim_{t\to+0}\psi(t) = +\infty$. Arguing similarly as in the proof of Theorem 2.3, we shall show first that the tensor product operator T_ψ is bounded from X into $X(I \times I)$, and then make use of Lemma 4.13(a).

Let $z_n := n^{-1/2}\sum_{i=1}^{n} r_i$, $n = 1, 2, \ldots$ By the central limit theorem, for all $u > 0$ we have

$$\lim_{n\to\infty} m\{t \in [0, 1] : |z_n(t)| > u\} = \Psi(u).$$

In consequence, Lemma 2.1 implies that $\lim_{n\to\infty} z_n^*(t) = \psi(t)$ a.e. on $[0, 1]$.

For any function $f \in X''$ and every $m = 1, 2, \ldots$ we define the truncation $f_m(s) := \min\{m, |f(s)|\}$. Then, $f_m \in L_\infty \subset X$ for each m and $f_m(s) \uparrow |f(s)|$ a.e. as $m \to \infty$. Moreover, for any $n, m = 1, 2, \ldots$ we set $x^{n,m} = \{x_k^{n,m}\}_{k=1}^\infty$, where $x_k^{n,m} := f_m/\sqrt{n}$ if $1 \le k \le n$, and $x_k^{n,m} := 0$ if $k > n$. As it can be easily seen, $x^{n,m} \in X(\ell_2)$ and $\|x^{n,m}\|_{X(\ell_2)} = \|f_m\|_X$ for all $n, m = 1, 2, \ldots$ Therefore, by assumption, we have

$$\|f_m(s)z_n^*(t)\|_{X(I\times I)} = \Big\|\sum_{i=1}^{n} \frac{1}{\sqrt{n}} f_m(s)r_i(t)\Big\|_{X(I\times I)} \le M\|x^{n,m}\|_{X(\ell_2)} = M\|f_m\|_X.$$

Since the norm of X is order semi-continuous, this space is isometrically embedded into its second Köthe dual X'' (see Appendix C or [151, Theorem VI.1.6]). Therefore, from the last inequality for all $n, m = 1, 2, \ldots$ it follows that

$$\|f_m(s)z_n^*(t)\|_{X''(I\times I)} \le M\|f_m\|_{X''} \le M\|f\|_{X''}.$$

Thanks to the fact that the space X'' has the Fatou property, tending m and n to ∞, we get

$$\|T_\psi f\|_{X''(I\times I)} = \|f(s)\psi(t)\|_{X''(I\times I)} \le M\|f\|_{X''} \quad \text{for all} \quad f \in X''.$$

Since $\alpha_X = \alpha_{X''}$ (see Appendix C or [168, Theorem II.4.11]), by Lemma 4.13, this implies that $\alpha_X > 0$, and the proof is completed. □

In conclusion, we consider the complementability problem of the subspace Rad X in a s.s. $X(I \times I)$. Recall that in the scalar case the similar task has been solved in Chap. 2 (see Theorem 2.5). Namely, it has been proved there that the Rademacher subspace $\mathcal{R}(X)$ is complemented in a s.s. X if and only if $G \subset X \subset G'$, where G is the separable part of the Orlicz space L_{N_2}, $N_2(t) = e^{t^2} - 1$. It turns out that in the case of Rad X this embedding condition should be replaced with the more restrictive condition of nontriviality of the Boyd indices of X.

Theorem 4.9 *Let X be a s.s. on $I = [0, 1]$ with an order semi-continuous norm. Then the following conditions are equivalent:*

1) the subspace Rad X *is complemented in $X(I \times I)$;*
2) the Boyd indices of X are nontrivial, i.e., $0 < \alpha_X \le \beta_X < 1$.

A crucial role in the proof of this theorem will be played by the orthogonal projection

$$Pf(s,t) = \sum_{i=1}^{\infty} \int_0^1 f(s,v) r_i(v)\,dv \cdot r_i(t). \tag{4.36}$$

We start with studying some properties of this operator.

Proposition 4.2 *The projection P is bounded in a s.s. $X(I \times I)$ if and only if $0 < \alpha_X \le \beta_X < 1$. Moreover, the image of P coincides with* Rad X.

Proof Let $0 < \alpha_X \le \beta_X < 1$. Assuming that $f(s,t) \in L_p(I \times I)$ for some $1 < p < \infty$, by Fubini's theorem, we have $f(s,\cdot) \in L_p(I)$, $s \in E$, where $m(E) = 1$. Therefore (see the proof of Theorem 2.5),

$$\sum_{i=1}^{\infty} \Big(\int_0^1 f(s,v) r_i(v)\,dv \Big)^2 < \infty, \ s \in E,$$

and for every $1 < p < \infty$ there exists a constant $C_p > 0$ such that for all $s \in E$ we have

$$\int_0^1 |Pf(s,t)|^p\,dt \le C_p \int_0^1 |f(s,t)|^p\,dt.$$

According to Lemma 4.12, the first of these relations assures that the series from the right-hand side of (4.36) converges a.e. on the square $I \times I$. Moreover, integrating the last inequality over E, we conclude that the operator P is bounded in $L_p(I \times I)$ for all $1 < p < \infty$. Therefore, by the condition $0 < \alpha_X \le \beta_X < 1$ and the Boyd interpolation Theorem D.3, we have that $X \subset \bigcup_{1<p<\infty} L_p$ and P is bounded on the space $X(I \times I)$. This implies, in particular, that the series from the right-hand side of (4.36) converges a.e. on $I \times I$ for every $f \in X(I \times I)$, whence the image of the projection P is contained in Rad X. Conversely, if $f(s,t) \in$ Rad X, then this function is representable in the form (4.27) and from Lemma 4.12 it follows that $\sum_{i=1}^{\infty} x_i^2(s) < \infty$ a.e. on $[0, 1]$. Hence, $f(s,\cdot) \in L_2(I)$ for almost all $s \in I$, and the direct substitution of formula (4.27) into equation (4.36) shows that $Pf = f$. Thus, the image of P is Rad X.

Let us prove the converse. Assuming that P is bounded in the space $X(I \times I)$, we denote $A := \|P\|_{X(I\times I)\to X(I\times I)}$. Also, for any $f \in X(I \times I)$ and $g \in X'(I \times I)$ we set

$$\langle f, g\rangle := \int_0^1\int_0^1 f(s,t)g(s,t)\,ds\,dt.$$

One can easily check that P is a self-adjoint operator, i.e., $\langle Pf, g\rangle = \langle f, Pg\rangle$. Moreover, since the norm of X is order semi-continuous, it follows that X is isometrically embedded into the second Köthe dual space X''. Hence, by definition of the Köthe dual norm, we get

$$\begin{aligned} A &= \sup_{\|f\|_{X(I\times I)}\le 1} \|Pf\|_{X''(I\times I)} = \sup_{\|f\|_{X(I\times I)}\le 1}\ \sup_{\|g\|_{X'(I\times I)}\le 1} \langle Pf, g\rangle \\ &= \sup_{\|g\|_{X'(I\times I)}\le 1}\ \sup_{\|f\|_{X(I\times I)}\le 1} \langle f, Pg\rangle = \sup_{\|g\|_{X'(I\times I)}\le 1} \|Pg\|_{X'(I\times I)}. \end{aligned}$$

Thus, P is bounded in $X'(I \times I)$ and

$$\|P\|_{X'(I\times I)\to X'(I\times I)} = A. \tag{4.37}$$

Furthermore, if a function $f \in$ Rad X and f is representable in the form (4.27), then, by Fubini's theorem, for arbitrary $g \in X'(I \times I)$ we have

$$\begin{aligned} \langle f, g\rangle &= \int_0^1\int_0^1 \sum_{i=1}^{\infty} x_i(s)r_i(t)\cdot g(s,t)\,ds\,dt = \int_0^1 \sum_{i=1}^{\infty} x_i(s)\int_0^1 g(s,t)r_i(t)\,dt\,ds \\ &= \int_0^1\int_0^1 \sum_{i=1}^{\infty} x_i(s)r_i(t)\cdot \sum_{i=1}^{\infty}\int_0^1 g(s,v)r_i(v)\,dv\; r_i(t)\,dt\,ds \;=\; \langle f, Pg\rangle. \end{aligned}$$

Therefore, since the norm of X is order semi-continuous, from (4.37) it follows that

$$\begin{aligned}\|f\|_{X(I\times I)} &= \sup\{\langle f, g\rangle : \|g\|_{X'(I\times I)} \le 1\} \\ &= \sup\{\langle f, Pg\rangle : \|g\|_{X'(I\times I)} \le 1\} \\ &\le \sup\{\langle f, h\rangle : h \in \mathrm{Rad}\ X, \|h\|_{X'(I\times I)} \le A\}. \qquad (4.38)\end{aligned}$$

Let us introduce one more notation: for any $x = \{x_i\} \in X(\ell_2)$ and $y = \{y_i\} \in X'(\ell_2)$ we put

$$\langle\langle x, y\rangle\rangle := \sum_{i=1}^{\infty} \int_0^1 x_i(s) y_i(s)\, ds.$$

Recall that the Rademacher system is orthonormal on $[0, 1]$. Therefore, if $f \in \mathrm{Rad}\ X$ and $g \in \mathrm{Rad}\ X'$, that is,

$$f(s,t) = \sum_{i=1}^{\infty} x_i(s) r_i(t) \text{ and } g(s,t) = \sum_{i=1}^{\infty} y_i(s) r_i(t),$$

where $x_i \in X$ and $y_i \in X'$, then, again by Fubini's theorem, we have $\langle f, g\rangle = \langle\langle x, y\rangle\rangle$, $x = \{x_i\}$, $y = \{y_i\}$. Moreover, from the Cauchy-Schwarz-Bunyakovskii inequality, it follows $|\langle\langle x, y\rangle\rangle| \le \|x\|_{X(\ell_2)}\|y\|_{X'(\ell_2)}$, whence $X'(\ell_2) \subset (X(\ell_2))^*$ and $\|y\|_{(X(\ell_2))^*} \le \|y\|_{X'(\ell_2)}$. Finally, applying inequality (4.35) for X', we get

$$\Big\|\sum_{i=1}^{\infty} y_i(s) r_i(t)\Big\|_{X'(I\times I)} \ge \frac{1}{\sqrt{3}}\|y\|_{X'(\ell_2)}.$$

The last relations, combined with inequality (4.38), imply that

$$\begin{aligned}\Big\|\sum_{i=1}^{\infty} x_i(s) r_i(t)\Big\|_{X(I\times I)} &\le \sup\{< x, y >: y \in X'(\ell_2), \|y\|_{X'(\ell_2)} \le \sqrt{3}A\} \\ &\le \sup\{< x, y >: y \in X'(\ell_2), \|y\|_{(X(\ell_2))^*} \le \sqrt{3}A\} \\ &\le \sqrt{3}A\|x\|_{X(\ell_2)}.\end{aligned}$$

Thus, by Theorem 4.8 and its proof, we obtain that $\alpha_X > 0$.

In addition, according to (4.37), the projection P is bounded in the Köthe dual $X'(I \times I)$. Therefore, as above, $\alpha_{X'} > 0$, which implies that $\beta_X = 1 - \alpha_{X'} < 1$ (see Appendix C or [168, Theorem II.4.11]). This completes the proof of the proposition.

□

Proof of Theorem 4.9 Taking into account Proposition 4.2, it suffices only to prove that the complementability of Rad X in the space $X(I \times I)$ implies the boundedness of the projection P in the latter space. Further, we shall argue largely in the same way as in the proof of Theorem 2.5.

Let $t = \sum_{i=1}^{\infty} \alpha_i 2^{-i}$, $u = \sum_{i=1}^{\infty} \beta_i 2^{-i}$ $(\alpha_i, \beta_i = 0, 1)$ be the binary expansions of numbers $t, u \in [0, 1]$. One can readily see that the interval $I = [0, 1]$, equipped with the operation

$$t \oplus u := \sum_{i=1}^{\infty} 2^{-i}[(\alpha_i + \beta_i) mod 2],$$

is a compact Abelian group. Moreover, for any $(s, t) \in I \times I, u \in I$ we set $w_u(s, t) = (s, t \oplus u)$. Since the transformation w_u, $u \in I$, preserves the Lebesgue measure on $I \times I$, the operators T_u, $0 \le u \le 1$, defined by

$$T_u f(s, t) := f \circ w_u(s, t) = f(s, t \oplus u),$$

form a family of isometries in $X(I \times I)$. It is obvious that the subspace Rad X is invariant with respect to the action of these operators. Thus, according to the well-known Rudin theorem [255, Theorem 5.18], there exists a bounded in $X(I \times I)$ projection Q, commuting with all the operators T_u, $0 \le u \le 1$, such that the image of Q is Rad X. It remains to show that $Q = P$.

First of all, we represent the projection Q as follows:

$$Qf(s, t) = \sum_{i=1}^{\infty} (Q_i f)(s) r_i(t), \tag{4.39}$$

where Q_i are linear operators, $Q_i : X(I \times I) \to X$. Since Q is a projection onto Rad X, we have

$$Q_i(x(s) r_j(t)) = \begin{cases} x(s), i = j, \\ 0, i \ne j. \end{cases} \tag{4.40}$$

Let us consider the sets

$$E_i := \left\{ u \in [0, 1] : u = \sum_{i=1}^{\infty} \alpha_j 2^{-j}, \quad \alpha_i = 0 \right\} \quad \text{and} \quad E_i^c := [0, 1] \backslash E_i.$$

One can check that

$$r_i(t \oplus u) = \begin{cases} r_i(t), u \in E_i, \\ -r_i(t), u \in E_i^c. \end{cases}$$

Combining this together with the equation $w_u^2 = I$ (I is the identity operator) and the commutativity conditions $T_u Q = QT_u$, $0 \le u \le 1$, we get

$$Q_i(f \circ w_u) = \begin{cases} Q_i f, u \in E_i, \\ -Q_i f, u \in E_i^c. \end{cases}$$

Therefore, since $m(E_i) = m(E_i^c) = 1/2$, we have

$$\int_{E_i} Q_i(f \circ w_u)\,du = \frac{1}{2} Q_i f, \quad \int_{E_i^c} Q_i(f \circ w_u)\,du = -\frac{1}{2} Q_i f.$$

Further, recall that each sequence $\{x_i(s)r_i(t)\}_{i=1}^{\infty}$, $x_i \in X$, is unconditional in $X(I \times I)$ with constant 1 (see (4.33)). This implies that

$$\|Q_i f\|_X \le \|Qf\|_{X(I\times I)} \le \|Q\| \|f\|_{X(I\times I)},$$

which means that the operators Q_i, $i = 1, 2, \ldots$, are bounded. Therefore, in the preceding equations Q_i can be taken out of the integral sign, whence

$$Q_i f = Q_i \left(\int_{E_i} (f \circ w_u) du - \int_{E_i^c} (f \circ w_u) du \right). \tag{4.41}$$

Clearly, for each $u \in [0, 1]$ the mapping $t \mapsto t \oplus u$ is a measure-preserving transformation of $[0, 1]$. Moreover, for every $t \in [0, 1]$ we have

$$\{v \in [0, 1] : v = t \oplus u, \quad u \in E_i\} = \begin{cases} E_i, t \in E_i, \\ E_i^c, t \in E_i^c \end{cases}$$

and

$$\{v \in [0, 1] : v = t \oplus u, \quad u \in E_i^c\} = \begin{cases} E_i^c, t \in E_i, \\ E_i, t \in E_i^c. \end{cases}$$

Consequently,

$$\int_{E_i} (f \circ w_u)(s,t)\,du = \int_{E_i} f(s,v)\,dv \cdot \chi_{E_i}(t) + \int_{E_i^c} f(s,v)\,dv \cdot \chi_{E_i^c}(t)$$

and

$$\int_{E_i^c} (f \circ w_u)(s,t)\,du = \int_{E_i^c} f(s,v)\,dv \cdot \chi_{E_i}(t) + \int_{E_i} f(s,v)\,dv \cdot \chi_{E_i^c}(t).$$

Since $r_i = \chi_{E_i} - \chi_{E_i^c}$ for all $i = 1, 2, \ldots$, this yields

$$\int_{E_i} (f \circ w_u)(s,t)\,du - \int_{E_i^c} (f \circ w_u)(s,t)\,du = \int_0^1 f(s,v) r_i(v)\,dv \cdot r_i(t).$$

Thus, taking into account (4.40) and (4.41), we get

$$(Q_i f)(s) = \int_0^1 f(s,v) r_i(v)\,dv, \quad i = 1, 2, \ldots$$

Finally, comparing (4.36) and (4.39), we deduce that the projections Q and P coincide, which finishes the proof. □

Comments and References The principal result of this chapter is the Kahane–Khintchine inequality (4.1), showing the equivalence of the L_p-norms of Rademacher sums with vector coefficients for all values $1 \le p < \infty$. Its original version was proved by Kahane (see Theorem 2.4 in [148]). The proof given here is due to Borell [80] (his work remained unpublished) and is based on applying Lemma 4.1 from [66]. In our presentation we follow [182, Theorem 1.e.13 and Lemma 1.e.14]). For the first time, Corollary 4.3 was proved by Kwapień in [171] by using a version of the Kahane–Khintchine inequality from [148]. Here, this result is obtained as a consequence of Corollary 4.2.

Theorems 4.2 and 4.3, which are used in the proof of estimates for the norms of vector-valued Rademacher sums, have been proved by Talagrand in [275]. Inequalities (4.16), (4.17), and (4.24) can be treated as vector-valued counterparts of Montgomery-Smith's and Hitczenko's inequalities from Chap. 1. The only substantial difference from the scalar case is that, in general, the expectation

$$\mathbb{E}\Big\|\sum_{n=1}^{\infty} x_n r_n\Big\|_F,$$

where x_n, $n = 1, 2, \ldots$, are vectors from a Banach space F, cannot be replaced (up to equivalence) with a quantity, depending only on the norms of x_n, $n = 1, 2, \ldots$. The above results (Theorems 4.4, 4.5 and Corollary 4.7) have been proved by Dilworth and Montgomery-Smith in the paper [108].

Theorem 4.6 related to exponential Orlicz spaces has been obtained also in the paper [108]. Moment and tail bounds for the vector-valued Rademacher sums in terms of weak parameters make up the content of the work by Oleszkiewicz [221]; in particular, Theorem 4.7 is borrowed from this paper. The classical estimate from Lemma 4.7 is due to Kahane [148, Theorem 2.3].

The subspace Rad X of a s.s. $X(I \times I)$ turned out to be useful when studying geometrical properties of s.s.'s. In particular, in Chapter 2.d of the monograph [182], by using Rad X, it is proved that arbitrary s.s. X with the nontrivial Boyd indices is primary. The latter means the following: if X is represented as a direct sum of its closed subspaces, then at least one of them is isomorphic to X itself. The definitions of the spaces Rad X and $X(\ell_2)$ are somewhat different from those used in [182], but coincide with them when X is separable. The sufficiency of the condition $\alpha_X > 0$ for (4.32) to hold (see Theorem 4.8) is proved in [182, Proposition 2.d.1], while the necessity in [36]. The tensor product and operators connected with this notion (see Lemma 4.13) had been studied very intensively (see e.g. [9, 11, 14, 52, 53, 202–205, 222]). Sharp conditions, under which the subspace Rad X is complemented in a s.s. $X(I \times I)$ (see Theorem 4.9), have been found in [36]. Earlier, in [182, Proposition 2.d.2], it was obtained a weaker result, asserting that, in the case of a separable s.s. X, the condition $0 < \alpha_X \le \beta < 1$ implies that the subspace Rad X is complemented in $X(I \times I)$.

Chapter 5
Optimal Constants in the Khintchine Type Inequalities

According to Khintchine's inequality (see Theorem 1.4), for every $p > 0$ and all $n \in \mathbb{N}$ there exist constants $c > 0$ and $C > 0$, which depend on p, such that

$$c\Big(\sum_{i=1}^{n} a_i^2\Big)^{1/2} \le \left(\int_0^1 \Big|\sum_{i=1}^{n} a_i r_i(t)\Big|^p dt\right)^{1/p} \le C\Big(\sum_{i=1}^{n} a_i^2\Big)^{1/2}, \tag{5.1}$$

where $a_i \in \mathbb{R}$, $i = 1, 2, \ldots, n$, are arbitrary.

In this chapter, we shall investigate the problem of finding the optimal constants for both sides of this inequality. Note that there are two possible ways of setting the task depending on whether $n \in \mathbb{N}$ is fixed or not. Further, by $k_p(n)$ and $K_p(n)$ (resp. k_p and K_p) we denote the best (or optimal) constants, for which inequality (5.1) holds for the fixed $n \in \mathbb{N}$ (resp. for all $n \in \mathbb{N}$). More precisely, $k_p(n) := \max c$, $K_p(n) := \min C$, where c and C are constants for which (5.1) is fulfilled for a *given* $p > 0$ and a *given* $n \in \mathbb{N}$ (resp. $k_p := \max c$, $K_p := \min C$, where c and C are constants for which this inequality holds for a *given* $p > 0$ and *all* $n \in \mathbb{N}$). One of the most effective methods to deal with these problems is based on using the so-called Schur-concavity notion.

5.1 The Schur-Concavity and Majorization

Let $n \in \mathbb{N}$, $a = (a_i)_{i=1}^n \in \mathbb{R}^n$. By $a^\sharp = (a_i^\sharp)_{i=1}^n$ will be denoted the vector, having the same components as a, but which are taken in the decreasing order, i.e., $a_1^\sharp \ge a_2^\sharp \ge \cdots \ge a_n^\sharp$.

S. V. Astashkin, *The Rademacher System in Function Spaces*,
https://doi.org/10.1007/978-3-030-47890-2_5

We say that a vector $b = (b_i)_{i=1}^n$ *majorizes* (or *dominates*) a vector $a = (a_i)_{i=1}^n$ (write: $a \prec\prec b$) if

$$\sum_{i=1}^{k} a_i^\sharp \le \sum_{i=1}^{k} b_i^\sharp, \quad k = 1, 2, \dots, n-1,$$

and

$$\sum_{i=1}^{n} a_i = \sum_{i=1}^{n} b_i.$$

In particular, we have

$$\left(\frac{1}{n}, \dots, \frac{1}{n}\right) \prec\prec \left(\frac{1}{n-1}, \dots, \frac{1}{n-1}, 0\right) \prec\prec \left(\frac{1}{2}, \frac{1}{2}, 0, \dots 0\right) \prec\prec (1, 0, \dots, 0), \tag{5.2}$$

and if $a_i \ge 0$, $\sum_{i=1}^n a_i = 1$, then

$$\left(\frac{1}{n}, \dots, \frac{1}{n}\right) \prec\prec (a_1, \dots, a_n) \prec\prec (1, 0, \dots, 0). \tag{5.3}$$

Let $F(t)$ be a real-valued function defined on a set $D \subset \mathbb{R}^n$. It is said to be *Schur-convex* on the set D whenever $F(a) \le F(b)$ provided that $a, b \in D$ and $a \prec\prec b$. A function $F(t)$ is said to be *Schur-concave* if $-F$ is Schur-convex.

Let $n \in \mathbb{N}$, $n \ge 2$. A set $D \subset \mathbb{R}^n$ is *symmetric* if for each $(t_1, t_2, \dots, t_n) \in D$ and every permutation $\pi : \{1, 2, \dots, n\} \to \{1, 2, \dots, n\}$ the vector $(t_{\pi(1)}, t_{\pi(2)}, \dots, t_{\pi(n)})$ belongs to D as well. Similarly, we say that a function $F : D \to \mathbb{R}$ is *symmetric* if for each $(t_1, t_2, \dots, t_n) \in D$ and every permutation $\pi : \{1, 2, \dots, n\} \to \{1, 2, \dots, n\}$ $F(t_1, t_2, \dots, t_n) = F(t_{\pi(1)}, t_{\pi(2)}, \dots, t_{\pi(n)})$.

Theorem 5.1 *Let $D \subset \mathbb{R}^n$ be an open symmetric convex set. Suppose a symmetric, continuously differentiable function $F : D \to \mathbb{R}$ satisfies the following condition:*

$$\frac{\partial F}{\partial t_i} - \frac{\partial F}{\partial t_j} \ge 0 \tag{5.4}$$

for all $(t_k)_{k=1}^n \in D$ such that $t_i > t_j$, $i, j = 1, \dots, n$. Then, F is a Schur-convex function on D.

Proof Taking into account that D and F are symmetric, it suffices to prove the Schur-convexity of F on the intersection $D' := D \cap \{t \in \mathbb{R}^n : t_1 \ge t_2 \ge \cdots \ge t_n\}$.

Assume that $a, b \in D', a \prec\prec b$. Then if $a = (a_k)_{k=1}^n, b = (b_k)_{k=1}^n$, by definition, we have

$$\sum_{i=1}^{k} a_i \leq \sum_{i=1}^{k} b_i, \quad k = 1, 2, \ldots, n-1,$$

and

$$\sum_{i=1}^{n} a_i = \sum_{i=1}^{n} b_i.$$

Hence, setting $\tilde{a}_k := \sum_{i=1}^k a_i$ and $\tilde{b}_k := \sum_{i=1}^k b_i$, $k = 1, 2, \ldots, n$, we get

$$\tilde{a}_k \leq \tilde{b}_k, \quad k = 1, 2, \ldots n-1, \quad \text{and} \quad \tilde{a}_n = \tilde{b}_n.$$

Therefore, the theorem will be proved once we show that

$$\tilde{F}(u_1, u_2, \ldots, u_n) := F(u_1, u_2 - u_1, \ldots, u_n - u_{n-1})$$

is an increasing function for each of its arguments $u_1, u_2, \ldots, u_{n-1}$ provided if $t = (t_k) \in D'$, where $t_1 = u_1$, $t_k = u_k - u_{k-1}$, $k = 2, \ldots n$. Since for every $k = 1, 2, \ldots n-1$ we have

$$\frac{\partial \tilde{F}}{\partial u_k} = \frac{\partial F}{\partial t_k} - \frac{\partial F}{\partial t_{k+1}}$$

and $t_1 \geq t_2 \geq \cdots \geq t_n$, then, by hypothesis, this derivative is nonnegative. This finishes the proof. □

Denote by $\mathcal{F}$ the set of all continuously differentiable functions $f : \mathbb{R} \to \mathbb{R}$ such that for each $h \geq 0$ the mapping

$$t \mapsto t^{-1}\left(f'(t+h) - f'(-t+h) + f'(t-h) - f'(-t-h)\right)$$

is an increasing function on the semi-axis $(0, \infty)$. In particular, it is clear that $f \in \mathcal{F}$ whenever the function

$$t \mapsto t^{-1}\left(f'(t+h) - f'(-t+h)\right) \tag{5.5}$$

is increasing on $(0, \infty)$ for every $h \in \mathbb{R}$.

Further, we need the following simple result.

Lemma 5.1 *Let $f \in \mathcal{F}$ and let ξ be a bounded symmetrically distributed r.v. on a probability space* $(\Omega, \Sigma, \mathbb{P})$. *Then, the mapping*

$$t \mapsto t^{-1}\mathbb{E}\left(f'(t+\xi) - f'(-t+\xi)\right)$$

is an increasing function on the semi-axis $(0, \infty)$.

Proof Since ξ is symmetrically distributed, then denoting

$$n_\xi(\tau) := \mathbb{P}\{\omega \in \Omega : \xi(\omega) > \tau\}, \quad \tau \in \mathbb{R},$$

we have

$$t^{-1}\mathbb{E}(f'(t+\xi) - f'(-t+\xi)) = \frac{1}{t}\int_{-\infty}^{\infty}(f'(t+\tau) - f'(-t+\tau))\,dn_\xi(\tau)$$

$$= \int_0^\infty \frac{1}{t}\left(f'(t+\tau) - f'(-t+\tau) + f'(t-\tau) - f'(-t-\tau)\right)dn_\xi(\tau).$$

Observe that, by condition, for each $\tau > 0$, the function standing under the last integral sign is increasing on $(0, \infty)$. Therefore, the result, which we wished, follows. □

For arbitrary $n \in \mathbb{N}$ and $a = (a_k)_{k=1}^n \in \mathbb{R}^n$, we set $S_n(a) := \sum_{k=1}^n a_k r_k$ (as usual, r_k, $k = 1, 2, \ldots$, are the Rademacher functions defined on $[0, 1]$).

Theorem 5.2 *If $f \in \mathcal{F}$ and $(a_1^2, \ldots, a_n^2) \prec\prec (b_1^2, \ldots, b_n^2)$, then we have*

$$\mathbb{E}f(S_n(b)) \le \mathbb{E}f(S_n(a)). \tag{5.6}$$

Proof Let the function F be defined on the set $D := \{u = (u_1, \ldots, u_n) : u_i > 0,\ i = 1, \ldots, n\}$ by $F(u) := \mathbb{E}f(S_n(\sqrt{u_1}, \ldots, \sqrt{u_n}))$. Show that F is Schur-concave.

Since $f \in \mathcal{F}$ and the Rademacher functions are independent and symmetrically distributed, then F is a symmetric, continuously differentiable function on the symmetric set D. Therefore, by Theorem 5.1, it is sufficient to check that $-F$ satisfies inequality (5.4) for $i = 1$ and $j = 2$.

Let $n \ge 3$. Then, by using the linearity of the expectation, we get

$$\frac{\partial F}{\partial u_1} - \frac{\partial F}{\partial u_2} = (2(u_1u_2)^{1/2})^{-1}\mathbb{E}\left((r_1u_2^{1/2} - r_2u_1^{1/2})f'(u_1^{1/2}r_1 + u_2^{1/2}r_2 + \xi)\right),$$

where $\xi := \sum_{k=3}^n u_k^{1/2}r_k$ is a bounded symmetrically distributed r.v. on $[0, 1]$. In view of the independence of Rademacher functions and Fubini's theorem,

$$\frac{\partial F}{\partial u_1} - \frac{\partial F}{\partial u_2} = (2(u_1u_2)^{1/2})^{-1}\mathbb{E}_\xi\mathbb{E}_{r_1,r_2}\left((r_1u_2^{1/2} - r_2u_1^{1/2})f'(u_1^{1/2}r_1 + u_2^{1/2}r_2 + \xi)\right)$$

$$= \frac{(u_1^{1/2} + u_2^{1/2})(u_2^{1/2} - u_1^{1/2})}{8(u_1 u_2)^{1/2}}$$

$$\times \mathbb{E}_\xi \Bigg(\frac{f'(u_1^{1/2} + u_2^{1/2} + \xi) - f'(-u_1^{1/2} - u_2^{1/2} + \xi)}{u_1^{1/2} + u_2^{1/2}}$$

$$- \frac{f'(u_2^{1/2} - u_1^{1/2} + \xi) - f'(-u_2^{1/2} + u_1^{1/2} + \xi)}{u_2^{1/2} - u_1^{1/2}} \Bigg).$$

Since by condition $f \in \mathcal{F}$, Lemma 5.1 implies that the expectation $\mathbb{E}_\xi(\cdot)$ from the right-hand side of this equation is nonnegative if $u_2 > u_1$. Thus, the latter condition assures that $\partial F/\partial u_1 - \partial F/\partial u_2 \geq 0$. As said above, then the function F is Schur-concave. For $n = 1$ and $n = 2$ the reasoning is only simplified.

By Proposition 2.2, Rademacher sums $\sum_{k=1}^n u_k r_k$ and $\sum_{k=1}^n |u_k| r_k$ are identically distributed. Hence, the theorem is proved in the case when all components of the vectors $(a_1^2, \dots, a_n^2)$ and $(b_1^2, \dots, b_n^2)$ are nonzero. If some of the numbers a_i and b_i are equal to zero, the desired result follows by continuity. □

We proceed now with consideration of functions from $\mathcal{F}$, which are the most important in this context.

Proposition 5.1 *If $r = 2$ or $r \geq 3$, then the function $f(t) = |t|^r$ belongs to the set $\mathcal{F}$.*

Proof We need to prove only that in this case, for each $h \in \mathbb{R}$, the function defined by (5.5) is increasing for $t > 0$. Since it is obvious for $r = 2$, we can assume that $r \geq 3$.

Throughout the proof we set $\langle t \rangle^p := |t|^p \operatorname{sign} t$, where $t \in \mathbb{R}$ and $p > 0$. Because $f'(t) = r \cdot \langle t \rangle^{r-1}$, for all $t \geq 0$ and $h \in \mathbb{R}$ we have

$$t^{-1} \big(f'(t+h) - f'(-t+h) \big) = r \cdot g_h(t),$$

where

$$g_h(t) := \frac{\langle t+h \rangle^{r-1} + \langle t-h \rangle^{r-1}}{t}.$$

One can easily check that g_h is continuously differentiable for $t \geq 0$. Show that, for arbitrary $h \in \mathbb{R}$, we have $g_h'(t) \geq 0$, $t \geq 0$.

After differentiation we get

$$g_h'(t) = \frac{(r-1)(|t+h|^{r-2} + |t-h|^{r-2})t - (\langle t+h \rangle^{r-1} + \langle t-h \rangle^{r-1})}{t^2}$$

$$= \frac{|t+h|^{r-2}((r-2)t - h) + |t-h|^{r-2}((r-2)t + h)}{t^2}.$$

Derivative of the numerator of the last fraction is equal to

$$(r-2)\left(\langle t+h\rangle^{r-3}((r-2)t-h)+|t+h|^{r-2}+\langle t-h\rangle^{r-3}((r-2)t+h)+|t-h|^{r-2}\right)$$
$$=(r-2)(r-1)t\left(\langle t+h\rangle^{r-3}+\langle t-h\rangle^{r-3}\right).$$

Since the latter expression is nonnegative for all $r\geq 3$, $t\geq 0$ and $h\in\mathbb{R}$, the numerator equals zero for $t=0$, we have $g_h'(t)\geq 0$. Thus, the proof is completed. □

From Proposition 5.1, Theorem 5.2 and relation (5.3) it follows

Theorem 5.3 *For arbitrary $n\in\mathbb{N}$ and $p\in\{2\}\cup[3,\infty)$ we have*

$$\max\left\{\left\|\sum_{k=1}^{n}a_k r_k\right\|_p : \sum_{k=1}^{n}a_k^2=1\right\}=\left(\int_0^1\left|\frac{1}{\sqrt{n}}\sum_{k=1}^{n}r_k(t)\right|^p dt\right)^{1/p}.$$

In particular, this implies that for every $p\in\{2\}\cup[3,\infty)$ the sequence

$$\left(\int_0^1\left|\frac{1}{\sqrt{n}}\sum_{k=1}^{n}r_k(t)\right|^p dt\right)^{1/p},\quad n\in\mathbb{N},$$

is increasing.

As a consequence of Theorem 5.3, we conclude that, if $p\in\{2\}\cup[3,\infty)$, the best constant $K_p(n)$ in inequality (5.1) is defined as follows:

$$K_p(n)=\left(\int_0^1\left|\frac{1}{\sqrt{n}}\sum_{k=1}^{n}r_k(t)\right|^p dt\right)^{1/p}. \tag{5.7}$$

Equivalently, by definition of the Rademacher functions and easy combinatorial arguments (see [284]), we have

$$K_p(n)=\left(2^{-n}n^{-p/2}\sum_{k=0}^{n}C_n^k|n-2k|^p\right)^{1/p}, \tag{5.8}$$

where $C_n^k:=n!/(k!(n-k)!)$.

In this respect the case $0<p<2$ is quite trivial; namely, $K_p(n)=1$. Indeed, firstly,

$$\left\|\sum_{k=1}^{n}a_k r_k\right\|_p\leq\left\|\sum_{k=1}^{n}a_k r_k\right\|_2=\left(\sum_{k=1}^{n}a_k^2\right)^{1/2}.$$

Secondly, taking $a_1 = 1$, $a_2 = \cdots = a_n = 0$, we get $\| \sum_{k=1}^n a_k r_k \|_p = 1$ for all $0 < p \le \infty$; thus, in this case, the last inequality becomes an equality.

As for the lower bound in (5.1), then for $p \ge 2$, clearly, we have $k_p(n) = 1$, because then

$$\Big(\sum_{k=1}^n a_k^2\Big)^{1/2} = \Big\| \sum_{k=1}^n a_k r_k \Big\|_2 \le \Big\| \sum_{k=1}^n a_k r_k \Big\|_p.$$

On the other hand, the same choice of coefficients: $a_1 = 1$, $a_2 = \cdots = a_n = 0$ shows that the latter inequality cannot be improved.

The Schur-concavity turns out useful also when one tries to find the best constants k_p and K_p (i.e., when Khintchine's inequality (5.1) is considered for all $n \in \mathbb{N}$). First of all, by Theorem A.1, the sequence of sums

$$S_n := \frac{1}{\sqrt{n}} \sum_{k=1}^n r_k, \quad n = 1, 2, \ldots$$

converges in distribution as $n \to \infty$ to a standard Gaussian r.v. $\mathfrak{g}$. Combining this fact together with Theorem 5.3, we infer that, for each $p \in \{2\} \cup [3, \infty)$, the moments $\mathbb{E}|S_n|^p$ form an increasing sequence, converging to $\mathbb{E}|\mathfrak{g}|^p$. Thus, by using an easy calculation, we get

$$K_p = \lim_{n\to\infty} K_p(n) = \lim_{n\to\infty} \|S_n\|_p = \|\mathfrak{g}\|_p = \sqrt{2}\left(\frac{\Gamma(\frac{p+1}{2})}{\sqrt{\pi}}\right)^{1/p}, \tag{5.9}$$

where $\Gamma(x)$ is the gamma-function. Specifically, from the well-known equality

$$\Gamma\left(\frac{2m+1}{2}\right) = \frac{1 \cdot 3 \cdot \cdots \cdot (2m-1)}{2^m} \cdot \sqrt{\pi} \tag{5.10}$$

it follows that

$$K_{2m}^{2m} = \mathbb{E}|\mathfrak{g}|^{2m} = \frac{(2m)!}{2^m m!} = (2m-1)!! \ \text{ for all } \ m = 1, 2, \ldots, \tag{5.11}$$

where $N!!$ denotes the product of all positive integers that have the same parity as N and do not exceed N. Moreover, since $\Gamma(x) \sim \sqrt{2\pi} e^{-x} x^{x-1/2}$ as $x \to \infty$ [118, II.12, Problem 22, p. 66], then (5.9) implies

$$\lim_{p\to\infty} \frac{K_p\sqrt{e}}{\sqrt{p}} = \lim_{p\to\infty} \frac{\|\mathfrak{g}\|_p\sqrt{e}}{\sqrt{p}} = 1. \tag{5.12}$$

Historically, the left-hand side of inequality (5.1) had caused the greatest interest in the case $p = 1$. In the next section, we shall pay a special attention to this important issue.

5.2 The Comparison of L_1- and L_2-Norms of Rademacher Sums and Littlewood Conjecture

As far back as in the 1930-s, Littlewood stated the following conjecture, directly connected with the optimality problem of constants in inequality (5.1).

Let K be a cube in $\mathbb{R}^n$, whose side length is equal to 2, and let H be a $(n-1)$-dimensional plane passing through its center. The conjecture asserts that the mean of distances from 2^n vertices of the cube K to H is not less than $1/\sqrt{2}$.

It is obvious that it can be assumed that the vertices of K are the sign arrangements, i.e., the vectors θ of the form $\theta = (\theta_1, \theta_2, \ldots, \theta_n)$, where $\theta_i = \pm 1$, $i = 1, 2, \ldots, n$. Then, if $a = (a_1, a_2, \ldots, a_n)$ is the unit vector normal to the plane H, then the answer to the above conjecture is "yes" if and only if

$$2^{-n} \sum_{\theta=\pm 1} \left| \sum_{i=1}^{n} \theta_i a_i \right| \geq \frac{1}{\sqrt{2}}$$

for all $a = (a_1, a_2, \ldots, a_n)$ such that $\sum_{k=1}^{n} a_k^2 = 1$. As it can be easily seen, the positive answer is equivalent to the fact that the left-hand side of inequality (5.1) holds for $p = 1$ and $c = 1/\sqrt{2}$. Observe that the simplest situation when $n = 2$ and $a_1 = a_2 = 1$ indicates that the constant c in this inequality cannot be larger than $1/\sqrt{2}$.

In 1976, the above conjecture was solved by Szarek in affirmative. Moreover, later on it was shown that a similar result holds even for Rademacher sums with vector coefficients. Before to state and prove this interesting result, we introduce somewhat other, more convenient for us, notation of the optimal constants in the inequalities under consideration.

Let $1 \leq p \leq q < \infty$. According to the Kahane–Khintchine inequality (see Theorem 4.1 or Corollary 4.2), there exists $C > 0$, depending only on p and q, such that for any $n \in \mathbb{N}$, arbitrary Banach space F and all $x_1, x_2, \ldots, x_n \in F$ we have

$$\left(\int_0^1 \left\| \sum_{i=1}^{n} r_i(t) x_i \right\|_F^q dt \right)^{1/q} \leq C \left(\int_0^1 \left\| \sum_{i=1}^{n} r_i(t) x_i \right\|_F^p dt \right)^{1/p}. \tag{5.13}$$

Let $K_{p,q}$ be the best (i.e., the least) constant C, for which (5.13) holds. By $K^{\mathbb{R}}_{p,q}$ we denote the analogous constant for the scalar case, i.e., $K^{\mathbb{R}}_{p,q} = \min C$ such that

$$\Big(\int_0^1 \Big|\sum_{i=1}^n r_i(t)a_i\Big|^q dt\Big)^{1/q} \le C\Big(\int_0^1 \Big|\sum_{i=1}^n r_i(t)a_i\Big|^p dt\Big)^{1/p}, \tag{5.14}$$

where $a_1, a_2, \dots, a_n \in \mathbb{R}$. Clearly, $K^{\mathbb{R}}_{p,q} \le K_{p,q}$.

It is well known that the Rademacher functions form a subsystem of the classical Walsh system. Recall its definition: $w_0(t) = r_0(t) = 1$ and if $n \in \mathbb{N}$ is representable as $n = \sum_{i=0}^k \varepsilon_i 2^i$, where $\varepsilon_k = 1$ and $\varepsilon_i = 0$ or $1, i = 0, 1, \dots, k-1$, then according to Paley's enumeration we set

$$w_n(t) = \prod_{i=0}^k \big(r_{i+1}(t)\big)^{\varepsilon_i}.$$

If $w_n = r_{m_1} \dots r_{m_l}$, $1 \le m_1 < \dots < m_l$, then we shall refer to l as the *order* of w_n and denote it by $\operatorname{ord} w_n$. It is well known (see e.g. [126, Claims 1.1.5 and 2.6.3]) that $\{w_n\}_{n=0}^\infty$ is a complete orthonormal system in $L_p[0, 1]$ for every $1 \le p < \infty$.

Theorem 5.4 *For arbitrary linear normed space X, every $n \in \mathbb{N}$, and all $x_1, x_2, \dots, x_n$ from X it holds*

$$\int_0^1 \Big\|\sum_{k=1}^n x_k r_k(t)\Big\|_X^2 dt \le 2\Big(\int_0^1 \Big\|\sum_{k=1}^n x_k r_k(t)\Big\|_X dt\Big)^2. \tag{5.15}$$

Thus, $K^{\mathbb{R}}_{1,2} = K_{1,2} = \sqrt{2}$.

Proof For any fixed $x_1, x_2, \dots, x_n \in X$ we define the function f_x by

$$f_x(t) := \Big\|\sum_{k=1}^n x_k r_k(t)\Big\|_X.$$

Then, if E is the linear space of all functions $f : [0, 1] \to \mathbb{R}$, which are constant on each of the dyadic intervals $((j-1)2^{-n}, j2^{-n})$, $j = 1, 2, \dots, 2^n$, then $f_x \in E$. Let us equip E with the usual L_2-norm, i.e.,

$$\|f\|_2 := \Big(\int_0^1 f(t)^2\, dt\Big)^{1/2}.$$

Because the linear dimension of E equals 2^n, the set W_n of all Walsh functions of the form $w = r_{m_1} \dots r_{m_l}$, $1 \le m_1 < \dots < m_l \le n$, is an orthonormal basis in E.

Therefore, for every function $f \in E$ we can write

$$f = \sum_{w \in W_n} \langle f, w \rangle w, \tag{5.16}$$

where $\langle f, w \rangle := \int_0^1 f(t) w(t)\, dt,\ w \in W_n$. Since $r_j(1-t) = -r_j(t)$ for all $j = 1, 2, \dots$ and $t \in [0, 1]$, we have $f_x(1-t) = f_x(t)$ for arbitrary $x_k \in X,\ k = 1, 2, \dots, n$. Hence, $\langle f_x, w \rangle = 0$ if $\operatorname{ord} w = 1$, i.e., if w is a Rademacher function.

Let us define on E the following "differentiation operation": for $f \in E$, representable in the form (5.16), we set

$$\mathcal{D} f := \sum_{w \in W_n} \operatorname{ord} w \cdot \langle f, w \rangle w.$$

It is clear that every function f from E can be uniquely written as $F(r_1(t), \dots, r_n(t))$, where F is a certain function defined on the set of all vectors from $\mathbb{R}^n$ of the form $(\theta_1, \dots, \theta_n)$, $\theta_i = \pm 1$. Show that

$$\begin{aligned} 2\mathcal{D} f = \sum_{j=1}^{n} \big(& F(r_1, \dots, r_{j-1}, r_j, r_{j+1}, \dots, r_n) \\ & - F(r_1, \dots, r_{j-1}, -r_j, r_{j+1}, \dots, r_n) \big). \end{aligned} \tag{5.17}$$

Since each $f \in E$ has expansion (5.16) with respect to the basis W_n, it suffices to check equation (5.17) only for Walsh functions. It is obvious for $w = w_0$. Let $w \in W_n$ and $\operatorname{ord} w \geq 1$, that is, $w = r_{m_1} \dots r_{m_l}$, $1 \leq m_1 < \dots < m_l \leq n$. Observe that this element of W_n is generated by the function $F(u_1, \dots, u_n) = u_{m_1} \dots u_{m_l}$. Consequently,

$$F(r_1, \dots, r_n) - F(\dots, r_{j-1}, -r_j, r_{j+1}, \dots) = \begin{cases} 2F(r_1, \dots, r_n) & \text{if } j \in \{m_1, \dots, m_l\} \\ 0, & \text{otherwise.} \end{cases}$$

Summing these equations over all $j = 1, \dots, n$, we obtain that the sum of increments of F over all coordinates from the right-hand side of (5.17) is equal to

$$2l F(r_1, \dots, r_n) = 2 \cdot \operatorname{ord} w \cdot w = 2\mathcal{D} w.$$

Therefore, (5.17) is proved.

Further, by (5.17) and the triangle inequality, we get the following pointwise inequality:

$$
\begin{aligned}
2\mathcal{D}f_x &= n\|x_1r_1 + \cdots + x_nr_n\| - \sum_{j=1}^{n} \| - x_jr_j + \sum_{i\neq j} x_ir_i\|_X \\
&\leq n\|x_1r_1 + \cdots + x_nr_n\| - (n-2)\|x_1r_1 + \cdots + x_nr_n\| = 2f_x,
\end{aligned}
$$

which holds for arbitrary $x_1, x_2, \ldots, x_n \in X$ and all $t \in [0, 1]$. It remains to check that (5.15) is a consequence of the latter inequality. Indeed, recalling that $\langle f_x, w\rangle = 0$ if $\operatorname{ord} w = 1$, by Parseval's Identity, we get

$$
\begin{aligned}
\|f_x\|_2^2 &= \int_0^1 f_x^2(t)\,dt \geq \int_0^1 f_x(t)\mathcal{D}f_x(t)\,dt \\
&= \sum_{w\in W_n} \operatorname{ord} w \cdot \langle f_x, w\rangle^2 \geq 2 \sum_{w\in W_n, w\neq w_0} \langle f_x, w\rangle^2.
\end{aligned}
$$

Hence,

$$
\|f_x\|_2^2 \geq 2(\|f_x\|_2^2 - \langle f_x, w_0\rangle^2),
$$

and therefore,

$$
\int_0^1 \Big\|\sum_{k=1}^{n} x_kr_k(t)\Big\|_X^2 dt = \|f_x\|_2^2 \leq 2\langle f_x, w_0\rangle^2 = 2\Big(\int_0^1 \Big\|\sum_{k=1}^{n} x_kr_k(t)\Big\|_X dt\Big)^2.
$$

Since the optimality of the constant $\sqrt{2}$ both in the scalar and vector cases follows from the reasoning given in the very beginning of the section, the proof is completed. □

5.3 Identification of the Constant $K_{2,4}$

Using a completely other technique based on studying a certain differential operator and corresponding inequalities of Poincaré type, the optimal constants in inequalities (5.13) and (5.14) can be found in the case $p = 2, q = 4$. Namely, in this section we show that $K_{2,4} = K_{2,4}^{\mathbb{R}} = \sqrt[4]{3}$.

Let M be a probability distribution on the interval $I = (-a, a)$, $0 < a \leq \infty$, with a continuous, symmetric and positive density function m on I. We shall assume that M has the finite second moment, i.e., $\int_I s^2 m(s)\,ds < \infty$. Moreover, let the functions $p, u : I \to \mathbb{R}$ be defined as follows:

$$p(x) := \int_x^a sm(s)\,ds \quad \text{and} \quad u(x) := \frac{p(x)}{m(x)}.$$

Obviously, p and u are continuous, symmetric and positive on I.

Denote by $C_0^\infty(I)$ the space of smooth (infinitely differentiable) functions on I with compact supports in I. For $f \in C_0^\infty(I)$ we define the operator L by

$$Lf(x) := xf'(x) - u(x)f''(x) = -\frac{(p(x)f'(x))'}{m(x)}.$$

Observe that L is nonnegative and formally symmetric as an operator defined on the space $L_2(I, M)$. Indeed, integrating by parts, for any $f, g \in C_0^\infty(I)$ we get

$$\int_I fLg\,dM = -\int_I f(pg')'\,dx = \int_I pf'g'\,dx = \int_I uf'g'\,dM.$$

Therefore, according to [113, § XIII.6] (see also [283, Theorems 5.8 and 6.3]), L can be extended to a self-adjoint operator on $L_2(I, M)$ (which will be denoted further still by L) with the domain $D(L)$, consisting of all functions $f \in L_2(I, M)$ such that f' is absolutely continuous on I, $xf'(x) - u(x)f''(x) \in L_2(I, M)$, and

$$\lim_{x \to a-} p(x)f'(x) = \lim_{x \to -a+} p(x)f'(x) = 0.$$

One can easily check that 0 and 1 are eigenvalues of the operator L. Moreover, the only eigenvectors corresponding to 0 (resp. 1) are multiples of the function $f(x) \equiv 1$ (resp. $f(x) = x$). Also, it is known [113, Theorem XIII.7.40] that the remaining part of the spectrum $\sigma(L)$ of L is contained in the interval $(1, \infty)$. Let us define

$$\lambda_2(L) := \inf\{\lambda > 1 : \lambda \in \sigma(L)\}.$$

Taking into account the above observation and using standard properties of the spectrum of a self-adjoint operator, it is easy to see that $\lambda_2(L)$ is the largest constant c, for which the conditions $f \in D(L)$ and $\int_I xf(x)\,dM = 0$ imply the following inequality of Poincaré type:

$$\int_I f^2\,dM \le \Big(\int_I f\,dM\Big)^2 + \frac{1}{c}\int_I fLf\,dM. \tag{5.18}$$

Next, assume that we have I_i, M_i, p_i, u_i, L_i for $i = 1, 2, \ldots, n$ as above, and set $\mathbf{I} = I_1 \times I_2 \times \cdots \times I_n$, $\mathbf{M} = M_1 \otimes M_2 \otimes \cdots \otimes M_n$ (here, by $\otimes$ we denote the tensor product). Then,

$$\mathbf{L} := L_1 \otimes Id_2 \otimes \cdots \otimes Id_n + Id_1 \otimes L_2 \otimes Id_3 \otimes \cdots \otimes Id_n + \cdots + Id_1 \otimes \cdots \otimes Id_{n-1} \otimes L_n,$$

where Id is the identity operator, is a nonnegative and self-adjoint operator on the space $L_2(\mathbf{I}, \mathbf{M})$. Moreover, for each $f \in C_0^\infty(\mathbf{I})$ we have

$$\mathbf{L}f(x_1, x_2, \dots, x_n) = \sum_{i=1}^{n} \Big(x_i \frac{\partial f}{\partial x_i} - u_i(x_i) \frac{\partial^2 f}{\partial x_i^2} \Big).$$

Also, it is easy to see that $\sigma(\mathbf{L}) = \sigma(L_1) + \sigma(L_2) + \cdots + \sigma(L_n)$ and if ϕ_i is an eigenvector, corresponding to $\lambda_i \in \sigma(L_i)$, then $\phi_1 \otimes \phi_2 \otimes \cdots \otimes \phi_n$ is an eigenvector of the operator $\mathbf{L}$, corresponding to the eigenvalue $\lambda_1 + \lambda_2 + \cdots + \lambda_n$. Hence, it is immediately follows that $0, 1 \in \sigma(\mathbf{L})$ and all eigenvectors, corresponding to 0, are multiples of the function $f(x_1, \dots, x_n) \equiv 1$, while the eigenspace, corresponding to 1, is n-dimensional and is spanned by the coordinate functions $f_i(x_1, \dots, x_n) = x_i$, $i = 1, 2, \dots, n$. The remaining part of $\sigma(\mathbf{L})$ is contained in $(1, \infty)$ and

$$\lambda_2(\mathbf{L}) = \min\{\lambda > 1 : \lambda \in \sigma(\mathbf{L})\} = \min\{2, \lambda_2(L_1), \lambda_2(L_2), \dots, \lambda_2(L_n)\}.$$

Thus, in the same way as (5.18), we get

$$\int_{\mathbf{I}} f^2 \, d\mathbf{M} \le \Big(\int_{\mathbf{I}} f \, d\mathbf{M} \Big)^2 + \frac{1}{\lambda_2(\mathbf{L})} \int_{\mathbf{I}} f \mathbf{L} f \, d\mathbf{M} \tag{5.19}$$

for every $f \in D(\mathbf{L})$ such that $\int_{\mathbf{I}} x_i f(x) \, d\mathbf{M} = 0$, $i = 1, 2, \dots, n$.

Consider now a special functional class from $D(\mathbf{L})$. Let $g : \mathbb{R}^n \to [0, \infty)$ be a smooth norm on $\mathbb{R}^n$, that is, a positively 1-homogeneous, convex, symmetric and smooth function on the set $\mathbb{R}^n \setminus \{0\}$. As it is easy to see, then there is a convex, symmetric and smooth function $h : \mathbb{R}^n \to [0, \infty)$, which coincides with g outside the unit ball in $\mathbb{R}^n$. For every $r > 0$ we define the function $g_r(x) := rh(x/r)$. Then, if $p \ge 1$ and each of the distributions M_i has finite $2p$-moment, we have $g_r^p \in D(\mathbf{L})$ and from the definition of $\mathbf{L}$ it follows that

$$\begin{aligned} \int_{\mathbf{I}} g_r^p \mathbf{L} g_r^p \, d\mathbf{M} &= p \int_{\mathbf{I}} g_r^{2p-1} \Big(\sum_{i=1}^{n} x_i \frac{\partial g_r}{\partial x_i} \Big) \, d\mathbf{M} \\ &\quad - p(p-1) \int_{\mathbf{I}} g_r^{2p-2} \Big[\sum_{i=1}^{n} u_i(x_i) \Big(\frac{\partial g_r}{\partial x_i} \Big)^2 \Big] \, d\mathbf{M} \\ &\quad - p \int_{\mathbf{I}} g_r^{2p-1} \Big[\sum_{i=1}^{n} u_i(x_i) \frac{\partial^2 g_r}{\partial x_i^2} \Big] \, d\mathbf{M}. \end{aligned}$$

Integrating by parts and using the fact that $u_i(x_i)m_i(x_i) = \int_{x_i}^{a_i} sm_i(s)\,ds$, we can rewrite the second integral from the right-hand side of this equation in the form:

$$\int_{I_i} g_r^{2p-2} u_i(x_i)\Big(\frac{\partial g_r}{\partial x_i}\Big)^2\,dM_i = -\frac{1}{2p-1}\int_{I_i} g_r^{2p-1}\frac{\partial}{\partial x_i}\Big(\frac{\partial g_r}{\partial x_i}\cdot u_i(x_i)m_i(x_i)\Big)\,dx_i$$

$$= -\frac{1}{2p-1}\int_{I_i} g_r^{2p-1} u_i(x_i)\frac{\partial^2 g_r}{\partial x_i^2}\,dM_i$$

$$+\frac{1}{2p-1}\int_{I_i} g_r^{2p-1} x_i\frac{\partial g_r}{\partial x_i}\,dM_i.$$

Therefore, by Fubini's theorem, it holds

$$\int_{\mathbf{I}} g_r^p \mathbf{L} g_r^p\,d\mathbf{M} = \frac{p^2}{2p-1}\int_{\mathbf{I}} g_r^{2p-1}\Big(\sum_{i=1}^n x_i\frac{\partial g_r}{\partial x_i}\Big)\,d\mathbf{M}$$

$$-\frac{p^2}{2p-1}\int_{\mathbf{I}} g_r^{2p-1}\Big[\sum_{i=1}^n u_i(x_i)\frac{\partial^2 g_r}{\partial x_i^2}\Big]\,d\mathbf{M}.$$

Moreover, since the function g_r is convex, we have $\partial^2 g_r/\partial x_i^2 \geq 0$. Hence, we get

$$\int_{\mathbf{I}} g_r^p \mathbf{L} g_r^p\,d\mathbf{M} \leq \frac{p^2}{2p-1}\int_{\mathbf{I}} g_r^{2p-1}\Big(\sum_{i=1}^n x_i\frac{\partial g_r}{\partial x_i}\Big)\,d\mathbf{M}. \tag{5.20}$$

Also, from the symmetry of g_r^p on $\mathbb{R}^n$ it follows $\int_{\mathbf{I}} x_i g_r^p\,d\mathbf{M} = 0$, $i = 1, 2, \ldots, n$. As a result, inequalities (5.19) and (5.20) imply that for $\lambda_2 = \lambda_2(\mathbf{L})$ we have

$$\int_{\mathbf{I}} g_r^{2p}\,d\mathbf{M} \leq \Big(\int_{\mathbf{I}} g_r^p\,d\mathbf{M}\Big)^2 + \frac{p^2}{\lambda_2(2p-1)}\int_{\mathbf{I}} g_r^{2p-1}\Big(\sum_{i=1}^n x_i\frac{\partial g_r}{\partial x_i}\Big)\,d\mathbf{M}. \tag{5.21}$$

Let us check that 1-homogeneity of g implies that

$$\sum_{i=1}^n x_i\frac{\partial g}{\partial x_i} = g.$$

Indeed, since for all $i = 2, \ldots, n$

$$g'_{x_i}(x_1, x_2, \ldots, x_n) = (x_1 g(1, x_2/x_1, \ldots, x_n/x_1))'_{x_i} = g'_{x_i}(1, x_2/x_1, \ldots, x_n/x_1),$$

then

$$\begin{aligned} g'_{x_1}(x_1, x_2, \ldots, x_n) &= \big(x_1 g(1, x_2/x_1, \ldots, x_n/x_1)\big)'_{x_1} \\ &= g(1, x_2/x_1, \ldots, x_n/x_1) - \sum_{i=2}^{n} x_i {x_1}^{-1} g'_{x_i}(1, x_2/x_1, \ldots, x_n/x_1) \\ &= {x_1}^{-1}\Big(g(x_1, x_2, \ldots, x_n) - \sum_{i=2}^{n} x_i g'_{x_i}(x_1, x_2, \ldots, x_n)\Big), \end{aligned}$$

and the desired result follows. Moreover, g_r coincides with g outside the ball of radius r and the functions g_r and $\sum_{i=1}^{n} x_i(\partial g_r/\partial x_i)$ are uniformly bounded on each bounded set. Therefore, passing to the limit in (5.21) as $r \to 0$, we get

$$\int_{\mathbf{I}} g^{2p}\, d\mathbf{M} \le \Big(\int_{\mathbf{I}} g^p\, d\mathbf{M}\Big)^2 + \frac{p^2}{\lambda_2(2p-1)} \int_{\mathbf{I}} g^{2p}\, d\mathbf{M},$$

whence

$$\Big(\int_{\mathbf{I}} g^{2p}\, d\mathbf{M}\Big)^{1/(2p)} \le C_p \Big(\int_{\mathbf{I}} g^p\, d\mathbf{M}\Big)^{1/p}. \tag{5.22}$$

Here, $C_p = (\lambda_2(2p-1)/(\lambda_2(2p-1) - p^2))^{1/(2p)}$ for $p \le \lambda_2 + \sqrt{\lambda_2^2 - \lambda_2}$.

Thus, we come to the following result.

Theorem 5.5 *Let ξ_i, $i = 1, 2, \ldots, n$, be independent symmetrically distributed r.v.'s such that, for each $i = 1, 2, \ldots, n$, ξ_i has M_i-distribution satisfying the above conditions. If $\lambda_2(L_i) \ge \lambda > 1$ for all $i = 1, 2, \ldots, n$ and some $\lambda \le 2$, then for $1 \le p < \lambda + \sqrt{\lambda^2 - \lambda}$ and any vectors x_i, $i = 1, 2, \ldots, n$, from an arbitrary linear normed space F we have*

$$\Big(\int_{\mathbf{I}} \Big\| \sum_{i=1}^{n} x_i \xi_i \Big\|_F^{2p} d\mathbf{M}\Big)^{1/(2p)} \le C_p \Big(\int_{\mathbf{I}} \Big\| \sum_{i=1}^{n} x_i \xi_i \Big\|_F^{p} d\mathbf{M}\Big)^{1/p},$$

where $C_p = (\lambda(2p-1)/(\lambda(2p-1) - p^2))^{1/(2p)}$.

Proof By standard approximation arguments, we can assume that the norm $\|\cdot\|_F$ is smooth. Therefore, substituting the function $g(t_1, t_2, \ldots, t_n) = \|\sum_{i=1}^{n} x_i t_i\|_F$ into (5.22), we obtain what we wished. □

Proceed now with the special case when, for every $i = 1, 2, \ldots, n$, M_i coincides with the symmetric beta-distribution M_α on $I = (-1, 1)$ with the density function

m_α defined by

$$m_\alpha(x) = \frac{2^{2\alpha-1}}{B(\alpha,\alpha)}(1-x^2)^{\alpha-1},$$

where $\alpha > 0$ and $B(\mu,\nu) = \int_0^1 (1-t)^{\mu-1} t^{\nu-1}\,dt$, $\mu,\nu > 0$. Then, one can easily check that

$$u_\alpha(x) = \frac{1}{2\alpha}(1-x^2)\ ,\quad L_\alpha f = xf' - \frac{1}{2\alpha}(1-x^2)f'',$$

and

$$\sigma(L_\alpha) = \left\{\frac{n(n+2\alpha-1)}{2\alpha},\ n = 0,1,\ldots\right\},$$

with Jacoby polynomials as the corresponding eigenvectors. Hence, $\lambda_2(L_\alpha) = (2\alpha+1)/\alpha \geq 2$.

Assume that $0 < \alpha, \delta < 1$. Since

$$M_\alpha(-1+\delta, 1-\delta) = \int_{-1+\delta}^{1-\delta} m_\alpha(t)\,dt = \frac{2^{2\alpha-1}}{B(\alpha,\alpha)}\int_{-1+\delta}^{1-\delta}(1-t^2)^{\alpha-1}\,dt \leq \frac{2^{2\alpha-1}\delta^{\alpha-1}}{B(\alpha,\alpha)}$$

and $B(\alpha,\alpha) \to \infty$ as $\alpha \to 0$ and δ is fixed, we have $\lim_{\alpha\to 0} M_\alpha(-1+\delta, 1-\delta) = 0$. In consequence, from the symmetry of M_α and the equality $M_\alpha(-1,1) = 1$ it follows that

$$M_\alpha(-1,-1+\delta) = M_\alpha(1-\delta,1) \to \frac{1}{2} \quad \text{as} \quad \alpha \to 0.$$

Thus, M_α tends to the distribution of a Rademacher function on $[0,1]$ as $\alpha \to 0$. Moreover, for every α, $\lambda_2(L_\alpha) \geq 2$, and the function $(\lambda(2p-1)/(\lambda(2p-1)-p^2))^{1/(2p)}$ decreases with increasing λ. As a result, applying Theorem 5.5, we get the first assertion of the following main result of this section.

Theorem 5.6 *Let $1 \leq p < 2+\sqrt{2}$. For an arbitrary linear normed space F and any vectors $x_i \in F$, $i = 1,2,\ldots,n$, we have*

$$\Big(\int_0^1 \Big\|\sum_{i=1}^n x_i r_i(t)\Big\|_F^{2p} dt\Big)^{1/(2p)} \leq \Big(\frac{4p-2}{4p-2-p^2}\Big)^{1/(2p)}\Big(\int_0^1 \Big\|\sum_{i=1}^n x_i r_i(t)\Big\|_F^{p} dt\Big)^{1/p}.$$

In particular, for $p = 1$ and $p = 2$ it holds

$$\Big(\int_0^1 \Big\|\sum_{i=1}^n x_i r_i(t)\Big\|_F^{2} dt\Big)^{1/2} \leq \sqrt{2}\int_0^1 \Big\|\sum_{i=1}^n x_i r_i(t)\Big\|_F dt$$

and

$$\Big(\int_0^1 \Big\|\sum_{i=1}^n x_i r_i(t)\Big\|_F^4\,dt\Big)^{1/4} \le \sqrt[4]{3}\Big(\int_0^1 \Big\|\sum_{i=1}^n x_i r_i(t)\Big\|_F^2\,dt\Big)^{1/2},$$

respectively. Moreover, constants $\sqrt{2}$ *and* $\sqrt[4]{3}$ *are optimal in these inequalities even in the case when* $F = \mathbb{R}$.

Proof In view of the preceding observations, it suffices only to show the optimality of constant $\sqrt[4]{3}$ in the last inequality. By the central limit theorem, for every $p > 1$ we have

$$\Big(\int_0^1 \Big|\sum_{i=1}^n \frac{1}{\sqrt{n}} r_i(t)\Big|^p\,dt\Big)^{1/p} \to \|\mathfrak{g}\|_p \quad \text{as } n \to \infty, \tag{5.23}$$

where $\mathfrak{g}$ is a standard Gaussian r.v. Therefore, from (5.9) and (5.10) it follows that

$$\Big(\int_0^1 \Big|\sum_{i=1}^n \frac{1}{\sqrt{n}} r_i(t)\Big|^2\,dt\Big)^{1/2} \to 1 \quad \text{as } n \to \infty$$

and

$$\Big(\int_0^1 \Big|\sum_{i=1}^n \frac{1}{\sqrt{n}} r_i(t)\Big|^4\,dt\Big)^{1/4} \to \sqrt[4]{3} \quad \text{as } n \to \infty.$$

Clearly, this completes the proof. □

In the next section, we shall show that the best constants in inequality (5.14) can be found also for more general values p and q.

5.4 The Constants $K^{\mathbb{R}}_{p,q}$ for Even p and q

As above, $\mathfrak{g}$ is a standard Gaussian r.v. and $\Gamma(x)$ is the gamma-function.

Theorem 5.7 *For arbitrary even integers* $q > p \ge 2$ *we have*

$$K^{\mathbb{R}}_{p,q} = \frac{(\mathbb{E}|\mathfrak{g}|^q)^{1/q}}{(\mathbb{E}|\mathfrak{g}|^p)^{1/p}} = (\sqrt{\pi})^{\frac{1}{p}-\frac{1}{q}}\left(\Gamma\left(\frac{q+1}{2}\right)\right)^{1/q}\left(\Gamma\left(\frac{p+1}{2}\right)\right)^{-1/p}.$$

A crucial role in the proof of Theorem 5.7 is played by the fact that the binomial convolution of two log-concave sequences is also log-concave (see [282, Theorem 1] or [217]). Recall that a sequence $(a_i)_{i=0}^{\infty}$ of nonnegative real numbers is called *log-concave* if the set $\{i \geq 0 : a_i > 0\}$ is an interval of integers and $a_i^2 \geq a_{i-1}a_{i+1}$ for all $i \geq 1$.

Theorem 5.8 *Let $(a_j)_{j=0}^{\infty}$ and $(b_j)_{j=0}^{\infty}$ be two log-concave sequences of positive real numbers. Then the sequence $(c_n)_{n=0}^{\infty}$ defined by*

$$c_n := \sum_{i=0}^{n} C_n^i a_i b_{n-i}, \quad n = 0, 1, \ldots,$$

is also log-concave (as usual, $C_n^i = n!/(i!(n-i)!)$).

We shall say that an r.v. f is *ultra sub-Gaussian* if either $f = 0$ a.e., or f is symmetrically distributed, all moments of f are finite, and the sequence $(a_i)_{i=0}^{\infty}$, defined by $a_i = \mathbb{E}|f|^{2i}/\mathbb{E}|\mathfrak{g}|^{2i}$ for $i \geq 1$ and $a_0 = 1$, is log-concave.

Lemma 5.2 *Any Rademacher function r is an ultra sub-Gaussian r.v.*

Proof Clearly, r is symmetrically distributed and $\mathbb{E}|r|^k = 1$, $k = 1, 2, \ldots$. Therefore, by definition, we have $a_i = (\mathbb{E}|\mathfrak{g}|^{2i})^{-1}$, $i \geq 1$. Then, from the Cauchy-Schwarz-Bunyakovskii inequality it follows for each $i \geq 1$

$$\mathbb{E}|\mathfrak{g}|^{2i} = \int_0^1 |\mathfrak{g}|^{i-1}|\mathfrak{g}|^{i+1}\,dt \leq (\mathbb{E}|\mathfrak{g}|^{2(i-1)})^{1/2}(\mathbb{E}|\mathfrak{g}|^{2(i+1)})^{1/2}.$$

Thus, $a_i^2 \geq a_{i-1}a_{i+1}$ for all $i \geq 1$. □

Lemma 5.3 *If f_1 and f_2 are independent ultra sub-Gaussian r.v.'s, then their sum $f_1 + f_2$ is also ultra sub-Gaussian.*

Proof In the case when either f_1 or f_2 is equal to 0 a.e., the assertion is obvious. Otherwise, let

$$a_i := \mathbb{E}|f_1|^{2i}/\mathbb{E}|\mathfrak{g}|^{2i}, \quad b_i := \mathbb{E}|f_2|^{2i}/\mathbb{E}|\mathfrak{g}|^{2i}, \quad \text{and} \quad c_i := \mathbb{E}|f_1 + f_2|^{2i}/\mathbb{E}|\mathfrak{g}|^{2i}$$

for $i \geq 1$, and $a_0 = b_0 = c_0 = 1$. Recall that $\mathbb{E}|\mathfrak{g}|^{2k} = (2k-1)!!$, where the notation $N!!$ is understood as in equation (5.11). Then, since f_1 and f_2 are independent and symmetrically distributed, we infer

$$\begin{aligned} c_n &= \frac{1}{(2n-1)!!}\sum_{i=0}^{n} C_{2n}^{2i}\mathbb{E}|f_1|^{2i}\mathbb{E}|f_2|^{2n-2i} \\ &= \sum_{i=0}^{n} \frac{(2n)!!}{(2i)!!(2n-2i)!!}a_i b_{n-i} = \sum_{i=0}^{n} C_n^i a_i b_{n-i}, \end{aligned}$$

where $(-1)!! = 0!! = 1$. Hence, the desired result follows now from Theorem 5.8. □

Proof of Theorem 5.7 By Lemmas 5.2 and 5.3, the Rademacher sum $S_n := \sum_{k=1}^{n} x_k r_k$ is an ultra sub-Gaussian r.v. for every $n \in \mathbb{N}$ and $x_k \in \mathbb{R}$, $k = 1, 2, \ldots, n$. Therefore, the sequence $(a_i)_{i=0}^{\infty}$, where $a_i = \mathbb{E}|S_n|^{2i}/\mathbb{E}|\mathfrak{g}|^{2i}$, $i = 1, 2, \ldots$, and $a_0 = 1$, is log-concave. Multiplying inequalities $a_i^2 \geq a_{i-1}a_{i+1}$ over all $i = 1, 2, \ldots, s$, we get $a_s^{2s} \geq a_s^{s-1}a_{s+1}^s$, or equivalently $a_{s+1}^{1/(s+1)} \leq a_s^{1/s}$ for all $s = 1, 2, \ldots$ Then, in particular, we have $a_{q/2}^{2/q} \leq a_{p/2}^{2/p}$ for all even $q > p \geq 2$. This means that inequality (5.14) holds with the constant $C := (\mathbb{E}|\mathfrak{g}|^q)^{1/q}/(\mathbb{E}|\mathfrak{g}|^p)^{1/p}$. The fact that C is optimal follows immediately from limit relation (5.23). Finally, by (5.9), we can easily express the last constant via the gamma-function. □

5.5 Asymptotic Coincidence of "scalar" and "vector" Optimal Constants

Recall that $K_{p,q}$ (resp. $K_{p,q}^{\mathbb{R}}$) is the optimal constant in the Kahane–Khintchine inequality (5.13) (resp. in its scalar variant, Khintchine's inequality (5.14)).

While the inequality $K_{p,q}^{\mathbb{R}} \leq K_{p,q}$ is obvious, it can be easily checked also that ratio of these constants tends to 1 as $p, q \to \infty$. Indeed, substituting $a_i = n^{-1/2}$, $i = 1, 2, \ldots, n$, into inequality (5.14), we come to the estimate

$$\left(\int_0^1 \left|\sum_{i=1}^{n} \frac{1}{\sqrt{n}} r_i(t)\right|^q dt\right)^{1/q} \leq K_{p,q}^{\mathbb{R}} \left(\int_0^1 \left|\sum_{i=1}^{n} \frac{1}{\sqrt{n}} r_i(t)\right|^p dt\right)^{1/p}, \quad n = 1, 2, \ldots$$

Taking now $n \to \infty$ we get $K_{p,q}^{\mathbb{R}} \geq \|\mathfrak{g}\|_q/\|\mathfrak{g}\|_p$ by (5.23). On the other hand, from Corollary 4.2 it follows that $K_{p,q} \leq \sqrt{(q-1)/(p-1)}$ if $1 < p \leq q < \infty$. Therefore,

$$1 \leq \frac{K_{p,q}}{K_{p,q}^{\mathbb{R}}} \leq \sqrt{\frac{q-1}{p-1}} \cdot \frac{\|\mathfrak{g}\|_p}{\|\mathfrak{g}\|_q}. \tag{5.24}$$

Since $\|\mathfrak{g}\|_r \sim \sqrt{r/e}$ as $r \to \infty$ (see (5.12)), then $\lim_{r\to\infty} \sqrt{r-1}/\|\mathfrak{g}\|_r = \sqrt{e}$. As a result, (5.24) implies that $K_{p,q}/K_{p,q}^{\mathbb{R}} \to 1$ as $p, q \to \infty$.

Here, we shall show that the above asymptotic equality of the constants $K_{p,q}$ and $K_{p,q}^{\mathbb{R}}$ holds, in fact, in a quite stronger sense: uniformly with respect to $p \in [1, q]$ as $q \to \infty$. The proof of this assertion will be based on the use of an L_p-estimate of deviation of the norm of a Rademacher sum with vector coefficients from its expectation by the "weak" L_p-norm of this sum.

Throughout this section, F is an arbitrary Banach space. For every $p > 0$ and a random vector ξ with values in F, we set $\Sigma_p(\xi) := (\mathbb{E}\|\xi\|_F^p)^{1/p}$ and

$$\sigma_p(\xi) := \sup_{x^* \in F^*: \|x^*\| \le 1} (\mathbb{E}|x^*(\xi)|^p)^{1/p}.$$

Clearly, $\Sigma_p(\xi) \ge \sigma_p(\xi)$. Moreover, as usual, we denote $a_+ := \max(a, 0)$ for any $a \in \mathbb{R}$.

Theorem 5.9 *Let* $S := \sum_{k=1}^n x_k r_k$, *where* $n \in \mathbb{N}$ *and* $x_k \in F$, $k = 1, 2, \ldots, n$, *are arbitrary. Then, for all* $\delta \in (0, 1]$ *and* $p \ge 1$ *we have*

$$\left\| \left(\|S\| - \frac{(1+\delta)^2}{\delta} \mathbb{E}\|S\| \right)_+ \right\|_p \le (1+\delta)\sigma_p(S).$$

Having postponed the proof of Theorem 5.9 for later, we show first that the above-mentioned property of the constants $K_{p,q}^{\mathbb{R}}$ and $K_{p,q}$ is a consequence of this result.

Theorem 5.10 *The following equation holds:*

$$\lim_{q\to\infty} \sup_{p\in[1,q]} \frac{K_{p,q}}{K_{p,q}^{\mathbb{R}}} = 1.$$

Proof Let $1 \le p \le q < \infty$. By Theorem 5.9, for every $\delta \in (0, 1]$ we have

$$\Sigma_q(S) \le \frac{(1+\delta)^2}{\delta}\mathbb{E}\|S\| + \left\| \left(\|S\| - \frac{(1+\delta)^2}{\delta} \mathbb{E}\|S\| \right)_+ \right\|_q \le \frac{(1+\delta)^2}{\delta}\mathbb{E}\|S\| + (1+\delta)\sigma_q(S).$$

Since the minimum of the right-hand side of this inequality is attained at

$$\delta = (\mathbb{E}\|S\|)^{1/2}(\mathbb{E}\|S\| + \sigma_q(S))^{-1/2},$$

then by inserting this value, we get

$$\Sigma_q(S) \le \sigma_q(S) + 2(\mathbb{E}\|S\|)^{1/2}(\mathbb{E}\|S\| + \sigma_q(S))^{1/2} + 2\mathbb{E}\|S\|.$$

Hence, in view of the definition of $\sigma_p(S)$ and $K_{p,q}^{\mathbb{R}}$, from the inequalities $\sigma_p(S) \le \Sigma_p(S)$ and $\mathbb{E}\|S\| \le \Sigma_p(S)$ it follows that

$$\begin{aligned}\Sigma_q(S) &\le K_{p,q}^{\mathbb{R}} \cdot \sigma_p(S) + 2\Sigma_p(S)^{1/2}(\Sigma_p(S) + K_{p,q}^{\mathbb{R}}\sigma_p(S))^{1/2} + 2\Sigma_p(S) \\ &\le (K_{p,q}^{\mathbb{R}} + 2(1 + K_{p,q}^{\mathbb{R}})^{1/2} + 2) \cdot \Sigma_p(S),\end{aligned}$$

whence

$$K_{p,q} \leq K^{\mathbb{R}}_{p,q} + 2(1 + K^{\mathbb{R}}_{p,q})^{1/2} + 2.$$

Dividing this inequality by $K^{\mathbb{R}}_{p,q}$, we obtain

$$\frac{K_{p,q}}{K^{\mathbb{R}}_{p,q}} \leq 1 + \frac{2(1 + K^{\mathbb{R}}_{p,q})^{1/2} + 2}{K^{\mathbb{R}}_{p,q}}. \tag{5.25}$$

Assume now that $p \in [1, q^{1/5}]$. Then, since $K^{\mathbb{R}}_{p,q} \geq \|\mathfrak{g}\|_q/\|\mathfrak{g}\|_p$ and $\|\mathfrak{g}\|_r \sim \sqrt{r/e}$ as $r \to \infty$, we have $K^{\mathbb{R}}_{p,q} \to \infty$ as $q \to \infty$. So, in this case the assertion is a consequence of inequality (5.25). On the other hand, if $p \in [q^{1/5}, q]$ we come to the desired result by (5.24) as well, because then both p and q tend to infinity. □

Proof of Theorem 5.9 We begin with the same reasoning as in the beginning of the proof of Theorem 4.7 from the previous chapter.

Let $\{r'_k\}_{k=1}^n$ be a collection of the Rademacher functions independent with respect to the functions r_k, $k = 1, 2, \ldots, n$, and let $S' := \sum_{k=1}^n x_k r'_k$, where $x_k \in F$. For each $x \in F$ we find a norm-one functional $x^* \in F^*$ such that $x^*(x) = \|x\|$. Then, we have the following pointwise estimate:

$$\|x\| - \|x - \delta S'\| = x^*(x) - \|x - \delta S'\| \leq x^*(x) - x^*(x - \delta S') = \delta x^*(S').$$

Hence, taking into account the definition of $\sigma_p(S')$, we easily deduce that

$$\mathbb{E}(\|x\| - \|x - \delta S'\|)_+^p \leq \delta^p \cdot \mathbb{E}|x^*(S')|^p \leq \delta^p \sigma_p(S')^p.$$

Consequently, since the random vectors $S' = \sum_{k=1}^n x_k r'_k$ and $S = \sum_{k=1}^n x_k r_k$ are independent and identically distributed, by Fubini's theorem, we get

$$\mathbb{E}(\|S\| - \|S - \delta S'\|)_+^p \leq \delta^p \sigma_p(S)^p. \tag{5.26}$$

Further, as in the previous chapter, just before Lemma 4.11, denote by (f_1, g_1), $(f_2, g_2), \ldots, (f_n, g_n)$ independent $\mathbb{R}^2$-valued vectors satisfying the following conditions:

$$\mathbb{P}\{(f_k, g_k) = (1, 1 - \delta)\} = \frac{1}{2},$$

$$\mathbb{P}\{(f_k, g_k) = (1, 1 + \delta)\} = \frac{\delta}{2(1 + \delta)},$$

and

$$\mathbb{P}\{(f_k, g_k) = (-1, -1-\delta)\} = \frac{1}{2(1+\delta)}.$$

Suppose also that $\{\epsilon_k\}_{k=1}^n$ is a collection, which is similar to the collection $\{r_k\}_{k=1}^n$ and independent with respect to $\{(f_k, g_k)\}_{k=1}^n$.

Recall that, by Lemma 4.11, the vectors $(S, S - \delta S')$ and $\left(\sum_{k=1}^n \epsilon_k x_k f_k, \sum_{k=1}^n \epsilon_k x_k g_k\right)$ are identically distributed. Therefore, inequality (5.26) can be rewritten as follows:

$$.\mathbb{E}\left(\left\|\sum_{k=1}^n \epsilon_k x_k f_k\right\| - \left\|\sum_{k=1}^n \epsilon_k x_k g_k\right\|\right)_+^p \le \delta^p \sigma_p(S)^p. \tag{5.27}$$

Since the function $x \mapsto x_+^p$ is increasing and convex, by the Jensen inequality (see Appendix A), we get

$$\mathbb{E}\left(\left\|\sum_{k=1}^n \epsilon_k x_k f_k\right\| - \left\|\sum_{k=1}^n \epsilon_k x_k g_k\right\|\right)_+^p \ge \mathbb{E}_\epsilon\left(\mathbb{E}_{f,g}\left\|\sum_{k=1}^n \epsilon_k x_k f_k\right\| - \mathbb{E}_{f,g}\left\|\sum_{k=1}^n \epsilon_k x_k g_k\right\|\right)_+^p, \tag{5.28}$$

where $\mathbb{E}_{f,g}$ (resp. $\mathbb{E}_\epsilon$) is the expectation with respect to the σ-algebra, generated by the collection $\{(f_k, g_k)\}_{k=1}^n$ (resp. $\{\epsilon_k\}_{k=1}^n$) (recall that $\{(f_k, g_k)\}_{k=1}^n$ and $\{\epsilon_k\}_{k=1}^n$ are independent).

Next, let each of r.v.'s η_k, $k = 1, 2, \ldots, n$, have the same distribution as r_1, and let $\eta_1, \eta_2, \ldots, \eta_n, \epsilon_1, \epsilon_2, \ldots, \epsilon_n$ are independent. Then, clearly, for every fixed $\omega \in \Omega$, the collections $\{\epsilon_k(\omega)\eta_k\}_{k=1}^n$, $\{\eta_k\}_{k=1}^n$, and $\{r_k\}_{k=1}^n$ are similar. Observe also that $\mathbb{E} g_k = 0$ and $|g_k| \le 1 + \delta$ a.e. Consequently, applying Lemma 4.10 to the function

$$\Phi(u_1, \ldots, u_n) := \left\|\sum_{k=1}^n \epsilon_k u_k x_k\right\|,$$

where $(u_1, \ldots, u_n) \in [-1-\delta, 1+\delta]^n$, we get

$$\mathbb{E}_{f,g}\left\|\sum_{k=1}^n \epsilon_k x_k g_k\right\| \le \mathbb{E}_\eta\left\|\sum_{k=1}^n \epsilon_k (1+\delta) x_k \eta_k\right\| = (1+\delta)\mathbb{E}\|S\|$$

($\mathbb{E}_\eta$ is the expectation with respect to the σ-algebra, generated by the collection $\{\eta_k\}_{k=1}^n$). On the other hand, from Minkowski's inequality and the fact that $\mathbb{E} f_k = \delta/(1+\delta)$ it follows

$$\mathbb{E}_{f,g}\left\|\sum_{k=1}^n \epsilon_k x_k f_k\right\| \ge \left\|\sum_{k=1}^n \epsilon_k x_k \mathbb{E} f_k\right\| = \frac{\delta}{1+\delta}\left\|\sum_{k=1}^n \epsilon_k x_k\right\| = \frac{\delta}{1+\delta}\|S\|.$$

The last two inequalities, combined together with estimates (5.27) and (5.28), imply that

$$\mathbb{E}\Big(\frac{\delta}{1+\delta}\|S\| - (1+\delta)\mathbb{E}\|S\|\Big)_+^p \le \delta^p \cdot \sigma_p(S)^p.$$

Hence, finally we have

$$\left\|\left(\|S\| - \frac{(1+\delta)^2}{\delta}\mathbb{E}\|S\|\right)_+\right\|_p \le (1+\delta)\sigma_p(S),$$

as we wished. □

5.6 The Optimal Constant in the Khintchine Type Inequality for L_{N_2}

As we know (see Theorem 2.3), a Khintchine type inequality holds in the exponential Orlicz space L_{N_α} if and only if $\alpha \ge 2$. Show that the knowledge of the optimal constants in Khintchine's L_p-inequality for $p = 2m$, $m = 1, 2, \dots$, allows us to find the optimal constant in a similar inequality for the space L_{N_2}.

Let $n \in \mathbb{N}$ and $a_i \in \mathbb{R}$, $i = 1, 2, \dots, n$. Denote

$$S_n := \sum_{i=1}^n a_i r_i \quad \text{and} \quad A_n := \sum_{i=1}^n a_i^2.$$

By Theorem 1.3, we have

$$\mathbb{E}S_n^{2m} \le K_{2m}^{2m} A_n^m, \quad m = 1, 2, \dots,$$

where $K_{2m}^{2m} = (2m-1)!!$ (see equation (5.11)). Therefore, using the Taylor expansions of the functions e^x and $(1-x)^{-1/2}$, for any $C > 0$ such that $2C^{-2} < 1$ we get

$$\begin{aligned}\mathbb{E}\exp\left(\frac{S_n^2}{C^2 A_n}\right) &= \sum_{m=0}^\infty \frac{\mathbb{E}S_n^{2m}}{C^{2m}A_n^m m!} \le \sum_{m=0}^\infty \frac{K_{2m}^{2m}}{C^{2m} m!} \\ &= \sum_{m=0}^\infty \frac{(2m-1)!!}{C^{2m} m!} = \sum_{m=0}^\infty \frac{(2m-1)!!}{2^m m!}\left(\frac{2}{C^2}\right)^m = \left(1 - \frac{2}{C^2}\right)^{-1/2}.\end{aligned}$$

By solving the equation

$$\left(1-\frac{2}{C^2}\right)^{-1/2}=2,$$

we find $C=\sqrt{8/3}$. Thus, since

$$\|f\|_{L_{N_2}}:=\inf\left\{\lambda>0:\ \mathbb{E}(\exp(f/\lambda)^2-1)\le 1\right\},$$

we come to the following Khintchine type inequality for the space L_{N_2}:

$$\Big\|\sum_{i=1}^n a_i r_i\Big\|_{L_{N_2}}\le\sqrt{\frac{8}{3}}\Big(\sum_{i=1}^n a_i^2\Big)^{1/2}. \tag{5.29}$$

Let us check that the constant $\sqrt{8/3}$ in (5.29) is optimal. To this end, because a direct calculation gives $\|\mathfrak{g}\|_{L_{N_2}}=\sqrt{8/3}$, it suffices to show that

$$\Big\|\frac{1}{\sqrt{n}}\sum_{i=1}^n r_i\Big\|_{L_{N_2}}\to\|\mathfrak{g}\|_{L_{N_2}}\ \text{ as } n\to\infty$$

(here, as above, $\mathfrak{g}$ is a standard Gaussian r.v.). In turn, one can easily see that, in view of definition of the L_{N_2}-norm, the latter relation follows from the equation

$$\lim_{n\to\infty}\mathbb{E}\exp(\bar{S}_n/\lambda)^2=\mathbb{E}\exp(\mathfrak{g}/\lambda)^2 \tag{5.30}$$

where $\bar{S}_n:=n^{-1/2}\sum_{i=1}^n r_i$, $n=1,2,\dots$, and $\lambda>0$.

Let us prove (5.30). First, according to Theorem 5.3 (see also (5.23)), for every $m=1,2,\dots$ the moments $\mathbb{E}\bar{S}_n^{2m}$ increase and converge to $\mathbb{E}\mathfrak{g}^{2m}$ as $n\to\infty$. Hence, by using the Taylor expansion of the function e^x in the same way as in the proof of inequality (5.29), we get

$$\limsup_{n\to\infty}\mathbb{E}\exp(\bar{S}_n/\lambda)^2\le\mathbb{E}\exp(\mathfrak{g}/\lambda)^2.$$

On the other hand, for every $k=1,2,\dots$

$$\sum_{m=1}^k\frac{\mathbb{E}\mathfrak{g}^{2m}}{\lambda^{2m}m!}=\lim_{n\to\infty}\left(\sum_{m=1}^k\frac{\mathbb{E}\bar{S}_n^{2m}}{\lambda^{2m}m!}\right)\le\liminf_{n\to\infty}\mathbb{E}\exp(\bar{S}_n/\lambda)^2,$$

whence taking $k\to\infty$ we infer

$$\mathbb{E}\exp(\mathfrak{g}/\lambda)^2\le\liminf_{n\to\infty}\mathbb{E}\exp(\bar{S}_n/\lambda)^2.$$

Thus, (5.30) is proved. As was above observed, this finishes the proof.

Comments and References The Schur-concavity notion was introduced by Schur [262] in 1923, precisely in the same year, when the work [154] with "Khintchine's inequality" was published. Theorem 5.1, which allows to identify the Schur-concavity of a function by calculating its first partial derivatives, was proved by Schur in [262] for functions defined on $\mathbb{R}_+^n$ and then was extended by Ostrovsky in [225] to the case of $\mathbb{R}^n$.

The line of reasoning, based on using the Schur-concavity property for finding the optimal constants in Khintchine's inequality, has been known already a long time. So, Theorem 5.2 is borrowed from the note [114] by Eaton published as early as in 1970. Rather close results and proofs one can find also in the work of Komorowski [162], which appeared much later, in 1988. Ten years earlier than Eaton, similar results were obtained by Whittle [284], though without of explicit using the Schur-concavity property. In the latter work, in particular, it was erroneously stated that the function

$$\varphi_r(x_1, x_2, \dots, x_n) := \mathbb{E}\left|\sum_{k=1}^{n} \sqrt{x_k} r_k\right|^r$$

is Schur-concave for $x_k \geq 0$, $k = 1, 2, \dots, n$, not only if $r = 2$ or $3 \leq r < \infty$, but also when $2 < r < 3$. It can be easily shown that this is not the case, and for this reason the Schur-concavity notion is not so useful for finding the best constant $K_r(n)$ if $2 < r < 3$. As far as we know, its value in the latter case has remained unknown. The same can be said also about the constant $k_r(n)$ for the values $0 < r < 2$.

The value of the best constant K_p for all $p > 0$ was obtained by direct calculations in the prominent work of Haagerup [127] in 1982 (it is determined by formula (5.9)). Much earlier, in 1961, Stechkin proved the so-called theorem on best lacunary systems and found the value of K_{2m}, $m \in \mathbb{N}$, without using the Schur-concavity. Comparing the results of his work [270] with the Haagerup theorem, one can come to the conclusion that, for every $p > 2$, the Rademacher sequence is one of the best infinite lacunary systems in the following sense. For arbitrary $p > 2$ and $n \in \mathbb{N}$ we define $M = M_p(n)$ as the least constant, for which there is a collection of linearly independent functions $\{f_k\}_{k=1}^n \subset L_p[0, 1]$ satisfying the inequality

$$\left\|\sum_{k=1}^{n} a_k f_k\right\|_p \leq M \left\|\sum_{k=1}^{n} a_k f_k\right\|_2 \tag{5.31}$$

for all real numbers $a_1, a_2, \dots, a_n$. In [270], it is proved that

$$M_p(n) = \sqrt{n}\left[\frac{\Gamma\left(\frac{p+1}{2}\right)\Gamma\left(\frac{n}{2}\right)}{\sqrt{\pi}\,\Gamma\left(\frac{n+p}{2}\right)}\right]^{1/p}.$$

Moreover, Stechkin has constructed an infinite orthonormal system $\{f_k\}_{k=1}^{\infty}$ in $L_2[0, 1]$ such that inequality (5.31) holds with $M = M_p(n)$ for every $n \in \mathbb{N}$. Observe that from the Stirling formula it follows

$$M_p := \lim_{n\to\infty} M_p(n) = \sqrt{2}\left[\frac{\Gamma\left(\frac{p+1}{2}\right)}{\sqrt{\pi}}\right]^{1/p}.$$

Therefore, if a system $\{f_k\}_{k=1}^{\infty}$ satisfies inequality (5.31) with some M for all $n \in \mathbb{N}$ and real numbers $a_1, a_2, \dots, a_n$, then $M \geq M_p$. Since by (5.9) we have $K_p = M_p$, the Rademacher functions form a lacunary system, for which (5.31) is fulfilled (for a *given* $p > 2$ and *all* $n \in \mathbb{N}$) with the least possible constant. Note that Stechkin himself, not having Haagerup's results, concluded that the Rademacher sequence has the above property only in the case when $p = 2m$, $m \in \mathbb{N}$.

The optimal constant k_p in inequality (5.1) is determined by the formula

$$k_p = \begin{cases} 2^{\frac{1}{2}-\frac{1}{p}}, & 0 < p \leq p_0, \\ \sqrt{2}\left(\frac{\Gamma(\frac{p+1}{2})}{\sqrt{\pi}}\right)^{1/p}, & p_0 \leq p \leq 2, \\ 1, & 2 \leq p < \infty, \end{cases}$$

where $p_0 = 1,84742\dots$ is the root of the equation

$$\Gamma\left(\frac{p+1}{2}\right) = \frac{\sqrt{\pi}}{2}.$$

The non-trivial case of this problem, when $p \in (0, 1) \cup (1, 2)$, has been investigated by Haagerup [127] as well. Also, we should mention here the interesting work [218] by Nazarov and Podkorytov, which have developed a quite different, rather ingenious approach to finding the optimal constants in Khintchine's inequality.

According to Theorem 5.3 (see also the preceding discussion), a special role in finding optimal constants in Khintchine's inequality in L_p-spaces is played by the norms $\|\frac{1}{\sqrt{n}}\sum_{k=1}^{n} r_k\|_p$, $n = 1, 2, \dots$ At the same time, in [235], Pinelis disproved a well-known longstanding conjecture, apparently due to Edelman [235, p. 2], that the "maximal" tail Rademacher function $M(\tau)$ defined by

$$M(\tau) := \sup\left\{m\left\{t \in [0, 1] : \sum_{i=1}^{\infty} a_i r_i(t) \geq \tau\right\} : a_i \in \mathbb{R}, \sum_{i=1}^{\infty} a_i^2 = 1\right\}, \quad \tau > 0,$$

is equal to the function

$$M^{=}(\tau) := \sup_{n\in\mathbb{N}} m\left\{t \in [0, 1] : \frac{1}{\sqrt{n}}\sum_{i=1}^{n} r_i(t) \geq \tau\right\}, \quad \tau > 0.$$

A formulation of Littlewood's conjecture from the beginning of Sect. 5.2 is taken from the paper [130]. It was resolved in affirmative by Szarek [273] in 1976. As it turned out later, the same result holds even in the vector-valued direction. We present here the proof of this remarkable Latała–Oleszkiewicz theorem [176], following to its exposition in the paper [247] (see Theorem 5.4 there). Estimate (5.15) was somewhat improved in the paper [236] for Banach spaces X whose norm admits an additive decomposition, i.e., there exists a Borel measure μ on the dual space X^* such that

$$\|x\|_X = \int_{X^*} |x^*(x)|\, d\mu \quad \text{for all } \; x \in X.$$

Theorem 5.5 and Corollary 5.6 have been proved in the paper [173]. The main result of Sect. 5.4 is Theorem 5.7 (see [217, Theorem 3]), which is being a direct consequence of a certain property of the binomial convolution of log-concave sequences of real numbers (the proof of Theorem 5.8 see in [282, Theorem 1] or [217]). Theorems 5.9 and 5.10 about asymptotic equality of the best constants in scalar and vector-valued Khintchine's inequalities have been proved by Oleszkiewicz in [221]. In the same paper, one can find the following conjecture due to Kwapień, which looks as rather plausible because of the statements of Theorems 5.4, 5.6 and 5.10: for arbitrary $0 < p \le q < \infty$ it holds $K_{p,q}^{\mathbb{R}} = K_{p,q}$. Finally, the optimal constant in the Khintchine type inequality for the exponential Orlicz space L_{N_2} has been found in [232] (in the account of this result we follow the survey [233]). In conclusion, note that the value of optimal constant C in the inequality

$$\Big\|\sum_{i=1}^{n} a_i r_i\Big\|_{L_{N_\alpha}} \le C\Big(\sum_{i=1}^{n} a_i^2\Big)^{1/2}$$

in the case when $\alpha > 2$, apparently, remains still unknown.

Chapter 6
Rademacher Chaos in Symmetric Spaces

Definition 6.1 Let $d \in \mathbb{N}$. The *(homogeneous) Rademacher chaos of order d* consists of all d-fold products of distinct Rademacher functions, i.e., this is the system

$$\{r_{i_1} r_{i_2} \dots r_{i_d}\}_{1 \le i_1 < i_2 < \dots < i_d < \infty}. \tag{6.1}$$

It is clear that the first-order Rademacher chaos is just the Rademacher system, and in the case of arbitrary order, this is a subsequence of the classical sequence of Walsh functions. In particular, system (6.1) is orthonormal on $[0, 1]$ (see e.g. Corollary 1.2). At the same time, unlike the Rademacher system, functions from the Rademacher chaos of order $d \ge 2$ are not independent. Moreover, the latter system is not even multiplicative, that is, the products of pairwise distinct functions of this system, in general, fail to be mean zero (see Definition B.7). For instance, if $d = 2$, multiplying the functions $r_1 r_2$, $r_1 r_3$ and $r_2 r_3$, we have

$$\int_0^1 r_1(t) r_2(t) r_1(t) r_3(t) r_2(t) r_3(t)\, dt = 1.$$

However, as we shall see, many properties of the Rademacher chaos are rather analogous to those of systems of independent uniformly bounded functions. Our focus will be on studying some geometrical properties of the subspaces spanned by these systems in s.s.'s.

Next, we confine ourselves to considering the second-order chaos, i.e., the system $\{r_i r_j\}_{1 \le i < j < \infty}$, firstly, in order to avoid cumbersome formulas, and secondly, due to the fact that corresponding results in the case of the chaos of an arbitrary order can be proved similarly (see Remarks 6.3 and 6.4 below).

S. V. Astashkin, *The Rademacher System in Function Spaces*,
https://doi.org/10.1007/978-3-030-47890-2_6

6.1 A.e. Convergence of Rademacher Chaos Series

As in the case of the usual Rademacher system, we begin with the a.e. convergence of Rademacher chaos series. The convergence of a series of the form

$$\sum_{1\le i<j<\infty} a_{i,j} r_i(t) r_j(t), \quad \text{where } a_{i,j} \in \mathbb{R},\ t \in [0,1], \tag{6.2}$$

will be meant as the convergence of the sequence of "triangular" sums

$$S_n(t) := \sum_{1\le i<j\le n} a_{i,j} r_i(t) r_j(t), \quad n \in \mathbb{N}. \tag{6.3}$$

According to Theorem 1.1, the sequence $\{r_k\}_{k=1}^\infty$ is a convergence system, i.e., a series $\sum_{k=1}^\infty a_k r_k(t)$ converges a.e. on $[0,1]$ provided if $a = (a_k)_{k=1}^\infty \in \ell_2$. Show that the Rademacher chaos possesses the same property.

Theorem 6.1 *If a sequence* $(a_{i,j})_{1\le i<j<\infty} \in \ell_2$, *then series* (6.2) *converges a.e. on* $[0,1]$.

Proof As was said above, $\{r_i r_j\}_{1\le i<j<\infty}$ is an orthonormal system on $[0,1]$. Therefore, the hypothesis $(a_{i,j})_{1\le i<j<\infty} \in \ell_2$ implies that the sequence of partial sums $\{S_n\}$, defined by (6.3), is Cauchy in the space $L_2[0,1]$ and so $\|S_n - f\|_2 \to 0$ as $n \to \infty$ for some $f \in L_2[0,1]$. Consequently, for arbitrary $0 \le a < b \le 1$ we have

$$\lim_{n\to\infty} \int_a^b (f(t) - S_n(t))^2\, dt = 0.$$

Hence, by the Cauchy-Schwarz-Bunyakovskii inequality,

$$\lim_{n\to\infty} \int_a^b (f(t) - S_n(t))\, dt = 0. \tag{6.4}$$

Let now $I = (a,b)$ be an interval of constancy of the function r_k, so is of S_k. Since each term of the sum $S_n - S_k$, $n > k$, contains as a factor the function r_j, with some $j > k$, we get

$$\int_I (S_n(t) - S_k(t))\, dt = 0.$$

This equation, combined together with (6.4), implies

$$\int_I f(t)\, dt = \int_I S_k(t)\, dt.$$

Next, one can use exactly the same reasoning as in the concluding part of the proof of Theorem 1.1. □

As we know (see Theorem 1.2), the Rademacher system satisfies the following *enhanced* converse statement: if a series $\sum_{k=1}^{\infty} a_k r_k(t)$ converges on some subset of the interval [0, 1] of positive measure, then $a = (a_k)_{k=1}^{\infty} \in \ell_2$. For the Rademacher chaos of order $d \geq 2$, this is no longer the case. Indeed, let us consider a series of the form:

$$\sum_{k=2d}^{\infty} c_k f_{d-1}(t) r_k(t), \quad \text{with} \quad f_{d-1}(t) := \prod_{i=1}^{d-1} \big(r_{2i-1}(t) + r_{2i}(t)\big),$$

where $c_k \in \mathbb{R}$, $k \geq 2d$, are arbitrary. One can readily see that this series vanishes on the set $E := \{t \in [0, 1] : f_{d-1}(t) = 0\}$. Moreover, since the Rademacher functions are independent, we have

$$m(E) = 1 - m\{f_{d-1}(t) \neq 0\} = 1 - \prod_{i=1}^{d-1} m\big\{r_{2i-1}(t) + r_{2i}(t) \neq 0\big\} = 1 - 2^{-d+1}.$$

Thus, independently of coefficients c_k, $k \geq 2d$, each of the above Rademacher chaos series of order d (see (6.1)) converges on a subset of [0, 1] of measure $1 - 2^{-d+1}$.

However, the converse result to Theorem 6.1 holds. In the proof of this and other results of the section we shall repeatedly make use of the so-called decoupling method. A key idea of this method is the replacement of a chaos with its decoupled version. In the case of the second-order Rademacher chaos, this decoupled chaos can be identified with the system of functions $\{r_i(s) r_j(t)\}_{i,j=1}^{\infty}$ defined on the square $I \times I$, where $I = [0, 1]$. Once some assertion is proved for functions of the form

$$\sum_{1 \leq i < j < \infty} a_{i,j} r_i(s) r_j(t), \quad s, t \in I$$

(as a rule, to do this is simpler than for functions defined by (6.2)), then by using "decoupling machinery" the result may be extended to the initial chaos.

As above, in the next theorem the convergence of series is understood as that of the sequence of "triangular" sums (6.3).

Theorem 6.2 *If a series* (6.2) *converges in measure on* [0, 1], *then we have* $(a_{i,j})_{1 \leq i < j < \infty} \in \ell_2$.

Proof First of all, according to the known decoupling result (see e.g. [174, Theorems 6.9.2 and 6.4.1]), there exist universal positive constants k_1, k_2, K_1, K_2

such that for any $z > 0$ and arbitrary positive integers $n < l$ the following inequalities hold:

$$K_1 m\Big\{t \in [0,1] : \Big|\sum_{j=n}^{l}\sum_{i=1}^{j-1} a_{i,j} r_i(t) r_j(t)\Big| > k_1 z\Big\}$$

$$\leq m\Big\{(s,t) \in I \times I : \Big|\sum_{j=n}^{l}\sum_{i=1}^{j-1} a_{i,j} r_i(s) r_j(t)\Big| > z\Big\}$$

$$\leq K_2 m\Big\{t \in [0,1] : \Big|\sum_{j=n}^{l}\sum_{i=1}^{j-1} a_{i,j} r_i(t) r_j(t)\Big| > k_2 z\Big\}. \tag{6.5}$$

Since (6.2) converges in measure on $[0, 1]$, this implies that the series

$$\sum_{1 \leq i < j < \infty} a_{i,j} r_i(s) r_j(t) \tag{6.6}$$

converges in measure on the square $I \times I$. By Fubini's theorem, the latter is equivalent to the fact that for every $\alpha > 0$

$$\lim_{\substack{n\to\infty \\ l>n}} \int_0^1 m\Big\{t \in [0,1] : \Big|\sum_{j=n}^{l}\sum_{i=1}^{j-1} a_{i,j} r_i(s) r_j(t)\Big| > \alpha\Big\}\, ds = 0. \tag{6.7}$$

Let us prove that, for almost all $s \in [0, 1]$, series (6.6) converges in measure on $[0, 1]$.

Denote $v_j(s) := \sum_{i=1}^{j-1} a_{i,j} r_i(s)$, $j = 2, 3, \dots$. Assuming the contrary, we find a set $E \subset [0, 1]$, $m(E) > 0$, such that for each $s \in E$ there are $p = p(s) \in \mathbb{N}$ and $q = q(s) \in \mathbb{N}$ satisfying

$$\limsup_{\substack{n\to\infty \\ l>n}} m\Big\{t \in [0,1] : \Big|\sum_{j=n}^{l} v_j(s) r_j(t)\Big| > \frac{1}{p}\Big\} > \frac{1}{q}.$$

Then, a standard reasoning shows that there exist a set $F \subset [0, 1]$, $m(F) > 0$, and positive constants β, ε such that for all $s \in F$ we have

$$\limsup_{\substack{n\to\infty \\ l>n}} m\Big\{t \in [0,1] : \Big|\sum_{j=n}^{l} v_j(s) r_j(t)\Big| > \beta\Big\} > \varepsilon. \tag{6.8}$$

Then, for every $s \in F$ we can find $N_1(s) \in \mathbb{N}$ satisfying the estimate

$$m\left\{t \in [0,1] : \left|\sum_{j=2}^{N_1(s)} v_j(s)r_j(t)\right| > \frac{\beta}{2}\right\} > \frac{\varepsilon}{2}. \tag{6.9}$$

Indeed, from (6.8) it follows that there are positive integers $n_1 < l_1$ such that

$$m\left\{t \in [0,1] : \left|\sum_{j=n_1}^{l_1} v_j(s)r_j(t)\right| > \beta\right\} > \varepsilon.$$

Since

$$m\left\{t \in [0,1] : \left|\sum_{j=n_1}^{l_1} v_j(s)r_j(t)\right| > \beta\right\} \le m\left\{t \in [0,1] : \left|\sum_{j=2}^{l_1} v_j(s)r_j(t)\right| > \frac{\beta}{2}\right\}$$
$$+ m\left\{t \in [0,1] : \left|\sum_{j=2}^{n_1-1} v_j(s)r_j(t)\right| > \frac{\beta}{2}\right\},$$

then at least one of the terms from the right-hand side of this inequality exceeds $\varepsilon/2$. So, choosing $N_1(s)$ appropriately, we get (6.9).

Further, let $N_1 \in \mathbb{N}$ satisfy the inequalities $N_1 > 2$ and $m\{s \in F : N_1(s) > N_1\} \le m(F)2^{-2}$. Then, setting $F_1 := \{s \in F : N_1(s) \le N_1\}$, we have $m(F_1) \ge \frac{3}{2^2}m(F)$. Moreover, it is clear that inequality (6.9) holds for every $s \in F_1$ and some $N_1(s) \le N_1$. Similarly, using (6.8) once more, for any $s \in F_1$ and some $N_2(s) \ge N_1$ we get

$$m\left\{t \in [0,1] : \left|\sum_{j=N_1}^{N_2(s)} v_j(s)r_j(t)\right| > \frac{\beta}{2}\right\} > \frac{\varepsilon}{2}. \tag{6.10}$$

Choose now $N_2 > N_1$ so that $m\{s \in F_1 : N_2(s) > N_2\} \le m(F)2^{-3}$. Then, denoting $F_2 := \{s \in F_1 : N_2(s) \le N_2\}$, we have $m(F_2) \ge \frac{5}{2^3}m(F)$. In addition, for every $s \in F_2$ one can find $N_1(s)$ and $N_2(s)$, satisfying inequalities $1 \le N_1(s) \le N_1 \le N_2(s) \le N_2$, such that estimates (6.9) and (6.10) hold.

Continuing a quite similar reasoning further, we obtain the set $F_\infty := \bigcap_{i=1}^\infty F_i$ of positive measure (because $m(F_\infty) \ge m(F)/2 > 0$) and positive integers $N_0 := 2 < N_1 < N_2 < N_3 < \dots$ such that for every $s \in F_\infty$ there is a sequence $\{N_i(s)\}_{i=1}^\infty$, satisfying the conditions: $N_{i-1} \le N_i(s) \le N_i$, $i = 1, 2, \dots$, and

$$m\left\{t \in [0,1] : \left|\sum_{j=N_i}^{N_{i+1}(s)} v_j(s)r_j(t)\right| > \frac{\beta}{2}\right\} > \frac{\varepsilon}{2}, \quad i = 0, 1, \dots$$

From the latter estimate, by Proposition 2.1, it follows that

$$m\Big\{t \in [0,1] : \Big|\sum_{j=N_i}^{N_{i+1}} v_j(s) r_j(t)\Big| > \frac{\beta}{2}\Big\} > \frac{\varepsilon}{4} \tag{6.11}$$

for all $s \in F_\infty$ and $i = 0, 1, \ldots$ Integrating this inequality, we get

$$\int_0^1 m\Big\{t \in [0,1] : \Big|\sum_{j=N_i}^{N_{i+1}} v_j(s) r_j(t)\Big| > \frac{\beta}{2}\Big\}\, ds$$

$$\geq \int_{F_\infty} m\Big\{t \in [0,1] : \Big|\sum_{j=N_i}^{N_{i+1}} v_j(s) r_j(t)\Big| > \frac{\beta}{2}\Big\}\, ds \geq \frac{\varepsilon}{4} m(F_\infty) \geq \frac{\varepsilon}{8} m(F) > 0, \quad i = 0, 1, 2, \ldots$$

Clearly, this contradicts equation (6.7) if we take $\alpha = \beta/2$. As a result, series (6.6) converges in measure on $[0, 1]$ for almost all $s \in [0, 1]$.

Next, by Corollary 1.8, for almost all $s \in [0, 1]$ we have

$$\sum_{j=2}^{\infty} \Big(\sum_{i=1}^{j-1} a_{i,j} r_i(s)\Big)^2 < \infty.$$

Hence, for arbitrarily small $\delta > 0$ there is a set $G \subset [0, 1]$, with $m(G) \geq 1 - \delta$, such that for some $K > 0$ and all $s \in G$ it holds

$$\sum_{j=2}^{\infty} \Big(\sum_{i=1}^{j-1} a_{i,j} r_i(s)\Big)^2 \leq K.$$

Then, integration over G gives us

$$\sum_{j=2}^{\infty} \int_G \Big(\sum_{i=1}^{j-1} a_{i,j} r_i(s)\Big)^2 ds \leq K m(G).$$

On the other hand, if $\delta < 1/32$, from Corollary 1.7 it follows for all $j = 2, 3, \ldots$ that

$$\int_G \Big(\sum_{i=1}^{j-1} a_{i,j} r_i(s)\Big)^2 ds \geq \frac{1}{64^2} \sum_{i=1}^{j-1} a_{i,j}^2.$$

Thus, combining two last inequalities, we conclude that

$$\sum_{1\le i<j<\infty} a_{i,j}^2 = \sum_{j=2}^{\infty}\sum_{i=1}^{j-1} a_{i,j}^2 \le 64^2 Km(G) < \infty,$$

and so the proof is completed. □

6.2 Rademacher Chaos as a Basic Sequence

In this section, we shall identify a class of s.s.'s, in which the functions $r_i r_j$, $1 \le i < j < \infty$, form a basis in their closed linear span. In other words, we are interested in finding the conditions on a s.s. X, under which $\{r_i r_j\}_{1\le i<j<\infty}$ is a basic sequence in X (see Definition B.1). Before considering this problem we clearly need to equip the above sequence with some ordering. As was mentioned, both sequences $\{r_k\}_{k=1}^{\infty}$ and $\{r_i r_j\}_{1\le i<j<\infty}$ are included into the Walsh system $\{w_n\}_{n=0}^{\infty}$, and therefore the order produced by positions of these functions in the sequence $\{w_n\}_{n=0}^{\infty}$ seems to be the most natural. Then, if the Walsh system is considered with the Paley enumeration (see [153, § IV.5, p. 134] or [126, § 1.1]), we have $w_0 = 1$, $w_{2^k} = r_{k+1}$, $k = 0, 1, ..$ According to this, the second-order Rademacher chaos should be rewritten as follows:

$$\varphi_1 = r_1 r_2,\ \varphi_2 = r_1 r_3,\ \varphi_3 = r_2 r_3, \dots,$$
$$\varphi_{(k-2)(k-1)/2+1} = r_1 r_k, \dots, \varphi_{(k-1)k/2} = r_{k-1} r_k, \dots \tag{6.12}$$

Show that the Rademacher chaos, ordered as above, is a basic sequence in a quite wide class of s.s.'s.

Theorem 6.3 *The system* $\{r_i r_j\}_{1\le i<j<\infty}$*, ordered according to* (6.12)*, is a basic sequence in every interpolation s.s.* X *between the spaces* L_1 *and* L_∞.

Proof Here, we shall make use of the following property of the Walsh system $\{w_n\}_{n=0}^{\infty}$ [126, § 2.1, Formula (2.1.11), p. 39]. Let

$$S_p x(t) = \sum_{i=0}^{p} \int_0^1 x(s) w_i(s)\, ds \cdot w_i(t), \quad p = 0, 1, 2, \dots,$$

be the partial Fourier-Walsh sum operators. Then, denoting $\sigma_k := S_{2^k}$, $k = 0, 1, 2, \dots$, for $t \in \Delta_k^j = \left((j-1)2^{-k}, j2^{-k}\right)$, $j = 1, 2, \dots, 2^k$, we have

$$\sigma_k x(t) = 2^k \int_{\Delta_k^j} x(u)\, du.$$

This means that the operator σ_k coincides with the averaging operator with respect to the system of dyadic intervals $\{\Delta_k^j\}_{j=1}^{2^k}$. Hence, one can easily check (see also [168, § II.3, p. 80–81]) that σ_k is a one-norm operator in both spaces L_1 and L_∞ for each $k = 0, 1, \ldots$ Combining this with the interpolation property of X between L_1 and L_∞, we immediately deduce that the operators σ_k are uniformly bounded in X. More formally, there is a constant $B = B(X) > 0$ such that for all $k = 0, 1, \ldots$ and $x \in X$ we have

$$\|\sigma_k x\| \leq B\, \|x\|. \tag{6.13}$$

By the definition of basic sequences (see Definition B.1), the proof will be finished once we find a constant $K > 0$, depending only on B, such that for arbitrary positive integers $m < n$ and $a_i \in \mathbb{R}$ it holds

$$\Big\| \sum_{i=1}^{m} a_i \varphi_i \Big\|_X \leq K \Big\| \sum_{i=1}^{n} a_i \varphi_i \Big\|_X. \tag{6.14}$$

Since the Walsh system is orthonormal, by the definition of the operators σ_k, the functions $z := \sum_{i=1}^m a_i\varphi_i$ and $y := \sum_{i=1}^n a_i\varphi_i$, where $m < n$ and $a_i \in \mathbb{R}$ are arbitrary, can be represented as follows:

$$z = \sigma_k y + \sum_{j=1}^{p} b_j r_j \cdot r_{k+1}, \quad y = \sigma_l y + \sum_{j=1}^{q} c_j r_j \cdot r_{l+1},$$

where $0 \leq k \leq l$, $1 \leq p < k+1$ and $1 \leq q < l+1$. Assume first that $k < l$. Then,

$$\sigma_{k+1} y - \sigma_k y = \sum_{j=1}^{k} d_j r_j \cdot r_{k+1},$$

where $d_j = b_j$ if $1 \leq j \leq p$. In consequence, from Proposition 2.1 and estimate (6.13) it follows

$$\begin{aligned} \|z\| &\leq \|\sigma_k y\| + \Big\| \sum_{j=1}^{p} b_j r_j \Big\| \leq B\|y\| + \Big\| \sum_{j=1}^{k} d_j r_j \Big\| \\ &= B\|y\| + \|\sigma_{k+1} y - \sigma_k y\| \leq 3B\|y\|, \end{aligned}$$

and hence (6.14) is proved.

If $k = l$, we have $p \leq q$ and $c_j = b_j$, $1 \leq j \leq p$. We can write

$$z - \sigma_k y = \sum_{j=1}^{p} b_j r_j \cdot r_{k+1} \ \text{ and } \ y - \sigma_k y = \sum_{j=1}^{q} c_j r_j \cdot r_{k+1}.$$

Therefore, as above, we get

$$\begin{aligned}\|z\| &\le \|\sigma_k y\| + \Big\|\sum_{j=1}^{p} b_j r_j\Big\| \le B\|y\| + \Big\|\sum_{j=1}^{q} c_j r_j\Big\| \\ &= B\|y\| + \|y - \sigma_k y\| \le (1+2B)\|y\|.\end{aligned}$$

Thus, in this case inequality (6.14) holds as well, which finishes the proof. □

6.3 Unconditionality of the Rademacher Chaos in Symmetric Spaces

Recall (see Proposition 2.2) that the Rademacher system is a symmetric basic sequence in an arbitrary s.s. X. Moreover, for any sequence $(a_j)_{j=1}^\infty \in \ell_2$ we have

$$\Big\|\sum_{j=1}^{\infty} a_j r_j\Big\|_X = \Big\|\sum_{j=1}^{\infty} a_j^* r_j\Big\|_X,$$

where $(a_j^*)_{j=1}^\infty$ is the decreasing permutation of the sequence $(|a_j|)_{j=1}^\infty$. Clearly, the Rademacher chaos does not have so perfect basic properties. In this chapter, we shall find sharp conditions on a s.s. X, under which this system (as in the previous section, denote it by $\{\varphi_i\}_{i=1}^\infty$) is symmetric in X, i.e.,

$$C^{-1}\Big\|\sum_{i=1}^{\infty} a_i\varphi_i\Big\|_X \le \Big\|\sum_{i=1}^{\infty} a_i\varphi_{\pi(i)}\Big\|_X \le C\Big\|\sum_{i=1}^{\infty} a_i\varphi_i\Big\|_X,$$

where $C > 0$ does not depend on a permutation π of the set of positive integers. It is worth to emphasize that these conditions are both necessary and sufficient for the Rademacher chaos to be unconditional. This means (see Appendix B) that there exists a constant $K_0 > 0$ such that the inequality

$$\Big\|\sum_{i=1}^{n} \theta_i a_i\varphi_i\Big\|_X \le K_0\Big\|\sum_{i=1}^{n} a_i\varphi_i\Big\|_X.$$

holds for arbitrary $n \in \mathbb{N}$, $\theta_i = \pm 1$ and $a_i \in \mathbb{R}$, $i = 1, 2, \ldots, n$. In fact, we shall establish even the following much stronger result: the unconditionality of the Rademacher chaos in a s.s. X implies its equivalence in X to the unit vector basis in ℓ_2.

As above, L_{N_1} is the Orlicz space on [0, 1], generated by the function $N_1(t) = e^t - 1$, and $H := L_{N_1}^\circ$ is the closure of L_∞ in L_{N_1}. We are ready now to state the main result of this section.

Theorem 6.4 *Let X be a s.s. on $[0, 1]$. The following conditions are equivalent:*

(1) the system $\{r_i r_j\}_{1\le i<j<\infty}$ is equivalent in X to the unit vector basis in ℓ_2, i.e., there exists a constant $C > 0$, which depends only on the space X, such that for all $(a_{i,j})_{1\le i<j<\infty} \in \ell_2$ we have

$$C^{-1}\,\|(a_{i,j})\|_{\ell_2} \le \Big\| \sum_{1\le i<j<\infty} a_{i,j} r_i r_j \Big\|_X \le C\,\|(a_{i,j})\|_{\ell_2}; \tag{6.15}$$

(2) the continuous embedding $X \supset H$ holds;
(3) $\{r_i r_j\}_{1\le i<j<\infty}$ is an unconditional basic sequence in X.

The proof of Theorem 6.4 consists of two main parts. First, we prove the equivalence (1) $\Leftrightarrow$ (2) and, second, the implication (3) $\Rightarrow$ (1) (the converse statement is obvious). Here, we again shall make use of reasoning based on the decoupling techniques. Namely, we begin with proving the equivalence (1) $\Leftrightarrow$ (2) for the auxiliary decoupled chaos $\{r_i(s)r_j(t)\}_{i,j=1}^{\infty}$ and then extend this result to the chaos itself.

As above, we associate to every s.s. X on the interval $I = [0, 1]$ the s.s. $X(I\times I)$, consisting of all Lebesgue measurable functions $x = x(s, t)$ on the square $I\times I$ such that $x^* \in X$. This space is equipped with the natural norm $\|x\|_{X(I\times I)} = \|x^*\|_X$.

Theorem 6.5 *Let X be a s.s. on I. The system $\{r_i(s)r_j(t)\}_{i,j=1}^{\infty}$ is equivalent in the space $X(I \times I)$ to the unit vector basis in ℓ_2 if and only if $X \supset H$.*

Proof Suppose first $X \supset H$. Denote

$$y(s,t) := \sum_{i,j=1}^{\infty} b_{i,j} r_i(s) r_j(t), \quad s, t \in I, \tag{6.16}$$

where $b = (b_{i,j})_{i,j=1}^{\infty} \in \ell_2$. Then, applying subsequently Fubini's theorem, Minkowski's inequality for the $L_{q/2}$-norm and twice Khintchine's inequality (see Theorem 1.3), we get

$$\begin{aligned}
\|y\|_q &= \left(\int_{I\times I} |y(s,t)|^q\,ds\,dt\right)^{1/q} = \left(\int_0^1\int_0^1 |y(s,t)|^q\,ds\,dt\right)^{1/q}\\
&\le \sqrt{q}\left\{\int_0^1 \left(\sum_{i=1}^{\infty}\Big|\sum_{j=1}^{\infty} b_{i,j} r_j(t)\Big|^2\right)^{q/2} dt\right\}^{1/q}\\
&\le \sqrt{q}\left\{\sum_{i=1}^{\infty}\left(\int_0^1 \Big|\sum_{j=1}^{\infty} b_{i,j} r_j(t)\Big|^q dt\right)^{2/q}\right\}^{1/2} \le q\cdot\|b\|_{\ell_2}.
\end{aligned}$$

Thus, we have $\|y\|_q \le q\|b\|_{\ell_2}$. Obviously, this inequality holds also for $1 \le q < 2$. Therefore, by using the Taylor expansion

$$\exp(uz) = \sum_{k=0}^{\infty} \frac{u^k}{k!} z^k, \quad u > 0,$$

we come to the estimate

$$\int_0^1 \int_0^1 \{\exp(u|y(s,t)|) - 1\}\, dsdt = \sum_{k=1}^{\infty} \frac{u^k}{k!} \int_0^1 \int_0^1 |y(s,t)|^k dsdt \le \sum_{k=1}^{\infty} \frac{u^k}{k!} k^k \|b\|_{\ell_2}^k.$$

Then, if $u < (2e\|b\|_{\ell_2})^{-1}$, according to the Stirling formula (1.13), the sum in the right-hand side of this inequality does not exceed 1. Therefore, by the definition of H, we have $y \in H$ and $\|y\|_{H(I\times I)} \le 2e\|b\|_{\ell_2}$. So, by hypothesis,

$$\Big\| \sum_{i,j=1}^{\infty} b_{i,j} r_i(s) r_j(t) \Big\|_{X(I\times I)} \le 2Ce\|b\|_{\ell_2},$$

where C is the embedding constant of $X \supset H$.

On the other hand, from the embedding $X \subset L_1$, Khintchine's L_1-inequality (see Theorem 5.4), and Minkowski's inequality it follows

$$\begin{aligned} \|y\|_{X(I\times I)} \ge c_X \|y\|_1 &\ge \frac{c_X}{\sqrt{2}} \int_0^1 \left[\sum_{i=1}^{\infty} \Big(\sum_{j=1}^{\infty} b_{i,j} r_j(t) \Big)^2 \right]^{1/2} dt \\ &\ge \frac{c_X}{\sqrt{2}} \left\{ \sum_{i=1}^{\infty} \left[\int_0^1 \Big| \sum_{j=1}^{\infty} b_{i,j} r_j(t) \Big| dt \right]^2 \right\}^{1/2} \ge \frac{c_X}{2} \|b\|_{\ell_2}. \end{aligned} \tag{6.17}$$

Thus, the equivalence of the system $\{r_i(s) r_j(t)\}_{i,j=1}^{\infty}$ in $X(I \times I)$ to the unit vector basis in ℓ_2 is established.

Conversely, setting $b_{i,j} = c_i d_j$, by condition, we have

$$\Big\| \sum_{i,j=1}^{\infty} b_{i,j} r_i(s) r_j(t) \Big\|_{X(I\times I)} \le C\|(b_{i,j})\|_{\ell_2} = C\|(c_i)\|_{\ell_2} \|(d_j)\|_{\ell_2}.$$

In particular, since X is symmetric, for $c_i = d_i = n^{-1/2}$ if $1 \le i \le n$, and $c_i = d_i = 0$ if $i > n$, $n = 1, 2, \ldots$, in view of the last inequality, it follows

$$\|v_n^*(s) v_n^*(t)\|_{X(I\times I)} \le C, \quad n = 1, 2, \ldots, \tag{6.18}$$

where $v_n := n^{-1/2} \sum_{i=1}^{n} r_i$. As above, by Theorem A.1, we have

$$\lim_{n\to\infty} m\{s \in I : v_n^*(s) > z\} = \Psi(z),$$

with

$$\Psi(z) = \frac{2}{\sqrt{2\pi}} \int_z^\infty \exp(-u^2/2)du, \quad z > 0.$$

Therefore, due to Lemma 2.1,

$$\lim_{n\to\infty} v_n^*(s) = \Psi^{-1}(s), \quad 0 < s \le 1,$$

and, taking into account (6.18) and the fact that the space X'' has the Fatou property, we get

$$\Psi^{-1}(s)\Psi^{-1}(t) \in X''(I \times I).$$

Further, since $\Psi(z) \ge (1/\sqrt{2\pi}) \cdot e^{-z^2}$ for $z \ge 0$ (see the proof of Theorem 2.3), the function $\ln^{1/2}(e/s)\ln^{1/2}(e/t)$ belongs to $X''(I \times I)$. Moreover, observe that for every $0 < u \le 1$ it holds

$$\ln^{1/2}(e/s)\ln^{1/2}(e/t) \ge \ln(e/\sqrt{u}) \ge \frac{1}{2}\ln(e/u), \quad 0 < s, t \le \sqrt{u}.$$

Thus, by definition of the decreasing rearrangement of a measurable function, we have

$$\left(\ln^{1/2}(e/\cdot)\ln^{1/2}(e/\cdot)\right)^*(u) \ge \frac{1}{2}\ln(e/u), \quad 0 < u \le 1,$$

whence $\ln(e/u) \in X''$.

Recall that the function $\ln(e/u)$ has the largest (up to a constant) rearrangement in the space L_{N_1}; more precisely, for any function $x \in L_{N_1}$ there is $C > 0$ such that $x^*(u) \le C\ln(e/u)$, $0 < u \le 1$ (see Lemma 2.4). Therefore, the fact that $\ln(e/u) \in X''$ implies $L_{N_1} \subset X''$. Then, by Lemma 2.2, $H = L_{N_1}^\circ \subset (X'')^\circ = X^\circ \subset X$, and Theorem 6.5 is proved. □

To establish the equivalence (1) ⇔ (2) in Theorem 6.4, we need two more auxiliary statements, the first of which is an immediate consequence of Fubini's theorem and Proposition 2.2.

Lemma 6.1 *Let A and B be finite subsets of $\mathbb{N}$. Then, for every permutation π of $\mathbb{N}$ the functions*

$$\sum_{i\in A, j\in B} b_{i,j} r_i(s) r_j(t) \quad \textit{and} \quad \sum_{i\in A, j\in B} b_{i,j} r_i(s) r_{\pi(j)}(t)$$

are identically distributed on the square $I \times I$.

Lemma 6.2 *There are universal constants $A > 0$ and $B > 0$ such that for arbitrary $n \in \mathbb{N}$ and $c_{i,j} \in \mathbb{R}$, $1 \le i < j \le n$, we have*

$$\int_0^1 \Big(\sum_{1\le i<j\le n} c_{i,j} r_i(t) r_j(t) \Big)^4 dt \le A \Big(\sum_{1\le i<j\le n} c_{i,j}^2 \Big)^2 \tag{6.19}$$

and

$$\Big(\sum_{1\le i<j\le n} c_{i,j}^2 \Big)^{1/2} \le B \int_0^1 \Big| \sum_{1\le i<j\le n} c_{i,j} r_i(t) r_j(t) \Big| dt. \tag{6.20}$$

Proof First of all, applying decoupling inequality (6.5) once more, with a universal constant $C > 0$ we get

$$\int_0^1 \Big(\sum_{1\le i<j\le n} c_{i,j} r_i(t) r_j(t) \Big)^4 dt \le C \int_0^1 \int_0^1 \Big(\sum_{1\le i<j\le n} c_{i,j} r_i(u) r_j(v) \Big)^4 du\, dv.$$

Hence, by Fubini's theorem and Theorem 5.6,

$$\int_0^1 \Big(\sum_{1\le i<j\le n} c_{i,j} r_i(t) r_j(t) \Big)^4 dt \le C \int_0^1 \int_0^1 \Big(\sum_{j=2}^n \Big(\sum_{i=1}^{j-1} c_{i,j} r_i(u) \Big) r_j(v) \Big)^4 du\, dv$$

$$\le 3C \int_0^1 \Big(\sum_{j=2}^n \Big(\sum_{i=1}^{j-1} c_{i,j} r_i(u) \Big)^2 \Big)^2 du$$

$$= 3C \int_0^1 \Big(\sum_{1\le i<j\le n} c_{i,j}^2 + 2 \sum_{j=3}^n \sum_{1\le i<k<j} c_{i,j} r_i(u) c_{k,j} r_k(u) \Big)^2 du$$

$$\le 3C \Big\{ \sum_{1\le i<j\le n} c_{i,j}^2 + 2 \Big(\int_0^1 \Big(\sum_{1\le i<k<n-1} \Big(\sum_{j=k+1}^n c_{i,j} c_{k,j} \Big) r_i(u) r_k(u) \Big)^2 du \Big)^{1/2} \Big\}^2$$

Since the system $\{r_i r_k\}_{1\le i<k<\infty}$ is orthonormal on [0, 1], then from Minkowski's inequality it follows

$$\Big(\int_0^1 \Big(\sum_{1\le i<k<n-1}\Big(\sum_{j=k+1}^{n} c_{i,j}c_{k,j}\Big)r_i(u)r_k(u)\Big)^2 du\Big)^{1/2} = \Big(\sum_{1\le i<k<n-1}\Big(\sum_{j=k+1}^{n} c_{i,j}c_{k,j}\Big)^2\Big)^{1/2}$$

$$\le \sum_{j=3}^{n}\Big(\sum_{1\le i<k<j}(c_{i,j}c_{k,j})^2\Big)^{1/2} = \sum_{j=3}^{n}\Big(\sum_{k=2}^{j-1} c_{k,j}^2 \sum_{i=1}^{k-1} c_{i,j}^2\Big)^{1/2} \le \sum_{1\le i<j\le n} c_{i,j}^2.$$

As a result, two last inequalities imply (6.19) with constant $A = 27C$.

Further, denote $S_n := \sum_{1\le i<j\le n} c_{i,j} r_i r_j$. Arguing in the same way as in the second part of the proof of Theorem 1.4 and applying (6.19), we come to the inequality

$$\|(c_{i,j})\|_{\ell_2}^2 \le A^{1/3}\|S_n\|_1^{2/3}\|(c_{i,j})\|_{\ell_2}^{4/3},$$

whence (6.20) follows immediately. □

***Proof of Theorem 6.4 (The Equivalence* (1) ⇒ (2))** The proof is based on the passage to the corresponding decoupled chaos and the subsequent application of Theorem 6.5.

Assume first that $X \supset H$. Denote

$$x(t) := \sum_{1\le i<j<\infty} a_{i,j} r_i(t) r_j(t) \ \text{ and } \ x_m(t) := \sum_{1\le i<j\le m} a_{i,j} r_i(t) r_j(t), \quad m = 2, 3, \dots,$$

where $a = (a_{i,j})_{1\le i<j<\infty} \in \ell_2$. Simple combinatorial arguments show that

$$x_m(t) = 2^{2-m} \sum_{D\subset\{1,\dots,m\}} \Big(\sum_{i\in D,\, j\notin D} a_{i,j} r_i(t) r_j(t)\Big), \tag{6.21}$$

where the external summation is taken over all subsets D of $\{1, \dots, m\}$. As it is easy to see, the functions

$$\sum_{i\in D,\, j\notin D} a_{i,j} r_i(t) r_j(t), \quad t \in I, \quad \text{and} \quad \sum_{i\in D,\, j\notin D} a_{i,j} r_i(s) r_j(t), \quad (s, t) \in I \times I,$$

are identically distributed. Therefore, as $X \supset H$, from Theorem 6.5, equation (6.21) and the Cauchy-Schwarz-Bunyakovskii inequality it follows

$$\|x_m\|_X \le 2^{2-m} C \sum_{D\subset\{1,\dots,m\}} \Big(\sum_{i\in D,\, j\notin D} a_{i,j}^2\Big)^{1/2}$$

$$\leq 2^{2-m} C \left(\sum_{D \subset \{1,\ldots,m\}} 1 \right)^{1/2} \left(\sum_{D \subset \{1,\ldots,m\}} \sum_{i \in D, j \notin D} a_{i,j}^2 \right)^{1/2}$$

$$\leq 2^{2-m/2} C \left(2^{m-2} \sum_{1 \leq i < j \leq m} a_{i,j}^2 \right)^{1/2} = 2C \|(a_{i,j})_{1 \leq i < j \leq m}\|_{\ell_2}.$$

A similar inequality for the norms $\|x_l - x_m\|_X$, where $1 \leq m < l < \infty$, shows that $\{x_m\}$ is a Cauchy sequence in X, and hence, passing to the limit in the last inequality as $m \to \infty$, we get $\|x\|_X \leq 2C\|a\|_{\ell_2}$. Thus, the right-hand side inequality in (6.15) is proved. As to the opposite estimate, then it holds in every s.s. X provided that the series $\sum_{1 \leq i < j < \infty} a_{i,j} r_i r_j$ converges in X. Indeed, since $X \subset L_1$, then, applying Lemma 6.2 (see inequality (6.20)), for any $m \in \mathbb{N}$ we get

$$\|(a_{i,j})_{1 \leq i < j \leq m}\|_{\ell_2} \leq B \Big\| \sum_{1 \leq i < j \leq m} a_{i,j} r_i r_j \Big\|_1 \leq B' \Big\| \sum_{1 \leq i < j \leq m} a_{i,j} r_i r_j \Big\|_X.$$

Taking now $m \to \infty$, we come to the estimate

$$\|(a_{i,j})\|_{\ell_2} \leq B' \Big\| \sum_{1 \leq i < j < \infty} a_{i,j} r_i r_j \Big\|_X. \tag{6.22}$$

In the converse setting, assume that the system $\{r_i r_j\}_{1 \leq i < j < \infty}$ is equivalent in a s.s. X to the unit vector basis in ℓ_2. Consider the partial sums of series (6.16), that is, the functions

$$y_m(s,t) := \sum_{i,j=1}^{m} b_{i,j} r_i(s) r_j(t), \quad m = 1, 2, \ldots$$

Let π be a permutation of $\mathbb{N}$ such that

$$\pi(\{1, 2, \ldots, m\}) = \{m+1, m+2, \ldots, 2m\}.$$

Then, it is clear that the functions

$$y'_m(t) := \sum_{i,j=1}^{m} b_{i,j} r_i(t) r_{\pi(j)}(t), \ t \in I, \quad \text{and} \quad y''_m(t) := \sum_{i,j=1}^{m} b_{i,j} r_i(s) r_{\pi(j)}(t), \ (s,t) \in I \times I,$$

are identically distributed. Since the same holds also for y_m and y''_m by Lemma 6.1, then the functions y_m and y'_m are identically distributed. Therefore, by hypothesis, for some $C > 0$ we have

$$\|y_m\|_{X(I\times I)} = \|y'_m\|_X \le C\|(b_{i,j})_{i,j=1}^m\|_{\ell_2}.$$

From a similar inequality for the norms $\|y_l - y_m\|_{X(I\times I)}$, $1 \le m < l < \infty$, it follows that $\{y_m\}$ is a Cauchy sequence in $X(I\times I)$. So, passing to the limit in the last inequality as $m \to \infty$, we infer

$$\|y\|_{X(I\times I)} \le C\|(b_{i,j})\|_{\ell_2}.$$

Since the opposite estimate is fulfilled in every s.s. (see (6.17)), then the system $\{r_i(s)r_j(t)\}_{i,j=1}^\infty$ is equivalent in $X(I\times I)$ to the unit vector basis in ℓ_2. As a result, by Theorem 6.5, we have $X \supset H$, and so the equivalence (1) $\Leftrightarrow$ (2) is proved. □

We proceed now with the proof of the implication (3) $\Rightarrow$ (1), which is more complicated. First, we prove this result under the additional condition that $X \supset G$. As above, G is the closure of L_∞ in the Orlicz space L_{N_2}, generated by the function $N_2(t) = e^{t^2} - 1$. Clearly, $G \subset H$.

Proposition 6.1 *Let X be a s.s. on $[0,1]$, $X \supset G$. If $\{r_ir_j\}_{1\le i<j<\infty}$ is an unconditional basic sequence in X, then it is equivalent in this space to the unit vector basis in ℓ_2.*

For the proof of this statement, we need the following property of Orlicz spaces.

Lemma 6.3 *Let M be an Orlicz function, and let $z = z(u,t)$ be a measurable function on the square $I\times I$. Suppose that the function $z(\cdot,t)$ belongs to the space L_M for almost all $t \in I$ and moreover*

$$\operatorname*{ess\,sup}_{t\in I} \|z(\cdot,t)\|_{L_M} < \infty.$$

Then, $z(u,\cdot) \in L_M$ for almost all $u \in I$ and

$$\int_0^1 \|z(u,\cdot)\|_{L_M}\,du \leqslant 2\operatorname*{ess\,sup}_{t\in I} \|z(\cdot,t)\|_{L_M}.$$

Proof First of all, setting $a := \operatorname{ess\,sup}_{t\in I}\|z(\cdot,t)\|_{L_M}$, for almost all $t \in I$ we have

$$\int_0^1 M\left(\frac{|z(u,t)|}{a}\right)du \leqslant 1.$$

Hence, by Fubini's theorem,

$$\int_0^1\int_0^1 M\left(\frac{|z(u,t)|}{a}\right)dt\,du = \int_0^1\int_0^1 M\left(\frac{|z(u,t)|}{a}\right)du\,dt \leqslant 1.$$

On the other hand, let

$$l(u) := 1 + \int_0^1 M\left(\frac{|z(u,t)|}{a}\right)dt.$$

Then, since $l(u) \geqslant 1$, the function M is convex, and

$$\int_0^1 M\left(\frac{|z(u,t)|}{a}\right)dt \leqslant l(u),$$

we have

$$\int_0^1 M\left(\frac{|z(u,t)|}{al(u)}\right)dt \leqslant 1,$$

whence $\|z(u,\cdot)\|_{\mathrm{L}_M} \leqslant al(u)$. Summarizing all, we get

$$\int_0^1 \|z(u,\cdot)\|_{\mathrm{L}_M}\,du \leqslant a\int_0^1 l(u)\,du \leqslant a\left(1 + \int_0^1\int_0^1 M\left(\frac{|z(u,t)|}{a}\right)dt\,du\right) \leqslant 2a,$$

as we wished. □

Proof of Proposition 6.1. We apply here the so-called randomization against the sequence of usual Rademacher functions $\mathbf{r}_{i,j}(u)$, $1 \leq i < j < \infty$, which are enumerated in an arbitrary order.

Since the norms of G and L_{N_2} coincide on L_∞, the embedding $X \supset G$ and Lemma 6.3 imply that

$$\begin{aligned}\int_0^1 \Big\|\sum_{1\leq i<j\leq n} a_{i,j}\mathbf{r}_{i,j}(u)r_i(\cdot)r_j(\cdot)\Big\|_X du &\leq C\int_0^1 \Big\|\sum_{1\leq i<j\leq n} a_{i,j}\mathbf{r}_{i,j}(u)r_i(\cdot)r_j(\cdot)\Big\|_G du \\ &\leq 2C \operatorname*{ess\,sup}_{t\in I}\Big\|\sum_{1\leq i<j\leq n} a_{i,j}\mathbf{r}_{i,j}(\cdot)r_i(t)r_j(t)\Big\|_G\end{aligned}$$

for any $n \in \mathbb{N}$ and $a_{i,j} \in \mathbb{R}$, $1 \le i < j \le n$. Hence, from the Khintchine type inequality for G (see Theorem 2.2) it follows

$$\int_0^1 \Big\| \sum_{1\le i<j\le n} a_{i,j}\mathbf{r}_{i,j}(u) r_i r_j \Big\|_X du \le C' \, \|(a_{i,j})\|_{\ell_2}. \tag{6.23}$$

According to (6.22), the opposite inequality

$$\int_0^1 \Big\| \sum_{1\le i<j\le n} a_{i,j}\mathbf{r}_{i,j}(u) r_i r_j \Big\|_X du \ge c \, \|(a_{i,j})\|_{\ell_2} \tag{6.24}$$

holds as well.

Finally, by hypothesis, the system $\{r_i r_j\}_{1\le i<j<\infty}$ is unconditional in X. Therefore, with a constant, depending only on X, we have

$$\Big\| \sum_{1\le i<j\le n} a_{i,j} r_i r_j \Big\|_X \asymp \int_0^1 \Big\| \sum_{1\le i<j\le n} a_{i,j}\mathbf{r}_{i,j}(u) r_i r_j \Big\|_X du$$

for arbitrary $n \in \mathbb{N}$ and $a_{i,j} \in \mathbb{R}$, $1 \le i < j \le n$. Thus, the desired result follows from this equivalence, combined with estimates (6.23) and (6.24). □

From inequalities (6.22) and (6.23) we immediately obtain the following statement (the definition of a system of random unconditional convergence (RUC-system) see in Appendix B).

Corollary 6.1 *If X is a s.s. such that $X \supset G$, then the Rademacher chaos $\{r_i r_j\}_{1\le i<j<\infty}$ is an RUC-system in X.*

Corollary 6.2 *For every s.s. X the following conditions are equivalent:*

(i) $X \supset G$;

(ii) with a constant, independent of $(a_{i,j})_{1\le i<j<\infty} \in \ell_2$, it holds

$$\int_0^1 \Big\| \sum_{1\le i<j<\infty} a_{i,j}\mathbf{r}_{i,j}(u) r_i r_j \Big\|_X du \asymp \|(a_{i,j})_{1\le i<j<\infty}\|_{\ell_2}.$$

Proof The implication $(i) \Rightarrow (ii)$ follows from Proposition 6.1 and its proof (see inequalities (6.23) and (6.24)).

Conversely, assume that (ii) holds. Putting $a_{1,j} = b_j$, $j = 2, 3, \dots$, where $(b_j)_{j=2}^{\infty} \in \ell_2$ is arbitrary, while $a_{i,j} = 0$ if $i \ne 1$, by Proposition 2.2 and the hypothesis of the corollary, we get

$$\Big\| \sum_{j=2}^{\infty} b_j r_j \Big\|_X = \int_0^1 \Big\| \sum_{1\le i<j<\infty} a_{i,j}\mathbf{r}_{i,j}(u) r_i r_j \Big\|_X du \le C \Big(\sum_{j=2}^{\infty} b_j^2 \Big)^{1/2}.$$

Observing that this estimate holds for any sequence $(b_j)_{j=2}^{\infty} \in \ell_2$, by Theorem 2.3, we conclude that $X \supset G$. □

Corollary 6.3 *Suppose X is a s.s. on $[0, 1]$, $X \supset G$. For each sequence $\{a_{i,j}\}_{1 \le i < j < \infty} \in \ell_2$ there is a sign arrangement $\{\theta_{i,j}\}_{1 \le i < j < \infty}$ such that*

$$\Big\| \sum_{1 \le i < j < \infty} \theta_{i,j} a_{i,j} r_i r_j \Big\|_X \asymp \|(a_{i,j})\|_{\ell_2}$$

with a constant that depends only on X.

Proof In view of Corollary 6.2, we have

$$\inf\Big\{\Big\| \sum_{1 \le i < j < \infty} \theta_{i,j} a_{i,j} r_i r_j \Big\|_X : \theta_{i,j} = \pm 1\Big\} \le \int_0^1 \Big\| \sum_{1 \le i < j < \infty} a_{i,j} \mathbf{r}_{i,j}(u) r_i r_j \Big\|_X du \le C \|(a_{i,j})\|_{\ell_2}.$$

Since the opposite inequality holds as well (see (6.22)), everything is done. □

In the general situation, the implication (3) ⇒ (1) of Theorem 6.4 will be obtained as a consequence of the above special case and some geometrical properties of the Rademacher chaos in Marcinkiewicz spaces that are "closely" located to L_∞.

Recall that the Rademacher sequence is a Sidon system (see Remark 1.4 and equation (1.17)). As we shall see, the behaviour of the Rademacher chaos in L_∞ is completely different. We begin with the classical lower estimate for the decoupled Rademacher chaos.

Proposition 6.2 *Suppose that*

$$F(s, t) := \sum_{i,j=1}^{\infty} a_{i,j} r_i(s) r_j(t) \in L_\infty(I \times I).$$

Then, we have

$$\|(a_{i,j})\|_{\ell_{4/3}} \le \sqrt{2} \|F\|_\infty. \tag{6.25}$$

Proof Without loss of generality, we can assume that $(a_{i,j})_{1 \le i < j < \infty}$ is an eventually vanishing sequence, i.e., for some N and all $i, j \ge N$ we have $a_{i,j} = 0$. Clearly,

$$\sum_{i,j=1}^{\infty} |a_{i,j}|^{4/3} = \sum_{i=1}^{\infty} \Big(\sum_{j=1}^{\infty} |a_{i,j}|^{2/3} |a_{i,j}|^{2/3} \Big).$$

Applying now twice Hölder's inequality, first, to the inner sum with exponents 3 and $3/2$, and then to the external sum with exponents $3/2$ and 3, we get

$$\sum_{i,j=1}^{\infty} |a_{i,j}|^{4/3} \le \Big(\sum_{i=1}^{\infty}\Big(\sum_{j=1}^{\infty} a_{i,j}^2\Big)^{1/2}\Big)^{2/3} \cdot \Big(\sum_{i=1}^{\infty}\Big(\sum_{j=1}^{\infty} |a_{i,j}|\Big)^2\Big)^{1/3}.$$

Hence, estimating the second factor from the right-hand side by Minkowski's inequality, we obtain that

$$\sum_{i,j=1}^{\infty} |a_{i,j}|^{4/3} \le \Big(\sum_{i=1}^{\infty}\Big(\sum_{j=1}^{\infty} a_{i,j}^2\Big)^{1/2}\Big)^{2/3} \Big(\sum_{j=1}^{\infty}\Big(\sum_{i=1}^{\infty} a_{i,j}^2\Big)^{1/2}\Big)^{2/3}. \tag{6.26}$$

On the other hand, according to Khintchine's L_1-inequality (see Theorem 5.4),

$$\sum_{i=1}^{\infty}\Big(\sum_{j=1}^{\infty} a_{i,j}^2\Big)^{1/2} \le \sqrt{2}\sum_{i=1}^{\infty}\int_0^1 \Big|\sum_{j=1}^{\infty} a_{i,j} r_j(t)\Big|\,dt$$

$$= \sqrt{2}\int_0^1 \sup_{0<s\le 1} |F(s,t)|\,dt \le \sqrt{2}\|F\|_\infty.$$

Similarly,

$$\sum_{j=1}^{\infty}\Big(\sum_{i=1}^{\infty} a_{i,j}^2\Big)^{1/2} \le \sqrt{2}\|F\|_\infty.$$

It remains to observe that (6.25) is an immediate consequence of two last inequalities and (6.26). □

Further, we shall consider the "random" L_∞-norms of linear combinations of the functions $r_i(s)r_j(t)$, $1 \le i < j < \infty$. Namely, for any integer $n \ge 2$ and $\theta = (\theta_{i,j})_{1\le i<j\le n}$, $\theta_{i,j} = \pm 1$, we set

$$\psi_n(\theta) := \Big\| \sum_{1\le i<j\le n} \theta_{i,j} r_i(s) r_j(t) \Big\|_{L_\infty(I\times I)}. \tag{6.27}$$

From the definition of the Rademacher functions it follows immediately that $\max_\theta \psi_n(\theta)$ is attained at $\theta_{i,j} = 1$, $1 \le i < j \le n$, and it is equal to the number of terms in the corresponding sum, i.e., $n(n-1)/2$. At the same time, as the following result shows, the average of the quantity $\psi_n(\theta)$ over sign arrangements is equivalent to $\min_\theta \psi_n(\theta)$, $n \in \mathbb{N}$. Meanwhile, this indicates that the exponent $4/3$ in inequality (6.25) is sharp.

Proposition 6.3 *For every $n \geq 2$ it holds*

$$\frac{1}{4}n^{3/2} \leq \min_{\theta} \psi_n(\theta) \leq 2^{-(n-1)n/2} \sum_{\theta} \psi_n(\theta) < 6n^{3/2}. \tag{6.28}$$

Proof First, by Proposition 6.2, for arbitrary $\theta_{i,j} = \pm 1$, $1 \leq i < j \leq n$, we have

$$\psi_n(\theta) \geq \frac{1}{\sqrt{2}} \left(\frac{n(n-1)}{2} \right)^{3/4} \geq \frac{1}{4} n^{3/2}.$$

Hence, the left-hand side inequality in (6.28) is proved.

Next, note that $2^{-(n-1)n/2} \sum_{\theta} \psi_n(\theta)$ is the average of the numbers $\psi_n(\theta)$, taken over all sign arrangements θ. Equivalently, we can write

$$2^{-(n-1)n/2} \sum_{\theta} \psi_n(\theta) = \int_0^1 \Big\| \sum_{1 \leq i < j \leq n} \mathbf{r}_{i,j}(u) r_i(s) r_j(t) \Big\|_\infty du, \tag{6.29}$$

where $\{\mathbf{r}_{i,j}\}_{1 \leq i < j \leq n}$ are the first $n(n-1)/2$ Rademacher functions, enumerated in an arbitrary way. This observation, clearly, implies the middle inequality in (6.28).

The proof of the right-hand side inequality in (6.28) will be based on applying the well-known theorem on uniform norms of "random" sums [148, Theorem 6.1, p. 68–69]. If "random" coefficients are Rademacher functions, this result reads as follows: Let (Ω, μ) be a measure space, $\mu(\Omega) < \infty$, and let B be a linear space of real-valued functions, $B \subset L_\infty(\Omega, \mu)$. Assume that there exists $\rho > 0$ with the following property: for each $f \in B$ there is a measurable set $E = E(f) \subset \Omega$ such that $\mu(E) \geq (1/\rho)\mu(\Omega)$ and $|f(\omega)| \geq 1/2\|f\|_\infty$ for all $\omega \in E$. Then, for every "random" sum

$$P(u, \omega) = \sum_{k=1}^n r_k(u) f_k(\omega), \quad \text{where} \quad f_k \in B,$$

and all $z > 2$ we have

$$m\left\{ u \in [0,1] : \|P(u, \cdot)\|_{L_\infty(\Omega)} \geq 3 \left(\sum_{k=1}^n \|f_k\|_\infty^2 \ln(2\rho z) \right)^{1/2} \right\} \leq \frac{2}{z}. \tag{6.30}$$

Let now $n \in \mathbb{N}$ and B_n be the space, consisting of all functions of the form:

$$f(s, t) = \sum_{1 \leq i < j \leq n} a_{i,j} r_i(s) r_j(t), \quad a_{i,j} \in \mathbb{R},$$

defined on the square $I \times I$. It is clear that for each $f \in B_n$ the equality $|f(s,t)| = \|f\|_\infty$ is valid on a set $E \subset I \times I$ of measure $m(E) \geq 2^{-2n+1}$. Hence, by (6.30), for all $z > 2$ we have

$$m\Big\{u \in I : \Big\| \sum_{1 \le i < j \le n} \mathbf{r}_{i,j}(u) r_i(s) r_j(t) \Big\|_\infty \ge 3\Big(\sum_{1 \le i < j \le n} \ln(2^{2n} z)\Big)^{1/2}\Big\} \le \frac{2}{z}.$$

Denote

$$F(u) := \Big\| \sum_{1 \le i < j \le n} \mathbf{r}_{i,j}(u) r_i(s) r_j(t) \Big\|_\infty, \quad u \in I.$$

Since

$$\Big(\sum_{1 \le i < j \le n} \ln(2^{2n} z)\Big)^{1/2} \le \Big(\frac{n(n-1)(2n+1)}{2} \cdot \ln z \Big)^{1/2} \le n^{3/2} \ln^{1/2} z,$$

from the preceding inequality it follows that

$$m\{u \in I : F(u) \ge 3n^{3/2}\tau\} \le 2e^{-\tau^2}, \quad \tau \ge 1.$$

Finally, combining this together with (6.29), we get

$$\begin{aligned} 2^{-(n-1)n/2} \sum_{\theta} \psi_n(\theta) = \|F\|_1 &= 3n^{3/2} \int_0^\infty m\{u \in I : X(u) \ge 3n^{3/2}\tau\}\, d\tau \\ &\le 3n^{3/2}\Big(1 + 2\int_1^\infty e^{-\tau^2}\, d\tau\Big) < 6n^{3/2}, \end{aligned}$$

which completes the proof. □

Analogous result holds also for the Rademacher chaos.

Corollary 6.4 *For every integer $n \ge 2$ there exists a sign arrangement $\theta = (\theta_{i,j})_{1 \le i < j \le n}$ such that*

$$\Big\| \sum_{1 \le i < j \le n} \theta_{i,j} r_i r_j \Big\|_{L_\infty(I)} < 6n^{3/2}. \tag{6.31}$$

Proof This statement is a direct consequence of Proposition 6.3 and the following obvious inequality

$$\Big\| \sum_{1 \le i < j \le n} a_{i,j} r_i r_j \Big\|_{L_\infty(I)} \le \Big\| \sum_{1 \le i < j \le n} a_{i,j} r_i(s) r_j(t) \Big\|_{L_\infty(I \times I)},$$

which holds for arbitrary $n \in \mathbb{N}$ and $a_{i,j} \in \mathbb{R}$. □

From Proposition 6.3 and Corollary 6.4 we get the following result (for the definition of a Sidon system see Definition B.8).

Corollary 6.5 *Each of the sequences* $\{r_i(t)r_j(s)\}_{1\le i<j<\infty}$ *and* $\{r_i(t)r_j(t)\}_{1\le i<j<\infty}$ *fails to be a Sidon system on* $I \times I$ *and* I*, respectively.*

The probabilistic proof of Proposition 6.3 does not provide any specific sign arrangement, which the minimum (up to order) of the quantity $\psi_n(\theta)$ (see (6.27)) is attained at. For this reason, we prove one more assertion of such a sort for $\psi_n(\theta)$ in the case when $n = 2^k$, $k = 1, 2, \ldots$

Recall that the first 2^k Walsh functions are defined by

$$w_0 = 1 \;\text{ and }\; w_{2^i+j} = r_{i+1}w_j, \;\; i = 0, 1, \ldots, k-1, \;\; j = 0, \ldots, 2^i - 1$$

(see [153, § IV.5, p. 134] or [126, § 1.1, Formula (1.1.5), p. 2]). Since such a function is equal to $+1$ or -1 on each of the intervals $\Delta_k^i = ((i-1)2^{-k}, i2^{-k})$, $1 \le i \le 2^k$, we can introduce the numbers

$$\varepsilon_{i,j} := \operatorname{sign} w_{j-1}(t)\ ,\ t \in \Delta_k^i \;\; (i, j = 1, 2, \ldots, 2^k),$$

which form the so-called *Walsh matrix* $\varepsilon = (\varepsilon_{i,j})_{i,j=1}^{2^k}$.

Proposition 6.4 $\psi_{2^k}(\varepsilon) \le 2^{3k/2}$, $k = 0, 1, 2, \ldots$

Proof For a fixed $s \in I$ we consider the function

$$x_s(u) = \sum_{i=1}^{2^k} r_i(s)\chi_{\Delta_k^i}(u), \;\; 0 \le u \le 1.$$

Because $|x_s(u)| = 1$ a.e., we have $\|x_s\|_2 = 1$. Moreover, the Fourier-Walsh coefficients of the function $x_s(u)$ can be represented by the Walsh matrix as follows:

$$c_{j-1}(x_s) = \sum_{i=1}^{2^k} \int_{\Delta_k^i} w_{j-1}(u)\,du \cdot r_i(s) = 2^{-k}\sum_{i=1}^{2^k} \varepsilon_{i,j}r_i(s), \;\; 1 \le j \le 2^k.$$

Therefore, by using the Cauchy-Schwarz-Bunyakovskii and Hölder inequalities, we get

$$\begin{aligned}\psi_{2^k}(\varepsilon) &= \sup_{0<s\le 1}\sum_{j=1}^{2^k}\left|\sum_{i=1}^{2^k}\varepsilon_{i,j}r_i(s)\right| = 2^k \sup_{0<s\le 1}\sum_{j=1}^{2^k}|c_{j-1}(x_s)| \\ &\le 2^{3k/2}\sup_{0<s\le 1}\left(\sum_{j=1}^{2^k}c_{j-1}(x_s)^2\right)^{1/2} \le 2^{3k/2}\|x_s\|_2 = 2^{3k/2},\end{aligned}$$

completing the proof. □

Before to proceed with the concluding part of the proof of Theorem 6.4, we introduce some additional notation. If X is a s.s. on $[0,1]$, then $\mathcal{R}_{\mathrm{ch}}(X)$ is the subspace of X, consisting of all functions of the form

$$x(t) = \sum_{1\le i<j<\infty} a_{i,j} r_i(t) r_j(t), \quad \text{where } (a_{i,j})_{1\le i<j<\infty} \in \ell_2. \tag{6.32}$$

Moreover, for an arbitrary sign arrangement $\theta = (\theta_{i,j})_{1\le i<j<\infty}$ we define the operator T_θ on $\mathcal{R}_{\mathrm{ch}}(X)$ as follows: if $x(t) \in \mathcal{R}_{\mathrm{ch}}(X)$ is representable as in (6.32), then

$$T_\theta x(t) := \sum_{1\le i<j<\infty} \theta_{i,j} a_{i,j} r_i(t) r_j(t), \quad 0 \le t \le 1.$$

Obviously, if $\{r_i r_j\}_{1\le i<j<\infty}$ is an unconditional basic sequence in X, then for every θ the operator T_θ acts boundedly on $\mathcal{R}_{\mathrm{ch}}(X)$.

Further, we need one more lemma.

Lemma 6.4 *Let $\{n_k\}_{k=1}^\infty$ be an increasing sequence of positive integers, $t_k := 2^{-2n_{k+1}-1}$, $m_k := (n_{k+1} - n_k)(n_{k+1} - n_k - 1)/2$, and*

$$y_k(t) := \sum_{n_k<i<j\le n_{k+1}} r_i(t) r_j(t), \quad k = 1, 2, \ldots$$

Suppose that a series $S := \sum_{k=1}^\infty c_k y_k$, where $c_k > 0$, $k = 1, 2, \ldots$, converges in measure on $[0, 1]$. Then, there is a universal constant $\alpha_0 \in (0, 1)$ such that

$$S^*(\alpha_0 t_k) \ge \alpha_0 \sum_{l=1}^k m_l c_l, \quad k = 1, 2, \ldots$$

Proof Precisely as in the beginning of the proof of Theorem 6.2, the hypothesis of the lemma implies that the series

$$\bar{S}(u, v) := \sum_{k=1}^\infty c_k \bar{y}_k(u, v), \quad \text{where } \bar{y}_k(u, v) := \sum_{n_k<i<j\le n_{k+1}} r_i(u) r_j(v), \quad (u, v) \in I \times I,$$

converges in measure on the square $I \times I$. Observe that, if $1 \le l \le k$, then by definition of the Rademacher functions, for all $(u, v) \in Q_k := (0, 2^{-n_{k+1}}) \times (0, 2^{-n_{k+1}})$ we have $\bar{y}_l(u, v) = m_l$. Consequently,

$$\bar{S}(u, v) = \sum_{l=1}^k m_l c_l + \sum_{l=k+1}^\infty c_l \bar{y}_l(u, v), \quad (u, v) \in Q_k.$$

Furthermore, as the second sum from the right-hand side of this equation is symmetrically distributed on the square Q_k, then there exists a set $E_k \subset Q_k$, $m(E_k) = m(Q_k)/2 = 2^{-2n_{k+1}-1}$, where this sum is nonnegative. Therefore, we have

$$\bar{S}(u, v) \geq \sum_{l=1}^{k} m_l c_l \quad \text{for all} \quad (u, v) \in E_k.$$

Note that the function $\bar{S}(u, v)$ is the decoupled version of S. Hence, precisely as in the proof of Theorem 6.2, according to [174, Theorems 6.9.2 and 6.4.1], there is a universal constant $\alpha_0 \in (0, 1)$ such that for every $z > 0$

$$\alpha_0 m\{(u, v) \in I \times I : |\bar{S}(u, v)| > z\} \leq m\{t \in [0, 1] : |S(t)| > \alpha_0 z\}.$$

Combining this with the preceding estimate, we get

$$m\Big\{t \in [0, 1] : |S(t)| > \alpha_0 \sum_{l=1}^{k} m_l c_l\Big\} \geq \alpha_0 m\{(u, v) \in I \times I : |\bar{S}(u, v)| > \sum_{l=1}^{k} m_l c_l\} \geq \alpha_0 t_k,$$

and the desired inequality follows from definition of the decreasing rearrangement of a measurable function. □

Let X be either L_∞ or a Marcinkiewicz space that is "closely" located to L_∞. The next two assertions show that the appropriate choice of signs $\theta_{i,j}$, $1 \leq i < j < \infty$, assures that the image of the operator T_θ, defined on $\mathcal{R}_{\text{ch}}(X)$, contains functions having a certain "logarithmic" distribution.

Proposition 6.5 *There exists a sign arrangement $\theta = (\theta_{i,j})_{1 \leq i < j < \infty}$ such that for every $\gamma \in (0, 1/2)$ we can find a function $x \in \mathcal{R}_{\text{ch}}(L_\infty)$ satisfying the estimate:*

$$(T_\theta x)^*(t) \geq b \log_2^\gamma(2/t), \quad 0 < t \leq \alpha_0 2^{-9},$$

where $b > 0$ depends only on γ.

Proof Observe that the functions

$$\sum_{2^k < i < j \leq 2^{k+1}} a_{i,j} r_i r_j \quad \text{and} \quad \sum_{1 \leq i < j \leq 2^k} a_{i+2^k, j+2^k} r_i r_j$$

are identically distributed for arbitrary $a_{i,j} \in \mathbb{R}$, $2^k < i < j \leq 2^{k+1}$. Hence, according to Corollary 6.4, for each $k \in \mathbb{N}$ there are $\theta_{i,j} = \pm 1$, $2^k < i < j \leq 2^{k+1}$ such that the function $z_k := \sum_{2^k < i < j \leq 2^{k+1}} \theta_{i,j} r_i r_j$ satisfies the following inequality:

$$\|z_k\|_\infty \le 6 \cdot 2^{3k/2}. \tag{6.33}$$

We set $x_k := 2^{(\gamma-2)k} z_k$ and

$$x := \sum_{k=1}^{\infty} x_k = \sum_{1\le i<j<\infty} a_{i,j} r_i r_j,$$

where $a_{i,j} = 2^{(\gamma-2)k}\theta_{i,j}$ if $2^k < i < j \le 2^{k+1}$, $k = 1, 2, \dots$, and $a_{i,j} = 0$ otherwise. Then, from (6.33) it follows that

$$\|x\|_\infty \le 6 \cdot \sum_{k=1}^{\infty} 2^{(\gamma-1/2)k} = \frac{6}{2^{1/2-\gamma} - 1},$$

whence $x \in \mathcal{R}_{\rm ch}(L_\infty)$.

Let now a sign arrangement θ consist of the chosen $\theta_{i,j}$, $2^k < i < j \le 2^{k+1}$, $k = 1, 2, \dots$, and arbitrary $\theta_{i,j}$ for the remaining pairs (i, j), $i < j$. For such a sign arrangement θ we have

$$T_\theta x = \sum_{k=1}^{\infty} 2^{(\gamma-2)k}\, y_k,$$

where

$$y_k := \sum_{2^k<i<j\le 2^{k+1}} r_i r_j, \quad k = 1, 2, \dots$$

We apply Lemma 6.4 for $n_k = 2^k$ and $c_k = 2^{(\gamma-2)k}$. Then, $t_k = 2^{-2^{k+2}-1}$ and $m_k \ge 2^{2k-2}$. Therefore, for all $k = 1, 2, \dots$ we get

$$(T_\theta x)^*(\alpha_0 t_k) \ge \frac{\alpha_0}{4} \sum_{i=1}^{k} 2^{2i} 2^{(\gamma-2)i} \ge \alpha_0 2^{\gamma k-2} \ge b' \log_2^\gamma (2/t_k),$$

where $b' > 0$ depends only on γ. Clearly, for each $t \in (0, \alpha_0 2^{-9}]$ there is $k = 1, 2, \dots$ such that $\alpha_0 t_{k+1} < t \le \alpha_0 t_k$. Combining this observation with the last inequality, we conclude

$$(T_\theta x)^*(t) \ge (T_\theta x)^*(\alpha_0 t_k) \ge b' \log_2^\gamma (2/t_k) \ge 2^{-\gamma} b' \log_2^\gamma (2/t_{k+1}) \ge b \log_2^\gamma (2/t),$$

where $b = 2^{-\gamma} b'$, and the proof is finished. □

Further, $M(\varphi_\gamma)$ is the Marcinkiewicz space, generated by the function $\varphi_\gamma(t) = t\log_2^\gamma(2/t)$, $\gamma > 0$.

Corollary 6.6 *Suppose that a s.s. X on $[0, 1]$ has the following property: for an arbitrary sign arrangement $\theta = (\theta_{i,j})_{1\le i<j<\infty}$ the operator T_θ is bounded in the subspace $\mathcal{R}_{\text{ch}}(X)$. Then,*

$$X \supset \bigcup_{\gamma\in(0,1/2)} M(\varphi_\gamma).$$

Proof According to Proposition 6.5, we can find a sign arrangement θ such that for a given $\gamma \in (0, 1/2)$ and a certain function $x \in \mathcal{R}_{\text{ch}}(L_\infty)$ (depending on γ) it holds

$$(T_\theta x)^*(t) \ge b\,\log_2^\gamma(2/t),\quad 0 < t \le \alpha_0 2^{-9}.$$

Since $X \supset L_\infty$ and X is symmetric, from this inequality and the hypothesis it follows that $\log_2^\gamma(2/t) \in X$. Consequently, by Proposition 2.4, $X \supset M(\varphi_\gamma)$. Because $\gamma \in (0, 1/2)$ is arbitrary, the proof is completed. □

Theorem 6.6 *There exists a sign arrangement $\theta = (\theta_{i,j})_{1\le i<j<\infty}$ such that for every $\gamma \in (0, 1)$ and $\delta \in (0, \gamma/2)$ we can find a function $x \in \mathcal{R}_{\text{ch}}(M(\varphi_\gamma))$ that satisfies the inequality*

$$(T_\theta x)^*(t) \ge d\,\log_2^{1/2+\delta}(2/t),\quad 0 < t \le \alpha_0 2^{-9}, \tag{6.34}$$

where $d > 0$ depends only on γ and δ.

Proof Let us choose a sign arrangement $\theta = (\theta_{i,j})_{1\le i<j<\infty}$ and functions z_k, y_k, $k = 1, 2, \dots$, in the same way as in the proof of Proposition 6.5. By Proposition 2.4, the Marcinkiewicz space $M(\varphi_1)$, with $\varphi_1(t) = t\log_2(2/t)$, coincides with the Orlicz space L_{N_1}, $N_1(t) = e^t - 1$. Therefore, applying the equivalence (1) ⇔ (2), which is already proved, we have

$$\|z_k\|_{M(\varphi_1)} \asymp \Big(\sum_{2^k<i<j\le 2^{k+1}} |\theta_{i,j}|^2\Big)^{1/2} \asymp 2^k,\quad k = 1, 2, \dots \tag{6.35}$$

Next, for every $\gamma \in (0, 1)$ there is a constant $C > 0$ such that for all $x \in L_\infty$

$$\|x\|_{M(\varphi_\gamma)} \le C\,\|x\|_\infty^{1-\gamma}\,\|x\|_{M(\varphi_1)}^\gamma. \tag{6.36}$$

Indeed, by using Proposition 2.4 once more, we get

$$\|x\|_{M(\varphi_\gamma)} \le C \sup_{0<t\le 1} (x^*(t) \log_2^{-\gamma}(2/t))$$
$$\le C \left(\sup_{0<t\le 1} x^*(t) \right)^{1-\gamma} \cdot \left(\sup_{0<t\le 1} (x^*(t) \log_2^{-1}(2/t)) \right)^{\gamma}$$
$$\le C \, \|x\|_\infty^{1-\gamma} \, \|x\|_{M(\varphi_1)}^{\gamma}.$$

Thus, from (6.33), (6.35), and (6.36) it follows that for some $D > 0$

$$\|z_k\|_{M(\varphi_\gamma)} \le D\, 2^{(3-\gamma)k/2}, \quad k = 1, 2, \dots \tag{6.37}$$

Let us define for any $\upsilon > 0$ the functions

$$x_k := 2^{(\gamma-3-2\upsilon)k/2} z_k \ \text{ and } \ x := \sum_{k=1}^{\infty} x_k.$$

Then, according to (6.37), we have $x \in \mathcal{R}_{\mathrm{ch}}(M(\varphi_\gamma))$. Moreover, applying Lemma 6.4 to the function

$$T_\theta x(t) = \sum_{k=1}^{\infty} 2^{(\gamma-3-2\upsilon)k/2} y_k,$$

in the case when $n_k = 2^k$, $t_k = 2^{-2^{k+2}-1}$, and $c_k = 2^{(\gamma-3-2\upsilon)k/2}$, we infer that for every $k = 1, 2, \dots$

$$(T_\theta x)^*(\alpha_0 t_k) \ge \frac{\alpha_0}{4} \sum_{i=1}^{k} 2^{2i} \, 2^{(\gamma-2\upsilon-3)i/2} \ge \alpha_0 2^{(1+\gamma-2\upsilon)k/2-2} \ge d' \log_2^{(1+\gamma-2\upsilon)/2}(2/t_k),$$

where $d' > 0$. Hence, using the same reasoning as in the proof of Proposition 6.5, we deduce

$$(T_\theta x)^*(t) \ge d \log_2^{(1+\gamma-2\upsilon)/2}(2/t), \quad 0 < t \le \alpha_0 2^{-9}, \tag{6.38}$$

with some $d > 0$, depending only on γ and υ. By condition, $\delta < \gamma/2$; therefore, choosing υ so that $0 < \upsilon < \gamma/2 - \delta$, we get $\gamma - 2\upsilon > 2\delta$. Thus, (6.34) is an immediate consequence of inequality (6.38), and so the theorem is proved. □

In the same way as Corollary 6.6 from Proposition 6.5, from Theorem 6.6 it follows

Corollary 6.7 *Let X be a s.s. such that $X \supset M(\varphi_\gamma)$ for some $\gamma \in (0, 1)$. Suppose that the operator T_θ is bounded in $\mathcal{R}_{\mathrm{ch}}(X)$ for every sign arrangement*

$\theta = (\theta_{i,j})_{1\le i<j<\infty}$. *Then, we have*

$$X \supset \bigcup_{0<\delta<\gamma/2} M(\varphi_{1/2+\delta}).$$

Now everything is ready to complete the proof of Theorem 6.4.

Proof of the Implication (3) $\Rightarrow$ (1). Since the system $\{r_i r_j\}_{1\le i<j<\infty}$ is unconditional in X, the operator T_θ is bounded in the subspace $\mathcal{R}_{\mathrm{ch}}(X)$ for every sign arrangement θ. Hence, according to Corollary 6.6, we have $X \supset M(\varphi_\gamma)$ for each $\gamma \in (0, 1/2)$. Consequently, applying Corollary 6.7, we get $X \supset M(\varphi_{1/2+\delta})$ for each $\delta \in (0, \gamma/2)$. As $M(\varphi_{1/2+\delta}) \supset G$ for $\delta > 0$, we have, a fortiori, $X \supset G$. Thus, by Proposition 6.1, the system $\{r_i r_j\}_{1\le i<j<\infty}$ is equivalent in X to the unit vector basis in ℓ_2. □

By Theorem 3.2, for every interpolation sequence space E between the spaces ℓ_1 and ℓ_2 there is a function s.s. X on $[0, 1]$ such that the Rademacher system spans in X a sequence space, which coincides with E, that is,

$$\Big\| \sum_{j=1}^{\infty} c_j r_j \Big\|_X \asymp \|(c_j)\|_E.$$

In contrast to that, for the Rademacher chaos we have

Corollary 6.8 *Let E be a Banach sequence lattice, and let X be a s.s. on* $[0, 1]$. *Then, the equivalence*

$$\Big\| \sum_{1\le i<j<\infty} a_{i,j} r_i r_j \Big\|_X \asymp \|(a_{i,j})\|_E \tag{6.39}$$

holds if and only if $E = \ell_2$ and $X \supset H$.

Proof If $E = \ell_2$ and $X \supset H$, the desired result follows from Theorem 6.4.

Conversely, from (6.39) it follows that $\{r_i r_j\}_{1\le i<j<\infty}$ is an unconditional basic sequence in X. Hence, applying Theorem 6.4 once more, we have $E = \ell_2$ and $X \supset H$. □

Remark 6.1 An inspection of the proofs of the above results (first of all, Theorems 6.3 and 6.4) shows that similar statements hold also for the decoupled Rademacher chaos (or the multiple Rademacher system) $\{r_i(s) r_j(t)\}_{i,j=1}^{\infty}$ on the square $I \times I$.

Remark 6.2 From Theorem 6.4, in particular, it follows that exponential integrability of the Rademacher chaos, i.e., the fact that a function

$$\exp\Big(\alpha\Big| \sum_{1\le i<j<\infty} b_{i,j} r_i r_j \Big|\Big)$$

belongs to $L_1[0,1]$ for every $\alpha > 0$ provided that $\sum_{1\le i<j<\infty} b_{i,j}^2 < \infty$, is sharp. It is instructive also to compare the equivalence (1) $\Leftrightarrow$ (2) from the latter theorem with Theorem 2.3 for the Rademacher sequence, where the role of H is played by the space G, the closure of L_∞ in the Orlicz space L_{N_2}, with $N_2(t) = e^{t^2} - 1$.

Remark 6.3 Let $d \in \mathbb{N}$, $d \ge 2$, and let H_d be the closure of L_∞ in the Orlicz space $L_{N_{2/d}}$, generated by an Orlicz function equivalent to the function $N_{2/d}(t) = \exp(t^{2/d}) - 1$. Then, arguing similarly as in the proof of Theorem 6.4, one can show that for an arbitrary s.s. X on $[0,1]$ the following conditions are equivalent:

(i) the Rademacher chaos of order d, defined by (6.1), is equivalent in X to the unit vector basis in ℓ_2;
(ii) the continuous embedding $X \supset H_d$ holds;
(iii) the Rademacher chaos of order d is an unconditional basic sequence in X.

6.4 Complementability of Subspaces Spanned by the Rademacher Chaos

In the concluding section of this chapter we consider the complementability problem of the subspace $\mathcal{R}_{\text{ch}}(X)$ in a s.s. X. It is not hard to guess that the result will be stated by using the space H (precisely in the same way, as in Theorem 2.5 we use the space G).

Theorem 6.7 *Let X be a s.s. on* $[0,1]$. *The subspace $\mathcal{R}_{\text{ch}}(X)$, consisting of all functions representable in the form* (6.2)*, is complemented in X if and only if $H \subset X \subset H'$.*

Proof Since the reasoning, used in the first part of the proof, is similar to that given in the proof of Theorem 2.5, we restrict ourselves in this part only to a few remarks.

First of all, by using Theorem 6.4 for both spaces X and X', one can easily show that the embeddings $H \subset X \subset H'$ guarantee the boundedness in X of the orthogonal projection

$$P_{\text{ch}} f(t) = \sum_{1\le i<j<\infty} c_{i,j}(f) r_i(t) r_j(t), \tag{6.40}$$

where

$$c_{i,j}(f) = \int_0^1 f(s) r_i(s) r_j(s)\, ds, \quad 1 \le i < j < \infty. \tag{6.41}$$

In the converse setting, as the projection P_{ch} is self-adjoint, the embeddings $H \subset X \subset H'$ follow from the boundedness of P_{ch} in the space X. Thus, it remains only to prove that complementability of $\mathcal{R}_{\rm ch}(X)$ in X implies that $P_{\rm ch}$ is bounded in this space.

Recall (see the proof of Theorem 2.5) that the interval [0, 1], equipped with the operation

$$t \oplus u = \sum_{i=1}^{\infty} 2^{-i}[(\alpha_i + \beta_i) mod 2],$$

where $t = \sum_{i=1}^{\infty} \alpha_i 2^{-i}$, $u = \sum_{i=1}^{\infty} \beta_i 2^{-i}$, $\alpha_i, \beta_i = 0, 1$, are the binary expansions of numbers $t, u \in [0, 1]$, is a compact Abelian group. Since for every $u \in [0, 1]$ the transformation $w_u(s) = s \oplus u$ of [0, 1] preserves the Lebesgue measure, then the operators $T_u f := f \circ w_u$, $0 \le u \le 1$, are isometries in X. Moreover, one can easily check that the subspace $\mathcal{R}_{\rm ch}(X)$ is invariant with respect to their action. Therefore, again by the Rudin theorem [255, Theorem 5.18], there exists a bounded projection $Q : X \to X$, $Q(X) = \mathcal{R}_{\rm ch}(X)$, which commutes with all operators T_u, $0 \le u \le 1$. We need only to check that Q coincides with the above orthogonal projection $P_{\rm ch}$ (see (6.40) and (6.41)).

Let us begin with the following representation of the projection Q:

$$Qf(t) = \sum_{1\le i<j<\infty} Q_{i,j}(f) r_i(t) r_j(t), \quad f \in X, \tag{6.42}$$

where the coordinate functionals $Q_{i,j}$, $1 \le i < j < \infty$, are bounded on X. Indeed, from Lemma 6.2 and the embedding $X \subset L_1$ it follows

$$|Q_{i,j}(f)| \le \|(Q_{i,j}(f))\|_{\ell_2} \le B\|Q(f)\|_1 \le B'\|Q(f)\|_X \le B'\|Q\|\|f\|_X.$$

Moreover, observe that for arbitrary $1 \le k < m$ we have

$$Q_{i,j}(r_k r_m) = \begin{cases} 1, k = i \text{ and } m = j, \\ 0, k \ne i \text{ or } m \ne j. \end{cases} \tag{6.43}$$

Next, let us consider the sets

$$E_i = \left\{u \in [0, 1] : u = \sum_{j=1}^{\infty} \alpha_j 2^{-j}, \ \ \alpha_i = 0\right\}, \quad E_i^c = [0, 1] \backslash E_i.$$

It is easy to check that

$$r_i(t \oplus u) = \begin{cases} r_i(t), u \in E_i, \\ -r_i(t), u \in E_i^c. \end{cases}$$

Therefore, taking into account commutativity equations $T_u Q = Q T_u$, $u \in [0,1]$, we get

$$Q_{i,j}(T_u f) = \begin{cases} Q_{i,j}(f), u \in E_i \cap E_j \text{ or } u \in E_i^c \cap E_j^c \\ -Q_{i,j}(f), u \in E_i \cap E_j^c \text{ or } u \in E_i^c \cap E_j. \end{cases}$$

Since for arbitrary $1 \le i < j < \infty$

$$m(E_i \cap E_j) = m(E_i^c \cap E_j) = m(E_i \cap E_j^c) = m(E_i^c \cap E_j^c) = \frac{1}{4},$$

then

$$\int_{E_i \cap E_j} Q_{i,j}(T_u f)\, du = \int_{E_i^c \cap E_j^c} Q_{i,j}(T_u f)\, du = \frac{1}{4} Q_{i,j}(f)$$

and

$$\int_{E_i \cap E_j^c} Q_{i,j}(T_u f)\, du = \int_{E_i^c \cap E_j} Q_{i,j}(T_u f)\, du = -\frac{1}{4} Q_{i,j}(f).$$

Thanks to the boundedness of the functionals $Q_{i,j}$ on X, they can be taken out of the integral sign in the last equations. Hence, we have

$$\begin{aligned} Q_{i,j}(f) = Q_{i,j}\Big(&\int_{E_i \cap E_j} T_u f\, du + \int_{E_i^c \cap E_j^c} T_u f\, du \\ &- \int_{E_i \cap E_j^c} T_u f\, du - \int_{E_i^c \cap E_j} T_u f\, du \Big). \end{aligned} \tag{6.44}$$

Furthermore, one can easily see that

$$\{s \in [0,1] : s = t \oplus u, \quad u \in E_i\} = \begin{cases} E_i, t \in E_i, \\ E_i^c, t \in E_i^c \end{cases}$$

$$\{s \in [0,1] : s = t \oplus u, \quad u \in E_i^c\} = \begin{cases} E_i^c, t \in E_i, \\ E_i, t \in E_i^c. \end{cases}$$

Thus, since the transformation ω_u preserves the Lebesgue measure, we get

$$\begin{aligned} \int_{E_i \cap E_j} T_u f(t)\, du = &\int_{E_i \cap E_j} f(s)\, ds \cdot \chi_{E_i \cap E_j}(t) + \int_{E_i^c \cap E_j^c} f(s)\, ds \cdot \chi_{E_i^c \cap E_j^c}(t) \\ &+ \int_{E_i \cap E_j^c} f(s)\, ds \cdot \chi_{E_i \cap E_j^c}(t) + \int_{E_i^c \cap E_j} f(s)\, ds \cdot \chi_{E_i^c \cap E_j}(t), \end{aligned}$$

$$\int_{E_i^c\cap E_j} T_u f(t)\,du = \int_{E_i^c\cap E_j} f(s)\,ds\cdot\chi_{E_i\cap E_j}(t) + \int_{E_i\cap E_j^c} f(s)\,ds\cdot\chi_{E_i^c\cap E_j^c}(t)$$
$$+ \int_{E_i^c\cap E_j^c} f(s)\,ds\cdot\chi_{E_i\cap E_j^c}(t) + \int_{E_i\cap E_j} f(s)\,ds\cdot\chi_{E_i^c\cap E_j}(t),$$

$$\int_{E_i\cap E_j^c} T_u f(t)\,du = \int_{E_i\cap E_j^c} f(s)\,ds\cdot\chi_{E_i\cap E_j}(t) + \int_{E_i^c\cap E_j} f(s)\,ds\cdot\chi_{E_i^c\cap E_j^c}(t)$$
$$+ \int_{E_i\cap E_j} f(s)\,ds\cdot\chi_{E_i\cap E_j^c}(t) + \int_{E_i^c\cap E_j^c} f(s)\,ds\cdot\chi_{E_i^c\cap E_j}(t),$$

and

$$\int_{E_i^c\cap E_j^c} T_u f(t)\,du = \int_{E_i^c\cap E_j^c} f(s)\,ds\cdot\chi_{E_i\cap E_j}(t) + \int_{E_i\cap E_j} f(s)\,ds\cdot\chi_{E_i^c\cap E_j^c}(t)$$
$$+ \int_{E_i^c\cap E_j} f(s)\,ds\cdot\chi_{E_i\cap E_j^c}(t) + \int_{E_i\cap E_j^c} f(s)\,ds\cdot\chi_{E_i^c\cap E_j}(t).$$

Combining these relations together with (6.44) and the obvious equation $r_i(t) = \chi_{E_i}(t) - \chi_{E_i^c}(t)$, we have

$$Q_{i,j}(f) = Q_{i,j}\Big(\int_0^1 f(s)r_i(s)r_j(s)\,ds\cdot r_i(t)r_j(t)\Big) = \int_0^1 f(s)r_i(s)r_j(s)\,ds,$$

that is, $Q_{i,j}(f) = c_{i,j}(f)$ for all $1 \le 1 < j < \infty$ (see (6.41)). As a result, $Q = P_{\text{ch}}$, and the proof is completed. □

Analogous assertion holds also for the multiple Rademacher system $\{r_i(s)r_j(t)\}_{i,j=1}^{\infty}$. Since the proof of this fact is similar to that of Theorem 6.7, we skip it. Note only that instead of the projection P_{ch} (resp. the transformations w_u of $I = [0, 1]$) it needs to consider the projection P'_{ch} defined by

$$P'_{\text{ch}} f(s,t) = \sum_{i,j=1}^{\infty} \int_0^1\int_0^1 x(u,v)r_i(u)r_j(v)\,du\,dv\cdot r_i(s)r_j(t)$$

(resp. the transformations $\psi_{u,v}(s,t) = (s\oplus u, t\oplus v)$, $0 \le u, v \le 1$, of the square $I\times I$) (see [17]).

Theorem 6.8 *The subspace of all functions from a s.s. $X(I\times I)$, representable in the form* (6.16)*, is complemented in $X(I\times I)$ if and only if $H \subset X \subset H'$.*

Remark 6.4 In a analogous way it may be shown that the subspace, generated by the Rademacher chaos of an arbitrary integer order d (see (6.1)) is complemented in a s.s. X if and only if $H_d \subset X \subset H_d'$. Here, as above (see Remark 6.3), H_d is the closure of L_∞ in the Orlicz space $L_{N_{2/d}}$, corresponding to an Orlicz function equivalent to the function $N_{2/d}(t) = \exp(t^{2/d}) - 1$, and H_d' is the Köthe dual space for H_d.

Comments and References The origins of the theory of chaoses, generated by systems of independent r.v.'s, go back to the work of Wiener [285], published as long as in 1941 (see also [286]). Since then the random polylinear forms and polynomial chaos, introduced in [285], became important objects of research reflected in vast literature. We mention here, first of all, the monographs devoted, wholly or in part, to this topic: [177, Chapter 4] by Ledoux and Talagrand, [174, Chapter 6] by Kwapień and Woyczyński, [104] by De la Peña and Giňe (see also bibliography therein). On the one hand, these books contain general methods of studying chaoses (maximal inequalities, contraction principles, decoupling methods, moment and distribution inequalities, convergence theorems), on the other hand, the reader can meet here various properties of some specific chaoses (first of all, Rademacher and Gaussian). And finally, subsystems of the Walsh system, indexed by sets of a certain combinatorial dimension, integer or fractional (in particular, the Rademacher chaos considered here) are the main object of the remarkable monograph [76] by Blei.

The purpose of this chapter, whose contents is taken mainly from the papers [15] and [20] due to Astashkin, is the study of properties of the Rademacher chaos in general s.s.'s.

Theorem 6.2 has been proved in [58]. For a close result, see [174, Corollary 6.7.2]; moreover, Chapter 6 of the monograph [174] contains some results of such a sort for more general chaoses. The condition of X to be an interpolation space with respect to the couple (L_1, L_∞) in Theorem 6.3 is not too restrictive. Indeed, in view of the well-known results (see e.g. [168, § II.4]) all the most important in applications s.s.'s (Orlicz, Lorentz, Marcinkiewicz, etc.) possess the above interpolation property. At the same time, as was recently shown in [47], this interpolation assumption is in fact superfluous: the Rademacher chaos of arbitrary order is a basic sequence in *every* s.s.

The heart of this chapter is Theorem 6.4, which gives necessary and sufficient conditions, under which the Rademacher chaos is an unconditional basic sequence in a s.s. This criterion, proved in [20], is formulated in fairly comprehensible terms of embeddings similar to as the Rodin–Semenov Theorem 2.3 is. Its proof, presented here, is essentially based on using the decoupling method, the point of which is the passage from the Rademacher chaos to the Rademacher multiple system and vice versa. As being an important tool in research, this method is widely represented in literature. In addition to the above-cited monographs, we mention here, without pretending to completeness, the papers [105–107, 172], where various decoupling moment and distribution inequalities in vector-valued setting have been obtained, the paper [149], where the possibility of applying the decoupling techniques to studying the Rademacher chaos in quasi-normed spaces has been examined, as well as the papers [84] and [186], containing interesting applications of the latter method to Banach space theory and harmonic analysis. Regarding to the behaviour of the Rademacher chaos, as well as of the multiple Rademacher system in the L_p-spaces, see, for instance, [182, Proposition 2.d.6], [246], and [77]. In the last of these papers, Blei and Janson have revealed deep connections between combinatorial dimension of a subset $\mathcal{M}$ of $\mathbb{N}^d := \mathbb{N} \times \mathbb{N} \times \ldots \times \mathbb{N}$ (d factors) and a version of the central limit theorem, which is satisfied by the sequence of Walsh functions, indexed by $\mathcal{M}$. An extension of Theorem 6.4 to a Rademacher chaos of an arbitrary integer combinatorial dimension has been obtained recently in [47]. There is a number of results close to Lemma 6.3 (see e.g. [90], [271], and [78]). In the same papers, the reader can find examples of s.s.'s, in which a similar assertion fails to hold.

The concept of a system of random unconditional convergence (RUC-system), which is a natural generalization of the notion of an unconditional basic sequence, has been introduced in the paper [74]. Last time it is intensively studied, in particular, within the framework of the theory of s.s.'s [42, 111, 263, 271].

Inequality (6.25) from Proposition 6.2 was proved by Littlewood as early as in 1930. Accordingly, it is usually called the 4/3 Littlewood inequality. One can say also that the sequence $\{r_i(s)r_j(t)\}_{1 \le i < j < \infty}$ is a Sidon system of order $4/3$ (from this point of view, a classical Sidon system is a Sidon system of order 1). The proof of this inequality, presented here, is borrowed from Section 2.5 of the monograph [76]. There is also its extension to systems of arbitrary order; the d-fold multiple Rademacher sequence is a Sidon system of order $2d/(d+1)$ (see [76, § 7.10, Theorem 34] or [116]). Moreover, from the above-mentioned decoupling inequalities (see e.g. [105, 106]) it follows that a similar result holds also for the Rademacher chaos of arbitrary order d. Corollary 6.4, obtained in [15], gives a quantitative estimation of the lack of classical sidonicity (i.e., of order 1) of the Rademacher chaos, in a certain sense answering to a question from the paper [246] concerning to geometrical properties of the subspace, spanned by this chaos in L_∞. In a different way, the result of Corollary 6.5 has been proved in [246, Theorem 4].

Finally, Theorem 6.7, which gives a criterion of the complementability of the closed linear span of the Rademacher chaos in a s.s., has been obtained in [15].

Chapter 7
The Comparison of Systems of Random Variables

The main purpose of this chapter is comparing systems of measurable functions (or r.v.'s) with the "model" Rademacher system. To clarify the meaning of the word "comparison", we introduce the following definitions.

Definition 7.1 We shall say that a system of r.v.'s $\{f_i\}_{i=1}^{\infty}$ defined on a probability space $(\Omega, \Sigma, \mathbb{P})$ *is dominated in distribution* by a system of r.v.'s $\{g_i\}_{i=1}^{\infty}$ defined on a probability space $(\Omega', \Sigma', \mathbb{P}')$ ($\{f_i\} \overset{\mathbb{P}}{\prec} \{g_i\}$) if there exists $C > 0$ such that for arbitrary $m \in \mathbb{N}$, $a_i \in \mathbb{R}$, $i = 1, 2, \ldots, m$, and $z > 0$ we have

$$\mathbb{P}\Big\{\omega \in \Omega : \Big|\sum_{i=1}^{m} a_i f_i(\omega)\Big| > z\Big\} \le C\mathbb{P}'\Big\{\omega' \in \Omega' : \Big|\sum_{i=1}^{m} a_i g_i(\omega')\Big| > C^{-1}z\Big\}.$$

Definition 7.2 Systems of random variables $\{f_i\}$ and $\{g_i\}$ are *equivalent in distribution* (we write $\{f_i\} \overset{\mathbb{P}}{\sim} \{g_i\}$) if both $\{f_i\} \overset{\mathbb{P}}{\prec} \{g_i\}$ and $\{g_i\} \overset{\mathbb{P}}{\prec} \{f_i\}$. More explicitly, $\{f_i\} \overset{\mathbb{P}}{\sim} \{g_i\}$ if for some $C > 0$ and for any $m \in \mathbb{N}$, $a_i \in \mathbb{R}$, $i = 1, 2, \ldots, m$, and $z > 0$

$$\begin{aligned} C^{-1}\mathbb{P}\Big\{\omega \in \Omega : \Big|\sum_{i=1}^{m} a_i f_i(\omega)\Big| > Cz\Big\} &\le \mathbb{P}'\Big\{\omega' \in \Omega' : \Big|\sum_{i=1}^{m} a_i g_i(\omega')\Big| > z\Big\} \\ &\le C\mathbb{P}\Big\{\omega \in \Omega : \Big|\sum_{i=1}^{m} a_i f_i(\omega)\Big| > C^{-1}z\Big\}. \end{aligned}$$

Later on, we shall need similar definitions also in the case of vector-valued sums.

Definition 7.3 We say that a system of r.v.'s $\{f_i\}_{i=1}^{\infty}$ defined on a probability space $(\Omega, \Sigma, \mathbb{P})$ *is dominated in distribution in the vector sense* by a system of r.v.'s

S. V. Astashkin, *The Rademacher System in Function Spaces*,
https://doi.org/10.1007/978-3-030-47890-2_7

$\{g_i\}_{i=1}^{\infty}$ defined on $(\Omega', \Sigma', \mathbb{P}')$ ($\{f_i\} \overset{\vec{\mathbb{P}}}{\prec} \{g_i\}$) whenever there is a constant $C > 0$ such that for every Banach space F, arbitrary $m \in \mathbb{N}$, $x_i \in F$, $i = 1, 2, \ldots, m$, and $z > 0$ it holds

$$\mathbb{P}\Big\{\omega \in \Omega : \Big\|\sum_{i=1}^{m} x_i f_i(\omega)\Big\|_F > z\Big\} \leq C\mathbb{P}'\Big\{\omega' \in \Omega' : \Big\|\sum_{i=1}^{m} x_i g_i(\omega')\Big\|_F > C^{-1}z\Big\}.$$

Systems of r.v.'s $\{f_i\}$ and $\{g_i\}$ will be called *equivalent in distribution in the vector sense* ($\{f_i\} \overset{\vec{\mathbb{P}}}{\sim} \{g_i\}$) if simultaneously $\{f_i\} \overset{\vec{\mathbb{P}}}{\prec} \{g_i\}$ and $\{g_i\} \overset{\vec{\mathbb{P}}}{\prec} \{f_i\}$.

Here, we find sharp conditions, under which a general system of r.v.'s is dominated in distribution by the Rademacher system or equivalent to the latter one. In particular, we shall show that a uniformly bounded system $\{f_i\}$ satisfies the relation $\{f_i\} \overset{\mathbb{P}}{\prec} \{r_i\}$ if and only if any series with respect to $\{f_i\}$ with coefficients from ℓ_2 is exponentially square integrable. More formally, the latter condition means that from $(a_i)_{i=1}^{\infty} \in \ell_2$ it follows

$$\exp\Big(\sum_{i=1}^{\infty} a_i f_i\Big)^2 \in L_1(\Omega)$$

(cf. Theorem 2.2). In the proof of this and other results of the chapter we still make use of the notions and tools of the interpolation theory of operators. In particular, by exploiting Peetre's $\mathcal{K}$-functional for some special Banach couples, we shall find useful necessary and sufficient conditions for the equivalence in distribution of an arbitrary system of r.v.'s to the Rademacher system both in the scalar and vector sense. In the final sections, we apply the results obtained to studying multiplicative systems and sequences of characters defined on a compact Abelian group.

We start with some general properties of vector-valued sums of r.v.'s, as well as moment and distribution comparison inequalities for them.

7.1 A Contraction Principle for the Rademacher Sequence

As we know (see Proposition 2.2), the Rademacher functions form a symmetric and hence an unconditional (with constant 1) sequence in every symmetric function space. This means that the inequalities $|a_i| \leq |b_i|$, $i = 1, 2, \ldots$, $a_i, b_i \in \mathbb{R}$, imply the similar estimate for the norms $\|\sum_{i=1}^{n} a_i r_i\|_X$ and $\|\sum_{i=1}^{n} b_i r_i\|_X$ in every s.s. X. It turns out that vector-valued Rademacher sums possess a similar property.

Theorem 7.1 (Contraction Principle) *Let $\Phi : \mathbb{R}_+ \to \mathbb{R}_+$ be a convex function. Then, for every Banach space F, arbitrary $n = 1, 2, \ldots$, $x_i \in F$, and $\alpha_i \in \mathbb{R}$ such* *that* $|\alpha_i| \le 1$, $i = 1, 2, \ldots, n$, *it holds*

$$\mathbb{E}\Phi\Big(\Big\|\sum_{i=1}^{n} \alpha_i x_i r_i\Big\|_F\Big) \le \mathbb{E}\Phi\Big(\Big\|\sum_{i=1}^{n} x_i r_i\Big\|_F\Big). \tag{7.1}$$

Moreover, for each $\tau > 0$ *we have*

$$m\Big\{\Big\|\sum_{i=1}^{n} \alpha_i x_i r_i\Big\|_F > \tau\Big\} \le 2m\Big\{\Big\|\sum_{i=1}^{n} x_i r_i\Big\|_F > \tau\Big\}. \tag{7.2}$$

Proof According to the hypothesis of the theorem, the mapping

$$(\alpha, \ldots, \alpha_n) \mapsto \mathbb{E}\Phi\Big(\Big\|\sum_{i=1}^{n} \alpha_i x_i r_i\Big\|_F\Big),$$

defined on the cube

$$[-1, +1]^n := \{(\alpha_1, \ldots, \alpha_n) \in \mathbb{R}^n : |\alpha_i| \le 1, i = 1, \ldots, n\},$$

is convex. Therefore, it attains the maximal value at one of the extreme points of the set $[-1, +1]^n$, i.e., when $\alpha_i = \pm 1$, $i = 1, \ldots, n$. Since the Rademacher functions are independent, symmetrically and identically distributed, the left-hand side of inequality (7.1) for all such values of α_i is equal exactly to its right-hand side. Thus, (7.1) is proved.

Next, in view of the above-mentioned properties of the Rademacher functions, we can assume that $\alpha_1 \ge \cdots \ge \alpha_n \ge 0$. Then, setting $S_k := \sum_{i=1}^{k} x_i r_i$, $k = 1, \ldots, n$, and $\alpha_{n+1} = 0$, we have

$$\sum_{i=1}^{n} \alpha_i x_i r_i = \sum_{k=1}^{n} \alpha_k (S_k - S_{k-1}) = \sum_{k=1}^{n} (\alpha_k - \alpha_{k+1}) S_k.$$

Hence,

$$\Big\|\sum_{i=1}^{n} \alpha_i x_i r_i\Big\|_F \le \max_{k=1,\ldots,n} \|S_k\|_F,$$

and applying Levy's maximal inequality (see Theorem A.2), we get (7.2). □

Now, combining Theorem 7.1 with the randomization against a Rademacher sequence, we are in position to prove the following comparison test for sequences of independent r.v.'s, which will be used repeatedly later.

Theorem 7.2 (Kwapień–Rychlik Inequality) *Let $\{f_i\}_{i=1}^{\infty}$ and $\{g_i\}_{i=1}^{\infty}$ be two sequences of independent symmetrically distributed r.v.'s such that*

$$\mathbb{P}\{|g_i| > z\} \leq \mathbb{P}\{|f_i| > z\} \tag{7.3}$$

for all $z > 0$ and $i \in \mathbb{N}$.

Then, for every Banach space F, all $n \in \mathbb{N}$ and $x_i \in F$, $i = 1, 2, \ldots, n$, we have

$$\mathbb{P}\Big\{\Big\|\sum_{i=1}^{n} x_i g_i\Big\|_F > z\Big\} \leq 2\mathbb{P}\Big\{\Big\|\sum_{i=1}^{n} x_i f_i\Big\|_F > z\Big\}.$$

Moreover, if $\Phi : \mathbb{R}_+ \to \mathbb{R}_+$ is a convex function, then

$$\mathbb{E}\Phi\Big(\Big\|\sum_{i=1}^{n} x_i g_i\Big\|_F\Big) \leq \mathbb{E}\Phi\Big(\Big\|\sum_{i=1}^{n} x_i f_i\Big\|_F\Big).$$

Proof Observe that from inequality (7.3) and definition of the decreasing rearrangement it follows $g_i^*(t) \leq f_i^*(t)$, $i = 1, 2, \ldots$, for all $t \in (0, 1]$. Let us define on the product probability space $([0, 1]^{\infty}, m_{\infty}) := \prod_{i=1}^{\infty}([0, 1], m_i)$, where m_i is the usual Lebesgue measure on $[0, 1]$, $i = 1, 2, \ldots$, two sequences of r.v.'s $\{\tilde{f}_i\}_{i=1}^{\infty}$ and $\{\tilde{g}_i\}_{i=1}^{\infty}$ as follows: if $\bar{t} := (t_1, t_2, \ldots) \in \prod_{i=1}^{\infty}[0, 1]$, then $\tilde{f}_i(\bar{t}) := f_i^*(t_i)$ and $\tilde{g}_i(\bar{t}) := g_i^*(t_i)$. It is easy to see that we have the following: (1) each of these sequences consists of independent r.v.'s; (2) $\tilde{f}_i$ (resp. $\tilde{g}_i$) has the same distribution as $|f_i|$ (resp. $|g_i|$); (3) $\tilde{g}_i \leq \tilde{f}_i$ a.e., $i = 1, 2, \ldots$

Suppose that $\{r_i\}$ is the Rademacher sequence independent both with respect to $\{f_i\}$ and $\{g_i\}$. Since the sequences $\{f_i\}$ and $\{g_i\}$ are symmetrically distributed, they have the same distribution as the sequences $\{r_i|f_i|\}$ and $\{r_i|g_i|\}$, respectively. Therefore, by Fubini's theorem and Theorem 7.1 (see inequality (7.2)), for every $z > 0$ we have

$$\begin{aligned}
\mathbb{P}\Big\{\Big\|\sum_{i=1}^{n} x_i g_i\Big\|_F > z\Big\} &= (m \times \mathbb{P})\Big\{(s, \omega) \in [0, 1] \times \Omega : \Big\|\sum_{i=1}^{n} x_i r_i(s)|g_i(\omega)|\Big\|_F > z\Big\} \\
&= (m \times m_{\infty})\Big\{(s, \bar{t}) \in [0, 1] \times [0, 1]^{\infty} : \Big\|\sum_{i=1}^{n} x_i r_i(s)\tilde{g}_i(\bar{t})\Big\|_F > z\Big\} \\
&= \int_{[0,1]^{\infty}} m\Big\{s \in [0, 1] : \Big\|\sum_{i=1}^{n} x_i r_i(s)\frac{\tilde{g}_i(\bar{t})}{\tilde{f}_i(\bar{t})}\tilde{f}_i(\bar{t})\Big\|_F > z\Big\}\, dm_{\infty} \\
&\leq 2\int_{[0,1]^{\infty}} m\Big\{s \in [0, 1] : \Big\|\sum_{i=1}^{n} x_i r_i(s)\tilde{f}_i(\bar{t})\Big\|_F > z\Big\}\, dm_{\infty} \\
&= 2\mathbb{P}\Big\{\Big\|\sum_{i=1}^{n} x_i f_i\Big\|_F > z\Big\}.
\end{aligned}$$

Thus, the first inequality is proved. The second one can be obtained similarly by using inequality (7.1) from Theorem 7.1. □

In a similar way, one can prove also the next comparison theorem, where one of the sequences formed by the Rademacher functions.

Theorem 7.3 *Let $\Phi : \mathbb{R}_+ \to \mathbb{R}_+$ be a convex function, and let $\{\eta_i\}_{i=1}^{\infty}$ be an arbitrary sequence of independent symmetrically distributed r.v.'s such that $\mathbb{E}|\eta_i| < \infty$ for each $i = 1, 2, \ldots$ Then, for every Banach space F, any $n = 1, 2, \ldots$ and $x_i \in F$, $i = 1, 2, \ldots, n$, we have*

$$\mathbb{E}\Phi\Big(\inf_{i=1,2,\ldots} \mathbb{E}|\eta_i| \cdot \Big\| \sum_{i=1}^{n} x_i r_i \Big\|_F\Big) \le \mathbb{E}\Phi\Big(\Big\| \sum_{i=1}^{n} x_i \eta_i \Big\|_F\Big). \tag{7.4}$$

Proof Without loss of generality, we can assume that the sequences $\{\eta_i\}$ and $\{r_i\}$ are independent. Then, by condition, $\{\eta_i\}$ has the same distribution as the sequence $\{r_i \cdot |\eta_i|\}$. Hence, from Fubini's theorem, the Jensen inequality (see Appendix A), and Theorem 7.1 it follows

$$\begin{aligned}
\mathbb{E}\Phi\Big(\Big\| \sum_{i=1}^{n} x_i \eta_i \Big\|_F\Big) &= \mathbb{E}\Phi\Big(\Big\| \sum_{i=1}^{n} x_i r_i |\eta_i| \Big\|_F\Big) \\
&\ge \mathbb{E}\Phi\Big(\Big\| \sum_{i=1}^{n} x_i r_i \mathbb{E}|\eta_i| \Big\|_F\Big) \\
&\ge \mathbb{E}\Phi\Big(\inf_{i=1,2,\ldots} \mathbb{E}|\eta_i| \Big\| \sum_{i=1}^{n} x_i r_i \Big\|_F\Big).
\end{aligned}$$
□

In particular, the latter result implies that vector-valued Gaussian sums dominate (up to a constant) the corresponding Rademacher ones.

Corollary 7.1 *If $\Phi : \mathbb{R}_+ \to \mathbb{R}_+$ is a convex function and $\{\mathfrak{g}_i\}_{i=1}^{\infty}$ is a sequence of independent standard Gaussian r.v.'s, then for every Banach space F and any $x_i \in F$ we have*

$$\mathbb{E}\Phi\Big(\Big\| \sum_{i=1}^{n} x_i r_i \Big\|_F\Big) \le \mathbb{E}\Phi\Big(\Big(\frac{\pi}{2}\Big)^{1/2} \Big\| \sum_{i=1}^{n} x_i \mathfrak{g}_i \Big\|_F\Big). \tag{7.5}$$

Proof It suffices to apply Theorem 7.3, taking into account that $\mathbb{E}|\mathfrak{g}_i| = (2/\pi)^{1/2}$ (see Appendix A). □

Remark 7.1 In connection with Theorem 7.3, one can ask whether the opposite inequality to (7.5) holds? In order to find out this, we shall apply the contraction principle stated in Theorem 7.1 for $\Phi(u) = u$, assuming that the sequence $\{\mathfrak{g}_i\}$ is independent with respect to $\{r_i\}$. We have

$$\mathbb{E}_r \left\| \sum_{i=1}^{n} x_i \mathfrak{g}_i r_i \right\|_F \leq \max_{i=1,\dots,n} |\mathfrak{g}_i| \mathbb{E}_r \left\| \sum_{i=1}^{n} x_i r_i \right\|_F \tag{7.6}$$

(here, the symbol $\mathbb{E}_r$ denotes integration against $\{r_i\}$). Show that

$$K^{-1}(\ln(n+1))^{1/2} \leq \mathbb{E} \max_{i=1,\dots,n} |\mathfrak{g}_i| \leq K(\ln(n+1))^{1/2} \tag{7.7}$$

for some $K > 0$ and all $n \in \mathbb{N}$. Indeed, first from Fubini's theorem for every $u > 0$ it follows

$$\begin{aligned}\mathbb{E} \max_{i=1,\dots,n} |\mathfrak{g}_i| &\leq u + n \int_u^\infty \mathbb{P}\{|\mathfrak{g}| > t\}\, dt = u + \frac{2n}{\sqrt{2\pi}} \int_u^\infty \int_t^\infty e^{-z^2/2}\, dz\, dt \\ &\leq u + \frac{2n}{\sqrt{2\pi}} \int_u^\infty z e^{-z^2/2}\, dz = u + \frac{2n}{\sqrt{2\pi}} \exp(-u^2/2).\end{aligned}$$

Hence, choosing $u = (2 \ln n)^{1/2}$, we come to the upper estimate in (7.7).

As to the lower estimate, since $\mathbb{E}|\mathfrak{g}_i| = (2/\pi)^{1/2}$, we can assume that n is sufficiently large. Thanks to the fact that $\mathfrak{g}_i$, $i = 1, 2, \dots, n$, are independent and identically distributed, for each $M > 0$ we have

$$\mathbb{E} \max_{i=1,\dots,n} |\mathfrak{g}_i| \geq \int_0^M \left(1 - (1 - m\{|\mathfrak{g}| > z\})^n\right) dz \geq M \left(1 - (1 - m\{|\mathfrak{g}| > M\})^n\right).$$

Moreover, taking into account that

$$m\{|\mathfrak{g}| > M\} = \frac{2}{\sqrt{2\pi}} \int_M^\infty e^{-z^2/2}\, dz \geq \frac{2}{\sqrt{2\pi}} e^{-(M+1)^2/2}$$

and choosing $M = (\ln(n+1))^{1/2}$, for all n sufficiently large we get

$$m\{|\mathfrak{g}| > M\} \geq \frac{2}{\sqrt{2\pi}} \cdot \frac{1}{n+1} \geq \frac{1}{3n}.$$

Thus,

$$\mathbb{E} \max_{i=1,\dots,n} |\mathfrak{g}_i| \geq M \left(1 - \left(1 - \frac{1}{3n}\right)^n\right) \geq \left(1 - e^{-1/3}\right) (\ln(n+1))^{1/2},$$

and so estimate (7.7) is proved.

Let us return to inequality (7.6). Since the sequences $\{\mathfrak{g}_i r_i\}$ and $\{\mathfrak{g}_i\}$ have the same distribution, integrating (7.6) against $\{\mathfrak{g}_i\}$ and again applying Fubini's theorem, by (7.7), we obtain

$$\mathbb{E}\Big\|\sum_{i=1}^{n} x_i \mathfrak{g}_i\Big\|_F \le K(\ln(n+1))^{1/2}\mathbb{E}\Big\|\sum_{i=1}^{n} x_i r_i\Big\|_F. \tag{7.8}$$

Of course, it is *wrong* to say that the latter inequality is completely analogous to inequality (7.5), because the constant in its right-hand side depends on $n \in \mathbb{N}$. It turns out that in the case of an arbitrary Banach space, inequality (7.8) cannot be significantly improved. In order to show this, it suffices to assume that x_i, $i = 1, 2, \dots$, in (7.8) form the unit vector basis of the space c_0. Indeed, on the one hand, we obviously have

$$\mathbb{E}\Big\|\sum_{i=1}^{n} x_i r_i\Big\|_{\ell_\infty} = \mathbb{E}\max_{i=1,\dots,n} |r_i| = 1.$$

On the other hand, according to (7.7),

$$\mathbb{E}\Big\|\sum_{i=1}^{n} x_i \mathfrak{g}_i\Big\|_{\ell_\infty} = \mathbb{E}\max_{i=1,\dots,n} |\mathfrak{g}_i| \ge K^{-1}(\ln(n+1))^{1/2}, \quad n = 1, 2, \dots$$

Thus, this negative result holds for every Banach space, which contains subspaces isomorphic to finite-dimensional spaces ℓ_∞^n with arbitrarily large n. However, if F does not have the latter property, inequality (7.8) is true with a constant independent of n (but depending on the space F) (see [177, Theorem 9.16, Proposition 9.14, and Formula (9.12)]).

Next, we shall need also a comparison principle, which shows that, under certain conditions, moment comparison inequalities for r.v.'s imply the corresponding ones for their distributions.

7.2 A Comparison Principle for Distribution Functions of r.v.'s

Let f be a nonnegative r.v. defined on a probability space $(\Omega, \Sigma, \mathbb{P})$. Then the mapping $p \mapsto (\mathbb{E}f^p)^{1/p}$ is an increasing continuous function on the interval $[1, \infty)$. Therefore, we can put

$$\alpha_f := \mathbb{E}f \quad \text{and} \quad \beta_f := \lim_{p\to\infty} (\mathbb{E}f^p)^{1/p}.$$

Moreover, if $\alpha_f < \beta_f$, then for every $z \in [\alpha_f, \beta_f)$ there is $p \in [1, \infty)$ such that $(\mathbb{E}f^p)^{1/p} = z$.

Proposition 7.1 *Suppose that f is a nonnegative r.v. defined on a probability space $(\Omega, \Sigma, \mathbb{P})$ such that $\mathbb{E}f^p < \infty$ for each $p < \infty$. Assume that there exist constants $A > 0$ and $B > 0$ such that for all $p \geq 1$ we have*

$$(\mathbb{E}f^{2p})^{1/(2p)} \leq A(\mathbb{E}f^p)^{1/p} \tag{7.9}$$

and

$$(\mathbb{E}g^p)^{1/p} \leq B(\mathbb{E}f^p)^{1/p} \tag{7.10}$$

for some nonnegative r.v. g defined on a probability space $(\Omega', \Sigma', \mathbb{P}')$. Then, for some $C > 0$, depending only on A and B, and all $z > 0$ it holds

$$\mathbb{P}'\{g > z\} \leq C\mathbb{P}\{f > z/C\}. \tag{7.11}$$

Proof First of all, by the Paley–Zygmund inequality (see Proposition 1.4) and hypothesis (7.9), for any $p \geq 1$ we have

$$\begin{aligned}\mathbb{P}\Big\{f > \frac{1}{3}(\mathbb{E}f^p)^{1/p}\Big\} &\geq \mathbb{P}\{f^p \geq 2^{-p}\mathbb{E}f^p\} \geq (1-2^{-p})^2\frac{(\mathbb{E}f^p)^2}{\mathbb{E}f^{2p}} \\ &\geq (1-2^{-p})^2 A^{-2p} \geq (2A)^{-2p},\end{aligned} \tag{7.12}$$

where the last estimate follows from the elementary inequality $(1-2^{-p})^2 \geq 2^{-2p}$, $p \geq 1$. Hence, from the definition of α_f it follows

$$\mathbb{P}\{f > z\} \geq (2A)^{-2} \text{ if } 0 < z \leq \frac{\alpha_f}{3}. \tag{7.13}$$

On the other hand, applying Chebyshev's inequality and hypothesis (7.10), for arbitrary $p \geq 1$ and $\lambda > 0$ we obtain that

$$\mathbb{P}'\{g \geq \lambda(\mathbb{E}f^p)^{1/p}\} \leq \lambda^{-p}\frac{\mathbb{E}g^p}{\mathbb{E}f^p} \leq \left(\frac{B}{\lambda}\right)^p.$$

In particular, for $\lambda = Be$, this implies

$$\mathbb{P}'\{g \geq Be(\mathbb{E}f^p)^{1/p}\} \leq e^{-p}, \quad p \geq 1. \tag{7.14}$$

Moreover, from (7.14) and the definition of β_f it follows that

$$\mathbb{P}'\{g \geq z\} = 0 \text{ if } z \geq Be\beta_f. \tag{7.15}$$

Next, let us introduce the functions $E(s)$ and $G(s)$ by

$$E(s) := \sup\{p \geq 1 : (\mathbb{E}f^p)^{1/p} \leq s\} \text{ and } G(s) := \inf\{p \geq 1 : (\mathbb{E}f^p)^{1/p} \geq s\}, \quad s > 0$$

(with the convention that $\sup \varnothing = 1$ and $\inf \varnothing = +\infty$). One can readily check that these functions increase and $G(s) \le E(s)$, $s > 0$.

Assume that

$$\frac{\alpha_f}{3} < z < \frac{\beta_f}{3}. \tag{7.16}$$

Then, by the definition of β_f, we have

$$U := \{p \ge 1 : \mathbb{E} f^p \ge (3z)^p\} \ne \varnothing$$

and, according to (7.12), for every $p \in U$

$$\mathbb{P}\{f > z\} \ge \mathbb{P}\{f > \frac{1}{3}(\mathbb{E} f^p)^{1/p}\} \ge (2A)^{-2p}.$$

Consequently, because $G(3z) = \inf U$, we get

$$\mathbb{P}\{f > z\} \ge (2A)^{-2G(3z)},$$

which implies that

$$\mathbb{P}\{f > z\} \ge e^{-C_1 G(3z)}, \tag{7.17}$$

where $C_1 := 2\ln(2A) \ge 1$.

Furthermore, let us show that condition (7.16) assures that

$$C_1 G(3z) \le E(3A^{[C_1]+1} z)$$

(as usual, $[a]$ denotes the integer part of a real number a). Indeed, by (7.16) and the definition of α_f and β_f, we can find $p \ge 1$ such that $(\mathbb{E} f^p)^{1/p} = 3z$, whence $(\mathbb{E} f^{G(3z)})^{1/G(3z)} = 3z$. Therefore, taking into account that $M := 2^{[C_1]+1} \ge C_1$, by (7.9), we get

$$\begin{aligned}(\mathbb{E} f^{C_1 G(3z)})^{1/(C_1 G(3z))} &\le (\mathbb{E} f^{M G(3z)})^{1/(M G(3z))} \\ &\le A^{[C_1]+1} (\mathbb{E} f^{G(3z)})^{1/G(3z)} \\ &= 3z A^{[C_1]+1}.\end{aligned}$$

Combining this inequality with the definition of $E(s)$, we obtain the desired estimate. Thus, from (7.17) it follows that

$$\mathbb{P}\{f > z\} \ge e^{-E(C_2 z)} \text{ if } \frac{\alpha_f}{3} < z < \frac{\beta_f}{3}, \tag{7.18}$$

with $C_2 := 3A^{[C_1]+1}$.

Next, if $z > Be\alpha_f$, then by the definition of α_f, we have

$$V := \left\{p \geq 1 : (\mathbb{E}f^p)^{1/p} \leq \frac{z}{Be}\right\} \neq \varnothing,$$

and so, according to (7.14), for every $p \in V$

$$\mathbb{P}'\{g \geq z\} \leq \mathbb{P}'\{g \geq Be(\mathbb{E}f^p)^{1/p}\} \leq e^{-p}.$$

Since $E(B^{-1}e^{-1}z) = \sup V$, the latter estimate implies

$$\mathbb{P}'\{g \geq z\} \leq e^{-E(B^{-1}e^{-1}z)} \text{ if } z > Be\alpha_f,$$

and hence

$$\mathbb{P}'\{g > z\} \leq e^{-E(B^{-1}e^{-1}z)} \text{ if } z > 2Be\alpha_f. \tag{7.19}$$

Let us check now that inequality (7.11) is fulfilled with $C := \max(2BeC_2, (2A)^2)$. At first, if $z/C \leq \alpha_f/3$, then from (7.13) and the choice of C it follows

$$\mathbb{P}'\{g > z\} \leq 1 \leq (2A)^2\mathbb{P}\left\{f > \frac{z}{C}\right\} \leq C\mathbb{P}\left\{f > \frac{z}{C}\right\}.$$

Suppose next that $\alpha_f/3 < z/C < \beta_f/3$. Since $A \geq 1$, then $C_2 \geq 3$ and $C \geq 6Be$. So, $z > C\alpha_f/3 \geq 2Be\alpha_f$. Therefore, by (7.18), (7.19) and the inequality $Be \leq C/C_2$, we have again

$$\mathbb{P}'\{g > z\} \leq e^{-E(B^{-1}e^{-1}z)} \leq e^{-E(C_2z/C)} \leq C\mathbb{P}\left\{f > \frac{z}{C}\right\}.$$

Finally, in the case when $z/C \geq \beta_f/3$, we get $z \geq 2Be\beta_f$. This inequality, combined with (7.15), implies $\mathbb{P}'\{g > z\} = 0$. Thus, inequality (7.11) holds for all $z > 0$, which completes the proof. □

7.3 Comparison of Systems of r.v.'s with the Rademacher Sequence: The Scalar Case

The main result of this section is a criterion of domination in distribution of an arbitrary system of r.v.'s by the Rademacher sequence. Further, as above, we use the notation

$$\kappa(t, a) := \mathcal{K}(t, a; \ell_1, \ell_2),$$

where $a = (a_i)_{i=1}^{\infty} \in \ell_2$ and $t > 0$. If $a = (a_i)_{i=1}^{m}$ for some $m \in \mathbb{N}$, then we set $\kappa(t, a) = \kappa(t, b)$, where $b = (b_i)_{i=1}^{\infty}$, $b_i = a_i$ if $1 \le i \le m$, and $b_i = 0$ if $i > m$.

Theorem 7.4 *For an arbitrary system $\{g_i\}_{i=1}^{\infty}$ of r.v.'s defined on a probability space $(\Omega, \Sigma, \mathbb{P})$, the following conditions are equivalent:*

(*a*) *there exists $C > 0$, which does not depend on $m \in \mathbb{N}$, $a = (a_i)_{i=1}^{m}$ and $t \in [0, 1]$, such that*

$$\int_0^t \Big(\sum_{i=1}^{m} a_i g_i\Big)^*(s)\, ds \le C\, t\, \kappa(\ln^{1/2}(e/t), a); \tag{7.20}$$

(*b*) *there exists $C_1 > 0$, which does not depend on $m \in \mathbb{N}$, $a = (a_i)_{i=1}^{m}$ and $p \ge 1$, such that*

$$\Big\| \sum_{i=1}^{m} a_i g_i \Big\|_p \le C_1\, \kappa(\sqrt{p}, a); \tag{7.21}$$

(*c*) $\{g_i\} \overset{\mathbb{P}}{\prec} \{r_i\}$.

Moreover, if $\{g_i\}_{i=1}^{\infty}$ is a uniformly bounded system, that is, $|g_i(\omega)| \le D$, $\omega \in \Omega$, $i = 1, 2, \dots$, then each of the conditions (a), (b), (c) is equivalent to the following:

(*d*) *for every sequence $a = (a_i)_{i=1}^{\infty} \in \ell_2$ the function $g = \sum_{i=1}^{\infty} a_i g_i$ belongs to the Orlicz space $L_{N_2}(\Omega)$.*

Recall that, according to Theorem 3.3, with a constant not depending on $a = (a_i)_{i=1}^{\infty} \in \ell_2$, we have

$$\mathcal{K}\Big(t, \sum_{i=1}^{\infty} a_i r_i; L_\infty, G\Big) \asymp \kappa(t, a), \quad t > 0,$$

where G is the closure of L_∞ in the Orlicz space L_{N_2}. In the proof of Theorem 7.4 we need a similar equivalent expression for the $\mathcal{K}$-functional of a Rademacher sum with respect to the Banach couple (L_p, L_∞), $1 \le p < \infty$.

Proposition 7.2 *For every $1 \le p < \infty$, with a constant independent of sequences of coefficients $a = (a_i)_{i=1}^{\infty} \in \ell_2$ and $0 < u \le 1$, we have*

$$\mathcal{K}\Big(u, \sum_{i=1}^{\infty} a_i r_i; L_p, L_\infty\Big) \asymp u^{1/p} \cdot \kappa(\ln^{1/2}(e/u), a). \tag{7.22}$$

Proof First, by equivalence (D.1),

$$\mathcal{K}(u, x; L_p, L_\infty) \asymp \Big(\int_0^{u^p} x^*(s)^p\, ds\Big)^{1/p}, \quad 0 \le u \le 1.$$

Moreover, denoting $x_a = \sum_{i=1}^{\infty} a_i r_i$ and applying Corollary 1.9, we have

$$x_a^*(s) \le \beta^{-1}\kappa(\ln^{1/2}(e/s), a), \quad 0 < s \le 1,$$

and

$$x_a^*(\beta s) \ge \beta\kappa(\ln^{1/2}(e/s), a), \quad 0 < s \le 1,$$

with a universal constant $\beta \in (0, 1)$. From these relations it follows that

$$\int_0^u x_a^*(s)^p\,ds \asymp \int_0^u \kappa(\ln^{1/2}(e/s), a)^p\,ds, \quad 0 < u \le 1,$$

with a constant, depending only on p. Therefore, since $t \mapsto \kappa(t, a)$ is an increasing and concave function, the proof will be completed once we show that, with some constant C_p, it holds

$$\int_0^u \kappa(\ln^{1/2}(e/s), a)^p\,ds \le C_p u \cdot \kappa(\ln^{1/2}(e/u), a)^p, \quad 0 < u \le 1. \tag{7.23}$$

Suppose first that $u = e^{-k}$, $k \in \mathbb{N}$. By using the concavity of $\kappa(t, a)$, we get

$$\begin{aligned}\int_0^{e^{-k}} \kappa(\ln^{1/2}(e/s), a)^p\,ds &= \sum_{i=k}^{\infty}\int_{e^{-i-1}}^{e^{-i}} \kappa(\ln^{1/2}(e/s), a)^p\,ds\\ &\le 2\sum_{i=k}^{\infty}\kappa(i^{1/2}, a)^p\,e^{-i} \le \frac{2\kappa(k^{1/2}, a)^p}{k^{p/2}}\sum_{i=k}^{\infty} i^{p/2}e^{-i}\\ &\le \frac{2e\kappa(k^{1/2}, a)^p}{k^{p/2}}\int_k^{\infty} t^{p/2}\cdot e^{-t}\,dt.\end{aligned} \tag{7.24}$$

Let us estimate the integral from the right-hand side of this inequality. Integrating by parts, we get

$$\begin{aligned}\int_k^{\infty} t^{p/2}\cdot e^{-t}\,dt &= k^{p/2}e^{-k} + \frac{p}{2}\int_k^{\infty} t^{p/2-1}\cdot e^{-t}\,dt\\ &\le k^{p/2}e^{-k} + \frac{p}{2k}\int_k^{\infty} t^{p/2}\cdot e^{-t}\,dt.\end{aligned}$$

whence for all $k > p$

$$\int_k^{\infty} t^{p/2}\cdot e^{-t}\,dt \le \frac{2k}{2k-p}k^{p/2}e^{-k} \le 2k^{p/2}e^{-k}.$$

Otherwise, if $k \le p$, we have

$$\int_k^\infty t^{p/2} \cdot e^{-t}\,dt \le C'_p k^{p/2} e^{-k},$$

where $C'_p := e^p \int_1^\infty t^{p/2} \cdot e^{-t}\,dt > 2$. Combining the last estimates with (7.24), we obtain (7.23) for $u = e^{-k}$, $k \in \mathbb{N}$, with the constant $2eC'_p$. To extend this inequality to the whole interval $(0, 1]$, it suffices to take into account once more that the function $t \mapsto \kappa(t, a)$ is increasing and concave. □

Applying the last proposition in the case $p = 1$ together with the formula for the $\mathcal{K}$-functional $\kappa(t, a)$ from Lemma 1.5, we get

Corollary 7.2 *With a constant, not depending on $a = (a_i)_{i=1}^\infty \in \ell_2$ and $0 < u \le 1$, it holds*

$$\int_0^u \Big(\sum_{i=1}^\infty a_i r_i\Big)^*(t)\,dt \asymp u \left\{ \sum_{i=1}^{[\ln(e/u)]} a_i^* + \ln^{1/2}(e/u) \Big[\sum_{i=[\ln(e/u)]+1}^{\infty} (a_i^*)^2 \Big]^{1/2} \right\}.$$

Proof of Theorem 7.4 **(*a*) ⇒ (*b*)** By Proposition 7.2, inequality (7.20) can be rewritten as follows:

$$\int_0^u \Big(\sum_{i=1}^m a_i g_i\Big)^*(s)\,ds \le C' \int_0^u \Big(\sum_{i=1}^m a_i r_i\Big)^*(s)\,ds, \quad 0 < u \le 1. \tag{7.25}$$

Recall that L_p, $1 \le p \le \infty$, is an interpolation space with constant 2 between the spaces L_1 and L_∞ (see Theorem D.1 and Remark D.1). Hence, by Theorems D.6 and 1.6, (7.21) is an immediate consequence of inequality (7.25).

(*b*) ⇒ (*c*). Applying once more Theorem 1.6, we conclude that inequality (7.21) is equivalent to the following:

$$\Big\| \sum_{i=1}^m a_i g_i \Big\|_p \le B \Big\| \sum_{i=1}^m a_i r_i \Big\|_p,$$

where B does not depend on $p \ge 1$, $m \in \mathbb{N}$, and $a_i \in \mathbb{R}$, $i = 1, 2, \ldots, m$. Therefore, the r.v.'s

$$g(\omega) := \Big| \sum_{i=1}^m a_i g_i(\omega) \Big|, \quad \omega \in \Omega, \quad \text{and} \quad f(t) := \Big| \sum_{i=1}^m a_i r_i(t) \Big|, \quad 0 \le t \le 1,$$

satisfy inequality (7.10) of Proposition 7.1 with constant B. As to (7.9), then this inequality is fulfilled for f with constant $2\sqrt{2}$. Indeed, if $p \geq 2$, by Corollary 4.2,

$$\|f\|_{2p} \leq \sqrt{\frac{2p-1}{p-1}}\|f\|_p \leq 2\|f\|_p.$$

In the remaining case $1 \leq p < 2$, according to Theorems 1.4 and 5.4, we have

$$\|f\|_{2p} \leq \sqrt{2p}\|(a_i)\|_{\ell_2} \leq 2\sqrt{p}\|f\|_1 \leq 2\sqrt{2}\|f\|_p.$$

Finally, applying Proposition 7.1, we obtain (c).

$(c) \Rightarrow (a)$. By condition, for some $A > 0$, not depending on $m \in \mathbb{N}$, $a_i \in \mathbb{R}$, $i = 1, 2, \ldots, m$, and $z > 0$, we have

$$\mathbb{P}\left\{\omega \in \Omega : \left|\sum_{i=1}^{m} a_i g_i(\omega)\right| > z\right\} \leq Am\left\{t \in [0,1] : \left|\sum_{i=1}^{m} a_i r_i(t)\right| > A^{-1}z\right\}. \tag{7.26}$$

Note that the space $L_1[0,1]$, equipped with the norm $\|x\|_u := \int_0^u x^*(s)\,ds$, is symmetric. Therefore, according to Theorem C.3, from inequality (7.26) it follows (7.25) with a constant, which depends only on A. Since by Proposition 7.2 inequality (7.25) implies (7.20), the proof of the equivalence of conditions $(a) - (c)$ is completed.

$(c) \Rightarrow (d)$. First, by condition, we have (7.26). Consequently, since the space L_{N_2} is symmetric, applying Theorems C.3 and 2.2, we get that for all $m \in \mathbb{N}$ and $a = (a_i)_{i=1}^{\infty} \in \ell_2$

$$\left\|\sum_{i=1}^{m} a_i g_i\right\|_{L_{N_2}} \leq C\left\|\sum_{i=1}^{m} a_i r_i\right\|_{L_{N_2}} \leq \sqrt{2e}C\|a\|_{\ell_2}$$

with a constant C depending only on A. This clearly implies that the series $\sum_{i=1}^{\infty} a_i g_i$ converges in $L_{N_2}(\Omega)$ and hence (d) is proved.

Observe that, in the proof of this implication, we did not use the uniform boundedness of the functions g_i, $i = 1, 2\ldots$ (as it is easy to see, the last property follows directly from condition (c)).

$(d) \Rightarrow (a)$. We first check that the linear operator $T(a_i) := \sum_{i=1}^{\infty} a_i g_i$ is bounded from ℓ_2 into $L_{N_2}(\Omega)$.

To this end, we consider the corresponding maximal operator

$$M(a_i)(\omega) := \sup_{n=1,2,..} |T_n(a_i)(\omega)|, \text{ where } T_n(a_i) := \sum_{i=1}^{n} a_i g_i.$$

Obviously, each of the operators T_n, $n = 1, 2, \ldots$, is linear and bounded from ℓ_2 into the space $S(\Omega) := S(\Omega, \Sigma, \mathbb{P})$ of all measurable $\mathbb{P}$-a.e. finite functions on Ω, equipped with the topology of convergence in probability $\mathbb{P}$. Moreover, observe that, by condition, we have $M(a_i) < \infty$ $\mathbb{P}$-a.e. on Ω for every sequence $(a_i) \in \ell_2$. Therefore, by the well-known Banach continuity principle (see e.g. [6, Theorem 2.7]), M is boundedly acts from ℓ_2 in $S(\Omega)$. Clearly, the same is true also for the operator T.

Next, we apply the Closed Graph theorem. Suppose that $(a_{m,i}) \to (a_i)$ in ℓ_2 and $T(a_{m,i}) \to g$ in $L_{N_2}(\Omega)$ as $m \to \infty$. Since L_{N_2} is continuously embedded into $S(\Omega)$, we have also that $T(a_{m,i}) \to g$ in $S(\Omega)$. On the other hand, we know that $T(a_{m,i}) \to T(a_i)$ in $S(\Omega)$. Hence, $g = T(a_i)$, i.e., the graph of T is closed. As a result, T is bounded from ℓ_2 into $L_{N_2}(\Omega)$.

Recall now that, according to Proposition 2.4, the Orlicz space L_{N_2} coincides with the Marcinkiewicz space $M(\varphi)$, where $\varphi(t) = t \ln^{1/2}(e/t)$. Therefore, the boundedness of T from ℓ_2 into $L_{N_2}(\Omega)$ means that

$$\int_0^t \Big(\sum_{i=1}^{\infty} a_i g_i\Big)^*(s)\,ds \le C \cdot t \ln^{1/2}(e/t)\|a\|_{\ell_2},$$

where $C > 0$ does not depend on $a = (a_i)_{i=1}^{\infty} \in \ell_2$ and $0 < t \le 1$.

As usual, $(a_k^*)_{k=1}^{\infty}$ denotes decreasing permutation of the sequence $(|a_i|)$. Let $a_k^* = |a_{i_k}|$, $k = 1, 2, \ldots$ As was already said above, the functional $\int_0^u x^*(s)\,ds$ is a norm on the space $L_1[0, 1]$ for every $0 < u \le 1$. Therefore, by using the last inequality and the uniform boundedness of g_i, $i = 1, 2, \ldots$, for each $j \in \mathbb{N}$ we get

$$\begin{aligned}
\int_0^{e^{-j}} \Big(\sum_{i=1}^{\infty} a_i g_i\Big)^*(s)\,ds &\le \int_0^{e^{-j}} \Big(\sum_{k=1}^{j} a_{i_k} g_{i_k}\Big)^*(s)\,ds + \int_0^{e^{-j}} \Big(\sum_{k=j+1}^{\infty} a_{i_k} g_{i_k}\Big)^*(s)\,ds \\
&\le e^{-j} D \sum_{k=1}^{j} |a_{i_k}| + C e^{-j} \ln^{1/2}(e^{j+1}) \Big(\sum_{k=j+1}^{\infty} |a_{i_k}|^2\Big)^{1/2} \\
&\le \max(D, 2C) e^{-j} \left\{ \sum_{k=1}^{j} a_k^* + \sqrt{j} \Big[\sum_{k=j+1}^{\infty} (a_k^*)^2\Big]^{1/2} \right\} \\
&\le 4 \max(D, 2C) e^{-j} \mathcal{K}(\sqrt{j}, a; \ell_1, \ell_2),
\end{aligned}$$

where the last estimate is a consequence of the Holmstedt inequality, proved in Lemma 1.5.

Thus, (7.20) is fulfilled for $u = e^{-j}$, $j \in \mathbb{N}$. Since the standard reasons based on the concavity of the functions under consideration allow us to extend this inequality to the whole interval [0, 1], the proof is completed. □

Remark 7.2 One can easily see that the condition of the uniform boundedness of r.v.'s g_i, $i = 1, 2, \ldots$, in (*d*) cannot be skipped. Indeed, consider on [0, 1] the functions

$$g_i(t) = \ln^{1/2}(e/t) \cdot \chi_{[2^{-i},2^{-i+1})}(t), \quad i = 1, 2, \dots$$

Clearly, $\{g_i\} \overset{\mathbb{P}}{\not\prec} \{r_i\}$. At the same time, for each sequence $a = (a_i)_{i=1}^{\infty} \in \ell_2$ we have

$$\Big\| \sum_{i=1}^{\infty} a_i g_i \Big\|_{L_{N_2}} \le \| \ln^{1/2}(e/t) \|_{L_{N_2}} \cdot \max_{i=1,2,\dots} |a_i| \le C \|a\|_{\ell_2},$$

that is, $\sum_{i=1}^{\infty} a_i g_i \in L_{N_2}$.

With Theorem 7.4 in hand, we are in position to prove a criterion of the equivalence in distribution of an arbitrary sequence of r.v.'s to the Rademacher sequence. Later on, this result will allow us to give meaningful examples of systems $\{f_n\}_{n=1}^{\infty}$ such that $\{f_n\} \overset{\mathbb{P}}{\sim} \{r_n\}$ and also to find necessary and sufficient conditions, under which a sequence of r.v.'s contains a subsequence equivalent in distribution to the sequence $\{r_n\}$.

Theorem 7.5 *Let $\{f_n\}_{n=1}^{\infty}$ be a sequence of r.v.'s defined on a probability space $(\Omega, \Sigma, \mathbb{P})$. The following conditions are equivalent:*

(i) $\{f_n\} \overset{\mathbb{P}}{\sim} \{r_n\}$;

(ii) *with a constant, which does not depend on $p \ge 1$, $m \in \mathbb{N}$ and $a_n \in \mathbb{R}$, $n = 1, 2, \dots, m$, we have*

$$\Big\| \sum_{n=1}^{m} a_n f_n \Big\|_p \asymp \Big\| \sum_{n=1}^{m} a_n r_n \Big\|_p;$$

(iii) *with a constant, which does not depend on $p \ge 1$, $m \in \mathbb{N}$ and $a = (a_n)_{n=1}^{m} \in \mathbb{R}^m$, it holds*

$$\Big\| \sum_{n=1}^{m} a_n f_n \Big\|_p \asymp \kappa(\sqrt{p}, a).$$

Proof Obviously, the equivalence $(ii) \Leftrightarrow (iii)$ is an immediate consequence of Theorem 1.6. Observe also that the implication $(i) \Rightarrow (ii)$ is straightforward, because of the definition of the equivalence in distribution. Thus, it remains to prove the implication $(ii) \Rightarrow (i)$. Moreover, in view of the preceding theorem, we need to show only that $\{r_n\} \overset{\mathbb{P}}{\prec} \{f_n\}$.

We put

$$f(\omega) := \Big|\sum_{i=1}^{m} a_i f_i(\omega)\Big|, \quad \omega \in \Omega, \quad \text{and} \quad g(t) := \Big|\sum_{i=1}^{m} a_i r_i(t)\Big|, \quad 0 \le t \le 1.$$

It is easy to see that f and g satisfy inequality (7.10) from Proposition 7.1 with the constant B implied in the equivalence from condition (ii). Furthermore, similarly as in the proof of the implication $(b) \Rightarrow (c)$ of Theorem 7.4, we obtain that

$$\|f\|_{2p} \le B\|g\|_{2p} \le 2\sqrt{2}B\|g\|_p \le 2\sqrt{2}B^2\|f\|_p \quad \text{for all} \quad p \ge 1.$$

Therefore, inequality (7.9) of Proposition 7.1 also holds. So, applying this proposition, we conclude that $\{r_n\} \overset{\mathbb{P}}{\prec} \{f_n\}$. □

Remark 7.3 For the future, let us note that a similar result is true also for finite collections of r.v.'s. Moreover, an inspection of the given proof shows that the constants implied in the equivalences from conditions (i)–(iii) of a version of Theorem 7.5 in this case will depend only on each other.

7.4 Comparison of Systems of r.v.'s with the Rademacher Sequence: The Vector Case

In this section we briefly consider some results related to comparing systems of r.v.'s with the Rademacher system in the vector case.

Let $1 \leqslant p < \infty$, and let F be a Banach space. Recall that $\ell_p^w(F)$ denotes the Banach space of all sequences $\{x_n\}_{n=1}^\infty \subset F$ such that $(x^*(x_n))_{n=1}^\infty \in \ell_p$ for every $x^* \in F^*$, equipped with the following natural norm:

$$\|\{x_n\}\|_{\ell_p^w(F)} := \sup_{\|x^*\| \le 1} \|(x^*(x_n))\|_{\ell_p}.$$

Moreover, as above, for every $\{x_n\} \in \ell_2^w(F)$ we set

$$\kappa_w(t, \{x_n\}) := \sup_{\|x^*\| \leqslant 1} \kappa(t, (x^*(x_n))), \quad t > 0,$$

where $\kappa(t, a) = \mathcal{K}(t, a; \ell_1, \ell_2)$, $a = (a_n)_{n=1}^\infty \in \ell_2$.

Theorem 7.6 *Suppose $\{f_n\}_{n=1}^\infty$ is a sequence of r.v.'s defined on a probability space $(\Omega, \Sigma, \mathbb{P})$ such that*

$$\Big(\mathbb{E}\Big\|\sum_{n=1}^{m} x_n f_n\Big\|_F^p\Big)^{1/p} \le C\Big(\int_0^1 \Big\|\sum_{n=1}^{m} x_n r_n(t)\Big\|_F^p dt\Big)^{1/p},$$

where C does not depend on $p \geq 1$, $m \in \mathbb{N}$, *a Banach space* F *and vectors* $x_n \in F$, $n = 1, 2, \ldots, m$.

Then, $\{f_n\} \overset{\vec{\mathbb{P}}}{\prec} \{r_n\}$.

Proof As in the case of the implication $(b) \Rightarrow (c)$ of Theorem 7.4, the proof will be based on using Proposition 7.1.

For a fixed $m \in \mathbb{N}$, any Banach space F and $x_n \in F$, $n = 1, 2, \ldots, m$, we define the r.v.'s

$$r_x(t) := \Big\| \sum_{n=1}^{m} x_n r_n(t) \Big\|_F, \quad t \in [0, 1], \quad \text{and} \quad f_x(\omega) := \Big\| \sum_{n=1}^{m} x_n f_n(\omega) \Big\|_F, \quad \omega \in \Omega.$$

First, according to Theorem 4.5, with a constant, not depending on $p \geq 1$, $m \in \mathbb{N}$, a Banach space F and $x_n \in F$, $n = 1, 2, \ldots, m$, we have

$$(\mathbb{E}{r_x}^p)^{1/p} \asymp W(p, \{x_n\}), \tag{7.27}$$

where for every $p > 0$ and $\{x_n\} \in \ell_2^w(F)$ we put

$$W(p, \{x_n\}) := \mathbb{E}r_x + \kappa_w(\sqrt{p}, \{x_n\}),$$

Moreover, by condition, for some $C_1 > 0$, it holds

$$(\mathbb{E}{f_x}^p)^{1/p} \leq C_1(\mathbb{E}{r_x}^p)^{1/p}, \quad p \geq 1.$$

Therefore, if C_2 is the equivalence constant of (7.27), by the concavity of the function $p \mapsto W(p, \{x_n\})$, we get

$$(\mathbb{E}r_x^{2p})^{1/(2p)} \leq C_2 W(2p, \{x_n\}) \leq \sqrt{2}C_2 W(p, \{x_n\}) \leq \sqrt{2}C_2^2(\mathbb{E}{r_x}^p)^{1/p}, \quad p \geq 1.$$

Thus, the r.v.'s $r_x(t)$ and $f_x(\omega)$ satisfy all the conditions of Proposition 7.1. Hence, with some $C > 0$,

$$\mathbb{P}\{\omega \in \Omega : f_x(\omega) > z\} \leq Cm\left\{t \in [0, 1] : r_x(t) > \frac{z}{C}\right\}, \quad z > 0.$$

Since C is independent of $m \in \mathbb{N}$, a Banach space F and $x_n \in F$, $n = 1, 2, \ldots, m$, the proof is finished. □

Now, the following criterion of the equivalence in distribution in the vector sense to the Rademacher system can be proved exactly in the same way as Theorem 7.5.

Theorem 7.7 *Let* $\{f_n\}_{n=1}^{\infty}$ *be a sequence of r.v.'s defined on a probability space* $(\Omega, \Sigma, \mathbb{P})$. *The following conditions are equivalent:*

(1) $\{f_n\} \overset{\vec{\mathbb{P}}}{\sim} \{r_n\}$;

(2) *with a constant, which does not depend on* $p \geq 1$, $m \in \mathbb{N}$, *a Banach space* F *and* $x_n \in F$, $n = 1, 2, \ldots, m$, *it holds*

$$\left(\mathbb{E}\Big\|\sum_{n=1}^{m} x_n f_n\Big\|_F^p\right)^{1/p} \asymp \left(\int_0^1 \Big\|\sum_{n=1}^{m} x_n r_n(t)\Big\|_F^p dt\right)^{1/p};$$

(3) *with a constant, which does not depend on* $p \geq 1$, $m \in \mathbb{N}$, *a Banach space* F *and* $x_n \in F$, $n = 1, 2, \ldots, m$, *we have*

$$\left(\mathbb{E}\Big\|\sum_{n=1}^{m} x_n f_n\Big\|_F^p\right)^{1/p} \asymp \int_0^1 \Big\|\sum_{n=1}^{m} x_n r_n(t)\Big\|_F dt + \kappa_w(\sqrt{p}, \{x_n\}).$$

In the concluding part of the chapter, we shall apply the results obtained to the study of some special classes of systems of r.v.'s. We start with considering multiplicative systems.

7.5 Multiplicative Systems

Recall that a finite or countable system of r.v.'s $\{f_i\}$ defined on a probability space $(\Omega, \Sigma, \mathbb{P})$ is called *multiplicative* if for any pairwise distinct $i_1, i_2, \ldots, i_k$, $k \in \mathbb{N}$, it holds

$$\mathbb{E}(f_{i_1} \cdot f_{i_2} \cdot \cdots \cdot f_{i_k}) = 0.$$

In the case when we additionally have

$$\mathbb{E}(f_{i_1} \cdot f_{i_2} \cdot \cdots \cdot f_{i_k} \cdot f_i^2) = 0$$

for any pairwise distinct $i_1, i_2, \ldots, i_k$ and $i \neq i_j$, $j = 1, 2, \ldots, k$, a system $\{f_i\}$ is said to be *strongly multiplicative* (see also Definition B.7).

Theorem 7.8 *Suppose that* $\{f_i\}_{i=1}^{\infty}$ *(resp.* $\{g_i\}_{i=1}^{\infty}$*) is a multiplicative (resp. strongly multiplicative) system of r.v.'s defined on a probability space* $(\Omega, \Sigma, \mathbb{P})$ *(resp.* $(\Omega', \Sigma', \mathbb{P}')$*) such that* $g_i \neq 0$ *for each* $i = 1, 2, \ldots$ *and*

$$\|f_i\|_\infty \cdot \|g_i\|_\infty \leq \mathbb{E}g_i^2, \quad i = 1, 2, \ldots \tag{7.28}$$

Then, for arbitrary $p \geq 1$, $n \in \mathbb{N}$, *a Banach space* F *and* $x_i \in F$, $i = 1, 2, \ldots, n$, *we have*

$$\left(\mathbb{E}\Big\|\sum_{i=1}^{n} x_i f_i\Big\|_F^p\right)^{1/p} \leq 2\left(\mathbb{E}\Big\|\sum_{i=1}^{n} x_i g_i\Big\|_F^p\right)^{1/p}. \tag{7.29}$$

Proof Let us consider the integral operator

$$(Tu)(\omega) = \int_{\Omega'} K(\omega, \omega')u(\omega')\, d\mathbb{P}', \quad \omega \in \Omega,$$

with the kernel

$$K(\omega, \omega') = \prod_{i=1}^{n} \{1 + \alpha_i f_i(\omega) g_i(\omega')\}, \quad (\omega, \omega') \in \Omega \times \Omega',$$

where

$$\alpha_i = \frac{1}{\mathbb{E} g_i^2}, \quad i = 1, 2, \ldots .$$

We list some properties of the kernel $K(\omega, \omega')$. At first, from (7.28) it follows that

$$K(\omega, \omega') \geq 0 \quad \text{for almost all } (\omega, \omega') \in \Omega \times \Omega'. \tag{7.30}$$

Furthermore, since $\{f_i\}_{i=1}^{\infty}$ and $\{g_i\}_{i=1}^{\infty}$ are multiplicative systems, we have

$$\int_{\Omega'} K(\omega, \omega')\, d\mathbb{P}' = 1 \quad \text{for almost all } \omega \in \Omega \tag{7.31}$$

and

$$\int_{\Omega} K(\omega, \omega')\, d\mathbb{P} = 1 \quad \text{for almost all } \omega' \in \Omega'. \tag{7.32}$$

The relations (7.30) and (7.31) imply that the operator T is bounded from $L_\infty(\Omega')$ into $L_\infty(\Omega)$. Indeed, for all $u \in L_\infty(\Omega')$ and $\omega \in \Omega$

$$|Tu(\omega)| \leq \int_{\Omega} K(\omega, \omega')|u(\omega')|\, d\mathbb{P}' \leq \int_{\Omega} K(\omega, \omega')\, d\mathbb{P}' \cdot \|u\|_\infty = \|u\|_\infty,$$

whence

$$\|T\|_{L_\infty(\Omega') \to L_\infty(\Omega)} \leq 1. \tag{7.33}$$

We show now that

$$\|T\|_{L_1(\Omega') \to L_1(\Omega)} \leq 1. \tag{7.34}$$

To this end, observe that, by Fubini's theorem, for all $v \in L_\infty(\Omega)$ and $u \in L_1(\Omega')$ we have

$$\int_{\Omega'} (T'v)(\omega')u(\omega')\,d\mathbb{P}' = \int_{\Omega} (Tu)(\omega)v(\omega)\,d\mathbb{P},$$

where

$$(T'v)(\omega') := \int_{\Omega} K(\omega, \omega')v(\omega)\,d\mathbb{P}, \quad \omega' \in \Omega'.$$

Hence, it follows that

$$\|T\|_{L_1(\Omega')\to L_1(\Omega)} = \|T'\|_{L_\infty(\Omega)\to L_\infty(\Omega')}.$$

On the other hand, using (7.30) and (7.32), instead of (7.30) and (7.31), precisely in the same way as in the case of T, one can check that

$$\|T'\|_{L_\infty(\Omega)\to L_\infty(\Omega')} \le 1.$$

Combining these facts, we infer that T is bounded from $L_1(\Omega')$ into $L_1(\Omega)$ and inequality (7.34) holds.

According to the Riesz–Thorin interpolation Theorem D.1 (see also Remark D.1), estimates (7.33) and (7.34) imply that T is bounded from $L_p(\Omega')$ into $L_p(\Omega)$ for every $1 \le p \le \infty$ and

$$\|T\|_{L_p(\Omega')\to L_p(\Omega)} \le 2\max\left(\|T\|_{L_\infty(\Omega')\to L_\infty(\Omega)}, \|T\|_{L_1(\Omega')\to L_1(\Omega)}\right) \le 2. \tag{7.35}$$

We observe next that

$$f_j = Tg_j \quad \text{for all} \quad j = 1, 2, \ldots, n.$$

Indeed, since $\{g_i\}_{i=1}^\infty$ is a strongly multiplicative system, taking into account the definition of α_i, we have

$$\begin{aligned}(Tg_j)(\omega) &= \int_{\Omega'} \prod_{i=1}^{n} \{1 + \alpha_i f_i(\omega) g_i(\omega')\} \cdot g_j(\omega')\,d\mathbb{P}' \\ &= \int_{\Omega'} \Big\{ g_j(\omega') + \alpha_j f_j(\omega) g_j^2(\omega') + g_j(\omega') \sum_{i\neq j} \alpha_i f_i(\omega) g_i(\omega') \\ &\quad + g_j(\omega') \sum_{k=2}^{n} \sum_{1\le i_1<\ldots<i_k\le n} \alpha_{i_1} \cdot \ldots \cdot \alpha_{i_k} f_{i_1}(\omega) g_{i_1}(\omega') \ldots f_{i_k}(\omega) g_{i_k}(\omega') \Big\}\,d\mathbb{P}' = f_j(\omega).\end{aligned}$$

Therefore, for any $x_j \in F$, $j = 1, 2, \ldots, n$,

$$\sum_{j=1}^{n} x_j f_j = (T \otimes I)\Big(\sum_{j=1}^{n} x_j g_j\Big),$$

where $T \otimes I$ is the tensor product of T and the identity $I : F \to F$. Hence, for all $\omega \in \Omega$ we get

$$\Big\|\sum_{j=1}^{n} x_j f_j(\omega)\Big\|_F = \Big\|(T \otimes I)\Big(\sum_{j=1}^{n} x_j g_j\Big)(\omega)\Big\|_F = \Big\|\int_{\Omega'} K(\omega, \omega') \sum_{j=1}^{n} x_j g_j(\omega')\, d\mathbb{P}'\Big\|_F$$

$$\le \int_{\Omega'} K(\omega, \omega') \Big\|\sum_{j=1}^{n} x_j g_j(\omega')\Big\|_F d\mathbb{P}' = T\Big(\Big\|\sum_{j=1}^{n} x_j g_j\Big\|_F\Big)(\omega).$$

Since inequality (7.29) is an immediate consequence of this inequality and estimate (7.35), the proof is completed. □

By using Theorems 7.6–7.8, we are able to prove the following result on the comparison of a multiplicative (strongly multiplicative) system of r.v.'s with the Rademacher system.

Theorem 7.9 *Suppose that a multiplicative system $\{\xi_i\}_{i=1}^{\infty}$ of r.v.'s, defined on a probability space $(\Omega, \Sigma, \mathbb{P})$, is uniformly bounded, i.e.,*

$$|\xi_i(\omega)| \le D \text{ for all } i = 1, 2, \ldots \text{ and } \omega \in \Omega. \tag{7.36}$$

Then, $\{\xi_i\} \overset{\vec{\mathbb{P}}}{\prec} \{r_i\}$ with a constant, which depends only on D.

If additionally $\{\xi_i\}_{i=1}^{\infty}$ is strongly multiplicative and

$$d := \inf_{i=1,2,\ldots} \mathbb{E}\xi_i^2 > 0, \tag{7.37}$$

we have that $\{\xi_i\} \overset{\vec{\mathbb{P}}}{\sim} \{r_i\}$, where the implied equivalence constant depends only on D and d.

Proof Suppose first $\{\xi_i\}$ is multiplicative. Since this system is uniformly bounded by D, then condition (7.28) of Theorem 7.8 is satisfied by the r.v.'s $f_i = \xi_i/D$ and $g_i = r_i$, $i = 1, 2, \ldots$ Therefore, in view of the fact that the Rademacher system is strongly multiplicative (see Corollary 1.2) and Theorem 7.8, for any $p \ge 1$, $n \in \mathbb{N}$, arbitrary Banach space F and all $x_i \in F$, $i = 1, 2, \ldots, n$, we have

$$\Big(\mathbb{E}\Big\|\sum_{i=1}^{n} x_i \xi_i\Big\|_F^p\Big)^{1/p} \le 2D\Big(\mathbb{E}\Big\|\sum_{i=1}^{n} x_i r_i\Big\|_F^p\Big)^{1/p}.$$

Hence, by Theorem 7.6, we get the first assertion.

Let now $\{\xi_i\}_{i=1}^{\infty}$ be strongly multiplicative and let this system satisfy condition (7.37). Observe that in this case (7.28) holds also for the r.v.'s $f_i = (d/D)r_i$ and $g_i = \xi_i$. So, exploiting Theorem 7.8 once more, we obtain for any $p \geq 1$, $n \in \mathbb{N}$, arbitrary Banach space F and all $x_i \in F$, $i = 1, 2, \ldots, n$, the inequality

$$\left(\mathbb{E}\Big\|\sum_{i=1}^{n} x_i r_i\Big\|_F^p\right)^{1/p} \leq \frac{2D}{d}\left(\mathbb{E}\Big\|\sum_{i=1}^{n} x_i \xi_i\Big\|_F^p\right)^{1/p}.$$

To complete the proof, it is left to refer to Theorem 7.7. □

Applying the latter theorem in the partial case of systems of independent r.v.'s (see Example B.2), we come to the following rather elegant statement.

Corollary 7.3 *Every sequence of independent r.v.'s* $\{\xi_i\}_{i=1}^{\infty}$, $\mathbb{E}\xi_i = 0$, $i = 1, 2, \ldots$, *satisfying conditions* (7.36) *and* (7.37), *is equivalent in distribution in the vector sense to the Rademacher system.*

Moreover, from Theorem 7.9 and the definition of s.s.'s (see also Theorem C.3) it follows

Corollary 7.4 *Let* $\{\xi_i\}_{i=1}^{\infty}$ *be a multiplicative system of r.v.'s, defined on a probability space* $(\Omega, \Sigma, \mathbb{P})$ *and satisfying condition* (7.36). *Then, for an arbitrary s.s.* X *on* $[0, 1]$, *all* $n \in \mathbb{N}$ *and any real numbers* $a_1, a_2, \ldots, a_n$ *it holds*

$$\Big\|\sum_{i=1}^{n} a_i \xi_i\Big\|_X \leq C\Big\|\sum_{i=1}^{n} a_i r_i\Big\|_X,$$

where $C > 0$ *depends only on* D.

If, in addition, $\{\xi_i\}_{i=1}^{\infty}$ *is strongly multiplicative and also* (7.37) *holds, then for an arbitrary s.s.* X *on* $[0, 1]$, *all* $n \in \mathbb{N}$ *and any real numbers* $a_1, a_2, \ldots, a_n$ *we have*

$$\Big\|\sum_{i=1}^{n} a_i \xi_i\Big\|_X \asymp \Big\|\sum_{i=1}^{n} a_i r_i\Big\|_X,$$

with the implied equivalence constant, depending only on D *and* d.

7.6 Sequences of Characters on Compact Abelian Groups

Recall that the Rademacher system can be treated as a sequence of characters, defined on the Cantor group $\{-1, 1\}^{\mathbb{N}}$ (see Comments and References concluding Chap. 1). Thus, the idea of comparing systems of characters, defined on a compact Abelian group, with the Rademacher system looks very natural.

We start with a result, which is analogous in a certain sense to Theorem 7.8. If $\mathcal{S}$ is a compact Abelian group, then $C(\mathcal{S})$ will denote the space of all continuous complex-valued functions f, defined on $\mathcal{S}$, with the norm $\|f\|_{C(\mathcal{S})} := \max_{s\in\mathcal{S}} |f(s)|$.

Theorem 7.10 *Let $\mathcal{S}$ and $\mathcal{T}$ be two compact Abelian groups with Haar probability measures μ and ν, respectively. Suppose that $\{f_j\}_{j=1}^{\infty}$ and $\{g_j\}_{j=1}^{\infty}$ are sequences of characters defined on $\mathcal{S}$ and $\mathcal{T}$, respectively, satisfying the following condition: there exist constants $C > 0$ and $c > 0$ such that for all $n \in \mathbb{N}$ and complex numbers $a_1, a_2, \dots, a_n$ we have*

$$c \cdot \max_{s\in\mathcal{S}} \Big|\sum_{i=1}^{n} a_i f_i(s)\Big| \le \max_{t\in\mathcal{T}} \Big|\sum_{i=1}^{n} a_i g_i(t)\Big| \le C \cdot \max_{s\in\mathcal{S}} \Big|\sum_{i=1}^{n} a_i f_i(s)\Big|. \tag{7.38}$$

Then, for arbitrary $1 \le p < \infty$, $n \in \mathbb{N}$, every Banach space F (over the complex field), and all vectors $x_1, x_2, \dots, x_n$ from F it holds

$$\frac{c}{C}\Big(\int_{\mathcal{S}} \Big\|\sum_{i=1}^{n} x_i f_i(s)\Big\|_F^p d\mu\Big)^{1/p} \le \Big(\int_{\mathcal{T}} \Big\|\sum_{i=1}^{n} x_i g_i(t)\Big\|_F^p d\nu\Big)^{1/p}$$
$$\le \frac{C}{c}\Big(\int_{\mathcal{S}} \Big\|\sum_{i=1}^{n} x_i f_i(s)\Big\|_F^p d\mu\Big)^{1/p}. \tag{7.39}$$

Proof First of all, we observe that, in view of (7.38), the mapping U defined by $Uf_j = g_j$, $j = 1, 2, \dots$, extends to an isomorphism of the closed linear span $[f_j]$ in $C(\mathcal{S})$ onto the closed linear span $[g_j]$ in $C(\mathcal{T})$. Moreover, $\|U\| \le C$ and $\|U^{-1}\| \le c^{-1}$.

For a fixed $t \in \mathcal{T}$ denote by δ_t the point mass (or the Dirac measure) at t. Let also φ_t^* denote the restriction of δ_t to $[g_j]$. Clearly, φ_t^* is a bounded linear functional on $[g_j]$ with $\|\varphi_t^*\| \le 1$. Moreover, for each $j = 1, 2, \dots$ we have

$$g_j(t) = \varphi_t^*(g_j) = \varphi_t^*(Uf_j) = (U^*\varphi_t^*)(f_j). \tag{7.40}$$

Let μ_t be a complex Borel measure (that is, a countably additive function with finite variation, defined on the σ-algebra of Borel subsets of $\mathcal{S}$), which according to the Riesz Representation theorem (see e.g. [79, Theorem 6.5.5]) corresponds to the Hahn-Banach extension of the linear functional $U^*\varphi_t^*$ onto $C(\mathcal{S})$. Define also the complex measures μ_t' and $\overline{\mu_t}$ on Borel sets $V \subset \mathcal{S}$ by

$$\mu_t'(V) = \mu_t(-V), \quad \overline{\mu}_t(V) = \overline{\mu_t(V)}.$$

Note that their total variations can be estimated as follows:

$$\|\overline{\mu}_t\| = \|\mu_t'\| = \|\mu_t\| \le \|U^*\| = \|U\| \le C. \tag{7.41}$$

Summing up, we conclude that for every $p \in \mathbb{N}$ and an arbitrary partition $\{A_j\}_{j=1}^p$ of the set of positive integers there are $N \in \mathbb{N}$ and real numbers b_i, $i = 1, 2, \ldots, N$, satisfying (8.18), such that

$$I_N = \int_\Omega R_N(\omega)\varphi(\omega)\, d\mathbb{P} \geq \frac{1}{2D} \sum_{j=1}^p \Big(\sum_{i \in A_j^N} a_i^2 \Big)^{1/2}. \tag{8.20}$$

The next stage of the proof is estimation of the integral I_N from the above (up to a constant) by the norm $\|\varphi\|_p$. To this end, in view of Hölder's inequality

$$|I_N| \leq \|R_N\|_{p'} \|\varphi\|_p, \quad \text{where } p' := \frac{p}{p-1}, \tag{8.21}$$

and the fact that the Riesz products R_N are nonnegative (because of $|b_i\varphi_i(\omega)| \leq 1$, $i = 1, 2, \ldots$; see (8.18)), it suffices to find an upper estimate for the integral L_N defined by

$$L_N := \|R_N\|_{p'}^{p'} = \int_\Omega R_N(\omega)^{p'}\, d\mathbb{P}. \tag{8.22}$$

Let $p \in \mathbb{N}$, $p \geq 3$. As above, $\{A_j\}_{j=1}^p$ is a partition of $\mathbb{N}$, $A_j^N = \{i = 1, \ldots, N : i \in A_j\}$, and b_i are real numbers satisfying inequalities (8.18). From the elementary estimate $(1+x)^y \leq 1 + yx$ $(x \geq 0, 0 < y \leq 1)$ and the inequality $|\varphi_i(\omega)| \leq D$ it follows that

$$\begin{aligned} R_N(\omega)^{p'} &= \prod_{i=1}^N (1 + b_i\varphi_i(\omega))^{1+1/(p-1)} \\ &\leq \prod_{i=1}^N (1 + b_i\varphi_i(\omega)) \left(1 + \frac{1}{p-1} b_i\varphi_i(\omega)\right) \\ &\leq \prod_{i=1}^N \left(1 + \frac{D^2}{p-1} b_i^2 + \frac{p}{p-1} b_i\varphi_i(\omega)\right) \\ &= \prod_{j=1}^p \prod_{i \in A_j^N} \left(1 + \frac{D^2}{p-1} b_i^2 + \frac{p}{p-1} b_i\varphi_i(\omega)\right). \end{aligned}$$

Denoting $|B| := \text{card}\, B$, we see that the inner product in the last expression is equal to the sum

$$1+\frac{D^2}{p-1}\sum_{i\in A_j^N} b_i^2+\left(\frac{D^2}{p-1}\right)^2\sum_{\substack{i_1<i_2\\ i_1,i_2\in A_j^N}} b_{i_1}^2 b_{i_2}^2+\cdots+\left(\frac{D^2}{p-1}\right)^{|A_j^N|}\prod_{i\in A_j^N} b_i^2+$$

$$+\sum_{B_j\overset{\neq}{\subset} A_j^N,\, B_j\neq\varnothing}\left\{1+\left(\frac{D^2}{p-1}\right)^{|A_j^N|-|B_j|}\prod_{i\in A_j^N\setminus B_j} b_i^2\right\}\left(\frac{p}{p-1}\right)^{|B_j|}\prod_{i\in B_j} b_i\varphi_i(\omega)+$$

$$+\left(\frac{p}{p-1}\right)^{|A_j^N|}\prod_{i\in A_j^N} b_i\varphi_i(\omega).$$

Observe that, according to inequalities (8.18),

$$1+\frac{D^2}{p-1}\sum_{i\in A_j^N} b_i^2+\left(\frac{D^2}{p-1}\right)^2\sum_{\substack{i_1<i_2\\ i_1,i_2\in A_j^N}} b_{i_1}^2 b_{i_2}^2+\cdots+\left(\frac{D^2}{p-1}\right)^{|A_j^N|}\prod_{i\in A_j^N} b_i^2\le\frac{p-1}{p-2}$$

and for any $j=1,2,\dots,p$ and $B_j\overset{\neq}{\subset} A_j^N$

$$1+\left(\frac{D^2}{p-1}\right)^{|A_j^N|-|B_j|}\prod_{i\in A_j^N\setminus B_j} b_i^2\le\frac{p}{p-1}.$$

Therefore, we come to the following upper estimate of the r.v. $R_N(\omega)^{p'}$:

$$R_N(\omega)^{p'}\le\prod_{j=1}^{p}\left\{\frac{p-1}{p-2}+\sum_{B_j\subset A_j^N,\, B_j\neq\varnothing}\left(\frac{p}{p-1}\right)^{|B_j|+1}\prod_{i\in B_j} b_i\varphi_i(\omega)\right\}.$$

Hence (see (8.22)),

$$L_N\le\left(\frac{p-1}{p-2}\right)^p+\left(\frac{p-1}{p-2}\right)^{p-1}\sum_{j=1}^{p}\sum_{B_j\subset A_j^N}\left(\frac{p}{p-1}\right)^{|B_j|+1}\left|\mathbb{E}\Big(\prod_{i\in B_j} b_i\varphi_i\Big)\right|$$

$$+\left(\frac{p-1}{p-2}\right)^{p-2}\sum_{1\le j_1<j_2\le p}\sum_{B_{j_1}\subset A_{j_1}^N}\sum_{B_{j_2}\subset A_{j_2}^N}\left(\frac{p}{p-1}\right)^{|B_{j_1}|+|B_{j_2}|+2}\left|\mathbb{E}\Big(\prod_{k=1}^{2}\prod_{i\in B_{j_k}} b_i\varphi_i\Big)\right|$$

$$\cdots+\left(\frac{p-1}{p-2}\right)^{p-s}\sum_{1\le j_1<\cdots<j_s\le p}\sum_{B_{j_1}\subset A_{j_1}^N}$$

$$\cdots\sum_{B_{j_s}\subset A_{j_s}^N}\left(\frac{p}{p-1}\right)^{\sum_{k=1}^{s}|B_{j_k}|+s}\left|\mathbb{E}\Big(\prod_{k=1}^{s}\prod_{i\in B_{j_k}} b_i\varphi_i\Big)\right|$$

$$\cdots+\sum_{B_1\subset A_1^N}\cdots\sum_{B_p\subset A_p^N}\left(\frac{p}{p-1}\right)^{\sum_{j=1}^{p}|B_j|+p}\left|\mathbb{E}\Big(\prod_{j=1}^{p}\prod_{i\in B_j} b_i\varphi_i\Big)\right|. \tag{8.23}$$

Note that for all $p\ge 3$ we have

$$\frac{p}{p-1}\le 2 \quad\text{and}\quad \left(\frac{p-1}{p-2}\right)^{p-s}\left(\frac{p}{p-1}\right)^{s}\le 4e,\quad s=0,1,\dots,p. \tag{8.24}$$

In addition, since the sets A_j^N, $j=1,2,\dots,p$, are nonoverlapping, then the integrands in the right-hand side of (8.23) are pairwise distinct products of the functions φ_i, $i=1,2,\dots,N$, and each of these integrands is a product of different functions φ_i. Observe also that the number of factors in integrands is equal to $|B_j|$, $j=1,2,\dots,p$, for terms of the first sum, $|B_{j_1}|+|B_{j_2}|$, $1\le j_1<j_2\le p$, for terms of the second one,..., $\sum_{j=1}^{p}|B_j|$ for terms of the last one. Hence, the maximal index of the functions φ_i in each integrand is not less than $|B_j|$, $|B_{j_1}|+|B_{j_2}|,\dots$, $\sum_{j=1}^{p}|B_j|$, respectively. As a result, from (8.23), (8.24) and (8.18) it follows

$$L_N\le 4e\left(1+\sum_{i=1}^{N}2^i\sum_{1\le j_1<\cdots<j_s\le i-1}D^{-s-1}\left|\mathbb{E}(\varphi_{j_1}\dots\varphi_{j_s}\varphi_i)\right|\right), \tag{8.25}$$

where the inner sum is taken over all sets of indices $1\le j_1<\cdots<j_s\le i-1$. Since $D\ge 1$, then combining this inequality with (8.10) and the choice of ε_i, we infer

$$L_N\le 4e\Big(1+\sum_{i=1}^{\infty}8^{-i}\Big)<13.$$

Therefore, taking into account (8.21) and (8.22), for all $N\in\mathbb{N}$ and $p=3,4,\dots$ we get

$$|I_N|\le 13\|\varphi\|_p.$$

In turn, this inequality and (8.20) yield

$$\sum_{i=1}^{p}\Big(\sum_{i\in A_j} a_i^2\Big)^{1/2} \le 26D\|\varphi\|_p,$$

whence, in view of (8.8) and (8.9) (see also Lemma 1.3), it follows that

$$\kappa(\sqrt{p}, a) \le 26\sqrt{2}\, D\Big\|\sum_{i=1}^{\infty} a_i\varphi_i\Big\|_p \tag{8.26}$$

for positive integers $p \ge 3$ and all $a = (a_i)_{i=1}^{\infty} \in \ell_2$.

Show that a similar inequality holds also for $p = 1$ and $p = 2$. Since $\|a\|_{\ell_2} \le \|a\|_{\ell_1}$, then we have $\kappa(1, a) = \|a\|_{\ell_2}$. Hence, as the function $t \mapsto \kappa(t, a)$ is concave, it suffices to prove that

$$\|a\|_{\ell_2} \le M\Big\|\sum_{i=1}^{\infty} a_i\varphi_i\Big\|_1 \tag{8.27}$$

for some $M = M(D)$ and all $a = (a_i)_{i=1}^{\infty} \in \ell_2$.

Taking into account that $\|\varphi_i\|_2 = 1$, by the Cauchy-Schwarz-Bunyakovskii inequality and (8.10), we have for any $m \in \mathbb{N}$

$$\begin{aligned}\Big\|\sum_{i=1}^{m} a_i\varphi_i\Big\|_2^2 &= \sum_{i=1}^{m} a_i^2 + 2\sum_{1\le i<j\le m} a_i a_j \mathbb{E}(\varphi_i\varphi_j)\\ &\ge \sum_{i=1}^{m} a_i^2\Big\{1 - 2\Big(\sum_{1\le i<j\le m}\mathbb{E}(\varphi_i\varphi_j)^2\Big)^{1/2}\Big\}\\ &\ge \sum_{i=1}^{m} a_i^2\Big\{1 - 2\sum_{j=1}^{\infty}\varepsilon_j\Big\} \ge \frac{1}{2}\sum_{i=1}^{m} a_i^2.\end{aligned}$$

Since a series $\sum_{i=1}^{\infty} a_i\varphi_i$ converges in $L_2(\Omega)$ for each $(a_i)_{i=1}^{\infty} \in \ell_2$, the last inequality yields

$$\Big\|\sum_{i=1}^{\infty} a_i\varphi_i\Big\|_2^2 \ge \frac{1}{2}\|(a_i)\|_{\ell_2}^2.$$

Thus, applying successively Hölder's inequality, (8.11), and Theorem 1.4, we get

$$\|a\|_{\ell_2}^2 \le 2\Big\|\sum_{i=1}^{\infty} a_i\varphi_i\Big\|_2^2 \le 2\Big(\int_0^1 \Big|\sum_{i=1}^{\infty} a_i\varphi_i(\omega)\Big|\,d\mathbb{P}\Big)^{2/3}\Big(\int_0^1 \Big(\sum_{i=1}^{\infty} a_i\varphi_i(\omega)\Big)^4 d\mathbb{P}\Big)^{1/3}$$

$$\le 2\cdot 2^{4/3}D^{4/3}\Big\|\sum_{i=1}^{\infty} a_i\varphi_i\Big\|_1^{2/3}\Big(\int_0^1 \Big(\sum_{i=1}^{\infty} a_i r_i(t)\Big)^4 dt\Big)^{1/3}$$

$$\le 16\cdot D^{4/3}\Big\|\sum_{i=1}^{\infty} a_i\varphi_i\Big\|_1^{2/3}\|a\|_{\ell_2}^{4/3},$$

which implies inequality (8.27) with the constant $M = 64D^2$.

Summarizing the estimates obtained, we see that inequality (8.7) is proved for positive integers p. Combining this together with the concavity of the function $t \mapsto \kappa(t, a)$, we conclude that (8.7) holds for all real numbers $p \ge 1$ with a constant that depends only on D and d. As was observed above, this finishes the proof. □

One can readily check that the conditions (α)–(γ) of Theorem 8.2 are fulfilled for uniformly bounded orthonormal systems of r.v.'s. Therefore, we obtain the following important result.

Theorem 8.4 *Every orthonormal system $\{f_n\}_{n=1}^{\infty}$ of r.v.'s on a probability space $(\Omega, \Sigma, \mathbb{P})$ such that $|f_n(\omega)| \le D$ $\mathbb{P}$-a.e. on Ω for all $n = 1, 2, \dots$ contains a subsystem $\{\varphi_i\}_{i=1}^{\infty}$, equivalent in distribution to the Rademacher system with a constant depending only on D.*

As was observed above, the Rademacher system is a perfect lacunary sequence. Indeed, by Theorem 1.3, $\{r_n\}$ is a S_∞-system with the constants $K_p \asymp \sqrt{p}$, $p \ge 1$ (see Definition B.10), and simultaneously, by Remark 1.4, it is a Sidon system. It is clearly that every sequence of r.v.'s, equivalent in distribution to the Rademacher system, possesses the same properties. Thus, we can say that the results obtained give sharp conditions, under which an uniformly bounded sequence of r.v.'s contains a subsequence with the "best" lacunary properties.

Show that every sequence $\{\varphi_i\}_{i=1}^{\infty}$, equivalent in distribution to the Rademacher system, has a strengthened sidonicity: measure of the set, on which the absolute value of a sum $\sum_{i=1}^{m} a_i\varphi_i$ exceeds (up to a constant independent of m) the sum $\sum_{i=1}^{m} |a_i|$, may be controlled by m.

Corollary 8.3 *Assume that a system $\{f_n\}_{n=1}^{\infty}$ of r.v.'s on a probability space $(\Omega, \Sigma, \mathbb{P})$ satisfies the conditions of Theorem 8.2. Then, $\{f_n\}_{n=1}^{\infty}$ contains a subsystem $\{\varphi_i\}_{i=1}^{\infty}$ with the following property: for some constants $\alpha_1, \alpha_2, \alpha_3 > 0$, depending only on D and d, and for each sum $\Phi = \sum_{i=1}^{m} a_i\varphi_i$, $m \in \mathbb{N}$, $a_i \in \mathbb{R}$, there exists a set $E = E(\Phi) \subset \Omega$, $\mathbb{P}(E) > \alpha_1 2^{-m}$, such that*

$$|\Phi(\omega)| \ge \alpha_2\,\|\Phi\|_\infty \ge \alpha_3 \sum_{i=1}^{m} |a_i|, \quad \omega \in E.$$

Proof By Theorem 8.2, we can find a subsequence $\{\varphi_i\} \subset \{f_n\}$ such that for some $C > 0$ (which depends only on D and d) and all $z > 0$ we have

$$C^{-1}m(\{|R(t)| > Cz\} \le \mathbb{P}\{|\Phi(\omega)| > z\} \le Cm\{|R(t)| > C^{-1}z\},$$

where $\Phi(\omega) = \sum_{i=1}^{m} a_i\varphi_i(\omega)$, $\omega \in \Omega$, and $R(t) = \sum_{i=1}^{m} a_i r_i(t)$, $0 \le t \le 1$. This implies, in particular, that

$$C^{-1}\|\Phi\|_\infty \le \|R\|_\infty \le C\|\Phi\|_\infty.$$

Moreover, because $\|R\|_\infty = \sum_{i=1}^{m} |a_i|$, we obtain

$$\begin{aligned}\mathbb{P}\{|\Phi(\omega)| > (2C^2)^{-1}\|\Phi\|_\infty\} &\ge \mathbb{P}\{|\Phi(\omega)| > (2C)^{-1}\|R\|_\infty\}\\ &\ge C^{-1}m\{|R(t)| > 2^{-1}\|R\|_\infty\} \ge C^{-1}2^{-m}.\end{aligned}$$

Denote now $\alpha_1 := C^{-1}$, $\alpha_2 := (2C^2)^{-1}$ and $E = E(\Phi) := \{\omega \in \Omega : |\Phi(\omega)| > \alpha_2\|\Phi\|_\infty\}$. Then, $\mathbb{P}(E) > \alpha_1 2^{-m}$ and for all $\omega \in E$

$$|\Phi(\omega)| > \alpha_2\|\Phi\|_\infty \ge \alpha_2 C^{-1}\|R\|_\infty = \alpha_3\sum_{i=1}^{m}|a_i|,$$

where $\alpha_3 = \alpha_2 C^{-1}$. □

According to Proposition 1.3 (see also Comments and References concluding Chap. 1), the Rademacher sequence is a strong Sidon–Zygmund system (see Definition B.9). The next result shows that, under certain conditions, a sequence of measurable functions on [0, 1] contains a subsequence, which is everywhere "locally" equivalent in distribution to the Rademacher system (and hence it is also a strong Sidon–Zygmund system).

Theorem 8.5 *Suppose a system $\{f_n\}_{n=1}^\infty$ of measurable functions on $[0, 1]$ contains a subsequence $\{f_{n_k}\}$, satisfying the conditions (α) and (β) of Theorem 8.2 as well having the following property: for each interval $I \subset [0, 1]$*

$$d(I) := \liminf_{k\to\infty}\int_I f_{n_k}^2(x)\,dx > 0. \tag{8.28}$$

Then, there exists a subsequence $\{\varphi_i\} \subset \{f_n\}$ such that for any interval $I \subset [0, 1]$ the sequence $\{\varphi_i\chi_I\}_{i=i_0}^\infty$, with some $i_0 = i_0(I)$, is equivalent in distribution to the Rademacher system on $[0, 1]$.

Proof Thanks to Theorem 7.5, it suffices to extract a subsystem $\{\varphi_i\} \subset \{f_n\}$ such that for each interval $I \subset [0, 1]$ and some $i_0 = i_0(I)$

$$\Big\|\sum_{i=1}^{\infty} a_i\varphi_{i_0+i-1}\chi_I\Big\|_p \asymp \mathcal{K}(\sqrt{p}, a) \tag{8.29}$$

for all $p \ge 1$ and $a = (a_i) \in \ell_2$ (with the implied constant depending on D and I).

First of all, by Theorem 8.1, there is $\{g_i\} \subset \{f_{n_k}\}$ satisfying for all $p \geq 1$ and $a = (a_i) \in \ell_2$ the inequality

$$\Big\| \sum_{i=1}^{\infty} a_i g_i \chi_I \Big\|_p \leq C\mathcal{K}(\sqrt{p}, a), \tag{8.30}$$

where a constant $C > 0$ depends only on D. Thus, it remains to find a subsystem $\{\varphi_i\} \subset \{g_i\}$ such that for any interval $I \subset [0, 1]$ the sequence $\{\varphi_i \chi_I\}_{i=i_0}^{\infty}$, with some $i_0 = i_0(I)$, satisfies the opposite inequality.

By condition, $g_i \xrightarrow{w} 0$ in L_2, which implies clearly that $\{g_i \chi_I\}$ is a weakly null sequence in L_2 for each interval $I \subset [0, 1]$. However, we shall need a somewhat stronger property: let us show that for every function $g \in L_2$ we have

$$\lim_{i\to\infty} \sup_{I\subset[0,1]} \Big| \int_I g(t) g_i(t)\,dt \Big| = 0. \tag{8.31}$$

To this end, given $g \in L_2$ and $\delta > 0$, by using the absolute continuity of the Lebesgue integral, choose a positive integer $l = l(g, \delta)$ satisfying the condition:

$$\omega(1/l) := \sup_{E\subset[0,1], m(E)\leq 1/l} \Big(\int_E g(t)^2\,dt \Big)^{1/2} < \frac{\delta}{3D}.$$

Next, let us split $[0, 1]$ into l intervals $J_1, \ldots, J_l$ of equal length. Because $\{g_i\}$ is weakly null in L_2, we can find i_1 such that

$$\sum_{k=1}^{l} \Big| \int_{J_k} g(t) g_i(t)\,dt \Big| < \frac{\delta}{3} \quad \text{for all } i \geq i_1.$$

Observe that any interval $I \subset [0, 1]$ is a union of some intervals J_k and no more than two intervals J' and J'' of length less than $1/l$. Hence, the latter estimates and the Cauchy-Schwarz-Bunyakovskii inequality yield

$$\Big| \int_I g(t) g_i(t)\,dt \Big| \leq \frac{\delta}{3} + \Big| \int_{J'} g(t) g_i(t)\,dt \Big| + \Big| \int_{J''} g(t) g_i(t)\,dt \Big| \leq \frac{\delta}{3} + 2\omega(1/l)D < \delta$$

if $i \geq i_1$. Since $\delta > 0$ is arbitrarily small, the proof of (8.31) is completed.

Further, let $\varepsilon_i = 16^{-i}$, $i = 1, 2, \ldots$, as in the proof of Theorem 8.2. In the same way as in this proof, using (8.31) and (8.28), we can extract, for every interval $I \subset [0, 1]$, a subsequence $\{\varphi_i\} \subset \{g_i\}$ that satisfies, starting from some index $i_0(I)$, inequality (8.10) adopted to I. Then, due to the fast convergence ε_i to zero as $i \to \infty$, repeating the reasoning from the proof of Theorem 8.2, we can show that for all

$p \geq 1$ and $a = (a_i) \in \ell_2$

$$\kappa(\sqrt{p}, a) \leq C_1 \Big\| \sum_{i=1}^{\infty} a_i \varphi_{i_0+i-1} \chi_I \Big\|_p,$$

where a constant $C_1 > 0$ depends only on D and I. Combining this inequality together with estimate (8.30), we get (8.29), and so the theorem is proved. □

8.3 Density of Subsystems of Finite Systems Equivalent in Distribution to Sets of Rademacher Functions

We now turn to considering the problem of extraction of lacunary subsystems from *finite* uniformly bounded orthonormal systems $\{f_n\}_{n=1}^N$ of functions, defined on the interval [0, 1]. We show that any such system contains a subsystem $\{f_{n_i}\}_{i=1}^s$ of "logarithmic density" that is equivalent in distribution to the set $\{r_i\}_{i=1}^s$ of Rademacher functions with a constant independent of N. Let us start with a formulation of the main result.

Theorem 8.6 *Suppose that $\{f_n\}_{n=1}^N$ is an orthonormal system of functions on $[0, 1]$, $|f_n(x)| \leq D$ for all $x \in [0, 1]$ and $n = 1, 2, \ldots, N$. Then, there exists a subsystem $\{f_{n_i}\}_{i=1}^s$, $1 \leq n_1 < n_2 < \cdots < n_s \leq N$, where $s \geq \max\{[\frac{1}{7}\log_2 N], 1\}$, such that for some $C > 0$, depending only on D, all $a_i \in \mathbb{R}$, $i = 1, 2, \ldots, s$, and $z > 0$ we have*

$$C^{-1} m\Big\{\Big|\sum_{i=1}^s a_i r_i(x)\Big| > Cz\Big\} \leq m\Big\{\Big|\sum_{i=1}^s a_i f_{n_i}(x)\Big| > z\Big\} \leq C m\Big\{\Big|\sum_{i=1}^s a_i r_i(x)\Big| > C^{-1} z\Big\}.$$

For the proof we shall need two lemmas. First of them shows that an appropriate thinning finite systems allows us, precisely as in the case of infinite sequences of r.v.'s, to get a subsystem with arbitrarily small mixed moments of its elements.

Lemma 8.2 *Let a system $\{f_n\}_{n=1}^N$, $N \geq 2^7$, satisfies all the conditions of Theorem 8.6. Then, there are functions f_{n_i}, $1 \leq n_1 < n_2 < \cdots < n_s \leq N$, where $s \geq [\frac{1}{7}\log_2 N]$, for which the following inequality holds:*

$$\sum_{\theta \in \mathcal{A}_s} \left\{ \mathbb{E}\left[\prod_{i=1}^s \left(\frac{f_{n_i}}{D}\right)^{\theta_i} \right] \right\}^2 \leq 10^{-s}, \tag{8.32}$$

where $\mathcal{A}_s$ is the family of all collections $\theta = (\theta_i)_{i=1}^s$ such that

$$a)\ \theta_i = 0, 1 \text{ or } 2\ ;\quad b)\ \sum_{i:\theta_i=1} 1 \geq 1\ ;\quad c)\ \sum_{i:\theta_i=2} 1 \leq 1.$$

Proof Let $s \in \mathbb{N}$, $s \le N$. We denote by $\mathcal{E}_N^s$ the family of all collections of positive integers $\{n_k\}_{k=1}^s$, satisfying the inequalities $1 \le n_1 < n_2 < \cdots < n_s \le N$. The combinatorial proof of the lemma will be based on estimation from above of the average value of sums of the form

$$I(\{n_k\}) := \sum_{\theta \in \mathcal{A}_s} \left\{ \mathbb{E}\left[\prod_{k=1}^{s} \left(\frac{f_{n_k}}{D} \right)^{\theta_i} \right] \right\}^2,$$

corresponding to all $\{n_k\}_{k=1}^s \in \mathcal{E}_N^s$. Since $\min_{\{n_k\} \in \mathcal{E}_N^s} I(\{n_k\})$ is less than or equal to this average, it suffices to prove that the last average does not exceed 10^{-s}.

First of all, observe that each of the terms of the sum

$$S := \sum_{\{n_k\} \in \mathcal{E}_N^s} I(\{n_k\}) \tag{8.33}$$

can be rewritten as

$$\left(\frac{1}{D^j} \int_0^1 \prod_{k=1}^{j} f_{n_k}(x)\, dx \right)^2, \quad 1 \le j \le s, \ \{n_k\} \in \mathcal{E}_N^j, \tag{8.34}$$

or

$$\left(\frac{1}{D^{j+1}} \int_0^1 f_{n_q}(x) \prod_{k=1}^{j} f_{n_k}(x)\, dx \right)^2, \quad 1 < j \le s, \ 1 \le q \le j, \ \{n_k\} \in \mathcal{E}_N^j. \tag{8.35}$$

Let us estimate the sums

$$S_j' := \sum_{\{n_k\} \in \mathcal{E}_N^j} \left(\int_0^1 \prod_{k=1}^{j} f_{n_k}(x)\, dx \right)^2, \quad j = 1, 2, \ldots, s,$$

and

$$S_j'' := \sum_{\{n_k\} \in \mathcal{E}_N^j} \sum_{q=1}^{j} \left(\int_0^1 f_{n_q}(x) \prod_{k=1}^{j} f_{n_k}(x)\, dx \right)^2, \quad j = 2, 3, \ldots, s$$

separately. Since $\{f_n\}_{n=1}^N$ is an orthonormal system, then, by the Bessel inequality, we have

$$S_1' = \sum_{n=1}^{N} \left(\int_0^1 f_n(x)\,dx \right)^2 \le \|\chi_{[0,1]}\|_2^2 = 1. \tag{8.36}$$

In the case $j > 1$ we can estimate S_j'' as follows:

$$S_j' \le \sum_{\{p_k\}\in\mathcal{E}_N^{j-1}} \sum_{\{n_k\}\in\mathcal{E}_N^{j}} \delta(\{p_k\},\{n_k\}) \left(\int_0^1 \prod_{k=1}^{j} f_{n_k}(x)\,dx \right)^2,$$

where $\delta(\{p_k\},\{n_k\})$ is equal to 1 if $\{p_k\} \subset \{n_k\}$, and 0, otherwise. Since for each set $\{p_k\} \in \mathcal{E}_N^{j-1}$ the Bessel inequality gives

$$\sum_{\{n_k\}\in\mathcal{E}_N^{j}:\{n_k\}\supset\{p_k\}} \left(\int_0^1 \prod_{k=1}^{j} f_{n_k}(x)\,dx \right)^2 \le \sum_{n=1}^{N} \left\{ \int_0^1 \left(\prod_{k=1}^{j-1} f_{p_k}(x) \right) f_n(x)\,dx \right\}^2$$
$$\le \left\| \prod_{k=1}^{j-1} f_{p_k} \right\|_2^2 \le D^{2(j-1)},$$

then from the preceding inequality it follows that

$$S_j' \le \sum_{\{p_k\}\in\mathcal{E}_N^{j-1}} D^{2(j-1)} = C_N^{j-1} D^{2(j-1)}, \quad j = 2, 3, \ldots, s, \tag{8.37}$$

where C_m^l, $m, l \in \mathbb{N}$, $0 \le l \le m$, are the binomial coefficients.

As to the sum S_j'', it can be represented in the form

$$S_j'' = \sum_{p=1}^{N} S_j''(p),$$

where

$$S_j''(p) := \sum_{\{n_k\}\in\mathcal{E}_N^{j-1}:\ p\notin\{n_k\}} \left(\int_0^1 f_p(x)^2 \prod_{k=1}^{j-1} f_{n_k}(x)\,dx \right)^2.$$

Estimating the sum $S''_j(p)$ in the same way as S'_j, we come to the inequality

$$S''_j(p) \le C_{N-1}^{j-2} D^{2j}, \quad j = 2, 3, \dots, s,$$

which implies that

$$S''_j \le N C_{N-1}^{j-2} D^{2j}, \quad j = 2, 3, \dots, s. \tag{8.38}$$

Note that for $s > 1$ each of the terms of the form (8.34) and (8.35) occurs in the sum (8.33) exactly C_{N-j}^{s-j} times. Therefore, taking into account that $D \ge 1$ and setting $S''_1 = 0$, thanks to inequalities (8.36)–(8.38), we get

$$S \le \sum_{j=1}^{s} \frac{1}{D^{2j}} C_{N-j}^{s-j}(S'_j + S''_j) \le \sum_{j=1}^{s} C_{N-j}^{s-j}(C_N^{j-1} + N C_{N-1}^{j-2}).$$

Combining this with the fact that the number of elements of the family $\mathcal{E}_N^s$ is C_N^s, we arrive at the following estimate for the average of sums (8.33):

$$\frac{S}{C_N^s} \le \sum_{j=1}^{s} \frac{C_{N-j}^{s-j}}{C_N^s}(C_N^{j-1} + N C_{N-1}^{j-2}).$$

One can readily check that $j C_N^{j-1} \ge N C_{N-1}^{j-2}$. Hence, the right-hand side in the preceding inequality does not exceed the quantity

$$\begin{aligned}(1+s)\sum_{j=1}^{s} \frac{C_{N-j}^{s-j} C_N^{j-1}}{C_N^s} &= (1+s)\sum_{j=1}^{s} \frac{1}{N-j+1} \cdot \frac{s!}{(s-j)!(j-1)!} \\ &\le \frac{(1+s)}{N-s} \sum_{j=1}^{s} j C_s^j \le \frac{2^s s(1+s)}{N-s} \le 10^{-s}\end{aligned}$$

if $N \ge (128)^s$, or equivalently $s \le \frac{1}{7}\log_2 N$. As a result, this yields

$$\frac{S}{C_N^s} \le 10^{-s},$$

which completes the proof. □

Let $\{f_n\}_{n=1}^N$ be a uniformly bounded system of functions defined on $[0, 1]$. The second lemma will allow us, under certain conditions, to extend the domain of functions f_n, $n = 1, 2, \dots, N$, i.e., $[0, 1]$, to a wider interval of the real line so that the new set of functions defined on the latter interval becomes multiplicative (see Definition B.7), being still uniformly bounded.

Lemma 8.3 *Let* $\{g_i\}_{i=1}^s$ *be a system of functions on* $[0, 1]$ *such that* $|g_i(x)| \le D$ *for all* $i = 1, 2, \dots, s$ *and* $x \in [0, 1]$. *Suppose that*

$$\max_{\theta \in \mathcal{A}_s{}'} \left| \mathbb{E}\left[\prod_{i=1}^{s} \left(\frac{g_i}{D} \right)^{\theta_i} \right] \right| < 2^{-s}, \tag{8.39}$$

where $\mathcal{A}_s{}'$ *is the family of all* $\theta = (\theta_i)_{i=1}^s \in \mathcal{A}_s$ *such that* $\theta_i = 0$ *or* 1.

Then, there exists a multiplicative system $\{h_i\}_{i=1}^s$ *of functions defined on the interval* $[0, 2]$, *satisfying the conditions:* $h_i(x)\chi_{[0,1]}(x) = g_i(x)$ *and* $|h_i(x)| \le D$ *for all* $0 \le x \le 2$ *and* $i = 1, 2, \dots, s$.

Proof Let us represent the interval $[1, 2)$ as a union of 2^s nonoverlapping intervals of the same length, i.e.,

$$[1, 2) = \bigcup_{k=1}^{2^s} \Delta_k, \quad \text{where} \quad \Delta_k = [a_{k-1}, a_k), \; a_k = 1 + k2^{-s}, \; k = 0, 1, 2, \dots, 2^s.$$

Since card $\mathcal{A}_s{}' = 2^s - 1$ (recall that $(0, \dots, 0) \notin \mathcal{A}_s{}'$), we can establish a one-to-one mapping between $\mathcal{A}_s{}'$ and the family of intervals $\{\Delta_k\}_{k=1}^{2^s-1}$. Choose and fix arbitrarily one of these mappings.

Let $k = 1, 2, \dots, 2^s - 1$ be fixed, and let $\theta = (\theta_i)_{i=1}^s$ be the element of $\mathcal{A}_s{}'$, corresponding to the interval Δ_k. Equivalently, we can rewrite this element as

$$\theta = \{i_j\}_{j=1}^m := \{i = 1, 2, \dots, s : \theta_i = 1\}, \quad \text{where} \; i_1 < i_2 < \dots < i_m,$$

for some $m = 1, 2, \dots, s$. We wish to find extensions h_i of the functions g_i to the interval $[0, 2]$ satisfying the following conditions:

$$\int_{\Delta_k} h_{i_1} h_{i_2} \dots h_{i_m} \, dx = -\mathbb{E}(g_{i_1} g_{i_2} \cdots g_{i_m}) \tag{8.40}$$

and

$$\int_{\Delta_k} h_{i'_1} h_{i'_2} \dots h_{i'_l} \, dx = 0 \tag{8.41}$$

for any collection $\{i'_j\}_{j=1}^l$, which does not correspond (as an element of $\mathcal{A}_s{}'$) to the interval Δ_k.

Before to define the functions h_i, $i = 1, 2, \dots, s$, we set for arbitrary $t \in \Delta_k$

$$u_t(x) := \chi_{[a_{k-1},t)}(x) - \chi_{[t,a_k]}(x), \quad x \in \Delta_k.$$

Then, the function

$$v(t) := \int_{\Delta_k} u_t(x)\,dx = 2t - a_{k-1} - a_k, \quad a_{k-1} \le t \le a_k,$$

takes all values from the interval $[-2^{-s}, 2^{-s}]$. Therefore, because $D^{-m}|\mathbb{E}(g_{i_1} g_{i_2} \ldots g_{i_m})| < 2^{-s}$ by condition (8.39), there is a value $t_k \in (a_{k-1}, a_k)$ such that

$$D^m v(t_k) = -\mathbb{E}(g_{i_1} g_{i_2} \ldots g_{i_m}). \tag{8.42}$$

Suppose first that $m \ge 2$. Next, we need a system of step-functions $d_j(x)$, $x \in [a_{k-1}, t_k)$, $j = 1, 2, \ldots, m-1$, with the following properties:

$$|d_j(x)| \equiv 1 \text{ and } \int_{a_{k-1}}^{t_k} d_{j_1} d_{j_2} \ldots d_{j_l}\,dx = 0 \tag{8.43}$$

for arbitrary sets $1 \le j_1 < j_2 < \cdots < j_l \le m-1$ (for d_j we may take, for instance, usual Rademacher functions but with the domain $[a_{k-1}, t_k)$ rather than $[0, 1]$). By $c_j(x)$, $j = 1, 2, \ldots, m-1$, we denote the analogous system of functions defined on the interval $[t_k, a_k]$.

Now everything is ready to define the functions $h_i(x)$, $i = 1, 2, \ldots, s$, for $x \in \Delta_k$. If $i \ne i_j$ for any $j = 1, 2, \ldots, m$ (recall that the collection $\{i_j\}_{j=1}^m$ is an element of the family $\mathcal{A}_s{}'$, corresponding to the interval Δ_k), we set: $h_i(x) \equiv 0$. Otherwise,

$$h_{i_j}(x) = D \cdot (d_j(x)\chi_{[a_{k-1}, t_k)}(x) + c_j(x)\chi_{[t_k, a_k]}(x)) \text{ if } j = 1, 2, \ldots, m-1$$

and

$$h_{i_m}(x) = D \cdot \Big(\prod_{j=1}^{m-1} d_j(x)\chi_{[a_{k-1}, t_k)}(x) - \prod_{j=1}^{m-1} c_j(x)\chi_{[t_k, a_k]}(x)\Big).$$

By (8.42) and (8.43), we have

$$\int_{\Delta_k} h_{i_1} h_{i_2} \ldots h_{i_m}\,dx = D^m \int_{\Delta_k} u_{t_k}(x)\,dx = D^m v(t_k) = -\mathbb{E}(g_{i_1} g_{i_2} \ldots g_{i_m}),$$

whence it follows (8.40). As to equation (8.41), it is a consequence of the definition of h_i, $i = 1, 2, \ldots, s$, and (8.43).

In the case $m = 1$ we define h_i on Δ_k in the following way:

$$h_i(x) \equiv 0 \text{ if } i \ne i_1, \text{ and } h_{i_1}(x) = D u_{t_k}(x).$$

One can easily see that (8.40) and (8.41) still are fulfilled (see (8.42)).

Finally, define the functions h_i on the remaining part of the interval $[0, 2]$ by setting $h_i(x) = g_i(x)$, $x \in [0, 1]$, and $h_i(x) = 0$, $x \in \Delta_{2^s}$. Then, it can be easily checked that from (8.40) and (8.41) it follows that the system $\{h_i\}_{i=1}^s$ is multiplicative on $[0, 2]$, and moreover $|h_i(x)| \le D$ for all $0 \le x \le 2$ and $i = 1, 2, \dots, s$. □

Proof of Theorem 8.6 Further, as above, $\kappa(t, a) = \mathcal{K}(t, a; \ell_1, \ell_2)$, $a = (a_i)_{i=1}^\infty \in \ell_2$ and $t > 0$. Moreover, as in the previous chapter, for $a = (a_i)_{i=1}^m$, $m \in \mathbb{N}$, we set $\kappa(t, a) = \kappa(t, b)$, where $b = (b_i)_{i=1}^\infty$, $b_i = a_i$ if $1 \le i \le m$, and $b_i = 0$ if $i > m$.

According to Theorem 7.5 and subsequent Remark 7.3 it suffices to find a subsystem $\{f_{n_i}\}_{i=1}^s$, with $s \ge \max\{[\frac{1}{7}\log_2 N], 1\}$, that satisfies the equivalence:

$$\Big\|\sum_{i=1}^s a_i f_{n_i}\Big\|_p \asymp \kappa(\sqrt{p}, a) \text{ for all } p \ge 1 \text{ and } a = (a_i)_{i=1}^s, a_i \in \mathbb{R}, \tag{8.44}$$

with a constant depending only on D.

First of all, by using Lemma 8.2, we extract a subsystem $\{f_{n_i}\}_{i=1}^s$, $s \ge \max\{[\frac{1}{7}\log_2 N], 1\}$ satisfying condition (8.32). Because $\mathcal{A}_s' \subset \mathcal{A}_s$, inequality (8.39) holds for the functions $g_i := f_{n_i}$, $i = 1, 2, \dots, s$, as well. Consequently, by Lemma 8.3, we can define a multiplicative set $\{h_i\}_{i=1}^s$ of functions, defined on the interval $[0, 2]$, such that $h_i(x)\chi_{[0,1]}(x) = g_i(x)$ and $|h_i(x)| \le D$ for all $0 \le x \le 2$ and $i = 1, 2, \dots, s$. One can readily check that the functions $h_i'(x) = h_i(2x)$ form then a multiplicative set on $[0, 1]$, $|h_i'(x)| \le D$. Hence, applying Corollary 7.4 and Theorem 1.6, we get that for arbitrary $p \ge 1$ and $a_i \in \mathbb{R}$, $i = 1, 2, \dots, s$,

$$\Big\|\sum_{i=1}^s a_i h_i'\Big\|_{L_p[0,1]} \le C\kappa(\sqrt{p}, a),$$

where $C > 0$ depends only on D. Since $f_{n_i}(2x) = h_i'(x)$, $0 \le x \le 1/2$, and $\|\sigma_2\|_{X\to X} \le 2$ for every s.s. X, where $\sigma_2 f(x) = f(x/2)$ (see Appendix C), the latter inequality yields

$$\Big\|\sum_{i=1}^s a_i f_{n_i}\Big\|_p \le 2C\kappa(\sqrt{p}, a). \tag{8.45}$$

Let us prove the opposite inequality. To this end, we set $f := \sum_{i=1}^s a_i f_{n_i}$, where a sequence $a = (a_i)_{i=1}^s$ is fixed.

Suppose first that $p \in \mathbb{N}$ and $p \ge 3$. Let $s \le 16D^2$. Then, using subsequently Parseval's Identity, the Cauchy-Schwarz-Bunyakovskii inequality, and the definition of Peetre's $\mathcal{K}$-functional, we immediately obtain

$$\|f\|_p \ge \|f\|_2 = \|a\|_{\ell_2} \ge \frac{1}{\sqrt{s}}\sum_{i=1}^s |a_i| \ge \frac{1}{4D}\kappa(\sqrt{p}, a), \tag{8.46}$$

as we wished. Therefore, we can assume that

$$s > 16D^2. \tag{8.47}$$

Further, we proceed with the same plan as in the proof of Theorem 8.3. Let $\{A_j\}_{j=1}^p$ be an arbitrary partition of the set $\{1, 2, \dots, s\}$ (possibly, some of the sets A_j are empty). We form the Riesz products

$$R_s := \prod_{i=1}^{s} \left(1 + b_i f_{n_i}\right),$$

where the coefficients b_i will be chosen later so that $|b_i| \le 1/D$, $i = 1, 2, \dots, s$, and estimate first the integral I_s, defined by

$$I_s := \int_0^1 f(x) R_s(x)\, dx,$$

from below.

Let $\mathcal{A}_s$ and $\mathcal{A}_s{}'$ be the families defined in Lemmas 8.2 and 8.3, respectively. Since $\| f_{n_k} \|_2 = 1$, $k = 1, 2, \dots, s$, we can represent I_s as follows:

$$I_s = \sum_{i=1}^{s} a_i b_i + \sum_{i=1}^{s} a_i \gamma_i, \tag{8.48}$$

where

$$\gamma_i := \int_0^1 f_{n_i}(x)\, dx + \int_0^1 f_{n_i}(x) \cdot \sum_{\theta \in \mathcal{A}_s{}'(i)} \prod_{k=1}^{s} \left(b_k f_{n_k}(x)\right)^{\theta_k} dx.$$

Here, $\mathcal{A}_s{}'(i) := \mathcal{A}_s{}' \setminus \{\theta^i\}$, $\theta^i = (\theta^i_j)$, where $\theta^i_i = 1$ and $\theta^i_j = 0$ if $j \ne i$. Taking into account that $|b_i| \le 1/D$, $i = 1, 2, \dots, s$, we infer

$$\sum_{i=1}^{s} |\gamma_i| \le D \sum_{\theta \in \mathcal{A}_s} \left| \int_0^1 \prod_{k=1}^{s} \left(\frac{f_{n_k}(x)}{D} \right)^{\theta_k} dx \right|.$$

Moreover, note that $\operatorname{card} \mathcal{A}_s = 2^s - 1 + (2^{s-1} - 1)s \le s 2^s$. Therefore, from the Cauchy-Schwarz-Bunyakovskii inequality and (8.32) it follows

$$\sum_{i=1}^{s} |\gamma_i| \le D (s 2^s)^{1/2} 10^{-s/2} = D (s 5^{-s})^{1/2}.$$

In consequence, since (8.47) implies $(s5^{-s})^{1/2} \le 8/s < 2^{-1}D^{-2}$, we get

$$\sum_{i=1}^{s} |\gamma_i| \le \frac{1}{2D}. \tag{8.49}$$

Next, for every $j = 1, 2, \dots, p$ we choose numbers b_i, $i \in A_j$, satisfying the inequality

$$\sum_{i \in A_j} b_i^2 \le D^{-2}, \quad j = 1, 2, \dots, p.$$

so that

$$\sum_{i \in A_j} a_i b_i = \frac{1}{D}\Big(\sum_{i \in A_j} a_i^2\Big)^{1/2}.$$

Then, by (8.48) and (8.49), we have

$$I_s \ge \frac{1}{D}\sum_{j=1}^{p}\Big(\sum_{i\in A_j} a_i^2\Big)^{1/2} - \sum_{j=1}^{p}\Big|\sum_{i\in A_j} a_i\gamma_i\Big| \ge \frac{1}{D}\sum_{j=1}^{p}\Big(\sum_{i\in A_j} a_i^2\Big)^{1/2}$$
$$-\sum_{j=1}^{p}\Big(\sum_{i\in A_j} a_i^2\Big)^{1/2}\Big(\sum_{i\in A_j}|\gamma_i|\Big) \ge \frac{1}{2D}\sum_{j=1}^{p}\Big(\sum_{i\in A_j} a_i^2\Big)^{1/2}.$$

Since $\{A_j\}_{j=1}^{p}$ is an arbitrary partition of the set $\{1, 2, \dots, s\}$, from this inequality and (8.8) it follows that

$$I_s \ge \frac{1}{2\sqrt{2}D}\kappa(\sqrt{p}, a) \;\text{ if }\; p \in \mathbb{N}, p \ge 3. \tag{8.50}$$

Let us estimate now the integral I_s from above. By Hölder's inequality, we have

$$|I_s| \le \|R_s\|_{p'}\|f\|_p \;, \quad p' = \frac{p}{p-1}. \tag{8.51}$$

Then, precisely in the same way as in the proof of Theorem 8.2 (see inequality (8.25)), we prove that

$$\|R_s\|_{p'}^{p'} \le 4e\Big(1 + 2^s \sum_{\theta \in \mathcal{A}_s'} \Big|\mathbb{E}\Big\{\prod_{i=1}^{s}\Big[\frac{f_{n_i}}{D}\Big]^{\theta_i}\Big\}\Big|\Big).$$

Hence, as $\mathcal{A}_s{}' \subset \mathcal{A}_s$, applying the Cauchy-Schwarz-Bunyakovskii inequality and (8.32), we get

$$\|R_s\|_{p'} \leq 4e(1 + 2^{3s/2} \cdot 10^{-s/2}) \leq 8e.$$

As a result, from (8.50) and (8.51) it follows that

$$\kappa(\sqrt{p}, a) \leq 16\sqrt{2}eD\|f\|_p \ \text{ if } \ p \in \mathbb{N}, p \geq 3.$$

Similar estimates hold also for $p = 1$ and $p = 2$. Indeed, it is clear that $\kappa(1, a) = \|a\|_{\ell_2}$. Therefore, by Khintchine's L_1-inequality (see Theorem 5.4), $\kappa(1, a) \leq \sqrt{2}\|f\|_1$. Furthermore, from the concavity of the function $t \mapsto \kappa(t, a)$ and Parseval's Identity, it follows $\kappa(\sqrt{2}, a) \leq \sqrt{2}\kappa(1, a) = \sqrt{2}\|f\|_2$.

Extending then in a standard way the estimates obtained to all real numbers $p \geq 1$, we get

$$\kappa(\sqrt{p}, a) \leq C\|f\|_p \ \text{ for all } \ p \geq 1,$$

with some $C = C(D)$. Combining this fact together with (8.45), we come to equivalence (8.44), whose constant depends only on D. As was said above, this completes the proof. □

As an inspection of the first part of the proof of Theorem 8.6 shows, reasoning in the same way as in the proof of Theorem 7.9, we can obtain the following "vector-valued" result.

Theorem 8.7 *Let $\{f_n\}_{n=1}^N$ be an orthonormal system of functions on $[0, 1]$, $|f_n(t)| \leq D$ for all $t \in [0, 1]$ and $n = 1, 2, \ldots, N$.*

Then, there exists a system $\{f_{n_i}\}_{i=1}^s$, $1 \leq n_1 < n_2 < \cdots < n_s \leq N$, where $s \geq \max\{[\frac{1}{7}\log_2 N], 1\}$, such that for some $C > 0$, depending only on D, every Banach space F, all $x_1, x_2, \ldots, x_n$ from F and $z > 0$ the following inequality holds:

$$m\Big\{t \in [0, 1] : \Big\|\sum_{i=1}^s x_i f_{n_i}(t)\Big\|_F > z\Big\} \leq Cm\Big\{t \in [0, 1] : \Big\|\sum_{i=1}^s x_i r_i(t)\Big\|_F > C^{-1}z\Big\}.$$

Remark 8.1 As in the case of infinite systems, a system of functions $\{f_{n_i}\}_{i=1}^s$ from Theorem 8.6 is both a S_∞- and a Sidon system with a constant that depends only on D (but independent of the number of functions N of the initial system). Moreover, the results of this section on extraction of both infinite and finite subsystems, equivalent in distribution to the Rademacher system, are the best possible in the sense that further thinning cannot improve distributions of sums of functions from selected subsequences. Indeed, by Corollary 2.2, for every subsequence $\{r_{n_i}\}_{i=1}^\infty$ of

Rademacher functions, any $m \in \mathbb{N}$, $a_i \in \mathbb{R}$, $i = 1, 2, \dots, m$, and $z > 0$ we have

$$m\Big\{t \in [0,1] : \Big|\sum_{i=1}^{m} a_i r_{n_i}(t)\Big| > z\Big\} = m\Big\{t \in [0,1] : \Big|\sum_{i=1}^{m} a_i r_i(t)\Big| > z\Big\}.$$

Hence, if $\{f_n\} \overset{\mathbb{P}}{\sim} \{r_n\}$, then clearly $\{f_{n_i}\} \overset{\mathbb{P}}{\sim} \{r_i\}$ for every subsequence $\{f_{n_i}\}_{i=1}^{\infty} \subset \{f_n\}$ as well.

In the conclusion of this section, we show that the statement of Theorem 8.6 cannot be improved up to order. More precisely, the "logarithmic density" of a subsystem, equivalent in distribution to the corresponding set of Rademacher functions, in general, is the best result. We shall prove this fact even for the sidonicity, which is clearly weaker than the equivalence in distribution to the Rademacher system.

Proposition 8.1 *Suppose that a sequence of the form* $\{\sqrt{2}\cos 2\pi n_k x\}_{k=1}^{\infty}$ *is a Sidon system on* $[0, 1]$, *i.e., for some* $C_1 > 0$, *all* $m \in \mathbb{N}$ *and* $a_n \in \mathbb{R}$, $n = 1, 2, \dots, m$, *we have*

$$\sum_{k=1}^{m} |a_k| \le C_1 \Big\|\sum_{k=1}^{m} a_k \cos 2\pi n_k x\Big\|_{\infty}. \tag{8.52}$$

Then,

$$\operatorname{card}\{k : n_k \le N\} \le C_2 \ln N, \quad N = 2, 3, \dots, \tag{8.53}$$

where $C_2 > 0$ *depends only on* C_1.

Proof Denote $f_n(x) := \sqrt{2}\cos 2\pi n x$, $n = 1, 2, \dots$ We prove the following distribution estimate for the L_∞-norms of "random" trigonometric sums:

$$m\Big\{t \in [0,1] : \Big\|\sum_{n=1}^{N} a_n f_n r_n(t)\Big\|_{\infty} > 8\Big(\sum_{n=1}^{N} a_n^2\Big)^{1/2}\sqrt{\ln N}\Big\} \le \frac{72}{N^2}. \tag{8.54}$$

Show first that for each sum $P := \sum_{n=1}^{N} a_n f_n$ we have

$$\|P\|_\infty \le 2 \max_{k=0,1,\dots,36N^2} \Big|P\Big(\frac{k}{36N^2}\Big)\Big|. \tag{8.55}$$

Indeed, since the system $\{f_n\}_{n=1}^{\infty}$ is orthonormal, by the Cauchy-Schwarz-Bunyakovskii inequality, for all $0 \le x \le 1$

$$|P'(x)| \le 2\sqrt{2}\pi N \sum_{n=1}^{N} |a_n| \le 2\sqrt{2}\pi N^{3/2}\Big(\sum_{n=1}^{N} a_n^2\Big)^{1/2} \le 2\sqrt{2}\pi N^{3/2}\|P\|_\infty.$$

Let $\|P\|_\infty = |P(x_0)|$, $x_0 \in [0,1]$. Then $x_0 \in [(k-1)/(36N^2), k/(36N^2)]$ for some $k = 1, 2, \ldots, 36N^2$, and hence

$$\left|P(x_0) - P\left(\frac{k}{36N^2}\right)\right| \le \frac{1}{36N^2}\|P'\|_\infty \le \frac{2\sqrt{2}\pi N^{3/2}}{36N^2}|P(x_0)| < \frac{1}{2}|P(x_0)|.$$

This gives $|P(k/(36N^2))| > |P(x_0)|/2 = \|P\|_\infty/2$, and so inequality (8.55) is proved.

Applying (8.55), we can estimate the distribution functions of the L_∞-norms of "random" trigonometric sums from the left-hand side of inequality (8.54) as follows:

$$m\left\{t \in [0,1] : \left\|\sum_{n=1}^{N} a_n f_n r_n(t)\right\|_\infty > 8\left(\sum_{n=1}^{N} a_n^2\right)^{1/2}\sqrt{\ln N}\right\} \le$$

$$m\left\{t \in [0,1] : \max_{k=0,1,\ldots,36N^2}\left|\sum_{n=1}^{N} a_n f_n\left(\frac{k}{36N^2}\right) r_n(t)\right| > 4\left(\sum_{n=1}^{N} a_n^2\right)^{1/2}\sqrt{\ln N}\right\} \le$$

$$36N^2 \cdot \max_{k=0,1,\ldots,36N^2} m\left\{t \in [0,1] : \left|\sum_{n=1}^{N} a_n f_n\left(\frac{k}{36N^2}\right) r_n(t)\right| > 4\left(\sum_{n=1}^{N} a_n^2\right)^{1/2}\sqrt{\ln N}\right\}.$$

On the other hand, from the exponential distribution estimate proved in Proposition 1.2 and the obvious inequality $|f_n(x)| \le \sqrt{2}$, $n = 1, 2, \ldots$, it follows that the last quantity in the preceding chain of inequalities is dominated by

$$72N^2 \max_{k=0,1,\ldots,36N^2} \exp\left(-\frac{8\ln N\sum_{n=1}^{N} a_n^2}{\sum_{n=1}^{N} a_n^2 f_n^2(k/(36N^2))}\right) \le 72N^2\exp(-4\ln N) = \frac{72}{N^2}.$$

Summing up, we obtain (8.54).

Set now $a_{n_k} = 1$, $k = 1, 2, \ldots$ and $a_n = 0$ if $n \ne n_k$ for all $k = 1, 2, \ldots$ Then, as one can easily see, inequality (8.52) does not contradict estimate (8.54) for this choice of coefficients if only

$$\sum_{k:n_k \le N} 1 \le 4\sqrt{2}C_1\sqrt{\ln N}\left(\sum_{k:n_k\le N} 1\right)^{1/2}.$$

Hence, we get the desired inequality (8.53) with the constant $C_2 = 32C_1^2$. □

According to the last result, we cannot hope that a general system of r.v.'s contains a Sidon subsystem, whose density is larger than "logarithmic". However, this restriction does not extend, in general, to other types of lacunarity. In the final section of this chapter, we show that, for each $p > 0$, every orthonormal uniformly bounded system of r.v.'s contains subsystems, whose density is equal to a certain (depending on p) power of the number of elements of the initial system, with the L_p-norms of sums dominated (up to a constant) by those of the corresponding Rademacher sums.

8.4 Extraction of Subsystems with "sub-Rademacherian" L_p-Norms of Sums

Theorem 8.8 *Suppose* $\{f_n\}_{n=1}^N$ *is an orthonormal system of r.v.'s on a probability space* $(\Omega, \Sigma, \mathbb{P})$, $|f_n(\omega)| \le D$ *for all* $\omega \in \Omega$ *and* $n = 1, 2, \dots, N$. *For each* $p > 0$ *let* $p^* = 4$ *if* $p < 2$, *and let* p^* *be the least even positive integer such that* $p^* \ge p$ *if* $p \ge 2$.

Then, if $k \le N^{1/p^*}$, *there exists a set* $I \subset \{1, 2, \dots, N\}$, $\operatorname{card} I = k$, *such that for some constant* $C > 0$, *which depends only on* D, *and all* $a_n \in \mathbb{R}$, $n \in I$, *we have*

$$\Big\| \sum_{n \in I} a_n f_n \Big\|_{L_p(\Omega)} \le C \Big\| \sum_{n \in I} a_n r_n \Big\|_p . \tag{8.56}$$

Let us first introduce some notations, which will be used both in a forthcoming auxiliary statement and in the proof of Theorem 8.8.

Given set J and positive integer r by J^r we denote the set of all collections $(j_1, j_2, \dots, j_r)$ such that $j_k \in J$, $k = 1, 2, \dots, r$. Moreover, for arbitrary $Q = (q_1, q_2, \dots, q_r) \in \{1, 2, \dots, N\}^r$, functions f_n and $a_n \in \mathbb{R}$, $n = 1, 2, \dots, N$, we set

$$f_Q := \prod_{i=1}^r f_{q_i} \quad \text{and} \quad a_Q := \prod_{i=1}^r a_{q_i} .$$

Finally, let $\langle f, g \rangle := \int_\Omega f(\omega) g(\omega)\, d\mathbb{P}$.

Lemma 8.4 *Let r.v.'s* f_n, $n = 1, 2, \dots, N$, *satisfy the conditions of Theorem 8.8,* $k \le N$. *For every* $r \in \mathbb{N}$ *there exists a set* $I \subset \{1, 2, \dots, N\}$, $\operatorname{card} I = k$, *such that*

$$\sum_{(Q, Q') \in \mathcal{I}_1(r)} \langle f_Q, f_{Q'} \rangle^2 \le \frac{2 r k^{2r}}{N} D^{4r-4}, \tag{8.57}$$

where $\mathcal{I}_1(r)$ *is the family of all pairs* $(Q, Q') \in I^r \times I^r$, *for which at least one of the elements occurs in the sequence* $(q_1, q_2, \dots, q_r, q_1', q_2', \dots, q_r')$ *exactly once.*

Proof Denote by $S_r(I)$ the sum from the left-hand side of inequality (8.57). As in the case of Lemma 8.2, the idea of the proof is to estimate the average of these sums over all sets $I \subset \{1, 2, \dots, N\}$, $\operatorname{card} I = k$.

If $I_0 := \{1, 2, \ldots, N\}$, we have

$$\sum_{I:\,\mathrm{card}\,I=k} S_r(I) \le \sum_{I:\,\mathrm{card}\,I=k} 2r \sum_{i\in I} \sum_{\substack{V\in(I\setminus\{i\})^{r-1},\\ Q'\in(I\setminus\{i\})^r}} \langle f_i f_V, f_{Q'}\rangle^2$$

$$\le 2rC_{N-2r}^{k-2r} \sum_{V\in I_0^{r-1},\,Q'\in I_0^r} \sum_{i=1}^{N} \langle f_i, f_V f_{Q'}\rangle^2$$

To clarify this chain of inequalities, let us observe that the factor $2r$ appears in the first of them because the element i ($\in I$), which occurs exactly once in the sequence

$$(q_1, q_2, \ldots, q_r, q_1', q_2', \ldots, q_r'),$$

may appear at any of $2r$ places. As to the factor C_{N-2r}^{k-2r}, then it estimates the number of ways, in which a given term $\langle f_i, f_V f_{Q'}\rangle^2$ enters this sum via different subsets I.

Next, since $\{f_n\}_{n=1}^N$ is a uniformly bounded orthonormal system, by the Bessel inequality, for any $V \in I_0^{r-1}$ and $Q' \in I_0^r$ we have

$$\sum_{i=1}^{N} \langle f_i, f_V f_{Q'}\rangle^2 \le \|f_V f_{Q'}\|_{L_2(\Omega)}^2 \le D^{4r-4}.$$

Hence,

$$\min_{I:\,\mathrm{card}\,I=k} S_r(I) \le \frac{1}{C_N^k} \sum_{I:\,\mathrm{card}\,I=k} S_r(I) \le \frac{2rC_{N-2r}^{k-2r} N^{2r-1}}{C_N^k} \cdot D^{4r-4}$$

$$\le \frac{2rk(k-1)\ldots(k-2r+1)N^{2r-2}}{(N-1)\ldots(N-2r+1)} \cdot D^{4r-4} \le \frac{2rk^{2r}}{N} \cdot D^{4r-4},$$

and the lemma is proved. □

Proof of Theorem 8.8 Clearly, it suffices to consider the case when p is an even positive integer, $p \ge 4$. Hence, we shall assume that $p = 2r$, where $r \ge 2$. Moreover, by Lemma 8.4, we need only to show that inequality (8.56) holds for every set $I \subset \{1, 2, \ldots, N\}$, $\mathrm{card}\,I = k$, that satisfies (8.57) with $k \le N^{1/p} = N^{1/(2r)}$.

We first represent

$$\Big\|\sum_{n\in I} a_n f_n\Big\|_{L_p(\Omega)}^p = \int_\Omega \Big(\sum_{n\in I} a_n f_n(\omega)\Big)^{2r} d\mathbb{P} = \sum_{Q,Q'\in I^r} a_Q a_{Q'} \langle f_Q, f_{Q'}\rangle$$

$$= \sum_{(Q,Q')\in\mathcal{I}_1(r)} a_Q a_{Q'}\langle f_Q, f_{Q'}\rangle + \sum_{(Q,Q')\notin\mathcal{I}_1(r)} a_Q a_{Q'}\langle f_Q, f_{Q'}\rangle$$
$$:= S_1 + S_2.$$

Now, estimate the sums S_1 and S_2 separately. Applying successively the Cauchy-Schwarz-Bunyakovskii inequality, (8.57), and the fact that $k \le N^{1/(2r)}$, we get

$$S_1 \le \Big(\sum_{(Q,Q')\in\mathcal{I}_1(r)} a_Q^2 a_{Q'}^2\Big)^{1/2}\Big(\sum_{(Q,Q')\in\mathcal{I}_1(r)} \langle f_Q, f_{Q'}\rangle^2\Big)^{1/2}$$
$$\le \Big(\sum_{Q\in I^r} a_Q^2\Big)\cdot\Big(2r\,D^{4r-4}\Big)^{1/2} = (2r)^{1/2}D^{2r-2}\Big(\sum_{n\in I} a_n^2\Big)^r$$
$$= p^{1/2}D^{p-2}\Big\|\sum_{n\in I} a_n f_n\Big\|_{L_2(\Omega)}^p \le 2^p D^{p-2}\Big\|\sum_{n\in I} a_n r_n\Big\|_p^p.$$

In turn, since the functions f_n, $n = 1, 2, \ldots N$, are normalized in $L_2(\Omega)$, the sum S_2 can be estimated as follows:

$$S_2 \le \Big(\sum_{(Q,Q')\notin\mathcal{I}_1(r)} |a_Q||a_{Q'}|\Big)\max_{(Q,Q')\notin\mathcal{I}_1(r)} |\langle f_Q, f_{Q'}\rangle| \le D^{2r-2}\cdot \sum_{(Q,Q')\notin\mathcal{I}_1(r)} |a_Q||a_{Q'}|. \tag{8.58}$$

Moreover, for every fixed sequence $(a_n)_{n\in I}$ we have

$$\sum_{(Q,Q')\notin\mathcal{I}_1(r)} |a_Q||a_{Q'}| \le 3^{2r}\sum_{V\in\mathcal{I}_2(r)} a_V, \tag{8.59}$$

where $\mathcal{I}_2(r)$ consists of those collections $(v_1, v_2, \ldots, v_{2r}) \in I^{2r}$, in which each element of I appears any even number of times. Indeed, for any pair $(Q, Q') \in I^r \times I^r$, clearly, we can find $V \in \mathcal{I}_2(r)$ such that $|a_Q|\cdot|a_{Q'}| \le a_V$. In addition, one can easily define a procedure of assigning V to (Q, Q'), under which a given $V \in \mathcal{I}_2(r)$ corresponds to at most 3^{2r} pairs $(Q, Q') \notin \mathcal{I}_1(r)$ (we just organize inside of V replacing elements with smaller ones).

On the other hand, since the Rademacher system is strongly multiplicative (see Corollary 1.2 and the proof of Theorem 1.3), it holds

$$\Big\|\sum_{n\in I} a_n r_n\Big\|_p^p = \int_0^1 \Big(\sum_{n\in I} a_n r_n(t)\Big)^{2r} dt = \sum_{V\in\mathcal{I}_2(r)} a_V.$$

Therefore, combining this equation with inequalities (8.58) and (8.59), we get

$$S_2 \le 3^p D^{p-2}\Big\|\sum_{n\in I} a_n r_n\Big\|_p^p.$$

As a result, summarizing the estimates for the sums S_1 and S_2, we arrive at inequality (8.56) with $C = 5D^{1-2/p}$. □

Comments and References A wide range of problems connected with the extraction of lacunary subsystems is quite fully represented in the survey "Lacunary series and independent functions" by Gaposhkin [122]. Recall here the most important results of [122] related to the integrability and absolute convergence of series corresponding to lacunary systems of functions.

As we know, by Khintchine's inequality, the sum $\sum_{n=1}^{\infty} a_n r_n(x)$, $0 \le x \le 1$, belongs to the space L_p, for every $p < \infty$, provided that a sequence of coefficients $(a_n)_{n=1}^{\infty} \in \ell_2$. For lacunary trigonometric series

$$\frac{a_0}{2} + \sum_{k=1}^{\infty} (a_k \cos n_k x + b_k \sin n_k x), \quad \text{where } \frac{n_{k+1}}{n_k} \ge q > 1, \tag{8.60}$$

a similar result was proved by Zygmund [290] (cf. Corollary 7.5 from the previous chapter). These results served as a motivation for Banach and Sidon to introduce the notions of lacunary system of order p (briefly, S_p-system), as well of S_∞-system (see Definition B.10). Banach has proved that every orthonormal sequence $\{f_n\}$ of functions on $[0, 1]$ such that $\limsup_{n\to\infty} \|f_n\|_p < \infty$ contains a S_p-subsystem (see [65] or [146, § VII.2]). Hence, in particular, it follows a possibility of the extraction of a S_∞-subsystem from arbitrary uniformly bounded orthonormal sequence of functions. A slightly more precise result was obtained later by Stechkin [122, Theorem 1.3.1]. The Stechkin theorem on extraction of S_p-subsystems reads as follows: Let $p > 2$. A sequence of measurable functions $\{f_n\}_{n=1}^{\infty}$ on $[0, 1]$ contains a S_p-subsystem if and only if there is a subsequence $\{f_{n_k}\} \subset \{f_n\}$ such that $\|f_{n_k}\|_p \le D$ for all $k = 1, 2, \ldots$ and $f_{n_k} \xrightarrow{w} 0$ in L_2.

The study of the absolute convergence of lacunary series has been started with the work [117] by Fatou, who proved that each pointwise converging series of the form (8.60), where $q > 3$, converges absolutely. Next, in [267], Sidon has established the same property for every lacunary trigonometric series (8.60) that defines a bounded function. Moreover, Zygmund has proved in [292] a local version of the latter result. Somewhat later, Kaczmarz and Steinhaus carried over Sidon's theorem to Rademacher series [145]. As a natural generalization of these developments, the notions of Sidon and Sidon-Zigmund systems have been introduced (see the definitions in Appendix B). The widest class of systems, containing Sidon and Sidon-Zigmund subsystems, has been identified by Gaposhkin [122, Theorems 1.4.1 and 1.4.2]. Let us formulate the Gaposhkin theorem. Suppose that a system $\{f_n\}_{n=1}^{\infty}$ of measurable functions on $[0, 1]$ possesses the following properties: (1) $\|f_n\|_2 = 1$, $n = 1, 2, \ldots$; (2) $|f_n(x)| \le D$ for all $n = 1, 2, \ldots$ and $x \in [0, 1]$; (3) there is a subsequence $\{f_{n_k}\} \subset \{f_n\}$ such that $f_{n_k} \xrightarrow{w} 0$ in L_2. Then, $\{f_n\}$ contains Sidon and Sidon-Zigmund subsystems. Moreover, if a subsequence $\{f_{n_k}\}$ in (3) can be chosen so that it satisfies the following additional condition:

$$\liminf_{k\to\infty}\int_F f_{n_k}^2(x)\,dx>0 \quad \text{for every } F\subset[0,1] \text{ such that } m(F)>0,$$

then $\{f_n\}$ contains a strong Sidon-Zigmund subsystem.

In this connection, the results of Sects. 8.1 and 8.2 show that, in fact, the same conditions as in the above classical results provide the existence of subsystems satisfying the same requirements, as well as having some additional properties. Thereby they strengthen and complement the above statements.

Theorem 8.1 and Corollary 8.2 have been proved in [18]. Lemma 8.1, essentially going back to Lemma A from [123], is obtained in the same paper. The statement of Corollary 8.1 about a possibility of the extraction of exponentially integrable subsystems from systems of r.v.'s, satisfying the conditions (a) and (b) of Theorem 8.1, was announced by Gaposhkin [122, Theorem 1.3.2]; a detailed proof of this fact has been published in [18]. All the results of Sect. 8.2 have been proved in the paper [21] (being announced, as well as the results of Sect. 8.3, in [19]). Note also that Theorem 8.2 has found interesting applications both in studying the Rademacher system itself (see Chap. 11) and as a useful tool by investigation of various geometrical properties of s.s.'s (see e.g. [112]).

We now turn to the issues related to extracting lacunary subsystems from finite systems of r.v.'s. Assume that $\{f_n\}_{n=1}^N$ is a system of r.v.'s. It is required, under some conditions imposed on $\{f_n\}_{n=1}^N$, to find $s=s(N)$ as large as possible such that there is a subset $\{f_{n_i}\}_{i=1}^s\subset\{f_n\}_{n=1}^N$, which satisfies a certain lacunarity property with a constant independent of N. Let us mention a result on the density of finite Sidon subsystems proved by Kashin (see [152] or [153, Theorem VIII.9]). The Kashin theorem reads: Let $\{f_n\}_{n=1}^N$ be an orthonormal system of functions on $[0,1]$, $|f_n(x)|\le D$ for all $n=1,2,\ldots,N$ and $x\in[0,1]$. Then, there exists a subsystem $\{f_{n_i}\}_{i=1}^s\subset\{f_n\}_{n=1}^N$, $1\le n_1<\cdots<n_s\le N$, such that $s\ge\max\{[\frac{1}{6}\log_2 N],1\}$ and for any $a_i\in\mathbb{R}$, $i=1,2,\ldots,s$, we have

$$D^{-1}\Big\|\sum_{i=1}^s a_i f_{n_i}\Big\|_\infty\le\sum_{i=1}^s|a_i|\le 4D\Big\|\sum_{i=1}^s a_i f_{n_i}\Big\|_\infty.$$

A somewhat earlier, a weaker result (with $s\ge c(D)\ln\ln N$) was obtained by Milman and Wolfson [207]. Theorem 8.6 shows that, in fact, every such a system contains a subsystem $\{f_{n_i}\}_{i=1}^s$ of the same "logarithmic density" (i.e., $s\ge c\log_2 N$) that is equivalent in distribution to the system $\{r_i\}_{i=1}^s$ of Rademacher functions with a constant independent of N. Already after that as the author proved this theorem in the paper [21] (this result was announced also in [19]), he found out that a similar assertion, with the replacement of the equivalence in distribution to the Rademacher system with the equivalence of L_p-norms of corresponding sums, was somewhat earlier proved by Szarek [274, Theorem 1]. Though the latter result, clearly, formally weaker than Theorem 8.6, however, being combined with

Theorem 7.5 and Remark 7.3, it implies Theorem 8.6. Note that the above Szarek's result holds in the vector-valued setting as well. Combinatorial Lemma 8.2, which plays a key role in the proof of Theorem 8.6, is due to Kashin (see [152] or [153, § VIII.4, Lemma 1]). Another important auxiliary result, Lemma 8.3, on extension of functions of an uniformly bounded system to a wider interval so that to obtain a multiplicative uniformly bounded system, has been proved in [21]. Proposition 8.1, which implies sharpness (up to order) of results of Sect. 8.3, has been obtained by Stechkin in [269]. In the proof of this result we follow the monograph [153] (see inequalities (71)–(73) in Chapter 4). Finally, Theorem 8.8 has been proved in [274, Theorem 2].

Chapter 9
Extreme Properties of the Rademacher System

The Rademacher system takes a special place among all sequences of r.v.'s because of its very wide scope of applications. This is caused, in particular, by the fact that it might be an "extreme point" in one sense or another of a certain class of sequences. Thanks to that, the validity of some properties for systems from this class often follows immediately from their validity for the Rademacher system.

9.1 Rademacher Functions and the Hardy–Littlewood–Pólya Submajorization

Recall that the notation $x \prec y$, where $x, y \in L_1[0, 1]$, means that

$$\int_0^t x^*(s)\, ds \le \int_0^t y^*(s)\, ds \quad \text{for all } \ 0 \le t \le 1.$$

This partial ordering, which is called usually the *Hardy–Littlewood–Pólya submajorization*, arises naturally when studying the class of interpolation spaces between L_1 and L_∞ (see the Calderón–Mityagin Theorem D.6). In this section, we prove an extreme property of the Rademacher system with respect to this ordering in the class of sequences of independent symmetrically distributed functions with equal L_1-norms.

Theorem 9.1 *Let* $\{g_i\}_{i=1}^\infty$ *be a sequence of independent symmetrically distributed functions on* $[0, 1]$ *such that* $\|g_i\|_1 = b > 0$, $i = 1, 2, \dots$

Then, for all $n \in \mathbb{N}$ *and* $c_i \in \mathbb{R}$ *we have*

$$\sum_{i=1}^n c_i r_i \prec \frac{1}{b} \sum_{i=1}^n c_i g_i.$$

S. V. Astashkin, *The Rademacher System in Function Spaces*,
https://doi.org/10.1007/978-3-030-47890-2_9

Proof The proof is based mainly on successive application of the conditional expectation operator to the given functions, which will allow us to get eventually a sequence, equivalent in distribution to the Rademacher sequence.

For any $i = 1, 2, \dots, n$ we set

$$E_i^+ := \{t \in [0, 1] : g_i(t) > 0\}, \quad E_i^- := \{t \in [0, 1] : g_i(t) < 0\}$$

and

$$E_i^0 := \{t \in [0, 1] : g_i(t) = 0\}.$$

Moreover, let $\mathcal{A}_i$ (resp. $\mathcal{A}$) be the σ-algebra generated by the family $\{E_i^+, E_i^-, E_i^0\}$ (resp. $\bigcup_{i=1}^n \mathcal{A}_i$). We show first that for all $i = 1, 2, \dots, n$

$$\mathbb{E}[g_i|\mathcal{A}] = \mathbb{E}[g_i|\mathcal{A}_i], \tag{9.1}$$

where $\mathbb{E}[f|\Sigma]$ denotes the conditional expectation of a function f with respect to a σ-subalgebra Σ of the σ-algebra of Lebesgue measurable subsets of $[0, 1]$ (see Appendix A).

Fix $i = 1, 2, \dots, n$. Since $\mathcal{A}_i \subset \mathcal{A}$, the r.v. $\mathbb{E}[g_i|\mathcal{A}_i]$ is $\mathcal{A}$-measurable. Therefore, by definition of the conditional expectation, equation (9.1) will be proved once we verify that for every set $A \in \mathcal{A}$ of positive measure it holds

$$\int_A \mathbb{E}[g_i|\mathcal{A}_i](t)\, dt = \int_A g_i(t)\, dt. \tag{9.2}$$

According to definition of the σ-algebra $\mathcal{A}$, the set A is a union of sets of the form $B_{j_1,\dots,j_m} = \bigcap_{p=1}^m C_{j_p}$, where $C_{j_p} \in \left\{E_{j_p}^+, E_{j_p}^-, E_{j_p}^0\right\}$, j_p are pairwise distinct and $1 \leqslant j_p \leqslant n$, $p = 1, \dots, m$. Assume that $j_p \neq i$ for each $p = 1, \dots, m$. Then, since $E_i^+ \cup E_i^- \cup E_i^0 = [0, 1]$, we have

$$B_{j_1,\dots,j_m} = (B_{j_1,\dots,j_m} \cap E_i^+) \cup (B_{j_1,\dots,j_m} \cap E_i^-) \cup (B_{j_1,\dots,j_m} \cap E_i^0).$$

This shows that the set A can be represented as follows:

$$A = (E_i^+ \cap F^+) \cup (E_i^- \cap F^-) \cup (E_i^0 \cap F^0), \tag{9.3}$$

where the sets F^+, F^- and F^0 belong to the σ-algebra, generated by the family $\bigcup_{j\neq i} \mathcal{A}_j$. Consequently, each of the functions $\mathbb{E}[g_i|\mathcal{A}_i]$ and g_i is (stochastically) independent with respect to the set $\{\chi_{F^+}, \chi_{F^-}, \chi_{F^0}\}$. So, since the intersections from the right-hand side of (9.3) are pairwise disjoint, by definition of the conditional expectation, we get

$$\int_A \mathbb{E}[g_i|\mathcal{A}_i](t)dt = \int_{E_i^+\cap F^+} \mathbb{E}[g_i|\mathcal{A}_i](t)dt + \int_{E_i^-\cap F^-} \mathbb{E}[g_i|\mathcal{A}_i](t)dt+$$

$$+ \int_{E_i^0\cap F^0} \mathbb{E}[g_i|\mathcal{A}_i](t)dt = \int_0^1 \mathbb{E}[g_i|\mathcal{A}_i](t)\chi_{E_i^+}(t)\chi_{F^+}(t)dt+$$

$$+ \int_0^1 \mathbb{E}[g_i|\mathcal{A}_i](t)\chi_{E_i^-}(t)\chi_{F^-}(t)dt + \int_0^1 \mathbb{E}[g_i|\mathcal{A}_i](t)\chi_{E_i^0}(t)\chi_{F^0}(t)dt =$$

$$= \int_{E_i^+} \mathbb{E}[g_i|\mathcal{A}_i](t)dt\; m(F^+) + \int_{E_i^-} \mathbb{E}[g_i|\mathcal{A}_i](t)dt\; m(F^-)+$$

$$+ \int_{E_i^0} \mathbb{E}[g_i|\mathcal{A}_i](t)dt\; m(F^0) = \int_{E_i^+} g_i(t)dt\; m(F^+) + \int_{E_i^-} g_i(t)dt\; m(F^-)+$$

$$+ \int_{E_i^0} g_i(t)dt\; m(F^0).$$

Similarly, we have

$$\int_A g_i(t)dt = \int_{E_i^+} g_i(t)dt\; m(F^+) + \int_{E_i^-} g_i(t)dt\; m(F^-) + \int_{E_i^0} g_i(t)dt\; m(F^0).$$

As a result, we obtain (9.2) and hence also (9.1).

Further, denote $f_i := \mathbb{E}[g_i|\mathcal{A}_i]$, $i = 1, 2, \ldots, n$. Then, from the equation

$$f_i(t) = \frac{1}{m(E_i^+)} \int_{E_i^+} g_i(s)\,ds \cdot \chi_{E_i^+}(t) + \frac{1}{m(E_i^-)} \int_{E_i^-} g_i(s)\,ds \cdot \chi_{E_i^-}(t), \quad 0 \le t \le 1, \tag{9.4}$$

it follows

$$\int_0^1 |f_i(t)|\,dt = \left|\int_{E_i^+} g_i(s)\,ds\right| + \left|\int_{E_i^-} g_i(s)\,ds\right|$$

$$= \int_{E_i^+} |g_i(s)|\,ds + \int_{E_i^-} |g_i(s)|\,ds = \int_0^1 |g_i(t)|\,dt,$$

whence

$$\|f_i\|_1 = \|g_i\|_1 = b, \quad i = 1, 2, \dots, n. \tag{9.5}$$

Moreover, note that, from the "interpolation" point of view, the quantity $\int_0^t x^*(s)\, ds$ is nothing but Peetre's $\mathcal{K}$-functional $\mathcal{K}(t, x; L_1, L_\infty)$ (see equivalence (D.1) and the comment following to it), i.e., for every $x \in L_1[0, 1]$ and all $t > 0$ we have

$$\int_0^t x^*(s)\, ds = \inf\{||x_0||_{L_1} + t||x_1||_{L_\infty} : x = x_0 + x_1,\ x_0 \in L_1, x_1 \in L_\infty\}.$$

Therefore, since the mapping $x \mapsto \mathbb{E}[x|\mathcal{A}]$ is a one-norm operator in each of the spaces L_1 and L_∞ (see Appendix A), for any representation $x = x_0 + x_1$, $x_0 \in L_1$, $x_1 \in L_\infty$ it holds

$$\int_0^t \mathbb{E}[x|\mathcal{A}]^*(s)\, ds \le \|\mathbb{E}[x_0|\mathcal{A}]\|_{L_1} + t\|\mathbb{E}[x_1|\mathcal{A}]\|_{L_\infty} \le \|x_0\|_{L_1} + t\|x_1\|_{L_\infty}.$$

Taking the infimum in the right-hand side over all above expansions of x, we obtain

$$\mathbb{E}[x|\mathcal{A}] \prec x. \tag{9.6}$$

Using now (9.1), (9.6) and the linearity of the conditional expectation, for any $c_i \in \mathbb{R}$ we have

$$\sum_{i=1}^n c_i f_i = \sum_{i=1}^n c_i \mathbb{E}[g_i|\mathcal{A}_i] = \sum_{i=1}^n c_i \mathbb{E}[g_i|\mathcal{A}] = \mathbb{E}\Big[\sum_{i=1}^n c_i g_i|\mathcal{A}\Big] \prec \sum_{i=1}^n c_i g_i. \tag{9.7}$$

Since g_i, $i = 1, 2, \dots, n$, are symmetrically distributed, then $m(E_i^+) = m(E_i^-)$, and hence the functions $g_i \chi_{E_i^+}$ and $-g_i \chi_{E_i^-}$ are identically distributed for every $i = 1, 2, \dots, n$. In consequence, the three-valued functions f_i, $i = 1, 2, \dots, n$, are symmetrically distributed as well. Moreover, they are independent as functions measurable with respect to the independent σ-algebras $\mathcal{A}_i$.

Let us denote

$$\alpha_1^i = \alpha_2^i := \frac{1}{2} m\{t \in [0, 1] : f_i(t) = 0\}$$

and

$$\alpha_3^i = \alpha_4^i := \frac{1}{2} m\{t \in [0, 1] : f_i(t) \neq 0\}.$$

Then, because f_i are independent and symmetrically distributed, one can easily construct a sequence of intervals $\Delta_{j_1,\dots,j_i} \subset [0,1]$, where $i = 1, 2, \dots, n$, $j_l = 1, 2, 3, 4$, with the following properties:

$$m(\Delta_{j_1,\dots,j_i}) = \alpha^1_{j_1} \cdot \alpha^2_{j_2} \cdot \dots \cdot \alpha^i_{j_i}, i = 1, 2, \dots, n, \tag{9.8}$$

and

$$\Delta_{j_1,\dots,j_i} = \bigcup_{k=1,2,3,4} \Delta_{j_1,\dots,j_i,k}, i = 1, 2, \dots, n-1. \tag{9.9}$$

Let $a^i_1 = a^i_2 = 0$ and let a^i_3 be (a unique) positive value of the function f_i, $a^i_4 = -a^i_3$. For every $i = 1, 2, \dots, n$ define the function $\bar{f}_i$ as follows: if $t \in \Delta_{j_1,\dots,j_i}$, then $\bar{f}_i(t) = a^i_{j_i}$. Since f_i, $i = 1, 2, \dots, n$, are independent, then (9.8) and (9.9) imply that $\{\bar{f}_i\}_{i=1}^n$ and $\{f_i\}_{i=1}^n$ are similar collections.

Setting

$$\bar{E}_i^+ := \bigcup_{\substack{j_l=1,2,3,4\\ 1\le l\le i-1,\ j_i=1,3}} \Delta_{j_1,\dots,j_{i-1},j_i}$$

and

$$\bar{E}_i^- := \bigcup_{\substack{j_l=1,2,3,4\\ 1\le l\le i-1,\ j_i=2,4}} \Delta_{j_1,\dots,j_{i-1},j_i},$$

we see that $m(\bar{E}_i^+) = m(\bar{E}_i^-) = 1/2$, $\bar{f}_i(t) \ge 0$ for $t \in \bar{E}_i^+$, and $\bar{f}_i(t) \le 0$ for $t \in \bar{E}_i^-$. Moreover, from definition of the intervals $\Delta_{j_1,\dots,j_i}$ it follows the independence of the σ-algebras $\bar{\mathcal{A}}_i$, $i = 1, 2, \dots, n$, generated by the two-element families $\{\bar{E}_i^+, \bar{E}_i^-\}$. Therefore, according to (9.5), we have

$$\mathbb{E}[\bar{f}_i|\bar{\mathcal{A}}_i](t) = \frac{b}{2m(\bar{E}_i^+)} \cdot \chi_{\bar{E}_i^+}(t) + \frac{b}{2m(\bar{E}_i^-)} \cdot \chi_{\bar{E}_i^-}(t), \quad 0 \le t \le 1,$$

Comparing this with definition of the Rademacher functions and taking into account the independence of the conditional expectations $\mathbb{E}[\bar{f}_i|\bar{\mathcal{A}}_i]$, $i = 1, 2, \dots, n$, we see that the collections $\{\mathbb{E}[\bar{f}_i|\bar{\mathcal{A}}_i]\}_{i=1}^n$ and $\{br_i\}_{i=1}^n$ are similar. Next, arguing precisely as in the proof of equation (9.1), one can show that

$$\mathbb{E}[\bar{f}_i|\bar{\mathcal{A}}] = \mathbb{E}[\bar{f}_i|\bar{\mathcal{A}}_i], \quad i = 1, 2, \dots, n,$$

where the σ-algebra $\bar{\mathcal{A}}$ is generated by the family $\bigcup_{i=1}^n \bar{\mathcal{A}}_i$. Therefore, summarizing all, we conclude that the collections $\{\mathbb{E}[\bar{f}_i|\bar{\mathcal{A}}]\}_{i=1}^n$ and $\{br_i\}_{i=1}^n$ are similar as well.

Hence, taking into account that these sequences are independent and any conditional expectation operator is linear, for any $c_i \in \mathbb{R}$ we have

$$\Big(\sum_{i=1}^{n} c_i r_i\Big)^* = \frac{1}{b}\Big(\mathbb{E}\Big[\sum_{i=1}^{n} c_i \bar{f}_i | \bar{\mathcal{A}}\Big]\Big)^*.$$

Moreover, since the collections $\{f_i\}_{i=1}^n$ and $\{\bar{f}_i\}_{i=1}^n$ are similar, relations (9.6) and (9.7) imply that

$$\sum_{i=1}^{n} c_i r_i \prec \frac{1}{b}\Big(\sum_{i=1}^{n} c_i \bar{f}_i\Big)^* = \frac{1}{b}\Big(\sum_{i=1}^{n} c_i f_i\Big)^* \prec \frac{1}{b}\Big(\sum_{i=1}^{n} c_i g_i\Big),$$

completing the proof. □

The following result is a direct consequence of Theorems 9.1 and D.6.

Theorem 9.2 *Let $\{g_i\}_{i=1}^\infty$ be a sequence of independent symmetrically distributed r.v.'s on $[0, 1]$ such that $\|g_i\|_1 = b > 0$, $i = 1, 2, \ldots$ If X is an interpolation s.s. between L_1 and L_∞, then for every $n \in \mathbb{N}$ and all $c_i \in \mathbb{R}$, $i = 1, 2, \ldots, n$, we have*

$$\Big\|\sum_{i=1}^{n} c_i r_i\Big\|_X \le \frac{C}{b}\Big\|\sum_{i=1}^{n} c_i g_i\Big\|_X,$$

where a constant C depends only on X.

9.2 Modular Inequalities for Sums of Independent Symmetrically Distributed r.v.'s

Here, we shall establish a certain minimality of the Rademacher system in the same class of systems of independent symmetrically distributed r.v.'s as in the previous section. However, now instead of the L_1-norms, we fix their L_2-norms. The main result has the form of a modular inequality, which involves an *Orlicz function* (see Appendix C). Further, it will be assumed that such an Orlicz function is extended to the whole real line to be even.

Theorem 9.3 *For any Orlicz function Φ the following conditions are equivalent:*

(a) the inequality

$$\int_0^1 \Phi\Big(\sum_{i=1}^{n} \sigma_i r_i(s)\Big)\, ds \le \mathbb{E}\Phi\Big(\sum_{i=1}^{n} f_i\Big), \qquad (9.10)$$

where $\sigma_i \geq 0$, $i = 1, 2, \ldots, n$, are arbitrary, holds for all systems $\{f_i\}_{i=1}^n$ of independent symmetrically distributed r.v.'s such that $\mathbb{E} f_i^2 = \sigma_i^2$;

(*b*) *for every $b > 0$ the function $\Psi_b(x) := \Phi(b + \sqrt{|x|}) + \Phi(b - \sqrt{|x|})$ is convex on $\mathbb{R}$;*

(*c*) *the function Φ is twice differentiable and Φ'' is convex on $\mathbb{R}$.*

Proof We prove firstly the equivalence of conditions (a) and (b).

Assume that a symmetrically distributed r.v. η and a Rademacher function r' are independent, and F is an even function on $\mathbb{R}$. Then, by Fubini's theorem, for arbitrary $b > 0$ and $\sigma \geq 0$ we have

$$\begin{aligned}\mathbb{E}_{r',\eta} F(br' + \sigma\eta) &= \frac{1}{2}\left(\mathbb{E}_\eta F(b + \sigma\eta) + \mathbb{E}_\eta F(-b + \sigma\eta)\right)\\ &= \frac{1}{2}\left(\mathbb{E}_\eta F(b + \sigma\eta) + \mathbb{E}_\eta F(b - \sigma\eta)\right) = \mathbb{E}_\eta F(b + \sigma\eta).\end{aligned}$$

Therefore, if condition (a) holds, then for any symmetrically distributed r.v. f, $\mathbb{E} f^2 = \sigma^2$, independent with respect to r', by (9.10), we get

$$\mathbb{E}_r \Phi\left(b + \sqrt{\mathbb{E} f^2} r\right) = \mathbb{E}_r \Phi(b + \sigma r) = \mathbb{E}_{r,r'} \Phi(br' + \sigma r) \leq \mathbb{E}_{r',f} \Phi(br' + f) = \mathbb{E}_f \Phi(b + f),$$

where r is a Rademacher function such that $r \neq r'$. In addition, suppose that f and r are independent. Then, since f is symmetrically distributed, it has the same distribution as the r.v. $\sqrt{f^2} r$. Consequently, again according to Fubini's theorem, the preceding inequality can be rewritten in the following way:

$$\mathbb{E}_r \Phi\left(b + \sqrt{\mathbb{E} f^2} r\right) \leq \mathbb{E}_{r,f^2} \Phi\left(b + \sqrt{f^2} r\right) = \mathbb{E}_r \mathbb{E}_{f^2} \Phi\left(b + \sqrt{f^2} r\right).$$

Observe that f^2 can be treated here as an arbitrary nonnegative r.v. In particular, inserting in this inequality, instead of f^2, a two-valued r.v., which takes each of its nonnegative values u and v with probability $1/2$, we get

$$\mathbb{E}_r \Phi\left(b + \sqrt{\frac{u + v}{2}} r\right) \leq \frac{1}{2}\left(\mathbb{E}_r \Phi(b + ur) + \mathbb{E}_r \Phi(b + vr)\right)$$

for all $u, v \geq 0$. Since Φ is convex on $\mathbb{R}$, it is continuous. Therefore, the latter inequality implies that the function

$$\Psi_b(t) := \mathbb{E}_r \Phi\left(b + \sqrt{t} r\right) = \frac{1}{2}\left(\Phi(b + \sqrt{t}) + \Phi(b - \sqrt{t})\right)$$

is convex if $t \geq 0$ for every $b > 0$. Thus, (b) is proved.

Let us prove the converse. If the function Ψ_b is convex on $[0, \infty)$ for some b, then for every nonnegative r.v. g such that $\mathbb{E}g < \infty$, by the Jensen inequality (see e.g. [216, Theorem X.5.6] or Appendix A), we have

$$\Psi_b(\mathbb{E}g) \leq \mathbb{E}\Psi_b(g).$$

Suppose f is a symmetrically distributed r.v. such that $\mathbb{E}f^2 = \sigma^2$. Then, assuming that f and r are independent and setting $g := f^2$, by Fubini's theorem and the last inequality, we obtain

$$\mathbb{E}_{g,r}\Phi(b+f) = \mathbb{E}_g\mathbb{E}_r\Phi(b+\sqrt{g}r) = \frac{1}{2}\mathbb{E}_g\Psi_b(g) \geq \frac{1}{2}\Psi_b(\sigma^2) = \mathbb{E}_r\Phi(b+\sigma r).$$

We can now prove (a). Indeed, let $\{f_i\}_{i=1}^n$ be an arbitrary system of independent symmetrically distributed r.v.'s such that $\mathbb{E}f_i^2 = \sigma_i^2$, $i = 1, \ldots, n$. Using successively the last estimate and Fubini's theorem, we have

$$\begin{aligned}
\mathbb{E}\Phi\Big(\sum_{i=1}^n f_i\Big) &= \mathbb{E}_{f_1,\ldots,f_{n-1}}\mathbb{E}_{f_n}\Phi\Big(\sum_{i=1}^{n-1} f_i + f_n\Big) \\
&\geq \mathbb{E}\Phi\Big(\sum_{i=1}^{n-1} f_i + \sigma_n r_n\Big) = \mathbb{E}\Phi\Big(\sum_{i=1}^{n-2} f_i + \sigma_n r_n + f_{n-1}\Big) \\
&\geq \mathbb{E}\Phi\Big(\sum_{i=1}^{n-2} f_i + \sigma_n r_n + \sigma_{n-1} r_{n-1}\Big) \geq \cdots \geq \int_0^1 \Phi\Big(\sum_{i=1}^n \sigma_i r_i(s)\Big)\, ds,
\end{aligned}$$

and so (9.10) is proved. □

It remains to prove the equivalence of (b) and (c), which is an immediate consequence of the following two lemmas.

Lemma 9.1 *Let Φ be an Orlicz function. For a fixed $b > 0$ we consider the conditions:*

(i) *the function $\Psi_b(x) := \Phi(b+\sqrt{|x|}) + \Phi(b-\sqrt{|x|})$ is convex on $\mathbb{R}$;*
(ii) *Φ is differentiable on $\mathbb{R}$ and the function $\Psi_{b,1}(x) := \frac{1}{x}(\Phi'(b+x) - \Phi'(b-x))$ increases on $(0, \infty)$;*
(iii) *Φ is twice differentiable on $\mathbb{R}$ and for every $x \in \mathbb{R}$ we have*

$$x\left(\Phi''(b+x) + \Phi''(b-x)\right) - \left(\Phi'(b+x) - \Phi'(b-x)\right) \geq 0.$$

(iv) *Φ is twice continuously differentiable and its second derivative Φ'' is convex on $\mathbb{R}$.*

Then, the following chain of implications holds:

$$(iv) \Rightarrow (iii) \Rightarrow (ii) \Rightarrow (i).$$

Proof If (iv) is fulfilled, then we have

$$\Phi'(b+x) - \Phi'(b-x) = \int_{b-x}^{b+x} \Phi''(u)\,du \le 2x \cdot \frac{\Phi''(b+x) + \Phi''(b-x)}{2},$$

whence it follows (iii). Thus, the implication $(iv) \Rightarrow (iii)$ is proved.

To show that (iii) implies (ii) (resp. (ii) implies (i)), it suffices to calculate the first derivative of the function $\Psi_{b,1}$ (resp. Ψ_b). □

Lemma 9.2 *Suppose an Orlicz function Φ satisfies condition (i) from the preceding lemma for every $b > 0$. Then, Φ satisfies condition (iv).*

Proof To simplify reasoning, let us assume that the second derivative Φ'' (and hence the function Ψ_b'' for every $b \in \mathbb{R}$) exists and it is continuous (the proof of the lemma without this additional assumption see in [119]).

Observe first that for all $x \ge 0$ and $b \in \mathbb{R}$ it holds

$$x\left(\Phi''(b+x) + \Phi''(b-x)\right) - \left(\Phi'(b+x) - \Phi'(b-x)\right) \ge 0. \tag{9.11}$$

Indeed, because Ψ_b is convex, we have

$$\begin{aligned}\Psi_b''(x) &= \left(\frac{\Phi'(b+\sqrt{x}) - \Phi'(b-\sqrt{x})}{2\sqrt{x}}\right)' \\ &= \frac{\sqrt{x}(\Phi''(b+\sqrt{x}) + \Phi''(b-\sqrt{x})) - (\Phi'(b+\sqrt{x}) - \Phi'(b-\sqrt{x}))}{4x\sqrt{x}} \ge 0,\end{aligned}$$

which is equivalent to (9.11).

On the contrary, assume that Φ'' is not convex. Then, since it is continuous, there are $x_0 > 0$ and $b_0 \in \mathbb{R}$ such that

$$\Phi''(b_0) > \frac{1}{2}\left(\Phi''(b_0+x_0) + \Phi''(b_0-x_0)\right).$$

Let l be the straight line passing through the points $(b_0 - x_0, \Phi''(b_0 - x_0))$ and $(b_0 + x_0, \Phi''(b_0 + x_0))$. Using the continuity of Φ'' once more, we can find $x_0' > 0$ and $b_0' \in \mathbb{R}$ such that $b_0' - x_0'$ and $b_0' + x_0'$ are the nearest points to b_0 from the left and from the right, respectively, in which the graph of the function Φ'' crosses l. Then, from obvious geometric considerations it follows that

$$\Phi''(u) > \frac{1}{2}\left(\Phi''(b_0'+x_0') + \Phi''(b_0'-x_0')\right)$$

for all $u \in (b_0' - x_0', b_0' + x_0')$. This inequality implies

$$\Phi'(b_0'+x_0') - \Phi'(b_0'-x_0') = \int_{b_0'-x_0'}^{b_0'+x_0'} \Phi''(u)\,du > x_0' \cdot (\Phi''(b_0'+x_0') + \Phi''(b_0'-x_0')).$$

Since this contradicts (9.11), the proof is completed. □

If we fix, instead of the second moments (as in Theorem 9.3), the (absolute) first ones, then using averaging over signs, one can prove a similar result, which holds for *every* Orlicz function and all integrable r.v.'s.

Theorem 9.4 *For an arbitrary Orlicz function Φ, any $n \in \mathbb{N}$ and $a_i \geq 0$, $i = 1, 2, \ldots, n$, we have*

$$\int_0^1 \Phi\Big(\sum_{i=1}^n a_i r_i(t)\Big)\,dt \leq \int_0^1 \mathbb{E}\Phi\Big(\sum_{i=1}^n f_i r_i(t)\Big)\,dt \tag{9.12}$$

for all r.v.'s $f_i, f_2, \ldots, f_n$ such that $\mathbb{E}|f_i| = a_i$, $i = 1, 2, \ldots, n$.

In particular, for any $f_i, f_2, \ldots, f_n$, $\mathbb{E}|f_i| = a_i$, $i = 1, 2, \ldots, n$, there exist $\theta_i = \pm 1$, $i = 1, 2, \ldots, n$, satisfying the following inequality:

$$\int_0^1 \Phi\Big(\sum_{i=1}^n a_i r_i(t)\Big)\,dt \leq \mathbb{E}\Phi\Big(\sum_{i=1}^n \theta_i f_i\Big). \tag{9.13}$$

Proof For every sign arrangement $\theta = \{\theta_i\}_{i=1}^n$ we put

$$F_\theta := \{s \in [0, 1] : \ \mathrm{sign}\, f_i(s) = \theta_i \ \text{ for all } \ i = 1, 2, \ldots, n\}.$$

Moreover, if $\theta = \{\theta_i\}_{i=1}^n$ and $\eta = \{\eta_i\}_{i=1}^n$ are two sign arrangements, we set $\theta\eta := \{\theta_i \cdot \eta_i\}_{i=1}^n$. Then, since Φ is convex, by the Jensen inequality, we have

$$\begin{aligned}
\int_0^1 \Phi\Big(\sum_{i=1}^n a_i r_i(t)\Big)\,dt &= 2^{-n} \sum_{\eta_i=\pm 1} \Phi\Big(\sum_{i=1}^n a_i \eta_i\Big) = 2^{-n} \sum_{\eta_i=\pm 1} \Phi\Big(\mathbb{E}\Big(\sum_{i=1}^n |f_i| \cdot \eta_i\Big)\Big) \\
&\leq 2^{-n} \sum_{\eta_i=\pm 1} \sum_{\theta_i=\pm 1} \mathbb{E}\Big(\chi_{F_{\eta\theta}} \Phi\Big(\sum_{i=1}^n \eta_i |f_i|\Big)\Big) \\
&= 2^{-n} \sum_{\eta_i=\pm 1} \sum_{\theta_i=\pm 1} \mathbb{E}\Big(\chi_{F_{\eta\theta}} \Phi\Big(\sum_{i=1}^n \theta_i f_i\Big)\Big) \\
&= 2^{-n} \sum_{\theta_i=\pm 1} \mathbb{E}\Phi\Big(\sum_{i=1}^n \theta_i f_i\Big) = \int_0^1 \mathbb{E}\Phi\Big(\sum_{i=1}^n f_i r_i(t)\Big)\,dt,
\end{aligned}$$

and inequality (9.12) is proved. As to the existence of $\theta_i = \pm 1$, $i = 1, 2, \ldots, n$, such that inequality (9.13) holds, this is a direct consequence of (9.12). □

9.3 Maximality of the Rademacher Sequence in the Class of Uniformly Bounded Systems

In addition to being independent and symmetrically distributed, the Rademacher functions are also uniformly bounded; more precisely, $|r_i(s)| = 1$ a.e. on $[0, 1]$ for all $i \in \mathbb{N}$. This naturally leads to arising the conjecture that they have a certain property of maximality in the class of all sequences of independent symmetrically distributed r.v.'s $\{\xi_i\}$ such that $|\xi_i| \leqslant 1$ a.e. on a probability space $(\Omega, \Sigma, \mathbb{P})$. This is the case, indeed: from the Kwapień–Rychlik inequality (see Theorem 7.2) it follows that every such a sequence, for all $n \in \mathbb{N}$ and $t > 0$, satisfies the following estimate:

$$\mathbb{P}\Big\{\omega \in \Omega : \Big|\sum_{i=1}^{n} \xi_i(\omega)\Big| > t\Big\} \leqslant 2 \cdot m\Big\{s \in [0, 1] : \Big|\sum_{i=1}^{n} r_i(s)\Big| > t\Big\}. \tag{9.14}$$

It turns out that, by appropriate averaging of distributions, a similar inequality holds with constant 1 in the right-hand side.

Theorem 9.5 *Let $\{\xi_i\}_{i=1}^n$ be a system of independent symmetrically distributed (real-valued) r.v.'s on a probability space $(\Omega, \Sigma, \mathbb{P})$ such that $|\xi_i(\omega)| \leqslant 1$ $\mathbb{P}$-a.e. on Ω for all $1 \leqslant i \leqslant n$.*

Then, for every $x \geqslant 0$ we have

$$\int_x^{x+2} \mathbb{P}\Big\{\omega \in \Omega : \Big|\sum_{i=1}^{n} \xi_i(\omega)\Big| > t\Big\}\, dt \leqslant \int_x^{x+2} m\Big\{s \in [0, 1] : \Big|\sum_{i=1}^{n} r_i(s)\Big| > t\Big\}\, dt. \tag{9.15}$$

Remark 9.1 In contrast to (9.15), the inequality (9.14) fails without the factor 2 in the right-hand side. Moreover, the inequality

$$\int_x^{x+\tau} \mathbb{P}\Big\{\omega \in \Omega : \Big|\sum_{i=1}^{n} \xi_i(\omega)\Big| > t\Big\}\, dt \leqslant \int_x^{x+\tau} m\Big\{s \in [0, 1] : \Big|\sum_{i=1}^{n} r_i(s)\Big| > t\Big\}\, dt$$

does not hold if $\tau < 2$. Indeed, let $\xi_1 = r_1$ and $\xi_2 = 0$. Then

$$\mathbb{P}\{|\xi_1 + \xi_2| > t\} = \begin{cases} 1 \text{ if } & 0 < t < 1 \\ 0 \text{ if } & t \geqslant 1 \end{cases}$$

and

$$m\{|r_1 + r_2| > t\} = \begin{cases} 1/2 \text{ if } & 0 < t < 2 \\ 0 \quad \text{if} & t \geqslant 2. \end{cases}$$

Therefore, inserting $x = 0$, we get

$$\int_0^\tau \mathbb{P}\{|\xi_1 + \xi_2| > t\}\,dt = \min(1, \tau) \text{ and } \int_0^\tau m\{|r_1(s) + r_2(s)| > t\}\,dt = \frac{\tau}{2}.$$

It remains to note that the inequality $\min(1, \tau) \leqslant \tau/2$ holds only for $\tau \geqslant 2$.

To prove Theorem 9.5, we need some auxiliary statements, as well as the following notion from probability theory.

Let $\{H(x, y)\}_{y\in\mathbb{R}}$ be a family of distribution functions, $G(y)$ be a distribution function (as they are defined in probability theory). Then, the function

$$F(x) := \int_{-\infty}^{\infty} H(x, y)\,dG(y)$$

is called the *mixture of* $\{H(x, y)\}_{y\in\mathbb{R}}$ *with respect to* G. $F(x)$ is a distribution function as well; in particular, if G has a density, i.e., $G(y) = \int_{-\infty}^{y} g(u)\,du$, then

$$F(x) = \int_{-\infty}^{\infty} H(x, y)g(y)\,dy.$$

Lemma 9.3 *The distribution function $F(x)$ of an arbitrary symmetrically distributed (real-valued) r.v. ξ is a mixture of some family of distribution functions of two-valued symmetrically distributed r.v.'s.*

Proof Note that, for any symmetrically distributed r.v. ξ, the r.v.'s $|\xi|$ and $\operatorname{sign}\xi$ are independent (slightly modifying the usual definition of $\operatorname{sign} a$, here we set $\operatorname{sign}\xi = 1$ on A_+ and $\operatorname{sign}\xi = -1$ on A_-, where A_+ and A_- are arbitrary nonoverlapping sets such that $A_+ \cup A_- = \{\xi = 0\}$ and $\mathbb{P}(A_+) = \mathbb{P}(A_-) = \frac{1}{2}\mathbb{P}\{\xi = 0\}$). Consequently, for every $x \in \mathbb{R}$ we have

$$F(x) = \mathbb{P}\{|\xi| \cdot \operatorname{sign}\xi < x\} = \int_0^\infty \mathbb{P}\{y \cdot \operatorname{sign}\xi < x\}\,dG(y) = \int_0^\infty H(x, y)\,dG(y),$$

where G is the distribution function of the r.v. $|\xi|$ and $H(\cdot, y)$ is the distribution function of the r.v. $\xi_y = y \cdot \operatorname{sign} \xi$, $y \geqslant 0$. Since ξ_y is two-valued and symmetrically distributed for each $y \geqslant 0$, everything is done. □

Recall that the *convolution* of distribution functions F_1 and F_2 is the function $F_1 * F_2$ defined by

$$F_1 * F_2(x) := \int_{-\infty}^{\infty} F_2(x - y)\, dF_1(y).$$

Lemma 9.4 *Let F_i be the mixture of a family of distribution functions $\{H_i(x, y)\}_{y \in \mathbb{R}}$ with respect to a distribution function G_i, $i = 1, 2, \ldots, n$. Then, for every $x \in \mathbb{R}$ we have*

$$\underbrace{\int_{-\infty}^{\infty} \cdots \int_{-\infty}^{\infty}}_{n} H_1(\cdot, y_1) * H_2(\cdot, y_2) * \cdots * H_n(\cdot, y_n)(x)\, dG_1(y_1) \ldots dG_n(y_n) = F_1 * F_2 * \cdots * F_n(x).$$

Proof It suffices to consider the case $n = 2$ (next, we can proceed by induction). Since

$$F_i(x) := \int_{-\infty}^{\infty} H_i(x, y)\, dG_i(y), \quad i = 1, 2,$$

then applying Fubini's theorem yields

$$\int_{-\infty}^{\infty} \int_{-\infty}^{\infty} H_1(\cdot, y_1) * H_2(\cdot, y_2)(x)\, dG_1(y_1)\, dG_2(y_2)$$

$$= \int_{-\infty}^{\infty} \int_{-\infty}^{\infty} \int_{-\infty}^{\infty} H_2(x - t, y_2)\, dH_1(t, y_1)(x)\, dG_1(y_1)\, dG_2(y_2)$$

$$= \int_{-\infty}^{\infty} \int_{-\infty}^{\infty} \int_{-\infty}^{\infty} H_2(x - t, y_2)\, dG_2(y_2)\, dH_1(t, y_1)(x)\, dG_1(y_1)$$

$$= \int_{-\infty}^{\infty} \int_{-\infty}^{\infty} F_2(x - t)\, dH_1(t, y_1)(x)\, dG_1(y_1)$$

$$= \int_{-\infty}^{\infty} \int_{-\infty}^{\infty} H_1(t, y_1)\, dF_2(x-t)\, dG_1(y_1)$$

$$= \int_{-\infty}^{\infty} F_1(t)\, dF_2(x-t) = F_1 * F_2(x).$$

□

Proof of Theorem 9.5 We assume first that the result is already proved for arbitrary two-valued independent symmetrically distributed r.v.'s ξ_i such that $|\xi_i| \le 1$, $i = 1, 2, \dots, n$.

Let $F_i(x)$ be the distribution functions of r.v.'s ξ_i, $i = 1, 2, \dots, n$, satisfying the conditions of Theorem 9.5. Then, by Lemma 9.3,

$$F_i(x) = \int_0^1 H_i(x, y)\, dG_i(y), \quad i = 1, 2, \dots, n, \tag{9.16}$$

where G_i (resp. $H_i(\cdot, y)$, $0 \le y \le 1$) is the distribution function of the r.v. $|\xi_i|$ (resp. $\xi_{i,y} := y \cdot \operatorname{sign} \xi_i$). Moreover, since ξ_i, $i = 1, 2, \dots, n$, are independent, then the systems $\{|\xi_i|\}_{i=1}^n$ and $\{\xi_{i,y_i}\}_{i=1}^n$, $0 \le y_i \le 1$, consist also of independent r.v.'s. In particular, this implies that the convolution

$$H_1(\cdot, y_1) * H_2(\cdot, y_2) * \cdots * H_n(\cdot, y_n)(x)$$

is the distribution function of the sum $\sum_{i=1}^n \xi_{i,y_i}$ for any $0 \le y_i \le 1$, $i = 1, 2, \dots, n$. Combining this with the fact that ξ_{i,y_i} are symmetrically distributed r.v.'s, we get

$$\int_x^{x+2} \mathbb{P}\Big\{\omega \in \Omega : \Big|\sum_{i=1}^n \xi_{i,y_i}(\omega)\Big| > t\Big\}\, dt = 2 \int_x^{x+2} \mathbb{P}\Big\{\omega \in \Omega : \sum_{i=1}^n \xi_{i,y_i}(\omega) < -t\Big\}\, dt$$

$$= 2 \int_x^{x+2} H_1(\cdot, y_1) * H_2(\cdot, y_2) * \cdots * H_n(\cdot, y_n)(-t)\, dt.$$

Therefore, according to our assumption, it holds

$$2 \int_x^{x+2} H_1(\cdot, y_1) * H_2(\cdot, y_2) * \cdots * H_n(\cdot, y_n)(-t)\, dt \le \int_x^{x+2} m\Big\{s \in [0, 1] : \Big|\sum_{i=1}^n r_i(s)\Big| > t\Big\}\, dt$$

for any $0 \le y_i \le 1$, $i = 1, 2, \dots, n$. Let us integrate this inequality against the probability measure $dG_1(y_1) dG_2(y_2) \dots dG_n(y_n)$ over the cube $[0, 1]^n$. Then, taking into account (9.16) and applying Fubini's theorem together with Lemma 9.4, we obtain

$$2\int_x^{x+2} F_1 * F_2 * \cdots * F_n(-t)\,dt \le \int_x^{x+2} m\Big\{s\in[0,1]:\Big|\sum_{i=1}^n r_i(s)\Big|>t\Big\}\,dt.$$

Since r.v.'s ξ_i, $i=1,2,\ldots,n$, are independent and symmetrically distributed, then the latter inequality coincides with (9.15), and hence in this case the theorem is proved.

Thus, it remains to check that inequality (9.15) holds for any independent r.v.'s of the form $\xi_i = a_i \cdot r_i$, where $0 \le a_i \le 1$ and r_i are the Rademacher functions on $[0,1]$. Equivalently, we need to prove for arbitrary $0 \le a_i \le 1$ that

$$\int_x^{x+2} m\Big\{s\in[0,1]:\Big|\sum_{i=1}^n a_i r_i(s)\Big|>t\Big\}\,dt \leqslant \int_x^{x+2} m\Big\{s\in[0,1]:\Big|\sum_{i=1}^n r_i(s)\Big|>t\Big\}\,dt. \tag{9.17}$$

Setting

$$I(a_1,a_2,\ldots,a_n) := \int_x^{x+2} m\Big\{s\in[0,1]:\Big|\sum_{i=1}^n a_i r_i(s)\Big|>t\Big\}\,dt,$$

we can rewrite (9.17) as follows:

$$I(a_1,a_2,\ldots,a_n) \leqslant I(1,1,\ldots,1). \tag{9.18}$$

First of all, we observe that

$$I(a_1,a_2,\ldots,a_n) \leqslant I(\lambda a_1,\lambda a_2,\ldots,\lambda a_n) \tag{9.19}$$

for each $\lambda \geqslant 1$. Indeed,

$$\begin{aligned} I(\lambda a_1,\lambda a_2,\ldots,\lambda a_n) &= \int_x^{x+2} m\Big\{s:\lambda\Big|\sum_{i=1}^n a_i r_i(s)\Big|>t\Big\}\,dt \\ &\ge \int_x^{x+2} m\Big\{s:\Big|\sum_{i=1}^n a_i r_i(s)\Big|>t\Big\}\,dt \\ &= I(a_1,a_2,\ldots,a_n). \end{aligned}$$

The proof of inequality (9.18) (or equivalently (9.17)) will be based on successive increase of the parameter $\lambda \geqslant 1$ in such a way that, first, one of the coefficients of the sum $\sum_{i=1}^{n} \lambda a_i r_i$ became equal to 1, then another one and so on. Namely, for example, if $\lambda a_1 = 1$ (and still $0 \leqslant \lambda a_i \leqslant 1$ for $i = 2, 3, \ldots, n$), then inequality (9.19) takes the form:

$$I(a_1, a_2, \ldots, a_n) \leqslant I(1, \lambda a_2, \ldots, \lambda a_n).$$

Applying a similar operation to this inequality another $n - 1$ times we shall come eventually to (9.18). Thus, since (9.19) is established and the Rademacher functions are identically distributed and independent, it suffices to check that for any $1 \leq k < n$ and $\lambda \geqslant 1$ we have

$$I(1, \ldots, 1, a_{k+1}, \ldots, a_n) \leqslant I(1, \ldots, 1, \lambda a_{k+1}, \ldots, \lambda a_n). \tag{9.20}$$

Given $x \geqslant 0$ and $\lambda \geqslant 1$, consider the function

$$\Phi(\lambda) := \int_{x}^{x+2} m\Big\{u \in [0, 1] : \Big|\sum_{i=1}^{k} r_i(u) + \lambda \sum_{i=k+1}^{n} a_i r_i(u)\Big| > t\Big\}\, dt. \tag{9.21}$$

We need to prove that Φ increases for $\lambda \geqslant 1$.

Let

$$f(u) := \sum_{i=1}^{k} r_i(u) \quad \text{and} \quad g(u) := \sum_{i=k+1}^{n} a_i r_i(u).$$

By definition, f is a step-function of the form

$$f(u) = \sum_{i=1}^{2^k} b_i \chi_{\Delta_k^i}(u), \quad \text{where } \Delta_k^i = [(i-1)2^{-k}, i2^{-k}), \; i = 1, \ldots, 2^k.$$

Moreover, the function $f + \lambda g$ is symmetrically distributed and the functions $g \cdot \chi_{\Delta_k^i}$ are identically distributed for all $i = 1, \ldots, 2^k$. Therefore, using the introduced notation, we get

$$\Phi(\lambda) = \int_{x}^{x+2} m\,\{u \in [0, 1] : |f(u) + \lambda g(u)| > t\}\, dt$$

$$= 2 \int_{x}^{x+2} m\,\{u \in [0, 1] : f(u) + \lambda g(u) > t\}\, dt$$

$$= 2\sum_{i=1}^{2^k} \int_x^{x+2} m\left\{u \in \Delta_k^i : b_i + \lambda g(u) > t\right\} dt$$

$$= 2\sum_{i=1}^{2^k} \int_x^{x+2} m\left\{u \in \left[0, 2^{-k}\right] : g(u) > \frac{t - b_i}{\lambda}\right\} dt.$$

In consequence, since

$$\int_x^{x+2} m\left\{u \in \left[0, 2^{-k}\right] : g(u) > \frac{t - b_i}{\lambda}\right\} dt = \lambda \int_{(x-b_i)/\lambda}^{(x+2-b_i)/\lambda} m\left\{u \in \left[0, 2^{-k}\right] : g(u) > s\right\} ds,$$

we have

$$\Phi(\lambda) = 2\lambda \sum_{i=1}^{2^k} \int_{(x-b_i)/\lambda}^{(x+2-b_i)/\lambda} G(s)\, ds,$$

where

$$G(s) := m\left\{u \in \left[0, 2^{-k}\right] : g(u) > s\right\}.$$

Clearly, the function $G(s)$ decreases. Hence,

$$G\left(\frac{x - b_i}{\lambda}\right) \geqslant G\left(\frac{x + 2 - b_i}{\lambda}\right)$$

and

$$\int_{(x-b_i)/\lambda}^{(x+2-b_i)/\lambda} G(s)\, ds \geqslant \frac{2}{\lambda} G\left(\frac{x + 2 - b_i}{\lambda}\right).$$

Thus, using the last expression for the function $\Phi(\lambda)$ and the well-known differentiation formula

$$\frac{d}{dy} \int_{\varphi(y)}^{\psi(y)} f(x)\, dx = f(\psi(y))\psi'(y) - f(\varphi(y))\varphi'(y),$$

we can estimate $\Phi'(\lambda)$ from below as follows:

$$\Phi'(\lambda) = \frac{2}{\lambda}\Big(\sum_{i=1}^{2^k} G\left(\lambda^{-1}(x-b_i)\right)(x-b_i) - \sum_{i=1}^{2^k} G\left(\lambda^{-1}(x+2-b_i)\right)(x+2-b_i)\Big)$$

$$+ 2\sum_{i=1}^{2^k} \int_{(x-b_i)/\lambda}^{(x+2-b_i)/\lambda} G(s)\,ds \geqslant -\frac{4}{\lambda}\sum_{i=1}^{2^k} G\left(\lambda^{-1}(x+2-b_i)\right)$$

$$+ \frac{4}{\lambda}\sum_{i=1}^{2^k} G\left(\lambda^{-1}(x+2-b_i)\right) = 0$$

Hence, $\Phi'(\lambda) \geqslant 0$, which implies that $\Phi(\lambda)$ increases for $\lambda \geq 1$. As a result, we get inequality (9.20), and so the proof is completed. □

9.4 An Extremal Problem

In Chap. 5, when seeking the optimal constants in Khintchine's inequality (see Theorem 5.3), we proved that, for arbitrary $n \in \mathbb{N}$ and $p \in \{2\} \cup [3, \infty)$, the maximum

$$\max\left\{\Big\|\sum_{i=1}^{n} t_i r_i\Big\|_p : t_i \in \mathbb{R}, \sum_{i=1}^{n} t_i^2 = 1\right\}$$

is attained at $t_i = 1/\sqrt{n}$, $i = 1, 2, \ldots, n$. Here, we consider another extreme problem related as well to the norm $\|\sum_{i=1}^{n} t_i r_i\|_p$, $1 \leq p < \infty$.

Theorem 9.6 *For any $n \in \mathbb{N}$ we have*

$$\min\left\{\Big\|\sum_{i=1}^{n} t_i r_i\Big\|_p : t_i \in \mathbb{R}, \sum_{i=1}^{n} |t_i|^p = 1\right\} = n^{-1/p}\Big\|\sum_{i=1}^{n} r_i\Big\|_p$$

if $1 < p < 2$, and

$$\max\left\{\Big\|\sum_{i=1}^{n} t_i r_i\Big\|_p : t_i \in \mathbb{R}, \sum_{i=1}^{n} |t_i|^p = 1\right\} = n^{-1/p}\Big\|\sum_{i=1}^{n} r_i\Big\|_p$$

if $2 < p < \infty$.

Moreover, each of these extrema is attained only at the sequences $(t_i)_{i=1}^n$, $t_i = \pm n^{-1/p}$, $i = 1, 2, \ldots, n$.

In fact, we prove here a more general statement related to the norm of the diagonal operator acting from the n-dimensional subspace of space $L_p(\Omega, \mu)$, spanned by a system of symmetrically and identically distributed independent r.v.'s, in ℓ_p^n. To formulate it, denote by Δ the operator, defined on the linear span $[f_i]$ of functions $f_i \in L_p(\Omega, \mu)$, $i = 1, 2, \ldots, n$ (μ is a probability measure on Ω) by

$$\Delta\Big(\sum_{i=1}^{n} t_i f_i\Big) = (t_i)_{i=1}^n .$$

Moreover, let Δ^{-1} be the inverse operator for Δ.

Theorem 9.7 *Let $1 \le p < \infty$. Suppose $f_i : \Omega \to \mathbb{R}$ are symmetrically and identically distributed independent r.v.'s, $\|f_i\|_p = 1$, $i = 1, 2, \ldots, n$.*

Then, if $1 < p < 2$, we have

$$\|\Delta\| = n^{1/p} \Big\| \sum_{i=1}^{n} f_i \Big\|_p^{-1},$$

and the norm $\|\Delta\|$ is attained in the unit ball of the subspace $[f_i]$ only at functions of the form

$$\Big(\sum_{i=1}^{n} \pm f_i\Big) \Big\| \sum_{i=1}^{n} f_i \Big\|_p^{-1}, \quad i = 1, 2, \ldots, n.$$

If $2 < p < \infty$, then

$$\|\Delta^{-1}\| = n^{-1/p} \Big\| \sum_{i=1}^{n} f_i \Big\|_p \tag{9.22}$$

and this norm is attained in the unit ball of the space ℓ_p^n only at the vectors $(\pm n^{-1/p})_{i=1}^n$, $i = 1, 2, \ldots, n$.

We shall present a detailed proof of this theorem in the case $1 < p < 2$, starting with the following simple observation related to the definition of the norm of a linear operator.

Let X and Y be arbitrary linear normed spaces, and let A be a one-to-one linear operator bounded from X onto Y. Then, we have

$$\|A\|^{-1} = \inf\{\|A^{-1}y\|_X : \|y\|_Y = 1\}.$$

In addition, let $x \in X$ and $y = \|A\|^{-1}Ax$. Then, the equations $\|x\|_X = 1$ and $\|A\| = \|Ax\|_Y$ hold if and only if $\|y\|_Y = 1$ and $\|A\|^{-1} = \|A^{-1}y\|_X$.

Hence, the first part of Theorem 9.7 will be obtained once we prove the following statement.

Proposition 9.1 *Let $1 < p < 2$. Suppose that r.v.'s f_i, $i = 1, 2, \ldots, n$, satisfy all the conditions of Theorem 9.7. Then, setting*

$$m := \min\left\{\Big\|\sum_{i=1}^{n} t_i f_i\Big\|_p : t_i \in \mathbb{R}, \sum_{i=1}^{n} |t_i|^p = 1\right\},$$

we have $m = n^{-1/p}\|\sum_{i=1}^{n} f_i\|_p$.

Moreover, if $\|\sum_{i=1}^{n} t_i f_i\|_p = m$ for some sequence $(t_i)_{i=1}^{n}$ such that $\sum_{i=1}^{n} |t_i|^p = 1$, then $t_i = \pm n^{-1/p}$, $i = 1, 2, \ldots, n$.

Proof First of all, as in the proof of Proposition 5.1, we put $\langle t\rangle^p := |t|^p \cdot \operatorname{sign} t$ for any $t \in \mathbb{R}$ and $p > 0$. Then, for the function $f(t) := |t|^p$, $t \in \mathbb{R}$, where $p > 1$, we clearly have

$$f'(t) = p \cdot \langle t\rangle^{p-1}. \tag{9.23}$$

Further, a standard compactness argument implies the existence of a sequence $(a_i)_{i=1}^{n}$ such that $\sum_{i=1}^{n} |a_i|^p = 1$ and

$$\Big\|\sum_{i=1}^{n} a_i f_i\Big\|_p = m. \tag{9.24}$$

To get the desired result, it suffices to show that $a_i = \pm n^{-1/p}$ for all $i = 1, 2, \ldots, n$. Taking into account that the given r.v.'s f_i, $i = 1, 2, \ldots, n$, are independent, symmetrically and identically distributed, we can assume that $1 \geq a_1 \geq a_2 \geq \cdots \geq a_n \geq 0$.

Being a solution of the above extremal problem, the sequence (a_i) has to satisfy the equations:

$$\frac{\partial\phi}{\partial t_j}(a_1, a_2, \ldots, a_n, \lambda_0) = 0, \quad j = 1, 2, \ldots, n, \quad \text{for some } \lambda_0 \in \mathbb{R}, \tag{9.25}$$

where ϕ is Lagrange's function for this problem, i.e.,

$$\phi(t_1, t_2, \ldots, t_n, \lambda) := \int_\Omega \Big|\sum_{j=1}^{n} t_j f_j\Big|^p d\mu - \lambda \cdot \Big(\sum_{j=1}^{n} |t_j|^p - 1\Big).$$

According to rule (9.23), differentiation shows that (9.25) is equivalent to the following system of equations

$$\int_\Omega f_j\Big(\sum_{j=1}^n a_j f_j\Big)^{p-1} d\mu = \lambda_0 a_j^{p-1}, \quad j = 1, 2, \ldots, n. \tag{9.26}$$

Multiplying the jth equation from (9.26) by a_j and then summing up the resulting equations over all j, we get

$$\int_\Omega \Big|\sum_{j=1}^n a_j f_j\Big|^p d\mu = \lambda_0 \sum_{j=1}^n a_j^p = \lambda_0.$$

Hence, by (9.24),

$$\lambda_0 = m^p. \tag{9.27}$$

Furthermore, since the sequence $(a_i)_{i=1}^n$ decreases, we have $na_1^p \geq \sum_{j=1}^n a_j^p = 1$, and so $a_1 > 0$. Denote

$$b_j := \frac{a_j}{a_1}, \quad j = 1, 2, \ldots, n.$$

Then, $1 = b_1 \geq b_2 \geq \cdots \geq b_n \geq 0$, and after dividing the first equation from (9.26) (i.e., for $j = 1$) by a_1^{p-1} we deduce

$$\lambda_0 = \int_\Omega f_1\Big(\sum_{j=1}^n b_j f_j\Big)^{p-1} d\mu. \tag{9.28}$$

Also, note that, by the definition of m,

$$m \leq n^{-1/p}\Big\|\sum_{i=1}^n f_i\Big\|_p.$$

Therefore, from (9.27) and (9.28) it follows

$$\int_\Omega f_1\Big(\sum_{j=1}^n b_j f_j\Big)^{p-1} d\mu \leq n^{-1}\int_\Omega \Big|\sum_{j=1}^n f_j\Big|^p d\mu. \tag{9.29}$$

Introduce now the function H defined on $\mathbb{R}^n$ by

$$H(x_1, x_2, \ldots, x_n) := \int_\Omega f_1\Big(\sum_{j=1}^n x_j f_j\Big)^{p-1} d\mu.$$

Let also

$$m_1 := \inf\{H(x_1, \dots, x_n) : 1 = x_1 \ge x_2 \ge \cdots \ge x_n \ge 0\}.$$

Since $f_i \in L_p(\Omega, \mu)$, $i = 1, 2, \dots, n$, by using Hölder's inequality, one can readily verify that the function H is well-defined. In addition, because f_i are independent and identically distributed, we have

$$H(1, 1, \dots, 1) = \int_\Omega f_k \Big\langle \sum_{j=1}^n f_j \Big\rangle^{p-1} d\mu \quad \text{for all } k = 1, 2, \dots, n.$$

Summing up these equations over all k and using that $t \cdot \langle t \rangle^{p-1} = |t|^p$, we get

$$H(1, 1, \dots, 1) = n^{-1} \int_\Omega \Big| \sum_{j=1}^n f_j \Big|^p d\mu.$$

Thus, according to the definition of m_1, the equation

$$m_1 = H(1, 1, \dots, 1), \tag{9.30}$$

would imply the inequality, opposite to (9.29). This, combined together with (9.27) and (9.28), yields the first statement of Proposition 9.1. Since f_i, $i = 1, \dots, n$, are independent r.v.'s, normalized in $L_p(\Omega, \mu)$, equation (9.30) and the second statement of the proposition will be proved once we verify that from the equation

$$H(b_1, b_2, \dots, b_n) = m_1, \tag{9.31}$$

with some

$$1 = b_1 \ge b_2 \ge \cdots \ge b_n \ge 0 \text{ and } b_n < 1, \tag{9.32}$$

it follows the pairwise disjointness of f_i, $i = 1, \dots, n$ (clearly, nonzero independent r.v.'s fail to be disjoint).

To this end, we need the following auxiliary assertion. □

Lemma 9.5 *Let $1 < p < 2$.*

(*i*) *If $\alpha, \beta \in \mathbb{R}$ and*

$$h(t) := \alpha[\langle \alpha + t\beta \rangle^{p-1} + \langle \alpha - t\beta \rangle^{p-1}] \text{ for } t \ge 0,$$

then either $\alpha\beta = 0$ and then $h(t) = 2|\alpha|^p$, or $h(t)$ is a strictly decreasing function for $t \ge 0$.

(ii) If $u, v \in L_p(\Omega, \mu)$ *and*

$$h(t) := \int_\Omega u[\langle u(\omega) + tv(\omega)\rangle^{p-1} + \langle u(\omega) - tv(\omega)\rangle^{p-1}]\, d\mu(\omega) \ \textit{for } t \ge 0,$$

then either $uv = 0$ μ*-a.e. and then* $h(t) = 2\|u\|_p^p$, *or* $h(t)$ *is a strictly decreasing function for* $t \ge 0$.

Proof If $\alpha\beta \neq 0$, then equation (9.23) implies that $h'(t) < 0$ for $t > 0$. Hence, we get (i). Part (ii) follows from (i) by integration. □

Let us proceed with the proof of the proposition. Assume that a sequence $(b_k)_{k=1}^n$ satisfies (9.31) and also there is $k \in \mathbb{N}$, $1 \le k < n$, for which $b_k = 1$ and $b_{k+1} < 1$.

We show first that

$$\Big(\sum_{j=1}^{k} f_j\Big)\Big(\sum_{j=k+1}^{n} b_j f_j\Big) = 0 \ \ \mu - \text{a.e.} \tag{9.33}$$

If $b_{k+1} = 0$, then $b_j = 0$ for all $k+1 \le j \le n$, which implies (9.33). Therefore, we can assume that $b_{k+1} \neq 0$. Define the functions

$$u := \sum_{j=1}^{k} b_j f_j = \sum_{j=1}^{k} f_j, \quad v := \sum_{j=k+1}^{n} b_j f_j,$$

and

$$h(t) := \int_\Omega u[\langle u + tv\rangle^{p-1} + \langle u - tv\rangle^{p-1}]\, d\mu, \quad t \ge 0.$$

Since $0 < b_{k+1} < 1$, then, by Lemma 9.5, we have $h(1) \ge h(1/b_{k+1})$. Moreover, in view of the definition of H, taking into account once more that f_i are independent, symmetrically and identically distributed, we get

$$h(t) = 2\int_\Omega u\langle u + tv\rangle^{p-1}\, d\mu, \quad t \ge 0,$$

and

$$H(b_1, b_2, \ldots, b_n) = \int_\Omega f_i \Big\langle \sum_{j=1}^{k} f_j + \sum_{j=k+1}^{n} b_j f_j \Big\rangle^{p-1} d\mu \ \text{ for } 1 \le i \le k.$$

From these equations it follows that

$$H(b_1, b_2, \ldots, b_n) = \frac{1}{k}\int_\Omega \Big(\sum_{j=1}^{k} f_j\Big)\Big\langle \sum_{j=1}^{k} f_j + \sum_{j=k+1}^{n} b_j f_j \Big\rangle^{p-1} d\mu = \frac{1}{2k} h(1).$$

Next, set $\bar{b}_j := b_j$ if $1 \le j \le k$, and $\bar{b}_j := b_j/b_{k+1}$ if $k+1 \le j \le n$. It is clear that $1 = \bar{b}_1 = \cdots = \bar{b}_{k+1} \ge \bar{b}_{k+2} \ge \cdots \ge \bar{b}_n \ge 0$. Therefore, precisely in the same way as in the case of the quantity $H(b_1, b_2, \ldots, b_n)$, we get

$$H(\bar{b}_1, \bar{b}_2, \ldots, \bar{b}_n) = \frac{1}{k}\int_\Omega \Big(\sum_{j=1}^k f_j\Big)\Big\langle \sum_{j=1}^k f_j + \sum_{j=k+1}^n (b_j/b_{k+1}) f_j\Big\rangle^{p-1} d\mu = \frac{1}{2k} h(1/b_{k+1}).$$

On the other hand, (9.31) implies $H(b_1, b_2, \ldots, b_n) = m_1 \le H(\bar{b}_1, \bar{b}_2, \ldots, \bar{b}_n)$. Combining this inequality with the last equation, we obtain that $h(1) \le h(1/b_{k+1})$. Thus, $h(1) = h(1/b_{k+1})$, and consequently, by Lemma 9.5, $u \cdot v = 0$ μ-a.e. As a result, (9.33) is proved.

Show that

$$\Big(\sum_{j=1}^k f_j\Big) f_{k+1} = 0 \quad \mu - \text{a.e.} \tag{9.34}$$

Assume first that $b_{k+1} \ne 0$. Then, again thanks to the properties of f_i, $i = 1, \ldots, n$, from equation (9.33) it follows that

$$\Big(\sum_{j=1}^k f_j\Big)(-b_{k+1} f_{k+1} + b_{k+2} f_{k+2} + \cdots + b_n f_n) = 0 \quad \mu - \text{a.e.}$$

Subtracting the last equation from (9.33), we get

$$2b_{k+1} f_{k+1}\Big(\sum_{j=1}^k f_j\Big) = 0 \quad \mu - \text{a.e.},$$

which is equivalent to equation (9.34) because $b_{k+1} \ne 0$.

Let now $b_{k+1} = 0$. Then $b_j = 0$ for $j > k$ and $b_j = 1$ for $j \le k$. We proceed here as in the proof of (9.33), setting $\bar{b}_j := 1$ if $1 \le j \le k+1$, and $\bar{b}_j := 0$ if $k+1 < j \le n$. Consider the function

$$h_1(t) := \int_\Omega u[\langle u + t f_{k+1}\rangle^{p-1} + \langle u - t f_{k+1}\rangle^{p-1}]\, d\mu, \quad t \ge 0,$$

where, as above, $u = \sum_{j=1}^k f_j$. Then, again

$$h_1(t) := 2\int_\Omega u\langle u + t f_{k+1}\rangle^{p-1}\, d\mu, \quad t \ge 0,$$

$$h_1(0) = 2kH(b_1, b_2, \ldots, b_n) \quad \text{and} \quad h_1(1) = 2kH(\bar{b}_1, \bar{b}_2, \ldots, \bar{b}_n).$$

Since $H(b_1, b_2, \ldots, b_n) = m_1 \leq H(\bar{b}_1, \bar{b}_2, \ldots, \bar{b}_n)$, this implies that $h(0) \leq h(1)$. Thus, by Lemma 9.5, $u \cdot f_{k+1} = 0$ μ-a.e., and so (9.34) is proved.

In turn, from (9.34) it follows that $(-f_1 + f_2 + \cdots + f_k) \cdot f_{k+1} = 0$ μ-a.e. Subtracting the last equation from (9.34), we get $f_1 \cdot f_{k+1} = 0$ μ-a.e. Due to the properties of f_i this implies that $f_l \cdot f_m = 0$ μ-a.e. for all $l \neq m$. As was said above, this contradicts the assumptions, and therefore the proof is completed.

Since in the case $2 < p < \infty$ Theorem 9.7 can be proved in a similar way, we shall give only a sketch of the proof. Again it suffices to obtain the following statement.

Proposition 9.2 *Let $2 < p < \infty$, and let $f_i : \Omega \to \mathbb{R}$ be symmetrically and identically distributed independent r.v.'s, $\|f_i\|_p = 1$, $i = 1, 2, \ldots, n$. Then, setting*

$$M := \sup\left\{\left\|\sum_{i=1}^{n} t_i f_i\right\|_p : t_i \in \mathbb{R}, \sum_{i=1}^{n} |t_i|^p = 1\right\},$$

we have $M = n^{-1/p}\|\sum_{i=1}^{n} f_i\|_p$.

Moreover, if $\|\sum_{i=1}^{n} t_i f_i\|_p = M$ for some sequence $(t_i)_{i=1}^{n}$ such that $\sum_{i=1}^{n} |t_i|^p = 1$, then $t_i = \pm n^{-1/p}$, $i = 1, 2, \ldots, n$.

Proof As in Proposition 9.1, we apply Lagrange's method and then, defining in the same way the function H, introduce the quantity

$$M_1 := \sup\{H(x_1, \ldots, x_n) : 1 = x_1 \geq x_2 \geq \cdots \geq x_n \geq 0\}.$$

Next, we use an analogue of Lemma 9.5, but this time we should prove that $h(t)$ defined there is a strictly increasing function. □

Since the Rademacher functions are symmetrically and identically distributed independent r.v.'s on $[0, 1]$ with the Lebesgue measure, Theorem 9.6 is an immediate consequence of Theorem 9.7.

Comments and References Moment inequalities (1.11) from Chap. 1, comparing Gaussian and Rademacher sums, can be treated as a starting point for the results of Sect. 9.1. Theorem 9.1, showing a certain minimality of the Rademacher system with respect to the Hardy–Littlewood–Pólya submajorization in the class of independent and symmetrically distributed r.v.'s with the same L_1-norm has been proved in the paper [86] (see also Chapter IV in the dissertation [220]). This property of the Rademacher functions has been used in [86] for proving the complementability of the subspaces, spanned by appropriate sequences of independent functions.

Theorem 9.3 is one of the main results of the deep work [119]. Observe that in the simplest case, when $\Phi(t) = |t|^r$, the condition (c) of this theorem holds if and only if $r = 2$ or $r \geq 3$, which is equivalent to the requirement that the latter function is Schur-concave. If Φ is a 2-convex Orlicz function (i.e., the function $t \mapsto \Phi(\sqrt{t})$ is convex) such that Φ'' is concave on $[0, \infty)$, then another result from [119] reads that, for every $b \in \mathbb{R}$, the mapping

$$(x_1, x_2, \ldots, x_n) \mapsto \mathbb{E}\Phi\Big(b + \sum_{i=1}^{n} \sqrt{x_i} r_i\Big) - \sum_{i=1}^{n} \Phi(\sqrt{x_i})$$

is Schur-concave on $\mathbb{R}^n$. This fact implies one more extreme property of the Rademacher functions. Assume that $f_1, f_2, \ldots, f_n$ are pairwise disjoint, symmetrically distributed three-valued r.v.'s. Let

$$\mathbb{P}\{f_i = x_i\} = \mathbb{P}\{f_i = -x_i\} = \frac{t_i}{2}, \quad i = 1, 2, \ldots, n,$$

where $x_i \neq 0$, and $\mathbb{E}(\sum_{i=1}^n f_i^2) = \sum_{i=1}^n a_i^2$ (equivalently, $\sum_{i=1}^n t_i x_i^2 = \sum_{i=1}^n a_i^2$). Then, for every $b \in \mathbb{R}$ we have

$$\mathbb{E}\Big\{\Phi\Big(b + \sum_{i=1}^{n} f_i\Big) - \sum_{i=1}^{n} \Phi(f_i)\Big\} \leq \mathbb{E}\Big\{\Phi\Big(b + \sum_{i=1}^{n} a_i r_i\Big) - \sum_{i=1}^{n} \Phi(a_i r_i)\Big\}.$$

Theorem 9.4 has been proved in [210] and it was used there to characterize those s.s.'s X, for which the canonical inclusion of X into L_1 is strictly singular. Theorem 9.5 belongs to Mushtari [212]. All results of Sect. 9.4 are borrowed from the paper [120] by Figiel, Iwaniec, and Pełczyński. In fact, [120] contains more general assertions, related to any 1-symmetric and 1-unconditional sequence of functions $\{f_i\}_{i=1}^n$ from L_p. The first of these properties means that for arbitrary real numbers $t_1, t_2, \ldots, t_n$ and every permutation π of the set $\{1, 2, \ldots, n\}$ we have

$$\Big\|\sum_{i=1}^{n} t_i f_i\Big\|_p = \Big\|\sum_{i=1}^{n} t_i f_{\pi(i)}\Big\|_p.$$

A sequence $\{f_i\}_{i=1}^n \subset L_p$ is 1-unconditional whenever

$$\Big\|\sum_{i=1}^{n} t_i f_i\Big\|_p = \Big\|\sum_{i=1}^{n} \theta_i t_i f_i\Big\|_p$$

for all $t_i \in \mathbb{R}$ and $\theta_i = \pm 1$, $i = 1, 2, \ldots, n$. In conclusion, we mention the following interesting Schechtman theorem [261]: if $3 \leq p < \infty$, then every 1-unconditional basic sequence $\{f_k\}_{k=1}^\infty$ in $L_p[0, 1]$ such that $\|f_k\|_p = 1$, $k = 1, 2, \ldots$, is dominated in this space by the Rademacher system in the sense that

$$\Big\|\sum_{k=1}^{n} a_k f_k\Big\|_p \leq \Big\|\sum_{k=1}^{n} a_k r_k\Big\|_p$$

for all $n \in \mathbb{N}$ and $a_k \in \mathbb{R}$, $k = 1, 2, \ldots, n$.

Chapter 10
Bernoulli Processes

When studying Rademacher sums with vector coefficients, often it is very useful to keep in mind that the norm of such a sum is nothing but as the supremum of a certain Bernoulli process. More precisely, if $x_1, \dots, x_n$ are elements of a Banach space F, then by the Hahn-Banach theorem for each $u \in [0, 1]$ we have

$$\Big\| \sum_{i=1}^{n} x_i r_i(u) \Big\|_F = \sup_{t \in T} \Big| \sum_{i=1}^{n} t_i r_i(u) \Big|, \tag{10.1}$$

where T is the (compact) subset of $\mathbb{R}^n$ defined by

$$T = \{t = (t_i)_{i=1}^n, \ \ t_i = f(x_i) : f \in F^*, \|f\|_{F^*} \le 1\}.$$

This implies, in particular, that the comparison of the norms of vector-valued Rademacher sums reduces to that of the suprema of the corresponding Bernoulli processes from the right-hand side of equation (10.1). This circumstance is only one of the motivations of studying suprema of Bernoulli processes. We begin with estimates of their expectations in the case when coefficients of a process are subjected to a contraction transform.

10.1 A Contraction Principle for Bernoulli Processes

A function $\phi : \mathbb{R} \to \mathbb{R}$ is called a *contraction* if $|\phi(s) - \phi(t)| \le |s - t|$ for all $s, t \in \mathbb{R}$.

Theorem 10.1 *Let $\Phi : \mathbb{R}_+ \to \mathbb{R}_+$ be an increasing convex function, and let $\phi_i : \mathbb{R} \to \mathbb{R}$ be a contraction such that $\phi_i(0) = 0$, $i = 1, \dots, n$. Then, for every*

S. V. Astashkin, *The Rademacher System in Function Spaces*,
https://doi.org/10.1007/978-3-030-47890-2_10

bounded set $T \subset \mathbb{R}^n$ we have

$$\mathbb{E}\Phi\Big(\frac{1}{2}\sup_{t\in T}\Big|\sum_{i=1}^{n}\phi_i(t_i)r_i\Big|\Big) \le \mathbb{E}\Phi\Big(\sup_{t\in T}\Big|\sum_{i=1}^{n}t_ir_i\Big|\Big).$$

Proof We show first that for each increasing convex function $\Psi : \mathbb{R} \to \mathbb{R}$ it holds

$$\mathbb{E}\Psi\Big(\sup_{t\in T}\sum_{i=1}^{n}\phi_i(t_i)r_i\Big) \le \mathbb{E}\Psi\Big(\sup_{t\in T}\sum_{i=1}^{n}t_ir_i\Big). \tag{10.2}$$

Similarly as in the proof of the equivalence of conditions (a) and (b) of Theorem 9.3, by using Fubini's theorem and successive integration, one can prove that (10.2) is a consequence of the inequality

$$\mathbb{E}\Psi\left(\sup_{t\in T}(t_1+\phi(t_2)r)\right) \le \mathbb{E}\Psi\left(\sup_{t\in T}(t_1+t_2r)\right), \tag{10.3}$$

where ϕ is a contraction such that $\phi(0) = 0$, r is a Rademacher function, and T is a subset of $\mathbb{R}^2$, $t = (t_1, t_2)$. In other words, it suffices to check that for all points $s = (s_1, s_2)$ and $t = (t_1, t_2)$ from T, satisfying the conditions

$$t_1 + \phi(t_2) \ge s_1 + \phi(s_2) \quad \text{and} \quad s_1 - \phi(s_2) \ge t_1 - \phi(t_2), \tag{10.4}$$

the quantity

$$I(s,t) := \frac{1}{2}\Psi(t_1+\phi(t_2)) + \frac{1}{2}\Psi(s_1-\phi(s_2))$$

does not exceed the right-hand side of inequality (10.3). We consider separately the following 4 cases.

(i) $t_2 \ge 0$, $s_2 \ge 0$. Suppose $s_2 \le t_2$. We need to prove that

$$2I(s,t) \le \Psi(t_1+t_2) + \Psi(s_1-s_2), \tag{10.5}$$

or equivalently

$$\Psi(a) - \Psi(b) \le \Psi(a') - \Psi(b'), \tag{10.6}$$

where $a := s_1 - \phi(s_2)$, $b := s_1 - s_2$, $a' := t_1 + t_2$ and $b' := t_1 + \phi(t_2)$.

Since ϕ is a contraction, $\phi(0) = 0$ and $s_2 \ge 0$, we have $|\phi(s_2)| \le s_2$. Hence, $a \ge b$ and, by (10.4), $b \le b'$. Moreover, from the inequality $s_2 \le t_2$ it follows $\phi(t_2) - \phi(s_2) \le t_2 - s_2$. Therefore,

$$a - b = s_2 - \phi(s_2) \le t_2 - \phi(t_2) = a' - b'. \tag{10.7}$$

By condition, Ψ is convex and increasing. Consequently, for each positive x the function $\Psi(\cdot + x) - \Psi(\cdot)$ increases. Combining this with the inequality $b \le b'$, we get

$$\Psi(a) - \Psi(b) = \Psi(b + (a - b)) - \Psi(b) \le \Psi(b' + (a - b)) - \Psi(b').$$

Furthermore, by (10.7), we have $b' + a - b \le a'$. Thus, inequality (10.6) is proved. The case when $s_2 \ge t_2$ can be treated quite similarly with the replacement s_k with t_k, $k = 1, 2$, and ϕ with $-\phi$.

(ii) $t_2 \le 0$, $s_2 \le 0$. This case can be considered precisely in the same way as the previous one.

(iii) $t_2 \ge 0$, $s_2 \le 0$. As ϕ is a contraction and $\phi(0) = 0$, we have $\phi(t_2) \le t_2$ and $-\phi(s_2) \le -s_2$. Therefore, taking into account that Ψ increases, we get (10.5) as a consequence of the definition of $I(s, t)$.

(iv) $t_2 \le 0$, $s_2 \ge 0$. Similarly as in (iii), we have $\phi(t_2) \le -t_2$ and $-\phi(s_2) \le s_2$. In consequence, we have

$$2I(s, t) \le \Psi(t_1 - t_2) + \Psi(s_1 + s_2).$$

It remains to observe that the right-hand side of the latter inequality does not exceed that of inequality (10.3). Summarizing all, we conclude that (10.2) is proved.

Now we are in position to finish the proof of the theorem. Next, as usual, $u_+ = \max(u, 0)$ and $u_- = \max(-u, 0)$. Since Φ is convex, the Rademacher functions are symmetrically distributed and $(-u)_- = u_+$, we get

$$\begin{aligned}
\mathbb{E}\Phi\Big(\frac{1}{2}\sup_{t\in T}\Big|\sum_{i=1}^{n}\phi_i(t_i)r_i\Big|\Big) &\le \frac{1}{2}\Big\{\mathbb{E}\Phi\Big(\sup_{t\in T}\Big(\sum_{i=1}^{n}\phi_i(t_i)r_i\Big)_+\Big) \\
&\quad + \mathbb{E}\Phi\Big(\sup_{t\in T}\Big(\sum_{i=1}^{n}\phi_i(t_i)r_i\Big)_-\Big)\Big\} \\
&\le \mathbb{E}\Phi\Big(\sup_{t\in T}\Big(\sum_{i=1}^{n}\phi_i(t_i)r_i\Big)_+\Big).
\end{aligned}$$

Note that the function $\Psi(u) := \Phi((u)_+)$ is convex and increasing on $\mathbb{R}$. Thus, the preceding inequality and (10.2) imply that

$$\mathbb{E}\Phi\Big(\frac{1}{2}\sup_{t\in T}\Big|\sum_{i=1}^{n}\phi_i(t_i)r_i\Big|\Big) \le \mathbb{E}\Phi\Big(\sup_{t\in T}\Big(\sum_{i=1}^{n}t_ir_i\Big)_+\Big) \le \mathbb{E}\Phi\Big(\sup_{t\in T}\Big|\sum_{i=1}^{n}t_ir_i\Big|\Big),$$

and the desired result follows. □

Remark 10.1 The constant $1/2$ in the left-hand side of the inequality, proved in Theorem 10.1, is optimal. Indeed, let T be the subset of $\mathbb{R}^2$, consisting of two points $(1, 1)$ and $(-1, -1)$. Then, if $\phi_1(t) = t$, $\phi_2(t) = -|t|$ and $\Phi(t) = t$, this inequality becomes

$$\mathbb{E}\Big(\frac{1}{2}\max\left(|r_1 - r_2|, |r_1 + r_2|\right)\Big) \le \mathbb{E}|r_1 + r_2|.$$

Simple calculations show now that both parts here are equal to 1.

Applying Theorem 10.1 in the case when $\phi_i(t) = |t|$ for all $i = 1, 2, \dots, n$ and taking into account equation (10.1), we get

Corollary 10.1 *If $\Phi : \mathbb{R}_+ \to \mathbb{R}_+$ is a increasing convex function, then for every bounded set $T \subset \mathbb{R}^n$ we have*

$$\mathbb{E}\Phi\Big(\frac{1}{2}\sup_{t\in T}\Big|\sum_{i=1}^{n}|t_i|r_i\Big|\Big) \le \mathbb{E}\Phi\Big(\sup_{t\in T}\Big|\sum_{i=1}^{n}t_ir_i\Big|\Big).$$

In particular, for each Banach space F and arbitrary $x_1, \dots, x_n$ from F the following inequality holds:

$$\mathbb{E}\Phi\Big(\frac{1}{2}\sup_{\|f\|_{F^*}\le 1}\Big|\sum_{i=1}^{n}|f(x_i)|r_i\Big|\Big) \le \mathbb{E}\Phi\Big(\Big\|\sum_{i=1}^{n}x_ir_i\Big\|_F\Big).$$

10.2 A Minorant Sudakov Type Estimate

Next, we shall be interested in estimating the expectation of the supremum of a Bernoulli process defined on a set $T \subset \ell_2$, i.e., the quantity

$$b(T) := \mathbb{E}\sup_{t=(t_k)\in T}\sum_{k=1}^{\infty}t_kr_k.$$

We begin with listing simple properties of $b(T)$, assuming that $\vec{0} := (0, 0, \dots) \in T$:

(a) $b(T) \ge 0$;
(b) if $T \subset T'$, then $b(T) \le b(T')$;
(c) $\mathbb{E}\sup_{t\in T}|\sum_{k=1}^{\infty}t_kr_k| \le 2b(T)$.

The first two properties are obvious. Let us check (c). Indeed, since the r.v. $\xi_t := \sum_{k=1}^{\infty}t_kr_k$ is symmetrically distributed for every $t = (t_k) \in \ell_2$ and $\vec{0} \in T$, we get

$$\mathbb{E}\sup_{t\in T}|\xi_t| \le \mathbb{E}\sup_{t,s\in T}|\xi_t - \xi_s| = \mathbb{E}\sup_{t,s\in T}(\xi_t - \xi_s) = \mathbb{E}\sup_{t\in T}\xi_t + \mathbb{E}\sup_{s\in T}\xi_s = 2b(T).$$

Properties of $b(T)$ are closely connected with some geometrical features of the domain T, and, in particular, with its ε-entropy. Recall that *ε-entropy* ($\varepsilon > 0$) of a set $T \subset \ell_2$ is the quantity $\ln \nu(T, \varepsilon)$, where

$$\nu(T, \varepsilon) := \min\{\operatorname{card} C : C \subset T, C \text{ is an } \varepsilon\text{-net of } T \text{ with respect to the } \ell_2\text{−metric}\}.$$

Theorem 10.2 *There exists a universal constant $K > 0$ such that for any $\varepsilon > 0$ and every set $T \subset \ell_2$, satisfying the condition: $\|t\|_{\ell_\infty} \le \varepsilon^2/(Kb(T))$ for all $t \in T$, we have*

$$\varepsilon \cdot (\ln \nu(T, \varepsilon))^{1/2} \le Kb(T). \tag{10.8}$$

Proof We make use further of the fact that analogous property is enjoyed by a Gaussian process (even without any control over the ℓ_∞-norm of elements from its domain). Namely, if

$$\mathfrak{g}(U) := \mathbb{E} \sup_{t \in U} \sum_{k=1}^{\infty} t_k \mathfrak{g}_k$$

(here, as usual, $\mathfrak{g}_k$ are independent standard Gaussian r.v.'s), then there is $K_1 > 0$ such that for all $\varepsilon > 0$ and $U \subset \ell_2$

$$\varepsilon \cdot (\ln \nu(U, \varepsilon))^{1/2} \le K_1 \mathfrak{g}(U) \tag{10.9}$$

(see e.g. [177, Theorem 3.18]).

The "hardest" part of the proof of Theorem 10.2 is to get inequality (10.8) for some ε, say, for $\varepsilon = 1/2$. So, we aim to find a constant $K_0 > 0$ with the following property: if $T \subset \ell_2$, $\|t\|_{\ell_\infty} \le 1/(K_0 b(T))$ for $t \in T$, then

$$(\ln \nu(T, 1/2))^{1/2} \le K_0 b(T). \tag{10.10}$$

Observe first that in the case when $\|t\|_{\ell_2} \le 1/4$ for all $t \in T$ inequality (10.10) is obvious (because its left-hand side is equal to zero). Therefore, we can assume that there exists $t^0 = (t_k^0) \in T$ such that $\|t^0\|_{\ell_2} \ge 1/4$, and hence, by property (c) and Khintchine's L_1-inequality (see Theorem 5.4), we have

$$b(T) \ge \frac{1}{2}\mathbb{E}\Big|\sum_{k=1}^{\infty} t_k^0 r_k\Big| \ge \frac{\|t^0\|_{\ell_2}}{2\sqrt{2}} \ge \frac{1}{8\sqrt{2}}. \tag{10.11}$$

Let $s > 0$. We put

$$f_k := \mathfrak{g}_k \chi_{\{|\mathfrak{g}_k| \le s\}} \text{ and } h_k := \mathfrak{g}_k \chi_{\{|\mathfrak{g}_k| > s\}},$$

where $\mathfrak{g}_k$ are independent standard Gaussian r.v.'s, $k = 1, 2, \ldots$. Then, for every $U \subset T$ we have

$$\mathbb{E}\sup_{t\in U}\sum_{k=1}^{\infty} t_k \mathfrak{g}_k \le \mathbb{E}\sup_{t\in U}\sum_{k=1}^{\infty} t_k f_k + \mathbb{E}\sup_{t\in U}\sum_{k=1}^{\infty} t_k h_k.$$

Now, we estimate the expectations from the right-hand side of this inequality separately. First of all, since f_k are symmetrically distributed, the sequences $\{f_k\}$ and $\{|f_k|r_k\}$ are similar whenever the Rademacher sequence $\{r_k\}$ is independent with respect to $\{|f_k|\}$. Therefore, taking into account that $\|f_k\|_\infty \le s$ and applying Theorem 7.2, we obtain

$$\mathbb{E}\sup_{t\in U}\sum_{k=1}^{\infty} t_k f_k \le s\mathbb{E}\sup_{t\in U}\sum_{k=1}^{\infty} t_k r_k \le sb(T).$$

Combining this with the preceding inequality, we have

$$\mathbb{E}\sup_{t\in U}\sum_{k=1}^{\infty} t_k \mathfrak{g}_k \le sb(T) + \mathbb{E}\sup_{t\in U}\sum_{k=1}^{\infty} t_k h_k. \tag{10.12}$$

Let $\lambda > 0$ be so far arbitrary (it will be chosen later). Note that the r.v. h_k, as well as $\mathfrak{g}_k$, is symmetrically distributed, and so

$$\mathbb{E}\exp(\lambda h_k) = 1 + \lambda^2\phi_s(\lambda) \le \exp(\lambda^2\phi_s(\lambda)), \tag{10.13}$$

where

$$\phi_s(\lambda) := \frac{1}{2\lambda^2}\mathbb{E}\left(\exp(\lambda h_k) + \exp(-\lambda h_k) - 2\right).$$

Furthermore, since $\mathbb{E}\exp(\mathfrak{g}_1) < \infty$, from Lebesgue's dominated convergence theorem it follows that $\lim_{s\to\infty}\phi_s(1) = 0$. Therefore, we can choose (and fix) s so that $\phi_s(1) \le (36K_1^2)^{-1}$, where K_1 is the constant from (10.9). We prove that inequality (10.10) holds with $K_0 = 24sK_1^2$.

Since the function $x^{-2}(e^x + e^{-x} - 2)$ increases for $x > 0$, so does $\phi_s(\lambda)$ for $\lambda > 0$. Consequently, by (10.13), we have

$$\mathbb{E}\exp(\lambda t_k h_k) \le \exp\left(\lambda^2 t_k^2\phi_s(t_k\lambda)\right) \le \exp\left(\lambda^2 t_k^2\phi_s(\|t\|_{\ell_\infty}\lambda)\right)$$

for any $t \in \ell_2$ and $k \in \mathbb{N}$. Thus, if $\|t\|_{\ell_2} \le 1$, then we have

$$\mathbb{E}\exp\left(\lambda\sum_{k=1}^{\infty} t_k h_k\right) = \prod_{k=1}^{\infty}\mathbb{E}\exp(\lambda t_k h_k) \le \exp\left(\lambda^2\phi_s(\|t\|_{\ell_\infty}\lambda)\right).$$

Hence, as for any r.v. h and each $x \in \mathbb{R}$ from Chebyshev's inequality it follows

$$\mathbb{P}\{h \geq x\} \leq \exp(-\lambda x)\mathbb{E}\exp(\lambda h),$$

we infer that

$$\mathbb{P}\Big\{\sum_{k=1}^{\infty} t_k h_k \geq x\Big\} \leq \exp\Big(-\lambda x + \lambda^2 \phi_s(\|t\|_{\ell_\infty}\lambda)\Big). \tag{10.14}$$

Consider now a finite set $U \subset T$ such that $\operatorname{card} U = N$. Assume that $N \geq 3$ and $\sqrt{\ln N} \leq 4K_1 sb(T)$. Then, if $\lambda := 6K_1\sqrt{\ln N}$, we have $\lambda \leq 24K_1^2 sb(T)$. Moreover, by condition, $\|t\|_{\ell_\infty} \leq (24K_1^2 sb(T))^{-1}$ for $t \in T$. Consequently, from (10.14) it follows

$$\mathbb{P}\Big\{\sum_{k=1}^{\infty} t_k h_k \geq x\Big\} \leq \exp\Big(-\lambda x + \lambda^2 \phi_s(1)\Big).$$

Note that the choice of s and λ assures that $\lambda^2\phi_s(1) \leq \ln N$. Therefore, since $\operatorname{card} U = N$, the preceding estimate implies that

$$\mathbb{P}\Big\{\sup_{t\in U}\sum_{k=1}^{\infty} t_k h_k \geq x\Big\} \leq N^2 \exp(-\lambda x),$$

whence for every $y \geq 0$

$$\begin{aligned}
\mathbb{E}\sup_{t\in U}\sum_{k=1}^{\infty} t_k h_k &\leq \int_0^\infty \mathbb{P}\Big\{\sup_{t\in U}\sum_{k=1}^{\infty} t_k h_k \geq x\Big\}\,dx \\
&\leq y + \int_y^\infty N^2\exp(-\lambda x)\,dx \\
&\leq y + \frac{N^2}{\lambda}\exp(-\lambda y).
\end{aligned}$$

Setting here $y := \sqrt{\ln N}/(3K_1)$, we have $\lambda y = 2\ln N$, and hence,

$$\mathbb{E}\sup_{t\in U}\sum_{k=1}^{\infty} t_k h_k \leq y + \frac{1}{\lambda} = \frac{\sqrt{\ln N}}{3K_1} + \frac{1}{6K_1\sqrt{\ln N}} \leq \frac{2\sqrt{\ln N}}{3K_1} \tag{10.15}$$

(because $\ln N \geq 1$).

Suppose now additionally that $\nu(U, 1/2) = N$. Then, by using (10.9) for $\varepsilon = 1/2$ as well as inequalities (10.12) and (10.15), we obtain

$$\frac{1}{K_1}\sqrt{\ln N} \leq sb(T) + \frac{2}{3K_1}\sqrt{\ln N},$$

which implies that $\sqrt{\ln N} \leq 3K_1 sb(T)$. As a result, the assumption that there exists a set $U \subset T$ satisfying the condition

$$\nu(U, 1/2) = \operatorname{card} U = N \tag{10.16}$$

implies that from the estimate $\sqrt{\ln N} \leq 4K_1 sb(T)$ it follows, in fact, the more restrictive inequality: $\sqrt{\ln N} \leq 3K_1 sb(T)$. In other words, the inequality

$$3K_1 sb(T) < \sqrt{\ln N} \leq 4K_1 sb(T) \tag{10.17}$$

fails for such N.

Further, let N be the least positive integer such that $\sqrt{\ln N} > 3K_1 sb(T)$. Choosing s sufficiently large, we can assume that $K_1 sb(T) \geq 1$ (see also (10.11)). Hence, $N \geq 3$ and so

$$\sqrt{\ln N} - \sqrt{\ln(N-1)} \leq K_1 sb(T).$$

Consequently, $\sqrt{\ln N} \leq 4K_1 sb(T)$, which implies that N satisfies inequality (10.17). We claim that $\nu(T, 1/2) < N$.

Indeed, on the contrary, by the definition of $\nu(T, 1/2)$, one can inductively construct a set $U = \{t_1, t_2, \ldots, t_N\} \subset T$ such that $\|t_i - t_j\|_{\ell_2} > 1/2$ for $i \neq j$ and hence (10.16) holds. On the other hand, it is impossible because for the same N inequality (10.17) is fulfilled. This contradiction proves the claim, whence

$$(\ln \nu(T, 1/2))^{1/2} \leq \sqrt{\ln N} \leq 4K_1 sb(T) \leq K_0 b(T)$$

(because $K_0 = 24sK_1^2$), and so inequality (10.10) is proved.

It remains to show that (10.10) implies (10.8). We shall use an iteration procedure. Let B_{ℓ_2} be the closed unit ball in ℓ_2. It can be readily checked that, for any $t \in T$ and $l \in \mathbb{Z}$, it holds

$$\nu\left(T \cap (t + 2^{-l}B_{\ell_2}), 2^{-l-1}\right) = \nu\left((T - t) \cap 2^{-l}B_{\ell_2}, 2^{-l-1}\right).$$

Furthermore, from the definition of $b(T)$ it follows that $b(T - t) \leq 2b(T)$. Therefore, in view of (10.10), taking into account homogeneity of the quantities involved, we get

$$\nu\left(T \cap (t + 2^{-l}B_{\ell_2}), 2^{-l-1}\right) \leq \exp\left(K_0^2 2^{2l+2} b(T)^2\right)$$

provided that $\|t\|_{\ell_\infty} \le \left(K_0 2^{2l+2} b(T)\right)^{-1}$ for all $t \in T$. Hence,

$$\nu(T, 2^{-k}) \le \prod_{l<k} \sup_{t \in T} \nu\left(T \cap (t + 2^{-l} B_{\ell_2}), 2^{-l-1}\right)$$
$$\le \exp\left(K_0^2 \sum_{l<k} 2^{2l+2} b(T)^2\right) \le \exp\left(K_0^2 2^{2k+3} b(T)^2\right)$$

if $\|t\|_{\ell_\infty} \le \left(K_0 2^{2k} b(T)\right)^{-1}$ for all $t \in T$.

Let $\varepsilon > 0$ be arbitrary and k be the least positive integer, satisfying the inequality $2^{-k} \le \varepsilon$. Then, $2^k \le 2/\varepsilon$ and, as we already proved,

$$\nu(T, \varepsilon) \le \nu(T, 2^{-k}) \le \exp\left(K_0^2 2^{2k+3} b(T)^2\right) \le \exp\left(K_0^2 2^5 \varepsilon^{-2} b(T)^2\right),$$

provided that for all $t \in T$ we have

$$\|t\|_{\ell_\infty} \le (4K_0 b(T))^{-1} \varepsilon^2.$$

Thus, the desired result holds with the constant $K = 4\sqrt{2}K_0$. □

10.3 The Bernoulli Conjecture

A given set $T \subset \ell_2$, as in the preceding section, we put

$$b(T) := \mathbb{E} \sup_{t \in T} \sum_{k=1}^{\infty} t_k r_k.$$

Let us investigate how $b(T)$ can be estimated from the above. There are two easy ways to do this. It is obvious first that

$$b(T) \le \sup_{t \in T} \|t\|_{\ell_1}. \tag{10.18}$$

Moreover, one can easily show that

$$b(T) \le \sqrt{\frac{\pi}{2}} \mathfrak{g}(T), \tag{10.19}$$

where

$$\mathfrak{g}(T) := \mathbb{E} \sup_{t \in T} \sum_{k=1}^{\infty} t_k \mathfrak{g}_k$$

($\mathfrak{g}_k$ are independent standard Gaussian r.v.'s). Indeed, we can assume that the sequences $\{r_k\}$ and $\{\mathfrak{g}_k\}$ are independent. Then, $\{r_k|\mathfrak{g}_k|\}$ has the same distribution as $\{\mathfrak{g}_k\}$, and therefore

$$\mathfrak{g}(T) = \mathbb{E} \sup_{t\in T} \sum_{k=1}^{\infty} t_k r_k |\mathfrak{g}_k|.$$

Using now Fubini's theorem and the equality $\mathbb{E}|\mathfrak{g}_k| = \sqrt{2/\pi}$, we get

$$\mathfrak{g}(T) \geq \sqrt{\frac{2}{\pi}} \mathbb{E} \sup_{t\in T} \sum_{k=1}^{\infty} t_k r_k = \sqrt{\frac{2}{\pi}} b(T),$$

which implies (10.19).

Inequalities (10.18) and (10.19) show that the expectation of the supremum of a Bernoulli process is estimated from the above by each of the quantities $\sup_{t\in T} \|t\|_{\ell_1}$ and $\mathfrak{g}(T)$. Clearly, a similar control can be carried out by their combination. Specifically, if $T \subset T_1 + T_2 := \{t^1 + t^2 : t^k \in T_k\}$, then $b(T) \leq b(T_1) + b(T_2)$, whence from (10.18) and (10.19) it follows that

$$b(T) \leq \inf \left\{ \sup_{t\in T_1} \|t\|_{\ell_1} + \sqrt{\frac{\pi}{2}} \mathfrak{g}(T_2) \right\},$$

where the infimum is taken over all decompositions $T = T_1 + T_2$. It was open for about 25 years (under the name of *Bernoulli conjecture*) whether the above estimate may be (up to a constant) reversed. The next Bednorz–Latała theorem provides an affirmative answer.

Theorem 10.3 (Bednorz–Latała) *There exists a universal constant D such that for any nonempty set $T \subset \ell_2$ with $b(T) < \infty$ one may find a decomposition $T \subset T_1 + T_2$, satisfying the conditions:*

$$\sup_{t\in T_1} \sum_{i=1}^{\infty} |t_i| \leq Db(T) \quad \textit{and} \quad \mathfrak{g}(T_2) \leq Db(T).$$

From this theorem, in particular, it follows a remarkable characterization of sets $T \subset \ell_2$, for which the quantity $b(T)$ is finite. In order to formulate this result, we introduce some additional notations. If $U \subset \ell_2$, then $\overline{\mathrm{conv}}\, U$ denotes the closed linear hull of U in ℓ_2. Also, a given $t = (t_k)_{k=1}^{\infty} \in \ell_2$ we set $\xi_t := \sum_{k=1}^{\infty} t_k r_k$. And finally, as usual, for every $p > 0$ and any r.v. ξ on a probability space we put $\|\xi\|_p := (\mathbb{E}|\xi|^p)^{1/p}$.

Theorem 10.4 (Bednorz–Latała) *For an arbitrary Bernoulli process* $(\xi_t)_{t\in T}$, *where* $T \subset \ell_2$, *the following conditions are equivalent:*

(*a*) $b(T) < \infty$;
(*b*) *there are elements* $t^i \in \ell_2$, $i = 1, 2, \ldots$, *satisfying the conditions:* $T - T \subset \overline{\mathrm{conv}}\,\{t^n : n \geq 1\}$ *and* $\|\xi_{t^n}\|_{\ln(n+2)} \leq M$ *for some* M *and all* $n = 1, 2, \ldots$

Moreover, if the condition (*a*) *is fulfilled, then for* M *in* (*b*) *we can take* $Lb(T)$, *where* L *is a universal constant.*

Proof (*a*) $\Rightarrow$ (*b*). By Theorem 10.3, there exists a decomposition $T \subset T_1 + T_2$, with $\sup_{t\in T_1} \|t\|_{\ell_1} \leq Db(T)$ and $\mathfrak{g}(T_2) \leq Db(T)$. Then, we have

$$T - T \subset (T_1 - T_1) + (T_2 - T_2) \subset \overline{\mathrm{conv}}\,\{2(T_1 - T_1), 2(T_2 - T_2)\}.$$

Firstly, it is clear that $T_1 - T_1 \subset 2Db(T)\overline{\mathrm{conv}}\,\{e^i : i = 1, 2, \ldots\}$, where $\{e^i\}$ is the unit vector basis in ℓ_2. Secondly, from [277, Theorem 2.1.8] it follows the existence of elements $s^k \in \ell_2$, $k = 1, 2, \ldots$, such that $T_2 - T_2 \subset \overline{\mathrm{conv}}\,\{s^k : k = 1, 2, \ldots\}$ and

$$\sqrt{\ln(k+1)}\|s^k\|_{\ell_2} \leq D\mathfrak{g}(T_2) \leq D^2 b(T).$$

Note that $\|\xi_{e^i}\|_p = \|r_i\|_p = 1$ for each $p > 0$ and, according to Khintchine's inequality (see Theorem 1.4), we have $\|\xi_{s^k}\|_p \leq \sqrt{p}\|s^k\|_{\ell_2}$, $p \geq 1$. Now, it is easy to see that the sequence $\{t^n\}_{n=1}^{\infty}$ suitably formed by the elements $4Db(T)e^i$, $i = 1, 2, \ldots$, and $2s^k$, $k = 1, 2, \ldots$ satisfies all the conditions of (*b*).

(*b*) $\Rightarrow$ (*a*). Assume that $T - T \subset \overline{\mathrm{conv}}\{t^n : n \geq 1\}$ and $\|\xi_{t^n}\|_{\ln(n+2)} \leq M$, $n = 1, 2, \ldots$, for some sequence $\{t^n\}_{n=1}^{\infty} \subset \ell_2$. Then, for every $u \geq 1$ we have

$$\mathbb{P}\Big\{\sup_{s\in T-T} \xi_s \geq uM\Big\} \leq \mathbb{P}\Big\{\sup_{n\geq 1} \xi_{t^n} \geq uM\Big\} \leq \sum_{n=1}^{\infty} \mathbb{P}\{\xi_{t^n} \geq u\|\xi_{t^n}\|_{\ln(n+2)}\}.$$

Furthermore, since from Chebyshev's inequality it follows

$$\mathbb{E}|\xi_{t^n}|^{\ln(n+2)} \geq \mathbb{P}\{\xi_{t^n} \geq u\|\xi_{t^n}\|_{\ln(n+2)}\}u^{\ln(n+2)}\mathbb{E}|\xi_{t^n}|^{\ln(n+2)},$$

then

$$\mathbb{P}\{\xi_{t^n} \geq u\|\xi_{t^n}\|_{\ln(n+2)}\} \leq u^{-\ln(n+2)}.$$

Combining this with the preceding estimate, we get

$$\mathbb{P}\Big\{\sup_{s\in T-T} \xi_s \geq uM\Big\} \leq \sum_{n=1}^{\infty} u^{-\ln(n+2)}.$$

Integration then gives

$$\mathbb{E}\sup_{s\in T-T}\xi_s = M\int_0^\infty \mathbb{P}\Big\{\sup_{s\in T-T}\xi_s \ge uM\Big\}\,du$$

$$\le M\Big(e^2+\sum_{n=1}^\infty\int_{e^2}^\infty u^{-\ln(n+2)}\,du\Big)$$

$$= Me^2\Big(1+\sum_{n=1}^\infty \frac{1}{(n+2)^2\ln((n+2)/e)}\Big) := M'.$$

Finally, for any $t^0 \in T$ we have

$$b(T) = \mathbb{E}\sup_{t\in T}(\xi_t - \xi_{t^0}) = \mathbb{E}\sup_{t\in T}\xi_{t-t^0} \le \mathbb{E}\sup_{s\in T-T}\xi_s \le M' < \infty. \qquad \square$$

The last result has found many interesting applications. Next, we consider some of them.

10.4 L-Regularity of Rademacher Sums and the Comparison of Distributions of Random Vectors

Let f and g be random vectors with values in a Banach space F (possibly, defined on different probability spaces). We say that g is *weakly dominated* by f if

$$\mathbb{P}'\{|x^*(g)| \ge z\} \le \mathbb{P}\{|x^*(f)| \ge z\}$$

for all $x^* \in F^*$ and $z > 0$.

The result of Theorem 10.4 from the previous section motivates introduction of the next notion, which enables passing from weak to strong comparison of distributions.

Let $M > 0$. A random vector f with values in F will be called *M-regular* for some $K > 0$ if there exists a sequence $\{x_n^*\} \subset F^*$ such that

$$\|x_n^*(f)\|_{\ln(n+2)} := \Big(\mathbb{E}|x_n^*(f)|^{\ln(n+2)}\Big)^{1/\ln(n+2)} \le M\mathbb{E}\|f\|_F, \quad n = 1, 2, \ldots, \tag{10.20}$$

and

$$B_{F^*} := \{x^* \in F^* : \|x^*\| \le 1\} \subset cl_f(\text{conv}\{\pm x_n^* : n \ge 1\}), \tag{10.21}$$

where $cl_f(A)$ denotes the closure of a set $A \subset F^*$ with respect to the L_2-distance $d_f(x^*, y^*) := \left(\mathbb{E}|x^*(f) - y^*(f)|^2\right)^{1/2}$.

Example 10.1 Let F be a Banach space, $x_i \in F$, $i = 1, 2, \ldots$ Show that the random vector $S := \sum_{i=1}^m x_i r_i$ is L-regular for every $m \in \mathbb{N}$, where L is the constant from Theorem 10.4.

To this end, we consider the Bernoulli process with the domain

$$T := \left\{(x^*(x_i))_{i=1}^m \,:\, \|x^*\|_{F^*} \le 1\right\}$$

($\mathbb{R}^m$ is considered here as a natural subspace of ℓ_2). Since

$$b(T) = \mathbb{E} \sup_{\|x^*\|_{F^*} \le 1} \left|\sum_{i=1}^m x^*(x_i) r_i\right| = \mathbb{E}\|S\|_F < \infty,$$

by Theorem 10.4, there are vectors $t^n = (t_i^n)_{i=1}^m \in \mathbb{R}^m$, $n = 1, 2, \ldots$, such that the set $\mathrm{conv}\{t^n \,:\, n \ge 1\}$ dense in $T - T$ (with respect to the ℓ_2-distance) and $\|\xi_{t^n}\|_{\ln(n+2)} \le L\mathbb{E}\|S\|_F$ for all $n = 1, 2, \ldots$ Clearly, we can assume that $t^n = (x_n^*(x_i))_{i=1}^m$, where $x_n^* \in F^*$, $n = 1, 2, \ldots$ Then, we have

$$\|x_n^*(S)\|_{\ln(n+2)} = \left\|\sum_{i=1}^m x_n^*(x_i) r_i\right\|_{\ln(n+2)} \le L\mathbb{E}\|S\|_F, \quad n = 1, 2, \ldots$$

Now, we need to prove only that the set $\mathrm{conv}\{x_n^* \,:\, n = 1, 2, \ldots\}$ is dense in the unit ball B_{F^*} with respect to the distance $d_S(x^*, y^*) := \left(\mathbb{E}|x^*(S) - y^*(S)|^2\right)^{1/2}$.

Since $\vec{0} \in T$, we have $T \subset T - T$. Therefore, the set $\mathrm{conv}\{(x_n^*(x_i))_{i=1}^m \,:\, n \ge 1\}$ is dense in T with respect to the ℓ_2-distance, and hence for arbitrary $x^* \in B_{F^*}$ and $\varepsilon > 0$ one can find $r \in \mathbb{N}$ and $\lambda_k \ge 0$, $k = 1, \ldots, r$, such that $\sum_{k=1}^r \lambda_k = 1$ and

$$\sum_{i=1}^m \Big(x^*(x_i) - \sum_{k=1}^r \lambda_k x_k^*(x_i)\Big)^2 < \varepsilon.$$

This inequality and the fact that the Rademacher system is orthonormal imply

$$\mathbb{E}\Big(x^*(S) - \sum_{k=1}^r \lambda_k x_k^*(S)\Big)^2 = \mathbb{E}\Big(\sum_{i=1}^m \Big(x^*(x_i) - \sum_{k=1}^r \lambda_k x_k^*(x_i)\Big) r_i\Big)^2$$
$$= \sum_{i=1}^m \Big(x^*(x_i) - \sum_{k=1}^r \lambda_k x_k^*(x_i)\Big)^2 < \varepsilon.$$

As $\varepsilon > 0$ is arbitrary, the proof is completed.

Proposition 10.1 *If f is a M-regular random vector, which weakly dominates a vector g, then*

$$\mathbb{E}\|g\|_F \leq 20M\mathbb{E}\|f\|_F.$$

Proof Thanks to the M-regularity of f, there are $x_n^* \in F^*$, $n = 1, 2, \ldots$, satisfying conditions (10.20) and (10.21). From Chebyshev's inequality and the hypothesis of the proposition for every $z > 0$ it follows that

$$\begin{aligned}
\mathbb{P}'\left\{\sup_{n=1,2,\ldots} |x_n^*(g)| \geq z\right\} &\leq \sum_{n=1}^{\infty} \mathbb{P}'\left\{|x_n^*(g)| \geq z\right\} \leq \sum_{n=1}^{\infty} z^{-\ln(n+2)}\mathbb{E}|x_n^*(g)|^{\ln(n+2)} \\
&\leq \sum_{n=1}^{\infty} z^{-\ln(n+2)}\mathbb{E}|x_n^*(f)|^{\ln(n+2)} \\
&\leq \sum_{n=1}^{\infty} \left(\frac{M\mathbb{E}\|f\|_F}{z}\right)^{\ln(n+2)}.
\end{aligned}$$

Moreover, by condition,

$$d_g(x^*, y^*) = \left(\mathbb{E}|x^*(g) - y^*(g)|^2\right)^{1/2} \leq \left(\mathbb{E}|x^*(f) - y^*(f)|^2\right)^{1/2} = d_f(x^*, y^*),$$

and so the unit ball B_{F^*} is contained in $cl_g(\mathrm{conv}\,\{\pm x_n^* : n \geq 1\})$. Therefore, from the preceding estimate we get

$$\begin{aligned}
\mathbb{E}\|g\|_F &\leq \mathbb{E}\sup_{n=1,2,\ldots} |x_n^*(g)| \leq M\mathbb{E}\|f\|_F\Big(e^2 + \int_{e^2}^{\infty} \mathbb{P}'\left\{\sup_{n=1,2,\ldots} |x_n^*(g)| \geq zM\mathbb{E}\|f\|_F\right\} dz\Big) \\
&\leq M\mathbb{E}\|f\|_F\Big(e^2 + \sum_{n=1}^{\infty}\int_{e^2}^{\infty} z^{-\ln(n+2)}\, dz\Big) \leq 20M\mathbb{E}\|f\|_F.
\end{aligned}$$ □

Under certain additional conditions, the moment inequality for the norms of random vectors, proved in the last proposition, can be strengthened to a tail distribution estimate.

Theorem 10.5 *Let $f_1, f_2, \ldots$ be independent copies of a symmetrically distributed random vector f with values in a Banach space F. Suppose that there exist constants $M, \alpha, \beta > 0$ such that for all $n = 1, 2, \ldots$*

(*i*) *the random vector $(f_1, \ldots, f_n)$ with values in $\ell_\infty^n(F)$ is M-regular;*
(*ii*) $\mathbb{P}\left\{\max_{i=1,\ldots,n} \|f_i\|_F \geq \alpha\mathbb{E}\max_{i=1,\ldots,n} \|f_i\|_F\right\} \geq \beta$.

Then, if a random vector g is weakly dominated by f, for all $z > 0$ we have

$$\mathbb{P}'\{\|g\|_F \geq z\} \leq \frac{2}{\beta}\mathbb{P}\left\{\|f\|_F \geq \frac{\alpha z}{80M}\right\}. \tag{10.22}$$

Proof First, without loss of generality, we can assume that g is a symmetrically distributed vector. Indeed, if it is not the case, we consider an independent r.v. θ with respect to both vectors f and g such that $\mathbb{P}\{\theta = 1\} = \mathbb{P}\{\theta = -1\} = 1/2$. Then, one can readily check that the symmetrically distributed vectors θf and θg satisfy all the conditions of the theorem. Therefore, assuming that in this case the result is proved, we get inequality (10.22) for θf and θg. But then, clearly, (10.22) holds for f and g as well.

Next, for every $z > 0$ we find an integer $n \geq 2$ such that

$$\frac{2}{n} \geq \mathbb{P}', \{\|g\|_F \geq z\} \geq \frac{1}{n}. \tag{10.23}$$

Then, if $g_1, g_2, \ldots$ are independent copies of g, we have

$$\mathbb{P}'\left\{\max_{i=1,\ldots,n} \|g_i\|_F \geq z\right\} = 1 - \mathbb{P}'\left(\bigcap_{i=1}^{n}\{\|g_i\|_F < z\}\right) \geq 1 - \left(1 - \frac{1}{n}\right)^n \geq \frac{1}{2},$$

and so, according to Chebyshev's inequality,

$$\mathbb{E}\max_{i=1,\ldots,n} \|g_i\|_F \geq \frac{z}{2}. \tag{10.24}$$

Let η be an r.v. independent with respect to the sequences $\{f_i\}$, $\{g_i\}$ and such that $\mathbb{P}\{\eta = 1\} = \mathbb{P}\{\eta = 0\} = 1/2$. Since g is weakly dominated by f, for any $x_i^* \in F^*$, $i = 1, 2, \ldots, n$, and all $z > 0$ we have

$$\mathbb{P}'\{|x_i^*(\eta g_i)| \geq z\} \leq \mathbb{P}\{|x_i^*(\eta f_i)| \geq z\}.$$

Observe that the sequences $\{x_i^*(\eta g_i)\}_{i=1}^n$ and $\{x_i^*(\eta f_i)\}_{i=1}^n$ consist of independent symmetrically distributed r.v.'s. Therefore, by Theorem 7.2, for all $z > 0$

$$\begin{aligned}
\mathbb{P}'\left\{\sum_{i=1}^{n} |x_i^*(\eta g_i)| > z\right\} &= \mathbb{P}'\left\{\|(x_1^*(\eta g_1), \ldots, x_n^*(\eta g_n))\|_{\ell_1^n} > z\right\} \\
&\leq 2\mathbb{P}\left\{\|(x_1^*(\eta f_1), \ldots, x_n^*(\eta f_n))\|_{\ell_1^n} > z\right\} \\
&= \mathbb{P}\left\{\sum_{i=1}^{n} |x_i^*(f_i)| > z\right\}.
\end{aligned}$$

This inequality means that the vector $\eta \cdot (g_1, \ldots, g_n)$ is weakly dominated by the vector $(f_1, \ldots, f_n)$ in the space $\ell_\infty^n(F)$. Consequently, from inequality (10.24) and Proposition 10.1 it follows that

$$\frac{z}{4} \le \mathbb{E} \max_{i=1,\ldots,n} \|\eta g_i\|_F = \mathbb{E}\|\eta(g_1, \ldots, g_n)\|_{\ell_\infty^n(F)}$$
$$\le 20M\mathbb{E}\|(f_1, \ldots, f_n)\|_{\ell_\infty^n(F)} = 20M\mathbb{E} \max_{i=1,\ldots,n} \|f_i\|_F.$$

Combining this estimate together with condition (ii), we get

$$\beta \le \mathbb{P}\left\{\max_{i=1,\ldots,n} \|f_i\|_F \ge \alpha\mathbb{E} \max_{i=1,\ldots,n} \|f_i\|_F\right\}$$
$$\le \mathbb{P}\left\{\max_{i=1,\ldots,n} \|f_i\|_F \ge \frac{\alpha z}{80M}\right\} \le n\mathbb{P}\left\{\|f\|_F \ge \frac{\alpha z}{80M}\right\}.$$

Hence, in view of (10.23), it follows

$$\mathbb{P}\left\{\|f\|_F \ge \frac{\alpha z}{80M}\right\} \ge \frac{\beta}{n} \ge \frac{\beta}{2}\mathbb{P}'\left\{\|g\|_F \ge z\right\},$$

and the proof is completed. □

Recall that, in Chap. 6, we have introduced the notions of domination in distribution of systems of r.v.'s in the scalar and vector sense. In the concluding section of this chapter, using the latter result, we shall show that these notions are equivalent if the role of a dominating system is played by the Rademacher sequence.

10.5 Comparison of Distributions of Scalar- and Vector-Valued Sums of r.v.'s with Those of Rademacher Functions

Theorem 10.6 *Suppose that r.v.'s g_i, $i = 1, 2, \ldots$, defined on a probability space $(\Omega, \Sigma, \mathbb{P})$ satisfy the condition: $\{g_i\} \overset{\mathbb{P}}{\prec} \{r_i\}$. Then, $\{g_i\} \overset{\vec{\mathbb{P}}}{\prec} \{r_i\}$.*

Proof We prove first that there exist constants $M, \alpha, \beta > 0$ such that for every Banach space F, any $x_i \in F$, $i = 1, 2, \ldots$ and $m \in \mathbb{N}$ the random vector $S := \sum_{i=1}^m x_i r_i$ satisfies conditions (i) and (ii) of Theorem 10.5.

To this end, we consider the vector $\tilde{S} := (S_1, \ldots, S_n)$ with values in $\ell_\infty^n(F)$, where $n \in \mathbb{N}$ and $S_1, \ldots, S_n$ are independent copies of S. Since the Rademacher functions are independent and identically distributed, we can set

$$S_k = \sum_{i=1}^m x_i r_{i+m(k-1)}, \quad k = 1, 2, \ldots, n.$$

Then, if $\{e_k\}_{k=1}^n$ is the unit vector basis of $\mathbb{R}^n$, we have

$$\tilde{S} = \sum_{k=1}^n S_k e_k = \sum_{k=1}^n \sum_{i=1}^m x_i r_{i+m(k-1)} \cdot e_k = \sum_{k=1}^n \sum_{j=(k-1)m+1}^{km} x_{j-(k-1)m} e_k \cdot r_j.$$

Thus, the vector $\tilde{S}$ is a linear combination of the Rademacher functions with coefficients from $\ell_\infty^n(F)$, and hence, as it was checked above (see Example 10.1), $\tilde{S}$ is L-regular, where L is the constant from Theorem 10.4. Therefore, (i) is proved. For the same reason, by Theorem 5.4, we have

$$\mathbb{E} \max_{i=1,\dots,n} \|S_i\|_F^2 = \mathbb{E}\|\tilde{S}\|_{\ell_\infty^n(F)}^2 \le 2\left(\mathbb{E}\|\tilde{S}\|_{\ell_\infty^n(F)}\right)^2 = 2\left(\mathbb{E} \max_{i=1,\dots,n} \|S_i\|_F\right)^2.$$

Applying now the Paley–Zygmund inequality (see Proposition 1.4) to the r.v. $\max_{i=1,\dots,n} \|S_i\|_F$, we infer that

$$\mathbb{P}\left\{\max_{i=1,\dots,n} \|S_i\|_F \ge \frac{1}{2}\mathbb{E} \max_{i=1,\dots,n} \|S_i\|_F\right\} \ge \frac{1}{8},$$

whence it follows that (ii) holds with the constants $\alpha = 1/2$ and $\beta = 1/8$.

Suppose now that $\{g_i\} \overset{\mathbb{P}}{\prec} \{r_i\}$ with constant 1, i.e., for any $m \in \mathbb{N}$, $a_i \in \mathbb{R}$, $i = 1, 2, \dots, m$, and $z > 0$ we have

$$\mathbb{P}\left\{\omega \in \Omega : \left|\sum_{i=1}^m a_i g_i(\omega)\right| > z\right\} \le m\left\{s \in [0, 1] : \left|\sum_{i=1}^m a_i r_i(s)\right| > z\right\}.$$

This implies, obviously, that the random vector $\sum_{i=1}^m x_i g_i$ is weakly dominated by the vector $S = \sum_{i=1}^m x_i r_i$ for every Banach space F, arbitrary $m \in \mathbb{N}$ and $x_i \in F$, $i = 1, 2, \dots, m$. Hence, the desired assertion follows immediately from Theorem 10.5.

In the case when the domination inequality has the form

$$\mathbb{P}\left\{\omega \in \Omega : \left|\sum_{i=1}^m a_i g_i(\omega)\right| > z\right\} \le m\left\{s \in [0, 1] : \left|\sum_{i=1}^m a_i r_i(s)\right| > \frac{z}{C}\right\}$$

for some C, it suffices to notice that $\{g_i/C\} \overset{\mathbb{P}}{\prec} \{r_i\}$ with constant 1 and then to use the result already established. Thus, it remains to consider the case when for some $C \in \mathbb{N}$ and arbitrary $m \in \mathbb{N}$, $a_i \in \mathbb{R}$, $i = 1, 2, \dots, m$, $z > 0$ it holds

$$\mathbb{P}\left\{\omega \in \Omega : \left|\sum_{i=1}^m a_i g_i(\omega)\right| > z\right\} \le Cm\left\{s \in [0, 1] : \left|\sum_{i=1}^m a_i r_i(s)\right| > z\right\}. \tag{10.25}$$

The proof will be carried again by reduction to the case when $C = 1$. Let $\{e_j\}_{j=1}^C$ be the unit vector basis of $\mathbb{R}^C$ and let $h_1, h_2, \dots, h_m$ be $\mathbb{R}^C$-valued random vectors on Ω such that $h_1, \dots, h_m, g_1, \dots, g_m$ are independent and $\mathbb{P}\{h_i = e_j\} = 1/C$ for all $i = 1, 2, \dots, m$ and $j = 1, 2, \dots, C$. It follows from this definition that $h_i = \sum_{j=1}^C \chi_{E_i^{(j)}} e_j$, where the sets $E_i^{(j)}$ are pairwise disjoint for each $i = 1, 2, \dots, m$ and $\mathbb{P}(E_i^{(j)}) = 1/C$ for all i and j.

Consider the r.v.'s $h_i^{(j)} := \chi_{E_i^{(j)}}$, $i = 1, 2, \dots, m$, $j = 1, 2, \dots, C$. Clearly, they possess the following properties:

(a) $h_1^{(j)}, \dots, h_m^{(j)}, g_1, \dots, g_m$ are independent for every $j = 1, 2, \dots, C$;
(b) $\mathbb{P}\{h_i^{(j)} = 1\} = 1 - \mathbb{P}\{h_i^{(j)} = 0\} = 1/C, i = 1, 2, \dots, m, j = 1, 2, \dots, C$;
(c) $\sum_{j=1}^C h_i^{(j)} = 1$ for each $i = 1, 2, \dots, m$.

Let now F be an arbitrary Banach space, $x_i \in F$, $i = 1, 2, \dots, m$. We obtain from the property (c) that

$$\Big\{\Big\|\sum_{i=1}^m x_i g_i\Big\|_F > z\Big\} = \Big\{\Big\|\sum_{j=1}^C \sum_{i=1}^m x_i g_i h_i^{(j)}\Big\|_F > z\Big\} \subset \bigcup_{j=1}^C \Big\{\Big\|\sum_{i=1}^m x_i g_i h_i^{(j)}\Big\|_F > \frac{z}{C}\Big\}.$$

Since $h_i^{(j)}$ are identically distributed, this implies that

$$\mathbb{P}\Big\{\Big\|\sum_{i=1}^m x_i g_i\Big\|_F > z\Big\} \le C\mathbb{P}\Big\{\Big\|\sum_{i=1}^m x_i g_i h_i^{(1)}\Big\|_F > \frac{z}{C}\Big\}. \tag{10.26}$$

On the other hand, by the properties (a) and (b), for all $a_i \in \mathbb{R}$, $i = 1, 2, \dots, m$ and every $z > 0$ we have

$$\mathbb{P}\Big\{\Big|\sum_{i=1}^m a_i g_i h_i^{(1)}\Big| > z\Big\} = \frac{1}{C}\mathbb{P}\Big\{\Big|\sum_{i=1}^m a_i g_i\Big| > z\Big\}.$$

Combining this together with inequality (10.25), we get

$$\mathbb{P}\Big\{\Big|\sum_{i=1}^m a_i g_i h_i^{(1)}\Big| > z\Big\} \le m\Big\{\Big|\sum_{i=1}^m a_i r_i\Big| > z\Big\},$$

As was established above, in the case when $C = 1$ the conditions of Theorem 10.5 are fulfilled with the constants L, $\alpha = 1/2$ and $\beta = 1/8$. Therefore, applying the latter theorem, for arbitrary $x_1, x_2, \dots, x_m$ from F we get

$$\mathbb{P}\Big\{\Big\|\sum_{i=1}^m x_i g_i h_i^{(1)}\Big\|_F > z\Big\} \le 16m\Big\{\Big\|\sum_{i=1}^m x_i r_i\Big\|_F > \frac{z}{160L}\Big\}.$$

Comparing this inequality with (10.26), we conclude finally that

$$\mathbb{P}\Big\{\Big\|\sum_{i=1}^{m} x_i g_i\Big\|_F > z\Big\} \le 16Cm\Big\{\Big\|\sum_{i=1}^{m} x_i r_i\Big\|_F > \frac{z}{160CL}\Big\},$$

i.e., $\{g_i\} \overset{\vec{\mathbb{P}}}{\prec} \{r_i\}$. □

From Theorems 10.6 and 8.1 it follows

Theorem 10.7 *Suppose that* $\{g_n\}_{n=1}^{\infty}$ *is a sequence of r.v.'s defined on a probability space* $(\Omega, \Sigma, \mathbb{P})$, $|g_n(\omega)| \le D$ *for all* $\omega \in \Omega$, $n \in N$, *and* $g_n \overset{w}{\to} 0$ *in* $L_2(\Omega)$. *Then one can extract a subsequence* $\{g_{n_k}\} \subset \{g_n\}$ *such that* $\{g_{n_k}\} \overset{\vec{\mathbb{P}}}{\prec} \{r_k\}$ *with a constant depending only on* D.

Comments and References Bernoulli processes are important almost to the same extent as Gaussian ones. One can find their detailed investigation with a number of related results, for instance, in the monographs [177] and [277].

The Bernoulli conjecture was posed in [177, Problem 12] (see also Chapter 4 in [277]). Only almost 25 years later, an affirmative answer was obtained by Polish mathematicians Bednorz and Latała in the paper [68] (a little bit earlier, they announced this result in [67]). For a detailed account of the proof of this remarkable result as well as for some its nice applications and further results and conjectures we refer to another Talagrands' monograph [278]. Theorem 10.4, being one of the most interesting consequences of the affirmative solution of the Bernoulli conjecture, has been established also in [68]. The results of Sect. 10.4 have been proved in another paper by Latała [175].

Theorems 10.6 and 10.7 related to comparing the scalar and vector domination of systems of r.v.'s have been obtained in [35]. Independently, in the paper [83] by Bourgain and Lewko, devoted to studying variants of the well-known Kaczmarz's sidonicity problem, a result very close to Theorem 10.6 has been proved. One says that a sequence $\{g_k\}_{k=1}^{\infty}$ of r.v.'s defined on a probability space $(\Omega, \Sigma, \mathbb{P})$ is a $\psi_2(C)$-*system* $(C > 0)$ if for arbitrary $a_k \in \mathbb{R}$ we have

$$\Big\|\sum_{k=1}^{\infty} a_k g_k\Big\|_{L_{N_2}(\Omega)} \le C\Big(\sum_{k=1}^{\infty} a_k^2\Big)^{1/2}.$$

Here, as usual, $L_{N_2}(\Omega)$ is the Orlicz space, $N_2(u) = e^{u^2} - 1$. Then, the Bourgain–Lewko theorem (see [83, Theorem 1.9]) reads as follows:

Let $\{g_k\}_{k=1}^{\infty}$ be a $\psi_2(C)$-system, uniformly bounded by a constant M. Then, for every normed vector space F, arbitrary $n \in \mathbb{N}$ and $x_k \in F$, $k = 1, 2, \ldots, n$, it holds

$$\mathbb{E}\Big\|\sum_{k=1}^{n} x_k g_k\Big\|_F \le MC\int_0^1 \Big\|\sum_{k=1}^{n} x_k r_k(t)\Big\|_F \, dt. \tag{10.27}$$

Observe that this result is equivalent, in fact, to Theorem 10.6. Indeed, it follows first from [18, Theorem 3] that a uniformly bounded sequence $\{g_k\}_{k=1}^{\infty}$ is a $\psi_2(C)$-system if and only if $\{g_k\} \overset{\mathbb{P}}{\prec} \{r_k\}$ (Theorem 3 in [18] deals with orthonormal sequences $\{g_k\}$ but the same proof works also without this assumption). Secondly, although inequality (10.27) for the expectations is formally weaker than the distribution estimate, proved in Theorem 10.6, but the latter one can be obtained from (10.27), for example, in the same way as in the proof of [175, Theorem 1].

It is worth to note that the paper [83] contains also interesting problems related to properties of Sidon sets of measurable functions. Some of them were solved recently by Pisier in [243]. In particular, he proved that every orthonormal $\psi_2(C)$-system $\{g_k\}_{k=1}^{\infty}$, uniformly bounded by a constant M, satisfies the following: there is a constant $\alpha = \alpha(C, M)$ such that for each $a = (a_k)_{k=1}^{\infty} \in \ell_1$ we have

$$\sum_{k=1}^{\infty} |a_k| \le \alpha \sup_{t_1, t_2 \in [0,1]} \left| \sum_{k=1}^{\infty} a_k g_k(t_1) g_k(t_2) \right|$$

(see [243, Theorem 1.1]).

In [35], it is shown also that the result similar to Theorem 10.6 fails in the case when the Rademacher sequence, conversely, is *dominated* by a system $\{g_i\}$. At the same time, we have the following [35, Theorem 5].

Let X be a Banach function lattice, which has some finite cotype q, and let $\{g_k\}_{k=1}^{\infty}$ be a sequence of r.v.'s defined on a probability space $(\Omega, \Sigma, \mathbb{P})$ such that $\{r_k\} \overset{\mathbb{P}}{\prec} \{g_k\}$. Then, for some constant $B > 0$ and all $1 \le p < \infty$, $n \in \mathbb{N}$, $x_k \in X$, $k = 1, 2, \ldots, n$, it holds

$$\left(\int_0^1 \left\| \sum_{k=1}^{n} x_k r_k(t) \right\|_X^p dt \right)^{1/p} \le B \left(\mathbb{E} \left\| \sum_{k=1}^{n} x_k g_k \right\|_X^p \right)^{1/p}.$$

Here, B depends only on q, the constant of q-cotype of X, and the constant of the relation $\{r_k\} \overset{\mathbb{R}}{\prec} \{g_k\}$.

Finally, Theorems 10.1, 10.2 and their proofs are taken from [177, Theorem 4.12] and [276, Proposition 2.2], respectively.

Chapter 11
Rademacher Multiplicator Spaces

In this chapter, we study multiplicator spaces, generated by the Rademacher system in s.s.'s. They are called so because the elements of such a space $\mathcal{M}(X)$ can be regarded as multiplicators acting from the Rademacher subspace of a given s.s. X into the whole space X.

We shall focus mainly on issues related to the problem when there exists a symmetric norm that is equivalent on the Banach lattice $\mathcal{M}(X)$ to the initial $\mathcal{M}(X)$-norm. By abuse of terminology, we shall say in this case that $\mathcal{M}(X)$ is a s.s.

First of all, we show that for s.s.'s X, located sufficiently "far away" from L_∞, it is not the case. In this regard, we shall introduce the notion of symmetric kernel Sym (X) as the largest s.s. contained in $\mathcal{M}(X)$; we shall find its description and consider various concrete examples.

At the same time, if a s.s. X is located sufficiently "close" to L_∞, the space $\mathcal{M}(X)$ may be symmetric. Then, there are two different cases when this happens. Firstly, $\mathcal{M}(X)$ may coincide with the smallest s.s. on $[0, 1]$, i.e., L_∞; we shall establish that $\mathcal{M}(X) = L_\infty$ (with equivalence of norms) if and only if the separable part of X does not contain the function $\ln^{1/2}(e/t)$. Secondly, if X is somewhat more "remote" from L_∞, the space $\mathcal{M}(X)$ does not coincide with L_∞, but nevertheless it may be symmetric. We shall obtain sufficient and as well necessary conditions, under which this holds.

Along with the largest s.s. contained in the space $\mathcal{M}(X)$ (that is, the symmetric kernel Sym (X)), there is also the smallest one among s.s.'s with the Fatou property, which is, conversely, contains $\mathcal{M}(X)$. For proving this fact we introduce a construction of the so-called tail Rademacher multiplicator space $\mathcal{M}_T(X)$ that is somewhat more complicated than the notion of symmetric kernel. Properties of the latter space are studied in detail in the final section of the chapter and will be used further in the proof of a local version of Khintchine's inequality in s.s.'s.

S. V. Astashkin, *The Rademacher System in Function Spaces*,
https://doi.org/10.1007/978-3-030-47890-2_11

11.1 Definition and Basic Properties of $\mathcal{M}(X)$

Let X be a s.s. on $[0, 1]$. Recall that $\mathcal{R}(X)$ is the closed subspace in X, consisting of all $g \in X$ representable in the form: $g(t) = \sum_{k=1}^{\infty} c_k r_k(t)$, $0 \le t \le 1$, where $c_k \in \mathbb{R}$, $k = 1, 2, \dots$

Definition 11.1 If X is a s.s. on $[0, 1]$, then the *Rademacher multiplicator space* $\mathcal{M}(X) = \mathcal{M}(\mathcal{R}, X)$ is the set of all measurable functions $f : [0, 1] \to \mathbb{R}$ such that $f \cdot g \in X$ for every $g \in \mathcal{R}(X)$.

Proposition 11.1 *For every s.s.* X *on* $[0, 1]$ *the set* $\mathcal{M}(X)$ *is a Banach function lattice equipped with the norm*

$$\|f\|_{\mathcal{M}} := \sup\{\|f \cdot g\|_X : g \in \mathcal{R}(X), \|g\|_X \le 1\}.$$

Moreover, $\mathcal{M}(X) \subset X$ *and* $\|f\|_X \le \|f\|_{\mathcal{M}}$ *for* $f \in \mathcal{M}(X)$.

Proof We show first that $\|f\|_{\mathcal{M}} < \infty$ for each function $f \in \mathcal{M}(X)$. Let us check that the graph of the linear operator $T_f g := f \cdot g$, acting from $\mathcal{R}(X)$ into X, is closed provided if $f \in \mathcal{M}(X)$. Suppose that $g_n \in \mathcal{R}(X)$, $g \in \mathcal{R}(X)$, $g_n \to g$ and $T_f g_n \to h$ in X as $n \to \infty$, where $h \in X$. Then, $g_n \to g$ in measure, and hence, by the Riesz theorem, there is a subsequence $\{g_{n_k}\} \subset \{g_n\}$ that converges to g a.e. on $[0, 1]$. Consequently, $T_f g_{n_k} = f \cdot g_{n_k} \to f \cdot g$ a.e., whence $h = f \cdot g = T_f g$. Thus, the graph of T_f is closed. Since $\mathcal{R}(X)$ and X are Banach spaces, then, by the Closed Graph theorem (see e.g. [79, Theorem 6.2.7]), the operator T_f is bounded. This means that $\|f\|_{\mathcal{M}} = \|T_f\| < \infty$, as we desired.

The facts that the functional $f \mapsto \|f\|_{\mathcal{M}}$ is a norm on the linear space $\mathcal{M}(X)$, as well as that $\mathcal{M}(X)$ is a function lattice are immediate. Let us show the completeness of $\mathcal{M}(X)$.

Suppose that $\{f_n\}$ is a Cauchy sequence in $\mathcal{M}(X)$. From definition of the $\mathcal{M}(X)$-norm it follows that for every $g \in \mathcal{R}(X)$ the sequence $f_n \cdot g$ is Cauchy in the space X. Since X is complete, then there exists a function $f_g \in X$ such that $\|f_n g - f_g\|_X \to 0$ as $n \to \infty$. In particular, taking $g = r_1$, we get that $\|f_n - f\|_X \to 0$ for $f := f_{r_1} \cdot r_1 \in X$. Therefore, using the fact that the norm convergence in an arbitrary s.s. implies the convergence in measure, we conclude that $f_g = fg$ for each $g \in \mathcal{R}(X)$. Taking into account once more that $\{f_n\}$ is a Cauchy sequence in $\mathcal{M}(X)$, for any $\varepsilon > 0$ we can find N so that for all $m \ge n \ge N$ and $g \in \mathcal{R}(X)$, $\|g\|_X \le 1$, we have $\|f_n g - f_m g\|_X \le \varepsilon$. Passing to the limit then as $m \to \infty$, we get $\|f_n g - fg\|_X \le \varepsilon$ if $g \in \mathcal{R}(X)$, $\|g\|_X \le 1$. Equivalently, we have $\|f_n - f\|_{\mathcal{M}} \le \varepsilon$, i.e., $f \in \mathcal{M}(X)$ and $f_n \to f$ in $\mathcal{M}(X)$.

Since the final assertion of the proposition is obvious, the proof is completed. □

Next, we need the following property of Rademacher multiplicator spaces.

Proposition 11.2 *If a s.s. X has the Fatou property, then so has $\mathcal{M}(X)$.*

Proof Suppose that $\{x_n\}_{n=1}^{\infty} \subset \mathcal{M}(X)$, $0 \le x_n \uparrow x$, and

$$C := \sup_{n=1,2,\dots} \|x_n\|_{\mathcal{M}} < \infty.$$

Then, for every $z \in \mathcal{R}(X)$, $\|z\|_X \le 1$, we have $x_n|z| \uparrow x|z|$ and $\|zx_n\|_X \le \|x_n\|_{\mathcal{M}} \le C$ for all $n = 1, 2, \dots$ Therefore, since X has the Fatou property, we get $xz \in X$ and $\|xz\|_X \le C$, $n = 1, 2, \dots$ This implies that $x \in \mathcal{M}(X)$ and $\|x\|_{\mathcal{M}} \le C$, which completes the proof. □

The main aim of this section is to prove that the Rademacher multiplicator space $\mathcal{M}(X)$ is not symmetric if a s.s. X is located sufficiently "far away" from the space L_∞.

Recall first that, according to Proposition 2.4, we have

$$\|x\|_{L_{N_2}} \asymp \sup_{0<t\le 1} x^*(t) \ln^{-1/2}(e/t), \tag{11.1}$$

where, as above, L_{N_2} is the Orlicz space with $N_2(u) = e^{u^2} - 1$. Hence, there is $C > 0$ such that for all $x \in L_{N_2}$ and $0 < t \le 1$ the following inequality holds:

$$x^*(t) \le C \, \|x\|_{L_{N_2}} \cdot \ln^{1/2}(e/t). \tag{11.2}$$

By Theorem 2.2, we have

$$\Big\| \sum_{n=1}^{\infty} a_n r_n \Big\|_{L_{N_2}} \asymp \|(a_n)\|_{\ell_2}$$

and $\sum_{n=1}^{\infty} a_n r_n \in G$ if $(a_n) \in \ell_2$. Therefore, from (11.2) it follows

$$\Big(\sum_{n=1}^{\infty} a_n r_n\Big)^*(t) \le K \cdot \|(a_n)\|_{\ell_2} \cdot \ln^{1/2}(e/t), \quad 0 < t \le 1, \tag{11.3}$$

with a universal constant $K > 0$.

We proceed with the following key result related to comparing Rademacher multiplicator norms of characteristic functions.

Lemma 11.1 *Let X be a s.s. on $[0, 1]$, $X \supset G$. Then, there exists a constant $C > 0$ such that for every positive integer n of the form $n = 2^m$, $m \in \mathbb{N}$, we can find measurable sets B_n and D_n of the same measure, equal to $n2^{-n}$, satisfying the* *inequality*

$$\frac{\|\chi_{B_n}\|_{\mathcal{M}}}{\|\chi_{D_n}\|_{\mathcal{M}}} \le \frac{C}{n^{1/2}}\left(1+\frac{1}{\phi_X(n2^{-n})}\int_0^{n/2^n}\frac{\phi_X(t)}{t}\,dt\right)+\left(\frac{\phi_X(2^{-n})}{\phi_X(n2^{-n})}\right)^{1/2} \tag{11.4}$$

where ϕ_X is the fundamental function of X.

Proof First of all, since $X \supset G$, then $\mathcal{R}(X) = \ell_2$ and

$$\Big\|\sum_{n=1}^{\infty} a_n r_n\Big\|_X \asymp \|(a_n)\|_{\ell_2}$$

(see Remark 2.1). Therefore, we have

$$\|f\|_{\mathcal{M}} \asymp \sup\left\{\|fg\|_X : g = \sum_{k=1}^{\infty} c_k r_k \in X,\ \|(c_k)\|_{\ell_2} \le 1\right\}. \tag{11.5}$$

Let $J \subset \{1, 2, \ldots, 2^n\}$ be an arbitrary n-element set (it will be chosen later) and let $A = \bigcup_{j\in J}\Delta_n^j$ be the corresponding union of dyadic intervals $\Delta_n^j = ((j-1)2^{-n}, j2^{-n})$ of rank n. Note that $m(A) = n2^{-n}$. If χ_A is the characteristic function of this set, then from (11.5) it follows

$$\begin{aligned}
\|\chi_A\|_{\mathcal{M}} &\le C_1 \sup\left\{\Big\|\chi_A\sum_{i=1}^{\infty} b_i r_i\Big\|_X : \|(b_i)\|_{\ell_2} \le 1\right\}\\
&\le C_1 \sup\left\{\Big\|\chi_A\sum_{i=1}^{n} b_i r_i\Big\|_X : \|(b_i)\|_{\ell_2} \le 1\right\}\\
&+ C_1 \sup\left\{\Big\|\chi_A\sum_{i=n+1}^{\infty} b_i r_i\Big\|_X : \|(b_i)\|_{\ell_2} \le 1\right\}.
\end{aligned} \tag{11.6}$$

We estimate first the second term from the right-hand side of inequality (11.6). Observe that the functions

$$\chi_A(t)\cdot\sum_{i=n+1}^{\infty} b_i r_i(t) \quad\text{and}\quad \chi_{[0,n2^{-n}]}(t)\cdot\sum_{i=n+1}^{\infty} b_i r_i(t)$$

are identically distributed. Moreover, for each $\lambda > 0$ we have

$$m\left\{t\in[0,n2^{-n}] : \Big|\sum_{i=n+1}^{\infty} b_i r_i(t)\Big| > \lambda\right\} = \frac{n}{2^n}\, m\left\{t : \Big|\sum_{i=n+1}^{\infty} b_i r_i(t)\Big| > \lambda\right\},$$

whence

$$\Big(\chi_{[0,n2^{-n}]}\cdot\sum_{i=n+1}^{\infty}b_ir_i\Big)^*(t)=\Big(\sum_{i=n+1}^{\infty}b_ir_i\Big)^*(2^nt/n),\quad 0\le t\le n/2^n.$$

Consequently, since $\|x\|_X \le \|x\|_{\Lambda(\phi_X)}$, where $\Lambda(\phi_X)$ is the Lorentz space (see Appendix C), by (11.3), we get

$$\begin{aligned}\Big\|\chi_A\sum_{i=n+1}^{\infty}b_ir_i\Big\|_X&=\Big\|\chi_{[0,n2^{-n}]}\sum_{i=n+1}^{\infty}b_ir_i\Big\|_X\\&\le\Big\|\chi_{[0,n2^{-n}]}\sum_{i=n+1}^{\infty}b_ir_i\Big\|_{\Lambda(\phi_X)}\\&=\int_0^{n/2^n}\Big(\sum_{i=n+1}^{\infty}b_ir_i\Big)^*\Big(\frac{2^nt}{n}\Big)\phi_X{}'(t)\,dt\\&\le K\,\|(b_i)\|_{\ell_2}\int_0^{n/2^n}\ln^{1/2}\Big(\frac{ne}{2^nt}\Big)\phi_X{}'(t)\,dt.\end{aligned}$$

Integrating by parts, we can estimate the last integral as follows:

$$\begin{aligned}\int_0^{n/2^n}\ln^{1/2}\Big(\frac{ne}{2^nt}\Big)\phi_X{}'(t)\,dt&=\phi_X\Big(\frac{n}{2^n}\Big)-\lim_{t\to0}\ln^{1/2}\Big(\frac{ne}{2^nt}\Big)\phi_X(t)\\&\quad+\frac12\int_0^{n/2^n}\frac{\phi_X(t)}{t\ln^{1/2}(ne/2^nt)}\,dt\\&\le\phi_X\Big(\frac{n}{2^n}\Big)+\frac12\int_0^{n/2^n}\frac{\phi_X(t)}{t}\,dt.\end{aligned}$$

Thus,

$$\Big\|\chi_A\sum_{i=n+1}^{\infty}b_ir_i\Big\|_X\le K\,\|(b_i)\|_{\ell_2}\Big\{\phi_X\Big(\frac{n}{2^n}\Big)+\frac12\int_0^{n/2^n}\frac{\phi_X(t)}{t}\,dt\Big\}. \tag{11.7}$$

Proceed now with estimation of the first term from the right-hand side of inequality (11.6). Let a_{ij} be the value of the function r_i, $i = 1,2,\dots,n$, on the dyadic interval Δ_n^j, $1\le j\le 2^n$. Since $n=2^m$, then we can choose a set $J_1\subset\{1,2,\dots,2^n\}$, card $J_1=n$, such that the $n\times n$ matrix $n^{-1/2}\cdot(a_{ij})_{1\le i\le n,\,j\in J_1}$

is orthogonal. Then, denoting $B_n := \bigcup_{j\in J_1} \Delta_n^j$, we have

$$\sup\Big\{\Big\|\chi_{B_n}\sum_{i=1}^n b_i r_i\Big\|_X : \|(b_i)\|_{\ell_2}\le 1\Big\} = \sup\Big\{\Big\|\sum_{j\in J_1}\Big(\sum_{i=1}^n b_i r_i\Big)\chi_{\Delta_n^j}\Big\|_X : \|(b_i)\|_{\ell_2}\le 1\Big\}$$
$$= \sup\Big\{\Big\|\sum_{j\in J_1}\sum_{i=1}^n a_{ij}b_i\cdot\chi_{\Delta_n^j}\Big\|_X : \|(b_i)\|_{\ell_2}\le 1\Big\}$$
$$= n^{1/2}\sup\Big\{\Big\|\sum_{j\in J_1} c_j\chi_{\Delta_n^j}\Big\|_X : \|(c_j)_{j\in J_1}\|_{\ell_2}\le 1\Big\}.$$

Therefore, by (11.6) and (11.7),

$$\|\chi_{B_n}\|_{\mathcal{M}} \le C_2\Big(\phi_X\Big(\frac{n}{2^n}\Big) + \frac{1}{2}\int_0^{n/2^n}\frac{\phi_X(t)}{t}\,dt + n^{1/2}\sup\Big\{\Big\|\sum_{j\in J_1} c_j\chi_{\Delta_n^j}\Big\|_X : \|(c_j)_{j\in J_1}\|_{\ell_2}\le 1\Big\}\Big). \tag{11.8}$$

On the other hand, select a set $J_2 \subset \{1,2,\dots,2^n\}$, card $J_2 = n$, such that each column of the corresponding $n\times n$ matrix $(a_{ij})_{1\le i\le n, j\in J_2}$ has exactly one entry equal to -1 and the rest are equal to 1. Then, $\sum_{i=1}^n a_{ij} = n-2$ for every $j\in J_2$. Hence, for the set $D_n := \bigcup_{j\in J_2}\Delta_n^j$ it follows

$$\|\chi_{D_n}\|_{\mathcal{M}} \ge c\Big\|\sum_{j\in J_2}\Big(n^{-1/2}\sum_{i=1}^n r_i\Big)\chi_{\Delta_n^j}\Big\|_X$$
$$= c\Big\|\sum_{j\in J_2}\Big(n^{-1/2}\sum_{i=1}^n a_{ij}\Big)\chi_{\Delta_n^j}\Big\|_X$$
$$= c(n^{1/2} - 2n^{-1/2})\cdot\phi_X\Big(\frac{n}{2^n}\Big).$$

Combining this inequality with (11.8), we get

$$\frac{\|\chi_{B_n}\|_{\mathcal{M}}}{\|\chi_{D_n}\|_{\mathcal{M}}} \le \frac{C}{n^{1/2}}\Big(1 + \frac{1}{\phi_X(n2^{-n})}\int_0^{n/2^n}\frac{\phi_X(t)}{t}\,dt\Big) + \frac{C}{\phi_X(n2^{-n})}\sup\Big\{\Big\|\sum_{j\in J_1} c_j\chi_{\Delta_n^j}\Big\|_X : \|(c_j)_{j\in J_1}\|_{\ell_2}\le 1\Big\}, \tag{11.9}$$

with some $C>0$ independent of n. Finally, let us estimate the second term from the right-hand side of this inequality.

Let X_n be the n-dimensional subspace of X, spanned by the characteristic functions $\chi_{\Delta_n^j}$, $j \in J_1$. Consider the linear operator $T : \mathbb{R}^n \to X_n$ defined by $T(c_j) = \sum_{j\in J_1} c_j \chi_{\Delta_n^j}$. One can readily check that $\|T\|_{\ell_\infty^n \to X_n} = \phi_X(n2^{-n})$ and $\|T\|_{\ell_1^n \to X_n} = \phi_X(2^{-n})$. Since the conjugate operator $T^* : X_n^* \to \mathbb{R}^n$ is defined by $T^*x^* = (x^*(\chi_{\Delta_n^j}))$ and

$$\sum_{i=1}^n x_i^2 \le \max_{i=1,\dots,n} |x_i| \cdot \sum_{i=1}^n |x_i|,$$

we have

$$\|T\|_{\ell_2^n \to X_n} = \|T^*\|_{X_n^* \to \ell_2^n} \le \|T^*\|_{X_n^* \to \ell_1^n}^{1/2} \cdot \|T^*\|_{X_n^* \to \ell_\infty^n}^{1/2} = \|T\|_{\ell_\infty^n \to X_n}^{1/2} \cdot \|T\|_{\ell_1^n \to X_n}^{1/2}.$$

Thus,

$$\frac{1}{\phi_X(n2^{-n})} \cdot \sup\left\{\Big\|\sum_{j\in J_1} c_j \chi_{\Delta_n^j}\Big\|_X : \|(c_j)_{j\in J_1}\|_{\ell_2} \le 1\right\}$$
$$= \frac{\|T\|_{\ell_2^n \to X_n}}{\|T\|_{\ell_\infty^n \to X_n}} \le \frac{\|T\|_{\ell_1^n \to X_n}^{1/2}}{\|T\|_{\ell_\infty^n \to X_n}^{1/2}} = \left(\frac{\phi_X(2^{-n})}{\phi_X(n2^{-n})}\right)^{1/2}.$$

As a result, inequality (11.4) follows from the latter inequality and (11.9). □

By using Lemma 11.1, it is quite easy to show that, in many cases, Rademacher multiplicator spaces fail to be symmetric.

Recall that γ_ψ denotes the lower dilation exponent of a function ψ (see Appendix C).

Theorem 11.1 *If $\gamma_{\phi_X} > 0$, where ϕ_X is the fundamental function of a s.s. X, then the Banach lattice $\mathcal{M}(X)$ is not symmetric. In particular, this holds whenever the lower Boyd index α_X of X is positive.*

Proof According to the definition of γ_{ϕ_X}, for every $\alpha \in (0, \gamma_{\phi_X})$ there is a constant $C = C(\alpha) > 0$ such that for all $0 < s, t \le 1$ we have

$$\frac{\phi_X(st)}{\phi_X(s)} \le Ct^\alpha. \tag{11.10}$$

This implies, in particular, that $\phi_X(t) \le C\phi_X(1)t^\alpha$, $0 < t \le 1$, and hence $\Lambda(t^\alpha) \subset \Lambda(\phi_X)$, where $\Lambda(t^\alpha)$ and $\Lambda(\phi_X)$ are the Lorentz spaces with the fundamental functions t^α and ϕ_X, respectively. Moreover, $\Lambda(\phi_X) \subset X$ (see Appendix C), and from inequality (11.2), combined with the fact that

$$\int_0^1 \ln^{1/2}(e/t)t^{\alpha-1}\,dt < \infty,$$

it follows $L_{N_2} \subset \Lambda(t^\alpha)$. These embeddings imply that $L_{N_2} \subset X$. Thus, the hypothesis of Lemma 11.1 is satisfied, and therefore for every $n = 2^m$, $m \in \mathbb{N}$, there exist measurable sets B_n and D_n, $m(B_n) = m(D_n) = n2^{-n}$, such that inequality (11.4) holds.

Furthermore, from (11.10) we deduce that $\phi_X(2^{-n})/\phi_X(n2^{-n}) \le C n^{-\alpha}$ for all $n \in \mathbb{N}$. Thereby, in this case, the second term from the right-hand side of (11.4) converges to zero as $n \to \infty$. Moreover, the condition $\gamma_{\phi_X} > 0$ assures that for some $C' > 0$ we have

$$\int_0^{n/2^n} \frac{\phi_X(t)}{t}\,dt \le C'\phi_X(n2^{-n}), \quad n = 1, 2, \ldots$$

(see [168, § II.1, Corollary 2, p. 57]). Hence, the first term from the right-hand side of (11.4) is bounded from above by $(1 + C')n^{-1/2}$. Summing up, we see that the ratio

$$\frac{\|\chi_{B_n}\|_{\mathcal{M}}}{\|\chi_{D_n}\|_{\mathcal{M}}}$$

converges to zero as $n \to \infty$, and so $\mathcal{M}(X)$ fails to be symmetric (i.e., it is not symmetric after any equivalent renorming).

The last assertion of the theorem follows now at once from the inequality $\gamma_{\phi_X} \ge \alpha_X$ (see Appendix C). □

By Theorem 11.1, the Rademacher multiplicator spaces are not symmetric for many s.s.'s (for instance, for L_p, $1 \le p < \infty$). This makes rather natural for such a space X the problem of identifying the largest s.s., which is embedded into the Banach lattice $\mathcal{M}(X)$.

11.2 The Symmetric Kernel of $\mathcal{M}(X)$

We write $x \simeq y$ if the functions $x = x(t)$ and $y = y(t)$ are equimeasurable on $[0, 1]$, i.e.,

$$m\{t \in [0, 1] : |x(t)| > \tau\} = m\{t \in [0, 1] : |y(t)| > \tau\} \text{ for all } \tau > 0.$$

Definition 11.2 Let E be a Banach function lattice on $[0, 1]$. The *symmetric kernel* of E is the set $\mathrm{Sym}\,(E)$ defined by

$$\mathrm{Sym}\,(E) := \{x \in E : \text{ if } y \simeq x, \text{ then } y \in E\}.$$

Further, we consider only symmetric kernels of function lattices of the form $\mathcal{M}(X)$, where X is a s.s. Therefore, in what follows, to simplify the notation we shall write Sym (X) instead of Sym $(\mathcal{M}(X))$.

Proposition 11.3 *Let X be a s.s. on $[0, 1]$. The symmetric kernel* Sym (X) *of the Rademacher multiplicator space $\mathcal{M}(X)$, equipped with the norm*

$$\|x\|_{\text{Sym}} := \sup_{z:\, z\simeq x} \|z\|_{\mathcal{M}},$$

is the largest s.s. that is continuously embedded into $\mathcal{M}(X)$.

Proof Show first that $\|x\|_{\text{Sym}} < \infty$ if $x \in \text{Sym}(X)$. For any z such that $z \simeq x$ we consider the operator T_z defined as follows: $T_z y(t) := z(t)y(t)$, where $y \in \mathcal{R}(X)$. Since $z \in \mathcal{M}(X)$, then T_z is bounded from $\mathcal{R}(X)$ into X. For every function $y \in \mathcal{R}(X)$ we can find a measure-preserving mapping $\omega\colon [0, 1] \to [0, 1]$ such that $|y| = y^*(\omega)$ [70, Corollary 2.7.6]. Note that $x^*(\omega) \simeq x$. Consequently, $x^*(\omega) \in \mathcal{M}(X)$, whence $x^*(\omega)y \in X$. Since X is a s.s., it follows

$$\|T_{z^*} y^*\|_X = \|z^*(\omega)y^*(\omega)\|_X = \|z^*(\omega)y\|_X = \|x^*(\omega)y\|_X.$$

Moreover, by a known property of the decreasing rearrangement [168, § II.2, Property 10°, p. 67], we have

$$(zy)^*(2t) \le z^*(t)y^*(t), \quad 0 < t \le 1/2.$$

Combining this with the inequality $\|\sigma_2\|_{X\to X} \le 2$, where $\sigma_2 x(t) = x(t/2)$ (see Appendix C), we get

$$\sup_{z\simeq x} \|T_z y\|_X \le 2 \sup_{z\simeq x} \|T_{z^*} y^*\|_X \le 2\|x^*(\omega)y\|_X < \infty.$$

Thus, the family of operators $\{T_z : z \simeq x\}$, acting from $\mathcal{R}(X)$ into X, is pointwise bounded. Therefore, by the Uniform Boundedness Principle,

$$\|x\|_{\text{Sym}} = \sup_{z\simeq x} \|z\|_{\mathcal{M}} = \sup_{z\simeq x} \|T_z\|_{\mathcal{R}(X)\to X} < \infty.$$

Next, we check that the functional $x \mapsto \|x\|_{\text{Sym}}$ satisfies the triangle inequality. Let x_1 and x_2 belong to the set Sym (X), and let $z \simeq x_1 + x_2$. For arbitrary $\varepsilon > 0$ there exists a measure-preserving mapping $\tau\colon [0, 1] \to [0, 1]$ such that

$$\|\, |z| - |x_1(\tau) + x_2(\tau)|\, \|_\infty \le \frac{\varepsilon}{\|\chi_{[0,1]}\|_X}$$

[168, Lemma II.2.1, p. 60]. Furthermore, since $x_i(\tau) \simeq x_i$, then, by definition of the Sym(X)-norm, we have $\|x_i(\tau)\|_{\text{Sym}} = \|x_i\|_{\text{Sym}}$, $i = 1, 2$. Therefore, for every

$y \in \mathcal{R}(X)$ such that $\|y\|_X \le 1$ we get

$$\|zy\|_X \le \|(|z| - |x_1(\tau) + x_2(\tau)|)y\|_\infty \|\chi_{[0,1]}\|_X + \|(x_1(\tau) + x_2(\tau))y\|_X$$
$$\le \varepsilon + \|x_1(\tau)y\|_X + \|x_2(\tau)y\|_X \le \varepsilon + \|x_1\|_{\mathrm{Sym}} + \|x_2\|_{\mathrm{Sym}}.$$

Thus, $x_1 + x_2 \in \mathrm{Sym}\,(X)$ and

$$\|x_1 + x_2\|_{\mathrm{Sym}} \le \|x_1\|_{\mathrm{Sym}} + \|x_2\|_{\mathrm{Sym}} + \varepsilon.$$

Since $\varepsilon > 0$ is arbitrary, the desired inequality is proved. One can easily see that the rest of required properties is immediate. Consequently, the functional $x \mapsto \|x\|_{\mathrm{Sym}}$ is a norm, and hence $\mathrm{Sym}\,(X)$ is a linear normed space.

Let us prove the completeness of the space $\mathrm{Sym}\,(X)$. If $\{x_n\}$ is a Cauchy sequence in $\mathrm{Sym}\,(X)$, then it has the same property also in the space $\mathcal{M}(X)$. Hence, according to Proposition 11.1, we have $\|x_n - x\|_{\mathcal{M}} \to 0$ for some $x \in \mathcal{M}(X)$. Obviously, for any measure-preserving mapping $\lambda\colon [0,1] \to [0,1]$ the sequence $\{x_n(\lambda)\}$ lies in $\mathrm{Sym}\,(X)$ and it is Cauchy as well. Therefore, again there is a function $x_\lambda \in \mathcal{M}(X)$, for which $\|x_n(\lambda) - x_\lambda\|_{\mathcal{M}} \to 0$. Since the convergence in a Banach function lattice implies the convergence in measure, then $x_\lambda(t) = x(\lambda(t))$ a.e. on $[0,1]$. Using once more the fact that $\{x_n\}$ is a Cauchy sequence in $\mathrm{Sym}\,(X)$, a given $\varepsilon > 0$ we can find N such that for all $m \ge n \ge N$ and any measure-preserving mapping $\lambda\colon [0,1] \to [0,1]$ it holds:

$$\|x_n(\lambda) - x_m(\lambda)\|_{\mathcal{M}} \le \varepsilon.$$

Hence, taking $m \to \infty$, we get

$$\|x_n(\lambda) - x(\lambda)\|_{\mathcal{M}} \le \varepsilon. \tag{11.11}$$

Let now $z \simeq x_n - x$ for a fixed $n \ge N$. As above, there exists a measure-preserving mapping $\lambda_0\colon [0,1] \to [0,1]$ such that $\||z| - |x_n(\lambda_0) - x(\lambda_0)|\|_\infty \le \varepsilon$. Then, similarly as in the proof of the triangle inequality from (11.11) it follows for any $y \in \mathcal{R}(X)$, with $\|y\|_X \le 1$,

$$\|zy\|_X \le \varepsilon + \|(x_n(\lambda_0) - x(\lambda_0))y\|_X \le \varepsilon + \|x_n(\lambda_0) - x(\lambda_0)\|_{\mathcal{M}} \le 2\varepsilon,$$

and so, by the definition of z,

$$\|x_n - x\|_{\mathrm{Sym}} = \sup_{z\colon z \simeq x_n - x} \|z\|_{\mathcal{M}} \le 2\varepsilon, \quad n \ge N.$$

Thus, $x_n \to x$ in $\mathrm{Sym}\,(X)$, and the completeness of $\mathrm{Sym}\,(X)$ is proved.

Since the fact that Sym (X) is the largest s.s. continuously embedded into $\mathcal{M}(X)$ (more precisely, we have Sym $(X) \overset{1}{\subset} \mathcal{M}(X)$) follows directly from definition of the symmetric kernel, the proof is completed. □

Next, we present a useful description of the symmetric kernel of a Rademacher multiplicator space.

Theorem 11.2 *For every s.s. X the space* Sym (X) *consists of all functions $x = x(t)$ satisfying the condition: if $\sum_{n=1}^{\infty} a_n r_n \in \mathcal{R}(X)$, then $x^*(\sum_{n=1}^{\infty} a_n r_n)^* \in X$. Moreover, we have*

$$\frac{1}{2}\|x\|_{\mathrm{Sym}} \le \sup\left\{\left\|x^*\Big(\sum_{n=1}^{\infty} a_n r_n\Big)^*\right\|_X : \left\|\sum_{n=1}^{\infty} a_n r_n\right\|_X \le 1\right\} \le \|x\|_{\mathrm{Sym}}. \quad (11.12)$$

Proof Let $y := \sum_{n=1}^{\infty} a_n r_n \in \mathcal{R}(X)$, $\|y\|_X \le 1$. Then, as well as in the proof of Proposition 11.3, there is a measure-preserving mapping $\omega : [0, 1] \to [0, 1]$ such that $|y(t)| = y^*(\omega(t))$ a.e. Since X is a s.s., we get

$$\|x\|_{\mathrm{Sym}} \ge \|x^*(\omega)\|_{\mathcal{M}} \ge \|x^*(\omega)y\|_X = \|x^*(\omega)y^*(\omega)\|_X = \left\|x^*\Big(\sum_{n=1}^{\infty} a_n r_n\Big)^*\right\|_X,$$

and so the right-hand side inequality in (11.12) is proved.

In the converse way, by definition of the Sym (X)-norm, we have

$$\|x\|_{\mathrm{Sym}} = \sup_{z:\, z \simeq x} \|z\|_{\mathcal{M}} = \sup_{z:\, z \simeq x} \sup\left\{\left\|z \cdot \sum_{n=1}^{\infty} a_n r_n\right\|_X : \left\|\sum_{n=1}^{\infty} a_n r_n\right\|_X \le 1\right\}.$$

On the other hand, according to [168, § II.2, Property 10°, p. 67],

$$\Big(z \cdot \sum_{n=1}^{\infty} a_n r_n\Big)^*(2t) \le z^*(t) \cdot \Big(\sum_{n=1}^{\infty} a_n r_n\Big)^*(t), \quad 0 < t \le 1/2.$$

Since again $\|\sigma_2\|_{X \to X} \le 2$ (see Appendix C), this yields

$$\left\|z \cdot \sum_{n=1}^{\infty} a_n r_n\right\|_X \le 2\left\|z^*\Big(\sum_{n=1}^{\infty} a_n r_n\Big)^*\right\|_X = 2\left\|x^*\Big(\sum_{n=1}^{\infty} a_n r_n\Big)^*\right\|_X.$$

Thus, the left-hand side inequality in (11.12) is proved as well. □

In contrast to a Rademacher multiplicator space, its symmetric kernel admits a quite simple and transparent description provided that a given s.s. is located sufficiently "far" from L_∞.

Let X be a s.s. on [0, 1] with $\ln^{1/2}(e/t) \in X$. We denote by $X_{\ln^{1/2}}$ the set of all measurable functions $x = x(t)$ on [0, 1] such that

$$\|x\|_{\ln^{1/2}} := \|x^*(t)\ln^{1/2}(e/t)\|_X < \infty.$$

Firstly, thanks to the hypothesis, $X_{\ln^{1/2}}$ contains the space L_∞. Secondly, though the functional $x \mapsto \|x\|_{\ln^{1/2}}$, in general, is only a quasi-norm, which does not satisfy the triangle inequality, the additional condition $X \in \mathcal{I}(L_1, L_\infty)$ assures that it is equivalent to some symmetric norm. Indeed, the couple (L_1, L_∞) is $\mathcal{K}$-monotone (see Definition D.1), and therefore, by the Brudnyi–Kruglyak Theorem D.7, there exists a Banach lattice of two-sided sequences F such that $X = (L_1, L_\infty)^{\mathcal{K}}_F$. Combining this together with the equation

$$\mathcal{K}(t, x; L_1, L_\infty) = \int_0^t x^*(s)\,ds$$

(see (D.1)) and the fact that $\ln^{1/2}(e/t)$ is a decreasing function, we get

$$\|x\|_{\ln^{1/2}} \asymp \Big\|\Big(\int_0^{2^k} x^*(t)\ln^{1/2}(e/t)\,dt\Big)_k\Big\|_F.$$

Since the functional from the right-hand side of this equivalence is subadditive [168, § II.2, Corollary following Property 14°, p. 70], it defines a symmetric norm. Thus, if $X \in \mathcal{I}(L_1, L_\infty)$, then $X_{\ln^{1/2}}$ is a s.s.

In what follows, we shall often use the following characterization of the symmetric kernel of a Rademacher multiplicator space.

Theorem 11.3 *Let X be a s.s. such that $\ln^{1/2}(e/t) \in X$. Then,* $\mathrm{Sym}(X) = \mathrm{Sym}(X'') = (X'')_{\ln^{1/2}}$. *In particular, the space* $\mathrm{Sym}(X)$ *has the Fatou property.*

Proof By the hypothesis $\ln^{1/2}(e/t) \in X$ and Remark 2.1, we have

$$\Big\|\sum_{n=1}^\infty a_n r_n\Big\|_{X''} \asymp \Big\|\sum_{n=1}^\infty a_n r_n\Big\|_X \asymp \|(a_n)\|_{\ell_2}. \tag{11.13}$$

Let us assume first that X has the Fatou property and prove that

$$\mathrm{Sym}(X) = X_{\ln^{1/2}}. \tag{11.14}$$

In view of (11.3), for any sequence $(a_n) \in \ell_2$ and all $0 < t \le 1$ we have

$$x^*(t)\Big(\sum_{n=1}^\infty a_n r_n\Big)^*(t) \le K\|(a_n)\|_{\ell_2} x^*(t)\ln^{1/2}(e/t). \tag{11.15}$$

Suppose that $x \in X_{\ln^{1/2}}$, i.e., $x^*(t)\ln^{1/2}(e/t) \in X$. From (11.15) then it follows that $x^*\big(\sum_{n=1}^{\infty} a_n r_n\big)^* \in X$ and so, by Theorem 11.2, $x \in \mathrm{Sym}\,(X)$ and

$$\|x\|_{\mathrm{Sym}} \le 2\sup\Big\{\Big\|x^*\Big(\sum_{n=1}^{\infty} a_n r_n\Big)^*\Big\|_X : \Big\|\sum_{n=1}^{\infty} a_n r_n\Big\|_X \le 1\Big\}.$$

As a result, from (11.13) and (11.15) we get

$$\|x\|_{\mathrm{Sym}} \le C_1 \sup_{\|(a_n)\|_{\ell_2}\le 1} \Big\|x^*\Big(\sum_{n=1}^{\infty} a_n r_n\Big)^*\Big\|_X \le C_1 K \|x\|_{X_{\ln^{1/2}}},$$

which implies that $x \in \mathrm{Sym}\,(X)$.

Conversely, let $x \in \mathrm{Sym}\,(X)$. Denote $v_n = n^{-1/2}\sum_{i=1}^{n} r_i$, $n = 1, 2, \dots$ Observe that equivalence (11.13) yields $\sup_{n=1,2,\dots}\|v_n\|_X < \infty$. Therefore, by the central limit theorem (see Theorem A.1), there is $\lim_{n\to\infty} v_n^*(t) := v(t)$ a.e. on [0, 1] and

$$v(t) \ge \ln^{1/2}\Big(\frac{1}{t\sqrt{2\pi}}\Big), \quad 0 < t \le \frac{1}{\sqrt{2\pi}}.$$

Since X is a s.s. with the Fatou property, from this inequality and Theorem 11.2 it follows

$$\|x^*(t)\ln^{1/2}(e/t)\|_X \le C_2\|x^* v\|_X \le C_2 \liminf_{n\to\infty}\|x^* v_n^*\|_X \le C_3\|x\|_{\mathrm{Sym}}.$$

Thus, $x \in X_{\ln^{1/2}}$, and so equation (11.14) is proved.

Let now X be an arbitrary s.s. From (11.13) and the definitions it follows that $\mathcal{M}(X) \subset \mathcal{M}(X'')$, and hence $\mathrm{Sym}\,(X) \subset \mathrm{Sym}\,(X'')$. It remains to establish the opposite embedding.

Assume that $x \in \mathrm{Sym}\,(X'')$. Since the space X'' has the Fatou property, equation (11.14) is fulfilled for X'', i.e., $x^*(t)\ln^{1/2}(e/t) \in X''$. Moreover, by Theorem 2.2, for each sequence $a = (a_k) \in \ell_2$ the function $f_a := \sum_{k=1}^{\infty} a_k r_k$ belongs to the separable part of the Orlicz space L_{N_2}. Consequently, $\lim_{h\to 0}\|f_a^*\chi_{(0,h]}\|_{L_{N_2}} = 0$ (see Appendix C). Since

$$f_a^*(t)\ln^{-1/2}(e/t)\chi_{(0,h]}(t) \le C\|f_a^*\chi_{(0,h]}\|_{L_{N_2}}, \quad 0 < t \le 1,$$

by inequality (11.2), then

$$\lim_{t\to 0+} f_a^*(t)\ln^{-1/2}(e/t) = 0. \tag{11.16}$$

On the other hand, for any $0 < h, t \le 1$ we have

$$x^*(t) f_a^*(t)\chi_{[0,h]}(t) \le \sup_{0<s\le h}\big(f_a^*(s)\ln^{-1/2}(e/s)\big)\cdot x^*(t)\ln^{1/2}(e/t),$$

and so

$$\|x^* f_a^* \chi_{[0,h]}\|_{X''} \le \sup_{0<s\le h} \big(f_a^*(s) \ln^{-1/2}(e/s)\big) \cdot \|x^*(t) \ln^{1/2}(e/t)\|_{X''}.$$

Therefore, from (11.16) and the fact that $x^*(t) \ln^{1/2}(e/t) \in X''$ it follows

$$\|x^* f_a^* \chi_{[0,h]}\|_{X''} \to 0 \text{ as } h \to 0+.$$

Combining this relation with Lemma 2.2, we obtain $x^* f_a^* \in (X'')^\circ = X^\circ$. Since the latter holds for every $a \in \ell_2$, by Theorem 11.2, $x \in \mathrm{Sym}\,(X^\circ) \subset \mathrm{Sym}\,(X)$. As a result, $\mathrm{Sym}\,(X'') = (X'')_{\ln^{1/2}} \subset \mathrm{Sym}\,(X)$. The second assertion of the theorem is a consequence of the proved first one and the fact that the space $(X'')_{\ln^{1/2}}$ has the Fatou property. □

Corollary 11.1 *If X is a s.s. such that $\ln^{1/2}(e/t) \in X$, then the fundamental function of the symmetric kernel* $\mathrm{Sym}\,(X)$ *satisfies the following relation:*

$$\phi_{\mathrm{Sym}\,(X)}(t) \asymp \|\ln^{1/2}(e/s) \cdot \chi_{[0,t]}\|_X, \quad 0 < t \le 1.$$

Applying Theorem 11.3, we can identify the symmetric kernels of Rademacher multiplicator spaces for various classes of s.s.'s.

Example 11.1 Suppose that either $1 < p < \infty$, $1 \le q < \infty$ and $\alpha \in \mathbb{R}$, or $p = q = 1$ and $\alpha \ge 0$ (resp. $1 < p < \infty$ and $\alpha \in \mathbb{R}$). The Lorentz–Zygmund space $X = L_{p,q} \log^\alpha L$ consists of all measurable functions $x = x(t)$ on $[0, 1]$ such that

$$\|x\|_{p,q,\alpha} := \left(\int_0^1 (t^{1/p} \ln^\alpha(e/t) x^*(t))^q \frac{dt}{t}\right)^{1/q} < \infty$$

(resp.

$$\|x\|_{p,\infty,\alpha} := \sup_{0<t\le 1} \big(t^{1/p} \ln^\alpha(e/t) x^*(t)\big) < \infty\big).$$

Since $\ln^{1/2}(e/t) \in X$ for the above p, q and α, then Theorem 11.3 implies that

$$\|x\|_{\mathrm{Sym}} \asymp \left(\int_0^1 (t^{1/p} \ln^{1/2+\alpha}(e/t) x^*(t))^q \frac{dt}{t}\right)^{1/q}$$

(resp.

$$\|x\|_{\mathrm{Sym}} \asymp \sup_{0<t\le 1} (t^{1/p} \ln^{1/2+\alpha}(e/t) x^*(t))\big).$$

Thus, in both cases Sym $(L_{p,q}\log^{\alpha}L)$ is the Lorentz–Zygmund space $L_{p,q}\log^{1/2+\alpha}L$. In particular, if $1 \le p = q < \infty$ and $\alpha = 0$, Sym (L_p) is the Zygmund space $L_p \log^{1/2} L$ and

$$\|x\|_{\mathrm{Sym}\,(L_p)} \asymp \left(\int_0^1 \left(x^*(t) \ln^{1/2}(e/t)\right)^p dt \right)^{1/p}.$$

In the case when $1 < p < \infty$, $q = \infty$ and $\alpha = 0$, it follows that X is the Marcinkiewicz space $M(t^{1-1/p}) := L_{p,\infty}$ with the fundamental function $t^{1/p}$, and then Sym $(L_{p,\infty}) = L_{p,\infty} \log^{1/2} L$.

Example 11.2 Let $\Lambda_p(\varphi)$ be a Lorentz space (φ is an increasing concave function on $[0, 1]$, $\varphi(0) = 0$, $1 \le p < \infty$). Recall that this space consists of all measurable functions $x = x(t)$ on $[0, 1]$ such that

$$\|x\|_{\Lambda_p(\varphi)} := \left(\int_0^1 x^*(t)^p \, d\varphi(t) \right)^{1/p} < \infty.$$

Observe that $\ln^{1/2}(e/t) \in \Lambda_p(\varphi)$ if and only if $\int_0^1 \ln^{p/2}(e/t)\, d\varphi(t) < \infty$. In this case we have

$$\|x\|_{\mathrm{Sym}(\Lambda_p(\varphi))} \asymp \left(\int_0^1 x^*(t)^p \ln^{p/2}(e/t)\, d\varphi(t) \right)^{1/p}.$$

Since $\varphi'(t) \ln^{p/2}(e/t)$ is a decreasing, nonnegative and integrable function, this equivalence implies that Sym $(\Lambda_p(\varphi))$ is the Lorentz space $\Lambda_p(\psi_p)$, where

$$\psi_p(t) := \int_0^t \varphi'(s) \ln^{p/2}(e/s)\, ds.$$

In particular, for $p = 1$ we get Sym $(\Lambda(\varphi)) = \Lambda(\psi_1)$.

Example 11.1 shows that for each $1 < p < \infty$ the space L_p can be identified as the symmetric kernel Sym (X), where X is the Zygmund space $L_p(\log L)^{-1/2}$. Consider now a more general question about conditions under which a s.s. Y can be obtained as the symmetric kernel of the Rademacher multiplicator space for some s.s. Since Sym $(L_1) = L \log^{1/2} L$, then the embedding $Y \subset L \log^{1/2} L$ is an obvious necessary condition for that.

Theorem 11.4 *Let Y be an interpolation s.s. between the spaces $L\log^{1/2}L$ and L_∞, and let Y have the Fatou property. Then, there exists a s.s. X with the Fatou property such that* Sym $(X) = Y$.

Further, we need the fact that $(L\log^{1/2}L, L_\infty)$ is a $\mathcal{K}$-monotone couple (see Definition D.1). One can readily check that $L\log^{1/2}L = \Lambda(\varphi_1)$, where $\varphi_1(t) = t \ln^{1/2}(e/t)$. Therefore, the needed result is an immediate consequence of the following statement, the proof of which is based again on using the techniques of partial retracts (see the proofs of Lemma 3.2 and Proposition 3.3).

Proposition 11.4 *Suppose φ is a continuous increasing concave function on* $[0, 1]$, $\varphi(0) = 0$, *and* $1 \le p < \infty$. *Then, the Banach couple* $\vec{X} = (\Lambda_p(\varphi), L_\infty)$ *is* $\mathcal{K}$-*monotone.*

Proof Due to the well-known Sparr theorem [268] and Lemma 3.2 it suffices to show that $\vec{X}$ is a partial retract of the couple $\vec{Y} = (L_p(\varphi'), L_\infty)$, where $L_p(\varphi')$ denotes the weighted L_p-space with the norm

$$\|x\|_{L_p(\varphi')} := \Big(\int_0^1 |x(t)|^p \varphi'(t)\, dt\Big)^{1/p}.$$

So, let $x \in \Lambda_p(\varphi)$ (since $L_\infty \subset \Lambda_p(\varphi)$, we have $L_\infty + \Lambda_p(\varphi) = \Lambda_p(\varphi)$). By Lemma 3.3, there exists a linear operator T_1 bounded in the couple $\vec{X}$ such that $x^* = T_1 x$. Furthermore, one can check that the identity I acts boundedly from the couple $\vec{X}$ into the couple $\vec{Y}$. Indeed, the derivative $\varphi'(t)$ decreases and, therefore, in view of a property of the rearrangement of a measurable function (see [168, § II.2, Property 13°, p. 68]), it follows

$$\|y\|_{L_p(\varphi')} = \Big(\int_0^1 |y(t)|^p \varphi'(t)\, dt\Big)^{1/p} \le \Big(\int_0^1 y^*(t)^p\, d\varphi(t)\Big)^{1/p} = \|y\|_{\Lambda_p(\varphi)}.$$

Thus, the operator $U := IT_1$ is bounded from the couple $\vec{X}$ into the couple $\vec{Y}$ and $Ux = x^*$.

Let us construct now a linear operator V bounded from $\vec{Y}$ into $\vec{X}$ such that $Vx^* = x$. Without loss of generality, we can assume that $\varphi(1) = 1$. Let us select real numbers $1 = b_0 > b_1 > b_2 > \ldots$ such that $\varphi(b_n) = 2^{-n}$, $n = 0, 1, \ldots$ Then, the linear operator V_1, defined on the space $L_p(\varphi') = L_\infty + L_p(\varphi')$ by

$$V_1 y(s) := \sum_{n=1}^{\infty} 2^{n+1} \Big(\int_{b_{n+1}}^{b_n} y(t)\varphi'(t)\, dt\Big) \chi_{[b_n, b_{n-1})}(s),$$

is bounded in L_∞. Moreover, using Hölder's inequality, we have

$$|V_1 y(s)|^p \le \sum_{n=1}^{\infty} 2^{(n+1)p} \Big(\int_{b_{n+1}}^{b_n} |y(t)|\varphi'(t)\, dt\Big)^p \chi_{[0, b_{n-1})}(s)$$
$$\le \sum_{n=1}^{\infty} 2^{n+1} \Big(\int_{b_{n+1}}^{b_n} |y(t)|^p \varphi'(t)\, dt\Big) \chi_{[0, b_{n-1})}(s).$$

Since the latter sum in this inequality is a decreasing function on [0, 1], we get

$$\|V_1 y\|_{\Lambda_p(\varphi)}^p \le 4 \sum_{n=1}^{\infty} \int_{b_{n+1}}^{b_n} |y(t)|^p \varphi'(t)\, dt = 4\|y\|_{L_p(\varphi')}^p,$$

which implies that V_1 is a bounded operator from $L_p(\varphi')$ into $\Lambda_p(\varphi)$. Moreover, $x^*(t)$ decreases, and hence $V_1 x^*(s) \ge x^*(s)$ for all $0 < s \le 1$. Consequently, the linear operator

$$V_2 y(s) := \frac{x^*(s)}{V_1 x^*(s)} V_1 y(s), \quad 0 < s \le 1,$$

is bounded from the couple $\vec{Y}$ into the couple $\vec{X}$, and $V_2 x^*(t) = x^*(t)$. Applying now Lemma 3.3 once more, we find a linear operator T_2 bounded in the couple $\vec{X}$ such that $x = T_2 x^*$. Then, the operator $V := T_2 V_2$ is bounded from $\vec{Y}$ into $\vec{X}$, and $V x^* = x$.

Summing up, we see that an arbitrary function $x \in \Lambda_p(\varphi)$ as an element of the couple $\vec{X}$ is orbitally equivalent to its decreasing rearrangement x^* in the couple $\vec{Y}$. Thus, $\vec{X}$ is a partial retract of $\vec{Y}$, and the result follows. □

In the proof of Theorem 11.4 we shall make use also of the following technical lemma.

Lemma 11.2 *Let $\alpha > 0$ and $\gamma \in \mathbb{R}$. Then, we have*

$$\int_0^1 \left(x^*(\cdot) \ln^\gamma(e/\cdot)\right)^*(t) \ln^\alpha(e/t)\, dt \asymp \int_0^1 x^*(t) \ln^{\gamma+\alpha}(e/t)\, dt,$$

with a constant that depends only on α and γ.

Proof Obviously, it is sufficient to consider the case when $\gamma < 0$. Since $\alpha > 0$, the function $\ln^\alpha(e/t)$ decreases. Consequently,

$$\int_0^1 x^*(t) \ln^{\gamma+\alpha}(e/t)\, dt \le \int_0^1 \left(x^*(\cdot) \ln^\gamma(e/\cdot)\right)^*(t) \ln^\alpha(e/t)\, dt.$$

It remains to prove the opposite inequality.

Let first $x(t) = \chi_{(0,h]}(t)$, $0 < h < 1$. Since $\int_0^t \ln^\beta(e/s)\, ds \asymp t \ln^\beta(e/t)$ for any $\beta \in \mathbb{R}$, then

$$\int_0^h \left(\ln^\gamma(e/\cdot)\right)^*(t) \ln^\alpha(e/t)\, dt = \int_0^h \ln^\gamma\left(\frac{e}{h-t}\right) \ln^\alpha(e/t)\, dt \le \ln^\gamma(e/h) \int_0^h \ln^\alpha(e/t)\, dt$$
$$\asymp h \ln^{\gamma+\alpha}(e/h) \asymp \int_0^h \ln^{\gamma+\alpha}(e/t)\, dt.$$

Thus, for characteristic functions everything is proved.

If a function of the form $x^*(t)$ is finite-valued, it can be represented as follows:

$$x^*(t) = \sum_{k=1}^{N} b_k \chi_{(0,h_k]}(t), \quad b_k > 0,\ h_1 > h_2 > \ldots > h_N \geq 0. \tag{11.17}$$

Then, by [168, § II.2, Property 14°, p. 69–70], we have

$$\int_0^1 \left(x^*(\cdot) \ln^\gamma(e/\cdot)\right)^*(t) \ln^\alpha(e/t)\, dt \;=\; \sup_\omega \int_0^1 x^*(\omega(t)) \ln^\gamma(e/\omega(t)) \ln^\alpha(e/t)\, dt,$$

where the supremum is taken over all measure-preserving mappings $\omega : [0,1] \to [0,1]$. Hence, by using already considered case, we get

$$\begin{aligned}
\int_0^1 \left(x^*(\cdot) \ln^\gamma(e/\cdot)\right)^*(t) \ln^\alpha(e/t)\, dt &\leq \sum_{k=1}^{N} b_k \sup_\omega \int_0^1 \chi_{(0,h_k]}(\omega(t)) \ln^\gamma(e/\omega(t)) \ln^\alpha(e/t)\, dt \\
&\leq \sum_{k=1}^{N} b_k \int_0^{h_k} \left(\ln^\gamma(e/\cdot)\right)^*(t) \ln^\alpha(e/t)\, dt \\
&\leq C \sum_{k=1}^{N} b_k \int_0^{h_k} \ln^{\gamma+\alpha}(e/t)\, dt \\
&= C \int_0^1 x^*(t) \ln^{\gamma+\alpha}(e/t)\, dt.
\end{aligned}$$

Next, for arbitrary decreasing rearrangement $x^*(t)$ there is an increasing sequence of functions $x_n^*(t)$ of the form (11.17), converging to $x^*(t)$ a.e. on $[0,1]$. As it was proved,

$$\int_0^1 \left(x_n^*(\cdot) \ln^\gamma(e/\cdot)\right)^*(t) \ln^\alpha(e/t)\, dt \leq C \int_0^1 x_n^*(t) \ln^{\gamma+\alpha}(e/t)\, dt \leq C \int_0^1 x^*(t) \ln^{\gamma+\alpha}(e/t)\, dt.$$

Finally, applying Lemma 2.1, after taking $n \to \infty$ in the left-hand side of this inequality we come to the desired result. □

Proof of Theorem 11.4 Given s.s. Y we define the space $Y_{\ln^{-1/2}}$ as the set of all measurable functions $x = x(t)$ on $[0,1]$ such that

$$\|x\|_{\ln^{-1/2}} = \|x^*(t) \ln^{-1/2}(e/t)\|_Y < \infty.$$

Show that the functional $x \mapsto \|x\|_{\ln^{-1/2}}$ is equivalent to some symmetric norm. To this end, we first compute the $\mathcal{K}$-functional for the couple $(L\log^{1/2}L, L_\infty)$. Note that $L_\infty = \Lambda(\varphi_0)$ and, as was said above, $L\log^{1/2}L = \Lambda(\varphi_1)$, where $\varphi_0(s) = 1$ and $\varphi_1(s) = s \ln^{1/2}(e/s)$. Since the function $\varphi_0(s)/\varphi_1(s)$ is decreasing, then taking into account equation (D.2), we have for all $t \in (0, 1]$

$$\mathcal{K}(t, x; L\log^{1/2}L, L_\infty) = t\mathcal{K}(1/t, x; L_\infty, L\log^{1/2}L)$$
$$= \int_0^{s(t)} x^*(u)\, d(u \ln^{1/2}(e/u)),$$

where the function $s = s(t)$ is defined by the equation $s \ln^{1/2}(e/s) = t$. Hence, in view of the equivalence $(u \ln^{1/2}(e/u))' \asymp \ln^{1/2}(e/u)$, we get

$$\mathcal{K}(t, x; L\log^{1/2}L, L_\infty) \asymp \int_0^{s(t)} x^*(u) \ln^{1/2}(e/u)\, du, \quad 0 < t \le 1. \tag{11.18}$$

Further, according to Proposition 11.4, the Banach couple $(L\log^{1/2}L, L_\infty)$ is $\mathcal{K}$-monotone. Therefore, since Y is an interpolation space with respect to this couple, by Theorem D.7, there exists a Banach lattice of two-sided sequences E such that $(\min(1, 2^k))_{k=-\infty}^{\infty} \in E$ and

$$\|x\|_Y \asymp \|(\mathcal{K}(2^k, x; L\log^{1/2}L, L_\infty))_{k=-\infty}^{\infty}\|_E. \tag{11.19}$$

Denote by $\{e_k\}_{k=-\infty}^{\infty}$ the canonical basic sequence in a space of two-sided sequences. Noting that $L_\infty \subset L\log^{1/2}L$, for all $t \ge 1$ we have

$$\mathcal{K}(t, x; L\log^{1/2}L, L_\infty) \le C'\|x\|_{L\log^{1/2}L},$$

with a constant $C' > 0$ independent of $x \in L\log^{1/2}L$ and $t \ge 1$. Consequently, using the concavity of the function $t \mapsto \mathcal{K}(t, x; L\log^{1/2}L, L_\infty)$, we get

$$\Big\|\sum_{k=0}^{\infty} \mathcal{K}(2^k, x; L\log^{1/2}L, L_\infty)e_k\Big\|_E \le C'\|x\|_{L\log^{1/2}L} \Big\|\sum_{k=0}^{\infty} e_k\Big\|_E$$
$$\le C'\|x\|_{L\log^{1/2}L} \|(\min(1, 2^k))_{-\infty}^{\infty}\|_E$$
$$\le C\|x\|_{L\log^{1/2}L} \|e_0\|_E$$
$$\le C\Big\|\sum_{k=0}^{\infty} \mathcal{K}(2^{-k}, x; L\log^{1/2}L, L_\infty)e_{-k}\Big\|_E,$$

where $C > 0$ depends only on E. Thus, from equivalence (11.19) it follows that

$$\|x\|_Y \asymp \Big\|\sum_{k=0}^{\infty} \mathcal{K}(2^{-k}, x; L\log^{1/2}L, L_\infty)e_{-k}\Big\|_E.$$

Combining this with (11.18) and Lemma 11.2, we have

$$\|x\|_{\ln^{-1/2}} \asymp \Big\|\sum_{k=0}^{\infty}\int_0^{s(2^{-k})}\big(x^*(\cdot)\ln^{-1/2}(e/\cdot)\big)^*(u)\ln^{1/2}(e/u)\,du\; e_{-k}\Big\|_E$$
$$\asymp \Big\|\sum_{k=0}^{\infty}\int_0^{s(2^{-k})}x^*(u)\,du\; e_{-k}\Big\|_E.$$

This equivalence shows that the functional $x \mapsto \|x\|_{\ln^{-1/2}}$ is equivalent to some symmetric norm [168, § II.2, Property 8°, p. 64–65]. Moreover, $Y_{\ln^{-1/2}} \in \mathcal{I}(L_1, L_\infty)$.

Let us check now that the space $X := Y_{\ln^{-1/2}}$ has the Fatou property. Indeed, if $0 \le x_n(t) \uparrow x(t)$ a.e. and $\sup_n \|x_n\|_{\ln^{-1/2}} < \infty$, then Lemma 2.1 yields

$$x_n^*(t)\ln^{-1/2}(e/t) \uparrow x^*(t)\ln^{-1/2}(e/t) \text{ a.e. and } \sup_{n=1,2,\dots}\|x_n^*(t)\ln^{-1/2}(e/t)\|_Y < \infty.$$

Since Y has the Fatou property, then $x^*(t)\ln^{-1/2}(e/t) \in Y$, i.e., $x \in X$. Therefore, X has the Fatou property as well and so, by Theorem 11.3, we have

$$\|x\|_{\mathrm{Sym}} \asymp \|x^*\ln^{1/2}(e/t)\|_X$$
$$= \|\big(x^*(\cdot)\ln^{1/2}(e/\cdot)\big)^*(t)\ln^{-1/2}(e/t)\|_Y = \|x\|_Y.$$

As a result, $Y = \mathrm{Sym}\,(X)$, and the proof is completed. □

In the next two sections, we return to the question of when the Rademacher multiplicator space $\mathcal{M}(X)$ is symmetric. Theorem 11.1 gives a necessary condition: the lower dilation exponent of the fundamental function of such a s.s. X should be equal to zero, which means that X is located sufficiently "close" to the space L_∞. We shall consider separately the following two cases: (a) $\mathcal{M}(X)$ coincides with L_∞; (b) $\mathcal{M}(X)$ is a s.s. though $\mathcal{M}(X) \ne L_\infty$.

11.3 A Characterization of Symmetric Spaces X Such That $\mathcal{M}(X) = L_\infty$

In the main theorem of this section (see Theorem 11.5) we identify those s.s.'s X, for which the Rademacher multiplicator space $\mathcal{M}(X)$ is equal to L_∞ (with equivalence of norms). Let us emphasize that exactly the same conditions are necessary and sufficient both for the coincidence of the symmetric kernel $\mathrm{Sym}\,(X)$ and L_∞. First, an inspection of these conditions in connection with the Rodin–Semenov theorem (see Theorem 2.3) shows that $\mathcal{M}(X) = L_\infty$ if the Rademacher system is not equivalent in X to the unit vector basis in ℓ_2. On the other hand, the space $\mathcal{M}(X)$ may be equal to L_∞ when the latter equivalence holds (for example, for $X = G$).

Recall that X° is the separable part of a s.s. X, i.e., the closure of L_∞ in X.

Theorem 11.5 *Let X be a s.s. on* $[0, 1]$. *The following conditions are equivalent:*

(i) $\mathcal{M}(X) = L_\infty$;
(ii) $\mathrm{Sym}\,(X) = L_\infty$;
(iii) $\ln^{1/2}(e/t) \notin X^\circ$.

To prove this result, we need two lemmas.

Lemma 11.3 *Suppose X is a s.s. on* $[0, 1]$ *such that* $\ln^{1/2}(e/t) \notin X^\circ$. *There exists a constant $c_0 > 0$, which depends only on X, such that for every $\eta \in (0, 1]$ we can find $\delta_0 \in (0, \eta)$ satisfying the inequality:*

$$\frac{\| \ln^{1/2}(e/t)\chi_{[\delta,\eta]}\|_X}{\| \ln^{1/2}(e/t)\chi_{[\delta,1]}\|_X} \geq c_0 \quad \text{if } \delta \in (0, \delta_0).$$

Proof Show first that

$$\gamma := \inf_{0<\eta\leq 1} \lim_{\delta\to 0^+} \| \ln^{1/2}(e/t)\chi_{[\delta,\eta]}\|_X > 0. \tag{11.20}$$

Indeed, assuming that $\gamma = 0$, we can construct a decreasing sequence $\{\eta_n\}$ of positive real numbers such that $\lim_{n\to\infty} \eta_n = 0$ and

$$\| \ln^{1/2}(e/t)\chi_{[\eta_{n+1},\eta_n]}\|_X \leq 2^{-n}, \quad n = 1, 2, \ldots$$

Since X is a Banach space, this implies that $\ln^{1/2}(e/t) \in X$ and

$$\lim_{\eta\to 0+} \| \ln^{1/2}(e/t)\chi_{[0,\eta]}\|_X = 0.$$

Therefore, $\ln^{1/2}(e/t) \in X^\circ$, which contradicts our hypothesis. So, (11.20) is established.

Set now $\alpha := \lim_{\delta\to 0+} \| \ln^{1/2}(e/t)\chi_{[\delta,1]}\|_X$. Suppose $\alpha < \infty$. By (11.20), a given $\eta \in (0, 1]$, we have $\| \ln^{1/2}(e/t)\chi_{[\delta,\eta]}\|_X \geq \gamma/2$ for all sufficiently small $\delta > 0$. Hence, for such a number δ

$$\frac{\| \ln^{1/2}(e/t)\chi_{[\delta,\eta]}\|_X}{\| \ln^{1/2}(e/t)\chi_{[\delta,1]}\|_X} \geq \frac{\gamma}{2\alpha}.$$

In the case when $\alpha = \infty$, for each fixed $\eta \in (0, 1]$ and sufficiently small $\delta > 0$ such that $\delta < \eta$ we have

$$\frac{\| \ln^{1/2}(e/t)\chi_{[\delta,\eta]}\|_X}{\| \ln^{1/2}(e/t)\chi_{[\delta,1]}\|_X} \geq \frac{\| \ln^{1/2}(e/t)\chi_{[\delta,1]}\|_X - \| \ln^{1/2}(e/t)\chi_{[\eta,1]}\|_X}{\| \ln^{1/2}(e/t)\chi_{[\delta,1]}\|_X}$$
$$= 1 - \frac{\| \ln^{1/2}(e/t)\chi_{[\eta,1]}\|_X}{\| \ln^{1/2}(e/t)\chi_{[\delta,1]}\|_X} \geq \frac{1}{2}.$$

Thus, the desired inequality holds with the constant $c_0 := \min(1/2, \gamma/(2\alpha))$.

□

In the next lemma, we shall use again the $\mathcal{K}$-functional $\kappa(t, a) := \mathcal{K}(t, a; \ell_1, \ell_2)$, $t > 0$. Recall that, by the Holmstedt inequality from Theorem 1.5,

$$\frac{1}{4}\left\{\sum_{i=1}^{[t^2]} a_i^* + t\left[\sum_{i=[t^2]+1}^{\infty} (a_i^*)^2\right]^{1/2}\right\} \leq \kappa(t, a)$$
$$\leq \left\{\sum_{i=1}^{[t^2]} a_i^* + t\left[\sum_{i=[t^2]+1}^{\infty} (a_i^*)^2\right]^{1/2}\right\}, \tag{11.21}$$

where $(a_i^*)_{i=1}^{\infty}$ is the decreasing permutation of the sequence $(|a_n|)_{n=1}^{\infty}$.

Lemma 11.4 *Let $b^n := 1/n \sum_{k=1}^{n} e_k$, $n \in \mathbb{N}$, where e_k are the canonical unit vectors in a (one-sided) sequence space. Then, with a constant independent of $t > 0$ and $n \in \mathbb{N}$ we have*

$$\kappa(t, b^n) \asymp \min\left(1, \frac{t}{\sqrt{n}}\right). \tag{11.22}$$

Proof Suppose first that $t \in \mathbb{N}$. In the case when $t^2 \geq n$ equivalence (11.22) is a direct consequence of inequality (11.21). Let now $t^2 < n$. Then, again according to (11.21), we get

$$\frac{1}{4}\left(\frac{t^2}{n} + \frac{t}{n} \cdot (n - t^2)^{1/2}\right) \leq \kappa(t, b^n) \leq \frac{t^2}{n} + \frac{t}{n} \cdot (n - t^2)^{1/2},$$

or equivalently

$$\frac{1}{4}(v^2 + v(1 - v^2)^{1/2}) \leq \kappa(t, b^n) \leq v^2 + v(1 - v^2)^{1/2},$$

where $v := t/\sqrt{n} < 1$. Since

$$v \le v^2 + v(1-v^2)^{1/2} \le \sqrt{2}v, \quad 0 \le v \le 1,$$

the preceding inequality implies that

$$\frac{v}{4} \le \kappa(t, b^n) \le \sqrt{2}v.$$

Thus, equivalence (11.22) is established for positive integers t. Taking into account the concavity of the function $t \mapsto \kappa(t, b^n)$, this relation can be easily extended to all $t \ge 1$. It remains to observe that for $0 < t \le 1$, by definition of the $\mathcal{K}$-functional, we have $\kappa(t, b^n) = t\|b^n\|_2 = t/\sqrt{n}$. □

Proof of Theorem 11.5 The implication $(i) \Rightarrow (ii)$ follows immediately from the embeddings

$$L_\infty \subset \mathrm{Sym}\,(X) \subset \mathcal{M}(X),$$

which hold for every s.s. X.

$(ii) \Rightarrow (iii)$. If $\|(a_n)\|_{\ell_2} \le 1$, then from (11.3) it follows that for all $0 < t, h \le 1$

$$\chi_{[0,h]}(t)\Big(\sum_{n=1}^{\infty} a_n r_n\Big)^*(t) \le K\, \chi_{[0,h]}(t) \ln^{1/2}(e/t).$$

On the contrary, assume that $\ln^{1/2}(e/t) \in X^\circ$. Then we have equivalence (11.13). Therefore, from the last inequality and Theorem 11.2 for all $0 < h \le 1$ we get

$$\|\chi_{[0,h]}\|_{\mathrm{Sym}\,(X)} \asymp \sup_{\|(a_n)\|_{\ell_2} \le 1} \Big\|\chi_{[0,h]}\Big(\sum_{n=1}^{\infty} a_n r_n\Big)^*\Big\|_X \le C\, \|\chi_{[0,h]} \ln^{1/2}(e/t)\|_X.$$

Since (iii) is obvious for $X = L_\infty$, we can assume that $X \ne L_\infty$. Then X° is separable, and so $\|\chi_{[0,h]} \ln^{1/2}(e/t)\|_X \to 0$ as $h \to 0$. Thus, from the last inequality it follows that $\lim_{h\to 0} \|\chi_{[0,h]}\|_{\mathrm{Sym}\,(X)} = 0$, whence $\mathrm{Sym}\,(\mathcal{R}, X) \ne L_\infty$. This contradicts the assumption.

$(iii) \Rightarrow (i)$. We first prove that there exists a constant $c_1 > 0$, depending only on X, with the following property: for every integer $m \ge 0$ there is $n_0 \ge 1$ such that for all $n \ge n_0$ and any dyadic interval Δ of rank m we have

$$\Big\|\chi_\Delta \sum_{i=m+1}^{m+n} r_i\Big\|_X \ge c_1 \Big\|\sum_{i=m+1}^{m+n} r_i\Big\|_X. \tag{11.23}$$

Since the functions

$$\chi_\Delta \cdot \sum_{i=m+1}^{m+n} r_i \ \text{ and } \ \chi_{[0,2^{-m}]} \cdot \sum_{i=m+1}^{m+n} r_i$$

are equimeasurable, it suffices to prove (11.23) only for $\Delta = [0, 2^{-m}]$.

For arbitrary $m \geq 0$ and $n \geq 1$, we set

$$x_{m,n} := \frac{1}{n} \sum_{i=m+1}^{m+n} r_i \quad \text{and} \quad y_{m,n} := x_{m,n} \cdot \chi_{[0,2^{-m}]}.$$

Then, by definition of the Rademacher functions, we have

$$y_{m,n}^*(t) = x_{m,n}^*(2^m t), \quad 0 < t \leq 2^{-m}. \tag{11.24}$$

Moreover, according to Corollary 1.9, there exists a universal constant $\beta \in (0, 1)$ such that for every sequence $b = (b_n) \in \ell_2$ the function $f_b := \sum_{k=1}^{\infty} b_k r_k$ satisfies the conditions

$$f_b^*(t) \leq \beta^{-1} \kappa(\ln^{1/2}(e/t), b), \ \ 0 < t \leq 1,$$

and

$$f_b^*(\beta t) \geq \beta \kappa(\ln^{1/2}(e/t), b), \ \ 0 < t \leq 1.$$

In particular, we have

$$x_{m,n}^*(t) \leq \beta^{-1} \kappa(\ln^{1/2}(e/t), b_m^n), \ \ 0 < t \leq 1, \tag{11.25}$$

and

$$x_{m,n}^*(\beta t) \geq \beta \kappa(\ln^{1/2}(e/t), b_m^n), \ \ 0 < t \leq 1, \tag{11.26}$$

where $b_m^n := 1/n \sum_{k=m+1}^{m+n} e_k$ (as above, e_k are the canonical unit vectors in sequence spaces). Since $\kappa(t, b_m^n) = \kappa(t, b_0^n)$, $t > 0$, then, by Lemma 11.4, we get

$$\kappa(t, b_m^n) \asymp \min\left(1, \frac{t}{\sqrt{n}}\right)$$

with a constant independent of $t > 0$, $m = 0, 1, \ldots$, and $n = 1, 2, \ldots$ In consequence, from (11.24)–(11.26) it follows that for $0 < t \leq 1$

$$x_{m,n}^*(t) \leq C_1 G_n(t) \tag{11.27}$$

and

$$y_{m,n}^*(t) \geq C_2^{-1} F_{m,n}(t), \tag{11.28}$$

where

$$G_n(t) := \min\left\{1, \frac{\ln^{1/2}(e/t)}{\sqrt{n}}\right\}, \quad 0 < t \leq 1,$$

and

$$F_{m,n}(t) := \min\left\{1, \frac{\ln^{1/2}(2^{1-m}\beta/t)}{\sqrt{n}}\right\}, \quad 0 < t \leq 2^{-m}\beta.$$

We prove now the existence of a constant $c_2 > 0$ such that for every integer $m \geq 0$ the inequality

$$\|F_{m,n}\|_X \geq c_2\|G_n\|_X \tag{11.29}$$

holds for all sufficiently large $n \in \mathbb{N}$. To this end, we first show that for every integer $m \geq 0$

$$F_{m,n}(t) \geq \frac{1}{2}\, G_n(t) \quad \text{if } n \geq \frac{4m}{3} + \frac{4}{3}\log_2(e/\beta) \quad \text{and} \quad 0 < t < 2^{-4m/3}\beta^{4/3}. \tag{11.30}$$

In the case $0 < t \leq 2^{1-m}e^{-n}\beta$ (11.30) is obvious, because $F_{m,n}(t) = G_n(t) = 1$. Let now $2^{1-m}e^{-n}\beta < t \leq e^{1-n}$. Then, the first inequality from (11.30) turns into the estimate

$$\ln^{1/2}(2^{1-m}\beta/t) \geq \frac{1}{2}\sqrt{n}.$$

Taking into account that the function in the left-hand side of this relation decreases, this is valid for $2^{1-m}e^{-n}\beta < t \leq e^{1-n}$ whenever

$$\ln(2^{1-m} \cdot e^{n-1} \cdot \beta) \geq n/4.$$

One can readily check that the latter inequality holds if $n \geq 4m/3 + 4\log_2(e/\beta)/3$. Finally, in the case when $t > e^{1-n}$ the first inequality from (11.30) is equivalent to the following:

$$\ln^{1/2}(2^{1-m}\beta/t) \geq \frac{1}{2}\ln^{1/2}(e/t),$$

which, as it is easy to make sure, is satisfied for $0 < t < 2^{-4m/3}\beta^{4/3}$.

Thus, (11.30) is established and hence for $n \geq 4m/3 + 4\log_2(e/\beta)/3$ we get

$$\|F_{m,n}\|_X \geq \frac{1}{2}\|G_n\chi_{[0,c_3 2^{-4m/3}]}\|_X, \tag{11.31}$$

where $c_3 := \beta^{4/3}$. Combining this together with the definition of G_n, we conclude that for $n \geq 4m/3 + 4\log_2(e/\beta)/3$

$$\|F_{m,n}\|_X \geq \frac{1}{2\sqrt{n}}\|\ln^{1/2}(e/t)\chi_{[e^{1-n},c_3 2^{-4m/3}]}\|_X.$$

On the other hand, in view of Lemma 11.3, there exist a constant $c_0 \in (0, 1)$ (independent of m) and $n_1(m)$ such that for all $n \geq n_1(m)$ we have

$$\|\ln^{1/2}(e/t)\chi_{[e^{1-n},c_3 2^{-4m/3}]}\|_X \geq c_0\|\ln^{1/2}(e/t)\chi_{[e^{1-n},1]}\|_X.$$

Let $n \geq n_2(m) := \max\{4m/3 + 4\log_2(e/\beta)/3, n_1(m)\}$. From the last two inequalities it follows that $e^{1-n} < c_3 2^{-4m/3}$ and

$$\|F_{m,n}\|_X \geq \frac{c_0}{2}\|G_n\chi_{[e^{1-n},1]}\|_X.$$

Therefore, by (11.31), inequality (11.29) holds for $c_2 := c_0/4$ and $n \geq n_2(m)$.

Summarizing inequalities (11.27)–(11.29) and taking into account definition of the functions $x_{m,n}$ and $y_{m,n}$, for every $m \geq 0$ and all $n \geq n_2(m)$ we obtain

$$\frac{\|\chi_{[0,2^{-m}]}\sum_{i=m+1}^{m+n} r_i\|_X}{\|\sum_{i=m+1}^{m+n} r_i\|_X} = \frac{\|y_{m,n}\|_X}{\|x_{m,n}\|_X} \geq \frac{\|F_{m,n}\|_X}{C_1C_2\|G_n\|_X} \geq c_2C_1^{-1}C_2^{-1}.$$

As a result, (11.23) holds with the constant $c_1 := c_2C_1^{-1}C_2^{-1}$. We show next that a similar estimate is valid if we replace a dyadic interval with an arbitrary subset of $[0, 1]$ of positive measure.

Let $D \subset [0, 1]$, $m(D) > 0$. First of all, by the Lebesgue density theorem (see e.g. [215, Theorem IX.5.1]), for sufficiently large $m \in \mathbb{N}$ we can find a dyadic interval $\Delta := \Delta_m^{k_0} = [(k_0 - 1)2^{-m}, k_0 2^{-m}]$ such that

$$2^{-m} = m(\Delta) \geq m(\Delta \cap D) > 2^{-m-1}.$$

Consider the set $E = \bigcup_{k=1}^{2^m} E_m^k$, where E_m^k is obtained by shifting the set $\Delta \cap D$ to the interval Δ_m^k, $k = 1, 2, \ldots, 2^m$ (in particular, $E_m^{k_0} = \Delta \cap D$). Denote $f_i = r_i \cdot \chi_E$, $i \in \mathbb{N}$. It follows easily that $|f_i(t)| \leq 1$, $t \in [0, 1]$, and $\|f_i\|_{L_2} \geq 1/\sqrt{2}$. Moreover, since $\{r_i\}$ is a weakly null sequence in $L_2[0, 1]$, then the equation

$$\int_0^1 f_i(t)g(t)\,dt = \int_0^1 r_i(t)\chi_E(t)g(t)\,dt,$$

where $g \in L_2$ is arbitrary, shows that the sequence $\{f_i\}$ is weakly null in $L_2[0, 1]$ as well. Thus, this sequence satisfies all the conditions of Theorem 8.2 and hence it contains a subsequence $\{f_{i_j}\}$, equivalent in distribution to the Rademacher system. This means that there exists a constant $C > 0$ such that

$$C^{-1}m\Big\{t \in [0,1] : \Big|\sum_{j=1}^{l} a_j r_j(t)\Big| > Cz\Big\} \le m\Big\{t \in [0,1] : \Big|\sum_{j=1}^{l} a_j f_{i_j}(t)\Big| > z\Big\}$$
$$\le Cm\Big\{t \in [0,1] : \Big|\sum_{j=1}^{l} a_j r_j(t)\Big| > C^{-1}z\Big\}$$

for all $l \in \mathbb{N}$, $a_j \in \mathbb{R}$, $j = 1, 2, \dots, l$, and $z > 0$. Therefore, by the definition of r_j and f_i, for each $n \in \mathbb{N}$, we have

$$C^{-1}m\Big\{t \in [0,1] : \Big|\sum_{j=m+1}^{m+n} r_j(t)\chi_{[0,2^{-m}]}(t)\Big| > Cz\Big\}$$
$$\le m\Big\{t \in [0,1] : \Big|\sum_{j=m+1}^{m+n} f_{i_j}(t)\chi_\Delta(t)\Big| > z\Big\}$$
$$\le Cm\Big\{t \in [0,1] : \Big|\sum_{j=m+1}^{m+n} r_j(t)\chi_{[0,2^{-m}]}(t)\Big| > C^{-1}z\Big\}.$$

In particular, this implies that

$$\Big\|\sum_{j=m+1}^{m+n} r_{i_j}\chi_{\Delta\cap D}\Big\|_X \ge c_4\Big\|\sum_{j=m+1}^{m+n} r_j\chi_{[0,2^{-m}]}\Big\|_X,$$

where $c_4 > 0$ depends only on C and the space X. Thus, applying the assertion from the beginning of this part of the proof (see inequality (11.23)) to the interval $[0, 2^{-m}]$ and taking into account that $\{r_j\}$ is a symmetric basic sequence in X with constant 1 (see Proposition 2.2), for sufficiently large n we get

$$\Big\|\sum_{j=m+1}^{m+n} r_{i_j}\chi_{\Delta\cap D}\Big\|_X \ge c_1c_4\Big\|\sum_{j=m+1}^{m+n} r_j\Big\|_X = c_1c_4\Big\|\sum_{j=m+1}^{m+n} r_{i_j}\Big\|_X.$$

Consequently, by the definition of the space $\mathcal{M}(X)$, it holds

$$\|\chi_D\|_{\mathcal{M}(X)} \ge \|\chi_{\Delta\cap D}\|_{\mathcal{M}(X)} \ge c_1c_4.$$

Since $\mathcal{M}(X)$ is a Banach lattice, this inequality implies that $\mathcal{M}(X) \subset L_\infty$. We always have the opposite embedding. Hence, $\mathrm{Sym}\,(X) = L_\infty$, and so the proof is completed. □

Corollary 11.2 *If X is a separable s.s. on $[0, 1]$, then the following conditions are equivalent:*

(i) $\mathcal{M}(X) = L_\infty$;
(ii) $\mathrm{Sym}\,(X) = L_\infty$;
(iii) $\ln^{1/2}(e/t) \notin X$.

Corollary 11.3 *Suppose X and Y are s.s.'s such that $X \subset Y$. Then, we have $\mathcal{M}(X) \subset \mathcal{M}(Y)$ and $\mathrm{Sym}\,(X) \subset \mathrm{Sym}\,(Y)$.*

Proof If $\ln^{1/2}(e/t) \notin X^\circ$, then $\mathcal{M}(X) = L_\infty$ by Theorem 11.5, and so $\mathcal{M}(X) \subset \mathcal{M}(Y)$. Assume now that $\ln^{1/2}(e/t) \in X^\circ$. Then, $\ln^{1/2}(e/t) \in X \subset Y$, and therefore, from Corollary 2.1 it follows that the space G (the separable part of L_{N_2}) is contained both in X and in Y. Thus, appealing to Theorem 2.3, we have

$$\begin{aligned} \|x\|_{\mathcal{M}(Y)} &\asymp \sup\left\{\left\|x\sum_{n=1}^{\infty} a_n r_n\right\|_Y : \|(a_n)\|_{\ell_2} \le 1\right\} \\ &\le C\cdot \sup\left\{\left\|x\sum_{n=1}^{\infty} a_n r_n\right\|_X : \|(a_n)\|_{\ell_2} \le 1\right\} \\ &\asymp \|x\|_{\mathcal{M}(X)}, \end{aligned}$$

and the first embedding is proved. As to the second one, it is an immediate consequence of the first embedding and the definition of $\mathrm{Sym}\,(X)$. □

In the next corollary we provide examples of s.s.'s X such that $\mathcal{M}(X) = L_\infty$.

Corollary 11.4 *Let $p \ge 2$ and let X be either the Lorentz space $\Lambda(\varphi_p)$ with $\varphi_p(t) = \ln^{-1/p}(e/t)$, or the exponential Orlicz space L_{N_p}. Then, $\mathcal{M}(X) = L_\infty$.*

Proof By Theorem 11.5, it suffices to check that in both cases $\ln^{1/2}(e/t) \notin X^\circ$. If $X = \Lambda(\varphi_p)$, this follows at once from the definition of Lorentz spaces. Indeed, since $p \ge 2$, we have

$$\|\ln^{1/2}(e/t)\|_{\Lambda(\varphi_p)} = \frac{1}{p}\int_0^1 \ln^{1/2-1/p-1}(e/t)\,\frac{dt}{t} = \infty.$$

In the case when $X = L_{N_p}$ we can apply Corollary 2.1, which implies that X° is the separable part of the Marcinkiewicz space with the fundamental function φ_p. □

In view of Theorem 11.5, one might expect that L_{N_2} is the largest space among all s.s.'s X satisfying $\mathcal{M}(X) = \mathrm{Sym}\,(X) = L_\infty$. The following example indicates that this is not the case.

Example 11.3 Let $t_k \in [0, 1]$, $t_0 = 1 > t_1 > t_2 > \ldots, t_k \downarrow 0$, and let ψ be the piecewise linear function with knots at the points $(t_k, \ln^{-1/2}(e/t_k))$. Then, ψ is quasiconcave and satisfies the following conditions:

$$(1)\, 0 \le \psi(t) \le \ln^{-1/2}(e/t) \ ; \quad (2)\, \psi(t_k) = \ln^{-1/2}(e/t_k),\ k \ge 0.$$

Moreover, t_k, $k = 1, 2, \ldots$, may be chosen in such a way that additionally

$$(3) \inf_{0<t\le 1} \psi(t) \ln^{1/2}(e/t) = 0.$$

Let us consider the Marcinkiewicz space $M(t/\psi)$. From (1) and (3) (see also Proposition 2.4) it follows that $L_{N_2} \overset{\neq}{\subset} M(t/\psi)$. Furthermore, (2) implies

$$\limsup_{t\to 0} \psi(t) \ln^{1/2}(e/t) \ne 0.$$

Hence, because $tx^*(t) \le \int_0^t x^*(s)\, ds$, we deduce that $\ln^{1/2}(e/t) \notin (M(t/\psi))^\circ$ (see Appendix C or [168, Lemma II.5.4]). Thus, by Theorem 11.5, we have

$$\mathcal{M}(M(t/\psi)) = \mathrm{Sym}\,(M(t/\psi)) = L_\infty.$$

In the next section of the chapter, we show that there is a fairly wide class of s.s.'s such that the Rademacher multiplicator space $\mathcal{M}(X)$ is symmetric though $\mathcal{M}(X) \ne L_\infty$.

11.4 When $\mathcal{M}(X)$ Is Symmetric and Different from L_∞

Let us begin with some definitions and auxiliary results.

We shall say that a function φ on $[0, 1]$ satisfies the Δ^2-*condition* (briefly, $\varphi \in \Delta^2$) if there is a constant $D > 0$ such that

$$\varphi(t) \le D\varphi(t^2), \quad 0 < t \le 1. \tag{11.32}$$

Define also the operator Q by $Qx(t) := x(t^2)$, $0 \le t \le 1$.

Here, for some technical reasons, we prefer to use the logarithms with base 2. Then, inequality (11.3) will take the following form (see the first inequality from Remark 3.4):

$$\Big(\sum_{n=1}^{\infty} a_n r_n\Big)^*(t) \le K' \cdot \|(a_n)\|_{\ell_2} \cdot \log_2^{1/2}(2/t), \quad 0 < t \le 1. \tag{11.33}$$

Proposition 11.5 *Let X be a s.s. on* $[0, 1]$ *with the fundamental function* φ.

(1) If the operator Q is bounded in X, then $\varphi \in \Delta^2$.

(2) Suppose that $\varphi \in \Delta^2$ and X is an interpolation space between the Lorentz space $\Lambda(\varphi)$ and the Marcinkiewicz space $M(\tilde{\varphi})$, where $\tilde{\varphi}(t) := t/\varphi(t)$. Then, Q is bounded in X.

(3) If $\ln^{1/2}(e/t) \in X$ and $\varphi \in \Delta^2$, then there is a constant $C > 0$ such that for all $u \in (0, 1]$ we have

$$\| \log_2^{1/2}(2/t)\chi_{(0,u]}\|_X \le C\| \log_2^{1/2}(2u/t)\chi_{(0,u]}\|_X. \tag{11.34}$$

Proof

(1) From the boundedness Q in X it follows that

$$\|Q\chi_{[0,t]}\|_X \le \|Q\|\|\chi_{[0,t]}\|_X, \quad 0 < t \le 1, \tag{11.35}$$

which gives inequality (11.32) with $D = \|Q\|$.

(2) By Theorem C.1, the boundedness of a linear operator in a Lorentz space is a consequence of its boundedness on the set of all characteristic functions. Hence, (11.32) implies that Q is bounded in the space $\Lambda(\varphi)$. So, because $X \in \mathcal{I}(\Lambda(\varphi), M(\tilde{\varphi}))$, it remains to prove that Q is bounded also in the Marcinkiewicz space $M(\tilde{\varphi})$.

One can readily check that from the condition $\varphi \in \Delta^2$ it follows $\delta_\varphi = 0$, whence $\gamma_{\tilde{\varphi}} = 1 - \delta_\varphi = 1$ (see Appendix C). Consequently, by Theorem C.2, we have

$$\begin{aligned}\|Qx\|_{M(\tilde{\varphi})} &\le C \sup_{0<t\le 1} \varphi(t)x^*(t^2) = C \sup_{0<t\le 1} \varphi(\sqrt{t})x^*(t)\\ &\le CD \sup_{0<t\le 1} \varphi(t)x^*(t) \le CD\|x\|_{M(\tilde{\varphi})}.\end{aligned}$$

(3) First, for each $0 < u \le 1$ it follows that

$$\| \log_2^{1/2}(2/t)\chi_{(0,u]}\|_X \le \| \log_2^{1/2}(2/t)\chi_{(0,u^2]}\|_X + \| \log_2^{1/2}(2/t)\chi_{(u^2,u]}\|_X.$$

We estimate now the terms from the right-hand side separately. If $0 < t \le u^2$, then $\sqrt{t} \ge t/u$. Therefore,

$$\| \log_2^{1/2}(2/t)\chi_{(0,u^2]}\|_X \le \sqrt{2}\| \log_2^{1/2}(2/\sqrt{t})\chi_{(0,u^2]}\|_X \le \sqrt{2}\| \log_2^{1/2}(2u/t)\chi_{(0,u]}\|_X.$$

As to the second term, then, in view of (11.32), we have

$$\begin{aligned}\| \log_2^{1/2}(2/t)\chi_{(u^2,u]}\|_X &\le \sqrt{2}\log_2^{1/2}(2/u)\|\chi_{(u^2,u]}\|_X \le \sqrt{2}D\log_2^{1/2}(2/u)\varphi(u^2)\\ &= \sqrt{2}D\log_2^{1/2}(2u/u^2)\varphi(u^2) \le \sqrt{2}D\| \log_2^{1/2}(2u/t)\chi_{(0,u]}\|_X.\end{aligned}$$

Summing up three last inequalities, we obtain (11.34) with a constant independent of $u \in (0, 1]$. □

Remark 11.1

(a) Let φ be a quasiconcave function. Then, the Δ^2-condition (11.32) is equivalent both to the fact that $\varphi(t) \le C\varphi(t^\beta)$ for some $\beta > 1$, $C > 0$ and all $0 < t \le 1$, and as well to the following discrete version of (11.32): $\varphi(2^{-n}) \le C\varphi(2^{-\beta n})$, $n \in \mathbb{N}$, with $\beta > 1$ and $C > 0$.
(b) In view of elementary properties of the logarithm, inequality (11.34) holds if and only if there is $C > 0$ such that for all $n \in \mathbb{N}$ we have

$$\|\log_2^{1/2}(2/t)\chi_{(0,2^{-n}]}\|_X \le C\|\log_2^{1/2}(2^{1-n}/t)\chi_{(0,2^{-n}]}\|_X.$$

(c) Let $E \subset [0, 1]$. An inspection of the proof of Proposition 11.5(3) shows that, under the condition $\ln^{1/2}(e/t) \in X$, inequality (11.34) is fulfilled for $u \in E$ whenever we have (11.32) for $t \in E$.

We proceed now with a statement that gives a condition, ensuring that a Rademacher multiplicator space is symmetric.

Further, $\mathcal{D}_n$, $n = 0, 1, 2, \ldots$, denotes the set of all step-functions of the form

$$f = \sum_{k=1}^{2^n} a_k \chi_{\Delta_n^k}, \quad a_k \in \mathbb{R}, \tag{11.36}$$

where $\Delta_n^k = [(k-1)2^{-n}, k2^{-n}]$, $1 \le k \le 2^n$, are dyadic intervals of rank n. Moreover, we put $\mathcal{D} = \bigcup_{n=0}^{\infty} \mathcal{D}_n$.

Proposition 11.6 *Let X be a s.s. on $[0, 1]$ such that $\ln^{1/2}(e/t) \in X^\circ$. Suppose that for some $A > 0$ and all $n \in \mathbb{N}$, $b_1 \ge b_2 \ge \cdots \ge b_{2^n} \ge 0$ it holds*

$$\left\|\sum_{k=1}^{2^n} b_k \chi_{\Delta_n^k} \cdot \log_2^{1/2}(2/t)\right\|_X \le A\left\|\sum_{k=1}^{2^n} b_k \chi_{\Delta_n^k} \cdot \log_2^{1/2}\left(\frac{2}{2^n t + 1 - k}\right)\right\|_X. \tag{11.37}$$

Then $\mathcal{M}(X)$ is a s.s.

Proof Since always $\mathrm{Sym}\,(X) \overset{1}{\subset} \mathcal{M}(X)$ and $\mathrm{Sym}\,(X)$ is a s.s., we just have to prove the opposite embedding. To this end, we first estimate the norm $\|f\|_{\mathcal{M}}$ from below for a function $f \in \mathcal{D}_n$ of the form (11.36).

From the hypothesis $\ln^{1/2}(e/t) \in X^\circ$ it follows equivalence (11.13). Therefore, for some $C_1 > 0$ and all $n = 0, 1, 2, \ldots, m = 1, 2, \ldots$ we have

$$\left\|\sum_{k=n+1}^{n+m} \frac{1}{\sqrt{m}} r_k\right\|_X \le C_1,$$

whence by definition of the $\mathcal{M}(X)$-norm,

$$\|f\|_{\mathcal{M}} \geq C_1^{-1} \sup_{n,m\in\mathbb{N}} \left\| f \cdot \sum_{k=n+1}^{n+m} \frac{1}{\sqrt{m}} r_k \right\|_X. \tag{11.38}$$

Furthermore, it is easy to see that for arbitrary $n = 0, 1, 2, \dots$ and $k = 1, 2, \dots, 2^n$

$$\left(\chi_{\Delta_n^k} \cdot \sum_{i=n+1}^{n+m} r_i\right)^*(t) = \left(\sum_{i=1}^{m} r_i\right)^*(2^n t), \quad 0 \leq t \leq 2^{-n}.$$

Hence, because X is a s.s., we have

$$\begin{aligned}\left\| f \cdot \frac{1}{\sqrt{m}} \sum_{i=n+1}^{n+m} r_i \right\|_X &= \left\| \sum_{k=1}^{2^n} a_k \chi_{\Delta_n^k} \cdot \frac{1}{\sqrt{m}} \sum_{i=n+1}^{n+m} r_i \right\|_X \\ &= \left\| \sum_{k=1}^{2^n} a_k \chi_{\Delta_n^k} \cdot \frac{1}{\sqrt{m}} \left(\sum_{i=1}^{m} r_i\right)^*(2^n t + 1 - k) \right\|_X.\end{aligned}$$

Moreover, by Theorem A.1 (see also the proof of Theorem 2.3 and Remark 3.4),

$$\lim_{m\to\infty} \left(\frac{1}{\sqrt{m}} \sum_{i=1}^{m} r_i\right)^*(t) \geq \ln^{1/2} 2 \cdot \log_2^{1/2}\left(\frac{2}{e\sqrt{2\pi}\, t}\right), \quad 0 < t \leq \frac{1}{e\sqrt{2\pi}}.$$

Combining this together with the preceding equation, inequality (11.38) and the fact that X'' is a s.s. with the Fatou property, we get

$$\begin{aligned}\|f\|_{\mathcal{M}} &\geq C_2^{-1} \left\| \sum_{k=1}^{2^n} a_k \chi_{\Delta_n^k} \cdot \log_2^{1/2}\left(\frac{2}{2^n t + 1 - k}\right) \right\|_{X''} \\ &= C_2^{-1} \left\| \sum_{k=1}^{2^n} a_k^* \chi_{\Delta_n^k} \cdot \log_2^{1/2}\left(\frac{2}{2^n t + 1 - k}\right) \right\|_{X''},\end{aligned}$$

where $(a_k^*)_{k=1}^{2^n}$ is the decreasing permutation of the sequence $(|a_k|)_{k=1}^{2^n}$. On the other hand, since

$$\left(\sum_{k=1}^{2^n} a_k \chi_{\Delta_n^k}\right)^*(t) = \sum_{k=1}^{2^n} a_k^* \chi_{\Delta_n^k}(t), \quad 0 < t \leq 1,$$

by Theorem 11.3, we have

$$\|f\|_{\text{Sym}} \asymp \Big\| \sum_{k=1}^{2^n} a_k^* \chi_{\Delta_n^k} \cdot \log_2^{1/2}(2/t) \Big\|_{X''}.$$

In view of the hypothesis $\ln^{1/2}(e/t) \in X^\circ$ and the equation $(X'')^\circ = X^\circ$ (see Lemma 2.2), we may replace the norm of X'' in the latter formulae with that of X. As a result, applying inequality (11.37) with $b_k = a_k^*$, $k = 1, 2, \ldots, 2^n$, for some constant $C > 0$ and all $f \in \mathcal{D}$ we get

$$\|f\|_{\text{Sym}} \le C\|f\|_{\mathcal{M}}. \tag{11.39}$$

Let us extend inequality (11.39) to the whole space $\mathcal{M}(X)$. By Theorem 11.5, we have $\text{Sym}\,(X) \neq L_\infty$. Therefore, the fundamental function of the space $\text{Sym}\,(X)$ tends to zero as $t \to 0+$. Consequently, according to Lusin's theorem (see e.g. [215, Theorem IV.4.4]), for each function $g \in L_\infty$ there is a sequence $\{f_n\} \subset \mathcal{D}$ such that $\|f_n - g\|_{\text{Sym}} \to 0$. Observe that the embedding $\text{Sym}\,(X) \subset \mathcal{M}(X)$ implies $\|f_n - g\|_{\mathcal{M}} \to 0$. Moreover, by (11.39), we have $\|f_n\|_{\text{Sym}} \le C\|f_n\|_{\mathcal{M}}$, $n = 1, 2, \ldots$. In consequence,

$$\|g\|_{\text{Sym}} \le C\|g\|_{\mathcal{M}} \quad \text{for all} \quad g \in L_\infty.$$

Lastly, let $h \in \mathcal{M}(X)$ be an arbitrary nonnegative function. We can find a sequence $\{g_n\} \subset L_\infty$ such that $0 \le g_n \uparrow h$ a.e. on $[0, 1]$. From the preceding inequality then it follows that

$$\|g_n\|_{\text{Sym}} \le C\|g_n\|_{\mathcal{M}} \le C\|h\|_{\mathcal{M}}.$$

By Theorem 11.3, the space $\text{Sym}\,(X)$ has the Fatou property. Therefore, taking here $n \to \infty$, we obtain that $h \in \text{Sym}\,(X)$ and $\|h\|_{\text{Sym}} \le C\|h\|_{\mathcal{M}}$. This finishes the proof. □

We state now the first main result of this chapter, which gives a sufficient condition for a Rademacher multiplicator space to be symmetric.

Theorem 11.6 *If the operator $Qx(s) = x(s^2)$ is bounded in a s.s. X, then the space $\mathcal{M}(X)$ is symmetric.*

Proof Observe that, according to Theorem 11.5, the assumption $\ln^{1/2}(e/t) \notin X^\circ$ implies that $\mathcal{M}(X) = \text{Sym}\,(X) = L_\infty$. Therefore, we can assume that $\ln^{1/2}(e/t) \in X^\circ$, and hence, in view of Proposition 11.6, it suffices to prove inequality (11.37).

We shall present two different proofs of (11.37): the first is based on using distribution estimates; in the second we apply some extrapolation relations, which hold for s.s.'s located sufficiently "close" to the space L_∞.

The first proof. Denote

$$g(t) := \sum_{k=1}^{2^n} b_k \chi_{\Delta_n^k} \cdot \log_2^{1/2}(2/t) \text{ and } h(t) := \sum_{k=1}^{2^n} b_k \chi_{\Delta_n^k} \cdot \log_2^{1/2}\Big(\frac{2}{2^n t + 1 - k}\Big),$$

where $b_1 \ge b_2 \ge \cdots \ge b_{2^n} \ge 0$. Let us represent the function g in the form $g = g_1 + g_2$, with

$$g_1(t) := b_1 \cdot \ln^{1/2}(e/t) \cdot \chi_{[0,2^{-n}]}(t) \text{ and } g_2(t) := g(t) - g_1(t).$$

Since Q is bounded in X, then from Proposition 11.5 (see assertions (1) and (3)) it follows that X satisfies condition (11.34), which implies that

$$\|g_1\|_X \le C \cdot b_1 \Big\| \log_2^{1/2}(2^{1-n}/t) \cdot \chi_{[0,2^{-n}]} \Big\|_X \le C\|h\|_X. \tag{11.40}$$

To estimate $\|g_2\|_X$, we introduce the function

$$g_2'(t) := \sum_{j=1}^{n} \sum_{i=2^{j-1}+1}^{2^j} b_i (n+1-j)^{1/2} \chi_{\Delta_n^i}(t).$$

Taking into account that $\log_2^{1/2}(2/t)\chi_{\Delta_n^k}(t) \le 2(n+1-j)^{1/2}$ if $2^{j-1} < k \le 2^j$, $j = 1, 2, \ldots, n$, we have

$$|g_2(t)| \le 2|g_2'(t)|, \quad 0 < t \le 1. \tag{11.41}$$

Denoting by Q^{-1} the inverse operator for Q (i.e., $Q^{-1}x(t) := x(\sqrt{t})$), we get

$$Q^{-1}g_2'(t) = \sum_{j=1}^{n} \sum_{i=2^{j-1}+1}^{2^j} b_i (n+1-j)^{1/2} \chi_{[(i-1)^2 2^{-2n}, i^2 2^{-2n}]}(t).$$

Hence, the inequality

$$(Q^{-1}g_2')^*(t) \le h^*(t), \quad 0 < t \le 1, \tag{11.42}$$

will be proved once we show that for all $j = 1, 2, \ldots, n$ and $2^{j-1} < i \le 2^j$ it holds:

$$m\{t \in [0,1] : |h(t)| \ge b_i (n+1-j)^{1/2}\} \ge i^2 2^{-2n}.$$

Indeed, since $b_1 \geq b_2 \geq \cdots \geq b_{2^n} \geq 0$ and $i \leq 2^j$, we have

$$\begin{aligned} m\{t \in [0,1] : |h(t)| \geq b_i(n+1-j)^{1/2}\} &\geq i \cdot m\big\{t \in (0, 2^{-n}) : \log_2^{1/2}(2^{1-n}/t) \geq (n+1-j)^{1/2}\big\} \\ &\geq i \cdot 2^{j-2n} \geq i^2 2^{-2n}, \end{aligned}$$

and so (11.42) is proved.

Inequalities (11.41) and (11.42) imply that

$$\|g_2\|_X \leq 2\|Q(Q^{-1}g_2')\|_X \leq 2\|Q\|\|Q^{-1}g_2'\|_X \leq 2\|Q\|\|h\|_X.$$

Combining this estimate with (11.40), we obtain (11.37), and so the proof is completed. □

In the second proof of Theorem 11.6 and also somewhere next in the book we shall need the following useful statement.

Lemma 11.5 *Let $1 \leq p < \infty$ and let $f \in L_p[0,1]$ be arbitrary. Then, for every $n \in \mathbb{N}$ there exists a sequence $(a_k) \in \ell_2$ such that $a_k = 0$ if $k \leq n$, $\|(a_k)\|_{\ell_2} = 1$, and*

$$\|f\|_p \leq 3p^{-1/2} \cdot \Big\| f \cdot \sum_{k=1}^{\infty} a_k r_k \Big\|_p.$$

Proof Clearly, we can find a finite-valued function g of the form (11.36) satisfying the conditions: $|g| \leq |f|$ and $\|f\|_p \leq 3/2\|g\|_p$. Denote by N' the largest rank of the constancy intervals of g. Moreover, let $N := \max\{n, N'\}$. We choose $m \in \mathbb{N}$ such that $m-1 \leq p < m$ and set $a_k := m^{-1/2}$ for $N+1 \leq k \leq N+m$, and $a_k := 0$ for $k \leq N$ or $k > N+m$. Then, $\|(a_k)\|_{\ell_2} = 1$. Also, from definition of the Rademacher functions it follows that for each $m \in \mathbb{N}$ there is a set $A_m \subset [0,1]$ of measure 2^{1-m} such that for all $t \in A_m$ we have $|\sum_{k=N+1}^{N+m} r_k(t)| = m$. Hence,

$$\Big\| \sum_{k=1}^{\infty} a_k r_k \Big\|_p \geq m^{1/2} 2^{(1-m)/p} \geq 2^{-1} p^{1/2}.$$

Note that, thanks to the choice of N, the functions g and $\sum_{k=1}^{\infty} a_k r_k = \sum_{k=N+1}^{N+m} a_k r_k$ are independent. Therefore, applying (A.1), we get

$$\int_0^1 \Big| g(t) \sum_{k=1}^{\infty} a_k r_k(t) \Big|^p dt = \int_0^1 |g(t)|^p \, dt \int_0^1 \Big| \sum_{k=1}^{\infty} a_k r_k(t) \Big|^p dt,$$

whence, in view of the last inequality,

$$\|f\|_p \le \frac{3}{2}\|g\|_p \le 3p^{-1/2}\|g\|_p \Big\| \sum_{k=1}^{\infty} a_k r_k \Big\|_p$$
$$= 3p^{-1/2}\Big\| g \sum_{k=1}^{\infty} a_k r_k \Big\|_p \le 3p^{-1/2}\Big\| f \sum_{k=1}^{\infty} a_k r_k \Big\|_p,$$

as we wished. □

The Second Proof of Theorem 11.6 Let $f \in L_p[0,1]$ for all $p < \infty$. Since $t^{1/\log_2(2/t)} \ge 1/2$ if $0 < t \le 1$, we have

$$\|f\|_{\log_2(2/t)} \ge \Big(\int_0^t (f^*(s))^{\log_2(2/t)}\, ds\Big)^{1/\log_2(2/t)} \ge f^*(t) t^{1/\log_2(2/t)} \ge \frac{1}{2} f^*(t), \quad 0 < t \le 1.$$

Hence,

$$\Big\| f^*(t) \cdot \log_2^{1/2}(2/t) \Big\|_X \le 2 \Big\| \, \|f\|_{\log_2(2/t)} \log_2^{1/2}(2/t) \Big\|_X . \tag{11.43}$$

Suppose now that $f := \sum_{k=1}^{2^n} b_k \chi_{\Delta_n^k}$, where $0 \le b_{2^n} \le \cdots \le b_1$. Then, by Lemma 11.5, for arbitrary $1 \le p < \infty$ and $n = 0, 1, \ldots$ there exists a Rademacher sum $h = h_{p,f} = \sum_{k=n+1}^{\infty} a_k r_k$ such that $\|(a_k)\|_{\ell_2} = 1$ and

$$\|f\|_p \le 3p^{-1/2} \|f \cdot h\|_p. \tag{11.44}$$

By Theorem 2.2, we have $\|h\|_{L_{N_2}} \le \sqrt{2e}$. Therefore, in view of (11.33), for all $k = 1, 2, \ldots, 2^n$ it holds

$$\big(\chi_{\Delta_n^k} h\big)^*(s) = \Big(\sum_{i=1}^{\infty} a_{n+i} r_i\Big)^*(2^n s) \le K' \log_2^{1/2}\Big(\frac{2}{2^n s}\Big), \quad 0 < s \le 2^{-n}.$$

Thus,

$$(f \cdot h)^*(s) = \Big(\sum_{k=1}^{2^n} b_k \chi_{\Delta_n^k} h\Big)^*(s) \le K' \Big(\sum_{k=1}^{2^n} b_k \chi_{\Delta_n^k} \log_2^{1/2}\Big(\frac{2}{2^n \cdot +1 - k}\Big)\Big)^*(s).$$

Combining this estimate with (11.44), we infer

$$\|f\|_p \le 3K' p^{-1/2} \Big\| \sum_{k=1}^{2^n} b_k \chi_{\Delta_n^k} \log_2^{1/2}\Big(\frac{2}{2^n s + 1 - k}\Big) \Big\|_p .$$

Inserting here $p = \log_2(2/t)$, $0 < t \le 1$, multiplying the both sides of the inequality by $\log_2^{1/2}(2/t)$ and taking then the X-norm, we get

$$\Big\| \|f\|_{\log_2(2/t)} \log_2^{1/2}(2/t)\Big\|_X \le 3K' \Big\| \Big\| \sum_{k=1}^{2^n} b_k \chi_{\Delta_n^k} \log_2^{1/2}\Big(\frac{2}{2^n s+1-k}\Big)\Big\|_{\log_2(2/t)}\Big\|_X.$$

According to [45, Theorem 2.3], the operator Q is bounded in a s.s. X if and only if there is a constant $C_2 > 0$ such that for all $y \in X$ we have

$$\big\| \|y\|_{\log_2(2/t)}\big\|_X \le C_2 \|y\|_X.$$

Hence, by condition, from the preceding inequality it follows that

$$\Big\| \|f\|_{\log_2(2/t)} \log_2^{1/2}(2/t)\Big\|_X \le C \Big\| \sum_{k=1}^{2^n} b_k \chi_{\Delta_n^k} \log_2^{1/2}\Big(\frac{2}{2^n s+1-k}\Big)\Big\|_X.$$

Since inequality (11.37) is an immediate consequence of this inequality and (11.43) in the case when $f = \sum_{k=1}^{2^n} b_k \chi_{\Delta_n^k}$, then, by Proposition 11.6, the proof is completed. □

From Theorem 11.6 and Proposition 11.5(2) it follows

Corollary 11.5 *If X is an interpolation s.s. between the spaces $\Lambda(\varphi)$ and $M(\tilde{\varphi})$ and $\varphi \in \Delta^2$, then $\mathcal{M}(X) = \mathrm{Sym}\,(X)$. In particular, this holds for the Lorentz space $\Lambda(\varphi)$ and the Marcinkiewicz space $M(\tilde{\varphi})$ whenever $\varphi \in \Delta^2$.*

Remark 11.2 The boundedness of the operator Q in a s.s. X is not a necessary condition for the equality $\mathcal{M}(X) = \mathrm{Sym}\,(X)$. Indeed, for every function $\varphi \in \Delta^2$ there is a subspace X of the Marcinkiewich space $M(\tilde{\varphi})$ with the following properties: (a) X is a s.s.; (b) $X'' = M(\tilde{\varphi})$; (c) the operator Q is not bounded in X [45, Example 2.12]. Then, from (b), Theorem 11.3, and Corollaries 11.5, 11.3 it follows that

$$\mathrm{Sym}(X) = \mathrm{Sym}\,(X'') = \mathrm{Sym}\,(M(\tilde{\varphi})) = \mathcal{M}(M(\tilde{\varphi})) \supset \mathcal{M}(X).$$

Since the opposite embedding is obvious, as a result, we get $\mathcal{M}(X) = \mathrm{Sym}\,(X)$.

Remark 11.3 Assume that the operator Q is bounded in a s.s. X on $[0, 1]$. As it can be easily seen, the proofs of Proposition 11.6 and Theorem 11.6 reveal the following property of tails of Rademacher sums: for arbitrary $n \in \mathbb{N}$ and each function $f \in \mathcal{D}_n$ we have

$$\|f\|_{\mathcal{M}} \asymp \sup\Big\{\Big\|f \cdot \sum_{i=n+1}^{\infty} a_i r_i\Big\|_X : \Big\|\sum_{i=n+1}^{\infty} a_i r_i\Big\|_X \le 1\Big\} \tag{11.45}$$

with a constant independent of $n \in \mathbb{N}$ and $a_i \in \mathbb{N}$. Roughly speaking, this means that the $\mathcal{M}(X)$-norm of a function $f \in \mathcal{D}_n$ is attained (up to equivalence) at a certain tail Rademacher sum. Moreover, an inspection of the proof of Theorem 11.5 shows that in the case when $\mathcal{M}(X) = L_\infty$ an equivalence similar to (11.45) holds for the $\mathcal{M}(X)$-norms of characteristic functions of dyadic intervals.

In the concluding section of this chapter, we shall prove a result, which can be treated as an explanation of the above state of affairs. Namely, with a s.s. X we shall associate the so-called tail Rademacher multiplicator space $\mathcal{M}_T(X)$ such that $\mathcal{M}(X)$ is being a s.s. if and only $\mathcal{M}(X) = \mathcal{M}_T(X)$. Furthermore, it turns out that one of the equivalent norms in $\mathcal{M}_T(X)$ of a function $f \in \mathcal{D}_n$ is just exactly the right-hand side of equivalence (11.45).

We proceed now with the study of the problem of finding necessary conditions, under which a Rademacher multiplicator space is symmetric. First of all, let us observe that an obvious condition of such a sort is the existence of a constant $C > 0$ such that for all sets $E \subset [0, 1]$ we have

$$\|\chi_E\|_{\mathrm{Sym}} \le C\|\chi_E\|_{\mathcal{M}}.$$

In what follows, we consider separately two cases, when the norm $\|\chi_E\|_{\mathcal{M}}$ can be replaced here with a certain smaller quantity, subject to what is used by calculation of this quantity: Rademacher tails or Rademacher partial sums.

For an arbitrary function φ on $[0, 1]$ we set

$$\hat{\varphi}(t) := t^{-1/2}\varphi(2^{1-1/t}), \quad 0 < t \le 1.$$

Proposition 11.7 *Suppose that the fundamental function φ of a s.s. X satisfies the condition: $\gamma_{\hat{\varphi}} > 0$. Also, let $I \subset \mathbb{N}$ be such that for every $n \in I$*

$$\|\chi_{[0,2^{-n}]}\|_{\mathrm{Sym}} \le A \sup\left\{\Big\|\chi_{[0,2^{-n}]} \cdot \sum_{i=n+1}^{\infty} c_i r_i\Big\|_X : \Big\|\sum_{i=n+1}^{\infty} c_i r_i\Big\|_X \le 1\right\},$$

where $A > 0$ does not depend on n. Then, there exist $\alpha > 1$ and $C > 0$ such that for all $n \in I$ we have

$$\varphi(2^{-n}) \le C\varphi(2^{-\alpha n}).$$

For the proof we need the following technical result.

Lemma 11.6 *If φ is an increasing concave function on $[0, 1]$ satisfying the condition $\gamma_{\hat{\varphi}} > 0$, then the function $\ln^{1/2}(e/s)$ belongs to the Lorentz space $\Lambda(\varphi)$ and*

$$\sup_{n=1,2,\ldots} \frac{1}{\sqrt{n}\varphi(2^{-n})} \sum_{k=n}^{\infty} \frac{\varphi(2^{-k})}{\sqrt{k}} < \infty.$$

Proof According to the hypothesis, there are $\delta > 0$ and $C > 0$ such that for all $0 < u, t \le 1$ we have $\hat{\varphi}(tu) \le Ct^\delta \hat{\varphi}(u)$, that is,

$$\varphi(2^{1-1/(tu)}) \le Ct^{\delta+1/2}\varphi(2^{1-1/u}). \tag{11.46}$$

Putting here $u = 1$, after the change of variable $s = 2^{1-1/t}$ we obtain $\varphi(s) \le C_1 \log_2^{-1/2-\delta}(2/s)$, $0 < s \le 1$, where $C_1 = \varphi(1)C$. Therefore,

$$\|\log_2^{1/2}(2/s)\|_{\Lambda(\varphi)} = \int_0^1 \log_2^{1/2}(2/s)\, d\varphi(s) \le C_1(1/2+\delta)\int_0^1 \log_2^{-1-\delta}(2/s)\,\frac{ds}{s} < \infty,$$

which implies that $\log_2^{1/2}(2/s) \in \Lambda(\varphi)$, or equivalently $\ln^{1/2}(e/s) \in \Lambda(\varphi)$.

Let us prove the second assertion of the lemma. Setting $u = 1/n$, $n \in \mathbb{N}$, and $t = 2^{-j}$, $j = 0, 1, 2, \dots$ in inequality (11.46), by the concavity of φ, we get

$$\varphi(2^{-2^j n}) \le C2^{-(\delta+1/2)j}\varphi(2^{-n}).$$

Hence,

$$\sum_{k=n}^{\infty} \frac{\varphi(2^{-k})}{\sqrt{k}} = \sum_{j=0}^{\infty}\sum_{k=2^j n}^{2^{j+1}n-1} \frac{\varphi(2^{-k})}{\sqrt{k}} \le \sum_{j=0}^{\infty} 2^{j/2}\sqrt{n}\varphi(2^{-2^j n})$$

$$\le C\sqrt{n}\varphi(2^{-n})\sum_{j=0}^{\infty} 2^{-(\delta+1/2)j}2^{j/2} = C\sqrt{n}\varphi(2^{-n})\frac{2^\delta}{2^\delta - 1},$$

and the desired relation follows. □

The Proof of Proposition 11.7 Without loss of generality, we shall assume that the set I is infinite. Moreover, since the fundamental function of the space X° is equal to φ as well, then from Lemma 11.6 and the minimality of the Lorentz space among s.s.'s with the same fundamental function (see Appendix C) it follows that $\ln^{1/2}(e/t) \in X^\circ$. Consequently, by the hypothesis of the proposition, Corollary 11.1 and Theorem A.1, in the same way as in the proof of Proposition 11.6, we obtain that

$$\|\log_2^{1/2}(2/t)\chi_{[0,2^{-n}]}\|_X \le C_1\|\log_2^{1/2}(2^{1-n}/t)\chi_{[0,2^{-n}]}\|_X$$

for all $n \in I$. This inequality and the embeddings $\Lambda(\varphi) \subset X \subset M(\tilde{\varphi})$ (see Appendix C) imply that

$$\|\log_2^{1/2}(2/t)\chi_{[0,2^{-n}]}\|_{M(\tilde{\varphi})} \le C_1\|\log_2^{1/2}(2^{1-n}/t)\chi_{[0,2^{-n}]}\|_{\Lambda(\varphi)}, \quad n \in I. \tag{11.47}$$

Let us estimate the right-hand side of inequality (11.47) from above. First, we have

$$\| \log_2^{1/2}(2^{1-n}/t)\chi_{[0,2^{-n}]}\|_{\Lambda(\varphi)} = \sum_{k=n}^{\infty}\int_{2^{-k-1}}^{2^{-k}} \log_2^{1/2}(2^{1-n}/t)\,d\varphi(t)$$

$$\leq 2\sum_{k=n}^{\infty}(\varphi(2^{-k}) - \varphi(2^{-k-1}))\sqrt{k+1-n}.$$

Since for all $j > n$ the Abel transformation gives

$$\sum_{k=n}^{j}(\varphi(2^{-k}) - \varphi(2^{-k-1}))\sqrt{k+1-n}$$

$$= \sum_{k=n}^{j-1}\varphi(2^{-k-1})(\sqrt{k+2-n} - \sqrt{k+1-n}) + \varphi(2^{-n}) - \varphi(2^{-j-1})\sqrt{j+1-n}$$

$$\leq \sum_{k=n+1}^{j}\frac{\varphi(2^{-k})}{\sqrt{k-n}} + \varphi(2^{-n}),$$

then

$$\| \log_2^{1/2}(2^{1-n}/t)\chi_{[0,2^{-n}]}\|_{\Lambda(\varphi)} \leq 2\Big(\sum_{k=n+1}^{\infty}\frac{\varphi(2^{-k})}{\sqrt{k-n}} + \varphi(2^{-n})\Big).$$

On the other hand, from definition of the norm in a Marcinkiewicz space it follows that

$$\| \log_2^{1/2}(2/t)\chi_{[0,2^{-n}]}\|_{M(\tilde{\varphi})} \geq \sqrt{n}\varphi(2^{-n}), \quad n \in \mathbb{N}.$$

Thus, from (11.47) we get

$$\sqrt{n}\varphi(2^{-n}) \leq 2C_1\Big(\sum_{k=n+1}^{\infty}\frac{\varphi(2^{-k})}{\sqrt{k-n}} + \varphi(2^{-n})\Big), \quad n \in I.$$

Hence, for all $n \in I$ sufficiently large it holds

$$\sqrt{n}\varphi(2^{-n}) \leq 3C_1\sum_{k=n+1}^{\infty}\frac{\varphi(2^{-k})}{\sqrt{k-n}}. \tag{11.48}$$

It remains to estimate the right-hand side of this inequality.

Let $\varepsilon > 0$ be so far arbitrary (it will be chosen later). Then,

$$\sum_{k=n+1}^{[(1+\varepsilon)n]} \frac{\varphi(2^{-k})}{\sqrt{k-n}} \le \varphi(2^{-n}) \sum_{k=1}^{[(1+\varepsilon)n]-n} \frac{1}{\sqrt{k}}$$
$$\le 2\varphi(2^{-n})\sqrt{[(1+\varepsilon)n]-n}$$
$$\le 2\sqrt{\varepsilon}\varphi(2^{-n})\sqrt{n}.$$

Moreover, as $\gamma_{\hat{\varphi}} > 0$, by Lemma 11.6, we have

$$\sum_{k=2n}^{\infty} \frac{\varphi(2^{-k})}{\sqrt{k-n}} \le \sqrt{2}\sum_{k=2n}^{\infty} \frac{\varphi(2^{-k})}{\sqrt{k}} \le C_2\varphi(2^{-2n})\sqrt{n}. \tag{11.49}$$

Now, if $0 < \varepsilon \le (12C_1)^{-2}$, then from inequalities (11.48) and (11.49) it follows that

$$\sqrt{n}\varphi(2^{-n}) \le 6C_1\Big(\sum_{k=[(1+\varepsilon)n]+1}^{2n-1} \frac{\varphi(2^{-k})}{\sqrt{k-n}} + C_2\varphi(2^{-2n})\sqrt{n}\Big).$$

Combining this estimate with the inequality

$$\sum_{k=[(1+\varepsilon)n]+1}^{2n-1} \frac{\varphi(2^{-k})}{\sqrt{k-n}} \le \varphi(2^{-[(1+\varepsilon)n]})\sum_{k=1}^{n}\frac{1}{\sqrt{k}} \le 2\varphi(2^{-[(1+\varepsilon)n]})\sqrt{n},$$

we get

$$\varphi(2^{-n}) \le C\varphi(2^{-[(1+\varepsilon)n]}).$$

Thus, the desired inequality is established for sufficiently large $n \in I$ if $\alpha \in (1, 1+\varepsilon)$. Increasing C if it is necessary, one can make it true for all $n \in I$. □

To prove a similar result for partial Rademacher sums, we replace the interval $[0, 2^{-n}]$ with a certain special union of dyadic intervals of rank $2n$ of measure 2^{-n}.

Consider the matrix $\mathcal{A} = (\theta_{i,j})$, where $\theta_{i,j}$ denotes the value of the function r_j on the interval Δ_{2n}^i, $n \in \mathbb{N}$, $j = 1, 2, \dots, 2n$ and $i = 1, 2, \dots, 2^{2n}$. Let

$$\Omega_n := \big\{i \in \{1, 2, \dots, 2^{2n}\} : \theta_{i,j+n} = \theta_{i,j} \;\text{ for all }\; j = 1, 2, \dots, n\big\}$$

and

$$U_n := \bigcup_{i\in\Omega_n} \Delta_{2n}^i. \tag{11.50}$$

Since card $\Omega_n = 2^n$, then $m(U_n) = 2^{-n}$.

Proposition 11.8 *Let X be a s.s. such that $\ln^{1/2}(e/t) \in X^\circ$. Suppose that for all $n \in I \subset \mathbb{N}$ we have*

$$\|\chi_{U_n}\|_{\mathrm{Sym}} \le B \sup\left\{\left\|\chi_{U_n} \cdot \sum_{j=1}^{2n} c_j r_j\right\|_X : \left\|\sum_{j=1}^{2n} c_j r_j\right\|_X \le 1\right\},$$

where $B > 0$ does not depend on n. Then, there exist $\beta > 1$ and $C > 0$ such that for all $n \in I$ it holds

$$\varphi(2^{-n}) \le C\varphi(2^{-\beta n}).$$

Proof In view of the above definitions, for any $c_j \in \mathbb{R}$, $j = 1, 2, \ldots, 2n$,

$$\chi_{U_n} \cdot \sum_{j=1}^{2n} c_j r_j = \sum_{i \in \Omega_n} \Big(\sum_{j=1}^{2n} c_j \theta_{i,j}\Big) \cdot \chi_{\Delta^i_{2n}} = \sum_{i \in \Omega_n} b_i \chi_{\Delta^i_{2n}}, \tag{11.51}$$

where $b_i := \sum_{j=1}^{2n} c_j \theta_{i,j}$, $i \in \Omega_n$. Assuming that $\|(c_j)\|_{\ell_2} \le 1$, by the Cauchy-Schwarz-Bunyakovskii inequality, we have

$$|b_i| \le \|(c_j)\|_{\ell_2}\sqrt{2n} \le \sqrt{2n}, \quad i \in \Omega_n. \tag{11.52}$$

Furthermore, according to the choice of the set Ω_n, there is a sign arrangement $(\eta_j(i))_{j=1}^n$ such that

$$b_i = \sum_{j=1}^{2n} c_j \theta_{i,j} = \sum_{j=1}^{n} (c_j + c_{j+n})\eta_j(i), \quad i \in \Omega_n.$$

For $\delta > 0$, which will be chosen later, we consider the set B_n defined by

$$B_n := \{k \in \Omega_n : |b_k| \ge \delta\sqrt{n}\}.$$

Let us estimate card B_n from above. By the preceding expression for the coefficients b_i, $i \in \Omega_n$, the exponential distribution estimate of Rademacher sums from Proposition 1.2 and the inequality

$$\Big(\sum_{j=1}^{n} (c_j + c_{j+n})^2\Big)^{1/2} \le 2,$$

it follows that

$$\operatorname{card} B_n = \operatorname{card}\left\{\varepsilon_k = \pm 1 : \left|\sum_{k=1}^{n} (c_k + c_{k+n})\varepsilon_k\right| \ge \delta\sqrt{n}\right\}$$

$$= 2^n m\left\{t \in [0,1] : \left|\sum_{k=1}^{n}(c_k + c_{k+n})r_k(t)\right| \ge \delta\sqrt{n}\right\}$$
$$\le 2^{n+1}\cdot e^{-\delta^2 n/8} < 2^{n+1}\cdot 2^{-\delta^2 n/8}.$$

This estimate, combined with (11.51) and (11.52), yields the following inequality for the fundamental function φ of the space X:

$$\left\|\chi_{U_n}\cdot\sum_{j=1}^{2n} c_j r_j\right\|_X \le \left\|\sum_{i\in B_n} b_i\chi_{\Delta_n^i}\right\|_X + \left\|\sum_{i\in\Omega_n\setminus B_n} b_i\chi_{\Delta_n^i}\right\|_X$$
$$\le \sqrt{2n}\cdot\|\chi_{[0,\mathrm{card}B_n\cdot 2^{-2n}]}\|_X + \delta\sqrt{n}\|\chi_{U_n}\|_X$$
$$\le \sqrt{2n}\cdot\varphi(2\cdot 2^{-n(1+\delta^2/8)}) + \delta\sqrt{n}\varphi(2^{-n}).$$

Thus, from the condition $\ln^{1/2}(e/t)\in X^\circ$ and the concavity of φ we get

$$\sup\left\{\left\|\chi_{U_n}\cdot\sum_{j=1}^{2n} c_i r_i\right\|_X : \left\|\sum_{j=1}^{2n} c_i r_i\right\|_X \le 1\right\} \asymp \sup\left\{\left\|\chi_{U_n}\cdot\sum_{j=1}^{2n} c_i r_i\right\|_X : \|(c_j)\|_{\ell_2}\le 1\right\}$$
$$\le C_1\sqrt{n}\left(\varphi(2^{-n(1+\delta^2/8)}) + \delta\varphi(2^{-n})\right).$$

On the other hand, by Corollary 11.1,

$$\|\chi_{U_n}\|_{\mathrm{Sym}} \asymp \|\chi_{[0,2^{-n}]}\ln^{1/2}(e/t)\|_X \ge \sqrt{n}\,\varphi(2^{-n}),\quad n\in\mathbb{N}.$$

The last two inequalities and the hypothesis of the proposition imply that for every $n\in I$

$$\varphi(2^{-n}) \le C_2(\varphi(2^{-n(1+\delta^2/8)}) + \delta\varphi(2^{-n})).$$

Hence, choosing $\delta > 0$ sufficiently small, one can easily get

$$\varphi(2^{-n}) \le C\varphi(2^{-\beta n}),$$

where $C > 0$ does not depend on $n\in I$ and $\beta := 1 + \delta^2/8 > 1$. □

From Propositions 11.7 and 11.8 we deduce the following result.

Theorem 11.7 *Let X be a s.s. such that* $\ln^{1/2}(e/t)\in X^\circ$. *Suppose that for some* $D > 0$ *and all* $n\in\mathbb{N}$

$$\|\chi_{U_n}\|_{\mathrm{Sym}} \le D\|\chi_{U_n}\|_{\mathcal{M}}, \tag{11.53}$$

where the set U_n is defined by (11.50).

Then, X satisfies condition (11.34). *Moreover, if additionally $\gamma_{\hat{\varphi}} > 0$, where φ is the fundamental function of X, we have $\varphi \in \Delta^2$.*

Proof Denote by I_1 the set of all $n \in \mathbb{N}$ such that

$$\|\chi_{U_n}\|_{\mathrm{Sym}} \le 2D \sup\left\{\left\|\chi_{U_n}\cdot\sum_{i=2n+1}^{\infty} c_i r_i\right\|_X : \left\|\sum_{i=2n+1}^{\infty} c_i r_i\right\|_X \le 1\right\}. \tag{11.54}$$

Also, set $I_2 := \mathbb{N} \setminus I_1$. From the definition of $\mathcal{M}(X)$ and unconditionality of the Rademacher system in a s.s. (see Proposition 2.2) it follows that

$$\|\chi_{U_n}\|_{\mathcal{M}} \le \sup\left\{\left\|\chi_{U_n}\cdot\sum_{i=1}^{2n} c_i r_i\right\|_X : \left\|\sum_{i=1}^{2n} c_i r_i\right\|_X \le 1\right\}$$
$$+ \sup\left\{\left\|\chi_{U_n}\cdot\sum_{i=2n+1}^{\infty} c_i r_i\right\|_X : \left\|\sum_{i=2n+1}^{\infty} c_i r_i\right\|_X \le 1\right\}.$$

Hence, by inequalities (11.53) and (11.54), we get for each $n \in I_2$

$$\|\chi_{U_n}\|_{\mathrm{Sym}} \le 2D \sup\left\{\left\|\chi_{U_n}\cdot\sum_{i=1}^{2n} c_i r_i\right\|_X : \left\|\sum_{i=1}^{2n} c_i r_i\right\|_X \le 1\right\}. \tag{11.55}$$

Since $m(U_n) = 2^{-n}$ and Sym (X) is a s.s., we have $\|\chi_{U_n}\|_{\mathrm{Sym}} = \|\chi_{[0,2^{-n}]}\|_{\mathrm{Sym}}$. Furthermore, the functions $\chi_{U_n}\cdot\sum_{i=2n+1}^{\infty} c_i r_i$ and $\chi_{[0,2^{-n}]}\cdot\sum_{i=n+1}^{\infty} c_{n+i} r_i$ are identically distributed, which implies that

$$\sup\left\{\left\|\chi_{U_n}\cdot\sum_{i=2n+1}^{\infty} c_i r_i\right\|_X : \left\|\sum_{i=2n+1}^{\infty} c_i r_i\right\|_X \le 1\right\}$$
$$= \sup\left\{\left\|\chi_{[0,2^{-n}]}\cdot\sum_{i=n+1}^{\infty} c_{n+i} r_i\right\|_X : \left\|\sum_{i=n+1}^{\infty} c_{n+i} r_i\right\|_X \le 1\right\}.$$

In consequence, (11.54) is equivalent to the estimate

$$\|\chi_{[0,2^{-n}]}\|_{\mathrm{Sym}} \le 2D \sup\left\{\left\|\chi_{[0,2^{-n}]}\cdot\sum_{i=n+1}^{\infty} c_{n+i} r_i\right\|_X : \left\|\sum_{i=n+1}^{\infty} c_{n+i} r_i\right\|_X \le 1\right\}, \tag{11.56}$$

and hence, by Corollary 11.1 and Theorem A.1 (see also the proof of Proposition 11.6), to the inequality

$$\| \log_2^{1/2}(2/t)\chi_{(0,2^{-n}]}\|_X \le C_1 \| \log^{1/2}(2^{1-n}/t)\chi_{(0,2^{-n}]}\|_X. \tag{11.57}$$

As a result, (11.57) holds for all $n \in I_1$.

Moreover, according to Proposition 11.8, inequality (11.55) implies that there exist constants $\beta > 1$ and $C_2 > 0$ such that for all $n \in I_2$ we have

$$\varphi(2^{-n}) \le C_2 \varphi(2^{-\beta n}). \tag{11.58}$$

Therefore, taking into account Remark 11.1(c), we deduce that inequality (11.57) is valid also for $n \in I_2$. As a result, X satisfies condition (11.34) (see Remark 11.1(b)).

In the case when $\gamma_{\hat{\varphi}} > 0$, in view of (11.56) and Proposition 11.7, inequality (11.58) holds (possibly, with another constant) for $n \in I_1$ as well. Finally, by Remark 11.1(a), we get $\varphi \in \Delta^2$. □

Theorem 11.8 *Suppose that φ is an increasing concave function on* $[0, 1]$ *such that $\varphi(0) = 0$ and $\gamma_{\hat{\varphi}} > 0$, X is an interpolation s.s. between the spaces $\Lambda(\varphi)$ and $M(\tilde{\varphi})$. The following conditions are equivalent:*

(i) $\mathcal{M}(X)$ is symmetric;
(ii) there is $D > 0$ such that

$$\|\chi_{U_n}\|_{\mathrm{Sym}} \le D\|\chi_{U_n}\|_{\mathcal{M}}, \quad n \in \mathbb{N},$$

where the set U_n is defined by (11.50)*;*
(iii) $\varphi \in \Delta^2$;
(iv) for some $A > 0$ and all $n \in \mathbb{N}$, $f \in \mathcal{D}_n$ we have

$$\|f\|_{\mathrm{Sym}} \le A \sup\left\{\left\|f \cdot \sum_{k=n+1}^{\infty} a_k r_k\right\|_X : \left\|\sum_{k=n+1}^{\infty} a_k r_k\right\|_X \leqslant 1\right\};$$

(v) for some $A' > 0$ and all $n \in \mathbb{N}$ it holds

$$\|\chi_{[0,2^{-n}]}\|_{\mathrm{Sym}} \le A' \sup\left\{\left\|\chi_{[0,2^{-n}]} \cdot \sum_{k=n+1}^{\infty} a_k r_k\right\|_X : \left\|\sum_{k=n+1}^{\infty} a_k r_k\right\|_X \leqslant 1\right\}.$$

Proof The equivalence of conditions (i), (ii) and (iii) follows from Theorem 11.7 and Corollary 11.5. The implication $(iv) \Rightarrow (v)$ is obvious. An inspection of the proof of Theorem 11.7 shows that from (v) it follows (ii). Lastly, the implication $(iii) \Rightarrow (iv)$ is a consequence of Proposition 11.5, Remark 11.3 and the conditions of this theorem. □

The latter result can be improved for Marcinkiewicz spaces.

Theorem 11.9 *Let φ be a quasiconcave function on $[0, 1]$ such that $\delta_\varphi < 1$ and*

$$\lim_{t\to 0+} \ln^{1/2}(e/t)\varphi(t) = 0.$$

If X is the Marcinkiewicz space $M(\tilde{\varphi})$, then the following conditions are equivalent:

(a) $\mathcal{M}(X)$ is a s.s.;
(b) X satisfies condition (11.34);
(c) for some C_0 we have

$$\sqrt{n}\cdot\varphi(2^{-n}) \le C_0 \sup_{k\ge n} \sqrt{k-n}\cdot\varphi(2^{-k}), \quad n\in\mathbb{N}.$$

Moreover, if there exists $n_0 \in \mathbb{N}$ such that

$$c_\varphi := \sup_{n\ge n_0} \frac{\varphi(2^{-2n})}{\varphi(2^{-n})} \le \frac{1}{\sqrt{2}}, \tag{11.59}$$

then either of conditions (a)–(c) is equivalent to the following:

(d) $\varphi \in \Delta^2$.

Proof First of all, since $\delta_\varphi < 1$, then $\gamma_{\tilde{\varphi}} = 1 - \delta_\varphi > 0$ (see Appendix C) and so, by Theorem C.2, we have

$$\|x\|_X \asymp \sup_{0<t\le 1} x^*(t)\varphi(t). \tag{11.60}$$

Hence, from the quasiconcavity of φ, for each $n = 0, 1, \ldots$, it follows

$$\begin{aligned}\left\| \log_2^{1/2}(2^{1-n}/t)\chi_{[0,2^{-n}]}\right\|_X &\asymp \sup_{0<t\le 2^{-n}} \varphi(t)\log_2^{1/2}(2^{1-n}/t)\\ &\asymp \sup_{k\ge n} \sqrt{k-n}\cdot\varphi(2^{-k}).\end{aligned}$$

Therefore, by Remark 11.1(b), the space X satisfies (11.34) if and only if there is a constant $C > 0$ such that

$$\sup_{k\ge n} \sqrt{k}\cdot\varphi(2^{-k}) \le C \sup_{k\ge n} \sqrt{k-n}\cdot\varphi(2^{-k}), \quad n\in\mathbb{N}.$$

Since this estimate is equivalent to the inequality from condition (c), we get that $(b) \Leftrightarrow (c)$.

Further, from definition of the separable part X°, equivalence (11.60), and the hypothesis of the theorem it follows that $\ln^{1/2}(e/t) \in X^\circ$. Consequently, Theorem 11.7 yields the implication $(a) \Rightarrow (b)$.

Conversely, assume that (b) is fulfilled. Taking into account Proposition 11.6, one can see that for proving (a) it suffices to show that for some $C > 0$ and all functions f and g of the form

$$f(t) = \sum_{k=1}^{2^n} b_k \chi_{\Delta_n^k}(t) \log_2^{1/2}\left(\frac{2}{t}\right), \quad g(t) = \sum_{k=1}^{2^n} b_k \chi_{\Delta_n^k}(t) \log_2^{1/2}\left(\frac{2}{2^n t + 1 - k}\right),$$

where $n \in \mathbb{N}$ and $b_1 \geq b_2 \geq \cdots \geq b_{2^n} \geq 0$, we have

$$\|f\|_X \leq C\|g\|_X. \tag{11.61}$$

Indeed, on the one hand, using (11.60), the quasiconcavity of φ and condition (11.34), we get

$$\begin{aligned}\|f\|_X &\asymp \sup_{0\le j\le n} b_{2^j}\sqrt{n-j+1}\,\varphi(2^{j-n}) \leq \sup_{0\le j\le n} \|b_{2^j} \log_2^{1/2}(2/t)\chi_{[0,2^{j-n}]}\|_X \\ &\leq C \sup_{0\le j\le n} \|b_{2^j} \log_2^{1/2}(2^{j-n+1}/t)\chi_{[0,2^{j-n}]}\|_X.\end{aligned}$$

On the other hand, since

$$\log_2^{1/2}(2^{j-n+1}/t)\chi_{[0,2^{j-n}]}(t) = \sigma_{2^j}\left(\log_2^{1/2}(2^{1-n}/t)\chi_{[0,2^{-n}]}(t)\right)$$

(as usual, here σ_τ is the dilation operator), then the functions

$$\log_2^{1/2}(2^{j-n+1}/t)\chi_{[0,2^{j-n}]}(t) \text{ and } \sum_{i=1}^{2^j} \log_2^{1/2}\left(\frac{2}{2^n t + 1 - i}\right)\chi_{\Delta_n^i}(t)$$

are identically distributed. In consequence, from the inequalities $b_1 \geq \cdots \geq b_{2^j} \geq 0$ it follows

$$\|b_{2^j} \log_2^{1/2}(2^{j-n+1}/t)\chi_{[0,2^{j-n}]}\|_X \leq \left\| \sum_{i=1}^{2^j} b_i \log_2^{1/2}\left(\frac{2}{2^n t + 1 - i}\right)\chi_{\Delta_n^i}\right\|_X \leq \|g\|_X$$

for all $1 \leq j \leq n$. Clearly, the last estimates imply (11.61), and so (a) is proved.

Next, assuming that the function φ satisfies inequality (11.59) for $n \geq n_0$, let us prove that (c) implies (d).

By condition, for each $n \in \mathbb{N}$ we can find $k_n \geq n$ such that

$$\sqrt{n}\,\varphi(2^{-n}) \leq C_0\sqrt{k_n - n}\,\varphi(2^{-k_n}), \tag{11.62}$$

where C_0 does not depend on n. Hence, it follows that $\sqrt{n} \le C_0\sqrt{k_n - n}$, and so $k_n \ge \beta := 1 + C_0^{-2} \in (1, 2)$ (note that $C_0 > 1$). In the case when $\beta n \le k_n \le 2n$, by (11.62), we have

$$\varphi(2^{-n}) \le \sqrt{2}C_0\varphi(2^{-\beta n}).$$

Let now $k_n > 2n$. Then, $2^m n < k_n \le 2^{m+1}n$ for some $m \in \mathbb{N}$, and, in view of (11.59), for $n \ge n_0$ we have

$$\begin{aligned}\sqrt{k_n - n}\,\varphi(2^{-k_n}) &\le 2^{\frac{m+1}{2}}\sqrt{n}\,\varphi(2^{-2^m n})\\ &\le 2^{\frac{m+1}{2}}c_\varphi^{m-1}\sqrt{n}\,\varphi(2^{-2n})\\ &\le c_\varphi^{-1}\sqrt{2n}\,\varphi(2^{-2n}).\end{aligned}$$

Combining this inequality with (11.62), we get

$$\varphi(2^{-n}) \le \sqrt{2}C_0 c_\varphi^{-1}\varphi(2^{-2n})$$

if $k_n > 2n$ and $n \ge n_0$. Thus, the inequality

$$\varphi(2^{-n}) \le C\varphi(2^{-\beta n}),$$

where $C := c_\varphi^{-1}\sqrt{2}C_0$, holds for all $n \ge n_0$. Increasing, if it is necessary, the constant C, we arrive at the inequality, which is valid for all $n \in \mathbb{N}$. As a result, taking into account Remark 11.1(a), we get the condition (d).

Since from Proposition 11.5(3) it follows that (d) implies (b), everything is done. □

Example 11.4 Let us consider the exponential Orlicz space L_{N_p}, $p > 0$. As we already know (see Corollary 11.4), in the case when $p \ge 2$ we have $\mathcal{M}(L_{N_p}) = L_\infty$. Therefore, assume that $0 < p < 2$. Recall that, according to Proposition 2.4, L_{N_p} is simultaneously is the Marcinkiewicz space with the fundamental function $\varphi_p(t) = \ln^{-1/p}(e/t)$. Since $\delta_{\varphi_p} < 1$, $\lim_{t\to 0+}\ln^{1/2}(e/t)\varphi_p(t) = 0$ and φ_p satisfies condition (c) of Theorem 11.9, the Rademacher multiplicator space $\mathcal{M}(L_{N_p})$ is symmetric as well. In addition, by using Theorem 11.3, one can readily deduce that $\mathcal{M}(L_{N_p}) = L_{N_q}$ with $q = 2p/(2-p)$.

Example 11.5 Suppose that F is an increasing convex function on $[0, \infty)$ such that $F(0) = 0$. By $\operatorname{Exp} L^F$, as above (see Proposition 2.4), we denote the Orlicz space, corresponding to the function $e^{F(t)} - 1$. As was shown, this space coincides with the Marcinkiewicz space $M(\tilde{\varphi})$, where $\varphi(t) = 1/F^{-1}(\ln^{1/2}(e/t))$. If the function $F^{-1}(t)t^{-1/2}$ increases, then in the same way as in the preceding example, we can show, applying Theorems 11.3 and 11.9, that $\mathcal{M}(\operatorname{Exp} L^F) = \operatorname{Exp} L^\Psi$, where the function Ψ is defined by $\Psi^{-1}(t) = F^{-1}(t)t^{-1/2}$ (see details in [27, Theorem 10.6]).

In the final section of this chapter, we introduce and study the notion of tail Rademacher multiplicator space $\mathcal{M}_T(X)$ which, in a certain sense, complements the previously considered notion of symmetric kernel. Just like $\mathrm{Sym}(X)$ is being the largest s.s., *contained* in the Rademacher multiplicator space $\mathcal{M}(X)$, the space $\mathcal{M}_T(X)$ is the least s.s. with the Fatou property, *containing* $\mathcal{M}(X)$.

11.5 The Tail Rademacher Multiplicator Space and Its Properties

Let X be a s.s. on $[0, 1]$. Recall that

$$\Delta_n^k = [(k-1)2^{-n}, k2^{-n}), \quad n = 0, 1, \ldots, \quad k = 1, 2, \ldots, 2^n,$$

and $\mathcal{D}_n$ is the set of all step-functions of the form (11.36). It is clear that the union $\mathcal{D} := \bigcup_{n=1}^{\infty} \mathcal{D}_n$ is a linear subspace of X. As the first step to introduce the tail Rademacher multiplicator space, we define on $\mathcal{D}$ the following functional: if $f \in \mathcal{D}_n$, then

$$\|f\|_{\tilde{\mathcal{M}}_T} := \sup\left\{\left\|f \cdot \sum_{i=n+1}^{\infty} a_i r_i\right\|_X : \left\|\sum_{i=n+1}^{\infty} a_i r_i\right\|_X \le 1\right\}. \tag{11.63}$$

We claim that this quantity is well-defined, i.e., its value is independent of a representation f in the form (11.36). Indeed, if $f \in \mathcal{D}_n$, we have simultaneously $f \in \mathcal{D}_{n+j}$ for any $j = 1, 2, \ldots$ The function f is constant on each interval Δ_n^k, and the sums $\sum_{i=n+1}^{\infty} a_i r_i$ and $\sum_{i=n+j+1}^{\infty} a_{i-j} r_i$ restricted to such an interval are identically distributed. Since X is a s.s., this yields

$$\left\|f \cdot \sum_{i=n+1}^{\infty} a_i r_i\right\|_X = \left\|f \cdot \sum_{i=n+j+1}^{\infty} a_{i-j} r_i\right\|_X.$$

Therefore, the supremum in (11.63) may be taken equivalently over sums of the form $\sum_{i=n+j+1}^{\infty} b_i r_i$ rather than over sums $\sum_{i=n+1}^{\infty} a_i r_i$. This means that the quantity $\|f\|_{\tilde{\mathcal{M}}_T}$ is well-defined.

Applying a similar reasoning, one can check that the functional $f \mapsto \|f\|_{\tilde{\mathcal{M}}_T}$ is a norm on the linear space $\mathcal{D}$. Moreover, by definition, we get at once the inequality

$$\|f\|_{\tilde{\mathcal{M}}_T} \le \|f\|_{\mathcal{M}}, \quad f \in \mathcal{D}. \tag{11.64}$$

Let $\tilde{\mathcal{M}}_T(X)$ denote the completion of $\mathcal{D}$ with respect to the norm defined by (11.63).

Suppose first that $\ln^{1/2}(e/t) \notin X^\circ$. Then, from the assertion from the beginning of the proof of the implication $(iii) \Rightarrow (i)$ in Theorem 11.5 (see inequality (11.23)) it follows that there exists a constant $c_1 > 0$, which depends only on X, such that, for every $m \geq 0$, we can find $n_0 \geq 1$ with the following property: if $n \geq n_0$ and Δ is an arbitrary dyadic interval of rank m, then we have

$$\Big\|\chi_\Delta \sum_{i=m+1}^{m+n} r_i\Big\|_X \geq c_1 \Big\|\sum_{i=m+1}^{m+n} r_i\Big\|_X.$$

Hence, $\|\chi_\Delta\|_{\tilde{\mathcal{M}}_T} \geq c_1$ for each dyadic interval Δ. As a consequence, every Cauchy sequence $\{f_k\}$ of functions of the form (11.36) in the $\tilde{\mathcal{M}}_T(X)$-norm is a Cauchy sequence with respect to the uniform convergence on the interval $[0, 1]$ as well. In fact, if we suppose, on the contrary, that for some $\delta > 0$

$$\lim_{n\to\infty} \sup_{m>n} \|f_n - f_m\|_\infty \geq \delta,$$

then there exist $n_k \in \mathbb{N}$, $n_k \uparrow \infty$, $m_k > n_k$, and dyadic intervals Δ_k such that $|f_{n_k} - f_{m_k}| \geq \delta\chi_{\Delta_k}$ for all $k \in \mathbb{N}$. Consequently, $\|f_{n_k} - f_{m_k}\|_{\tilde{\mathcal{M}}_T} \geq c_1\delta$, $k = 1, 2, \ldots$, which contradicts the Cauchy condition of this sequence with respect to the $\tilde{\mathcal{M}}_T(X)$-norm. Thus, the space $\tilde{\mathcal{M}}_T(X)$ coincides, in this case, with the set of all uniform limits of sequences from $\mathcal{D}$. It is not hard to check that it consists of all functions f satisfying (possibly, after changing on a set of measure zero) the following property: for arbitrary $\varepsilon > 0$ there is $n \in \mathbb{N}$ such that the oscillation of f on each dyadic interval of rank n does not exceed ε, that is,

$$\sup_{t_1, t_2 \in \Delta_n^k} |f(t_1) - f(t_2)| \leq \varepsilon, \quad k = 1, 2, \ldots, 2^n.$$

It is obvious that then $\tilde{\mathcal{M}}_T(X)$ is not a Banach lattice and that

$$C[0, 1] \cup \mathcal{D} \overset{\neq}{\subset} \tilde{\mathcal{M}}_T(X) \overset{\neq}{\subset} L_\infty[0, 1],$$

where $C[0, 1]$ is the space of all continuous functions on $[0, 1]$.

Proceed now with considering the more interesting case when $\ln^{1/2}(e/t) \in X^\circ$. As we know (see Chap. 2), then the Rademacher sequence is equivalent in X to the unit vector basis in ℓ_2, and hence

$$\|f\|_{\tilde{\mathcal{M}}_T} \asymp \sup\Big\{\Big\|f \cdot \sum_{i=n+1}^{\infty} a_i r_i\Big\|_X : \sum_{i=n+1}^{\infty} a_i^2 \leq 1\Big\} \tag{11.65}$$

for all $f \in \mathcal{D}_n$. Thanks to this formula, there is another useful equivalent expression for the $\tilde{\mathcal{M}}_T$-norm on the space $\mathcal{D}$.

Lemma 11.7 *Let X be a s.s. on $[0,1]$ such that $\ln^{1/2}(e/t) \in X^{\circ}$. Then, for all $n \in \mathbb{N}$ and $f \in \mathcal{D}_n$ of the form* (11.36) *we have*

$$\|f\|_{\tilde{\mathcal{M}}_T} \asymp \left\| \sum_{k=1}^{2^n} a_k \log_2^{1/2}\left(\frac{2}{2^n t - k + 1}\right) \cdot \chi_{\Delta_n^k} \right\|_X$$

with a constant independent of n and f.

Proof Since $\ln^{1/2}(e/t) \in X^{\circ}$, then for any $n \in \mathbb{N}$, $k = 1, 2, \ldots, 2^n$, and every sequence $\{c_i\}_{i=n+1}^{\infty}$ such that $\sum_{i=n+1}^{\infty} c_i^2 \le 1$ from inequality (11.33) it follows

$$\left(\sum_{i=n+1}^{\infty} c_i r_i \cdot \chi_{\Delta_n^k}\right)^*(t) = \left(\sum_{i=1}^{\infty} c_{i+n} r_i\right)^*(2^n t)\chi_{[0,2^{-n}]}(t)$$
$$\le K' \log_2^{1/2}(2^{-n+1}/t)\chi_{[0,2^{-n}]}(t).$$

Let f be of the form (11.36). Taking into account that X is a s.s. and the intervals Δ_n^k, $k = 1, 2, \ldots, 2^n$, are pairwise disjoint, we deduce from the last inequality and (11.65) that

$$\|f\|_{\tilde{\mathcal{M}}_T} \le C \left\| \sum_{k=1}^{2^n} a_k \log_2^{1/2}\left(\frac{2}{2^n t - k + 1}\right) \cdot \chi_{\Delta_n^k} \right\|_X.$$

To prove the reverse inequality, we observe first that, in view of equivalence (11.65), for some constant $C_1 > 0$ it holds

$$\|f\|_{\tilde{\mathcal{M}}_T} \ge C_1^{-1} \sup_{m \in \mathbb{N}} \left\| \sum_{k=1}^{2^n} a_k \chi_{\Delta_n^k} \cdot \frac{1}{\sqrt{m}} \sum_{i=n+1}^{n+m} r_i \right\|_X$$
$$= C_1^{-1} \sup_{m \in \mathbb{N}} \left\| \sum_{k=1}^{2^n} a_k \chi_{\Delta_n^k} \cdot \frac{1}{\sqrt{m}} \left(\sum_{i=1}^{m} r_i\right)^*(2^n t - k + 1) \right\|_X.$$

Since X'' has the Fatou property, and, by Theorem A.1 (see also the proof of Theorem 2.3 and Remark 3.4), we have

$$\lim_{m\to\infty} \left(\frac{1}{\sqrt{m}} \sum_{i=1}^{m} r_i\right)^*(t) \ge \ln^{1/2} 2 \cdot \log_2^{1/2}\left(\frac{2}{e\sqrt{2\pi}\, t}\right), \quad 0 < t \le \frac{1}{e\sqrt{2\pi}},$$

then

$$\|f\|_{\tilde{\mathcal{M}}_T} \ge C_2^{-1} \left\| \sum_{k=1}^{2^n} a_k \log_2^{1/2}\left(\frac{2}{2^n t - k + 1}\right) \cdot \chi_{\Delta_n^k} \right\|_{X''}.$$

for some $C_2 > 0$. By assumption, $\ln^{1/2}(e/t) \in X^{\circ}$. Therefore, since $(X'')^{\circ} = X^{\circ}$ isometrically (see Lemma 2.2), we can replace here the norm of X'' with the norm of X, which finishes the proof. □

As before, X is a s.s. such that $\ln^{1/2}(e/t) \in X^{\circ}$. Using the last result, we can somewhat adjust the definition of the $\tilde{\mathcal{M}}_T(X)$-norm on the space $\mathcal{D}$.

For arbitrary $n \in \mathbb{N}$ and $f \in \mathcal{D}_n$ of the form (11.36), we set

$$\|f\|_{\mathcal{M}_T} := \Big\| \sum_{k=1}^{2^n} a_k \log_2^{1/2}\Big(\frac{2}{2^n t - k + 1}\Big) \cdot \chi_{\Delta_n^k} \Big\|_X. \tag{11.66}$$

As in the case of $\tilde{\mathcal{M}}_T(X)$-norm, this definition does not depend on a representation of f in the form (11.36). Moreover, one can readily check that the functional $f \mapsto \|f\|_{\mathcal{M}_T}$ is a norm on $\mathcal{D}$.

By Lemma 11.7, the $\tilde{\mathcal{M}}_T(X)$- and $\mathcal{M}_T(X)$-norms are equivalent on $\mathcal{D}$. Hence, the space $\tilde{\mathcal{M}}_T(X)$ may by treated as the completion of $\mathcal{D}$ with respect to the $\mathcal{M}_T(X)$-norm.

Lemma 11.8 *Let X be a s.s. on $[0, 1]$ such that $\ln^{1/2}(e/t) \in X^{\circ}$. Suppose that $E \subset [0, 1]$ is a measurable set of positive measure. Then, $\chi_E \in \tilde{\mathcal{M}}_T(X)$ and*

$$\|\chi_E\|_{\mathcal{M}_T} = \|\log_2^{1/2}(2m(E)/t)\,\chi_{[0,m(E)]}\|_X.$$

Proof Since E is measurable, there exists an increasing sequence of positive integers $\{n_k\}_{k=1}^{\infty}$ such that for some sequence of sets of the form

$$E_k = \bigcup_{j \in J_k} \Delta_{n_k}^j, \quad J_k \subset \{1, 2, \ldots, 2^{n_k}\},$$

we have $m(E \Delta E_k) \le 1/k$, $k \in \mathbb{N}$ (here, $A \Delta B = (A \setminus B) \cup (B \setminus A)$). Obviously, $\{\chi_{E_k}\} \subset \mathcal{D}$. Moreover, for any $k \ge l$

$$\chi_{E_k} - \chi_{E_l} = \sum_{j \in J_k} \chi_{\Delta_{n_k}^j} - \sum_{j \in J_l} \chi_{\Delta_{n_l}^j} = \sum_{j \in J_{k,l}} \pm \chi_{\Delta_{n_k}^j},$$

where $J_{k,l}$ is some subset of the set $\{1, 2, \ldots, 2^{n_k}\}$. We have

$$\sum_{j \in J_{k,l}} m(\Delta_{n_k}^j) = m(E_k \Delta E_l) \le m(E_k \Delta E) + m(E \Delta E_l) \le \frac{2}{l}.$$

Furthermore, from this inequality and (11.66) it follows that

$$\begin{aligned}\|\chi_{E_k} - \chi_{E_l}\|_{\mathcal{M}_T} &= \Big\| \sum_{j \in J_{k,l}} \pm \chi_{\Delta_{n_k}^j} \Big\|_{\mathcal{M}_T} \\ &= \Big\| \sum_{j \in J_{k,l}} \log_2^{1/2} \Big(\frac{2}{2^{n_k} t - j + 1} \Big) \cdot \chi_{\Delta_{n_k}^j} \Big\|_X \\ &\le \Big\| \log_2^{1/2} (4/(lt)) \cdot \chi_{[0,2/l]} \Big\|_X .\end{aligned}$$

Since X° is separable, the hypothesis $\ln^{1/2}(e/t) \in X^\circ$ implies that the right-hand side of the latter inequality tends to zero as $l \to \infty$. Therefore, $\{\chi_{E_k}\}$ is a Cauchy sequence in the $\mathcal{M}_T(X)$-norm. Taking into account that $\chi_{E_k} \to \chi_E$ in measure, we get $\chi_E \in \mathcal{M}_T(X)$ and $\|\chi_E - \chi_{E_k}\|_{\mathcal{M}_T} \to 0$. Moreover, using (11.66) once more, we have

$$\begin{aligned}\|\chi_{E_k}\|_{\mathcal{M}_T} &= \Big\| \sum_{j \in J_k} \log_2^{1/2} \Big(\frac{2}{2^{n_k} t - j + 1} \Big) \cdot \chi_{\Delta_{n_k}^j} \Big\|_X \\ &= \Big\| \log_2^{1/2} \Big(\frac{2 \operatorname{card}(J_k)}{2^{n_k} t} \Big) \cdot \chi_{[0,2^{-n_k} \operatorname{card}(J_k)]} \Big\|_X .\end{aligned} \tag{11.67}$$

Since $\operatorname{card}(J_k) \cdot 2^{-n_k} \to m(E)$, the sequence of functions

$$y_k(t) := \log_2^{1/2} \Big(\frac{2 \operatorname{card}(J_k)}{2^{n_k} t} \Big) \cdot \chi_{[0,2^{-n_k} \operatorname{card}(J_k)]}(t), \quad k = 1, 2, \ldots$$

converges pointwise to the function $y(t) := \log_2^{1/2} (2m(E)/t) \cdot \chi_{[0,m(E)]}(t)$. Therefore, from the inequality $|y_k(t) - y(t)| \le \log_2^{1/2}(2/t)$ and the hypothesis $\log_2^{1/2}(2/t) \in X^\circ$ it follows that $\|y_k - y\|_X \to 0$ as $k \to \infty$. Since $\|\chi_{E_k}\|_{\mathcal{M}_T} \to \|\chi_E\|_{\mathcal{M}_T}$, we obtain the desired result by taking $k \to \infty$ in (11.67). □

Lemma 11.9 *If X is a s.s. on $[0, 1]$ such that $\ln^{1/2}(e/t) \in X^\circ$, then $L_\infty \subset \tilde{\mathcal{M}}_T(X)$. Moreover, the $\mathcal{M}_T(X)$-norm is monotone and symmetric on L_∞, that is,*

(i) from $|f| \le |g|$ and $g \in L_\infty$ it follows $\|f\|_{\mathcal{M}_T} \le \|g\|_{\mathcal{M}_T}$;
(ii) $\|g\|_{\mathcal{M}_T} = \|f\|_{\mathcal{M}_T}$ for arbitrary equimeasurable functions $f, g \in L_\infty$.

In addition, if $0 \le f_n \uparrow f$, $f \in L_\infty$, then we have $\|f_n\|_{\mathcal{M}_T} \to \|f\|_{\mathcal{M}_T}$ as $n \to \infty$.

Proof Suppose first that f is a step-function, i.e.,

$$f = \sum_{j=1}^{m} \alpha_j \chi_{E_j},$$

where the sets E_j, $j = 1, 2, \ldots, m$, are pairwise disjoint. Since $\tilde{\mathcal{M}}_T(X)$ is a linear space, from Lemma 11.8 it follows $f \in \tilde{\mathcal{M}}_T(X)$. Let us prove that

$$\|f\|_{\mathcal{M}_T} = \left\| \sum_{j=1}^{m} \alpha_j \log_2^{1/2}\left(\frac{2m(E_j)}{t - \beta_{j-1}}\right)\chi_{(\beta_{j-1}, \beta_j]} \right\|_X, \tag{11.68}$$

where $\beta_j := \sum_{s=1}^{j} m(E_s)$, $j = 1, 2, \ldots, m$, and $\beta_0 = 0$.

As in the proof of Lemma 11.8, we apply an approximation argument. For every $j = 1, 2, \ldots, m$ we take a sequence of sets $\{E_j^k\}_{k=1}^{\infty}$ such that $\chi_{E_j^k} \in \mathcal{D}$ and $m(E_j \triangle E_j^k) \to 0$ as $k \to \infty$. Then, $h_k := \sum_{j=1}^{m} \alpha_j \chi_{E_j^k} \in \mathcal{D}$, $k = 1, 2, \ldots$, and from the convergence $\chi_{E_j^k} \to \chi_{E_j}$ in $\mathcal{M}_T(X)$ it follows that $\|h_k\|_{\mathcal{M}_T} \to \|f\|_{\mathcal{M}_T}$ as $k \to \infty$. Thus, to prove (11.68) it suffices to show that $\lim_{k\to\infty} \|h_k\|_{\mathcal{M}_T}$ coincides with the right-hand side of formula (11.68).

Taking into account that the sets $E_1^k, E_2^k, \ldots, E_m^k$ may be overlapped for a fixed $k \in \mathbb{N}$, we set

$$D_k := \bigcup_{1 \le j_1 < j_2 \le m} E_{j_1}^k \cap E_{j_2}^k.$$

Since E_j are pairwise disjoint and $m(E_j \triangle E_j^k) \to 0$ as $k \to \infty$ for every $j = 1, 2, \ldots, m$, we have $m(D_k) \to 0$. Moreover, for each $k \in \mathbb{N}$

$$h_k = \sum_{j=1}^{m} \alpha_j \chi_{E_j^k \setminus D_k} + \sum_{j=1}^{m} \alpha_j \chi_{E_j^k \cap D_k}.$$

Observe that the sets $E_j^k \setminus D_k$, $j = 1, 2, \ldots, m$, are pairwise disjoint and the functions $h_k' := \sum_{j=1}^{m} \alpha_j \chi_{E_j^k \setminus D_k}$, $k = 1, 2, \ldots$, belong to $\mathcal{D}$. Hence, by (11.66),we have

$$\|h_k'\|_{\mathcal{M}_T} = \left\| \sum_{j=1}^{m} \alpha_j \log_2^{1/2}\left(\frac{2m(E_j^k \setminus D_k)}{t - \beta_j^{k-1}}\right)\chi_{(\beta_{j-1}^k, \beta_j^k]} \right\|_X,$$

where $\beta_j^k := \sum_{s=1}^{j} m(E_s^k \setminus D_k)$ if $j = 1, 2, \ldots, m$, and $\beta_0^k = 0$. Since $m(D_k) \to 0$ and $m(E_j \triangle E_j^k) \to 0$, then $\beta_j^k \to \beta_j$ as $k \to \infty$ for every $j = 1, 2, \ldots, m$. Consequently, precisely as in the proof of Lemma 11.8, we get

$$\lim_{k\to\infty} \|h_k'\|_{\mathcal{M}_T} = \left\| \sum_{j=1}^{m} \alpha_j \log_2^{1/2}\left(\frac{2m(E_j)}{t - \beta_{j-1}}\right)\chi_{(\beta_{j-1}, \beta_j]} \right\|_X,$$

where $\beta_j = \sum_1^j m(E_s)$, $j = 1, 2, \ldots, m$, and $\beta_0 = 0$. On the other hand, from Lemma 11.8 it follows that

$$\Big\| \sum_{j=1}^{m} \alpha_j \chi_{E_j^k \cap D_k} \Big\|_{\mathcal{M}_T} \le \sum_{j=1}^{m} |\alpha_j| \cdot \|\chi_{E_j^k \cap D_k}\|_{\mathcal{M}_T}$$
$$\le \sum_{j=1}^{m} |\alpha_j| \cdot \Big\| \log_2^{1/2}(2m(D_k)/t)\, \chi_{[0,m(D_k)]} \Big\|_X .$$

Since $\log_2^{1/2}(2/t) \in X^\circ$ and $m(D_k) \to 0$, the right-hand side of this inequality tends to zero as $k \to \infty$. Therefore, $\lim_{k\to\infty} \|h_k\|_{\mathcal{M}_T} = \lim_{k\to\infty} \|h_k'\|_{\mathcal{M}_T}$, and so (11.68) is proved.

Suppose now that $f \in L_\infty$ is an arbitrary function and $\{f_k\}$ is a sequence of measurable step-functions uniformly converging to f. Setting $f_k := \sum_{j=1}^{m_k} \alpha_j^k \chi_{F_j^k}$, $k = 1, 2, \ldots$, without loss of generality, we can assume that, for every $j \in \{1, 2, \ldots, m_{k+1}\}$, there is $i \in \{1, 2, \ldots, m_k\}$ such that $F_j^{k+1} \subset F_i^k$. Then, taking into account (11.68), for all $k > l$ we get

$$\|f_k - f_l\|_{\mathcal{M}_T} \le \|f_k - f_l\|_\infty \Big\| \sum_{j=1}^{m_k} \log_2^{1/2}\Big(\frac{2m(F_j^k)}{t - \gamma_{j-1}^k}\Big) \chi_{[\gamma_{j-1}^k, \gamma_j^k)} \Big\|_X$$
$$\le \|f_k - f_l\|_\infty \| \log_2^{1/2}(2/t)\|_X,$$

where $\gamma_j^k := \sum_{s=1}^j m(F_s^k)$ if $j = 1, 2, \ldots, m_k$, and $\gamma_0^k = 0$. Therefore, $\{f_k\}$ is a Cauchy sequence in the $\mathcal{M}_T(X)$-norm, whence $f \in \tilde{\mathcal{M}}_T(X)$ and $\|f\|_{\mathcal{M}_T} = \lim_{k\to\infty} \|f_k\|_{\mathcal{M}_T}$.

Let $|f| \le |g|$, with $g \in L_\infty$. One can easily construct sequences of step-functions $\{f_k\}$ and $\{g_k\}$, converging uniformly to f and g, respectively, such that $|f_k| \le |g_k|$ for each $k = 1, 2, \ldots$ Then, according to (11.68), we have $\|f_k\|_{\mathcal{M}_T} \le \|g_k\|_{\mathcal{M}_T}$, $k = 1, 2, \ldots$ Since $\|f\|_{\mathcal{M}_T} = \lim_{k\to\infty} \|f_k\|_{\mathcal{M}_T}$ and $\|g\|_{\mathcal{M}_T} = \lim_{k\to\infty} \|g_k\|_{\mathcal{M}_T}$, this yields $\|f\|_{\mathcal{M}_T} \le \|g\|_{\mathcal{M}_T}$. In a similar way, one can prove also the last assertion of the lemma.

Finally, suppose that functions g and f are equimeasurable. It is not hard to find sequences of step-functions $\{f_k\}$ and $\{g_k\}$, converging uniformly to f and g, respectively, such that f_k and g_k are equimeasurable for every $k \in \mathbb{N}$. Moreover, by (11.68), we have $\|g_k\|_{\mathcal{M}_T} = \|f_k\|_{\mathcal{M}_T}$, $k = 1, 2, \ldots$, and also $\|f_k\|_{\mathcal{M}_T} \to \|f\|_{\mathcal{M}_T}$ and $\|g_k\|_{\mathcal{M}_T} \to \|g\|_{\mathcal{M}_T}$. Thus, $\|g\|_{\mathcal{M}_T} = \|f\|_{\mathcal{M}_T}$, and the proof is completed. □

Remark 11.4 As above, assume that X is a s.s. such that $\ln^{1/2}(e/t) \in X^\circ$. Then, by Theorem 11.3, we have

$$\|f\|_{\mathrm{Sym}} \asymp \|f^*(t) \ln^{1/2}(e/t)\|_{X''}, \tag{11.69}$$

where $\|f\|_{\mathrm{Sym}}$ is the norm of f in the symmetric kernel Sym (X) of the Rademacher multiplicator space $\mathcal{M}(X)$. Using (11.69), in the same way as in the proof of Lemmas 11.8 and 11.9, one can prove that the set $\mathcal{D}$ is dense in L_∞ with respect to the Sym (X)-norm.

Now, everything is ready in order to introduce definition of the tail Rademacher multiplicator space.

Let X be a s.s. on [0, 1] such that $\ln^{1/2}(e/t) \in X^\circ$. We define the linear space

$$\mathcal{M}_T(X) := \left\{ f \in X : \lim_{n\to\infty} \| f \cdot \chi_{\{t:|f(t)|\le n\}} \|_{\mathcal{M}_T} < \infty \right\},$$

equipped with the norm

$$\|f\|_{\mathcal{M}_T} := \lim_{n\to\infty} \| f \cdot \chi_{\{t:|f(t)|\le n\}} \|_{\mathcal{M}_T}. \tag{11.70}$$

Proposition 11.9 *Suppose that X is a s.s. with the Fatou property, $\ln^{1/2}(e/t) \in X^\circ$. Then, $\mathcal{M}_T(X)$ is a s.s. on [0, 1] with the Fatou property as well. Moreover, $\tilde{\mathcal{M}}_T(X)$ is the separable part of $\mathcal{M}_T(X)$, $\tilde{\mathcal{M}}_T(X) \subseteq X^\circ$ and $\mathcal{M}(X) \subseteq \mathcal{M}_T(X) \subseteq X$.*

Proof First of all, from Lemma 11.9 it follows that the $\mathcal{M}_T(X)$-norm has the lattice property and symmetric on L_∞. By definition, one can easily extend these properties to the whole space $\mathcal{M}_T(X)$. Let us verify now that the normed function lattice $\mathcal{M}_T(X)$ has the Fatou property.

To this end, we assume that $0 \le f_k \uparrow f$, $f_k \in \mathcal{M}_T(X)$ and $\sup_{k=1,2,\dots} \|f_k\|_{\mathcal{M}_T} < \infty$. We need to show that $f \in \mathcal{M}_T(X)$ and $\|f\|_{\mathcal{M}_T} = \lim_{k\to\infty} \|f_k\|_{\mathcal{M}_T}$. Indeed, since for every $n \in \mathbb{N}$ we have $f_k \cdot \chi_{\{f_k \le n\}} \uparrow f \cdot \chi_{\{f\le n\}}$ as $k \to \infty$, then Lemma 11.9 yields $f \cdot \chi_{\{f\le n\}} \in \mathcal{M}_T(X)$ and

$$\|f \cdot \chi_{\{f\le n\}}\|_{\mathcal{M}_T} = \lim_{k\to\infty} \|f_k \cdot \chi_{\{f_k\le n\}}\|_{\mathcal{M}_T} = \sup_{k=1,2,\dots} \|f_k \cdot \chi_{\{f_k\le n\}}\|_{\mathcal{M}_T}.$$

Therefore, using the above definition, we get

$$\begin{aligned}
\|f\|_{\mathcal{M}_T} &= \lim_{n\to\infty} \|f \cdot \chi_{\{f\le n\}}\|_{\mathcal{M}_T} = \sup_{n=1,2,\dots} \|f \cdot \chi_{\{f\le n\}}\|_{\mathcal{M}_T} \\
&= \sup_{n=1,2,\dots} \sup_{k=1,2,\dots} \|f_k \cdot \chi_{\{f_k\le n\}}\|_{\mathcal{M}_T} = \sup_{k=1,2,\dots} \sup_{n=1,2,\dots} \|f_k \cdot \chi_{\{f_k\le n\}}\|_{\mathcal{M}_T} \\
&= \sup_{k=1,2,\dots} \|f_k\|_{\mathcal{M}_T} = \lim_{k\to\infty} \|f_k\|_{\mathcal{M}_T} < \infty,
\end{aligned}$$

and so the desired property of $\mathcal{M}_T(X)$ is proved.

It is well-known that each normed function lattice with the Fatou property is complete (see e.g. [151, Theorems 6.1.2 and 6.1.7]). Consequently, $\mathcal{M}_T(X)$ is a s.s. with the Fatou property.

Next, according to Lemma 11.9 $L_\infty \subset \tilde{\mathcal{M}}_T(X)$, and hence the space $\tilde{\mathcal{M}}_T(X)$, as the completion of $\mathcal{D}$, coincides with the closure of L_∞ in $\mathcal{M}_T(X)$. In other words, $\tilde{\mathcal{M}}_T(X)$ is the separable part of $\mathcal{M}_T(X)$.

Furthermore, taking into account Remark 11.4, one can find, for every $f \in L_\infty$, a sequence $\{f_k\} \subset \mathcal{D}$ such that $\|f_k - f\|_{\mathrm{Sym}} \to 0$. In view of (11.64), the inequality $\|f\|_{\mathcal{M}_T} \le C\|f\|_{\mathcal{M}}$ holds with some constant $C > 0$ for any $f \in \mathcal{D}$. Therefore, since $\mathrm{Sym}(X) \subset \mathcal{M}(X)$, we have $\|f_k - f\|_{\mathcal{M}} \to 0$ and $\|f_k - f\|_{\mathcal{M}_T} \to 0$. Hence, $\|f\|_{\mathcal{M}_T} \le C\|f\|_{\mathcal{M}}$ for all $f \in L_\infty$. Suppose now that $f \in \mathcal{M}(X)$ is arbitrary. Writing the last inequality for the truncations $f^m := \min(|f|, m)$, $m = 1, 2, \ldots$, and taking into account the definition of the $\mathcal{M}_T(X)$-norm, we can extend the above inequality to the whole space $\mathcal{M}(X)$ by passing to the limit as $m \to \infty$. Thus, $\mathcal{M}(X) \subset \mathcal{M}_T(X)$.

Finally, equation (11.66) indicates that $\|f\|_X \le \|f\|_{\mathcal{M}_T}$ for $f \in \mathcal{D}$. Consequently, by definition, we get $\tilde{\mathcal{M}}_T(X) \subset X^\circ$. Since X possesses the Fatou property, the continuous embedding $\mathcal{M}_T(X) \subset X$ is a direct consequence of the definition of $\mathcal{M}_T(X)$. □

Remark 11.5 If $\ln^{1/2}(e/t) \in X^\circ$, then Lemma 11.8 implies the following expression for the fundamental function of the s.s. $\mathcal{M}_T(X)$:

$$\phi_{\mathcal{M}_T(X)}(u) = \|\log_2^{1/2}(2u/t) \cdot \chi_{[0,u]}\|_X, \quad 0 < u \le 1.$$

As we know, the largest s.s., contained in the multiplicator space $\mathcal{M}(X)$, is its symmetric kernel $\mathrm{Sym}(X)$. The next statement shows that the space $\mathcal{M}_T(X)$ is, conversely, the smallest s.s. with the Fatou property that contains the space $\mathcal{M}(X)$.

Theorem 11.10 *Let X be a s.s. on $[0, 1]$, $\ln^{1/2}(e/t) \in X^\circ$. If Y is a s.s. with the Fatou property such that $Y \supset \mathcal{M}(X)$, then $Y \supset \mathcal{M}_T(X)$.*

Proof Let $n \in \mathbb{N}$ and $f \in \mathcal{D}_n$ be arbitrary. The key point of the proof consists in finding a suitable function $f' \in \mathcal{D}_{2n}$, equimeasurable with f and satisfying the estimate:

$$\|f'\|_{\mathcal{M}} \le A\|f\|_{\mathcal{M}_T}, \tag{11.71}$$

with a constant $A > 0$ independent of $n \in \mathbb{N}$ and $f \in \mathcal{D}_n$.

Given $n \in \mathbb{N}$ we denote by θ_{ij} the (constant) value of the function r_j on the dyadic interval Δ_{2n}^i, $j = 1, \ldots, 2n$, $i = 1, \ldots, 2^{2n}$. Furthermore, for each sign arrangement $\varepsilon = (\varepsilon_j)_{j=1}^n \in \{1, -1\}^n$ we define the set

$$\Omega_n^\varepsilon := \big\{i \in \{1, \ldots, 2^{2n}\} : \theta_{i,j+n} = \varepsilon_j \cdot \theta_{i,j} \quad \text{for all} \quad j = 1, \ldots, n\big\}. \tag{11.72}$$

It is clear that $\Omega_n^{\varepsilon^1} \cap \Omega_n^{\varepsilon^2} = \varnothing$ whenever $\varepsilon^1, \varepsilon^2 \in \{1, -1\}^n$ are such that $\varepsilon^1 \neq \varepsilon^2$, and

$$\bigcup_{\varepsilon \in \{1,-1\}^n} \Omega_n^{\varepsilon} = \{1, 2, \ldots, 2^{2n}\}.$$

Moreover, for every $\varepsilon \in \{1, -1\}^n$ we put

$$U_n^{\varepsilon} := \bigcup_{i \in \Omega_n^{\varepsilon}} \Delta_{2n}^i.$$

Since for each ε the set Ω_n^{ε} contains 2^n elements, we have $m(U_n^{\varepsilon}) = 2^{-n}$. Observe also that $U_n^{\varepsilon_1} \cap U_n^{\varepsilon_2} = \varnothing$ if $\varepsilon^1 \neq \varepsilon^2$, and

$$\bigcup_{\varepsilon \in \{1,-1\}^n} U_n^{\varepsilon} = [0, 1].$$

Finally, given $f = \sum_{k=1}^{2^n} c_k \chi_{\Delta_n^k} \in \mathcal{D}_n$, where $c_k \geq 0$, $k = 1, 2, \ldots, 2^n$, we define the function f' by

$$f' = \sum_{k=1}^{2^n} c_k \chi_{U_n^{\varepsilon^k}}, \tag{11.73}$$

where $\{\varepsilon^k\}_{k=1}^{2^n}$ is an arbitrary enumeration of the set $\{1, -1\}^n$. Obviously, f' is equimeasurable with f and $f' \in \mathcal{D}_{2n}$. It remains to prove (11.71).

To this end, we need to estimate $\|f'\|_{\mathcal{M}}$ from above. Since $\ln^{1/2}(e/t) \in X$, then from Theorem 2.2 and Remark 2.1 it follows that the Rademacher system is equivalent in X to the unit vector basis in ℓ_2. Consequently, taking into account definition of the $\mathcal{M}(X)$-norm, for some $C' > 0$ we have

$$\begin{aligned}\|f'\|_{\mathcal{M}} &\leq C' \sup\left\{\left\|f' \cdot \sum_{j=1}^{2n} a_j r_j\right\|_X : \sum_{j=1}^{2n} a_j^2 \leq 1\right\} \\ &+ C' \sup\left\{\left\|f' \cdot \sum_{j=2n+1}^{\infty} a_j r_j\right\|_X : \sum_{j=2n+1}^{\infty} a_j^2 \leq 1\right\}. \end{aligned} \tag{11.74}$$

One can readily see that the functions $f' \cdot \sum_{j=2n+1}^{\infty} a_j r_j$ and $f \cdot \sum_{j=n+1}^{\infty} a_{j+n} r_j$ are identically distributed. Therefore, as above, by (11.33) and the inequality $\sum_{j=2n+1}^{\infty} a_j^2 \leq 1$, we get

$$\left(f' \cdot \sum_{j=2n+1}^{\infty} a_j r_j\right)^*(t) \leq K'\left(\sum_{k=1}^{2^n} c_k \chi_{\Delta_n^k} \log_2^{1/2} \frac{2}{2^n t - k + 1}\right)^*(t), \quad 0 < t \leq 1.$$

Hence, from definition of the $\mathcal{M}_T(X)$-norm it follows

$$\sup\left\{\left\|f'\cdot\sum_{j=2n+1}^{\infty}a_jr_j\right\|_X:\sum_{j=2n+1}^{\infty}a_j^2\le 1\right\}\le K'\|f\|_{\mathcal{M}_T}. \tag{11.75}$$

We prove now a similar estimate for the first term from the right-hand side of inequality (11.74). First of all, in view of the fact that θ_{ij} is the value of r_j on the interval Δ_{2n}^i, by (11.73), we have

$$g:=f'\cdot\sum_{j=1}^{2n}a_jr_j=\sum_{k=1}^{2^n}c_k\sum_{i\in\Omega_n^{\varepsilon^k}}b_i\chi_{\Delta_{2n}^i},$$

where for every $i=1,2,\ldots,2^{2n}$ we put

$$b_i=\sum_{j=1}^{2n}a_j\theta_{ij}.$$

According to definition (11.72), for $i\in\Omega_n^{\varepsilon^k}$, we have

$$b_i=\sum_{j=1}^{n}a_j\theta_{ij}+\sum_{j=1}^{n}a_{j+n}\varepsilon_j^k\theta_{ij}=\sum_{j=1}^{n}(a_j+a_{j+n}\varepsilon_j^k)\theta_{ij}.$$

Moreover, since $\sum_{j=1}^{2n}a_j^2\le 1$, for all $k=1,2,\ldots,2^n$ it holds

$$\left(\sum_{j=1}^{n}(a_j+a_{j+n}\varepsilon_j^k)^2\right)^{1/2}\le\sqrt{2}. \tag{11.76}$$

Therefore, applying the Cauchy-Schwarz-Bunyakovskii inequality, we get

$$|b_i|\le\sqrt{2n},\quad i=1,2,\ldots,2^{2n}. \tag{11.77}$$

Further, for arbitrary $k=1,2,\ldots,2^n$ and $j=1,2,\ldots,n$, we define the sets

$$B_n^k(j):=\left\{i\in\Omega_n^{\varepsilon^k}:|b_i|\ge\sqrt{n-j+1}\right\}.$$

Observe that for each $k=1,2,\ldots,2^n$ the set of sequences $\{\theta_{is}\}_{s=1}^n$, where $i\in\Omega_n^{\varepsilon^k}$, coincides with the family of all sign arrangements. In consequence, from the exponential distribution estimate from Proposition 1.2 and inequality (11.76) it follows that

$$\begin{aligned}\operatorname{card} B_n^k(j) &= \operatorname{card}\Big\{i \in \{1,\dots,2^n\} : \Big|\sum_{s=1}^n (a_s + a_{s+n}\varepsilon_s^k)\theta_{is}\Big| \ge \sqrt{n-j+1}\Big\} \\ &= 2^n m\Big(\Big\{t \in [0,1] : \Big|\sum_{s=1}^n (a_s + a_{s+n}\varepsilon_s^k) r_s(t)\Big| \ge \sqrt{n-j+1}\Big\}\Big) \\ &\le 2{\cdot}2^n{\cdot}e^{-(n-j+1)/4} < 2{\cdot}2^n{\cdot}2^{-(n-j+1)/4}.\end{aligned}$$

Note that the latter estimate holds also for $j = 0$ and $j = n+1$ if we put $B_n^k(0) := \varnothing$ and $B_n^k(n+1) := \Omega_n^{\varepsilon^k}$. Thus, for arbitrary $k = 1, 2, \dots, 2^n$ and $j = 1, 2, \dots, n+1$, we have

$$\begin{aligned}m\Big(\bigcup_{i\in B_n^k(j)\setminus B_n^k(j-1)} \Delta_{2n}^i\Big) &\le m\Big(\bigcup_{i\in B_n^k(j)} \Delta_{2n}^i\Big) \\ &= 2^{-2n}\operatorname{card} B_n^k(j) \\ &\le 2{\cdot}2^{-n}{\cdot}2^{-(n-j+1)/4}.\end{aligned} \tag{11.78}$$

Next, for every $j = 1, 2, \dots, n+1$ we consider the functions

$$g_j := \sum_{k=1}^{2^n} c_k \sum_{i\in B_n^k(j)\setminus B_n^k(j-1)} b_i \chi_{\Delta_{2n}^i}$$

and

$$h_j := \sum_{k=1}^{2^n} c_k \chi_{E_k^j}, \quad \text{where} \quad E_k^j = \bigcup_{i\in B_n^k(j)\setminus B_n^k(j-1)} \Delta_{2n}^i.$$

Moreover, let

$$\alpha_j := 2^{-5n/4}(2^{j/4} - 1), \quad j = 0, 1, 2, \dots, n, \quad \alpha_{n+1} = 2^{-n},$$

and

$$v_j := \sum_{k=1}^{2^n} c_k \log_2^{1/2}\Big(\frac{2}{2^n t - k + 1}\Big) \chi_{[\frac{k-1}{2^n}+\alpha_{j-1}, \frac{k-1}{2^n}+\alpha_j]}, \quad j = 1, 2, \dots, n.$$

Let us estimate first the function g_1^*. Noting that $\alpha_1 = 2^{-5n/4}(2^{1/4}-1)$, by the definition of v_1 and the monotonicity of the logarithm, one can readily see that

$$|v_1(t)| \geq \frac{\sqrt{n}}{2} \cdot \sum_{k=1}^{2^n} |c_k| \chi_{[\frac{k-1}{2^n}, \frac{k-1}{2^n}+\alpha_1]}(t).$$

Since the function h_1 takes the value c_k on the set $B_n^k(1)$, $k = 1, 2, \ldots, 2^n$, of measure not exceeding $2{\cdot}2^{-5n/4}$ (see (11.78)), the last inequality yields

$$h_1^*(t) \leq \frac{2}{\sqrt{n}} \cdot (\sigma_{2(2^{1/4}-1)^{-1}} v_1)^*(t), \quad 0 < t \leq 1,$$

where $\sigma_\tau x(t) := x(t/\tau)$. On the other hand, from (11.77) it follows

$$|g_1(t)| \leq \sqrt{2n}\, |h_1(t)|, \quad 0 < t \leq 1,$$

and hence

$$g_1^*(t) \leq 2\sqrt{2}(\sigma_{2(2^{1/4}-1)^{-1}} v_1)^*(t), \quad 0 < t \leq 1. \tag{11.79}$$

Proceed with estimation of the rearrangements g_j^*, $j = 2, 3, \ldots, n$. Again, taking into account the definition of v_j and the fact that

$$\log_2^{1/2}\left(2^{1-n}/\alpha_j\right) = \log_2^{1/2}\left(\frac{2 \cdot 2^{n/4}}{2^{j/4}-1}\right) \geq \left(1 + \frac{n}{4} - \frac{j}{4}\right)^{1/2} \geq \frac{1}{2}\sqrt{n-j+2},$$

we deduce

$$|v_j(t)| \geq \frac{1}{2}\sqrt{n-j+2} \sum_{k=1}^{2^n} |c_k| \chi_{[\frac{k-1}{2^n}+\alpha_{j-1}, \frac{k-1}{2^n}+\alpha_j]}(t), \quad 0 < t \leq 1.$$

Since $\alpha_j - \alpha_{j-1} = 2^{-n}{\cdot}2^{-(n-j+1)/4}(2^{1/4}-1)$, then combining this inequality with estimate (11.78) and the definition of h_j, we obtain

$$h_j^*(t) \leq \frac{2}{\sqrt{n-j+2}}(\sigma_{2(2^{1/4}-1)^{-1}} v_j)^*(t), \quad 0 < t \leq 1.$$

At the same time, by definition of the sets $B_n^k(j)$,

$$|b_i| \leq \sqrt{n-j+2} \quad \text{if} \quad i \notin B_n^k(j-1),$$

whence

$$|g_j(t)| \leq \sqrt{n-j+2}\,|h_j(t)|, \quad 0 < t \leq 1.$$

Consequently,

$$g_j^*(t) \leq 2(\sigma_{2(2^{1/4}-1)^{-1}} v_j)^*(t), \quad 0 < t \leq 1,$$

for all $j = 2, 3, \dots, n$. Observe that the functions g_j (resp. v_j) are pairwise disjoint. Therefore, from the last inequality and (11.79) it follows

$$\Big(\sum_{j=1}^{n} g_j\Big)^*(t) \leq 2\sqrt{2}(\sigma_{2(2^{1/4}-1)^{-1}} v)^*(t), \quad 0 < t \leq 1, \tag{11.80}$$

where

$$v := \sum_{k=1}^{2^n} c_k \log_2^{1/2}\left(\frac{2}{2^n t - k + 1}\right) \chi_{\Delta_n^k}.$$

It remains to estimate only the rearrangement g_{n+1}^*. For $i \in \Omega_n^{\varepsilon^k} \setminus B_n^k(n)$ we have $|b_i| \leq 1$, whence $|g_{n+1}(t)| \leq |h_{n+1}(t)|$, $0 < t \leq 1$. Therefore, the inequality $h_{n+1}^*(t) \leq v^*(t)$, $0 < t \leq 1$, implies that

$$g_{n+1}^*(t) \leq v^*(t), \quad 0 < t \leq 1.$$

Since $g = \sum_{j=1}^{n+1} g_j$, then from this inequality, estimate (11.80), the inequality $\|\sigma_\tau\|_{X\to X} \leq \max(1,\tau)$ (see Appendix C), and definition (11.66) it follows that for all sequences (a_j) such that $\sum_{j=1}^{2n} a_j^2 \leq 1$ it holds

$$\Big\| f' \cdot \sum_{j=1}^{2n} a_j r_j \Big\|_X = \|g\|_X \leq \Big\| \sum_{j=1}^{n} g_j \Big\|_X + \|g_{n+1}\|_X \leq C\|v\|_X = C\|f\|_{\mathcal{M}_T},$$

where $C = 1 + 4\sqrt{2}(2^{1/4}-1)^{-1}$. Summing up, we get

$$\sup\left\{ \Big\| f' \cdot \sum_{j=1}^{2n} a_j r_j \Big\|_X : \sum_{j=1}^{2n} a_j^2 \leq 1 \right\} \leq C\|f\|_{\mathcal{M}_T}.$$

Thus, by (11.74) and (11.75), inequality (11.71) is valid for all $f \in \mathcal{D}$ with the constant $A := C'(K' + C)$.

Now, one can easily complete the proof of the theorem. Indeed, assume that Y is a s.s. with the Fatou property such that $Y \supset \mathcal{M}(X)$. Then, since f and f' are equimeasurable and Y is a s.s., from (11.71) it follows that $\|f\|_Y \le A'\|f\|_{\mathcal{M}_T}$ for all $f \in \mathcal{D}$. By using Lemma 11.9, we can extend this inequality first to L_∞, and then, taking into account the Fatou property of Y and definition of the $\mathcal{M}_T(X)$-norm, to the whole space $\mathcal{M}_T(X)$. So, everything is done. □

The results obtained allow us to clarify the relationship between an initial s.s. X and the corresponding Rademacher multiplicator spaces.

Let X be a s.s. on $[0, 1]$ with the Fatou property such that $\ln^{1/2}(e/t) \in X^\circ$. The following chain of embeddings:

$$\mathrm{Sym}\,(X) \subset \mathcal{M}(X) \subset \mathcal{M}_T(X) \subset X \tag{11.81}$$

is an immediate consequence of Proposition 11.9 and definition of the symmetric kernel $\mathrm{Sym}\,(X)$.

In the next theorem we present the conditions, under which these embeddings turn into equalities.

Theorem 11.11 *Let X be a s.s. on $[0, 1]$ with the Fatou property such that $\ln^{1/2}(e/t) \in X^\circ$. The following conditions are equivalent:*

- *(i) $\mathcal{M}(X)$ is a s.s.;*
- *(ii) $\mathcal{M}(X) = \mathrm{Sym}\,(X)$;*
- *(iii) $\mathcal{M}(X) = \mathcal{M}_T(X)$;*
- *(iv) for all $n \in \mathbb{N}$ and arbitrary function $f \in \mathcal{D}_n$ we have*

$$\|f\|_{\mathcal{M}} \asymp \sup\left\{\left\|f \sum_{i=n+1}^{\infty} a_i r_i\right\|_X : \left\|\sum_{i=n+1}^{\infty} a_i r_i\right\|_X \le 1\right\}, \tag{11.82}$$

with a constant independent of $n \in \mathbb{N}$ and $f \in \mathcal{D}_n$.

Proof By definition, $\mathrm{Sym}\,(X)$ is the largest space among all s.s.'s Y such that $Y \subset \mathcal{M}(X)$. Hence, it follows that the conditions (i) and (ii) are equivalent.

Assume now that (ii) holds. Then, by Theorem 11.3, the s.s. $\mathcal{M}(X)$ has the Fatou property. Therefore, applying Theorem 11.10, we get $\mathcal{M}_T(X) \subset \mathcal{M}(X)$. Thus, in view of (11.81), we have (iii). The converse implication $(iii) \Rightarrow (ii)$ is a consequence of the fact that $\mathcal{M}_T(X)$ is a s.s.

Further, since the $\mathcal{M}_T(X)$- and $\tilde{\mathcal{M}}_T(X)$-norms are equivalent on the subspace $\mathcal{D}$ by Lemma 11.7, the implication $(iii) \Rightarrow (iv)$ is obvious. Conversely, let $f \in L_\infty$. By Lemma 11.9, there exists a sequence $\{f_k\} \subset \mathcal{D}$, converging to f in the $\mathcal{M}_T(X)$-norm. According to condition (iv), $\{f_k\}$ is a Cauchy sequence with respect to the $\mathcal{M}(X)$-norm. Hence,

$$\|f\|_{\mathcal{M}} = \lim_{k\to\infty} \|f_k\|_{\mathcal{M}} \le C \lim_{k\to\infty} \|f_k\|_{\mathcal{M}_T} = C\|f\|_{\mathcal{M}_T}.$$

Since X has the Fatou property, by Proposition 11.2, so has the space $\mathcal{M}(X)$. Therefore, the latter inequality may be extended to the whole space $\mathcal{M}_T(X)$, whence $\mathcal{M}_T(X) \subset \mathcal{M}(X)$. As a result, taking into account embeddings (11.81), we get (iii). □

Remark 11.6 Suppose that $\ln^{1/2}(e/t) \notin X^\circ$. Then, from Theorem 11.5 it follows $\mathrm{Sym}\,(X) = \mathcal{M}(X) = L_\infty$. Furthermore, according to Remark 11.3, we can conclude that in this case equivalence (11.82) holds as well.

An inspection of the results of this chapter shows that the equality $\mathcal{M}(X) = X$ is only possible if $X = L_\infty$. In contrast to that, the tail Rademacher multiplicator space $\mathcal{M}_T(X)$ coincides with X whenever the latter space is located sufficiently "far" from L_∞.

Theorem 11.12 *Suppose that a s.s. X on $[0, 1]$ has the Fatou property and $\ln^{1/2}(e/t) \in X^\circ$. Then, $\mathcal{M}_T(X) = X$ if and only if $\alpha_X > 0$, where α_X is the lower Boyd index of the space X.*

Proof Observe that for all $0 < t \le 1$ we have

$$\sum_{i=1}^{\infty} \sqrt{i} \cdot \chi_{[2^{-i},2^{-i+1}]}(t) \le \log_2^{1/2}(2/t) \le \sum_{i=1}^{\infty} \sqrt{i+1} \cdot \chi_{[2^{-i},2^{-i+1}]}(t),$$

Moreover. for every $k = 1, 2, \ldots, 2^n$

$$\log_2^{1/2}\left(\frac{2}{2^n t - k + 1}\right) \cdot \chi_{\Delta_n^k}(t) = \left(\sigma_{2^{-n}}\left(\log_2^{1/2}(2/\cdot)\right)\right)(t - (k-1)2^{-n}) \cdot \chi_{\Delta_n^k}(t)$$

(as usual, σ_τ is the dilation operator) and for all $n, i \in \mathbb{N}$

$$\sigma_{2^{-n}}\chi_{[2^{-i},2^{-i+1}]} = \sigma_{2^{-i}}\chi_{[2^{-n},2^{-n+1}]}.$$

Hence, for arbitrary $c_k \ge 0$, $k = 1, 2, \ldots, 2^n$, we get

$$\sum_{i=1}^{\infty} \sqrt{i} f_i(t) \le \sum_{k=1}^{2^n} c_k \log_2^{1/2}\left(\frac{2}{2^n t - k + 1}\right) \cdot \chi_{\Delta_n^k}(t) \le \sum_{i=1}^{\infty} \sqrt{i+1} f_i(t), \tag{11.83}$$

where the functions f_i, $i = 1, 2, \ldots,$ are defined by

$$f_i(t) := \sum_{k=1}^{2^n} c_k \cdot \left(\sigma_{2^{-i}}\chi_{[2^{-n},2^{-n+1}]}\right)(t - (k-1)2^{-n}) \cdot \chi_{\Delta_n^k}(t).$$

It is easy to see that f_i are pairwise disjoint and that for each $i \in \mathbb{N}$ the function f_i is equimeasurable with the function

$$\sigma_{2^{-i}} f = \sum_{k=1}^{2^n} c_k \sigma_{2^{-i}} \chi_{\Delta_n^k},$$

where $f = \sum_{k=1}^{2^n} c_k \chi_{\Delta_n^k}$.

Assume first that $\alpha_X > 0$. By definition of the Boyd indices, there is $\varepsilon > 0$ such that for some $M > 0$, all $\tau \in (0, 1]$ and $f \in X$ we have $\|\sigma_\tau f\|_X \le M \cdot \tau^\varepsilon \|f\|_X$. Consequently, from the right-hand side of (11.83) it follows

$$\begin{aligned} \|f\|_{\mathcal{M}_T} &= \left\| \sum_{k=1}^{2^n} c_k \log_2^{1/2}\left(\frac{2}{2^n t - k + 1}\right) \cdot \chi_{\Delta_n^k} \right\|_X \le \sum_{i=1}^{\infty} \sqrt{i+1} \cdot \|f_i\|_X \\ &= \sum_{i=1}^{\infty} \sqrt{i+1} \cdot \|\sigma_{2^{-i}} f\|_X \le M \sum_{i=1}^{\infty} \sqrt{i+1} \cdot 2^{-\varepsilon i} \|f\|_X. \end{aligned}$$

According to Lemma 11.9 and definition of the $\mathcal{M}_T(X)$-norm, this inequality can be extended to the whole of X, and so $X \subset \mathcal{M}_T(X)$. From this embedding and (11.81) it follows that $\mathcal{M}_T(X) = X$.

In the converse way, suppose that $\mathcal{M}_T(X) = X$. Then, from the left-hand side of inequality (11.83) for any $m \in \mathbb{N}$ we get

$$\begin{aligned} \|f\|_{\mathcal{M}_T} &= \left\| \sum_{k=1}^{2^n} c_k \log_2^{1/2}\left(\frac{2}{2^n t - k + 1}\right) \cdot \chi_{\Delta_n^k} \right\|_X \\ &\ge \left\| \sum_{i=1}^{\infty} \sqrt{i} \cdot f_i \right\|_X \ge \sqrt{m} \cdot \|f_m\|_X = \sqrt{m} \cdot \|\sigma_{2^{-m}} f\|_X. \end{aligned}$$

Therefore, by the hypothesis, for some constant $C > 0$, all $m \in \mathbb{N}$ and $f \in \mathcal{D}$ it holds

$$\|\sigma_{2^{-m}} f\|_X \le m^{-1/2} \cdot \|f\|_{\mathcal{M}_T} \le C \cdot m^{-1/2} \cdot \|f\|_X.$$

Since X has the Fatou property, one can readily check that

$$\|\sigma_\tau\|_{X \to X} = \sup_{f \in \mathcal{D}, \|f\|_X \le 1} \|\sigma_\tau f\|_X.$$

Thus, the latter inequality implies that

$$\|\sigma_{2^{-m}}\|_{X \to X} \le C \cdot m^{-1/2} \quad \text{for all } m \in \mathbb{N},$$

whence $\lim_{m\to\infty} \|\sigma_{2^{-m}}\|_{X\to X} = 0$. This means that $\alpha_X > 0$, which completes the proof. □

Let us return to embeddings (11.81). By Theorem 11.5, if a s.s. X is located very "close" to the space L_∞, or more precisely, when $\ln^{1/2}(e/t) \notin X^\circ$, we have $\mathrm{Sym}\,(X) = \mathcal{M}(X)$ (in this case the space $\mathcal{M}_T(X)$ is not defined). In the opposite situation when $\alpha_X > 0$ (for example, for $X = L^p$, $1 \le p < \infty$), according to Theorems 11.1 and 11.12,

$$\mathrm{Sym}\,(X) \overset{\neq}{\subset} \mathcal{M}(X) \overset{\neq}{\subset} \mathcal{M}_T(X) = X.$$

In conclusion of this section, we provide an example of a s.s. X, for which all the embeddings in (11.81) are strict, i.e.,

$$\mathrm{Sym}\,(X) \overset{\neq}{\subset} \mathcal{M}(X) \overset{\neq}{\subset} \mathcal{M}_T(X) \overset{\neq}{\subset} X.$$

Example 11.6 Let

$$\psi(t) := 2^{1-\log_2^{1/2}(2/t)}, \quad 0 < t \le 1.$$

Since for all sufficiently small $t > 0$ we have $\psi'(t) > 0$ and $\psi''(t) < 0$, then there exists an increasing concave function $\varphi(t)$ on $[0, 1]$ such that $\varphi(t) = \psi(t)$ if $t > 0$ is sufficiently small. We take for X the Marcinkiewicz space $M(\tilde{\varphi})$. Clearly, it has the Fatou property, or equivalently $X'' = X$. Moreover, since $\lim_{t\to 0^+} \psi(t)\ln^{1/2}(e/t) = 0$, we have $\ln^{1/2}(e/t) \in X^\circ$. Therefore, the Rademacher system is equivalent in X to the unit vector basis in ℓ_2, which implies that

$$\begin{aligned}
\|\chi_{[0,2^{-n}]}\|_{\mathcal{M}(X)} &\ge \sup\left\{\left\|\chi_{[0,2^{-n}]}\cdot\sum_{i=1}^n c_i r_i\right\|_X : \left\|\sum_{i=1}^n c_i r_i\right\|_X \le 1\right\} \\
&\asymp \sup\left\{\left\|\chi_{[0,2^{-n}]}\cdot\sum_{i=1}^n c_i r_i\right\|_X : \|(c_i)\|_2 \le 1\right\} \\
&\ge \left\{\left\|\chi_{[0,2^{-n}]}\cdot\sum_{i=1}^n \frac{1}{\sqrt{n}} r_i\right\|_X\right\} \asymp \sqrt{n}2^{-\sqrt{n}}.
\end{aligned}$$

On the other hand, because $\delta_\varphi = 0$, by Corollary 11.1 and Theorem C.2 (see also Remark 3.4), we have

$$\|\chi_{[0,2^{-n}]}\|_{\mathrm{Sym}} \asymp \sup_{0<t<2^{-n}} \ln^{1/2}(e/t)\cdot 2^{1-\log_2^{1/2}(2/t)} \asymp \sup_{0<t<2^{-n}} \log_2^{1/2}(2/t)\cdot 2^{1-\log_2^{1/2}(2/t)} \le C\sqrt{n}2^{-\sqrt{n}}.$$

In consequence,

$$\|\chi_{[0,2^{-n}]}\|_{\mathrm{Sym}} \asymp \|\chi_{[0,2^{-n}]}\|_{\mathcal{M}(X)} \asymp \sqrt{n}\cdot 2^{-\sqrt{n}}, \quad n \in \mathbb{N}.$$

At the same time, one can readily check that

$$\frac{\varphi(2^{-2n})}{\varphi(2^{-n})} \le 2^{-c\sqrt{n}}, \quad n = 1, 2, \ldots$$

where the constant $c > 0$ is sufficiently small. Hence, the function φ satisfies condition (11.59) and $\varphi \notin \Delta^2$. Thus, according to Theorem 11.9, the space $\mathcal{M}(X)$ fails to be symmetric. This implies that $\mathcal{M}(X) \neq \mathrm{Sym}\,(X)$ and also, by Proposition 11.9, $\mathcal{M}(X) \neq \mathcal{M}_T(X)$.

Finally, $\alpha_X = \gamma_\varphi = 0$ (see Appendix C). Therefore, from Theorem 11.12 it follows that $\mathcal{M}_T(X) \neq X$.

Comments and References The notion of Rademacher multiplicator space has arisen in connection with the theory of integration with respect to vector measures. Namely, in the paper [98], Curbera has proved that the set of integrable functions in the sense of Bartle, Dunford and Schwartz with respect to the vector measure ν with values in ℓ_2, defined by

$$\nu(A) = \Big(\int_A r_n(t)\,dt\Big)_{n=1}^{\infty} \quad (A \subset [0,1] \text{ is Lebesgue measurable}),$$

can be identified with the space $\mathcal{M}(L_1)$. Moreover, alternatively this space may be described as the set of all functions f such that for every measurable set $A \subset [0,1]$ the sequence of Fourier-Rademacher coefficients of the function $f\chi_A$ belongs to ℓ_2 (a description of Fourier-Rademacher coefficients of functions from general s.s.'s see in [22]). In the next paper [99], Curbera introduced explicitly definition of the Rademacher multiplicator space. He has obtained there also the first results related to properties such a space; in particular, he proved that the space $\mathcal{M}(X)$ is not symmetric provided that the lower dilation exponent of the fundamental function of X is positive (see Lemma 11.1 and Theorem 11.1).

The symmetric kernel of a Rademacher multiplicator space has been defined and studied in detail in [37]. Almost all results of Sect. 11.2 are taken from this paper; the only exception is the equation $\mathcal{M}(X'') = \mathcal{M}(X)$ from Theorem 11.3, proved in [39]. The proof of Proposition 11.4, based on using the partial retract techniques, is similar to that of Theorem 2 from the paper [102]. For a rather long time one could not provide an example of a function from the space $\mathcal{M}(L_1)$ that does not belong to the symmetric kernel $\mathrm{Sym}\,(L_1)$. Most recently, this was done in the paper [34].

One of the principal results of this chapter is Theorem 11.5, which identifies s.s.'s X such that $\mathcal{M}(X) = L_\infty$ and $\mathrm{Sym}\,(X) = L_\infty$. Before it was proved by Astashkin and Curbera in [37] (the equivalence $(ii) \Leftrightarrow (iii)$) and in [38] (the equivalence $(i) \Leftrightarrow (iii)$), a number of proofs of the first of these equalities, under some additional conditions, were appeared. So, in [100] and [101], the authors have imposed on a s.s. X the condition that the operator

$$x \longmapsto \left(\frac{1}{\varphi_1(t)}\int_0^t (x^*(s))^2\, d\varphi_1(s)\right)^{1/2}, \quad \text{where } \varphi_1(t) = \log_2^{-1}(2/t),$$

is bounded in X. Somewhat later, in [25], a stronger result, which reads that $\mathcal{M}(X) = L_\infty$ whenever X is an interpolation s.s. between the spaces L_∞ and L_{N_2}, was obtained. A useful Lemma 11.4 has been proved in the paper [25] as well.

Theorem 11.6 giving a sufficient condition, under which $\mathcal{M}(X)$ is a s.s., has been obtained in [39]. The proof of this result, based on using the extrapolation techniques, appeared for the first time in the paper [46] (note that the operator $Qx(s) = x(s^2)$ proved be very useful when studying extrapolation properties of s.s.'s [45]). An important Lemma 11.5 has been obtained in [99]. Theorems 11.7–11.9 have been proved in [39]. The construction of the tail Rademacher multiplicator space has been introduced and developed in the paper [41] for the purpose of proving a local version of Khintchine's inequality in general s.s.'s. All the main results of Sect. 11.5 have been obtained in [41] as well.

Chapter 12
Some Versions of Khintchine's Inequality

Let a s.s. X contain the separable part G of the Orlicz space L_{N_2}, $N_2(u) = e^{u^2} - 1$. According to Khintchine's inequality (see Theorem 2.2), there exists a constant $C = C(X) > 0$ such that for every sequence $a = (a_i)_{i=1}^{\infty} \in \ell_2$

$$C^{-1}\|a\|_{\ell_2} \le \Big\|\sum_{i=1}^{\infty} a_i r_i\Big\|_X \le C\|a\|_{\ell_2}. \tag{12.1}$$

Having possible applications in mind, it is of interest to get similar estimates in situations when the domain of the Rademacher functions is a certain subset $E \subset [0, 1]$ of positive measure rather than the whole interval $[0, 1]$. It is worth to observe, however, that non-trivial Rademacher sums may vanish on sets of positive measure. Say, the function $r_1 + r_2$ is equal to zero on the interval $(1/4, 3/4)$. Hence, if we deal with an arbitrary set E, a lower local estimate similar as in inequality (12.1) may hold only, at most, for tails of Rademacher sums. Moreover, note that the case when E is a dyadic interval, e.g., $\Delta_n^k = [(k-1)2^{-n}, k2^{-n}]$, with $n \in \mathbb{N}$ and $1 \le k \le 2^n$, really does not differ from that of the whole interval $[0, 1]$ due to the fact that the tail Rademacher sequence $(r_i)_{i=n+1}^{\infty}$ behaves itself on Δ_n^k completely in the same way as does the full sequence on $[0, 1]$. The question is so what happens when E is an *arbitrary* subset of $[0, 1]$ of positive measure.

As above, by ϕ_X we denote the fundamental function of a s.s. X, that is, $\phi_X(t) := \|\chi_{[0,t]}\|_X$. We shall say that *the local Khintchine inequality* holds in a s.s. X whenever for some constants $\alpha > 0$, $\beta > 0$ and any measurable set $E \subset [0, 1]$ with $m(E) > 0$ there exists a positive integer $N = N(E)$, depending only on E, such that

$$\alpha\phi_X(m(E))\Big(\sum_{i=N}^{\infty} a_i^2\Big)^{1/2} \le \Big\|\sum_{i=N}^{\infty} a_i r_i \chi_E\Big\|_X \le \beta\phi_X(m(E))\Big(\sum_{i=N}^{\infty} a_i^2\Big)^{1/2} \tag{12.2}$$

S. V. Astashkin, *The Rademacher System in Function Spaces*,
https://doi.org/10.1007/978-3-030-47890-2_12

for all sequences $(a_i)_{i=1}^{\infty} \in \ell_2$. By using some properties of the tail Rademacher multiplicator space $\mathcal{M}_T(X)$ obtained in the previous chapter, we show here that the local Khintchine inequality holds in a s.s. X if and only if the lower dilation exponent γ_{ϕ_X} of the function ϕ_X is positive. In particular, the latter condition is fulfilled for all separable Orlicz spaces and moreover for all s.s.'s with the positive Boyd index (the definitions see in Appendix C). In Sect. 12.2, we shall prove also for $X = L_2$ a sharp lower local estimate, where the choice of the domain of a given Rademacher sum $\sum_{i=1}^{\infty} a_i r_i$ depends on properties of a suitable tail of the sequence of coefficients.

Another opportunity to localize inequality (12.1) is to consider only some *special* classes of sets $E \subset [0, 1]$. More generally, in Sect. 12.3, we investigate, under which conditions on a function $w(t) \geq 0$, $0 \leq t \leq 1$, the following natural weighted version of inequality (12.1) holds:

$$C_{X,w}^{-1} \|a\|_{\ell_2} \leq \left\| w \cdot \sum_{i=1}^{\infty} a_i r_i \right\|_X \leq C_{X,w} \|a\|_{\ell_2}, \tag{12.3}$$

where $C_{X,w} > 0$ depends on X and w (but does not depend on a sequence $a = (a_i)_{i=1}^{\infty} \in \ell_2$). We refer (12.3) as to the *weighted Khintchine inequality*. As we shall see, the simpler right-hand side of this inequality is valid if and only if a weighted function w belongs to the Rademacher multiplicator space $\mathcal{M}(X)$ (see the previous chapter). As to the more delicate left-hand side inequality, we shall prove it first in the case when $m(\operatorname{supp} w) > 1/2$. The above example of the sum $r_1 + r_2$ shows that the latter inequality is necessary if one wants to have the left-hand side estimate in (12.3) as a consequence of a condition imposed on measure of the support of w. However, restricting to weighted functions, whose supports have a special structure, we shall be able to prove this estimate for those w that vanish on sets of measure arbitrarily close to 1. In particular, we shall establish the following lower local L_p-estimate: for every $p > 0$ and any set $E \subset [0, 1]$, with $m(E \cap (a, b)) > 0$ for each interval $(a, b) \subset [0, 1]$, there is a constant $\gamma = \gamma(E, p) > 0$ such that for all $a = (a_k)_{k=1}^{\infty} \in \ell_2$ it holds

$$\left(\int_E \left| \sum_{i=1}^{\infty} a_i r_i(t) \right|^p dt \right)^{1/p} \geq \gamma \|a\|_{\ell_2}.$$

In the concluding part of the chapter, we obtain some variants of Khintchine's L_1-inequality with sharp constants that refine and complement Theorem 5.4.

12.1 Local Khintchine's Inequality in Symmetric Spaces

We begin with the formulation of the main result of this section, which gives necessary and sufficient conditions, under which the local Khintchine's inequality holds in a s.s. X.

Recall that the lower dilation exponent γ_ψ of a function $\psi(s)$ on $[0, 1]$ is defined by

$$\gamma_\psi := \lim_{t\to 0+} \frac{\ln \mathcal{M}_\psi(t)}{\ln t},$$

where

$$\mathcal{M}_\psi(t) = \sup\left\{\frac{\psi(st)}{\psi(s)} : s \in (0, \min(1, 1/t))\right\}, \quad t > 0.$$

Theorem 12.1 *For an arbitrary s.s. X on $[0, 1]$ the following conditions are equivalent:*

(i) the local Khintchine inequality holds in X;
(ii) there exists a constant $M > 0$ such that

$$\left\| \log_2^{1/2}\left(\frac{2u}{t}\right)\chi_{[0,u]}\right\|_X \le M\phi_X(u), \quad 0 < u \le 1; \tag{12.4}$$

(iii) $\gamma_{\phi_X} > 0$.

We show first that the left-hand side of (12.2) is fulfilled in every s.s.

Proposition 12.1 *Let X be a s.s. on $[0, 1]$. For any measurable set $E \subset [0, 1]$ of positive measure there is a positive integer $N = N(E)$ such that*

$$\left\|\sum_{i=N}^{\infty} a_i r_i \chi_E\right\|_X \ge \frac{1}{48}\cdot\phi_X(m(E))\Big(\sum_{i=N}^{\infty} a_i^2\Big)^{1/2}$$

for each sequence $(a_i) \in \ell_2$ satisfying the condition $\sum_{i=1}^{\infty} a_i r_i \in X$.

Proof Let us prove first that for every $E \subset [0, 1]$, $m(E) > 0$, there is $N = N(E)$ with the following property:

$$m\left\{t \in E : \left|\sum_{i=N}^{\infty} a_i r_i(t)\right| \ge \frac{1}{2}\Big(\sum_{i=N}^{\infty} a_i^2\Big)^{1/2}\right\} \ge \frac{1}{24}\cdot m(E) \quad \text{for all} \quad (a_i) \in \ell_2. \tag{12.5}$$

Indeed, assuming that $0 < \varepsilon < m(E)/96$, for sufficiently large $N \in \mathbb{N}$ we can find a step-function $f = \sum_{k=1}^{2^{N-1}} c_k \chi_{\Delta_{N-1}^k}$, $c_k \in \mathbb{R}$, such that

$$\|\chi_E - f\|_1 < \varepsilon. \tag{12.6}$$

Next, let us define the set

$$A := \Big\{ t \in [0,1] : \Big| \sum_{i=N}^{\infty} a_i r_i(t) \Big| \le \frac{1}{2} \Big(\sum_{i=N}^{\infty} a_i^2 \Big)^{1/2} \Big\},$$

where the above number N and a sequence $(a_i) \in \ell_2$ are fixed. Applying equation (A.1) to the independent functions f and χ_A, by inequality (12.6) and the choice of ε, we get

$$\begin{aligned} m(E \cap A) &= \int_0^1 \chi_A(t)\chi_E(t)\, dt \le \int_0^1 \chi_A(t) \cdot |f(t)|\, dt + \varepsilon \\ &= \int_0^1 |f(t)|\, dt \cdot m(A) + \varepsilon \le m(E)m(A) + 2\varepsilon \le m(E)\Big(m(A) + \frac{1}{48} \Big). \end{aligned}$$

On the other hand, according to the Paley–Zygmund inequality (see Proposition 1.4), we have $m(A) \le 15/16$. Therefore, from the last estimate it follows that

$$m(E \cap A) \le \frac{23}{24} \cdot m(E),$$

which is equivalent to (12.5).

Suppose now that a set $E \subset [0,1]$ is of positive measure and a sequence $(a_i) \in \ell_2$ is such that $\sum_{i=1}^{\infty} a_i r_i \in X$. Then, we have (12.5) for some $N = N(E)$. Hence, putting $A^c := [0,1] \setminus A$, where the set A is defined as above, and taking into account the quasiconcavity of ϕ_X, we have

$$\begin{aligned} \Big\| \sum_{i=N}^{\infty} a_i r_i \chi_E \Big\|_X &\ge \Big\| \sum_{i=N}^{\infty} a_i r_i \chi_{E \cap A^c} \Big\|_X \ge \frac{1}{2} \Big(\sum_{i=N}^{\infty} a_i^2 \Big)^{1/2} \phi_X\big(m(E \cap A^c) \big) \\ &\ge \frac{1}{2} \Big(\sum_{i=N}^{\infty} a_i^2 \Big)^{1/2} \phi_X\Big(\frac{1}{24} \cdot m(E) \Big) \ge \frac{1}{48} \cdot \phi_X\big(m(E) \big) \Big(\sum_{i=N}^{\infty} a_i^2 \Big)^{1/2}, \end{aligned}$$

as we wished. □

Proof of Theorem 12.1 ***(i)*** $\Rightarrow$ ***(ii)*** By the hypothesis, inequality (12.2) holds, in particular, for the intervals $E = [0, 2^{-l}]$, $l = 0, 1, 2, \ldots$. Therefore, for every $l = 0, 1, 2, \ldots$ there is a positive integer $N(l)$ such that

$$\sup \Big\{ \Big\| \sum_{i=N(l)}^{\infty} a_i r_i \chi_{[0, 2^{-l}]} \Big\|_X : \sum_{i=N(l)}^{\infty} a_i^2 \le 1 \Big\} \le \beta \phi_X(2^{-l}). \qquad (12.7)$$

Observe that for any $N > l \geq 0$ the sequences $\{r_i\}_{i=l+1}^{\infty}$ and $\{r_i\}_{i=N}^{\infty}$ are similar when restricted to the interval $[0, 2^{-l}]$. This implies that we can take, in fact, in the last inequality $N(l) = l + 1$. Then, if $l = 0$ we come to the right-hand side of inequality (12.1), and so, by Theorem 2.3, $\ln^{1/2}(e/t) \in X^\circ$. Furthermore, applying Lemma 11.7, we get from (12.7) that for some constant $C > 0$

$$\phi_{\mathcal{M}_T(X)}(2^{-l}) := \|\chi_{[0,2^{-l}]}\|_{\mathcal{M}_T(X)} \leq C\phi_X(2^{-l}),$$

where $\mathcal{M}_T(X)$ is the tail Rademacher multiplicator space. Taking into account Remark 11.5, this inequality can be rewritten as follows:

$$\left\|\log_2^{1/2}\left(2^{1-l}/t\right)\chi_{[0,2^{-l}]}\right\|_X \leq C\,\phi_X(2^{-l}), \quad l \in \mathbb{N}.$$

Thus, (12.4) is proved if $u = 2^{-l}$, $l \in \mathbb{N}$. Since the quasiconcavity of the fundamental function ϕ_X allows easily to extend inequality (12.4) to the whole interval $(0, 1]$, the proof of (ii) is completed.

$(ii) \Leftrightarrow (iii)$. We assume first that condition (iii) holds. As above, it suffices to prove inequality (12.4) for $u = 2^{-l}$, $l \in \mathbb{N}$. From definition of the dilation function $\mathcal{M}_{\phi_X}$ it follows that

$$\begin{aligned}\left\|\log_2^{1/2}\left(2^{1-l}/t\right)\chi_{[0,2^{-l}]}\right\|_X &\leq \left\|\sum_{k=l}^{\infty}\sqrt{k-l+2}\,\chi_{[2^{-k-1},2^{-k}]}\right\|_X \\ &\leq \sum_{k=l}^{\infty}\sqrt{k-l+2}\,\phi_X(2^{-k}) \\ &\leq \sum_{k=l}^{\infty}\sqrt{k-l+2}\,\mathcal{M}_{\phi_X}(2^{-k+l})\,\phi_X(2^{-l}) \\ &= \sum_{j=0}^{\infty}\sqrt{j+2}\,\mathcal{M}_{\phi_X}(2^{-j})\,\phi_X(2^{-l}).\end{aligned}$$

Moreover, by condition, $\gamma_{\phi_X} > 0$, and hence there exist $\varepsilon > 0$ and $C' > 0$ such that

$$\mathcal{M}_{\phi_X}(t) = \sup_{0<s\leq 1}\frac{\phi_X(st)}{\phi_X(s)} \leq C't^{\varepsilon}, \quad 0 < t \leq 1.$$

Therefore,

$$\left\|\log_2^{1/2}\left(2^{1-l}/t\right)\chi_{[0,2^{-l}]}\right\|_X \leq C\,\phi_X(2^{-l}),$$

where

$$C := C' \sum_{j=0}^{\infty} \sqrt{j+2} \cdot 2^{-j\varepsilon} < \infty,$$

and (ii) is proved.

Let us prove the converse implication. From the hypothesis we obtain

$$\begin{aligned}\sqrt{k} \cdot \phi_X(2^{-l-k}) &\le \left\| \log_2^{1/2}\left(2^{1-l}/t\right) \chi_{[0,2^{-l-k}]} \right\|_X \\ &\le \left\| \log_2^{1/2}\left(2^{1-l}/t\right) \chi_{[0,2^{-l}]} \right\|_X \le M \cdot \phi_X(2^{-l})\end{aligned}$$

for all $l, k \in \mathbb{N}$. Consequently, because ϕ_X is quasiconcave, we can estimate the dilation function of ϕ_X as follows:

$$\mathcal{M}_{\phi_X}(2^{-k}) = \sup_{0<s\le 1} \frac{\phi_X(2^{-k}s)}{\phi_X(s)} \le \frac{2M}{\sqrt{k}}, \qquad k \in \mathbb{N}.$$

Thus, $\mathcal{M}_{\phi_X}(t) \to 0$ as $t \to 0$, which is equivalent to the inequality $\gamma_{\phi_X} > 0$.

$(iii) \Rightarrow (i)$. Since the left-hand side of inequality (12.2) holds by Proposition 12.1 in arbitrary s.s., we need only to prove its right-hand side.

Let us show first that condition $\gamma_{\phi_X} > 0$ guarantees the equivalence of the Rademacher system in X to the unit vector basis in ℓ_2. To achieve this, according to Theorem 2.2, it suffices to prove that $\ln^{1/2}(e/t) \in X^\circ$.

Take $p > 1/\gamma_{\phi_X}$. If $\Lambda(t^{1/p})$ is the Lorentz space with the fundamental function $t^{1/p}$, then we have

$$\| \ln^{1/2}(e/t) \|_{\Lambda(t^{1/p})} = \frac{1}{p} \int_0^1 \ln^{1/2}(e/t) t^{1/p-1}\, dt < \infty,$$

whence $\ln^{1/2}(e/t) \in \Lambda(t^{1/p})$. On the other hand, $p > 1/\gamma_{\phi_X}$, and so from the definition of γ_{ϕ_X} it follows that $\phi_X(t) \le Ct^{1/p}$, $0 \le t \le 1$, for some $C > 0$. Therefore, because $\Lambda(\phi_X)$ is the minimal space among all s.s.'s with the fundamental function ϕ_X (see Appendix C), $\Lambda(t^{1/p}) \subset \Lambda(\phi_X) \subset X$. Since the space $\Lambda(t^{1/p})$ is separable, the above embeddings imply that $\ln^{1/2}(e/t) \in X^\circ$. Thus, the Rademacher system is equivalent in X to the unit vector basis in ℓ_2.

Proceeding with the proof of the implication $(iii) \Rightarrow (i)$, we treat first the case when the set E is an union of pairwise disjoint dyadic intervals, that is,

$$E = \bigcup_{j=1}^{\infty} \Delta_{n_j}^{k_j}, \tag{12.8}$$

where $n_1 \le n_2 \le \dots$, $k_j \in \{1, 2, \dots, 2^{n_j}\}$, and $\Delta_{n_j}^{k_j} \cap \Delta_{n_i}^{k_i} = \varnothing$ if $n_j \ne n_i$ or $k_j \ne k_i$. Then, from the hypothesis $\gamma_{\phi_X} > 0$ it follows the existence of $\varepsilon > 0$ and $C > 0$ such that

$$\phi_X(2^{-k}) \le C2^{-\varepsilon k}, \quad k = 1, 2, \dots$$

Therefore, since $m(E) > 0$, there is $k_0 \in \mathbb{N}$ satisfying the inequality:

$$\sum_{k=k_0}^{\infty} \sqrt{k+2}\, \phi_X(2^{-k}) \le \phi_X(m(E)). \tag{12.9}$$

Let us represent E in the form: $E = E_1 \cup E_2$, where

$$E_1 := \bigcup_{j=1}^{j_0} \Delta_{n_j}^{k_j} \text{ and } E_2 := \bigcup_{j=j_0+1}^{\infty} \Delta_{n_j}^{k_j},$$

and let j_0 be chosen so that $m(E_2) < 2^{-k_0}$. Taking into account that the system $\{r_k\}$ is equivalent in X to the unit vector basis in ℓ_2, by definition of the $\mathcal{M}(X)$- and $\mathrm{Sym}(X)$-norms and Theorem 11.3, we get

$$\begin{aligned}\Big\|\sum_{i=1}^{\infty} a_i r_i \cdot \chi_{E_2}\Big\|_X &\le C_1\, \|\chi_{E_2}\|_{\mathcal{M}} \|(a_i)\|_{\ell_2} \le C_1\, \|\chi_{E_2}\|_{\mathrm{Sym}} \|(a_i)\|_{\ell_2} \\ &\asymp \|\log_2^{1/2}(2/t)\chi_{[0,m(E_2)]}\|_{X''} \|(a_i)\|_{\ell_2} \\ &\le \|\log_2^{1/2}(2/t)\chi_{[0,2^{-k_0}]}\|_X \|(a_i)\|_{\ell_2}.\end{aligned}$$

Moreover, from (12.9) it follows that

$$\begin{aligned}\|\log_2^{1/2}(2/t)\chi_{[0,2^{-k_0}]}\|_X &\le \Big\|\sum_{k=k_0}^{\infty} \sqrt{k+2}\ \chi_{[2^{-k-1},2^{-k}]}\Big\|_X \\ &\le \sum_{k=k_0}^{\infty} \sqrt{k+2}\, \phi_X(2^{-k}) \le \phi_X(m(E)).\end{aligned}$$

As a result, combining the last inequalities, we infer

$$\Big\|\sum_{i=1}^{\infty} a_i r_i \cdot \chi_{E_2}\Big\|_X \le C\|(a_i)\|_{\ell_2}\phi_X(m(E)). \tag{12.10}$$

Show that a similar estimate holds also for the set $E_1 = \bigcup_{j=1}^{j_0} \Delta_{n_j}^{k_j}$ if we replace full Rademacher sums with their suitable tails.

Denote $N := n_{j_0} + 1$. Then, clearly, for some set $I_{N-1} \subset \{1, 2, \dots, 2^{N-1}\}$ we have

$$E_1 = \bigcup_{k \in I_{N-1}} \Delta_{N-1}^k.$$

Obviously, for every function of the form $R_N := \sum_{i=N}^{\infty} a_i r_i \in X$ it follows that

$$\|R_N \cdot \chi_{E_1}\|_X = \Big\| \sum_{k \in I_{N-1}} \Big(\sum_{i=N}^{\infty} a_i r_i\Big)^* (2^{N-1}t - k + 1) \cdot \chi_{\Delta_{N-1}^k} \Big\|_X.$$

Hence, by inequality (11.33),

$$\|R_N \cdot \chi_{E_1}\|_X \le K' \Big\| \sum_{k \in I_{N-1}} \chi_{\Delta_{N-1}^k} \cdot \log_2^{1/2} \Big(\frac{2}{2^{N-1}t - k + 1}\Big) \Big\|_X \Big(\sum_{i=N}^{\infty} a_i^2\Big)^{1/2}.$$

Setting now $l := \operatorname{card} I_{N-1}$, we observe that the functions

$$\sum_{k \in I_{N-1}} \log_2^{1/2} \Big(\frac{2}{2^{N-1}t - k + 1}\Big) \cdot \chi_{\Delta_{N-1}^k} \quad \text{and} \quad \log_2^{1/2} \Big(\frac{2l}{2^{N-1}t}\Big) \cdot \chi_{[0, l2^{-N+1}]}$$

are identically distributed, and so the last inequality implies that

$$\|R_N \cdot \chi_{E_1}\|_X \le K' \Big\| \log_2^{1/2} \Big(\frac{2l}{2^{N-1}t}\Big) \cdot \chi_{[0, l2^{-N+1}]} \Big\|_X \Big(\sum_{i=N}^{\infty} a_i^2\Big)^{1/2}.$$

Thus, because $(ii) \Leftrightarrow (iii)$, we have

$$\|R_N \cdot \chi_{E_1}\|_X \le K' M \,\phi_X(l2^{-N+1}) \Big(\sum_{i=N}^{\infty} a_i^2\Big)^{1/2} \le K' M \,\phi_X(m(E)) \Big(\sum_{i=N}^{\infty} a_i^2\Big)^{1/2}.$$

Summing up, by (12.10), for the above $N = N(E)$ it holds

$$\|R_N \cdot \chi_E\|_X \le \|R_N \cdot \chi_{E_1}\|_X + \|R_N \cdot \chi_{E_2}\|_X \le (C + K'M) \cdot \phi_X(m(E)) \Big(\sum_{i=N}^{\infty} a_i^2\Big)^{1/2}.$$

This means that inequality (12.2) is proved in the case of sets of the form (12.8).

Suppose now that E is an arbitrary set of positive measure. We prove that this situation reduces to the above special case. To this end, let us construct the set F of the form (12.8) as follows. If $m(E \cap \Delta_0^1) = m(E) \ge \frac{1}{2} m(\Delta_0^1) = \frac{1}{2}$, we take as F just the dyadic interval Δ_0^1. In the case when $m(E \cap \Delta_0^1) < \frac{1}{2}$ we consider the sets

$E \cap \Delta_1^1$ and $E \cap \Delta_1^2$. We include the interval Δ_1^1 into F if $m(E \cap \Delta_1^1) \geq \frac{1}{2}m(\Delta_1^1)$. Note that then $m(E \cap \Delta_1^2) < \frac{1}{2}m(\Delta_1^2)$, and so Δ_1^2, conversely, is not included in F. On the next step we consider only those dyadic intervals of second rank that are contained in the interval Δ_1^2 and include them in the set F by using the same rule. As a result, we get as F a finite or countable union of disjoint dyadic intervals, and hence, as was already proved, the right-hand side of inequality (12.2) is fulfilled for F and for a suitable $N = N(F)$ with the constant $C + K'M$. Moreover, by construction, $m(F) \leq 2m(E)$ and the set F contains all density points of E, whence it follows that $\chi_E \leq \chi_F$ a.e. Thus, for the chosen N and all $(a_i) \in \ell_2$, in view of the quasiconcavity of ϕ_X, we have

$$\Big\| \sum_{i=N}^{\infty} a_i r_i \chi_E \Big\|_X \leq \Big\| \sum_{i=N}^{\infty} a_i r_i \chi_F \Big\|_X \leq (C + K'M)\phi_X(m(F))\Big(\sum_{i=N}^{\infty} a_i^2\Big)^{1/2}$$
$$\leq 2(C + K'M)\phi_X(m(E))\Big(\sum_{i=N}^{\infty} a_i^2\Big)^{1/2}.$$

This completes the proof. □

Remark 12.1 In the case when $X \neq L_\infty$ inequality (12.2) is equivalent to the following: there exist constants $\alpha' > 0$ and $\beta' > 0$ such that for any measurable set $E \subset [0, 1]$ of positive measure there is a positive integer $N = N(E)$ such that

$$\alpha'\phi_X(m(E))\Big\| \sum_{i=N}^{\infty} a_i r_i \Big\|_X \leq \Big\| \chi_E \cdot \sum_{i=N}^{\infty} a_i r_i \Big\|_X \leq \beta'\phi_X(m(E))\Big\| \sum_{i=N}^{\infty} a_i r_i \Big\|_X, \tag{12.11}$$

for arbitrary sequence $(a_i) \in \ell_2$ satisfying the condition: $\sum_{i=1}^{\infty} a_i r_i \in X$. To show this, it suffices to check that either of inequalities (12.11) and (12.2) implies that $\ln^{1/2}(e/t) \in X^\circ$. Indeed, then $\| \sum_{i=1}^{\infty} a_i r_i \|_X \asymp \|(a_i)\|_{\ell_2}$, and hence inequalities (12.2) and (12.11) are identical in this case.

On the contrary, let $\ln^{1/2}(e/t) \notin X^\circ$. Then, according to the assertion from the beginning of the proof of Theorem 11.5, there exists a constant $c_1 > 0$ which depends only on X such that for a given $m \geq 0$ we can find a positive integer $n_0 \geq 1$ such that for all $n \geq n_0$

$$\Big\| \sum_{i=m+1}^{m+n} r_i \chi_{[0,2^{-m}]} \Big\|_X \geq c_1 \Big\| \sum_{i=m+1}^{m+n} r_i \Big\|_X.$$

Assume now that (12.11) holds. Applying this inequality for $E = [0, 2^{-m}]$ and coefficients $a_i = 1$, we can take, clearly, $N = m + 1$. Therefore, for all $n \in \mathbb{N}$

we have

$$\Big\| \sum_{i=m+1}^{m+n} r_i \chi_{[0,2^{-m}]} \Big\|_X \le \beta' \phi_X(2^{-m}) \Big\| \sum_{i=m+1}^{m+n} r_i \Big\|_X.$$

Combining the last inequalities, we deduce that $\phi_X(2^{-m}) \ge c_1/\beta'$ for all $m \in \mathbb{N}$. Hence, as the function ϕ_X is quasiconcave, one can easily see that $X = L_\infty$, which contradicts the hypothesis. The alternative case when inequality (12.2) is fulfilled can be considered in a similar way. Thus, the equivalence of conditions (12.2) and (12.11) provided that $X \ne L_\infty$ is proved.

What does happen for $X = L_\infty$? As it is easy to see, then inequality (12.11) holds, while (12.2) does not.

Applying Theorem 12.1, we are able to find sharp conditions, under which the local Khintchine inequality holds in s.s.'s from the most important classes.

Example 12.1 Recall that the Lorentz–Zygmund space $X = L_{p,q}(\log L)^\sigma$ consists of all measurable functions $x(t)$ on $[0, 1]$ such that

$$\|x\|_{p,q,\sigma} := \left(\int_0^1 \left(t^{1/p} \ln^\sigma(e/t) x^*(t)\right)^q \frac{dt}{t} \right)^{1/q} < \infty$$

if either $1 < p < \infty$, $1 \le q < \infty$, $\sigma \in \mathbb{R}$, or $p = q = 1$, $\sigma \ge 0$, and

$$\|x\|_{p,\infty,\sigma} := \sup_{0<t\le 1} t^{1/p} \ln^\sigma(e/t) x^*(t) < \infty$$

if $1 < p < \infty$, $\sigma \in \mathbb{R}$. It is not hard to check that $\phi_X(t) \asymp t^{1/p} \ln^\sigma(e/t)$, and hence $\gamma_{\phi_X} = 1/p$. Therefore, according to Theorem 12.1, there exist constants $\alpha > 0$ and $\beta > 0$, depending on p, q, σ, such that for every set $E \subset [0, 1]$ of positive measure there is a positive integer $N = N(E)$ such that

$$\alpha \, \|\chi_E\|_{p,q,\sigma} \Big(\sum_{i=N}^\infty a_i^2 \Big)^{1/2} \le \Big\| \sum_{i=N}^\infty a_i r_i \chi_E \Big\|_{p,q,\sigma} \le \beta \, \|\chi_E\|_{p,q,\sigma} \Big(\sum_{i=N}^\infty a_i^2 \Big)^{1/2}$$

for every sequence $(a_i) \in \ell_2$. In particular, setting $\sigma = 0$, we get the local Khintchine inequality for the Lorentz spaces $L_{p,q}$ with $\|\chi_E\|_{p,q,0} = m(E)^{1/p}$. In the partial case when $p = q < \infty$ we arrive at the classical local Khintchine L_p-inequality.

Example 12.2 Let M be an Orlicz function satisfying the Δ_2-condition at infinity (i.e., for some $t_0 > 0$ and $C > 0$ we have $M(2t) \le CM(t)$ if $t \ge t_0$). This condition is equivalent in turn to the fact that the Orlicz space L_M is separable (see e.g. [151, Theorem IV.3.9]). One can also readily check that then the lower dilation exponent of the fundamental function $\phi_{L_M}(t) = 1/M^{-1}(1/t)$ is positive. Consequently, by

Theorem 12.1, the local Khintchine inequality holds in every separable Orlicz space L_M. Thus, there are constants $\alpha > 0$ and $\beta > 0$ such that for each set $E \subset [0,1]$ of positive measure we can find a positive integer $N = N(E)$ satisfying the following inequality

$$\alpha\Big(\sum_{i=N}^{\infty} a_i^2\Big)^{1/2} \le M^{-1}\Big(\frac{1}{m(E)}\Big)\Big\|\sum_{i=N}^{\infty} a_i r_i \chi_E\Big\|_{L_M} \le \beta\Big(\sum_{i=N}^{\infty} a_i^2\Big)^{1/2}$$

for any $(a_i) \in \ell_2$.

Example 12.3 Let φ be a quasiconcave function on $[0,1]$. Then, the fundamental function of the Marcinkiewicz space $M(\varphi)$ is equal to the function $\tilde{\varphi}(t) := t/\varphi(t)$. Therefore, if the upper dilation exponent $\delta_\varphi < 1$, we have $\gamma_{\tilde{\varphi}} = 1 - \delta_\varphi > 0$ (see Appendix C). Thus, in this case there exist constants $\alpha > 0$ and $\beta > 0$ such that for every set $E \subset [0,1]$ of positive measure there is a positive integer $N = N(E)$ such that for all $(a_i) \in \ell_2$ we have

$$\alpha\, m(E)\Big(\sum_{i=N}^{\infty} a_i^2\Big)^{1/2} \le \varphi(m(E))\Big\|\sum_{i=N}^{\infty} a_i r_i \chi_E\Big\|_{M(\varphi)} \le \beta\, m(E)\Big(\sum_{i=N}^{\infty} a_i^2\Big)^{1/2}.$$

Example 12.4 Suppose that φ is an increasing concave function on $[0,1]$ such that $\varphi(0) = 0$, $p \in [1,+\infty)$. Recall that the Lorentz space $\Lambda_p(\varphi)$ consists of all measurable functions f on $[0,1]$ such that

$$\|f\|_{\Lambda_p(\varphi)} := \Big(\int_0^1 f^*(s)^p\, d\varphi(s)\Big)^{1/p} < \infty.$$

The fundamental function of $\Lambda_p(\varphi)$ is equal to $\varphi(t)^{1/p}$. Hence, if the lower dilation exponent of the function φ is positive, there are constants $\alpha > 0$ and $\beta > 0$ such that for every set $E \subset [0,1]$ of positive measure there exists $N = N(E)$ with the property that

$$\begin{aligned}\alpha\,\varphi(m(E))^{1/p}\Big(\sum_{i=N}^{\infty} a_i^2\Big)^{1/2} &\le \Big\|\sum_{i=N}^{\infty} a_i r_i \chi_E\Big\|_{\Lambda_p(\varphi)}\\ &\le \beta\,\varphi(m(E))^{1/p}\Big(\sum_{i=N}^{\infty} a_i^2\Big)^{1/2}, \quad (a_i) \in \ell_2.\end{aligned}$$

12.2 A Lower Local L_2-Estimate for Rademacher Sums

Here, we prove a sharp lower local L_2-estimate, in which the domain of a given Rademacher sum $\sum_{i=0}^{\infty} a_i r_i$ is chosen depending in a sense on which a suitable tail of the sequence of coefficients is,

In this section, $r_0(t) \equiv 1, 0 \le t \le 1$.

Theorem 12.2 *For every measurable set $E \subset [0, 1]$ of positive measure and each $\lambda > 1$ there exists a dyadic interval $\Delta \subset [0, 1]$ (depending on E and λ) with the following property: for any sequence $(a_i)_{i=1}^{\infty} \in \ell_2$ we can find a partition $\Delta = J_1 \cup J_2$ that depends only on the tail $(a_i)_{i=N+1}^{\infty}$, where $N = -\log_2 m(\Delta)$, such that $m(J_1) = m(J_2) = m(\Delta)/2$ and*

$$\sum_{i=N+1}^{\infty} a_i^2 \le \max\left\{\frac{\lambda}{m(J_1)} \int_{J_1 \cap E} \left|\sum_{i=0}^{\infty} a_i r_i(t)\right|^2 dt\,,\ \frac{\lambda}{m(J_2)} \int_{J_2 \cap E} \left|\sum_{i=0}^{\infty} a_i r_i(t)\right|^2 dt\right\}.$$

For the proof of this result we need two lemmas.

Lemma 12.1 *For arbitrary $E \subset [0, 1]$ of positive measure and all $(a_i) \in \ell_2$ we have*

$$\int_E \left|\sum_{i=0}^{\infty} a_i r_i(t)\right|^2 dt \le \left(m(E) + \sqrt{m(E)}\right) \int_0^1 \left|\sum_{i=0}^{\infty} a_i r_i(t)\right|^2 dt.$$

Proof Since the system $\{r_j r_k\}_{j \ne k}$ is orthonormal on $[0, 1]$ (see e.g. Corollary 1.2), then for all $f \in L_2[0, 1]$

$$\sum_{j,k \ge 0, j \ne k} \left(\int_0^1 f r_k r_j \, dt\right)^2 \le \|f\|_2^2.$$

Therefore, from the Cauchy-Schwarz-Bunyakovskii inequality it follows that

$$\begin{aligned}
\int_E \left|\sum_{i=0}^{\infty} a_i r_i(t)\right|^2 dt &\le m(E) \sum_{i=0}^{\infty} a_i^2 + \sum_{j \ne k} a_j a_k \int_E r_j r_k \, dt \\
&\le m(E) \sum_{i=0}^{\infty} a_i^2 + \left(\sum_{j \ne k} (a_j a_k)^2\right)^{1/2} \left(\sum_{j \ne k} \left(\int_E r_j r_k \, dt\right)^2\right)^{1/2} \\
&\le m(E) \sum_{i=0}^{\infty} a_i^2 + \left(\sum_{j=0}^{\infty} a_j^2\right) \left(\sum_{j \ne k} \left(\int_E r_j r_k \, dt\right)^2\right)^{1/2} \\
&\le \left(m(E) + \sqrt{m(E)}\right) \sum_{i=0}^{\infty} a_i^2.
\end{aligned}$$

Furthermore, by Parseval's Identity,

$$\int_0^1 \Big|\sum_{i=0}^{\infty} a_i r_i(t)\Big|^2 dt = \sum_{i=0}^{\infty} a_i^2.$$

Combining these relations, we complete the proof. □

For a dyadic interval $\Delta_N = [m2^{-N}, (m+1)2^{-N})$, $m = 1, 2, \ldots, 2^N$, and a sequence $(a_i)_{i=1}^{\infty} \in \ell_2$ we define the sets:

$$\Delta_N^{++} := \Big\{t \in \Delta_N : \sum_{i=N+1}^{\infty} a_i r_i(t) > 0\Big\},$$

$$\Delta_N^{--} := \Big\{t \in \Delta_N : \sum_{i=N+1}^{\infty} a_i r_i(t) < 0\Big\},$$

$$\Delta_N^{0} := \Big\{t \in \Delta_N : \sum_{i=N+1}^{\infty} a_i r_i(t) = 0\Big\}.$$

It is straightforward to check that the sets Δ_N^{++} and Δ_N^{--} are pairwise disjoint and have equal measure. However, if the set Δ_N^0 is nonempty, the union $\Delta_N^{++} \cup \Delta_N^{--}$ is not equal to Δ_N. To arrange this, we find disjoint subsets $\Delta_N^{0,+}$ and $\Delta_N^{0,-}$ of Δ_N^0 of equal measure, whose union is Δ_N^0, and define $\Delta_N^{+} := \Delta_N^{++} \cup \Delta_N^{0,+}$ and $\Delta_N^{-} := \Delta_N^{--} \cup \Delta_N^{0,-}$. Then we have $\Delta_N^{+} \cup \Delta_N^{-} = \Delta_N$ and by construction $m(\Delta_N^+) = m(\Delta_N^-) = m(\Delta_N)/2$. Moreover, $\sum_{i=N+1}^{\infty} a_i r_i(t) \geq 0$ for $t \in \Delta_N^+$ and $\sum_{i=N+1}^{\infty} a_i r_i(t) \leq 0$ for $t \in \Delta_N^-$.

Further, as above, for any $E \subset [0, 1]$ we denote $E^c := [0, 1] \setminus E$.

Lemma 12.2 *Let $(a_i) \in \ell_2$, $N \in \mathbb{N}$, and let $\Delta_N \subset [0, 1]$ be a dyadic interval such that $m(\Delta_N) = 2^{-N}$. Then, for any set $E \subset [0, 1]$, satisfying the inequality*

$$\frac{m(E^c \cap \Delta_N)}{m(\Delta_N)} + \sqrt{\frac{m(E^c \cap \Delta_N)}{m(\Delta_N)}} < \frac{1}{2},$$

we have

$$\int_{\Delta_N} \Big|\sum_{i=N+1}^{\infty} a_i r_i(t)\Big|^2 dt \leq \left(\frac{1}{2} - \frac{m(E^c \cap \Delta_N)}{m(\Delta_N)} - \sqrt{\frac{m(E^c \cap \Delta_N)}{m(\Delta_N)}}\right)^{-1}$$
$$\times \int_{\Delta'_N \cap E} \Big|\sum_{i=N+1}^{\infty} a_i r_i(t)\Big|^2 dt,$$

where $\Delta'_N = \Delta_N^+$ or $\Delta'_N = \Delta_N^-$.

Proof For example, take $\Delta'_N = \Delta_N^+$. First, by the additivity of the integral,

$$\int_{\Delta_N} \Big| \sum_{i=N+1}^{\infty} a_i r_i(t) \Big|^2 dt = \int_{\Delta_N^+ \cap E} \Big| \sum_{i=N+1}^{\infty} a_i r_i(t) \Big|^2 dt + \int_{\Delta_N^- \cap E} \Big| \sum_{i=N+1}^{\infty} a_i r_i(t) \Big|^2 dt + \int_{\Delta_N \cap E^c} \Big| \sum_{i=N+1}^{\infty} a_i r_i(t) \Big|^2 dt.$$

Secondly, since the Rademacher functions are symmetrically distributed, from the definition of Δ_N^- it follows

$$\int_{\Delta_N^- \cap E} \Big| \sum_{i=N+1}^{\infty} a_i r_i(t) \Big|^2 dt \le \int_{\Delta_N^-} \Big| \sum_{i=N+1}^{\infty} a_i r_i(t) \Big|^2 dt = \frac{1}{2} \int_{\Delta_N} \Big| \sum_{i=N+1}^{\infty} a_i r_i(t) \Big|^2 dt.$$

Combining these inequalities, we have

$$\frac{1}{2} \int_{\Delta_N} \Big| \sum_{i=N+1}^{\infty} a_i r_i(t) \Big|^2 dt \le \int_{\Delta_N^+ \cap E} \Big| \sum_{i=N+1}^{\infty} a_i r_i(t) \Big|^2 dt + \int_{\Delta_N \cap E^c} \Big| \sum_{i=N+1}^{\infty} a_i r_i(t) \Big|^2 dt. \tag{12.12}$$

On the other hand, by an obvious change of variables,

$$\int_{\Delta_N \cap E^c} \Big| \sum_{i=N+1}^{\infty} a_i r_i(t) \Big|^2 dt = m(\Delta_N) \int_F \Big| \sum_{i=1}^{\infty} a_{i+N} r_i(t) \Big|^2 dt \tag{12.13}$$

for some measurable set $F \subset [0, 1]$ such that

$$m(F) = \frac{m(E^c \cap \Delta_N)}{m(\Delta_N)}. \tag{12.14}$$

Moreover, by Lemma 12.1,

$$\begin{aligned} \int_F \Big| \sum_{i=1}^{\infty} a_{i+N} r_i(t) \Big|^2 dt &\le \left(m(F) + \sqrt{m(F)} \right) \int_0^1 \Big| \sum_{i=1}^{\infty} a_{i+N} r_i(t) \Big|^2 dt \\ &= \left(m(F) + \sqrt{m(F)} \right) \frac{1}{m(\Delta_N)} \int_{\Delta_N} \Big| \sum_{i=N+1}^{\infty} a_i r_i(t) \Big|^2 dt. \end{aligned}$$

This estimate, (12.12), (12.13), and (12.14) yield

$$\left(\frac{1}{2} - \frac{m(E^c \cap \Delta_N)}{m(\Delta_N)} - \sqrt{\frac{m(E^c \cap \Delta_N)}{m(\Delta_N)}}\right) \int_{\Delta_N} \left| \sum_{i=N+1}^{\infty} a_i r_i(t) \right|^2 dt$$

$$\leq \int_{\Delta_N^+ \cap E} \left| \sum_{i=N+1}^{\infty} a_i r_i(t) \right|^2 dt,$$

and the required inequality is proved for $\Delta_N' = \Delta_N^+$. The case when $\Delta_N' = \Delta_N^-$ can be considered in the same way. □

Proof of Theorem 12.2 Given $\lambda > 1$, pick $\varepsilon > 0$ small enough such that

$$0 < \frac{1}{1/2 - \varepsilon - \sqrt{\varepsilon}} < 2\lambda.$$

Since almost all points of a set of positive measure are density points (see e.g. [215, Theorem IX.5.1]), then for every measurable set E, $m(E) > 0$, we can find a dyadic interval $\Delta_N \subset [0, 1]$ of length 2^{-N} such that

$$\frac{m(\Delta_N \cap E^c)}{m(\Delta_N)} < \varepsilon.$$

Furthermore, from the fact that $\{r_i\}_{i=1}^{\infty}$ is an orthonormal sequence on $[0, 1]$ it follows

$$\sum_{i=N+1}^{\infty} a_i^2 = \frac{1}{m(\Delta_N)} \int_{\Delta_N} \left| \sum_{i=N+1}^{\infty} a_i r_i(t) \right|^2 dt.$$

Hence, applying Lemma 12.2, we get

$$\sum_{i=N+1}^{\infty} a_i^2 \leq \frac{1}{m(\Delta_N)} \frac{1}{(1/2 - \varepsilon - \sqrt{\varepsilon})} \int_{\Delta_N' \cap E} \left| \sum_{i=N+1}^{\infty} a_i r_i(t) \right|^2 dt, \tag{12.15}$$

where $\Delta_N' = \Delta_N^+$ or $\Delta_N' = \Delta_N^-$.

Let us observe that the function $\sum_{i=0}^{N} a_i r_i$ is constant on the interval Δ_N. First, assume that it is positive. Then, for $t \in \Delta_N^+$

$$\left| \sum_{i=N+1}^{\infty} a_i r_i(t) \right| = \sum_{i=N+1}^{\infty} a_i r_i(t) \leq \sum_{i=0}^{\infty} a_i r_i(t) = \left| \sum_{i=0}^{\infty} a_i r_i(t) \right|.$$

Hence, choosing $\Delta'_N = \Delta_N^+$ in (12.15), we obtain

$$\sum_{i=N+1}^{\infty} a_i^2 \le \frac{1}{m(\Delta_N)} \frac{1}{(1/2 - \varepsilon - \sqrt{\varepsilon})} \int_{\Delta_N^+ \cap E} \Big| \sum_{i=N+1}^{\infty} a_i r_i(t) \Big|^2 dt$$
$$\le \frac{1}{m(\Delta_N)} \frac{1}{(1/2 - \varepsilon - \sqrt{\varepsilon})} \int_{\Delta_N^+ \cap E} \Big| \sum_{i=0}^{\infty} a_i r_i(t) \Big|^2 dt$$
$$\le \frac{2\lambda}{m(\Delta_N)} \int_{\Delta_N^+ \cap E} \Big| \sum_{i=0}^{\infty} a_i r_i(t) \Big|^2 dt$$
$$= \frac{\lambda}{m(J_1)} \int_{J_1 \cap E} \Big| \sum_{i=0}^{\infty} a_i r_i(t) \Big|^2 dt,$$

where $J_1 = \Delta_N^+$. We argue likewise when $\sum_{i=0}^{N} a_i r_i$ is a negative constant on Δ_N, in which case we pick $J_2 = \Delta_N^-$. Thus, the claim of the theorem is proved with $\Delta = \Delta_N$, $J_1 = \Delta_N^+$, and $J_2 = \Delta_N^-$. □

Corollary 12.1 *For every measurable set $E \subset [0, 1]$ of positive measure and each $\lambda > 1$ there exists a dyadic interval $\Delta \subset [0, 1]$ such that for all $(a_i) \in \ell_2$ we have*

$$\sum_{i=N+1}^{\infty} a_i^2 \le \inf_{a \in \mathbb{R}} \frac{2\lambda}{m(\Delta)} \int_{\Delta \cap E} \Big| \sum_{i=0}^{\infty} a_i r_i(t) - a \Big|^2 dt,$$

where $N = -\log_2 m(\Delta)$.

Proof Preserving the notation of Theorem 12.2, we get

$$\max \left\{ \frac{\lambda}{m(J_1)} \int_{J_1 \cap E} \Big| \sum_{i=0}^{\infty} a_i r_i(t) \Big|^2 dt \, , \, \frac{\lambda}{m(J_2)} \int_{J_2 \cap E} \Big| \sum_{i=0}^{\infty} a_i r_i(t) \Big|^2 dt \right\}$$
$$= \frac{2\lambda}{m(\Delta)} \max \left\{ \int_{J_1 \cap E} \Big| \sum_{i=0}^{\infty} a_i r_i(t) \Big|^2 dt \, , \, \int_{J_2 \cap E} \Big| \sum_{i=0}^{\infty} a_i r_i(t) \Big|^2 dt \right\}$$
$$\le \frac{2\lambda}{m(\Delta)} \int_{\Delta \cap E} \Big| \sum_{i=0}^{\infty} a_i r_i(t) \Big|^2 dt.$$

Since the choice of the interval Δ does not depend on a_0, we may replace here a_0 with $a_0 - a$, where $a \in \mathbb{R}$ is arbitrary. Then, one can easily see that the left-hand side of the last inequality does not exceed

$$\inf_{a \in \mathbb{R}} \frac{2\lambda}{m(\Delta)} \int_{\Delta \cap E} \Big| \sum_{i=0}^{\infty} a_i r_i(t) - a \Big|^2 dt.$$

Thus, the required result follows from Theorem 12.2. □

12.3 Weighted Khintchine's Inequality

As was already observed in the beginning of the chapter, there are close connections between the upper estimate in weighted Khintchine inequality (12.3) and the Rademacher multiplicator space $\mathcal{M}(X)$ (see the previous chapter). By Theorem 2.2, the condition $\ln^{1/2}(e/t) \in X''$ assures that the Rademacher sequence is equivalent in X to the unit vector basis in ℓ_2. Therefore, by definition of the $\mathcal{M}(X)$- and $\mathrm{Sym}(X)$-norms and Theorem 11.3, we get the following result.

Proposition 12.2 *Let X be a s.s. on $[0, 1]$ such that $\ln^{1/2}(e/t) \in X''$. Then, the inequality*

$$\left\| w \cdot \sum_{i=1}^{\infty} a_i r_i \right\|_X \le C \, \|w\|_{\mathcal{M}} \, \|a\|_{\ell_2} \tag{12.16}$$

holds for some constant $C > 0$, depending only on X, and all $a = (a_i) \in \ell_2$ if and only if $w \in \mathcal{M}(X)$. In particular, inequality (12.16) is fulfilled provided that $w \in (X'')_{\ln^{1/2}}$, i.e., when $\|w^*(t) \ln^{1/2}(e/t)\|_{X''} < \infty$.

Recall that, for each $1 \le p < \infty$, the symmetric kernel of L_p is the Zygmund space $L_p(\log L)^{1/2}$, consisting of all measurable functions f such that

$$\|f\|_{L_p(\log L)^{1/2}} := \left(\int_0^1 (f^*(t) \ln^{1/2}(e/t))^p \, dt \right)^{1/p}$$

(see Example 11.1). Consequently, since $\mathrm{Sym}(X) \subset \mathcal{M}(X)$, from Proposition 12.2 it follows

Corollary 12.2 *Assume that $1 \le p < \infty$ and $w \in L_p(\log L)^{1/2}$. There exists a constant $C > 0$, depending on p, such that for any $(a_i)_{i=1}^{\infty} \in \ell_2$ we have*

$$\left(\int_0^1 \left| w(t) \cdot \sum_{i=1}^{\infty} a_i r_i(t) \right|^p dt \right)^{1/p} \le C \|w\|_{L_p(\log L)^{1/2}} \|a\|_{\ell_2},$$

Now, we turn to the study of the much more delicate lower estimate in inequality (12.3).

Further, we repeatedly use the following version of the Paley–Zygmund inequality.

Lemma 12.3 *For arbitrary $0 \le b \le 1/\sqrt{2}$ and all $a = (a_i)_{i=1}^{\infty} \in \ell_2$ we have*

$$m\left\{ t \in [0, 1] : \left| \sum_{i=1}^{\infty} a_i r_i(t) \right| \ge b \|a\|_{\ell_2} \right\} \ge \left(\frac{1}{\sqrt{2}} - b \right)^2. \tag{12.17}$$

Proof First, from Khintchine's L_1-inequality with the optimal constant (see Theorem 5.4) it follows that for all $a = (a_i) \in \ell_2$

$$\int_0^1 \Big|\sum_{i=1}^{\infty} a_i r_i(t)\Big|\, dt \ge \frac{1}{\sqrt{2}} \|a\|_{\ell_2}. \tag{12.18}$$

Moreover, according to the Paley–Zygmund inequality from Proposition 1.4, for any $d \ge b \ge 0$ and every nonnegative r.v. f on a probability space $(\Omega, \mathbb{P})$, satisfying the conditions: $\int_\Omega f(\omega)\, d\mathbb{P} \ge d$ and $\int_\Omega f(\omega)^2\, d\mathbb{P} = 1$, we have

$$\mathbb{P}\{\omega \in \Omega : f(\omega) \ge b\} \ge (d-b)^2.$$

In view of (12.18), this estimate holds for each function $f := \|a\|_{\ell_2}{}^{-1} |\sum_{i=1}^{\infty} a_i r_i(t)|$, $0 \le t \le 1$, and $d = 1/\sqrt{2}$. As a result, we obtain (12.17) for every $0 \le b \le 1/\sqrt{2}$. □

By using the fact that the Rademacher functions are symmetrically distributed, we show that the lower estimate in inequality (12.3) holds provided if the location of the support of a weighted function w satisfies certain conditions.

For an arbitrary measurable function w and any $\eta > 0$ we set

$$M_\eta(w) := \big\{t \in [0,1] : |w(t)| \ge \eta\big\}.$$

Theorem 12.3 *Let X be a s.s. on* $[0, 1]$, $w \in \mathcal{M}(X)$. *Suppose that there exists* $\eta > 0$ *such that*

$$\alpha_\eta := \max\Big\{m\big(M_\eta(w) \cap \big[0, \tfrac{1}{2}\big]\big), m\big(M_\eta(w) \cap \big[\tfrac{1}{2}, 1\big]\big)\Big\} > \frac{1}{4}. \tag{12.19}$$

Then, for all $a = (a_i) \in \ell_2$ *we have*

$$\Big\| w \cdot \sum_{i=1}^{\infty} a_i r_i \Big\|_X \ge \frac{\eta}{4C_X} \Big(\alpha_\eta - \frac{1}{4}\Big)^2 \|a\|_{\ell_2},$$

where C_X is the embedding constant X into L_1.

We obtain Theorem 12.3 as an easy consequence of the following result.

Proposition 12.3 *Let a measurable set $E \subset [0, 1]$ satisfy the condition:*

$$\alpha := \max\Big\{m\big(E \cap \big[0, \tfrac{1}{2}\big]\big), m\big(E \cap \big[\tfrac{1}{2}, 1\big]\big)\Big\} > \frac{1}{4}.$$

If $0 < b \le \frac{1}{2}(\alpha - \frac{1}{4})$, *then for all* $a = (a_i) \in \ell_2$ *we have*

$$m\Big\{t \in E : \Big|\sum_{i=1}^{\infty} a_i r_i(t)\Big| \ge b\|a\|_{\ell_2}\Big\} > \frac{1}{2}\Big(\alpha - \frac{1}{4}\Big)$$

and

$$\Big\|\chi_E \cdot \sum_{i=1}^{\infty} a_i r_i\Big\|_1 \ge \frac{1}{4}\Big(\alpha - \frac{1}{4}\Big)^2 \|a\|_{\ell_2}.$$

Proof Assume that $\alpha = m(E \cap [0, 1/2])$. For every $b > 0$ we put

$$Q_b := \Big\{t \in [0, 1/2] : \Big|\sum_{i=1}^{\infty} a_i r_i(t)\Big| \ge b\|a\|_{\ell_2}\Big\}.$$

By definition of the Rademacher functions, the set

$$\Big\{t \in [0, 1] : \Big|\sum_{i=1}^{\infty} a_i r_i(t)\Big| \ge b\|a\|_{\ell_2}\Big\}$$

is symmetric with respect to the point $1/2$, and so, from Lemma 12.3 it follows that for all $b \in [0, 1/\sqrt{2}]$

$$m(Q_b) \ge \frac{1}{2}\Big(\frac{1}{\sqrt{2}} - b\Big)^2.$$

Moreover, since $E \cap Q_b \supset E \cap [0, 1/2] \setminus ([0, 1/2] \setminus Q_b)$, we have

$$m(E \cap Q_b) \ge m(E \cap [0, 1/2]) + m(Q_b) - \frac{1}{2}.$$

Therefore, if $0 < b \le \frac{1}{2}(\alpha - \frac{1}{4})$, then

$$\begin{aligned} m(E \cap Q_b) &\ge \alpha + \frac{1}{2}\Big(\frac{1}{\sqrt{2}} - b\Big)^2 - \frac{1}{2} \\ &\ge \frac{1}{2}\Big(\alpha - \frac{1}{4}\Big) + \Big\{\frac{1}{2}\Big(\alpha - \frac{1}{4}\Big) - \frac{1}{4} + \Big(\frac{1}{\sqrt{2}} - \frac{1}{2}\Big(\alpha - \frac{1}{4}\Big)\Big)^2\Big\}. \end{aligned}$$

Taking into account that the function $z \mapsto z - 1/4 + (1/\sqrt{2} - z)^2$ is positive for all $z \in \mathbb{R}$, from the latter inequality we get

$$m(E \cap Q_b) \ge \frac{1}{2}\Big(\alpha - \frac{1}{4}\Big).$$

Thus, the first desired estimate is proved if $\alpha = m(E \cap [0, 1/2])$. The case when $\alpha = m(E \cap [1/2, 1])$ can be considered in the same way. The second inequality is an immediate consequence of the first one and Chebyshev's inequality. □

Proof of Theorem 12.3 Recall that every s.s. X on $[0, 1]$ is continuously embedded into the space L_1 with some constant C_X, that is, $\|f\|_1 \le C_X \|f\|_X$ for all $f \in X$ (see Appendix C). Hence, from inequality (12.19) and Proposition 12.3 it follows

$$\begin{aligned}\Big\| w \cdot \sum_{i=1}^{\infty} a_i r_i \Big\|_X &\ge \Big\| w \chi_{M_\eta(w)} \cdot \sum_{i=1}^{\infty} a_i r_i \Big\|_X \\ &\ge \frac{\eta}{C_X} \Big\| \chi_{M_\eta(w)} \cdot \sum_{i=1}^{\infty} a_i r_i \Big\|_1 \\ &\ge \frac{\eta}{4C_X} \Big(\alpha_\eta - \frac{1}{4}\Big)^2 \|a\|_{\ell_2},\end{aligned}$$

as we wished. □

Corollary 12.3 *Let X be a s.s., $w \in \mathcal{M}(X)$. Suppose that*

$$\alpha_w := \max \Big\{ m\big(\operatorname{supp} w \cap \big[0, \frac{1}{2}\big]\big), m\big(\operatorname{supp} w \cap \big[\frac{1}{2}, 1\big]\big) \Big\} > \frac{1}{4}.$$

Then, for any $a = (a_i) \in \ell_2$ and every $\eta > 0$ small enough we have

$$\Big\| w \cdot \sum_{i=1}^{\infty} a_i r_i \Big\|_X \ge \frac{\eta}{4C_X} \Big(\alpha_w - \frac{1}{4}\Big)^2 \|a\|_{\ell_2}.$$

In particular, the result of Corollary 12.3 holds if $m(\operatorname{supp} w) > 1/2$. As was said above, if we restrict ourselves to conditions related to the *size of measure* of the support of w, the latter condition is sharp for the left-hand side estimate in (12.3) to be valid.

Next, we are going to examine conditions of a different kind on a weight w. In contrast to the above ones, they shall be linked with the *structure* of the support of w. As we shall see, the lower estimate in (12.3) may hold even for weights with supports of arbitrarily small measure. In more detail, we consider the case when w is the characteristic function of a suitable measurable subset of $[0, 1]$. More specifically, we prove that, for a certain class of measurable sets (which contains, in particular, some sets of arbitrarily small measure), we have the following: for each $p > 0$ and any set E from this class there exists a constant $\gamma = \gamma(E, p) > 0$ such that for all $a = (a_k)_{k=1}^{\infty} \in \ell_2$ we have

$$\Big(\int_E \Big| \sum_{i=1}^{\infty} a_i r_i(t) \Big|^p dt \Big)^{1/p} \ge \gamma \|a\|_{\ell_2}. \tag{12.20}$$

As above, $\Delta_n^k = ((k-1)/2^n, k/2^n)$, $n = 0, 1, 2, \dots, k = 1, 2, \dots, 2^n$.

Theorem 12.4 *Suppose that a measurable set $E \subset [0,1]$ possesses the property that $m(E \cap (a,b)) > 0$ for any interval $(a,b) \subset [0,1]$. Then for every $p > 0$ there exists a constant $\gamma = \gamma(E,p) > 0$ such that inequality* (12.20) *holds for all $a = (a_k)_{k=1}^\infty \in \ell_2$.*

Proof We again use the fact that almost all points of any subset of positive measure are its density points (see e.g. [215, Theorem IX.5.1]). Consequently, for each $n \in \mathbb{N}$ large enough we can find $k_0 = k_0(n)$ satisfying the inequality

$$m(E \cap \Delta_n^{k_0}) > \frac{5}{6} \cdot 2^{-n}. \tag{12.21}$$

Let us fix such a positive integer n and introduce some more notation. First, define the number $\delta = \delta(E)$ as follows:

$$\delta := \min_{k=1,\dots,2^n} 2^n m(E \cap \Delta_n^k).$$

Observing that $\delta > 0$ by condition, we set $A := 2^{3+6/p}\delta^{-1/p}$. And finally, for a fixed Rademacher sum $g = \sum_{i=1}^\infty a_i r_i$, with $a = (a_i) \in \ell_2$, and $b > 0$ we denote

$$Q_{E,b} := \left\{ t \in E : |g(t)| \geq b\|a\|_{\ell_2} \right\}.$$

Next, we consider two cases depending on how the norm $\|a\|_{\ell_2}$ is distributed over the given sequence $a = (a_k)_{k=1}^\infty$. If the norm $\|a\|_{\ell_2}$ is concentrated in the "tail" of a starting from $n+1$, then our argument will be based on the fact that a larger (in measure) part of the interval $\Delta_{k_0}^n$ lies in the set E (see inequality (12.21)). Otherwise, if the norm $\|a\|_{\ell_2}$ is concentrated in the corresponding "beginning" of this sequence, we shall employ the fact that a certain part of E is contained in each of the intervals Δ_k^n, $k = 1, 2, \dots, 2^n$, because $\delta > 0$.

So, we assume first that

$$\|a\|_{\ell_2} \leq A\Big(\sum_{i=n+1}^\infty a_i^2\Big)^{1/2}. \tag{12.22}$$

Then, since the sum $\sum_{i=1}^n a_i r_i$ is constant on the interval $\Delta_n^{k_0}$ (and equals, say, a_0), we obtain

$$\begin{aligned} m(Q_{E,b}) &\geq m\Big\{ t \in E : |g(t)| \geq Ab\Big(\sum_{i=n+1}^\infty a_i^2\Big)^{1/2}\Big\} \\ &\geq m\Big\{ t \in E \cap \Delta_n^{k_0} : \Big|a_0 + \sum_{i=n+1}^\infty a_i r_i(t)\Big| \geq Ab\Big(\sum_{i=n+1}^\infty a_i^2\Big)^{1/2}\Big\}. \end{aligned}$$

Let F denote the union of the subsets of the intervals Δ_n^k, $k = 1, 2, \dots, 2^n$, obtained by shifting the set $E \cap \Delta_n^{k_0}$ by suitable multiples of 2^{-n}. Then, $m(F) = 2^n m(E \cap \Delta_n^{k_0}) > 5/6$ by (12.21), and from definition of the Rademacher functions and the preceding inequality it follows that

$$m(Q_{E,b}) \geq 2^{-n} m\Big\{t \in F : \Big|a_0 + \sum_{i=n+1}^{\infty} a_i r_i(t)\Big| \geq Ab\Big(\sum_{i=n+1}^{\infty} a_i^2\Big)^{1/2}\Big\}. \tag{12.23}$$

On the other hand, by Lemma 12.3, assuming that $0 \leq b \leq 1/(\sqrt{2}A)$, we have

$$m\Big\{t \in [0,1] : \Big|\sum_{i=n+1}^{\infty} a_i r_i(t)\Big| \geq Ab\Big(\sum_{i=n+1}^{\infty} a_i^2\Big)^{1/2}\Big\} \geq \Big(\frac{1}{\sqrt{2}} - Ab\Big)^2,$$

and so, since Rademacher sums are symmetrically distributed on $[0, 1]$,

$$m\Big\{t \in [0,1] : \Big|a_0 + \sum_{i=n+1}^{\infty} a_i r_i(t)\Big| \geq Ab\Big(\sum_{i=n+1}^{\infty} a_i^2\Big)^{1/2}\Big\} \geq \frac{1}{2}\Big(\frac{1}{\sqrt{2}} - Ab\Big)^2.$$

Setting

$$b_0 := \frac{1}{A\sqrt{2}}\left(1 - \frac{2}{\sqrt{5}}\right), \tag{12.24}$$

we get $\left(1/\sqrt{2} - Ab_0\right)^2 = 2/5$; therefore, in view of the last inequality,

$$m\Big\{t \in F : \Big|a_0 + \sum_{i=n+1}^{\infty} a_i r_i(t)\Big| \geq Ab_0\Big(\sum_{i=n+1}^{\infty} a_i^2\Big)^{1/2}\Big\} \geq m(F) +$$

$$+ m\Big\{t \in [0,1] : \Big|a_0 + \sum_{i=n+1}^{\infty} a_i r_i(t)\Big| \geq Ab_0\Big(\sum_{i=n+1}^{\infty} a_i^2\Big)^{1/2}\Big\} - 1 > \frac{5}{6} + \frac{1}{5} - 1 = \frac{1}{30}.$$

Combining this with (12.23), we infer that $m(Q_{E,b_0}) \geq 2^{-n}/30$, and hence

$$\int_E |g(t)|^p \, dt \geq \int_{Q_{E,b_0}} |g(t)|^p \, dt \geq \frac{b_0^p 2^{-n}}{30} \|a\|_{\ell_2}^p.$$

As a result, we obtain (12.20) with $\gamma = \gamma_1 = b_0 2^{-n/p}/30^{1/p}$, where b_0, defined by (12.24), depends only on E.

Consider now the case when inequality (12.22) does not hold. Since $A \geq 2$, then we have

$$\sum_{i=n+1}^{\infty} a_i^2 \leq \sum_{i=1}^{n} a_i^2,$$

whence

$$m(Q_{E,b}) \geq m\Big\{t \in E : |g(t)| \geq \sqrt{2}b\Big(\sum_{i=1}^{n} a_i^2\Big)^{1/2}\Big\}. \tag{12.25}$$

Note that the sum $\sum_{i=1}^{n} a_i r_i$ is constant on each of the intervals Δ_n^k, $k = 1, 2, \ldots, 2^n$. In consequence, applying inequality (12.17) once more, we obtain that

$$m\Big\{t \in [0,1] : \Big|\sum_{i=1}^{n} a_i r_i(t)\Big| \geq \sqrt{2}b\Big(\sum_{i=1}^{n} a_i^2\Big)^{1/2}\Big\}$$
$$= \sum_{i \in I} m(\Delta_n^i) = \text{card}\, I \cdot 2^{-n} \geq \Big(\frac{1}{\sqrt{2}} - \sqrt{2}b\Big)^2$$

for some set of indices $I \subset \{1, 2, \ldots, 2^n\}$ and any $0 \leq b \leq 1/2$. Therefore, inserting $b = 1/4$ into this inequality, we arrive at the estimate

$$\text{card}\, I \geq 2^{n-3}.$$

Thus, if the sum $\sum_{i=1}^{n} a_i r_i$ is equal, say, to b_k on the interval Δ_n^k, $k = 1, 2, \ldots, 2^n$, then $|b_k| \geq \frac{\sqrt{2}}{4}\left(\sum_{i=1}^{n} a_i^2\right)^{1/2}$ for each $k \in I$. So, for any $t \in \Delta_n^k$, $k \in I$, at least one of the inequalities

$$\Big|b_k + \sum_{i=n+1}^{\infty} a_i r_i(t)\Big| \geq \frac{\sqrt{2}}{4}\Big(\sum_{i=1}^{n} a_i^2\Big)^{1/2} \quad \text{or} \quad \Big|b_k - \sum_{i=n+1}^{\infty} a_i r_i(t)\Big| \geq \frac{\sqrt{2}}{4}\Big(\sum_{i=1}^{n} a_i^2\Big)^{1/2}$$

holds. Further, for every $k \in I$ we introduce the sets

$$U_k := \Big\{t \in E \cap \Delta_n^k : \Big|b_k + \sum_{i=n+1}^{\infty} a_i r_i(t)\Big| \geq \frac{\sqrt{2}}{4}\Big(\sum_{i=1}^{n} a_i^2\Big)^{1/2}\Big\}$$

and

$$\tilde{U}_k := \Big\{t \in E \cap \Delta_n^k : \Big|b_k - \sum_{i=n+1}^{\infty} a_i r_i(t)\Big| \geq \frac{\sqrt{2}}{4}\Big(\sum_{i=1}^{n} a_i^2\Big)^{1/2}\Big\}.$$

Clearly, from the definition of δ and the above discussion it follows that

$$\max(m(U_k), m(\tilde{U}_k)) \geq \delta \cdot 2^{-n-1}$$

for arbitrary $k \in I$. Hence, $I = I_1 \cup I_2$, where

$$I_1 := \{k \in I : m(U_k) \geq \delta \cdot 2^{-n-1}\} \text{ and } I_2 := \{k \in I : m(\tilde{U}_k) \geq \delta \cdot 2^{-n-1}\}.$$

Thus, either card I_1 or card I_2 is greater than 2^{n-4}. In the former case, by (12.25), we have

$$\begin{aligned} m(Q_{E,1/4}) &\geq \sum_{k \in I_1} m\Big\{t \in E \cap \Delta_n^k : |g(t)| \geq \frac{\sqrt{2}}{4}\Big(\sum_{i=1}^n a_i^2\Big)^{1/2}\Big\} \\ &= \sum_{k \in I_1} m(U_k) \geq \delta \cdot 2^{-n-1} \text{card } I_1 \geq \delta \cdot 2^{-5}. \end{aligned} \tag{12.26}$$

In the latter one, denoting

$$\tilde{Q}_{E,1/4} := \Big\{t \in E : |\tilde{g}(t)| \geq \frac{1}{4}\|a\|_{\ell_2}\Big\},$$

where $\tilde{g} := \sum_{i=1}^n a_i r_i - \sum_{i=n+1}^\infty a_i r_i$, we similarly obtain

$$m(\tilde{Q}_{E,1/4}) > \delta \cdot 2^{-5}. \tag{12.27}$$

If (12.26) is valid, it follows immediately

$$\int_E |g(t)|^p \, dt \geq \int_{Q_{E,1/4}} |g(t)|^p \, dt \geq 4^{-p}\|a\|_{\ell_2}^p m(Q_{E,1/4}) \geq \delta \cdot 2^{-5-2p}\|a\|_{\ell_2}^p,$$

i.e., (12.20) is satisfied with the constant $\gamma = \delta^{1/p} \cdot 2^{-2-5/p}$.

Let now (12.27) hold. Then, as above,

$$\int_E |\tilde{g}(t)|^p \, dt \geq \delta \cdot 2^{-5-2p}\|a\|_{\ell_2}^p.$$

Moreover, since (12.22) fails in this case, we have

$$\Big(\sum_{i=n+1}^\infty a_i^2\Big)^{1/2} \leq A^{-1}\|a\|_{\ell_2}.$$

Without loss of generality, we may assume that $0 < p \leq 1$. From two last inequalities, the upper estimate in Khintchine's inequality (see Theorem 1.4), and definition of the number A it follows that

$$\begin{aligned}\int_E |g(t)|^p\,dt &\ge \int_E |\tilde{g}(t)|^p\,dt - 2^p \int_0^1 \Big|\sum_{i=n+1}^{\infty} a_i r_i(t)\Big|^p\,dt \\ &\ge \delta\cdot 2^{-5-2p}\|a\|_{\ell_2}^p - 2^p\Big(\sum_{i=n+1}^{\infty} a_i^2\Big)^{p/2} \\ &\ge (\delta\cdot 2^{-5-2p} - 2^p A^{-p})\|a\|_{\ell_2}^p = \delta 2^{-6-2p}\|a\|_{\ell_2}^p.\end{aligned}$$

Thus, inequality (12.20) holds in this case as well with the constant $\gamma = \gamma_2 = \delta^{1/p}\cdot 2^{-2-6/p}$. Taking into account the fact that $\gamma_1 < \gamma_2$, we set $\gamma = \gamma_1 = b_0 2^{-n/p}/30^{1/p}$, where b_0 is defined in (12.24). Since this constant depends only on E (because the choice of n is determined by this set), the theorem is proved. □

Remark 12.2 One can see that the value of the constant γ depends on $n \in \mathbb{N}$, which is chosen, to some extent, arbitrarily. Moreover, the function $\gamma(E, p, n) = b_0 2^{-n/p}/30^{1/p}$ decreases with increasing $n \in \mathbb{N}$. Therefore, to obtain inequality (12.20) with the greatest constant, we need to find the least positive integer n such that (12.21) holds for some $1 \le k_0 \le 2^n$. Then, this constant will depend both on the chosen n and on $\delta(n)$, i.e., on how uniformly the set E is distributed over the dyadic intervals of rank n.

Since every s.s. X on [0, 1] is continuously embedded into $L_1[0, 1]$ with some constant C_X, we obtain the following result.

Corollary 12.4 *Let X be a s.s. on* [0, 1]. *Suppose that a measurable set $E \subset [0, 1]$ is such that $m(E \cap (a, b)) > 0$ for each interval $(a, b) \subset [0, 1]$. Then for all $a = (a_k)_{k=1}^{\infty} \in \ell_2$ we have*

$$\Big\|\chi_E\cdot\sum_{i=1}^{\infty} a_i r_i\Big\|_X \ge \frac{\gamma(E,1)}{C_X}\|a\|_{\ell_2},$$

where $\gamma(E, 1)$ is the constant from Theorem 12.4.

Under some conditions, a result similar to Theorem 12.4 may be obtained also for more general weighted functions.

We say that a measurable set $E \subset [0, 1]$ belongs to the class $\mathcal{F}$ if there exist $n \in \mathbb{N}$, $\varepsilon \in (0, 2^{-n-2})$, $\delta \in (0, 1)$, and $\gamma \in (1/2, 1)$ such that the following conditions are satisfied:

(i) for some set of indices $I \subset \{1, 2, \ldots, 2^n\}$, with card $I > \gamma 2^n$, we have $m(E \cap \Delta_n^k) > \delta$;
(ii) there is k_0, $1 \le k_0 \le 2^n$, such that $m\big(E \cap \Delta_n^{k_0}\big) > \varepsilon + 3\cdot 2^{-n-2}$.

In particular, if a set E satisfies the condition $m(E \cap (a, b)) > 0$ for each interval $(a, b) \subset [0, 1]$ (as in Theorem 12.4), then $E \in \mathcal{F}$.

Let X be a s.s. on $[0, 1]$. For arbitrary $w \in \mathcal{M}(X)$ and $\eta > 0$ we define the set $M_\eta(w)$ in the same way as above (see Theorem 12.3). Then, reasoning in a similar way as in the proof of Theorem 12.4 and using Proposition 12.2, we arrive at the following result.

Theorem 12.5 *Suppose that* X *is a s.s. on* $[0, 1]$ *such that* $\ln^{1/2}(e/t) \in X''$ *and* $w \in \mathcal{M}(X)$. *If there exist* $\eta > 0$ *and a set* $F \in \mathcal{F}$ *such that* $M_\eta(w) \supset F$, *then for every sequence* $a = (a_i) \in \ell_2$ *we have*

$$c_w \|a\|_{\ell_2} \le \left\| w \cdot \sum_{i=1}^{\infty} a_i r_i \right\|_X \le C_X \|w\|_{\mathcal{M}} \|a\|_{\ell_2},$$

where the constant c_w *(resp.* C_X*) depends only on* w *(resp.* X*).*

12.4 Various Types of Khintchine's L_1-Inequality

In the concluding section of the chapter, we consider some L_1-estimates of Rademacher sums that refine and complement Khintchine's L_1-inequality from Theorem 5.4.

When studying various issues related to the Rademacher system, a substantial role is played by the sums $n^{-1/2}\sum_{k=1}^n r_k$, $n = 1, 2, \ldots$. In particular, as was proved in Theorem 5.3, for any $n \in \mathbb{N}$ and $p \in \{2\} \cup [3, \infty)$

$$\max\left\{ \left\| \sum_{k=1}^n a_k r_k \right\|_p : \sum_{k=1}^n a_k^2 = 1 \right\} = \left(\int_0^1 \left| \frac{1}{\sqrt{n}} \sum_{k=1}^n r_k(t) \right|^p dt \right)^{1/p}.$$

In the first theorem we establish some relations between the norms $\|\sum_{k=1}^n a_k r_k\|_1$ and $\|n^{-1/2}\sum_{k=1}^n r_k\|_1$ that turn into equalities for vectors $a = (a_1, \ldots, a_n)$, $\|a\|_{\ell_2} = 1$, with equal components.

Further, we adopt the following notation. For each vector $a = (a_1, \ldots, a_n) \in \mathbb{R}^n$, $n \ge 2$, let $a_1^\diamond, \ldots, a_n^\diamond$ be the sequence of the absolute values of the components of a in increasing order. Obviously, there exists a positive integer k, $1 \le k < n$, such that

$$a_k^\diamond \le \frac{\|a\|_{\ell_2}}{\sqrt{n}} \le a_{k+1}^\diamond. \tag{12.28}$$

Theorem 12.6 *For arbitrary* $n \in \mathbb{N}$, $n \ge 2$, *and* $a = (a_1, \ldots, a_n) \in \mathbb{R}^n$ *we have*

$$\left\| \sum_{i=1}^n a_i r_i \right\|_1 \ge \left\| \frac{1}{\sqrt{n}} \sum_{i=1}^n r_i \right\|_1 \|a\|_{\ell_2} + \frac{1}{2}\left(\|a\|_{\ell_1} - \sqrt{n}\|a\|_{\ell_2} \right), \tag{12.29}$$

$$\Big\| \sum_{i=1}^{n} a_i r_i \Big\|_1 \geq \Big\| \frac{1}{\sqrt{n}} \sum_{i=1}^{n} r_i \Big\|_1 \|a\|_{\ell_2} + \frac{1}{2} \left(a_1^{\diamond} + \cdots + a_k^{\diamond} - \frac{k}{\sqrt{n}} \|a\|_{\ell_2} \right), \tag{12.30}$$

and

$$\Big\| \sum_{i=1}^{n} a_i r_i \Big\|_1 \leq \Big\| \frac{1}{\sqrt{n}} \sum_{i=1}^{n} r_i \Big\|_1 \|a\|_{\ell_2} + \left(a_{k+1}^{\diamond} + \cdots + a_n^{\diamond} - \frac{n-k}{\sqrt{n}} \|a\|_{\ell_2} \right), \tag{12.31}$$

where k is defined by (12.28).

An equality in either of these relations is attained at $a_1 = a_2 = \cdots = a_n = \|a\|_{\ell_2} n^{-1/2}$.

We begin with an auxiliary assertion. Denote by D the subset of $\mathbb{R}^3$, consisting of vectors (a, b, c) such that $a \geq b \geq 0$ and $c \geq 0$. Also, let $f : D \to \mathbb{R}_+$ be the function defined by

$$f(a, b, c) := |a + b + c| + |a - b + c| + |a + b - c| + |a - b - c|.$$

Lemma 12.4 *Let*

$$g_1(a, b, c) := f(a, b, c) - 2a - 2b, \quad (a, b, c) \in D,$$

and

$$g_2(a, b, c) := -f(a, b, c) + 4a, \quad (a, b, c) \in D.$$

Then, for $i = 1, 2$ and any $a, b, a_1, b_1, c > 0$ such that $a \geq a_1 \geq b_1 \geq b$ and $a^2 + b^2 = a_1^2 + b_1^2$ we have $g_i(a, b, c) \geq g_i(a_1, b_1, c)$.

Proof Since the proof for both functions is quite similar, we restrict ourselves to the consideration of g_1. Clearly, it suffices to prove that the function $G_1(t) := g_1(r \cos t, r \sin t, c)$ decreases on the interval $[0, \pi/4]$ for arbitrary fixed $r, c > 0$. Moreover, thanks to the homogeneity of G_1, we can assume that $r = 1$. Therefore,

$$G_1(t) = -2 \sin t + 2c + |\cos t + \sin t - c| + |\cos t - \sin t - c|.$$

Let $0 < c < 1$. Then, if $t \in [0, \arccos(\sqrt{2}c/2) - \pi/4]$, we have $G_1(t) = 2 \cos t - 2 \sin t$ and, by calculation the derivative, one can easily see that G_1 decreases. If $t \in [\arccos(\sqrt{2}c/2) - \pi/4, \pi/4]$, then $G_1(t) = 2c$. Since G_1 is a continuous function, we can conclude that it is decreasing on the whole interval $[0, \pi/4]$. The cases when $1 \leq c < \sqrt{2}$ and $c \geq \sqrt{2}$ may be considered in a similar way. □

Proof of Theorem 12.6 Observe first that (12.30) is an easy consequence of inequality (12.29). Indeed, we need only to check that the right-hand side of (12.30) does not exceed the right-hand side of (12.29), that is,

$$a_1^{\diamondsuit} + \cdots + a_k^{\diamondsuit} - \frac{k}{\sqrt{n}}\|a\|_{\ell_2} \leq \|a\|_1 - \sqrt{n}\|a\|_{\ell_2}$$

But this inequality is equivalent to the estimate

$$\sqrt{n}(a_{k+1}^{\diamondsuit} + \cdots + a_n^{\diamondsuit}) \geq (n-k)\|a\|_{\ell_2},$$

which immediately follows from inequality (12.28).

Let us prove (12.29). Since $\{r_i\}_{i=1}^{\infty}$ is a symmetric basic sequence with constant 1 in every s.s. by Proposition 2.2, we can assume that $0 \leq a_1 \leq \cdots \leq a_n$. Furthermore, if $n = 2$, then (12.29) is equivalent to the obvious inequality $a_2 \geq (a_1 + a_2)/2$. Consequently, it suffices to consider the case when $n > 2$.

If $a_1 = a_2 = \cdots = a_n$, then (12.29) turns to an equality. So, let $a_1 < a_n$. Then, using the notation from Lemma 12.4, we get

$$\begin{aligned}
\Big\|\sum_{i=1}^{n} a_i r_i\Big\|_1 &= \frac{1}{2^n}\sum_{\varepsilon_i=\pm 1}\Big|\sum_{i=1}^{n} a_i\varepsilon_i\Big| \\
&= \frac{1}{2^{n-2}}\sum_{\varepsilon_i=\pm 1}\frac{1}{4}\Bigg(\Big|a_n + a_1 + \Big|\sum_{i=2}^{n-1} a_i\varepsilon_i\Big|\Big| \\
&\quad + \Big|a_n - a_1 + \Big|\sum_{i=2}^{n-1} a_i\varepsilon_i\Big|\Big| + \Big|a_n + a_1 - \Big|\sum_{i=2}^{n-1} a_i\varepsilon_i\Big|\Big| \\
&\quad + \Big|a_n - a_1 - \Big|\sum_{i=2}^{n-1} a_i\varepsilon_i\Big|\Big|\Bigg) \\
&= \frac{1}{2^{n-2}}\sum_{\varepsilon_i=\pm 1}\frac{1}{4}g_1\Big(a_n, a_1, \Big|\sum_{i=2}^{n-1} a_i\varepsilon_i\Big|\Big) + \frac{1}{2}(a_n + a_1).
\end{aligned}$$

Suppose that $\|a\|_{\ell_2} n^{-1/2} < \sqrt{(a_n^2 + a_1^2)/2}$. Then, if a_n' is chosen in such a way that $a_n^2 + a_1^2 = (a_n')^2 + \|a\|_{\ell_2}^2 n^{-1}$, we have $a_n > a_n' > \|a\|_{\ell_2} n^{-1/2} > a_1$. Therefore, from Lemma 12.4 (for g_1) it follows that

$$\Big\|\sum_{i=1}^{n} a_i r_i\Big\|_1 \geq \frac{1}{2^{n-2}}\sum_{\varepsilon_i=\pm 1}\frac{1}{4}g_1\Big(a_n', \frac{\|a\|_{\ell_2}}{\sqrt{n}}, \Big|\sum_{i=2}^{n-1} a_i\varepsilon_i\Big|\Big) + \frac{1}{2}(a_n + a_1)$$

$$= \left\| \frac{\|a\|_{\ell_2}}{\sqrt{n}} r_1 + \sum_{i=2}^{n-1} a_i r_i + a_n' r_n \right\|_1 + \frac{1}{2}\left(a_n + a_1 - a_n' - \frac{\|a\|_{\ell_2}}{\sqrt{n}} \right).$$

In the case when $\|a\|_{\ell_2} n^{-1/2} > \sqrt{(a_n^2 + a_1^2)/2}$, we define a_1' so that $a_n^2 + a_1^2 = (a_1')^2 + \|a\|_{\ell_2}^2 n^{-1}$. Then, applying Lemma 12.4 once more, we obtain

$$\left\| \sum_{i=1}^{n} a_i r_i \right\|_1 \geq \left\| a_1' r_1 + \sum_{i=2}^{n-1} a_i r_i + \frac{\|a\|_{\ell_2}}{\sqrt{n}} r_n \right\|_1 + \frac{1}{2}\left(a_n + a_1 - a_1' - \frac{\|a\|_{\ell_2}}{\sqrt{n}} \right).$$

Finally, if $\|a\|_{\ell_2} n^{-1/2} = \sqrt{(a_n^2 + a_1^2)/2}$, we have

$$\left\| \sum_{i=1}^{n} a_i r_i \right\|_1 \geq \left\| \frac{\|a\|_{\ell_2}}{\sqrt{n}} r_1 + \sum_{i=2}^{n-1} a_i r_i + \frac{\|a\|_{\ell_2}}{\sqrt{n}} r_n \right\|_1 + \frac{1}{2}\left(a_n + a_1 - 2\frac{\|a\|_{\ell_2}}{\sqrt{n}} \right).$$

Thus, by using the above procedure, one arrives eventually at some inequality of the form

$$\left\| \sum_{i=1}^{n} a_i r_i \right\|_1 \geq \left\| \sum_{i=1}^{n} b_i r_i \right\|_1 + \frac{1}{2}(a_1 + \cdots + a_n - b_1 - \cdots - b_n),$$

where $0 \leq b_1 \leq \cdots \leq b_n$, $b_1^2 + \cdots + b_n^2 = a_1^2 + \cdots + a_n^2$ and at least one of the numbers b_i is equal to $\|a\|_{\ell_2} n^{-1/2}$.

After at most $n-1$ similar steps, all the numbers b_i will be equal to $\|a\|_{\ell_2} n^{-1/2}$, and, as a result, we get

$$\left\| \sum_{i=1}^{n} a_i r_i \right\|_1 \geq \left\| \frac{1}{\sqrt{n}} \sum_{i=1}^{n} r_i \right\|_1 \cdot \|a\|_{\ell_2} + \frac{1}{2}(a_1 + \cdots + a_n - \sqrt{n}\|a\|_{\ell_2}).$$

Thus, the proof of (12.29) is completed.

Inequality (12.31) can be proved similarly by using the result of Lemma 12.4 for g_2. □

The next refinement of Khintchine's L_1-inequality gives an upper estimate for some special deviation of the L_1-norm of a Rademacher sum from a certain multiple of the ℓ_2-norm of the sequence of its coefficients.

Theorem 12.7 *For arbitrary $n \in \mathbb{N}$ and $a = (a_1, \dots, a_n) \in \mathbb{R}^n$ it holds*

$$\left| \Big\| \sum_{k=1}^{n} a_k r_k \Big\|_1 - \sqrt{\frac{2}{\pi}} \|a\|_{\ell_2} \right| \le \left(1 - \sqrt{\frac{2}{\pi}}\right) \|a\|_{\ell_\infty}. \tag{12.32}$$

Furthermore, inequality (12.32) *turns into an equality for* $a = e_1 = (1, 0, \dots, 0)$.

For proving this theorem we need some auxiliary assertions.

Lemma 12.5 *For every vector $a = (a_1, \dots, a_n) \in \mathbb{R}^n$ we have*

$$\Big\| \sum_{k=1}^{n} a_k r_k \Big\|_1 = \frac{2}{\pi} \int_0^\infty \frac{1 - \prod_{j=1}^{n} \cos(a_j t)}{t^2}\, dt. \tag{12.33}$$

Proof First of all, observe that for each $z \in \mathbb{R}$

$$|z| = \frac{2}{\pi} \int_0^\infty (1 - \cos(zt)) t^{-2}\, dt. \tag{12.34}$$

Indeed, after the change $u = |z|t$ and integration by parts we get

$$\int_0^\infty (1 - \cos(zt)) t^{-2}\, dt = |z| \int_0^\infty (1 - \cos u) u^{-2}\, du = |z| \int_0^\infty \frac{\sin u}{u}\, du = |z| \cdot \frac{\pi}{2},$$

which is equivalent to (12.34).

Further, denote

$$f := \sum_{k=1}^{n} a_k r_k \quad \text{and} \quad \phi(t) := \int_0^1 e^{itf(s)}\, ds.$$

Since f is symmetrically distributed, then

$$\phi(t) = \int_0^1 (\cos(tf(s)) + i \sin(tf(s)))\, ds = \int_0^1 \cos(tf(s))\, ds.$$

Therefore, from Fubini's theorem and equation (12.34) it follows that

$$\begin{aligned} \frac{2}{\pi} \int_0^\infty (1 - \phi(t)) t^{-2}\, dt &= \frac{2}{\pi} \int_0^\infty \int_0^1 (1 - \cos(tf(s)))\, ds \cdot t^{-2}\, dt \\ &= \frac{2}{\pi} \int_0^1 \int_0^\infty (1 - \cos(tf(s))) t^{-2}\, dt\, ds = \int_0^1 |f(s)|\, ds. \end{aligned}$$

By the independence of the Rademacher functions, we have

$$\phi(t) = \prod_{k=1}^{n} \int_0^1 e^{ita_k r_k(s)} = \prod_{k=1}^{n} \frac{1}{2}\left(e^{ita_k} + e^{-ita_k}\right) ds = \prod_{k=1}^{n} \cos(ta_k),$$

and hence (12.33) follows from the preceding equation. □

Lemma 12.6 *For every vector $a = (a_1, \ldots, a_n) \in \mathbb{R}^n$*

$$\|a\|_{\ell_2} = c_n \int_{\mathbb{R}^n} \frac{1 - \cos\langle a, b\rangle}{\|b\|_{\ell_2}^{n+1}}\, db, \tag{12.35}$$

where $\langle a, b\rangle$ is the usual inner product in $\mathbb{R}^n$ and $c_n := \Gamma((n+1)/2)\pi^{-\frac{n+1}{2}}$ (as above, $\Gamma(x)$ denotes the Γ-function).

Proof Since the integral in (12.35) is invariant with respect to rotations, we can assume that $a = ve_1$, with $v := \|a\|_{\ell_2}$. Then, introducing polar coordinates, we find

$$\begin{aligned} I := \int_{\mathbb{R}^n} \frac{1 - \cos\langle a, b\rangle}{\|b\|_{\ell_2}^{n+1}}\, db &= \int_{\mathbb{R}^n} \frac{1 - \cos(vb_1)}{\|b\|_{\ell_2}^{n+1}}\, db \\ &= m(S^{n-2}) \int_0^{\pi} (\sin\theta)^{n-2} \Big(\int_0^{\infty} \frac{s^{n-1}(1 - \cos(vs\cos\theta))}{s^{n+1}}\, ds \Big)\, d\theta, \end{aligned}$$

where $m(S^{n-2})$ is the Lebesgue measure of the unit euclidean sphere in $\mathbb{R}^{n-2}$. Hence, by equation (12.34),

$$I = m(S^{n-2}) \cdot \frac{\pi}{2} \cdot v \int_0^{\pi} (\sin\theta)^{n-2} |\cos\theta|\, d\theta.$$

Finally, rewriting the last integral in terms of the Γ-function and taking into account the equalities $m(S^{n-2}) = 2\pi^{\frac{n-1}{2}} / \Gamma((n-1)/2)$ and $z\Gamma(z) = \Gamma(z+1)$, we obtain that $I = c_n^{-1}\|a\|_{\ell_2}$. The proof is completed. □

Lemma 12.7 *For all $0 \le x \le 1$ and $0 \le t \le \pi/2$ the following inequality holds:*

$$\cos(xt) \ge (\cos t)^{x^2}.$$

Proof Let $g(s) := \cos s \cdot \exp(s^2/2)$. Then, $g(0) = 1$ and

$$g'(s) = (s\cos s - \sin s)\exp(s^2/2) \le 0,$$

because $\operatorname{tg} s \geq s$ for $s \in [0, \pi/2)$. Hence, $g(s) \leq 1$, which means that $\cos s \leq \exp(-s^2/2)$, or equivalently $\ln(1/\cos s) \geq s^2/2$. Thus,

$$2\ln(1/\cos s) \geq s^2 \geq \sin^2 s, \quad s \in [0, \pi/2). \tag{12.36}$$

Let us check that the function $f(s) := \sqrt{\ln(1/\cos s)}$ is convex on $[0, \pi/2)$. Indeed, we have

$$f'(s) = \frac{\sin s}{2\cos s\sqrt{\ln(1/\cos s)}}$$

and

$$f''(s) = \frac{1}{4} \cdot \frac{2\ln(1/\cos s) - \sin^2 s}{\cos^2 s(\ln(1/\cos s))^{3/2}}.$$

In view of (12.36), the numerator of the last fraction is positive for $s \in (0, \pi/2)$. Therefore, $f''(s) > 0$ on $(0, \pi/2)$, and so the claim is proved.

Since $f(0) = 0$, we have $f(xt) \leq xf(t)$ by convexity, whence $\cos(xt) \geq (\cos t)^{x^2}$ for arbitrary $0 \leq x \leq 1$ and $0 \leq t \leq \pi/2$. □

Lemma 12.8 *For each $\alpha \geq 1$ we have*

$$\sqrt{\alpha}\int_0^{\pi/2} (\cos t)^{\alpha} \ln\left(\frac{1}{\cos t}\right) t^{-2}\, dt < \sqrt{\frac{\pi}{8}}.$$

Proof We first prove that

$$\sqrt{\cos t}\ln\left(\frac{1}{\cos t}\right) \leq \frac{t^2}{2}, \quad 0 \leq t < \frac{\pi}{2}. \tag{12.37}$$

Setting $u(t) := \sqrt{\cos t}\ln(1/\cos t) - t^2/2$, we have

$$u'(t) = \frac{\sin t}{\sqrt{\cos t}}\left(\frac{1}{2}\ln(\cos t) + 1\right) - t,$$

and then, because $\ln(\cos t) \leq 0$ for $0 \leq t < \pi/2$,

$$u''(t) = \frac{1 + \cos^2 t}{4\cos^{3/2} t}\ln(\cos t) + \sqrt{\cos t} - 1 \leq 0.$$

Combining this inequality with the fact that $u(0) = u'(0) = 0$, we arrive at (12.37).

Next, multiplying (12.37) by a suitable expression, then integrating and passing to the Γ-function, we get

$$\sqrt{\frac{8\alpha}{\pi}}\int_0^{\pi/2} (\cos t)^{\alpha}\ln\left(\frac{1}{\cos t}\right) t^{-2}\, dt \leq \sqrt{\frac{2\alpha}{\pi}}\int_0^{\pi/2} (\cos t)^{\alpha - 1/2}\, dt = \sqrt{\frac{\alpha}{2}} \cdot \frac{\Gamma\left(\frac{\alpha}{2} + \frac{1}{4}\right)}{\Gamma\left(\frac{\alpha}{2} + \frac{3}{4}\right)}.$$

Since from definition of the Γ-function and easy calculations it follows that for all $x \geq 3/4$

$$\frac{\sqrt{x - 1/4} \cdot \Gamma(x)}{\Gamma(x + 1/2)} < 1,$$

the lemma is proved. □

Next, for any $x \geq 0$ we set

$$\Phi(x) := \frac{2}{\pi} \int_{\pi/2}^{\infty} \frac{\cos t \cos(xt)}{t^2}\, dt.$$

Clearly,

$$\Phi(x) := \frac{1}{\pi} \Big(\int_{\pi/2}^{\infty} \frac{\cos((x+1)t)}{t^2}\, dt + \int_{\pi/2}^{\infty} \frac{\cos(|x-1|t)}{t^2}\, dt \Big). \tag{12.38}$$

Then, integrating by parts, we obtain

$$\begin{aligned} \int_{\pi/2}^{\infty} \frac{\cos((x+1)t)}{t^2}\, dt &= \frac{2}{\pi} \cos((x+1)\pi/2) - (x+1) \int_{\pi/2}^{\infty} \frac{\sin((x+1)t)}{t}\, dt \\ &= \frac{2}{\pi} \cos((x+1)\pi/2) - (x+1) \int_{(x+1)\pi/2}^{\infty} \frac{\sin t}{t}\, dt. \end{aligned}$$

Denoting now

$$\mathrm{Si}(x) := \int_0^x \frac{\sin t}{t}\, dt,$$

by the equality $\mathrm{Si}(+\infty) = \pi/2$, we have

$$\int_{\pi/2}^{\infty} \frac{\cos((x+1)t)}{t^2}\, dt = \frac{2}{\pi} \cos((x+1)\pi/2) - (x+1)\pi/2 + (x+1)\mathrm{Si}((x+1)\pi/2).$$

Similarly,

$$\int_{\pi/2}^{\infty} \frac{\cos(|x-1|t)}{t^2}\, dt = \frac{2}{\pi} \cos((x-1)\pi/2) - |x-1|\pi/2 + |x-1|\mathrm{Si}(|x-1|\pi/2).$$

Consequently, (12.38) yields

$$\Phi(x) = \frac{1}{\pi}(x+1)\mathrm{Si}((x+1)\pi/2) + \frac{1}{\pi}|x-1|\mathrm{Si}(|x-1|\pi/2) - \max(1, x). \tag{12.39}$$

Hence, in particular, it follows that

$$\Phi(0) = \frac{2}{\pi}\mathrm{Si}(\pi/2) - 1 \approx -0,1273, \quad \Phi(1) = \frac{2}{\pi}\mathrm{Si}(\pi) - 1 \approx 0,1790. \qquad (12.40)$$

Lemma 12.9 *For any $x \geq 0$ it holds $\Phi(0) \leq \Phi(x) \leq \Phi(1)$.*

Proof Suppose first that $0 \leq x < 1$. Then Φ is differentiable and from equation (12.39) it follows that

$$\Phi'(x) = \frac{1}{\pi}\left\{\mathrm{Si}((x+1)\pi/2) - \mathrm{Si}((1-x)\pi/2)\right\}.$$

Observe that $0 < (1-x)\pi/2 \leq (1+x)\pi/2 < \pi$. Hence, because the function $\mathrm{Si}(x)$ increases on $[0,\pi]$, we conclude that $\Phi'(x) > 0$. Thus, Φ is an increasing function on $[0,1]$, and so in this case everything is done.

Now, let $x > 1$. Since $\mathrm{Si}(+\infty) = \pi/2$, then applying (12.39) once more, we have

$$\Phi(x) = -\frac{1}{\pi}(x+1)\int_{(x+1)\pi/2}^{\infty} \frac{\sin t}{t}\,dt - \frac{1}{\pi}(x-1)\int_{(x-1)\pi/2}^{\infty} \frac{\sin t}{t}\,dt.$$

Consequently, changing variables and integrating by parts, we get

$$\begin{aligned}
\Phi(x) &= -\frac{x}{\pi}\left(\int_{(x+1)\pi/2}^{\infty} \frac{\sin t}{t}\,dt + \int_{(x-1)\pi/2}^{\infty} \frac{\sin t}{t}\,dt\right) + \frac{1}{\pi}\int_{(x-1)\pi/2}^{(x+1)\pi/2} \frac{\sin t}{t}\,dt \\
&= -x\int_{(x-1)\pi/2}^{\infty} \frac{\sin t}{t(t+\pi)}\,dt + \frac{1}{\pi}\int_{(x-1)\pi/2}^{(x+1)\pi/2} \frac{\sin t}{t}\,dt \\
&= x\int_{(x-1)\pi/2}^{\infty} \frac{(\pi+2t)\cos t}{t^2(t+\pi)^2}\,dt + \frac{1}{\pi}\int_{(x-1)\pi/2}^{(x+1)\pi/2} \frac{\cos t}{t^2}\,dt.
\end{aligned}$$

This equation gives the estimate

$$|\Phi(x)| \leq x\int_{(x-1)\pi/2}^{\infty} \frac{\pi+2t}{t^2(t+\pi)^2}\,dt + \int_{(x-1)\pi/2}^{(x+1)\pi/2} \frac{dt}{t^2} = \frac{4}{\pi^2}\cdot\frac{1}{x-1}.$$

Therefore, for $x \geq 5$ we deduce that $|\Phi(x)| \leq \pi^{-2}$, and so, by (12.40), $\Phi(0) \leq \Phi(x) \leq \Phi(1)$.

It remains to consider the case when $1 < x < 5$. Direct calculations show (see equation (12.39)) that

$$\Phi''(x) = -\frac{2}{\pi}\frac{\cos(x\pi/2)}{x^2-1}.$$

Hence, Φ is convex in the range $1 < x < 3$, and it is concave if $3 < x < 5$. Since $\Phi'(x_0) = 0$, where $x_0 \approx 2,302$, then from (12.40) it follows

$$\Phi(x) \geq \Phi(x_0) \geq -0,035 \geq \Phi(0), \quad 1 < x < 3.$$

Moreover, because of the convexity, the graph of Φ is located below the line passing through the points $(1, \Phi(1))$ and $(3, \Phi(3))$. Therefore, taking into account that $\Phi(3) \approx -0,015 < \Phi(1)$, we get the inequality $\Phi(x) < \Phi(1)$ for $1 < x < 3$.

Analogously, $\Phi'(x_1) = 0$ for $x_1 \approx 4,180$. Consequently, in view of (12.40), we infer

$$\Phi(x) \leq \Phi(x_1) \leq 0,127 \leq \Phi(1), \quad 3 < x < 5.$$

Since the function Φ is concave on the interval $(3, 5)$, its graph is located above the line passing through the points $(3, \Phi(3))$ and $(5, \Phi(5))$. Therefore, from $\Phi(3) < \Phi(5)$ it follows that

$$\Phi(x) > \Phi(3) \approx -0,015 > \Phi(0), \quad 3 < x < 5.$$

Observe also that the left-side and right-side derivatives of Φ at the point $x = 1$ exist, but are not equal; namely,

$$\Phi'_-(1) = \frac{1}{\pi}\mathrm{Si}(\pi) \approx 0,59, \quad \Phi'_+(1) = \frac{1}{\pi}\mathrm{Si}(\pi) - 1 \approx -0,41,$$

Thus, all the cases are considered, and so the proof is completed. □

Proof of Theorem 12.7 Let $a = (a_j)_{j=1}^n \in \mathbb{R}^n$. We may assume, by homogeneity, that $\|a\|_{\ell_\infty} = 1$ and, by symmetry and independence of r_j, that $1 = a_1 \geq a_2 \geq \cdots \geq a_n \geq 0$. Denoting $\alpha := \|a\|_{\ell_2}^2$, we have $\alpha \geq 1$.

Then, by Lemma 12.5,

$$\Big\| \sum_{k=1}^n a_k r_k \Big\|_1 = \frac{2}{\pi} \int_0^\infty \frac{1 - \prod_{j=1}^n \cos(a_j t)}{t^2}\, dt = f(a) + g(a) + h(a),$$

where

$$f(a) := \frac{2}{\pi} \int_0^\infty \frac{1 - \exp(-\alpha t^2/2)}{t^2}\, dt,$$

$$g(a) := \frac{2}{\pi} \int_0^{\pi/2} \frac{\exp(-\alpha t^2/2) - \prod_{j=1}^n \cos(a_j t)}{t^2}\, dt + \frac{2}{\pi} \int_{\pi/2}^\infty \frac{\exp(-\alpha t^2/2)}{t^2}\, dt,$$

and

$$h(a) := -\frac{2}{\pi}\int_{\pi/2}^{\infty}\frac{\prod_{j=1}^{n}\cos(a_j t)}{t^2}\,dt.$$

Since $\Gamma(1/2) = \sqrt{\pi}$, then by substitution of $s = \alpha t^2/2$ and integration by parts we have

$$f(a) = \frac{\sqrt{\alpha}}{\sqrt{2\pi}}\int_0^{\infty}\frac{1-e^{-s}}{s^{3/2}}\,ds = \frac{\|a\|_{\ell_2}}{\sqrt{2\pi}}\cdot 2\int_0^{\infty} s^{-1/2}e^{-s}\,ds = \sqrt{\frac{2}{\pi}}\|a\|_{\ell_2},$$

whence

$$\Big\|\sum_{k=1}^{n} a_k r_k\Big\|_1 - \sqrt{\frac{2}{\pi}}\|a\|_{\ell_2} = g(a) + h(a). \tag{12.41}$$

To prove the theorem, it suffices to show that

$$0 \le g(a) \le g(e_1) \text{ and } h(e_1+e_2) \le h(a) \le h(e_1), \tag{12.42}$$

where $e_1 = (1,0,\ldots,0)$, $e_2 = (0,1,0,\ldots,0)$. Indeed, then for the function Φ defined before Lemma 12.9, in view of (12.41), we get

$$\begin{aligned}
-\Phi(1)\cdot\|a\|_{\ell_\infty} &= h(e_1+e_2) \le g(a) + h(a)\\
&= \Big\|\sum_{k=1}^{n} a_k r_k\Big\|_1 - \sqrt{\frac{2}{\pi}}\|a\|_{\ell_2} \le g(e_1) + h(e_1)\\
&= \|r_1\|_1 - \sqrt{\frac{2}{\pi}}\|e_1\|_{\ell_2} = \Big(1-\sqrt{\frac{2}{\pi}}\Big)\cdot\|a\|_{\ell_\infty}.
\end{aligned}$$

Since $\Phi(1) \approx 0,179 \le 1 - \sqrt{2/\pi} \approx 0,202$, one can see that (12.32) is proved.

We begin with proving the first inequality in (12.42). Note that the function $q(u) := \cos u \cdot \exp(u^2/2)$ is decreasing on $[0, \pi/2]$. This follows from the fact that

$$q'(u) = \exp(u^2/2)(u\cos u - \sin u) \le 0, \quad 0 \le u \le \pi/2.$$

Therefore, taking into account that $q(0) = 1$, we obtain $\cos u \le \exp(-u^2/2)$ for $0 \le u \le \pi/2$; moreover, since $0 \le a_j \le 1$, $j = 1,2,\ldots,n$, we have $\cos(a_j t) \le \exp(-a_j^2 t^2/2)$ for $0 \le t \le \pi/2$. Thus,

$$\exp(-\alpha t^2/2) - \prod_{j=1}^{n}\cos(a_j t) \ge 0,$$

whence $g(a) \geq 0$.

Further, by Lemma 12.7, we have

$$g(a) \leq \frac{2}{\pi} \int_0^{\pi/2} \frac{\exp(-\alpha t^2/2) - (\cos t)^\alpha}{t^2}\, dt + \frac{2}{\pi} \int_{\pi/2}^\infty \frac{\exp(-\alpha t^2/2)}{t^2}\, dt := k(\alpha).$$

Let us show that $k(\alpha)$ is a decreasing function for $\alpha \geq 1$. Differentiating under the integral sign, we obtain

$$\begin{aligned} k'(\alpha) &= -\frac{1}{\pi} \int_0^\infty \exp(-\alpha t^2/2)\, dt + \frac{2}{\pi} \int_0^{\pi/2} (\cos t)^\alpha \ln\Big(\frac{1}{\cos t}\Big) t^{-2}\, dt \\ &= -\frac{2}{\pi} \frac{1}{\sqrt{\alpha}} \Big(\sqrt{\frac{\pi}{8}} - \sqrt{\alpha} \int_0^{\pi/2} (\cos t)^\alpha \ln\Big(\frac{1}{\cos t}\Big) t^{-2}\, dt \Big). \end{aligned}$$

By Lemma 12.8, the last expression is negative for all $\alpha \geq 1$, which implies that $k'(\alpha) \leq 0$. Thus, $k(\alpha) \leq k(1) = g(e_1)$, and so, using the preceding estimate for $g(a)$, we arrive at the first inequality in (12.42).

It remains to establish the second inequality in (12.42). First, by the elementary equation

$$\cos y_1 \cos y_2 = \frac{1}{2}(\cos(y_1 + y_2) + \cos(y_1 - y_2))$$

and an induction argument, we get

$$\prod_{j=2}^n \cos(a_j t) = 2^{1-n} \sum_{\varepsilon_i = \pm 1} \cos((\varepsilon_2 a_2 + \varepsilon_3 a_3 + \cdots + \varepsilon_n a_n) t).$$

Therefore, since $a_1 = 1$, from definition of the functions h and Φ it follows

$$\begin{aligned} h(a) &= -\frac{2}{\pi} \cdot 2^{1-n} \sum_{\varepsilon_i = \pm 1} \int_{\pi/2}^\infty \frac{\cos t \cos((\varepsilon_2 a_2 + \varepsilon_3 a_3 + \cdots + \varepsilon_n a_n) t)}{t^2}\, dt \\ &= -2^{1-n} \sum_{\varepsilon_i = \pm 1} \Phi(|\varepsilon_2 a_2 + \varepsilon_3 a_3 + \cdots + \varepsilon_n a_n|). \end{aligned}$$

Hence, according to the definition of Φ and Lemma 12.9, we have

$$h(e_1 + e_2) = -\Phi(1) \leq h(a) \leq -\Phi(0) = h(e_1),$$

and, as a result, we come to (12.42), which finishes the proof. □

Comments and References A local version of Khintchine's inequality was given, for the first time, by Zygmund for the space $L_2[0, 1]$ [291] (see also [293,

Lemma V.8.3]). He has proved that there exist constants $\alpha_2 > 0$ and $\beta_2 > 0$ such that for any measurable set $E \subset [0, 1]$, with $m(E) > 0$, one can find a positive integer $N = N(E)$ satisfying the following property: for every sequence $(a_i) \in \ell_2$

$$\alpha_2\Big(\sum_{i=N}^{\infty} a_i^2\Big)^{1/2} \le \left(\frac{1}{m(E)}\int_E \Big|\sum_{i=N}^{\infty} a_i r_i(t)\Big|^2\,dt\right)^{1/2} \le \beta_2\Big(\sum_{i=N}^{\infty} a_i^2\Big)^{1/2}.$$

Later on, this Zygmund theorem was extended by Sagher and Zhou to the $L_p[0, 1]$-spaces, $0 < p < \infty$ [258]. It is worth to mention also that similar results have been obtained for some spaces of exponential integrability in [260] and much later in [96]. The main result of Sect. 12.1, Theorem 12.1, shows that the local Khintchine inequality holds in all s.s.'s that are located in a certain sense "far away" from the space L_∞ (cf. Rodin–Semenov's Theorem 2.3). This theorem appeared in the paper [41] as a result of studying the tail Rademacher multiplicator spaces $\mathcal{M}_T(X)$ (see the previous chapter). In connection with Proposition 12.1 and, in particular, with inequality (12.5) see the papers [92] by Burkholder and [259] by Sagher and Zhou. In the concluding part of the proof of Theorem 12.1 we use some reasoning from the proof of Theorem 1 in [258]. Theorem 12.2 is borrowed from [4] (see also [5]).

It is likely that the first version of weighted Khintchine's inequality (for the space L_2) has been obtained by Matuszewska and Orlicz [196, Lemma 4]. The Matuszewska–Orlicz theorem reads that for every measurable set $E \subset [0, 1]$ such that $m(E) > 3/4$ and for all $n = 1, 2, \ldots$ and $a_k \in \mathbb{R}$, $k = 1, 2, \ldots, n$, the following inequality holds

$$\sum_{k=1}^{n} a_k^2 \le 8\int_E \Big(\sum_{k=1}^{n} a_k r_k(t)\Big)^2\,dt.$$

Much later, a more general result of such a sort was obtained by Veraar in [281]. Let $p > 0$ and let a weighted function w satisfy the following conditions: (a) $w \in L_q[0, 1]$ for some $q > p$; (b) $m(\mathrm{supp} w) > 2/3$. Then, according to the Veraar theorem, there exists a constant $C > 0$, depending on p and w, such that for all $a = (a_i) \in \ell_2$ we have

$$C^{-1}\|a\|_{\ell_2} \le \left(\int_0^1 \Big|\sum_{i=1}^{\infty} a_i r_i(t)\Big|^p |w(t)|^p\,dt\right)^{1/p} \le C\|a\|_{\ell_2}.$$

This direction of research was continued in [40], where the latter inequality was extended to general s.s.'s. In the same paper, the authors have revealed close connections between the upper estimate in weighted Khintchine inequality (12.3) and the notion of Rademacher multiplicator space $\mathcal{M}(X)$. Moreover, they have found a sharp condition on measure of the support of a weighted function w, under which the more delicate lower estimate in (12.3) holds (see Corollary 12.3), and have shown that the latter estimate may be satisfied even for weights with supports of arbitrarily small measure (see Theorem 12.5). Theorem 12.4, proved in the paper [32], can be treated as a further development and strengthening of results obtained in [40].

In particular, if we take in Corollary 12.3 for w the characteristic function of a subset of [0, 1], then we immediately come to an affirmative answer to the following problem (the Scottish Book, No. 728, see [197, p. 315]) posed by Aronszajn in 1965:

Let $f(t) = \sum_{k=1}^{n} a_k r_k(t)$ *be a linear combination of Rademacher functions* r_k, $E \subset [0, 1]$ *a set of measure* $m(E) \in (0, 1]$. *Consider the inequality*

$$c \sum_{k=1}^{n} |a_k|^2 \leq \int_E |f(t)|^2 \, dt$$

with a constant c that depends on $m(E)$ *only. Whether or not this inequality holds for* $m(E) \in (1/2, 2/3]$?

Theorem 12.6 has been proved in [264]; in this paper the Schur convexity (see Chap. 5) is intensively used. Finally, Theorem 12.7 has been obtained in the paper [163] and successfully has been applied there for estimation of the projection constants of some finite dimensional spaces with a symmetric basis.

Chapter 13
Martingale Transforms of the Rademacher Sequence in Symmetric Spaces

Let $\{\mathcal{A}_k\}_{k=0}^{\infty}$ be a filtration of σ-algebras on a probability space $(\Omega, \mathcal{A}, \mathbb{P})$, i.e., $\mathcal{A}_0 \subset \mathcal{A}_1 \subset \dots \subset \mathcal{A}_k \subset \dots \subset \mathcal{A}$. A sequence of random variables (r.v.'s) $\{v_k\}_{k=1}^{\infty}$ is said to be *predictable with respect to this filtration* (or $\{\mathcal{A}_k\}$*-predictable*) if v_k is $\mathcal{A}_{k-1}$-measurable for every $k = 1, 2, \dots$.

Suppose we are given a sequence $\{\xi_k\}_{k=0}^{\infty}$ of independent r.v.'s on $(\Omega, \mathcal{A}, \mathbb{P})$ such that $\int_{\Omega} \xi_k \, d\mathbb{P} = 0$. Let $\mathcal{A}_k$ be the σ-algebra generated by the collection $\{\xi_0, \xi_1, \dots, \xi_k\}$, $k = 0, 1, \dots$. Then for arbitrary $\{\mathcal{A}_k\}$-predictable sequence $\{v_k\}_{k=1}^{\infty}$ the r.v.'s

$$\eta_n := \sum_{k=1}^{n} v_k \xi_k, \quad n = 1, 2, \dots$$

form a martingale with respect to the above filtration $\{\mathcal{A}_k\}_{k=1}^{\infty}$ (see Definition A.2). For this reason, such a sequence $\{\eta_n\}_{n=1}^{\infty}$ is called a *martingale transform* of the sequence $\{\xi_k\}_{k=0}^{\infty}$.

The main purpose of this not too long chapter is to extend the scope of Khintchine's inequalities to martingale transforms of the Rademacher sequence $\{r_k\}_{k=0}^{\infty}$, where $r_0(t) \equiv 1$, namely, to the class of martingales formed by the functions

$$f_n := \sum_{k=1}^{n} v_k r_k, \quad n = 1, 2, \dots \tag{13.1}$$

Here, $\{v_k\}_{k=1}^{\infty}$ is any $\{\Sigma_k\}$-predictable sequence, where the σ-algebra Σ_k generated by the dyadic intervals of rank k, i.e., by the intervals $\Delta_k^i = ((i-1)2^{-k}, i2^{-k})$, $i = 1, \dots, 2^k$, $k = 0, 1, \dots$. More explicitly, we shall be interested in the question, under which conditions on a s.s. X there exists a constant $C > 0$ such that for one class or another of $\{\Sigma_k\}$-predictable sequences $\{v_k\}_{k=1}^{\infty}$ from X the following

S. V. Astashkin, *The Rademacher System in Function Spaces*,
https://doi.org/10.1007/978-3-030-47890-2_13

inequality holds:

$$C^{-1}\Big\|\Big(\sum_{k=1}^{\infty} v_k^2\Big)^{1/2}\Big\|_X \le \Big\|\sum_{k=1}^{\infty} v_k r_k\Big\|_X \le C\Big\|\Big(\sum_{k=1}^{\infty} v_k^2\Big)^{1/2}\Big\|_X. \tag{13.2}$$

As we already know, in the simplest case when v_k are just any real numbers, according to the Rodin–Semenov Theorem 2.3, (13.2) is fulfilled in a s.s. X if and only if X contains the space G (recall that this is the closure of L_∞ in the Orlicz space L_{N_2} with $N_2(u) = e^{u^2} - 1$).

As we shall see soon, a necessary and sufficient condition, under which (13.2) holds in a s.s. X for *all* $\{\Sigma_k\}$-predictable sequences $\{v_k\}$, is the inequality $0 < \alpha_X \le \beta_X < 1$, where α_X and β_X are the Boyd indices of X. Moreover, we shall prove that there exists a special stopping time τ_0 with respect to the filtration $\{\Sigma_k\}$, which is critical in the following sense: if inequality (13.2) is valid for all sequences of the form $v_k = c_k \chi_{\{\tau_0 \ge k\}}$, where $c_k \in \mathbb{R}$, $k = 1, 2, \ldots$, are arbitrary, then the Boyd indices of X are nontrivial.

We shall consider also some other classes of $\{\Sigma_k\}$-predictable sequences. So, if v_k are arbitrary independent functions such that the sequences $\{v_k\}$ and $\{r_k\}$ are independent, then the inequality

$$C^{-1}\Big\|\Big(\sum_{k=1}^{\infty} v_k^2\Big)^{1/2}\Big\|_X \le \Big\|\sum_{k=1}^{\infty} v_k(s) r_k(t)\Big\|_{X([0,1]\times[0,1])} \le C\Big\|\Big(\sum_{k=1}^{\infty} v_k^2\Big)^{1/2}\Big\|_X, \tag{13.3}$$

where $C > 0$ depends only on X, holds if and only if the space X has the so-called Kruglov property ($X \in \mathbb{K}$). The definition of this property is given below; here, we note only that the condition $X \in \mathbb{K}$ is more restrictive than the embedding $X \supset G$, but however from $\alpha_X > 0$ it follows that $X \in \mathbb{K}$. The proof of the equivalence of (13.3) to the fact that $X \in \mathbb{K}$ will be based on applying of a version of Khintchine's inequality for independent symmetrically distributed functions.

And finally, when v_k are arbitrary linear combinations of the Rademacher functions, i.e., $v_k = \sum_{i=0}^{k-1} a_k^i r_i$, $a_k^i \in \mathbb{R}$, $k = 1, 2, \ldots$, inequality (13.2) is fulfilled (for all coefficients v_k of this form) if and only if $X \supset L_{N_1}^\circ$, where L_{N_1} is the Orlicz space, $N_1(u) = e^u - 1$, and, as above, Y° denotes the separable part of a s.s. Y.

We begin with a simple but useful observation related to a connection between martingale transforms (13.1) and Haar series.

13.1 Martingale Transforms and the Haar System

Recall that the Haar system is formed by the functions $h_{0,0}(t) = 1$,

$$h_{n,k}(t) = \begin{cases} 1, t \in \Delta_{n+1}^{2k-1} \\ -1, t \in \Delta_{n+1}^{2k} \\ 0, \text{ for all other } t \in [0,1], \end{cases}$$

where $n = 0, 1, \ldots, k = 1, \ldots, 2^n$. Define the Fourier–Haar coefficients of a function $f \in L_1 = L_1[0, 1]$ by the formula $c_{n,k} := \langle f, h_{n,k}\rangle$, where $\langle u, v\rangle := \int_0^1 u(t)v(t)\,dt$. It is well known (see e.g. [153, Theorem III.3]) that the Haar system is a basis in the space L_1. Therefore, every function $f \in L_1$ such that $\int_0^1 f(t)\,dt = 0$ admits the following Fourier–Haar expansion

$$f = \sum_{n=0}^{\infty}\sum_{k=1}^{2^n} c_{n,k}h_{n,k} \text{ (convergence in } L_1). \tag{13.4}$$

An important role in studying of various properties of the system $\{h_{n,k}\}$ is played by the Paley function $\mathcal{P}f$ defined for a function $f \in L_1$ as in (13.4) by

$$\mathcal{P}f := \Big(\sum_{n=0}^{\infty}\sum_{k=1}^{2^n} c_{n,k}^2 h_{n,k}^2\Big)^{1/2} = \Big(\sum_{n=0}^{\infty}\sum_{k=1}^{2^n} c_{n,k}^2 \chi_{\Delta_n^k}\Big)^{1/2}. \tag{13.5}$$

In particular, it is well known that the equivalence

$$\|f\|_X \asymp \|\mathcal{P}f\|_X, \ \ f \in X, \tag{13.6}$$

holds in a s.s. X if and only if the Boyd indices of X are nontrivial, i.e., $0 < \alpha_X \le \beta_X < 1$ (see e.g. [168, Theorem II.9.6] and [153, Theorem I.11]).

Let $\{v_k\}_{k=1}^{\infty}$ be a $\{\Sigma_k\}$-predictable sequence. By the classical Doob theorem (see e.g. [97, Theorem 4.4.14] or [265, Theorem 7.4.1]), the condition

$$\sup_{n=1,2,\ldots}\int_0^1 \Big|\sum_{k=1}^{n} v_k(t)r_k(t)\Big|\,dt < \infty$$

implies that the series $f_v := \sum_{k=1}^{\infty} v_k r_k$ converges on [0, 1] a.e. and $f_v \in L_1$. Observe that the terms v_k of any $\{\Sigma_k\}$-predictable sequence can be represented in the form

$$v_k := \sum_{i=1}^{2^{k-1}} c_{k-1,i}\chi_{\Delta_{k-1}^i}, \ \ k = 1, 2, \ldots,$$

with some $c_{k-1,i} \in \mathbb{R}$. Therefore, since

$$r_{k+1}(t) = \sum_{i=1}^{2^k} h_{k,i}(t), \ \ k = 0, 1, 2 \ldots,$$

we have

$$f_v := \sum_{j=1}^{\infty} v_j r_j = \sum_{k=1}^{\infty} \sum_{i=1}^{2^{k-1}} c_{k-1,i} h_{k-1,i} = \sum_{n=0}^{\infty} \sum_{k=1}^{2^n} c_{n,k} h_{n,k}.$$

Comparing this formula with (13.4), we see that $c_{n,k}$ are the Fourier–Haar coefficients of the function f_v. Thus, the Paley function $\mathcal{P} f_v$, defined by (13.5), coincides with the square function $\left(\sum_{k=1}^{\infty} v_k^2\right)^{1/2}$. Since $c_{n,k}$ are arbitrary in representation of v_k, $k = 1, 2, \ldots$, this observation yields, by the above-mentioned criterion related to equivalence (13.6), the following result.

Proposition 13.1 *Let X be a s.s. on* $[0, 1]$. *Then, there exists a constant $C > 0$ such that inequality* (13.2) *holds for arbitrary $\{\Sigma_k\}$-predictable sequence $\{v_k\} \subset X$ if and only if* $0 < \alpha_X \le \beta_X < 1$.

Now, we proceed with seeking conditions ensuring that inequality (13.2) holds in a s.s. X for various classes of $\{\Sigma_k\}$-predictable sequences.

13.2 Martingale Transforms Generated by Stopping Times

In this section, we consider martingale transforms (13.1) in the case when $v_k = c_k \chi_{\{\tau \ge k\}}$, where τ is a stopping time with respect to the filtration $\{\Sigma_k\}$ of σ-algebras, defined in the beginning of the chapter, and $c_k \in \mathbb{R}$, $k = 1, 2, \ldots$. Recall that an r.v. $\tau = \tau(\omega)$ with values in $\mathbb{N} \cup \{0\}$ is called a *stopping time with respect to a filtration* $\{\mathcal{A}_k\}_{k=0}^{\infty}$ of σ-algebras if $\{\omega : \tau(\omega) \le k\} \in \mathcal{A}_k$, $k = 0, 1, \ldots$.

Suppose that X is a s.s. such that $0 < \alpha_X \le \beta_X < 1$. Then, by Proposition 13.1, with some constants independent of a stopping time τ with respect to the filtration $\{\Sigma_k\}_{k=1}^{\infty}$ and any $c_i \in \mathbb{R}$, we have

$$\left\| \sum_{i=1}^{\infty} c_i \chi_{\{\tau \ge i\}} r_i \right\|_X \asymp \left\| \left(\sum_{i=1}^{\infty} c_i^2 \chi_{\{\tau \ge i\}} \right)^{1/2} \right\|_X.$$

Conversely, we show that, under some nonrestrictive conditions imposed on a s.s. X, the last equivalence for just one stopping time τ_0 defined by

$$\tau_0(t) := \sum_{k=1}^{\infty} k \chi_{(2^{-k}, 2^{-k+1}]}(t), \quad 0 < t \le 1, \tag{13.7}$$

implies that the Boyd indices of X are nontrivial.

Theorem 13.1 *Let X be a s.s. on* $[0, 1]$ *that is separable or has the Fatou property. Suppose that*

$$\Big\|\sum_{i=1}^{\infty} c_i \chi_{\{\tau_0 \geq i\}} r_i\Big\|_X \asymp \Big\|\Big(\sum_{i=1}^{\infty} c_i^2 \chi_{\{\tau_0 \geq i\}}\Big)^{1/2}\Big\|_X, \tag{13.8}$$

with a constant independent of $c_i \in \mathbb{R}$. *Then* $0 < \alpha_X \leq \beta_X < 1$.

In the proof we use some properties of the subspace $[h_{n,1}]$ generated in X by the subsequence $\{h_{n,1}\}_{n=0}^{\infty}$ of the first functions of the blocks forming the Haar system. Denote by Q the orthogonal projection onto this subspace defined for $f \in L_1 := L_1[0, 1]$ as follows:

$$Qf(t) := \sum_{n=0}^{\infty} \langle f, h_{n,1}\rangle h_{n,1}(t), \quad 0 < t \leq 1.$$

Proposition 13.2 *If X is a s.s. on* $[0, 1]$ *that is separable or has the Fatou property, then the projection Q is bounded on X.*

Proof By definition of the Haar functions, for any $n = 0, 1, \ldots, k = 2, 3, \ldots, 2^n$, and $j = 0, 1, \ldots$

$$\int_{2^{-j-1}}^{2^{-j}} h_{n,k}\, dt = 0.$$

Hence, because the Haar system is a basis in the space L_1 (see (13.4)), for every $f \in L_1$ such that $\int_0^1 f(t)\, dt = 0$ we have $Qf = \mathbb{E}[f|\Sigma]$, where $\mathbb{E}[\cdot|\Sigma]$ is the conditional expectation operator with respect to the σ-algebra Σ generated by the sequence of the intervals $\{(2^{-j-1}, 2^{-j})\}_{j=0}^{\infty}$. Since the operator $f \mapsto \mathbb{E}[f|\Sigma]$ is bounded on L_1 and on L_∞ with norm 1 (see Appendix A or [168, § II.3, p. 81]) and X is separable or has the Fatou property, we obtain

$$\|Qf\|_X = \|\mathbb{E}[f|\Sigma]\|_X \leq \|f\|_X$$

(see Appendix D or [168, Theorems II.4.9 and II.4.10]). This completes the proof. □

The next result will play a crucial role in the proof of Theorem 13.1.

Theorem 13.2 *Suppose that a s.s. X is separable or has the Fatou property. Then, the sequence $\{h_{n,1}\}_{n=0}^{\infty}$ is unconditional in X if and only if* $0 < \alpha_X \leq \beta_X < 1$.

Proof As was above-mentioned, if the Boyd indices of X are nontrivial, then we have equivalence (13.6), which implies that the Haar system $\{h_{n,k}\}$ (and hence its subsystem $\{h_{n,1}\}$) is unconditional in X. To prove the converse, we show that the operators

$$\mathcal{H}f(s) := \frac{1}{s}\int_0^s f(u)\,du \quad \text{and} \quad \mathcal{H}'f(s) = \int_s^1 \frac{f(u)}{u}\,du$$

are bounded on X. Then the result we need will follow from Theorem C.4.

Let $c_n \in \mathbb{R}$, $n = 0, 1, \ldots$, satisfy the condition:

$$c_{2n} = -c_{2n+1} > 0, \quad n = 0, 1, \ldots \tag{13.9}$$

For such a sequence $(c_n)_{n=0}^{\infty}$ we define on $[0, 1]$ the function $x(t)$ by

$$x(t) := \sum_{n=0}^{\infty} c_n h_{n,1}(t) = -c_0 \chi_{(1/2,1]}(t) + \sum_{n=1}^{\infty}\Big(- c_n + \sum_{k=0}^{n-1} c_k\Big)\chi_{(2^{-n-1},2^{-n}]}(t). \tag{13.10}$$

Then, from (13.9) it follows that

$$- c_n + \sum_{k=0}^{n-1} c_k = \begin{cases} -2c_{2j-1}, n = 2j-1 \\ -c_{2j}, n = 2j. \end{cases} \tag{13.11}$$

Moreover, setting

$$\tilde{x}(t) := \sum_{n=0}^{\infty} (-1)^n c_n h_{n,1}(t),$$

we get

$$\begin{aligned} \tilde{x}(t) &= -c_0 \chi_{(1/2,1]}(t) + \sum_{n=1}^{\infty}\Big((-1)^n c_n + \sum_{k=0}^{n-1}(-1)^k c_k\Big)\chi_{(2^{-n-1},2^{-n}]}(t) \\ &= -c_0 \chi_{(1/2,1]}(t) + \sum_{n=1}^{\infty}\Big((-1)^n c_n + \sum_{k=0}^{n-1}|c_k|\Big)\chi_{(2^{-n-1},2^{-n}]}(t). \end{aligned}$$

Thus, $\tilde{x}(t) = -c_0$ for $t \in (1/2, 1]$, and

$$\tilde{x}(t) = \sum_{k=0}^{2n} |c_k| \ \text{ if } \ t \in (2^{-2n-1}, 2^{-2n}], \quad n = 1, 2, \ldots \tag{13.12}$$

Next, for the same sequence $(c_n)_{n=0}^{\infty}$, satisfying condition (13.9), we define the step-function $y(t)$ by

$$y(t) := \sum_{k=0}^{\infty} |c_k| \chi_{(2^{-k-1}, 2^{-k}]}(t), \quad 0 < t \le 1. \tag{13.13}$$

We have

$$(\mathcal{H}' y)(2^{-n}) = \int_{2^{-n}}^{1} y(t) \frac{dt}{t} = \sum_{k=0}^{n-1} |c_k| \int_{2^{-k-1}}^{2^{-k}} \frac{dt}{t} = \ln 2 \sum_{k=0}^{n-1} |c_k|.$$

Since the function $(\mathcal{H}' y)(t)$ is decreasing, this implies that

$$(\mathcal{H}' y)(t) \le \ln 2 \sum_{n=1}^{\infty} \Big(\sum_{k=0}^{n-1} |c_k| \Big) \chi_{(2^{-n}, 2^{-n+1}]}(t), \quad 0 < t \le 1.$$

Comparing this inequality with (13.12) and taking into account definition of the dilation operator $\sigma_\tau f(t) = f(t/\tau)\chi_{[0,1]}(t)$, $\tau > 0$, we conclude that

$$(\mathcal{H}' y)(t) \le \ln 2 \cdot (|\tilde{x}(t)| + |\sigma_2 \tilde{x}(t)|), \quad 0 < t \le 1.$$

Hence (see Appendix C), it follows

$$\|\mathcal{H}' y\|_X \le 3 \ln 2 \cdot \|\tilde{x}\|_X. \tag{13.14}$$

By hypothesis, the sequence $\{h_{n,1}\}_{n=0}^{\infty}$ is unconditional in X; denote by K_0 the corresponding unconditional constant. Observe also that, in view of (13.10), (13.11), and (13.13), we have $|x| \le 2|y|$. Therefore, from (13.14) it follows that

$$\|\mathcal{H}' y\|_X \le 3K_0 \ln 2 \cdot \|x\|_X \le 6K_0 \ln 2 \cdot \|y\|_X. \tag{13.15}$$

Let $u = u^* \in X$. Then, setting

$$v(t) := \sum_{k=0}^{\infty} u(4^{-k}) \chi_{(4^{-k-1}, 4^{-k}]}(t), \quad 0 < t \le 1,$$

one can readily check that this function can be represented in the form (13.13), where the sequence $(c_n)_{n=0}^{\infty}$ of coefficients satisfies condition (13.9). Moreover, we have

$$v(t) \le u(t) \le \sigma_4 v(t), \quad 0 < t \le 1.$$

Consequently, $\|v\|_X \le \|u\|_X$ and (see again Appendix C)

$$\|\mathcal{H}'u\|_X \le \|\mathcal{H}'(\sigma_4 v)\|_X = \|\sigma_4(\mathcal{H}'v)\|_X \le 4\|\mathcal{H}'v\|_X.$$

Hence, applying (13.15) for $y = v$, we get

$$\|\mathcal{H}'u\|_X \le 24K_0 \ln 2 \cdot \|v\|_X \le 24K_0 \ln 2 \cdot \|u\|_X \tag{13.16}$$

for any function $u \in X$ such that $u = u^*$.

Further, we want to extend inequality (13.16) to the whole of X. To this end, we need the following property of the operator $\mathcal{H}'$.

Lemma 13.1 *For an arbitrary s.s. X we have*

$$\|\mathcal{H}'\|_{X\to X} = \sup_{\|u\|_X \le 1, u=u^*} \|\mathcal{H}'u\|_X.$$

According to this lemma (we postpone its proof), from inequality (13.16) it follows that the operator $\mathcal{H}'$ is bounded on X. So, it remains to prove the same property of the Hardy operator $\mathcal{H}$.

Let $\{\theta_n\}_{n=0}^\infty$ be an arbitrary sequence of signs. By Proposition 13.2 and the hypothesis of the theorem, the operator

$$Q_\theta f(t) := \sum_{n=0}^\infty \theta_n \langle f, h_{n,1}\rangle h_{n,1}(t), \quad 0 < t \le 1,$$

is bounded on X. Furthermore, it is easy to check that Q_θ is self-conjugate in the sense that

$$\int_0^1 Q_\theta f(t) g(t)\, dt = \int_0^1 f(t) Q_\theta g(t)\, dt$$

for all $f \in X$ and $g \in X'$. Hence, Q_θ is bounded on X' for any $\theta_n = \pm 1$, $n = 0, 1, \dots$. Thus, the sequence $\{h_{n,1}\}_{n=0}^\infty$ is unconditional in the space X' as well. But then, reasoning in the same way as for X, we deduce that the operator $\mathcal{H}'$ is bounded on X'.

Let us observe that the operators $\mathcal{H}$ and $\mathcal{H}'$ are conjugate to each other, i.e.,

$$\int_0^1 \mathcal{H}f(t) g(t)\, dt = \int_0^1 f(t)\mathcal{H}'g(t)\, dt \tag{13.17}$$

for all $f \in X$ and $g \in X'$. Moreover, since X is separable or has the Fatou property, then this space is embedded isometrically into its second Köthe dual X''

(see Appendix C). Therefore, from (13.17) and the boundedness of $\mathcal{H}'$ on X' it follows that

$$\|\mathcal{H}f\|_X = \|\mathcal{H}f\|_{X''} = \sup\left\{\int_0^1 \mathcal{H}f(t)g(t)\,dt : \|g\|_{X'} \le 1\right\}$$
$$= \sup\left\{\int_0^1 f(t)\mathcal{H}'g(t)\,dt : \|g\|_{X'} \le 1\right\} \le \|\mathcal{H}'\|_{X'\to X'}\|f\|_X.$$

So, the operator $\mathcal{H}$ is bounded on X, and hence the theorem is proved. □

Proof of Lemma 13.1 Since $\mathcal{H}'$ is a positive operator and X is a s.s., then we have

$$\|\mathcal{H}'\|_{X\to X} = \sup_{\|f\|_X\le 1,\, f\ge 0} \|\mathcal{H}'f\|_X.$$

Therefore, to prove the claim, it suffices to check that for every function $f \in X$, $f \ge 0$, we have

$$\|\mathcal{H}'f\|_X \le \|\mathcal{H}'f^*\|_X.$$

From definition of the decreasing rearrangement it follows that for each measurable function $g(t)$ the function $\mathcal{H}g^*(t)$ decreases on $[0, 1]$ and $(\mathcal{H}g)^*(t) \le \mathcal{H}g^*(t)$, $0 < t \le 1$. Moreover, similarly as in (13.17) we have

$$\int_0^1 g(t)\mathcal{H}'f(t)\,dt = \int_0^1 \mathcal{H}g(t)f(t)\,dt$$

for all $f \in X$ and $g \in X'$. In consequence, by a property of the decreasing rearrangement (see [168, § II.2, Property 13°, p. 68]), for any nonnegative $f \in X$ and $g \in X'$ we get

$$\int_0^1 g(t)\mathcal{H}'f(t)\,dt = \int_0^1 \mathcal{H}g(t)f(t)\,dt \le \int_0^1 \mathcal{H}g^*(t)f^*(t)\,dt = \int_0^1 g^*(t)\mathcal{H}'f^*(t)\,dt.$$

Recall that X is a s.s. isometrically embedded into X''. Hence, we have

$$\|\mathcal{H}'f\|_X = \|\mathcal{H}'f\|_{X''} = \sup\left\{\int_0^1 g(t)\mathcal{H}'f(t)\,dt : \|g\|_{X'} \le 1\right\}$$
$$\le \sup\left\{\int_0^1 g^*(t)\mathcal{H}'f^*(t)\,dt : \|g\|_{X'} \le 1\right\} \le \|\mathcal{H}'f^*\|_X,$$

as we wished. □

Now, everything is ready to complete the proof of the main result of this section.

Proof of Theorem 13.1 From (13.7) it follows that $\{\tau_0 \geq k\} = [0, 2^{-k+1}]$ for each $k = 1, 2, \ldots$. Therefore, we obtain that

$$\sum_{k=1}^{\infty} c_k \chi_{\{\tau_0 \geq k\}}(t) r_k(t) = \sum_{k=1}^{\infty} c_k h_{k-1,1}(t), \quad 0 < t \leq 1.$$

Combining this together with hypothesis (13.8), we see that the sequence $\{h_{n,1}\}_{n=0}^{\infty}$ is unconditional in X. Thus, applying Theorem 13.2, we arrive at the desired result. □

13.3 Vector-Valued Rademacher Series with Independent Coefficients

Here, we examine the class of martingale transforms of the Rademacher system generated by sequences of independent functions. By using a version of Khintchine's inequality for independent symmetrically distributed functions, we shall show that inequality (13.3) holds in this case if and only if a s.s. X possesses the so-called Kruglov property. Recall the definition of this property.

Let f be a measurable function (or an r.v.) on $[0, 1]$. Denote by $\pi(f)$ the r.v. $\sum_{i=1}^{N} f_i$, where f_i are independent copies of f (this means that f_i have the same distribution as f) and N is an r.v. that is independent with respect to the sequence $\{f_i\}$ and has the Poisson distribution with parameter 1, i.e.,

$$\mathbb{P}\{N = k\} = \frac{1}{e \cdot k!}, \quad k = 0, 1, 2, \ldots$$

Another (equivalent) definition of $\pi(f)$ can be given in terms of the characteristic functions (in probabilistic sense) of f and $\pi(f)$ by the equality:

$$\theta_{\pi(f)}(t) = \exp(\theta_f(t) - 1), \quad t \in \mathbb{R}$$

(see Appendix A).

Definition 13.1 A s.s. X on $[0, 1]$ is said to have the *Kruglov property* ($X \in \mathbb{K}$) if $\pi(f) \in X$ whenever $f \in X$.

Though the r.v. $\pi(f)$ looks, at first glance, rather artificial, the Kruglov property turns out very useful in studying various geometrical issues related to s.s.'s. Its importance is based mainly on the fact that the condition $X \in \mathbb{K}$ allows effectively to compare the X-norms of sums of independent mean zero (or positive) r.v.'s with the norms of sums of their pairwise disjoint copies in a certain s.s. on $[0, \infty)$. Roughly speaking, the assumption $X \in \mathbb{K}$ is in a sense a natural substitute of the weaker condition $X \supset G$ when one starts to consider the class of all sequences of independent symmetrically distributed (or positive) functions from a s.s. X rather than only the Rademacher system. So, the Kruglov property holds for s.s.'s sufficiently "remote" from the space L_∞. For instance, a s.s. X possesses this property provided that X contains L_p for some $p < \infty$ (in particular, if the lower Boyd index α_X is positive). However, the latter condition is not necessary: the exponential Orlicz space $L_{N_\alpha} \in \mathbb{K}$ if and only if $0 < \alpha \leq 1$.

Next, we make use of a generalized Khintchine inequality for sequences of independent symmetrically distributed functions proved in [26]. We present here this result without a proof.

Theorem 13.3 *Let X be a s.s. on $[0, 1]$ that is separable or has the Fatou property. Then $X \in \mathbb{K}$ if and only if for some $C > 0$ and any sequence of independent symmetrically distributed functions $\{f_k\}_{k=1}^\infty \subset X$ we have*

$$\Big\| \sum_{k=1}^\infty f_k \Big\|_X \leq C \Big\| \Big(\sum_{k=1}^\infty f_k^2 \Big)^{1/2} \Big\|_X. \tag{13.18}$$

As above, if X is a s.s. on $I = [0, 1]$, then $X(I \times I)$ is the corresponding s.s. on the square $I \times I$ equipped with the norm $\|f\|_{X(I\times I)} = \|f^*\|_X$, where f^* is the decreasing rearrangement of the function $f(s, t)$, $s, t \in I$.

Theorem 13.4 *If X is a s.s. on $[0, 1]$ that is separable or has the Fatou property, then the following properties are equivalent:*

(*i*) $X \in \mathbb{K}$;

(*ii*) *there exists $C > 0$ such that for any sequence of independent functions $\{f_k\}_{k=1}^\infty \subset X$*

$$C^{-1} \Big\| \Big(\sum_{k=1}^\infty f_k^2 \Big)^{1/2} \Big\|_X \leq \Big\| \sum_{k=1}^\infty f_k(s) r_k(t) \Big\|_{X(I\times I)} \leq C \Big\| \Big(\sum_{k=1}^\infty f_k^2 \Big)^{1/2} \Big\|_X; \tag{13.19}$$

(*iii*) *there exists $C > 0$ such that for any sequence of independent functions $\{f_k\}_{k=1}^\infty \subset X$*

$$C^{-1} \Big\| \Big(\sum_{k=1}^\infty f_k^2 \Big)^{1/2} \Big\|_X \leq \int_0^1 \Big\| \sum_{k=1}^\infty f_k(s) r_k(t) \Big\|_X dt \leq C \Big\| \Big(\sum_{k=1}^\infty f_k^2 \Big)^{1/2} \Big\|_X. \tag{13.20}$$

Proof First of all, we observe that the left-hand side inequalities in (13.19) and (13.20) are valid in every s.s. Let us prove this claim for (13.19).

The sequence $\{f_k(s)r_k(t)\}_{k=1}^\infty$ consists, clearly, of independent functions (because f_k, $k = 1, 2, \dots$, are independent by condition). Therefore, this sequence is unconditional in the space $X(I \times I)$ with constant 1 [87, Proposition 1.14] (alternatively, from Fubini's theorem and the contraction principle from Theorem 7.1 it follows that the above sequence is unconditional in $X(I \times I)$ with constant 2). Thus, by Minkowski's inequality and Khintchine's L_1-inequality (see Theorem 5.4), we get

$$\begin{aligned}\Big\|\sum_{k=1}^{\infty} f_k(s)r_k(t)\Big\|_{X(I\times I)} &= \int_0^1 \Big\|\sum_{k=1}^{\infty} f_k(s)r_k(t)r_k(u)\Big\|_{X(I\times I)} du \\ &\geq \Big\|\int_0^1 \Big|\sum_{k=1}^{\infty} f_k(s)r_k(t)r_k(u)\Big|\, du\Big\|_{X(I\times I)} \\ &\geq \frac{1}{\sqrt{2}}\Big\|\Big(\sum_{k=1}^{\infty} f_k^2\Big)^{1/2}\Big\|_X.\end{aligned}$$

The proof of the left-hand side inequality in (13.20) follows the same lines and we skip it.

$(i) \Rightarrow (ii)$. According to Theorem 13.3, we have inequality (13.18). Inserting independent and symmetrically distributed functions $f_k(s)r_k(t)$, $k = 1, 2, \ldots,$ into (13.18), we arrive at the right-hand side inequality in (13.19).

$(i) \Rightarrow (iii)$. As above, inequality (13.18) holds for every sequence of independent and symmetrically distributed functions. Using the standard symmetrization, we can extend this estimate to the class of all sequences of independent mean zero functions from X.

Indeed, let $\{f_k\}_{k=1}^{\infty} \subset X$ be a sequence of independent functions with $\int_0^1 f_k(t)\,dt = 0$, $k = 1, 2, \ldots$. Suppose that $\{f_k'\}_{k=1}^{\infty}$ is another sequence such that f_k' has the same distribution as f_k for each $k = 1, 2, \ldots$ and $\{f_k, f_m'\}_{k,m=1}^{\infty}$ is an independent family of functions. Then, the functions $h_k := f_k - f_k'$ are independent and symmetrically distributed. Consequently, by (13.18), we have

$$\Big\|\sum_{k=1}^{\infty} h_k\Big\|_X \leq C'\Big\|\Big(\sum_{k=1}^{\infty} h_k^2\Big)^{1/2}\Big\|_X \leq 2C'\Big\|\Big(\sum_{k=1}^{\infty} f_k^2\Big)^{1/2}\Big\|_X.$$

On the other hand, since $\int_0^1 f_k(t)\,dt = 0$ by condition for all $k = 1, 2, \ldots,$ then from [87, Proposition 1.11 and subsequent remark] it follows that

$$\Big\|\sum_{k=1}^{\infty} h_k\Big\|_X = \Big\|\sum_{k=1}^{\infty} f_k - \sum_{k=1}^{\infty} f_k'\Big\|_X \geq \Big\|\sum_{k=1}^{\infty} f_k\Big\|_X.$$

Combining two last inequalities, we get (13.18) for the sequence $\{f_k\}_{k=1}^{\infty}$.

Now, suppose that $\{f_k\}_{k=1}^{\infty} \subset X$ is an arbitrary sequence of independent functions. Then, the functions $g_k(s) := f_k(s) - \int_0^1 f_k(t)\,dt$, $k = 1, 2, \ldots,$ are independent as well, and moreover $\int_0^1 g_k(s)\,ds = 0$, $k = 1, 2, \ldots$. Hence, as was already proved, inequality (13.18) holds for the sequence $\{g_k(s)r_k(t)\}_{k=1}^{\infty}$. So, taking into account that $\|f\|_1 \le C_X\|f\|_X$ for $f \in X$ (see Appendix C) and using Minkowski's inequality, we obtain

$$\int_0^1 \Big\|\sum_{k=1}^{\infty} f_k(s)r_k(t)\Big\|_X dt = \int_0^1 \Big\|\sum_{k=1}^{\infty}\Big(g_k(s) + \int_0^1 f_k(s)\,ds\Big)\cdot r_k(t)\Big\|_X dt$$

$$\le \int_0^1 \Big\|\sum_{k=1}^{\infty} g_k(s)r_k(t)\Big\|_X dt + \int_0^1 \Big|\sum_{k=1}^{\infty}\int_0^1 f_k(s)\,ds\cdot r_k(t)\Big|\,dt$$

$$\le C\Big\|\Big(\sum_{k=1}^{\infty} g_k^2\Big)^{1/2}\Big\|_X + \Big(\sum_{k=1}^{\infty}\Big(\int_0^1 f_k(s)\,ds\Big)^2\Big)^{1/2} \le C\Big\|\Big(\sum_{k=1}^{\infty} f_k^2\Big)^{1/2}\Big\|_X$$

$$+ (C+1)\cdot\Big(\sum_{k=1}^{\infty}\Big(\int_0^1 f_k(s)\,ds\Big)^2\Big)^{1/2} \le C\Big\|\Big(\sum_{k=1}^{\infty} f_k^2\Big)^{1/2}\Big\|_X$$

$$+ (C+1)\cdot\int_0^1\Big(\sum_{k=1}^{\infty} f_k^2(s)\Big)^{1/2} ds \le (2C+1)C_X\Big\|\Big(\sum_{k=1}^{\infty} f_k^2\Big)^{1/2}\Big\|_X.$$

Thus, inequality (13.20) is proved.

$(iii) \Rightarrow (i)$. By Theorem 13.3, it suffices to prove inequality (13.18) for some $C > 0$ and any sequence of independent symmetrically distributed functions $\{f_k\}_{k=1}^{\infty} \subset X$. Since [87, Proposition 1.14] assures that every such a sequence is unconditional in X with constant 1, then from the hypothesis it follows that

$$\Big\|\sum_{k=1}^{\infty} f_k\Big\|_X = \int_0^1 \Big\|\sum_{k=1}^{\infty} f_k(s)r_k(t)\Big\|_X dt \le C\Big\|\Big(\sum_{k=1}^{\infty} f_k^2\Big)^{1/2}\Big\|_X.$$

$(ii) \Rightarrow (i)$. It is easy to check that, for each $k = 1, 2, \ldots$, the functions $f_k(s)$ on I and $f_k(s)r_k(t)$ on the square $I \times I$ are identically distributed. Indeed, since f_k are symmetrically distributed, for every $\tau \in \mathbb{R}$ we have

$$m(\{(s,t) \in I \times I : f_k(s)r_k(t) > \tau\}) = \frac{1}{2}[m(\{s \in I : f_k(s) > \tau\})$$
$$+ m(\{s \in I : -f_k(s) > \tau\})] = m(\{s \in I : f_k(s) > \tau\}).$$

Therefore, since each of the sequences $\{f_k\}_{k=1}^{\infty}$ and $\{f_k(s)r_k(t)\}_{k=1}^{\infty}$ consists of independent functions, it holds

$$\Big\|\sum_{k=1}^{\infty} f_k\Big\|_X = \Big\|\sum_{k=1}^{\infty} f_k(s)r_k(t)\Big\|_{X(I\times I)}.$$

As a result, from this observation it follows that (13.18) is a consequence of the right-hand side inequality in (13.19). This completes the proof. □

13.4 Martingale Transforms Generated by Linear Combinations of Rademacher Functions

In the concluding section of the chapter we consider the class of martingale transforms that are generated by $\{\Sigma_k\}$-predictable sequences of linear combinations of the Rademacher functions.

Theorem 13.5 *Let X be a s.s. on $[0, 1]$ and let $v_k = \sum_{i=0}^{k-1} a_k^i r_i$, where $r_0(t) \equiv 1$, $k = 1, 2, \ldots$. Then inequality (13.2) holds with some constant $C > 0$ independent of coefficients $a_k^i \in \mathbb{R}$ if and only if $X \supset L_{N_1}^{\circ}$, with $N_1(u) = e^u - 1$.*

In the proof of this result we need the fact that the exponential Orlicz space L_{N_2} is 2-convex. Recall that a Banach function lattice X is called *q-convex*, where $1 \leq q < \infty$, if there exists a constant $L > 0$ such that

$$\Big\| \Big(\sum_{k=1}^{n} |x_k|^q \Big)^{1/q} \Big\|_X \leq L \Big(\sum_{k=1}^{n} \|x_k\|_X^q \Big)^{1/q}$$

for every $n \in \mathbb{N}$ and any $x_1, x_2, \ldots, x_n$ from X. In the case when the last inequality is fulfilled for all pairwise disjoint elements of X, i.e.,

$$\Big\| \sum_{k=1}^{n} x_k \Big\|_X \leq L \Big(\sum_{k=1}^{n} ||x_k||_X^q \Big)^{1/q}, \tag{13.21}$$

one says that X *satisfies an upper q-estimate*. For example, for every $1 \leq q < \infty$ the space L_q is q-convex with constant 1. The next statement answers to the question, under which conditions an Orlicz space on $[0, 1]$ possesses these properties.

We say that an Orlicz function M is *q-convex*, $1 \leq q < \infty$, if the function $M(t^{1/q})$ is convex.

Proposition 13.3 *Suppose that M is an Orlicz function. For each $1 \leq q < \infty$ the following conditions are equivalent:*

(1) the Orlicz space L_M is q-convex;
(2) L_M satisfies an upper q-estimate;
(3) there are $c > 0$ and $t_0 > 0$ such that for arbitrary $s \geq 1$ and $t \geq t_0$ we have

$$M(st) \geq cs^q M(t). \tag{13.22}$$

Proof Since the implication (1) ⇒ (2) is trivial, it suffices to prove the implications (2) ⇒ (3) and (3) ⇒ (1).

So, let the space L_M satisfy an upper q-estimate, that is, inequality (13.21) holds for every $n \in \mathbb{N}$ and all pairwise disjoint functions $x_1, x_2, \dots, x_n$ from $X = L_M$. Without loss of generality, we can assume that $M(1) = 1$. Then, setting $v := 1/M(t)$ for a given $t \geq 1$, we see that $v \in (0, 1]$. Observe that $\|\chi_{[0,t]}\|_{L_M} = 1/M^{-1}(1/t)$, where M^{-1} is the inverse function for M. Therefore, inserting the functions $\chi_{[(k-1)v/n, kv/n]}$, instead of x_k, $k = 1, 2, \dots, n$, into inequality (13.21), we obtain that

$$M^{-1}(n/v) \leq Ln^{1/q} M^{-1}(1/v),$$

or after the change of variable $t = M^{-1}(1/v)$

$$M^{-1}(nM(t)) \leq Ln^{1/q} t.$$

Since M increases, this implies that

$$M(Ln^{1/q} t) \geq nM(t) \quad \text{for all} \quad t \geq 1 \text{ and } n \in \mathbb{N}. \tag{13.23}$$

For any $s \geq L$ we clearly can find $n \in \mathbb{N}$ such that

$$Ln^{1/q} \leq s < L(n + 1)^{1/q},$$

and hence, by (13.23), for $t \geq 1$ we have

$$M(st) \geq M(Ln^{1/q} t) \geq nM(t) \geq \frac{n+1}{2} M(t) \geq \frac{1}{2} L^{-q} s^q M(t).$$

Thus, inequality (13.22) is fulfilled for all $s \geq L$ if $c := L^{-q}/2$ and $t_0 := 1$. This estimate holds also for $1 \leq s \leq L$, because in this case from the definition of c it follows

$$cs^q M(t) \leq \frac{1}{2} M(t) \leq M(st).$$

As a result, the implication (2) $\Rightarrow$ (3) is proved.

We proceed now with proving the implication (3) $\Rightarrow$ (1). Let us show first that there exist a q-convex Orlicz function N and a constant $C > 0$ such that for all $t \geq 2t_0$

$$C^{-1} M(t/2) \leq N(t) \leq CM(t). \tag{13.24}$$

One can readily see that the function g defined by

$$g(t) := \begin{cases} M(t_0) t_0^{-q-1} t \text{ if } 0 \leq t \leq t_0 \\ \sup_{t_0 \leq s \leq t} M(s) s^{-q} \text{ if } t > t_0 \end{cases}$$

is increasing; moreover, in view of hypothesis (13.22), for $t \geq t_0$ we have

$$\frac{M(t)}{t^q} \leq g(t) \leq \frac{1}{c} \cdot \frac{M(t)}{t^q}. \tag{13.25}$$

We set $N(t) := \int_0^t g(u)u^{q-1}\,du$. Since $g(u)u^{q-1}$ increases, the function N is convex. Furthermore, changing variables, we get

$$N(t^{1/q}) = \frac{1}{q} \cdot \int_0^t g(u^{1/q})\,du,$$

and hence N is q-convex. Thus, taking into account inequality (13.25) and the above properties of g, we have for $t \geq t_0$

$$N(t) \leq g(t) \int_0^t u^{q-1}\,du \leq \frac{1}{q} \cdot g(t)t^q \leq \frac{1}{cq} \cdot M(t),$$

and for $t \geq 2t_0$

$$N(t) \geq \int_{t/2}^t g(u)u^{q-1}\,du \geq \frac{1-2^{-q}}{q} t^q g(t/2) \geq \frac{2^q - 1}{q} M(t/2).$$

Summing up, we arrive at inequality (13.24).

To complete the proof of the proposition, it remains to show that the Orlicz space L_N is q-convex. Indeed, it is well known that the (Luxemburg) norm of an Orlicz space on [0, 1] is determined (up to equivalence) by the behaviour of the corresponding Orlicz function for large arguments (see e.g. [167, Theorem 13.2]). Therefore, from inequality (13.24) and definition of the Luxemburg norm, it follows that the q-convexity of L_M is a consequence of this property of L_N.

To prove the q-convexity of L_N, we consider the Orlicz space $L_{N(u^{1/q})}$, which is well-defined because the function N is q-convex. One can easily check that

$$\|x\|_{L_{N(u^{1/q})}} = \|\,|x|^{1/q}\,\|_{L_N}^q.$$

Hence, from the triangle inequality in $L_{N(u^{1/q})}$ it follows

$$\Big\|\Big(\sum_{k=1}^n |x_k|^q\Big)^{1/q}\Big\|_{L_N} = \Big\|\sum_{k=1}^n |x_k|^q\Big\|_{L_{N(u^{1/q})}}^{1/q} \leq \Big(\sum_{k=1}^n \|\,|x_k|^q\,\|_{L_{N(u^{1/q})}}\Big)^{1/q}$$
$$= \Big(\sum_{k=1}^n \|x_k\|_{L_N}^q\Big)^{1/q}$$

for every $n \in \mathbb{N}$ and arbitrary $x_1, x_2, \ldots, x_n$ from L_N. Thus, L_N is q-convex, and so everything is done. □

In the proof of Theorem 13.5 we shall use also the following simple result.

Lemma 13.2 *For arbitrary Banach function lattice X and all $x, y \in X$, $x \geq 0$, $y \geq 0$, we have*

$$\|(xy)^{1/2}\|_X \leq \|x\|_X^{1/2}\|y\|_X^{1/2}.$$

Proof From the elementary inequality

$$(xy)^{1/2} \leq \frac{1}{2}(\theta x + \theta^{-1}y), \quad \theta > 0,$$

it follows

$$\|(xy)^{1/2}\|_X \leq \frac{1}{2}(\theta\|x\|_X + \theta^{-1}\|y\|_X), \quad \theta > 0,$$

Minimizing now the right-hand side of this inequality with respect to θ, we get the desired result. □

Proof of Theorem 13.5 As above, since $\|f\|_{L_1} \leq C_X\|f\|_X$ for $f \in X$ (see Appendix C), by Minkowski's inequality and Khintchine's L_1-inequality (see Theorem 5.4), we have

$$\Big\|\Big(\sum_{k=1}^{\infty} v_k^2\Big)^{1/2}\Big\|_X = \Big\|\Big(\sum_{k=1}^{\infty}\Big(\sum_{i=0}^{k-1} a_k^i r_i\Big)^2\Big)^{1/2}\Big\|_X \geq C_X^{-1}\Big\|\Big(\sum_{k=1}^{\infty}\Big(\sum_{i=0}^{k-1} a_k^i r_i\Big)^2\Big)^{1/2}\Big\|_{L_1}$$

$$\geq C_X^{-1}\Big(\sum_{k=1}^{\infty}\Big\|\sum_{i=0}^{k-1} a_k^i r_i\Big\|_{L_1}^2\Big)^{1/2} \geq \frac{1}{\sqrt{2}C_X}\Big(\sum_{k=1}^{\infty}\sum_{i=0}^{k-1}\big(a_k^i\big)^2\Big)^{1/2}. \quad (13.26)$$

Let us show that the opposite inequality is a consequence of the embedding $X \supset L_{N_1}^{\circ}$. To this end, representing

$$\Big(\sum_{k=1}^{\infty} v_k^2\Big)^{1/2} = \Big(\sum_{k=1}^{\infty}\sum_{i=0}^{k-1}\big(a_k^i\big)^2 + 2\sum_{k=1}^{\infty}\sum_{0\leq i<j<k} a_k^i a_k^j r_i r_j\Big)^{1/2},$$

we introduce the following notation:

$$A^2 := \sum_{k=1}^{\infty}\sum_{i=0}^{k-1}\big(a_k^i\big)^2 \quad \text{and} \quad U(t) := \sum_{k=1}^{\infty}\sum_{0\leq i<j<k} a_k^i a_k^j r_i(t) r_j(t).$$

Then, applying Lemma 13.2, we get

$$\Big\|\Big(\sum_{k=1}^{\infty} v_k^2\Big)^{1/2}\Big\|_X = \Big\|\Big(A^2+2U\Big)^{1/2}\Big\|_X \le B\|A^2+2U\|_X^{1/2} \le B\left(A^2+2\|U\|_X\right)^{1/2}, \tag{13.27}$$

where $B := \|\chi_{[0,1]}\|_X^{1/2}$. According to Theorem 6.4, the system $\{r_i r_k\}_{1\le i<k<\infty}$ is equivalent in a s.s. X to the unit vector basis in ℓ_2 if and only if $X \supset L_{N_1}^{\circ}$. Therefore, assuming that the latter embedding holds, by the equation

$$U(t) = \sum_{0\le i<j<\infty} \Big(\sum_{k=j+1}^{\infty} a_k^i a_k^j\Big) r_i(t) r_j(t)$$

and the Cauchy-Schwarz-Bunyakovskii inequality, we obtain that

$$\|U\|_X \asymp \Big(\sum_{0\le i<j<\infty}\Big(\sum_{k=j+1}^{\infty} a_k^i a_k^j\Big)^2\Big)^{1/2} \le \sum_{k=1}^{\infty}\Big(\sum_{0\le i<j<k}\left(a_k^i a_k^j\right)^2\Big)^{1/2}$$

$$= \sum_{k=1}^{\infty}\Big(\sum_{j=1}^{k-1}\left(a_k^j\right)^2 \sum_{i=0}^{j-1}\left(a_k^i\right)^2\Big)^{1/2} \le A^2.$$

Combining this estimate with (13.27), we infer that

$$\Big\|\Big(\sum_{k=1}^{\infty} v_k^2\Big)^{1/2}\Big\|_X \le \sqrt{3}BA = \sqrt{3}B\Big(\sum_{k=1}^{\infty}\sum_{i=0}^{k-1}\left(a_k^i\right)^2\Big)^{1/2}. \tag{13.28}$$

On the other hand, because $X \supset L_{N_1}^{\circ}$ and

$$\sum_{k=1}^{\infty} v_k r_k = \sum_{k=1}^{\infty}\sum_{i=0}^{k-1} a_k^i r_i r_k, \tag{13.29}$$

applying Theorem 6.4 once more, we conclude that

$$\Big\|\sum_{k=1}^{\infty} v_k r_k\Big\|_X \asymp \Big(\sum_{k=1}^{\infty}\sum_{i=0}^{k-1}\left(a_k^i\right)^2\Big)^{1/2}.$$

Hence, from inequalities (13.26) and (13.28) it follows (13.2).

Conversely, assume that (13.2) holds for $v_k = \sum_{i=0}^{k-1} a_k^i r_i$, $k = 1, 2, \ldots$. Then, in particular, with a constant independent of $c_k \in \mathbb{R}$, $k = 1, 2, \ldots$, we have $\|\sum_{k=1}^{\infty} c_k r_k\|_X \asymp (\sum_{k=1}^{\infty} c_k^2)^{1/2}$, and hence, by the Rodin–Semenov theorem

(Theorem 2.3), $X \supset G$. Furthermore, one may readily show that for any $\alpha > 0$ and $1 \le q < \infty$ there exist $c_{\alpha,q} > 0$ and $t_0 > 0$ such that for all $s \ge 1$ and $t \ge t_0$ it holds

$$\exp((st)^\alpha) - 1 \ge c_{\alpha,q} \cdot s^q \left(\exp(t^\alpha) - 1\right).$$

Consequently, in view of Proposition 13.3, the exponential Orlicz space L_{N_α} is q-convex for arbitrary $\alpha > 0$ and $1 \le q < \infty$. This implies that G as a subspace of L_{N_2} is 2-convex as well, and therefore, from the embedding $X \supset G$ and Theorem 2.2 it follows that

$$\begin{aligned}\Big\|\Big(\sum_{k=1}^{\infty} v_k^2\Big)^{1/2}\Big\|_X &\le \Big\|\Big(\sum_{k=1}^{\infty}\Big(\sum_{i=0}^{k-1} a_k^i r_i\Big)^2\Big)^{1/2}\Big\|_G \\ &\le \Big(\sum_{k=1}^{\infty}\Big\|\sum_{i=0}^{k-1} a_k^i r_i\Big\|_G^2\Big)^{1/2} \\ &\le C\Big(\sum_{k=1}^{\infty}\sum_{i=0}^{k-1}\left(a_k^i\right)^2\Big)^{1/2}.\end{aligned}$$

Combining this estimate together with (13.26), (13.29), and (13.2), we get

$$\Big\|\sum_{k=1}^{\infty}\sum_{i=0}^{k-1} a_k^i r_i r_k\Big\|_X \asymp \Big(\sum_{k=1}^{\infty}\sum_{i=0}^{k-1}\left(a_k^i\right)^2\Big)^{1/2}.$$

Hence, the system $\{r_i r_k\}_{1 \le i < k < \infty}$ is equivalent in X to the unit vector basis in ℓ_2. Thus, appealing to Theorem 6.4 once more, we infer that $X \supset L_{N_1}^\circ$. This completes the proof. □

Comments and References Let us recall the following classical Burkholder theorem [91] on comparing of a martingale with its square function in L_p-spaces proved in 1966: for every $1 < p < \infty$ there exists a constant $C_p > 0$ such that the inequality

$$C_p^{-1}\|f_\infty\|_p \le \|P(f)\|_p \le C_p\|f_\infty\|_p$$

holds for all uniformly integrable martingales $f = \{f_n\}_{n=0}^\infty$ (see the definition in Appendix A), where $f_\infty := \lim_{n\to\infty} f_n$ a.e. and

$$P(f) := \Big(f_0^2 + \sum_{n=1}^{\infty}(f_n - f_{n-1})^2\Big)^{1/2}.$$

Observe that this inequality fails for $p = 1$ or $p = \infty$. In [156], a similar problem has been considered in a more general setting of s.s.'s. Extending the above Burkholder's result, the Kikuchi theorem, established in [156], reads that the inequality

$$C^{-1}\|f_\infty\|_X \le \|P(f)\|_X \le C\|f_\infty\|_X$$

holds in a s.s. X for all uniformly integrable martingales $f = \{f_n\}_{n=0}^\infty \subset X$ if and only if the Boyd indices of X are nontrivial.

It is worth to mention interesting results related to the behaviour of martingale transforms of the Rademacher sequence in L_p-spaces, obtained by Hitczenko in the paper [135]. In particular, for every $1 \le p < \infty$ and arbitrary predictable sequence $\{v_k\}_{k=1}^\infty$ with respect to the filtration generated by the Rademacher system, he has proved the following decoupling-inequality:

$$\Big\|\sum_{k=1}^\infty v_k r_k\Big\|_p \le C\Big\|\sum_{k=1}^\infty v_k(s) r_k(t)\Big\|_{L_p(I\times I)},$$

where the constant C does not depend on p. Moreover, it turns out that this Hitczenko theorem gives in a sense the best comparison result, because of an example due to Talagrand [135] shows that the corresponding distribution estimate for sums, involved in the last inequality, fails.

The main results of this chapter have been obtained in the paper [33]. In the proof of Theorem 13.2 we apply an idea from the proof of Theorem 1 in [180]. A version of Khintchine's inequality (13.18) for sequences of independent symmetrically distributed functions, playing a key role in the proof of Theorem 13.4, has been proved in [26]. It is instructive to compare the result of Theorem 13.4, on the one hand, with Theorem 4.8 (see inequality (13.19)), on the other hand, with the well-known Maurey's inequality, showing that estimate (13.20) holds in an arbitrary q-concave Banach lattice X with some $q < \infty$ for each sequence of functions $\{f_k\}_{k=1}^\infty \subset X$ [182, Theorem 1.d.6(i)]. However, it is worth to highlight the essential difference between these results. Firstly, coefficients of sums in both mentioned statements are arbitrary functions, and secondly, the conditions ensuring inequalities (13.19) and (13.20) in these statements are much more restrictive than the condition $X \in \mathbb{K}$ in Theorem 13.4.

As to comparing conditions $X \in \mathbb{K}$ and $X \supset L_{N_1}^\circ$ (see Theorem 13.5), then from results proved in the paper [63] it follows that there are s.s.'s satisfying one of them and not satisfying another and vice versa. For a detailed account of the Kruglov property and its applications to the study of s.s.'s we refer to [31, 60, 62, 64, 87]. Regarding a characterization of q-convex Orlicz spaces in Proposition 13.3 see [150].

Chapter 14
Rademacher Functions in BMO and Paley Spaces

So far, we studied the behaviour of the Rademacher functions in symmetric function spaces. In contrast to that, in the next three chapters we shall focus on Banach function spaces that are not symmetric. We start with consideration of spaces that do not have even the lattice property.

14.1 Definition and Some Properties of BMO Type Spaces

For arbitrary function $f \in L_1 = L_1[0, 1]$ and any interval $I \subset [0, 1]$ we denote by f_I the average value of f over I, i.e.,

$$f_I := \frac{1}{m(I)} \int_I f(s)ds.$$

Then, a function $f \in L_1$ is said to have the *bounded mean oscillation* on $[0, 1]$ (briefly, $f \in BMO := BMO[0, 1]$) if

$$\|f\|_{BMO} := \sup_I \frac{1}{m(I)} \int_I |f(s) - f_I|\, ds < \infty, \tag{14.1}$$

where the supremum is taken over *all* intervals $I \subset [0, 1]$.

The functional $f \mapsto \|f\|_{BMO}$ is only a seminorm on the linear space BMO, because $\|f\|_{BMO} = 0$ if f is equal to a constant a.e. on $[0, 1]$. To turn BMO into a Banach space, we can either restrict this functional to the linear subspace of BMO of mean zero functions, or consider as a norm on BMO the functional

$$\|f\|'_{BMO} := \|f\|_{BMO} + \|f\|_1$$

(with the usual identification of functions equal a.e.).

S. V. Astashkin, *The Rademacher System in Function Spaces*,
https://doi.org/10.1007/978-3-030-47890-2_14

We also introduce the following dyadic counterpart of the BMO-space. Let $\mathfrak{D}$ be the set of all intervals of the form $\Delta_n^k = ((k-1)2^{-n}, k2^{-n})$, $k = 1, 2, \dots, 2^n$, $n = 0, 1, 2, \dots$. Then $BMO_d := BMO_d[0, 1]$ is the linear space of all $f \in L_1$ such that

$$\|f\|_d := \sup_{\Delta \in \mathfrak{D}} \frac{1}{m(\Delta)} \int_\Delta |f(s) - f_\Delta| \, ds < \infty.$$

Similarly as in the case of BMO, BMO_d becomes a Banach space when it is equipped with the norm

$$\|f\|'_d := \|f\|_d + \|f\|_1.$$

It is clear that

$$BMO \overset{1}{\subset} BMO_d. \tag{14.2}$$

Moreover, since

$$\begin{aligned} \frac{1}{m(I)} \int_I |f(s) - f_I| \, ds &\le \left(\frac{1}{m(I)} \int_I (f(s) - f_I)^2 \, ds \right)^{1/2} \\ &= \left(\frac{1}{m(I)} \int_I f(s)^2 \, ds - f_I^2 \right)^{1/2} \le \|f\|_\infty, \end{aligned}$$

then $L_\infty \overset{1}{\subset} BMO$. Simultaneously, $BMO \ne L_\infty$ and $BMO_d \ne BMO$. Indeed, by using the above definitions, it is not hard to show that

$$\ln|s - 1/2| \chi_{[0,1]}(s) \in BMO \setminus L_\infty \quad \text{and} \quad \ln|s - 1/2| \chi_{[1/2,1]}(s) \in BMO_d \setminus BMO$$

(see also [165, Examples 2.20 and 2.21]). These examples show also that BMO fails to be a function lattice.

Let us find a connection between the BMO- and BMO_d-norms. To this end, we define on L_1 the functional $A(f)$ by

$$A(f) := \sup_{\Delta_1, \Delta_2} |f_{\Delta_1} - f_{\Delta_2}|,$$

where the supremum is taken over all adjacent dyadic intervals Δ_1, Δ_2 of the same length.

Proposition 14.1 *For every function $f \in L_1$ we have*

$$\frac{1}{3} \left(\|f\|_d + A(f) \right) \le \|f\|_{BMO} \le 32 \left(\|f\|_d + A(f) \right). \tag{14.3}$$

Proof Clearly, the left-hand side inequality in (14.3) is an immediate consequence of (14.2) and the estimate

$$A(f) \leq 2 \, \|f\|_{BMO}. \tag{14.4}$$

To prove (14.4), we take two adjacent dyadic intervals Δ_1 and Δ_2, having the same length. Then

$$f_{\Delta_1 \cup \Delta_2} = \frac{1}{2}\left(\frac{1}{m(\Delta_1)}\int_{\Delta_1} f\,ds + \frac{1}{m(\Delta_2)}\int_{\Delta_2} f\,ds\right) = \frac{1}{2}(f_{\Delta_1} + f_{\Delta_2}). \tag{14.5}$$

Therefore, for the interval $I := \Delta_1 \cup \Delta_2$ we have

$$\begin{aligned}
\frac{1}{m(\Delta_1)}\int_{\Delta_1} |f - f_I|\,ds &= \frac{1}{2m(\Delta_1)}\int_{\Delta_1} |(f - f_{\Delta_1}) + (f - f_{\Delta_2})|\,ds \\
&\geq \frac{1}{2}\left|\frac{1}{m(\Delta_1)}\int_{\Delta_1} (f - f_{\Delta_1})\,ds + \frac{1}{m(\Delta_1)}\int_{\Delta_1} (f - f_{\Delta_2})\,ds\right| \\
&= \frac{1}{2}|f_{\Delta_1} - f_{\Delta_2}|,
\end{aligned}$$

and similarly

$$\frac{1}{m(\Delta_2)}\int_{\Delta_2} |f - f_I|\,ds \geq \frac{1}{2}|f_{\Delta_1} - f_{\Delta_2}|.$$

Thus,

$$\begin{aligned}
\frac{1}{m(I)}\int_I |f - f_I|\,ds &= \frac{1}{2}\left(\frac{1}{m(\Delta_1)}\int_{\Delta_1} |f - f_I|\,ds + \frac{1}{m(\Delta_2)}\int_{\Delta_2} |f - f_I)|\,ds\right) \\
&\geq \frac{1}{2}|f_{\Delta_1} - f_{\Delta_2}|,
\end{aligned}$$

and, by (14.1), we get (14.4).

Let us prove the right-hand side inequality in (14.3). For any interval $I \subset [0, 1]$ we can find adjacent dyadic intervals Δ_1 and Δ_2 of the same length such that $I \subset \Delta_1 \cup \Delta_2$ and

$$\frac{1}{2}m(\Delta_1) \leq m(I) \leq 2m(\Delta_1).$$

Then, we have

$$\begin{aligned}
\frac{1}{m(I)}\int_I |f(s)-f_I|\,ds &= \frac{1}{m(I)}\int_I \left|f(s)-\frac{1}{m(I)}\int_I f(t)\,dt\right| ds\\
&\le \frac{1}{m(I)^2}\int_I\int_I |f(s)-f(t)|\,dtds\\
&\le \frac{16}{m(\Delta_1\cup\Delta_2)^2}\int_{\Delta_1\cup\Delta_2}\int_{\Delta_1\cup\Delta_2} |f(s)-f(t)|\,dtds\\
&\le \frac{16}{m(\Delta_1\cup\Delta_2)^2}\int_{\Delta_1\cup\Delta_2}\int_{\Delta_1\cup\Delta_2} |f(s)-f_{\Delta_1\cup\Delta_2}|\,dsdt\\
&+ \frac{16}{m(\Delta_1\cup\Delta_2)^2}\int_{\Delta_1\cup\Delta_2}\int_{\Delta_1\cup\Delta_2} |f(t)-f_{\Delta_1\cup\Delta_2}|\,dsdt\\
&= \frac{32}{m(\Delta_1\cup\Delta_2)}\int_{\Delta_1\cup\Delta_2} |f(s)-f_{\Delta_1\cup\Delta_2}|\,ds.
\end{aligned}$$

This estimate, combined with equation (14.5), implies now that

$$\begin{aligned}
\frac{1}{m(I)}\int_I |f(s)-f_I|\,ds &\le \frac{16}{m(\Delta_1)}\int_{\Delta_1} |f(s)-\frac{1}{2}(f_{\Delta_1}+f_{\Delta_2})|\,ds\\
&+ \frac{16}{m(\Delta_2)}\int_{\Delta_2} |f(s)-\frac{1}{2}(f_{\Delta_1}+f_{\Delta_2})|\,ds\\
&\le \frac{16}{m(\Delta_1)}\int_{\Delta_1} |f(s)-f_{\Delta_1}|\,ds + \frac{16}{m(\Delta_2)}\int_{\Delta_2} |f(s)-f_{\Delta_2}|\,ds\\
&+ 16\,|f_{\Delta_1}-f_{\Delta_2}| \le 32\,\|f\|_d + 16\,A(f).
\end{aligned}$$

Hence, by definition of the BMO-seminorm, the right-hand side of inequality (14.3) is proved. □

14.2 Rademacher Sums in BMO Type Spaces

A key role in the study of the behaviour of Rademacher sums in the spaces BMO_d and BMO will be played by the following statement.

Proposition 14.2 *For arbitrary $n \in \mathbb{N}$ and all $a_k \in \mathbb{R}$, $k = 1, 2, \ldots, n$, we have*

$$\frac{1}{\sqrt{2}}\,\|(a_k)_{k=1}^n\|_{\ell_2} \le \left\|\sum_{k=1}^n a_k r_k\right\|_d \le \|(a_k)_{k=1}^n\|_{\ell_2} \tag{14.6}$$

and

$$\frac{2}{3}\max_{0\le j<m\le n}\Big|\sum_{k=j+1}^{m} a_k\Big| \le A\Big(\sum_{k=1}^{n} a_k r_k\Big) \le 4\max_{0\le j<m\le n}\Big|\sum_{k=j+1}^{m} a_k\Big|. \tag{14.7}$$

Proof Let $m = 0, 1, 2, \ldots$ For every dyadic interval Δ of length 2^{-m} it follows

$$(r_k)_\Delta = \begin{cases} \mathrm{sign}(r_k|_\Delta) & \text{if } k \le m, \\ 0 & \text{if } k > m. \end{cases}$$

Consequently, the average value of the sum $f = \sum_{k=1}^{n} a_k r_k$ on Δ is representable in the form

$$f_\Delta = \sum_{k=1}^{n} a_k \cdot (r_k)_\Delta = \sum_{k=1}^{\min(m,n)} a_k \cdot \mathrm{sign}(r_k|_\Delta).$$

Hence, if $m \ge n$, $f(x) = f_\Delta$ for all $x \in \Delta$, and so the mean oscillation of f on Δ vanishes. Moreover, by Proposition 2.2, $\{r_k\}_{k=1}^{\infty}$ is a symmetric basic sequence in L_1 with constant 1. Therefore, in the case when $m < n$ we have

$$\begin{aligned}
\frac{1}{m(\Delta)}\int_\Delta |f(x) - f_\Delta|\,dx &= \frac{1}{m(\Delta)}\int_\Delta \Big|\sum_{k=1}^{n} a_k\, r_k(x) - \sum_{k=1}^{m} a_k\,\mathrm{sign}(r_k|_\Delta)\Big|\,dx \\
&= \frac{1}{m(\Delta)}\int_\Delta \Big|\sum_{k=m+1}^{n} a_k\, r_k(x)\Big|\,dx = \int_0^1 \Big|\sum_{k=m+1}^{n} a_k\, r_{k-m}(x)\Big|\,dx \\
&= \Big\|\sum_{j=1}^{n-m} a_{m+j}\, r_j\Big\|_1 = \Big\|\sum_{k=m+1}^{n} a_k\, r_k\Big\|_1.
\end{aligned}$$

Hence, by the definition of $\|f\|_d$,

$$\|f\|_d = \Big\|\sum_{k=1}^{n} a_k\, r_k\Big\|_1,$$

and, as a result, (14.6) follows from the embedding $L_2[0,1] \overset{1}{\subset} L_1[0,1]$, Parseval's Identity, and Khintchine's L_1-inequality (see Theorem 5.4).

Proceed now with the proof of inequality (14.7). Let Δ_1 and Δ_2 be two adjacent dyadic intervals of length 2^{-m} each. Then, by the above observation,

$$f_{\Delta_1} - f_{\Delta_2} = \sum_{k=1}^{\min(m,n)} a_k \operatorname{sign}(r_k|_{\Delta_1}) - \sum_{k=1}^{\min(m,n)} a_k \operatorname{sign}(r_k|_{\Delta_2}).$$

Denote by Δ the smallest dyadic interval containing $\Delta_1 \cup \Delta_2$; let Δ have length 2^{-j}. Obviously, $j < m$ and $r_k|_{\Delta_1} = r_k|_{\Delta_2}$ if $k \le j$. Then for $j > n$ we have $f_{\Delta_1} = f_{\Delta_2}$, while for $j \le n$

$$f_{\Delta_1} - f_{\Delta_2} = \sum_{k=j+1}^{\min(m,n)} a_k \left[\operatorname{sign}(r_k|_{\Delta_1}) - \operatorname{sign}(r_k|_{\Delta_2})\right].$$

Moreover, from the definition of Δ it follows that the union $\Delta_1 \cup \Delta_2$ is located exactly in the middle of Δ. Assume that Δ_1 lies to the left of Δ_2. Then it is easy to see that $r_{j+1}|_{\Delta_1} = 1$, $r_{j+1}|_{\Delta_2} = -1$, and $r_k|_{\Delta_1} = -1$, $r_k|_{\Delta_2} = 1$ if $j+2 \le k \le m$. Thus, from the preceding equation we obtain

$$\left|f_{\Delta_1} - f_{\Delta_2}\right| = 2\left|\sum_{k=j+2}^{\min(m,n)} a_k - a_{j+1}\right|,$$

and so

$$A(f) = 2 \max_{0 \le j < m \le n} \left|\sum_{k=j+2}^{m} a_k - a_{j+1}\right|.$$

Also, one can readily check that

$$\frac{1}{3} \max_{0 \le j < m \le n} \left|\sum_{k=j+1}^{m} a_k\right| \le \max_{0 \le j < m \le n} \left|\sum_{k=j+2}^{m} a_k - a_{j+1}\right| \le 2 \max_{0 \le j < m \le n} \left|\sum_{k=j+1}^{m} a_k\right|.$$

Combining two last relations, we obtain inequality (14.7). □

Recall that $\mathcal{R}$ denotes the set of all functions f such that each of them can be represented as the sum of a Rademacher series converging a.e. on $[0, 1]$. Equivalently, $f \in \mathcal{R}$ if and only if $f = \sum_{k=1}^{\infty} a_k r_k$, where $(a_k)_{k=1}^{\infty} \in \ell_2$ (see Theorems 1.1 and 1.2).

The first of the next results follows immediately from inequality (14.6), showing that the closed linear span $[r_k]_{BMO_d}$ of the Rademacher system in BMO_d coincides with $\mathcal{R}$.

Theorem 14.1 *The sequence* $\{r_k\}_{k=1}^{\infty}$ *is equivalent in* BMO_d *to the unit vector basis in* ℓ_2. *More precisely, for any sequence* $a = (a_k)_{k=1}^{\infty} \in \ell_2$ *we have*

$$\frac{1}{\sqrt{2}} \|a\|_{\ell_2} \le \Big\| \sum_{k=1}^{\infty} a_k r_k \Big\|_d \le \|a\|_{\ell_2}.$$

From Propositions 14.1 and 14.2, combined with the elementary inequality

$$\max_{1\le m\le n} \Big| \sum_{k=1}^{m} a_k \Big| \le \max_{0\le j<m\le n} \Big| \sum_{k=j+1}^{m} a_k \Big| \le 2 \max_{1\le m\le n} \Big| \sum_{k=1}^{m} a_k \Big|,$$

it follows a similar statement on the behaviour of Rademacher sums in BMO.

Theorem 14.2 *For arbitrary* $n \in \mathbb{N}$ *and any* $a_k \in \mathbb{R}$, $k = 1, 2, \dots, n$, *we have*

$$\frac{2}{9}\Big[\Big(\sum_{k=1}^{n} a_k^2\Big)^{1/2} + \max_{1\le m\le n}\Big|\sum_{k=1}^{m} a_k\Big|\Big] \le \Big\|\sum_{k=1}^{n} a_k r_k\Big\|_{BMO}$$
$$\le 256\Big[\Big(\sum_{k=1}^{n} a_k^2\Big)^{1/2} + \max_{1\le m\le n}\Big|\sum_{k=1}^{m} a_k\Big|\Big].$$

Thus,

$$\Big\|\sum_{k=1}^{\infty} a_k r_k\Big\|_{BMO} \asymp \|(a_k)_{k=1}^{\infty}\|_{\ell_2} + \sup_{n\in\mathbb{N}}\Big|\sum_{k=1}^{n} a_k\Big|. \tag{14.8}$$

We present now some consequences of Theorem 14.2.

Corollary 14.1 $\{r_k\}_{k=1}^{\infty}$ *is a basic sequence in* BMO *equivalent in this space to each of its subsequences.*

Equivalence (14.8) shows that the Rademacher system fails to be unconditional in BMO. For example, $\sum_{k=1}^{\infty} (-1)^k r_k/k \in BMO$ but $\sum_{k=1}^{\infty} r_k/k \notin BMO$. Moreover, taking into account Corollary 14.1, we get

Corollary 14.2 *The system* $\{r_k\}_{k=1}^{\infty}$ *does not contain any unconditional basic subsequences in* BMO.

Corollary 14.3 L_∞ *is a unique (up to equivalence of norms) s.s. on* $[0, 1]$ *that is embedded into the space* BMO_d.

Proof For arbitrary $\delta \in (0, 1)$ we introduce the δ-"shift" of BMO_d, that is, the space $BMO_d(\delta)$ equipped with the norm $\|f\|_{d,\delta} := \|f_\delta\|_d$, where

$$f_\delta(s) := f(s-\delta)\chi_{[\delta,1]}(s) + f(s-\delta+1)\chi_{[0,\delta]}(s), \ s \in [0, 1]. \tag{14.9}$$

As was proved in [200], there exists $\delta_0 \in (0, 1)$ such that

$$BMO = BMO_d \cap BMO_d(\delta_0).$$

If the embedding $X \subset BMO_d$ holds for a s.s. X, then clearly $X \subset BMO_d(\delta_0)$. Hence, $X \subset BMO$. Next, since $\{r_k\}_{k=1}^{\infty}$ is an unconditional basic sequence with constant 1 in a s.s. (see Proposition 2.2), from Theorem 14.2 it follows that

$$\Big\|\sum_{k=1}^{n} a_k r_k\Big\|_X = \Big\|\sum_{k=1}^{n} |a_k| r_k\Big\|_X \geq c\Big\|\sum_{k=1}^{n} |a_k| r_k\Big\|_{BMO} \geq \frac{2c}{9}\|(a_k)\|_{\ell_1},$$

with some constant $c > 0$ independent of $n \in \mathbb{N}$ and $a_k \in \mathbb{R}$, $k = 1, 2, \ldots, n$. Thus, the sequence $\{r_k\}_{k=1}^{\infty}$ is equivalent in X to the unit vector basis in ℓ_1, and so, by Theorem 2.4, $X = L_\infty$ (with equivalence of norms). □

One can easily check that the Rademacher sequence is weakly null in every s.s. on [0, 1] (see also [1, Lemma 6.3.2]). In contrast to that, in BMO we have the following result.

Corollary 14.4 *The sequence $\{r_k\}_{k=1}^{\infty}$ is not weakly null in BMO.*

Proof Define a linear functional φ on the linear span of the Rademacher functions in BMO by

$$\varphi\Big(\sum_{k=1}^{n} a_k r_k\Big) := \sum_{k=1}^{n} a_k, \; a_k \in \mathbb{R}, \; k = 1, 2, \ldots, n, \; n \in \mathbb{N}.$$

According to Theorem 14.2, this functional is bounded on $[r_k]$ with respect to the BMO-norm. Therefore, by the Hahn-Banach theorem, it can be extended to a functional $\varphi_0 \in BMO^*$. Since $\varphi_0(r_n) = 1$ for all $n \in \mathbb{N}$, we have $r_n \overset{w}{\nrightarrow} 0$ in BMO. □

14.3 On Complementability of Rademacher Subspaces in BMO_d and BMO

We start with considering the simpler question on complementability of the subspace $[r_k]_{BMO_d} (= \mathcal{R})$ spanned by the Rademacher system in BMO_d.

Recall that a closed subspace F of a Banach space X is said to be *complemented* in X if there exists a bounded linear projection in X such that its image coincides with F.

Theorem 14.3 *The subspace $\mathcal{R}$ is complemented in BMO_d.*

Proof Let us check first that the embedding $BMO_d \subset L_p[0, 1]$ holds for every $1 \leq p < \infty$. Indeed, in view of a dyadic version of the classical John–Nirenberg theorem (see e.g. [165, Theorem 3.14]), for an arbitrary function $f \in BMO_d$, any dyadic interval Δ and all $\tau > 0$ we have

$$m\{x \in \Delta : |f(x) - f_\Delta| > \tau\} \le e \cdot m(\Delta) \cdot \exp\left(-\frac{\tau}{2e\,\|f\|_d}\right). \tag{14.10}$$

Passing to the decreasing rearrangements of functions involved in this inequality in the case $\Delta = [0, 1]$, we get

$$(f - f_{[0,1]})^*(t) \le 2e \cdot \|f\|_d \ln(e/t), \quad 0 < t \le 1.$$

Hence, for each $1 \le p < \infty$

$$\begin{aligned} \|f\|_p &\le \left(\int_0^1 |f(x) - f_{[0,1]}|^p\,dx\right)^{1/p} + |f_{[0,1]}| \\ &= \left(\int_0^1 [(f - f_{[0,1]})^*(t)]^p\,dt\right)^{1/p} + |f_{[0,1]}| \\ &\le 2e\|f\|_d\left(\int_0^1 \ln(e/t)^p\,dt\right)^{1/p} + |f_{[0,1]}| \\ &= C_p\|f\|_d + \left|\int_0^1 f(x)\,dx\right| \le C_p\|f\|_d', \end{aligned}$$

and the above claim is proved.

Further, recall that the orthogonal projection

$$Pf := \sum_{k=1}^\infty \int_0^1 f(u) r_k(u)\,du \cdot r_k$$

boundedly acts from the space $L_p[0,1]$ onto $[r_k]_{L_p} = \mathcal{R}$ if $1 < p < \infty$ (see the proof of Theorem 2.5). Therefore, from (14.6) and the last inequality it follows that for $1 < p < \infty$ and $f \in BMO_d$

$$\|Pf\|_d' \asymp \|Pf\|_p \le \|P\|_{L_p \to L_p}\|f\|_p \le C_p\|P\|_{L_p \to L_p}\|f\|_d'.$$

Thus, P is bounded in BMO_d, and so the proof of the theorem is completed. □

In contrast to Theorem 14.3, the subspace $\mathcal{R}_{BMO} := [r_k]_{BMO}$ spanned by the Rademacher system in BMO is not complemented. The subsequent proof will be based on the fact that $\mathcal{R}_{BMO}$ contains a further subspace isomorphic to the space c_0 of all vanishing sequences $\beta = (\beta_n)_{n=1}^\infty$ with the norm $\|\beta\|_{c_0} := \max_{n=1,2,\dots} |\beta_n|$.

Let $\mathcal{U} = \{u_n\}_{n=1}^\infty$ be an arbitrary block basic sequence of the Rademacher system, that is, a sequence of functions u_n, $n = 1, 2, \dots$, of the form

$$u_n = \sum_{k=m_n+1}^{m_{n+1}} a_k r_k, \quad n = 1, 2, \dots,$$

where $0 \le m_1 < m_2 < \ldots$ and $a_k \in \mathbb{R}$. From Proposition 14.2 it follows that $\|u_n\|_d \asymp \left(\sum_{k=m_n+1}^{m_{n+1}} a_k^2\right)^{1/2}$. For every such a block basic sequence we set

$$\gamma_n(\mathcal{U}) := \sum_{k=m_n+1}^{m_{n+1}} a_k, \; n = 1, 2, \ldots.$$

Proposition 14.3 *Assume that a block basic sequence $\mathcal{U} = \{u_n\}_{n=1}^{\infty}$ of the Rademacher system satisfies the following conditions:*

(*a*) $C^{-1} \le \|u_n\|_{BMO} \le C$ *for some* $C > 0$ *and all* $n = 1, 2, \ldots$.
(*b*) $\left(\sum_{k=m_n+1}^{m_{n+1}} a_k^2\right)^{1/2} \le 2^{-n}, \; n = 1, 2, \ldots$.
(*c*) $|\gamma_n(\mathcal{U})| \le 2^{-n}, \; n = 1, 2, \ldots$.

Then the sequence $\{u_n\}_{n=1}^{\infty}$ is equivalent in BMO to the unit vector basis in the space c_0.

Proof Let $u_n = \sum_{k=m_n+1}^{m_{n+1}} a_k r_k$, $n = 1, 2, \ldots$, where $0 \le m_1 < m_2 < \ldots$ and $a_k \in \mathbb{R}$. For brevity, we denote $\gamma_n := \gamma_n(\mathcal{U})$, $n = 1, 2, \ldots$. By Corollary 14.1, any block basic sequence $\mathcal{U}$ of the Rademacher system is a basic sequence in BMO. Therefore, if $f \in [u_n]_{BMO}$, then $f = \sum_{n=1}^{\infty} \beta_n u_n$, $\beta_n \in \mathbb{R}$, $n = 1, 2, \ldots$. We rewrite this sum in another way:

$$f = \sum_{n=1}^{\infty} \left(\sum_{k=m_n+1}^{m_{n+1}} \beta_n a_k r_k \right) = \sum_{k=1}^{\infty} \gamma_k r_k,$$

where $\gamma_k = \beta_n a_k$ if $m_n < k \le m_{n+1}$.

First, assuming that $m \in \mathbb{N}$ satisfies $m_i < m \le m_{i+1}$ with some $i \in \mathbb{N}$, we estimate the absolute value of the sum $\sum_{k=m_1+1}^{m} \gamma_k$ from above. By Proposition 14.2, inequality (14.4), and properties (*c*), (*a*), we have

$$\begin{aligned}
\left|\sum_{k=m_1+1}^{m} \gamma_k\right| &= \left|\sum_{k=m_1+1}^{m_i} \gamma_k + \sum_{k=m_i+1}^{m} \gamma_k\right| \le \sum_{n=1}^{i-1} |\beta_n| \left|\sum_{k=m_n+1}^{m_{n+1}} a_k\right| + |\beta_i| \left|\sum_{k=m_i+1}^{m} a_k\right| \\
&\le \max_n |\beta_n| \Big(\sum_{n=1}^{i-1} \Big| \sum_{k=m_n+1}^{m_{n+1}} a_k \Big| + \Big| \sum_{k=m_i+1}^{m} a_k \Big| \Big) \\
&\le \max_n |\beta_n| \Big(\sum_{n=1}^{i-1} 2^{-n} + 3A(u_i) \Big) \\
&\le (1 + 6\|u_i\|_{BMO}) \, \|(\beta_n)\|_{c_0} \le C' \|(\beta_n)\|_{c_0}
\end{aligned}$$

($A = A(f)$ is the functional defined in the previous section). Moreover, from (b) it follows that

$$\sum_{k=1}^{\infty} \gamma_k^2 = \sum_{n=1}^{\infty} \beta_n^2 \sum_{k=m_n+1}^{m_{n+1}} a_k^2 \leq \|(\beta_n)\|_{c_0}^2.$$

As a result, these inequalities, Theorems 1.3 and 14.2 imply that $\|f\|'_{BMO} \leq C''\|(\beta_n)\|_{c_0}$.

On the other hand, by (a), (b), Proposition 14.2, and Theorem 14.2, there is $\delta' > 0$ such that $A(u_n) \geq \delta'$ for all sufficiently large $n \in \mathbb{N}$. Also, according to Proposition 14.2, $A(u_n) > 0$ for all $n \in \mathbb{N}$. Therefore, $A(u_n) \geq \delta$ for some $\delta > 0$ and all $n \in \mathbb{N}$. Hence, from (14.4) and Proposition 14.2 it follows that

$$\|f\|_{BMO} \geq \frac{1}{2} A(f) \geq \frac{1}{3} \sup_{n \in \mathbb{N}} \sup_{m_n < p < q \leq m_{n+1}} \left| \sum_{k=p}^{q} \beta_n a_k \right|$$
$$\geq \frac{1}{12} \sup_{n \in \mathbb{N}} |\beta_n| \, A(u_n) \geq \frac{\delta}{12} \|(\beta_n)\|_{c_0}.$$

Summarizing all, we get

$$\left\| \sum_{n=1}^{\infty} \beta_n u_n \right\|'_{BMO} \asymp \|(\beta_n)\|_{c_0},$$

and the proof is completed. □

Corollary 14.5 *The space $\mathcal{R}_{BMO}$ contains a complemented subspace isomorphic to c_0.*

Proof Using Theorem 14.2, one can easily construct a block basic sequence $\mathcal{U} = \{u_n\}_{n=1}^{\infty}$ of the Rademacher system, satisfying the above conditions (a), (b), and (c). Then, by Proposition 14.3, the subspace $E := [u_n]_{BMO}$ of $\mathcal{R}_{BMO}$ is isomorphic to c_0. On the other hand, according to the well-known Sobczyk theorem (see e.g. [1, Corollary 2.5.9]), any subspace of a separable space X isomorphic to c_0 is complemented in X. Therefore, since $\mathcal{R}_{BMO}$ is separable, E is complemented in $\mathcal{R}_{BMO}$. □

Theorem 14.4 *The Rademacher subspace $\mathcal{R}_{BMO}$ is uncomplemented in BMO.*

Proof On the contrary, assume that $\mathcal{R}_{BMO}$ is a complemented subspace in BMO, and let P_1 be a bounded linear projection in BMO whose image is $\mathcal{R}_{BMO}$. By Corollary 14.5, there is a further subspace E that is isomorphic to c_0 and complemented in $\mathcal{R}_{BMO}$. Suppose that $P_2 : \mathcal{R}_{BMO} \to \mathcal{R}_{BMO}$ is a bounded linear projection such that its image coincides with E. Then the product $P := P_2 \cdot P_1$ is a linear projection, bounded in BMO, with the image E. Thus, BMO contains a complemented subspace isomorphic to c_0. On the other hand, it is well known that BMO is a dual space; more precisely, $BMO = (\mathrm{Re}H_1)^*$, where the space $\mathrm{Re}H_1$ consists of the real parts of complex-valued functions $f \in L_1$ such that their (trigonometric) conjugate function $\tilde{f} \in L_1$ (see, for example, [153, Theorem V.5]). However, this contradicts the Bessaga–Pełczyński theorem that a dual space cannot contain a complemented subspace isomorphic to c_0 (see [73, Corollary 4] and [72, Theorem 4 and its proof]). □

14.4 Geometrical Structure of the Rademacher Subspace in BMO

Here, we investigate the geometrical structure of the Rademacher space $\mathcal{R}_{BMO}$. More specifically, we shall prove that infinite-dimensional subspaces of $\mathcal{R}_{BMO}$ satisfy the following alternative.

Theorem 14.5 *Every infinite-dimensional subspace of the space $\mathcal{R}_{BMO}$ is either isomorphic to ℓ_2 and complemented in BMO, or contains a subspace isomorphic to c_0 and complemented in $\mathcal{R}_{BMO}$.*

To prove this result, we need the fact of the existence of a basis in the dual space $Z := \mathcal{R}^*_{BMO}$ for the Rademacher subspace in BMO. Clearly, a natural candidate for this role is the biorthogonal system for a basis of the space $\mathcal{R}_{BMO}$ itself. Recall (see e.g. [181, Proposition 1.b.1]) that the biorthogonal system of functionals $\{x_n^*\}_{n=1}^\infty$ for a basis $\{x_n\}_{n=1}^\infty$ of a Banach space X (that is, $x_n^*(x_n) = 1$, and $x_m^*(x_n) = 0$ for $m \neq n$) forms a basis in the dual space X^* if and only if $\{x_n\}_{n=1}^\infty$ is a *shrinking basis* (see Theorem B.1). This means that for each $x^* \in X^*$ the norm of the restriction $x^*|_{[x_n]_{n=m}^\infty}$ of this functional to the subspace $[x_n]_{n=m}^\infty$ vanishes as $m \to \infty$ (see Appendix B). It is worth to note that the biorthogonal system for the Rademacher system is a Rademacher system itself and it fails to be a basis in Z because, by Corollary 14.4, $\{r_n\}$ is not weakly null. However, the following result holds.

Proposition 14.4 *The functions*

$$s_n := r_n - r_{n-1},\ n = 1, 2, \dots,\ \textit{where } r_0 = 0,$$

form a shrinking basis in the space $\mathcal{R}_{BMO}$. Hence, the biorthogonal system $\{s_n^*\}_{n=1}^\infty$ is a basis in the dual space $Z = (\mathcal{R}_{BMO})^*$.

Proof If $f = \sum_{n=1}^\infty \beta_n s_n$, then

$$f = \sum_{n=1}^\infty \beta_n (r_n - r_{n-1}) = \sum_{n=0}^\infty (\beta_n - \beta_{n+1}) r_n, \text{ with } \beta_0 = 0.$$

Therefore, by Theorem 14.2, we have for some $C > 0$

$$C^{-1}\Big[\Big(\sum_{n=0}^{\infty}(\beta_n - \beta_{n+1})^2\Big)^{1/2} + \sup_{0\le m<n<\infty} |\beta_m - \beta_n|\Big] \le \|f\|_{BMO}$$
$$\le C\Big[\Big(\sum_{n=0}^{\infty}(\beta_n - \beta_{n+1})^2\Big)^{1/2} + \sup_{0\le m<n<\infty} |\beta_m - \beta_n|\Big]. \quad (14.11)$$

Hence, the sequence $\{s_n\}_{n=1}^{\infty}$ is a basis in the space $\mathcal{R}_{BMO}$ and also

$$|\beta_n| \le |\beta_n - \beta_1| + |\beta_1| \le C\,\|f\|_{BMO}, \quad n = 1, 2, \ldots \quad (14.12)$$

On the contrary, suppose that $\{s_n\}_{n=1}^{\infty}$ is not a shrinking sequence. Then, there exist $\varepsilon \in (0, 1)$, a functional $\varphi \in Z$ with $\|\varphi\|_Z = 1$, and a sequence of functions $f_n = \sum_{k=m_n}^{\infty} a_k^{m_n} s_k$, $1 \le m_1 < m_2 < \ldots$, such that $\|f_n\|_{BMO} = 1$ and $\varphi(f_n) \ge \varepsilon$, $n = 1, 2, \ldots$.

We construct two sequences of positive integers $\{q_i\}_{i=1}^{\infty}$ and $\{p_i\}_{i=1}^{\infty}$, $1 \le q_1 < p_1 < q_2 < p_2 < \ldots$, in the following way. Setting $q_1 = m_1$, we choose $p_1 > q_1$ so that

$$\Big\|\sum_{k=p_1+1}^{\infty} a_k^{q_1} s_k\Big\|_{BMO} < \frac{\varepsilon}{2}.$$

Suppose that $1 \le q_1 < p_1 < q_2 < p_2 < \ldots < q_{i-1} < p_{i-1}$, $i \ge 2$, are already chosen. As a next step, we take first for q_i the least number m_n that greater than p_{i-1}. Secondly, take for p_i a positive integer such that $p_i > q_i$ and

$$\Big\|\sum_{k=p_i+1}^{\infty} a_k^{q_i} s_k\Big\|_{BMO} < \frac{\varepsilon}{2}. \quad (14.13)$$

Let $\alpha_k^i = a_k^{q_i}$ if $q_i \le k \le p_i$, and $\alpha_k^i = 0$ if $p_i < k < q_{i+1}$, $i = 1, 2, \ldots$. Clearly, the functions $u_i := \sum_{k=q_i}^{q_{i+1}-1} \alpha_k^i s_k$, $i = 1, 2, \ldots$, form a block basic sequence of $\{s_n\}_{n=1}^{\infty}$. Moreover, from inequalities (14.11) and (14.12) it follows that

$$\Big(\sum_{k=q_i}^{q_{i+1}-1} (\alpha_k^i - \alpha_{k+1}^i)^2\Big)^{1/2} \le 2C, \quad i = 1, 2, \ldots \quad \text{and} \quad |\alpha_k^i| \le C, \quad k, i = 1, 2, \ldots \quad (14.14)$$

Furthermore, by the definition of u_i, we have

$$\sup_{i\in\mathbb{N}} \|u_i\|_{BMO} \le 2, \quad (14.15)$$

and, in view of the choice of the functional φ and inequality (14.13),

$$\varphi(u_i) = \varphi\Big(\sum_{k=q_i}^{p_i} a_k^{q_i} s_k\Big) = \varphi(f_i) - \varphi\Big(\sum_{k=p_i+1}^{\infty} a_k^{q_i} s_k\Big)$$

$$\geq \varphi(f_i) - \Big\| \sum_{k=p_i+1}^{\infty} a_k^{q_i} s_k \Big\|_{BMO} \geq \frac{\varepsilon}{2}. \tag{14.16}$$

Let us show that for every nonnegative sequence $\{\gamma_n\}_{n=1}^{\infty}$ satisfying the conditions

$$\sum_{n=1}^{\infty} \gamma_n^2 < \infty \text{ and } \sum_{n=1}^{\infty} \gamma_n = \infty,$$

the series $\sum_{n=1}^{\infty} \gamma_n u_n$ converges in BMO.

Indeed, let $b_k = \alpha_k^i \gamma_i$ for $q_i \leq k < q_{i+1}, i = 1, 2, \ldots$. Then, by (14.14), $|b_k| \leq C\gamma_i$, $q_i \leq k < q_{i+1}$, whence

$$\lim_{k\to\infty} b_k = 0. \tag{14.17}$$

Moreover,

$$\sum_{k=1}^{\infty} (b_k - b_{k+1})^2 = \sum_{i=1}^{\infty} \sum_{k=q_i}^{q_{i+1}-2} (\alpha_k^i \gamma_i - \alpha_{k+1}^i \gamma_i)^2$$

$$+ \sum_{i=1}^{\infty} (\alpha_{q_{i+1}-1}^i \gamma_i - \alpha_{q_{i+1}}^{i+1} \gamma_{i+1})^2 =: A_1 + A_2.$$

We shall estimate the terms A_1 and A_2 separately. Applying (14.14) once more, we get

$$A_1 = \sum_{i=1}^{\infty} \gamma_i^2 \sum_{k=q_i}^{q_{i+1}-2} (\alpha_k^i - \alpha_{k+1}^i)^2 \leq 4C^2 \sum_{i=1}^{\infty} \gamma_i^2$$

and

$$A_2 \leq 2 \sum_{i=1}^{\infty} \Big((\alpha_{q_{i+1}-1}^i)^2 \gamma_i^2 + (\alpha_{q_{i+1}}^{i+1})^2 \gamma_{i+1}^2 \Big)$$

$$\leq 4 \sup_{i\in\mathbb{N}} \max_{q_i \leq k \leq p_i} |a_k^{q_i}|^2 \sum_{i=1}^{\infty} \gamma_i^2 \leq 4\,C^2 \sum_{i=1}^{\infty} \gamma_i^2.$$

Therefore, according to (14.11) and (14.17), the series

$$\sum_{n=1}^{\infty} \gamma_n u_n = \sum_{k=1}^{\infty} b_k s_k$$

converges in BMO, and so, since the functional φ is continuous, from (14.16) it follows that

$$\varphi\Big(\sum_{n=1}^{\infty} \gamma_n u_n\Big) = \sum_{n=1}^{\infty} \gamma_n \varphi(u_n) \geq \frac{\varepsilon}{2} \sum_{n=1}^{\infty} \gamma_n = \infty.$$

This contradiction shows that $\{s_n\}_{n=1}^{\infty}$ is a shrinking basis in the space $\mathcal{R}_{BMO}$, which completes the proof of the proposition. □

Proof of Theorem 14.5 Let X be an infinite-dimensional subspace of the space $\mathcal{R}_{BMO}$. As above, by $\|f\|_d$ we denote the seminorm of a function $f \in BMO_d$ defined in the very beginning of the chapter.

Assume first that there is $c > 0$ such that

$$\|f\|_d \geq c\|f\|_{BMO} \quad \text{for all} \quad f \in X. \tag{14.18}$$

Then, by Theorem 14.1 and Parseval's Identity, for every function $f = \sum_{k=1}^{\infty} a_k r_k \in X$ it holds

$$\|f\|_{BMO} \leq c^{-1}\|f\|_d \leq c^{-1}\Big(\sum_{k=1}^{\infty} a_k^2\Big)^{1/2} = c^{-1}\|f\|_2. \tag{14.19}$$

From (14.2) and the embedding $BMO_d \subset L_2$ (see the proof of Theorem 14.3) it follows that $BMO \subset L_2$. Consequently, the BMO-norm is equivalent on X to the L_2-norm. Hence, X is isomorphic to ℓ_2 both in L_2 and in BMO. Clearly, X is complemented in L_2. Let Q denote the orthogonal projection in L_2 whose image is the subspace X. Then $\|Q\|_{L_2 \to L_2} = 1$ and, in view of inequality (14.19), for any $f \in BMO$ we have

$$\|Qf\|_{BMO} \leq c^{-1}\|Qf\|_2 \leq c^{-1}\|f\|_2 \leq C \cdot c^{-1}\|f\|_{BMO},$$

whence it follows that X is complemented in BMO as well.

Conversely, suppose that inequality (14.18) does not hold for every $c > 0$. Equivalently, there exists a sequence $\{f_n\}_{n=1}^{\infty} \subset X$, $\|f_n\|_{BMO} = 1$, $n = 1, 2, \dots$, such that $\|f_n\|_d \to 0$ as $n \to \infty$. Observe that $\{f_n\}_{n=1}^{\infty}$ does not contain subsequences converging in BMO. In fact, let $\lim_{k\to\infty} f_{n_k} := f \in BMO$ for some $\{f_{n_k}\} \subset \{f_n\}$. Then, according to Theorem 14.2, we would have simultaneously $\|f\|_{BMO} = 1$ and $\|f\|_d = 0$ (and hence $f = 0$), which is impossible. Therefore, passing to a subsequence if necessary (and preserving the notation), we may assume that for some $\delta > 0$

$$\|f_m - f_n\|_{BMO} \geq \delta, \quad m, n = 1, 2, \ldots, \quad m \neq n. \tag{14.20}$$

Let $\{s_k^*\}_{k=1}^{\infty} \subset Z$ be the sequence from Proposition 14.4. By using the diagonal process, we can select a subsequence $\{f_{n_i}\}_{i=1}^{\infty} \subset \{f_n\}$ such that for any $k \in \mathbb{N}$ we have

$$s_k^*(f_{n_{i+1}} - f_{n_i}) \to 0 \text{ as } i \to \infty.$$

Since $\{s_k^*\}_{k=1}^{\infty}$ is a basis in Z and $\|f_n\|_{BMO} = 1$, $n = 1, 2, \ldots$, this implies that $f_{n_{i+1}} - f_{n_i} \xrightarrow{w} 0$ in BMO. Then, by inequality (14.20), the sequence $\{f_{n_{i+1}} - f_{n_i}\}_{i=1}^{\infty}$ satisfies all the conditions of the Bessaga–Pełczyński selection principle (see Theorem B.3), and so there exist a block basic sequence $\mathcal{U} = \{u_k\}_{k=1}^{\infty}$ of the Rademacher system and a subsequence of $\{f_{n_{i+1}} - f_{n_i}\}$ (we denote it still by $\{f_{n_{i+1}} - f_{n_i}\}$) such that

$$\|u_k - (f_{n_{k+1}} - f_{n_k})\|_{BMO} \leq 2^{-k}, \ k = 1, 2, \ldots \tag{14.21}$$

Hence, in particular, $\|u_k\|_d \to 0$ and $u_k \xrightarrow{w} 0$ in BMO as $k \to \infty$. Consequently, if $u_k = \sum_{i=m_k+1}^{m_{k+1}} a_i r_i$, $1 \leq m_1 < m_2 < \ldots$, then by Proposition 14.2

$$\sum_{k=m_k+1}^{m_{k+1}} a_k^2 \to 0 \ \text{ as } \ k \to \infty.$$

Moreover, if φ_0 is the functional from the proof of Corollary 14.4, we get

$$\varphi_0(u_k) = \sum_{i=m_k+1}^{m_{k+1}} a_i \to 0 \text{ as } k \to \infty.$$

Finally, since $\|f_n\|_{BMO} = 1$, $n = 1, 2, \ldots$, then from (14.20) and (14.21) it follows that the sequence $\{u_k\}_{k=1}^{\infty}$ satisfies condition (a) from Proposition 14.3. These observations show that there is a subsequence $\{u_{k_i}\}_{i=1}^{\infty} \subset \{u_k\}$ (that is also a block basic sequence of $\{r_n\}$), satisfying conditions (a), (b), and (c) of the last proposition. Therefore, the subspace $[u_{k_i}]_{BMO}$ is complemented in $\mathcal{R}_{BMO}$ and isomorphic to c_0. Combining this together with inequality (14.21) and the principle of small perturbations (see Theorem B.2), we conclude that there is a subsequence of the sequence $\{f_{n_{i+1}} - f_{n_i}\}_{i=1}^{\infty}$ that possesses the above properties as well. Since this subsequence lies in X and, by Sobczyk theorem [1, Corollary 2.5.9], its span is complemented in $\mathcal{R}_{BMO}$, the proof is completed. □

Corollary 14.6 *Let $X \subset \mathcal{R}_{BMO}$ be an infinite-dimensional subspace of BMO. The following conditions are equivalent:*

(1) X does not contain subspaces isomorphic to c_0;
(2) X is isomorphic to a dual space;
(3) X is reflexive;
(4) X is isomorphic to ℓ_2;
(5) X is complemented in BMO;
(6) the BMO- and L_1-norms are equivalent on X.

Proof By Theorem 14.5 and Khintchine's L_1-inequality, each of the conditions (2)–(6) is a consequence of (1). Conversely, one can readily see that each of the conditions (3), (4), and (6) implies (1). Since X is separable, then again from the Sobczyk theorem it follows that every subspace isomorphic to c_0 is complemented in X. Therefore, the remaining implications (2) $\Rightarrow$ (1) and (5) $\Rightarrow$ (1) are direct consequences of the Bessaga–Pełczyński theorem [73, Corollary 4], which was used in the proof of Theorem 14.4. □

In conclusion, we present some examples of block basic sequences of the Rademacher system in BMO.

Example 14.1 (A Block Basic Sequence Equivalent to the Unit Vector Basis in ℓ_2) Let $u_k := r_{2k+1} - r_{2k}$ and $f = \sum_{k=1}^n a_k\, u_k$, $n = 1, 2, \ldots$. Then,

$$f = \sum_{k=1}^{n} a_k\,(r_{2k+1} - r_{2k}) = \sum_{k=1}^{n} a_k\, r_{2k+1} - \sum_{k=1}^{n} a_k\, r_{2k},$$

and, by (14.7),

$$A(f) \asymp \max_{1\le k\le n} |a_k| = \|(a_k)_{k=1}^n\|_{c_0}.$$

On the other hand, according to (14.6),

$$\frac{1}{\sqrt{2}}\|(a_k)_{k=1}^n\|_{\ell_2} \le \|f\|_d \le 2\|(a_k)_{k=1}^n\|_{\ell_2}.$$

Combining these relations with Theorem 14.2, we obtain

$$\left\|\sum_{k=1}^{\infty} a_k\, u_k\right\|_{BMO} \asymp \|(a_k)_{k=1}^{\infty}\|_{\ell_2}.$$

Example 14.2 (A Block Basic Sequence Equivalent to the Unit Vector Basis in c_0) Let us consider the block basic sequence $\mathcal{U} = \{u_n\}_{n=1}^{\infty}$ consisting of the functions

$$u_n = \sum_{k=m_n+1}^{m_{n+1}} a_k\, r_k \;\text{ with }\; m_{n+1} - m_n = 2^{2n},\; n = 1, 2, \ldots,$$

where

$$a_k = \begin{cases} 2^{-2n} & \text{if } m_n + 1 \le k \le \frac{m_n + m_{n+1}}{2}, \\ -2^{-2n} & \text{if } \frac{m_n+m_{n+1}}{2} + 1 \le k \le m_{n+1}. \end{cases}$$

Then, by (14.6) and (14.8), we have

$$\|u_n\|_d \asymp \Big(\sum_{k=m_n+1}^{m_{n+1}} 2^{-4n} \Big)^{1/2} = 2^{-n}, \ n = 1, 2, \ldots,$$

and

$$\|u_n\|_{BMO} \asymp 2^{-2n} \frac{m_{n+1} - m_n}{2} + 2^{-n} \asymp \frac{1}{2}, \ n = 1, 2, \ldots$$

Moreover,

$$\gamma_n(\mathcal{U}) = \sum_{k=m_n+1}^{(m_n+m_{n+1})/2} 2^{-2n} - \sum_{k=(m_n+m_{n+1})/2+1}^{m_{n+1}} 2^{-2n} = 0.$$

In consequence, $\mathcal{U}$ satisfies the conditions of Proposition 14.3, and so the sequence $\{u_n\}_{n=1}^{\infty}$ is equivalent in BMO to the unit vector basis in c_0.

14.5 Preliminaries Related to Paley Spaces

Here and in subsequent sections of this chapter, we consider another class of Banach function spaces closely related to the Fourier–Haar expansion of integrable functions.

As above, $\mathfrak{D}$ is the family of dyadic intervals from [0, 1], and $\{h_{n,k}\}_{(n,k)\in\Omega}$, where $\Omega := \{(0,0), (n,k), 1 \leqslant k \leqslant 2^n, n = 0, 1, \ldots\}$ is the Haar system normalized in L_∞ (see Sect. 13.1). Further, we shall use also other ways of indexing the Haar functions. Sometimes we shall write $\{1, h_\Delta\}_{\Delta\in\mathfrak{D}}$, where Δ is the support of the function h_Δ. Moreover, the formula $m = 2^n + k$ establishes a one-to-one correspondence between Ω and the set of positive integers, allowing us to represent this system using only one index: $\{h_m\}_{m\in\mathbb{N}}$.

Recall that the functions $2^{n/2}h_{n,k}$, $(n,k) \in \Omega$, form a complete orthonormal system in L_2, which is a monotone basis in any separable s.s. (see e.g. [168, Theorem II.9.4]). Furthermore, it is well known (see [182, 2.c.6] or [168, II.9.6]) that the Haar system is an unconditional basis in a separable s.s. X if and only if $0 < \alpha_X \leqslant \beta_X < 1$, where α_X and β_X are the Boyd indices of X. One more necessary and sufficient condition for unconditionality of the Haar system in a s.s.

X is the equivalence $\|x\|_X \asymp \|\mathcal{P}x\|_X$, where $\mathcal{P}x$ is the Paley function of a function $x \in L_1$ (see Sect. 13.1) [153, Theorem I.11].

Given a s.s. X on $[0, 1]$, we let $\mathcal{P}(X)$ denote the space of all functions $x \in L_1[0, 1]$ such that $\mathcal{P}x \in X$ and equip this space with the natural norm

$$\|x\|_{\mathcal{P}(X)} := \|\mathcal{P}x\|_X.$$

It is not hard to check that the space $\mathcal{P}(X)$ is complete for every s.s. X. Indeed, suppose that $x \in \mathcal{P}(X)$, $x = \int_0^1 x(s)\,ds + \sum_{\Delta\in\mathfrak{D}} a_\Delta h_\Delta$. We set

$$x_1 := \int_0^1 x(s)\,ds \ \text{ and } \ x_n := \sum_{\Delta\in\mathfrak{D}:m(\Delta)=2^{-n+2}} a_\Delta h_\Delta, \quad n = 2, 3, \dots$$

Then, the mapping $x \mapsto Ux := \{x_n\}_{n=1}^\infty$ is an isometric embedding of $\mathcal{P}(X)$ into the Banach space $\overline{X(\ell_2)}$ of all sequences $\{f_n\}_{n=1}^\infty$, $f_n \in X$, such that

$$\|\{f_n\}\|_{\overline{X(\ell_2)}} := \Big\|\Big(\sum_{n=1}^\infty f_n^2\Big)^{1/2}\Big\|_X < \infty$$

(one can find the subspace $X(\ell_2)$ of $\overline{X(\ell_2)}$ in Sect. 4.6). A standard reasoning shows that the image of the mapping U is closed in $\overline{X(\ell_2)}$. Therefore, $\mathcal{P}(X)$ is a Banach space. We shall say that $\mathcal{P}(X)$ is the *Paley space* corresponding to the s.s. X. In particular, in the case $X = L_1$ we get the well-known dyadic space H_1 [213, § 1.2]; note that the dual for H_1 is the space BMO_d (with equivalence of norms) [213, Theorems 1.2.2 and 1.2.3].

Let us consider the question when the space $\mathcal{P}(X)$ is being a normed function lattice with respect to the usual order on $[0, 1]$ (see Appendix C). As the following result shows, the answer is heavily depends on what the statement of this question is.

Theorem 14.6 *Let a s.s. X be separable or have the Fatou property. Then,*

(*i*) *$\mathcal{P}(X)$ is a normed function lattice if and only if $X = L_2$ and $\|x\|_X = \|x\|_2$ for all $x \in L_2$;*
(*ii*) *$\mathcal{P}(X)$ coincides (up to equivalence of norms) with a normed function lattice if and only if $0 < \alpha_X \leqslant \beta_X < 1$.*

Proof We show first that the space $\mathcal{P}(L_\infty)$ does not coincide (up to equivalence of norms) with any normed function lattice on $[0, 1]$.

To this end, observe that $\mathcal{P}\chi_{[0,1]} = 1$. Therefore, to prove the claim, it suffices to find a measurable function f such that $0 \le f \le \chi_{[0,1]}$ and the Paley function $\mathcal{P}f$ is unbounded on $[0, 1]$.

Let $f := \sum_{k=1}^\infty \chi_{I_k}$, where $I_k := [3 \cdot 4^{-k-1}, 4^{-k}]$, $k = 1, 2, \dots$ Then $0 \le f \le \chi_{[0,1]}$. On the other hand, if h_{Δ_k} is the Haar function supported on the interval

$\Delta_k = [0, 4^{-k}]$, then for all $k = 1, 2, \dots$ we have

$$\int_0^1 h_{\Delta_k} f\,dt = \int_0^{4^{-k}/2} f\,dt - \int_{4^{-k}/2}^{4^{-k}} f\,dt = \sum_{j=k+1}^{\infty} m(I_j) - m(I_k)$$
$$= -\frac{2}{3} \cdot 4^{-k-1} = -\frac{m(\Delta_k)}{6}.$$

Therefore, by definition, $(\mathcal{P}f)^2 \geq \frac{1}{36}\sum_{k=1}^{\infty} \chi_{\Delta_k}$, and so $\mathcal{P}f \notin L_\infty$.

Let us proceed with the proof of the theorem.

(i) By Parseval's Identity, for every $x = \sum_{m=1}^{\infty} c_m h_m \in L_2$ we have

$$\|\mathcal{P}x\|_2 = \Big(\sum_{m=1}^{\infty} c_m^2\Big)^{1/2} = \|x\|_2.$$

Hence, if $X = L_2$ and $\|x\|_X = \|x\|_2$, $x \in L_2$, then $\mathcal{P}(L_2) = L_2$ isometrically.

Conversely, let $x(t)$ be a step-function of the form

$$x(t) = \sum_{k=1}^{n} a_k \chi_{J_k}(t), \tag{14.22}$$

where the dyadic intervals J_k, $k = 1, 2, \dots, n$, are pairwise disjoint. If a function $\varepsilon(t)$, $|\varepsilon(t)| = 1$, $0 \leq t \leq 1$, takes the value $+1$ on the left half of the interval J_k and -1 on its right half, then $\varepsilon \cdot \chi_{J_k}$ is some Haar function $h_{i(k)}$ for each $k = 1, 2, \dots, n$. Therefore, $\varepsilon \cdot x = \sum_{k=1}^{n} a_k h_{i(k)}$ and

$$\mathcal{P}(\varepsilon \cdot x) = \Big(\sum_{k=1}^{n} (a_k h_{i(k)})^2\Big)^{1/2} = \sum_{k=1}^{n} |a_k| \chi_{J_k} = |x|. \tag{14.23}$$

Consequently, according to the hypothesis, we get

$$\|\mathcal{P}x\|_X = \|\mathcal{P}(\varepsilon \cdot x)\|_X = \|\,|x|\,\|_X = \|x\|_X. \tag{14.24}$$

Further, denote by S the orthogonal projection onto the set of constants, that is,

$$Sy(t) := \int_0^1 y(s)\,ds \cdot \chi_{[0,1]}(t).$$

Since any Haar sum $x = \sum_{k=1}^{n} c_k h_k$ is simultaneously a function of the form (14.22), then from (14.24) and the hypothesis it follows that

$$\|x\|_X = \Big\| \Big(\sum_{k=1}^{n} (c_k h_k)^2 \Big)^{1/2} \Big\|_X$$

and

$$\|(I - S)x\|_X = \Big\| \Big(\sum_{k=2}^{n} (c_k h_k)^2 \Big)^{1/2} \Big\|_X$$

(here, I is the identity). This implies that $\|(I - S)x\|_X \leq \|x\|_X$. If X is separable, then the set of all Haar sums is dense in this space, and hence $\|I - S\|_{X \to X} = 1$. Applying a characterization of the Hilbert space in the class of s.s.'s obtained in [121, Theorem 1], we get that $X = L_2$ isometrically, and thus the theorem is proved in the separable case.

Suppose X has the Fatou property, i.e., $X'' = X$. According to the observation from the beginning of the proof, we may assume that $X \neq L_\infty$. Then X° is a separable s.s. Moreover, since $\mathcal{P}(X^\circ) \subset \mathcal{P}(X)$ isometrically and $\mathcal{P}(X^\circ)$ is a sublattice of $\mathcal{P}(X)$, then the space $\mathcal{P}(X^\circ)$ satisfies the hypothesis of the part (i) of the theorem as well. Hence, as was already proved, $X^\circ = L_2$. As a result, by the definition of the Köthe duality, we have $X = X'' = (X^\circ)'' = L_2$.

(ii) If $0 < \alpha_X \leqslant \beta_X < 1$, then $\|\mathcal{P}x\|_X \asymp \|x\|_X$, and so immediately $\mathcal{P}(X) = X$ (with equivalence of norms).

For the converse, let x be an arbitrary function of the form (14.22). Then, by condition, there is a constant $C > 0$ such that

$$C^{-1}\|\mathcal{P}x\|_X \leq \|\mathcal{P}(\varepsilon \cdot x)\|_X \leq C\|\mathcal{P}x\|_X,$$

where $\varepsilon(t)$ is any function satisfying $|\varepsilon(t)| = 1$, $0 \leq t \leq 1$. On the other hand, choosing $\varepsilon(t)$ in the same way as in the proof of assertion (i) and taking into account equation (14.23), we get that $\|\mathcal{P}(\varepsilon \cdot x)\|_X = \|\,|x|\,\|_X \asymp \|x\|_X$ for any Haar sum x. Combining this with the preceding inequality, we conclude that $\|\mathcal{P}x\|_X \asymp \|x\|_X$ on the set of Haar sums. Now, in the case when X separable this equivalence may be easily extended to the whole space X, and hence the Boyd indices of X are nontrivial. Otherwise, if $X = X''$, we have the same result for the Boyd indices of the space X°, and then, using the equalities $\alpha_{X^\circ} = \alpha_{(X^\circ)''} = \alpha_X$ and $\beta_{X^\circ} = \beta_{(X^\circ)''} = \beta_X$ (see Appendix C), we complete the proof. □

14.6 Rademacher Projections in Paley Spaces

The behaviour of the Rademacher system in Paley spaces has quite interesting features. Recall that $\mathcal{R}$ is the set of all functions f representable in the form $f = \sum_{k=1}^{\infty} a_k r_k$, where $(a_k) \in \ell_2$. Since

$$r_{n+1} = \sum_{k=1}^{2^n} h_{n,k} = \sum_{\Delta \in \mathfrak{D}: m(\Delta)=2^{-n}} h_\Delta, \; n = 0, 1, 2 \ldots,$$

then $P\left(\sum_{n=1}^{\infty} a_n r_n\right)(t) = \left(\sum_{n=1}^{\infty} a_n^2\right)^{1/2}$. Hence, it follows that $\mathcal{R} \subset \mathcal{P}(X)$ and

$$\left\| \sum_{n=1}^{\infty} a_n r_n \right\|_{\mathcal{P}(X)} = \|(a_n)\|_{\ell_2} \tag{14.25}$$

for arbitrary s.s. X. Thus, every Paley space $\mathcal{P}(X)$ contains $\mathcal{R}$ as a closed linear subspace isometric to ℓ_2.

Theorem 14.7 *The subspace $\mathcal{R}$ is complemented in $\mathcal{P}(X)$ for any s.s. X.*

Proof Clearly, it suffices to show that the orthogonal projection

$$Pf(s) := \sum_{n=1}^{\infty} \int_0^1 f r_n \, dt \cdot r_n(s)$$

is bounded in $\mathcal{P}(X)$. Let us find first an expression of the norm $\|Pf\|_{\mathcal{P}(X)}$ via the Fourier–Haar coefficients of a given function $f \in \mathcal{P}(X)$.

If $f = \int_0^1 f\,dt + \sum_{\Delta \in \mathfrak{D}} a_\Delta h_\Delta$, then for each $n \in \mathbb{N}$ we have

$$\int_0^1 f r_n \, dt = \sum_{\Delta \in \mathfrak{D}: m(\Delta)=2^{-n+1}} a_\Delta \int_\Delta h_\Delta r_n \, dt = 2^{-n+1} \sum_{\Delta \in \mathfrak{D}: m(\Delta)=2^{-n+1}} a_\Delta .$$

Hence,

$$\begin{aligned} Pf &= \sum_{n=1}^{\infty} \Big(2^{-n+1} \sum_{\Delta \in \mathfrak{D}: m(\Delta)=2^{-n+1}} a_\Delta \Big) r_n \\ &= \sum_{n=1}^{\infty} \Big(2^{-n+1} \sum_{\Delta \in \mathfrak{D}: m(\Delta)=2^{-n+1}} a_\Delta \Big) \sum_{\Delta \in \mathfrak{D}: m(\Delta)=2^{-n+1}} h_\Delta \end{aligned}$$

(that is, the operator T averages the Fourier–Haar coefficients of f corresponding to each block of the Haar system). Thus, by definition of the norm in Paley spaces, we obtain

$$\|Pf\|^2_{\mathcal{P}(X)} = \sum_{n=1}^{\infty} \Big(2^{-n+1} \sum_{\Delta \in \mathfrak{D}: m(\Delta)=2^{-n+1}} a_\Delta\Big)^2. \tag{14.26}$$

Let us estimate now the norm $\|f\|_{\mathcal{P}(X)}$ from below. Fix $n \in \mathbb{N}$. Observe that for every $k = 1, 2, \ldots, 2^n$ there exists a unique sequence $\{i_j(k)\}_{j=0}^n$ such that $i_j(k) \in \{1, 2, \ldots, 2^j\}$ for all $j = 0, 1, \ldots, n$, $i_n(k) = k$ and $\Delta_0^{i_0(k)} \supset \Delta_1^{i_1(k)} \supset \cdots \supset \Delta_n^{i_n(k)}$.

Given $f \in \mathcal{P}(X)$, $f = \int_0^1 f\,dt + \sum_{\Delta \in \mathfrak{D}} a_\Delta h_\Delta$, the Paley function $\mathcal{P}f$ satisfies the inequality

$$(\mathcal{P}f)^2 \geq \sum_{k=1}^{2^n} \sum_{j=0}^{n} a^2_{\Delta_j^{i_j(k)}} \cdot \chi_{\Delta_n^k}.$$

Therefore, from Minkowski's inequality it follows that

$$\int_0^1 \mathcal{P}f(t)\,dt \geq 2^{-n} \sum_{k=1}^{2^n} \Big(\sum_{j=0}^{n} a^2_{\Delta_j^{i_j(k)}}\Big)^{1/2} \geq 2^{-n} \Big(\sum_{j=0}^{n} \Big(\sum_{k=1}^{2^n} |a_{\Delta_j^{i_j(k)}}|\Big)^2\Big)^{1/2}.$$

Noting that for any $j = 0, 1, \ldots, n$ and $m = 1, \ldots, 2^j$ it holds

$$\operatorname{card}\{k : i_j(k) = m\} = 2^{n-j}.$$

we see that this inequality yields

$$\begin{aligned}\int_0^1 \mathcal{P}f(t)\,dt &\geq 2^{-n}\Big(\sum_{j=0}^{n} \Big(2^{n-j} \sum_{m=1}^{2^j} |a_{\Delta_j^m}|\Big)^2\Big)^{1/2} \\ &= \Big(\sum_{j=0}^{n} \Big(2^{-j} \sum_{\Delta \in \mathfrak{D}: m(\Delta)=2^{-j}} |a_\Delta|\Big)^2\Big)^{1/2}.\end{aligned}$$

Letting here $n \to \infty$ and taking into account that X is continuously embedded into L_1, we get

$$\Big(\sum_{j=0}^{\infty} \Big(2^{-j} \sum_{\Delta \in \mathfrak{D}: m(\Delta)=2^{-j}} |a_\Delta|\Big)^2\Big)^{1/2} \leq C_X \|f\|_{\mathcal{P}(X)}.$$

Hence, by (14.26), $\|Pf\|_{\mathcal{P}(X)} \leq C_X \|f\|_{\mathcal{P}(X)}$, which proves the theorem. □

In view of equation (14.25), we obtain

Corollary 14.7 *For arbitrary s.s. X the Paley space $\mathcal{P}(X)$ contains a 1-complemented subspace isometric to ℓ_2.*

Remark 14.1 According to Theorem 2.5 (see also its proof) the orthogonal projection P is bounded in a s.s. X if and only if $G \subset X \subset G'$, where G is the separable part of the exponential Orlicz space L_{N_2}, and G' is the Köthe dual space for G. The comparison of this result with the proof of Theorem 14.7 reveals a substantial difference in the behaviour of the Rademacher system in s.s.'s and Paley spaces.

In the case when $X = L_1$, i.e., for the dyadic space H_1, Theorem 14.7 can be strengthened. To clarify this, we rewrite, by using (14.25) and definition of the orthogonal projection P (see also the proof of Theorem 14.7), the above result in the form:

$$\sum_{m=1}^{\infty} \langle g, r_m \rangle^2 \le \|g\|_{H_1}^2 \quad \text{for all} \quad g \in H_1,$$

where, as usual, $\langle f, g\rangle := \int_0^1 fg\,dt$. It is well known that $H_1 \overset{\neq}{\subset} L_1$ (see e.g. [153, Theorems III.4, III.10 and III.11]). It turns out that one of the H_1-norms of g in the right-hand side of this inequality may be replaced (up to a constant) with its smaller L_1-norm.

Theorem 14.8 *There exists a universal constant $\alpha > 0$ such that for all $g \in H_1$ we have*

$$\sum_{m=1}^{\infty} \langle g, r_m \rangle^2 \le \alpha \|g\|_{L_1} \|g\|_{H_1}. \tag{14.27}$$

In the proof of this result we make use of some properties of the dyadic space BMO_d. Recall that this space consists of all $f \in L_1[0,1]$ such that

$$\|f\|_d := \sup_{\Delta \in \mathfrak{D}} \frac{1}{m(\Delta)} \int_\Delta |f(s) - f_\Delta|\,ds < \infty,$$

where $f_\Delta := m(\Delta)^{-1} \int_\Delta f\,dt$. Observe first that for every $f = \sum_{J \in \mathfrak{D}} a_J h_J$ we have

$$\|f\|_d \le \sup_{\Delta \in \mathfrak{D}} \left(\frac{1}{m(\Delta)} \sum_{J \in \mathfrak{D}: J \subset \Delta} a_J^2 m(J) \right)^{1/2}. \tag{14.28}$$

Indeed, denoting

$$\|f\|_{d,2} := \sup_{\Delta \in \mathfrak{D}} \left(\frac{1}{m(\Delta)} \int_\Delta |f(s) - f_\Delta|^2\,ds \right)^{1/2},$$

from the Cauchy-Schwarz-Bunyakovskii inequality we immediately get

$$\|f\|_d \le \|f\|_{d,2}. \tag{14.29}$$

Moreover, for each interval $\Delta \in \mathfrak{D}$ it holds

$$f_\Delta = \Big(\sum_{J \in \mathfrak{D}: J \overset{\neq}{\supset} \Delta} a_J h_J \Big)\Big|_\Delta,$$

which implies that

$$\int_\Delta |f(s) - f_\Delta|^2 \, ds = \int_\Delta \Big| \sum_{J \in \mathfrak{D}} a_J h_J(s) - \sum_{J \in \mathfrak{D}: J \overset{\neq}{\supset} \Delta} a_J h_J(s) \Big|^2 ds$$
$$= \int_\Delta \Big| \sum_{J \in \mathfrak{D}: J \subset \Delta} a_J h_J(s) \Big|^2 ds = \sum_{J \in \mathfrak{D}: J \subset \Delta} a_J^2 m(J).$$

Hence,

$$\|f\|_{d,2} = \sup_{\Delta \in \mathfrak{D}} \left(\frac{1}{m(\Delta)} \sum_{J \in \mathfrak{D}: J \subset \Delta} a_J^2 m(J) \right)^{1/2}.$$

Combining this equation with (14.29) we arrive at inequality (14.28).

In the next lemma we establish a link between the spaces BMO_d and $\mathcal{P}(L_\infty)$.

Lemma 14.1 *The embedding $\mathcal{P}(L_\infty) \subset BMO_d$ holds. Moreover, $\|f\|_d \le \|f\|_{\mathcal{P}(L_\infty)}$ for all $f \in \mathcal{P}(L_\infty)$.*

Proof Let $f = \sum_{J \in \mathfrak{D}} a_J h_J$. For each interval $\Delta \in \mathfrak{D}$ we introduce the following local version of the Paley function:

$$\mathcal{P}_\Delta(f)(x) := \Big(\sum_{J \in \mathfrak{D}, J \subset \Delta} a_J^2 \chi_J(x) \Big)^{1/2}.$$

Clearly, $\mathcal{P}_\Delta(f)(x) \le \mathcal{P}f(x)$, $0 \le x \le 1$, and hence

$$\|\mathcal{P}_\Delta(f)\|_{L_\infty(\Delta, \frac{dx}{m(\Delta)})} \le \|\mathcal{P}f\|_{L_\infty[0,1]} = \|f\|_{\mathcal{P}(L_\infty)}.$$

On the other hand,

$$\|\mathcal{P}_\Delta(f)\|_{L_2(\Delta, \frac{dx}{m(\Delta)})} = \left(\frac{1}{m(\Delta)} \int_\Delta \mathcal{P}_\Delta(f)^2 \, dx \right)^{1/2}$$

$$= \left(\frac{1}{m(\Delta)} \sum_{J \in \mathfrak{D}: J \subset \Delta} a_J^2 \int_\Delta \chi_J(x)\, dx \right)^{1/2}$$

$$= \left(\frac{1}{m(\Delta)} \sum_{J \in \mathfrak{D}: J \subset \Delta} a_J^2 m(J) \right)^{1/2}.$$

Therefore, from the last inequality it follows that

$$\left(\frac{1}{m(\Delta)} \sum_{J \in \mathfrak{D}: J \subset \Delta} a_J^2 m(J) \right)^{1/2} \le \|\mathcal{P}_\Delta(f)\|_{L_\infty(\Delta, \frac{dx}{m(\Delta)})} \le \|f\|_{\mathcal{P}(L_\infty)}.$$

Since $\Delta \in \mathfrak{D}$ is arbitrary, this inequality, combined with (14.28), yields both assertions of the lemma. □

The subsequent proof of Theorem 14.8 is based on using two important results from the theory of BMO-spaces. The first one is a consequence of the well-known Garnett–Jones theorem on the BMO_d-distance of functions from the space $\mathcal{P}(L_\infty)$ to L_∞ [125]. According to this result, for every function $f \in \mathcal{P}(L_\infty)$ with $\|f\|_{\mathcal{P}(L_\infty)} \le 1$ and any $\varepsilon > 0$ there exists a function $u \in L_\infty$ such that

$$\|f - u\|_d \le \varepsilon \text{ and } \|u\|_{L_\infty} \le A\varepsilon^{-1}, \tag{14.30}$$

where $A > 0$ is a universal constant (by Lemma 14.1, $\mathcal{P}(L_\infty) \subset BMO_d$). Furthermore, we shall need the famous Fefferman inequality, which reads that for arbitrary $f \in BMO_d$ and $g \in H_1$ we have

$$\langle f, g \rangle \le 2\sqrt{2}\|f\|_d \|g\|_{H_1} \tag{14.31}$$

(see, for example, [213, Theorem 1.2.1]).

Proof of Theorem 14.8 Let $g \in H_1$. We set

$$f := \sum_{m=1}^\infty a_m r_m, \quad \text{where } a_m := \langle g, r_m \rangle \cdot \left(\sum_{n=1}^\infty \langle g, r_n \rangle^2 \right)^{-1/2}.$$

Note that

$$\langle g, f \rangle = \sum_{m=1}^\infty a_m \langle g, r_m \rangle = \left(\sum_{m=1}^\infty \langle g, r_m \rangle^2 \right)^{1/2}. \tag{14.32}$$

Moreover, since $\sum_{m=1}^\infty a_m^2 = 1$, we have $f \in \mathcal{P}(L_\infty)$ and $\mathcal{P}(f) = 1$. Therefore, applying the above-mentioned Garnett–Jones theorem to the function f with $\varepsilon :=$

$\|g\|_{L_1}^{1/2}\|g\|_{H_1}^{-1/2}$, we find a function $u \in L_\infty$ satisfying inequalities (14.30). As a result, from (14.30)–(14.32), and the choice of ε it follows that

$$\begin{aligned}\Big(\sum_{m=1}^{\infty}\langle g, r_m\rangle^2\Big)^{1/2} &= \langle g, f-u\rangle + \langle g, u\rangle \\ &\le 2\sqrt{2}\|f-u\|_d\|g\|_{H_1} + \|g\|_{L_1}\|u\|_{L_\infty} \\ &\le (2\sqrt{2}+A)\|g\|_{L_1}^{1/2}\|g\|_{H_1}^{1/2},\end{aligned}$$

and so inequality (14.27) is proved. □

14.7 The Subcouple of the Couple $(L_\infty, \mathcal{P}(L_\infty))$ Generated by the Rademacher System

As above, $R : \ell_2 \to \mathcal{R}$ is the operator defined by

$$Ra(t) := \sum_{k=1}^{\infty} a_k r_k(t), \quad a = (a_k)_{k=1}^{\infty} \in \ell_2.$$

In Chap. 3, we have investigated some properties of the Banach couple (L_∞, G) and its subcouple $(R(\ell_1), R(\ell_2))$. According to equation (14.25), R is an isometry from ℓ_2 into $\mathcal{P}(L_\infty)$. Therefore, $(R(\ell_1), R(\ell_2))$ is a subcouple of the Banach couple $(L_\infty, \mathcal{P}(L_\infty))$ as well (here, $R(\ell_1)$ and $R(\ell_2)(= \mathcal{R})$ are subspaces of L_∞ and $\mathcal{P}(L_\infty)$, respectively). Since the couples $(R(\ell_1), R(\ell_2))$ and (ℓ_1, ℓ_2) are isometric, then the definition of Peetre's $\mathcal{K}$-functional immediately yields

$$\mathcal{K}(t, Ra; R(\ell_1), R(\ell_2)) = \mathcal{K}(t, a; \ell_1, \ell_2). \tag{14.33}$$

Recall that a Banach couple (Y_0, Y_1) is said to be a *$\mathcal{K}$-closed subcouple* of a Banach couple (X_0, X_1) if

$$\mathcal{K}(t, y; Y_0, Y_1) \asymp \mathcal{K}(t, y; X_0, X_1)$$

with a constant independent of $y \in Y_0 + Y_1$ and $t > 0$.

For other definitions from the interpolation theory of operators that are used below, see Appendix D.

Theorem 14.9 *The Banach couple $(R(\ell_1), R(\ell_2))$ is a K-closed subcouple of the couple $(L_\infty, \mathcal{P}(L_\infty))$.*

Proof In view of equation (14.33), it suffices to show that

$$\mathcal{K}(t, Ra; L_\infty, \mathcal{P}(L_\infty)) \asymp \mathcal{K}(t, a; \ell_1, \ell_2),$$

with a constant independent of $a = (a_k)_{k=1}^\infty \in \ell_2$ and $t > 0$.

First of all, again by definition of the $\mathcal{K}$-functional, we have

$$\begin{aligned}\mathcal{K}(t, Ra; L_\infty, \mathcal{P}(L_\infty)) &\le \inf\{\|Rb\|_{L_\infty} + t\|Rc\|_{\mathcal{P}(L_\infty)} : a = b + c, b \in \ell_1, c \in \ell_2\} \\ &= \inf\{\|b\|_{\ell_1} + t\|c\|_{\ell_2} : a = b + c, b \in \ell_1, c \in \ell_2\} \\ &= \mathcal{K}(t, a; \ell_1, \ell_2).\end{aligned}$$

On the other hand, from [213, Corollary 1.3.2] it follows that $\mathcal{P}(L_\infty) \subset L_{N_2}$ and $\|x\|_{N_2} \le C\|x\|_{\mathcal{P}(L_\infty)}$ for all $x \in \mathcal{P}(L_\infty)$. Hence,

$$\mathcal{K}(t, Ra; L_\infty, G) = \mathcal{K}(t, Ra; L_\infty, L_{N_2}) \le C\mathcal{K}(t, Ra; L_\infty, \mathcal{P}(L_\infty)).$$

Also, according to Theorem 3.3,

$$\mathcal{K}(t, Ra; L_\infty, G) \asymp \mathcal{K}(t, a; \ell_1, \ell_2),$$

with a constant independent of $a = (a_k)_{k=1}^\infty \in \ell_2$ and $t > 0$. Thus,

$$\mathcal{K}(t, a; \ell_1, \ell_2) \le C'\mathcal{K}(t, Ra; L_\infty, \mathcal{P}(L_\infty)),$$

which finishes the proof. □

From Theorem 14.9 and the definition of interpolation spaces of the real $\mathcal{K}$-method we get the following result.

Corollary 14.8 *Let F be a Banach lattice of two-sided sequences such that $(\min(1, 2^k))_{k=-\infty}^\infty \in F$, and let $X := (L_\infty, \mathcal{P}(L_\infty))_F^{\mathcal{K}}$, $E := (\ell_1, \ell_2)_F^{\mathcal{K}}$. Then,*

$$\Big\| \sum_{n=1}^\infty a_n r_n \Big\|_X \asymp \|(a_n)\|_E,$$

with a constant independent of sequences (a_n).

By using Corollary 14.8, for any interpolation space E with respect to the couple (ℓ_1, ℓ_2), we are able to find a certain interpolation space X with respect to the couple $(L_\infty, \mathcal{P}(L_\infty))$ such that the norm of X is equivalent on $\mathcal{R} \cap X$ to the norm of E. In particular, arguing in the same way as in Example 3.1 from Chap. 3, one can show that for any $1 < p < 2$ and $\theta = 2(p-1)/p$ and the space $X_p := (L_\infty, \mathcal{P}(L_\infty))_{\theta,p}$

$$\Big\| \sum_{n=1}^{\infty} a_n r_n \Big\|_{X_p} \asymp \|(a_n)\|_{\ell_p},$$

with a constant independent of (a_n).

Comments and References The space of functions of bounded mean oscillation (BMO) was introduced by John and Nirenberg in 1961, when studying partial differential equations [141]. The definition of the functional $A(f)$ and Proposition 14.1 see in [124, Chapter VI, Exercises and Further Results, Problem 12(b), p. 266] or in [44]. The initial results on the behaviour of Rademacher sums in BMO have been obtained by Leibov in his Candidate's Dissertation [178]. In particular, he has proved equivalence (14.8), has described the properties of block basis sequences (Proposition 14.3) and the geometrical structure of the Rademacher subspace of BMO (Theorem 14.5). It is instructive to compare the last result with the well-known Kadec-Pełczyński alternative for subspaces of the L_p-spaces with $p > 2$ [147].

Theorems 14.3 and 14.4 on complementability of the Rademacher subspaces in BMO type spaces have been proved in the paper [44]. However, much earlier Müller and Schechtman established in [214, Theorem 1] the following more general result than Theorem 14.3. Let $1 \le p_1 < p_2 < \dots$ be an arbitrary increasing sequence of positive integers. Then, the Müller–Schechtman theorem reads that the Walsh functions

$$f_k = \prod_{i \in A_k} r_i, \quad \text{where } A_k \subset \{p_k + 1, p_k + 2, \dots, p_{k+1}\}, \tag{14.34}$$

form a sequence equivalent in the space BMO_d to the unit vector basis in ℓ_2, and moreover their span is complemented in BMO_d. Later on, the behaviour of sequences of functions (14.34) in BMO was studied in the paper [54]. In particular, in [54], it has been obtained the following generalization of Theorem 14.4: the subspace $[f_k]$ is complemented in BMO if and only if all the sets A_k, except possibly finitely many, contain an even number of elements.

It is likely that Paley (see [226]) was the first, who began to consider the square function $\mathcal{P}x$ corresponding to the Fourier–Haar expansion of a function $x \in L_1$. The class of Paley spaces $\mathcal{P}(X)$, where X is an arbitrary s.s. on $[0, 1]$, has been introduced and studied in the papers [56] and [57]. At the same time, in the partial case when $X = L_p$, $1 \le p \le \infty$, these spaces appeared much earlier (see, for example, the monograph [213]). In particular, if $X = L_1$, we get the well-known dyadic H_1-space [213, § 1.2]. Note also that § 1.3 of the latter book contains some results related to the space $\mathcal{P}(L_\infty)$, denoted there by SL_∞.

Theorems 14.6, 14.7, and 14.9 (cf. Theorem 3.3) have been proved in [56]. The more precise estimate for the Fourier-Rademacher coefficients of functions from H_1 (see Theorem 14.8) has been obtained by Bourgain in the paper [82].

Chapter 15
Rademacher Functions in the Cesàro Type Spaces

It is well known that the norms of many Banach spaces, playing a significant role in functional analysis and its applications, are generated by positive sublinear operators and the L_p-norms. In particular, rather simple and both important spaces of such a type are the Cesàro spaces Ces_p, $1 \le p \le \infty$. In this chapter, we investigate the behaviour of the Rademacher system in function spaces of the Cesàro type.

15.1 Definition and Some Properties of the Cesàro Type Spaces

The *Cesàro function spaces* $Ces_p := Ces_p[0, 1]$ consist of Lebesgue measurable functions f on $[0, 1]$ such that

$$\|f\|_{C(p)} := \left[\int_0^1 \left(\frac{1}{x}\int_0^x |f(t)|\,dt\right)^p dx\right]^{1/p} < \infty \quad \text{if} \quad 1 \le p < \infty,$$

and

$$\|f\|_{C(\infty)} := \sup_{0<x\le 1} \frac{1}{x}\int_0^x |f(t)|\,dt < \infty \quad \text{if} \quad p = \infty.$$

Clearly, Ces_p, $1 \le p \le \infty$, is a Banach function lattice and simultaneously this space fails to be symmetric with respect to any norm equivalent to the above introduced one. Indeed, for example, the functions $\chi_{[0,h]}$ and $\chi_{[1-h,1]}$ are identically distributed on $[0, 1]$ for each $0 < h \le 1$. However, an easy calculation shows that for all $0 < h \le 1$

$$\|\chi_{[0,h]}\|_{C(p)} \ge h^{1/p}, \quad \text{but} \quad \|\chi_{[1-h,1]}\|_{C(p)} = h^{1+1/p} \quad \text{for} \quad 1 \le p < \infty,$$

S. V. Astashkin, *The Rademacher System in Function Spaces*,
https://doi.org/10.1007/978-3-030-47890-2_15

and

$$\|\chi_{[0,h]}\|_{C(\infty)} = 1, \quad \text{but} \quad \|\chi_{[1-h,1]}\|_{C(\infty)} = h.$$

As we shall see further, the behaviour of the Rademacher functions in the spaces Ces_p differs substantially in the cases when $1 \le p < \infty$ and $p = \infty$. So, while in Ces_p, $1 \le p < \infty$, the system $\{r_n\}$ is equivalent to the unit vector basis in ℓ_2, in Ces_∞ it does not contain even unconditional basic sequences.

The space Ces_∞, known also as the *Korenblyum–Krein–Levin space*, belongs to the family of weighted $K_{w,q}$-spaces defined as follows. Let $1 \le q < \infty$, and let w be a quasiconcave function on $[0, 1]$, i.e., $w(0) = 0$, w is increasing, and $w(x)/x$ is a decreasing function on $(0, 1]$. The space $K_{w,q} := K_{w,q}[0, 1]$ consists of all measurable functions f on $[0, 1]$ such that

$$\|f\|_{K_{w,q}} := \sup_{0<x\le 1} \left(\frac{1}{w(x)} \int_0^x |f(t)|^q \, dt \right)^{1/q} < \infty. \tag{15.1}$$

Without loss of generality, we shall assume that $w(1) = 1$. Then it is easy to see that $K_{w,q} \overset{1}{\subset} L_q$ for every $1 \le q < \infty$. Moreover, $K_{1,q} = L_q$ isometrically. For brevity, in the case $w(x) = x$ the space $K_{w,q}$ will be denoted by K_q (it is also called often the Korenblyum–Krein–Levin space). In particular, we have $K_1 = Ces_\infty$. Using the Fatou lemma, it is not hard to show that the Banach lattice $K_{w,q}$ has the Fatou property for all w and q.

Later on, when studying the geometrical structure of the space K_1, we shall need the following result related to sequences of pairwise disjoint functions with supports "converging" to zero.

Proposition 15.1 *Suppose that* $\{x_n\}_{n=1}^\infty \subset K_1$, $\operatorname{supp} x_n \subset I_n := [a_n, b_n]$, $n = 1, 2, \ldots$, *where* $b_1 > a_1 > b_2 > a_2 > \cdots > 0$ *and* $b_n \to 0+$.

Then, there exists a subsequence $\{x_{n_k}\} \subset \{x_n\}$ *such that*

$$\sup_{k=1,2,\ldots} \|x_{n_k}\|_{K_1} \le \left\| \sum_{k=1}^\infty x_{n_k} \right\|_{K_1} \le 4 \sup_{k=1,2,\ldots} \|x_{n_k}\|_{K_1}. \tag{15.2}$$

Proof Clearly, we may assume (passing to a subsequence if necessary) that $x_n \ge 0$, $b_1 \le 1/2$, and

$$\sum_{k=n+1}^\infty b_k \le a_n, \quad n = 1, 2, \ldots \tag{15.3}$$

Moreover, since the left-hand side inequality in (15.2) holds for an arbitrary sequence of disjoint functions, it suffices only to prove the right-hand side inequality.

First of all, for any $n \in \mathbb{N}$ and $t \in (0, 1]$ we have

$$\begin{aligned}\frac{1}{t}\int_0^t \Big(\sum_{k=1}^n x_k(s)\Big)\,ds &= \frac{1}{t}\sum_{j=2}^n \sum_{i=n-j+2}^n \int_{a_i}^{b_i} x_i(s)\,ds\,\chi_{[a_{n-j+1},b_{n-j+1}]}(t)\\ &+ \frac{1}{t}\sum_{j=1}^n \int_{a_{n-j+1}}^t x_{n-j+1}(s)\,ds\,\chi_{[a_{n-j+1},b_{n-j+1}]}(t)\\ &+ \frac{1}{t}\sum_{j=1}^n \sum_{i=n-j+1}^n \int_{a_i}^{b_i} x_i(s)\,ds\,\chi_{[b_{n-j+1},a_{n-j}]}(t)\\ &= f_1(t)+f_2(t)+f_3(t)+f_4(t), \end{aligned} \tag{15.4}$$

where we set $a_0 = 1$,

$$f_1(t) := \frac{1}{t}\sum_{j=2}^n \sum_{i=n-j+2}^n \int_{a_i}^{b_i} x_i(s)\,ds\,\chi_{[a_{n-j+1},b_{n-j+1}]}(t),$$

$$f_2(t) := \frac{1}{t}\sum_{j=2}^n \sum_{i=n-j+2}^n \int_{a_i}^{b_i} x_i(s)\,ds\,\chi_{[b_{n-j+1},a_{n-j}]}(t),$$

$$f_3(t) := \frac{1}{t}\sum_{j=1}^n \int_{a_{n-j+1}}^{b_{n-j+1}} x_{n-j+1}(s)\,ds\,\chi_{[b_{n-j+1},a_{n-j}]}(t),$$

$$f_4(t) := \frac{1}{t}\sum_{j=1}^n \int_{a_{n-j+1}}^{t} x_{n-j+1}(s)\,ds\,\chi_{[a_{n-j+1},b_{n-j+1}]}(t).$$

Let now $0 \le a < b \le 1$. For each $x \in K_1$ we have

$$\int_0^1 x(s)\chi_{[a,b]}(s)\,ds \le b\cdot\frac{1}{b}\int_0^b |x(s)|\,ds \le b\|x\|_{K_1}.$$

Therefore, for all $0 \le a < b \le 1$ we have $\|\chi_{[a,b]}\|_{K_1'} \le b$, where K_1' is the Köthe dual space for K_1. In consequence,

$$\int_{a_i}^{b_i} x_i(s)\,ds = \int_0^1 x_i(s)\chi_{[a_i,b_i]}(s)\,ds \le \|\chi_{[a_i,b_i]}\|_{K_1'}\|x_i\|_{K_1} \le b_i\|x_i\|_{K_1},\quad i=1,2,\dots$$

Hence, using inequalities (15.3), we successively get

$$\|f_1\|_\infty = \max_{j=2,\dots,n}\left(\sum_{i=n-j+2}^{n}\int_{a_i}^{b_i} x_i(s)\,ds\cdot\frac{1}{a_{n-j+1}}\right)$$

$$\le \max_{j=2,\dots,n}\left(\sum_{i=n-j+2}^{n}\|x_i\|_{K_1}\cdot\frac{b_i}{a_{n-j+1}}\right)$$

$$\le \max_{j=2,\dots,n}\|x_j\|_{K_1}\sum_{i=n-j+2}^{n}\frac{b_i}{a_{n-j+1}} \le \max_{j=2,\dots,n}\|x_j\|_{K_1},$$

$$\|f_2\|_\infty = \max_{j=2,\dots,n}\left(\sum_{i=n-j+2}^{n}\int_{a_i}^{b_i} x_i(s)\,ds\cdot\frac{1}{b_{n-j+1}}\right)$$

$$\le \max_{j=2,\dots,n}\left(\sum_{i=n-j+2}^{n}\|x_i\|_{K_1}\cdot\frac{b_i}{b_{n-j+1}}\right) \le \max_{j=2,\dots,n}\|x_j\|_{K_1},$$

$$\|f_3\|_\infty = \max_{j=1,2,\dots,n}\int_{a_{n-j+1}}^{b_{n-j+1}} x_{n-j+1}(s)\,ds\cdot\frac{1}{b_{n-j+1}} \le \max_{j=1,2,\dots,n}\|x_j\|_{K_1},$$

and finally, by definition of the K_1-norm,

$$\|f_4\|_\infty = \max_{j=1,2,\dots,n}\ \sup_{a_{n-j+1}\le t\le b_{n-j+1}}\left(\frac{1}{t}\int_0^t x_{n-j+1}(s)\,ds\right) \le \max_{j=1,2,\dots,n}\|x_j\|_{K_1}.$$

Combining these estimates with inequality (15.4), we have for every $n = 1, 2, \dots$

$$\left\|\sum_{k=1}^{n} x_k\right\|_{K_1} \le \sum_{k=1}^{4}\|f_k\|_\infty \le 4\max_{j=1,2,\dots,n}\|x_j\|_{K_1}.$$

Since K_1 has the Fatou property, then passing in the last inequality to the limit as $n \to \infty$, we arrive at estimate (15.2). □

15.2 Rademacher Sums in the Cesàro Type Spaces

We start with studying the behaviour of Rademacher sums in the spaces Ces_p, $1 \le p < \infty$.

Theorem 15.1 *For every* $1 \le p < \infty$ *the sequence* $\{r_n\}_{n=1}^{\infty}$ *is equivalent in* Ces_p *to the unit vector basis in* ℓ_2.

Proof We need to prove that

$$\Big\| \sum_{k=1}^{\infty} a_k r_k \Big\|_{C(p)} \asymp \Big(\sum_{k=1}^{\infty} a_k^2 \Big)^{1/2} \tag{15.5}$$

with a constant independent of $a_k \in \mathbb{N}$, $k = 1, 2, \ldots$

First, let us show that the following continuous embeddings hold:

$$L_p \overset{p'}{\subset} Ces_p \overset{1}{\subset} Ces_1 = L_1(\ln 1/t) \overset{54}{\subset} L_{1/3} \text{ if } 1 < p < \infty, \tag{15.6}$$

where $p' = p/(p-1)$ and $L_1(\ln 1/t)$ is the weighted L_1-space with the norm

$$\|f\|_{L_1(\ln 1/t)} := \int_0^1 |f(t)| \ln(1/t)\, dt.$$

Indeed, the first embedding in (15.6) is a consequence of the classical Hardy inequality (see Appendix C); the second one follows directly from the embedding $L_p[0,1] \overset{1}{\subset} L_1[0,1]$. Further, since

$$\int_0^1 \left(\frac{1}{x} \int_0^x |f(t)|\, dt \right) dx = \int_0^1 \left(\int_t^1 \frac{1}{x}\, dx \right) |f(t)|\, dt = \int_0^1 |f(t)| \ln\left(\frac{1}{t}\right) dt,$$

then $Ces_1 = L_1(\ln 1/t)$ isometrically.

To prove the last embedding in (15.6), we assume that $\|f\|_{L_1(\ln 1/t)} = 1$. Then, from the elementary inequality $\ln(1/t) \ge 1 - t$, $0 < t \le 1$, and a property of the decreasing rearrangement [168, § II.2, Property 16, p. 70–71] it follows that

$$\begin{aligned} 1 = \int_0^1 |f(t)| \ln(1/t)\, dt &\ge \int_0^1 |f(t)|(1-t)\, dt \ge \int_0^1 f^*(1-t)(1-t)\, dt \\ &= \int_0^1 f^*(s) s\, ds \ge \int_0^t f^*(s) s\, ds \ge \frac{f^*(t) t^2}{2}, \end{aligned}$$

which implies $f^*(t) \le 2/t^2$. Thus,

$$\int_0^1 |f(t)|^{1/3}\, dt = \int_0^1 f^*(t)^{1/3}\, dt \le \int_0^1 (2t^{-2})^{1/3}\, dt = 3 \cdot \sqrt[3]{2},$$

whence $\|f\|_{1/3} = (\int_0^1 |f(t)|^{1/3}\, dt)^3 \le 54$.

Since equivalence (15.5) is an immediate consequence of embeddings (15.6) and Khintchine's L_p-inequality (see Theorem 1.4), the proof is completed. □

In the $K_{w,q}$-spaces the behaviour of the Rademacher system is more complicated.

Theorem 15.2 *For every $1 \le q < \infty$ and arbitrary quasiconcave function w we have*

$$\Big\| \sum_{k=1}^{\infty} a_k r_k \Big\|_{K_{w,q}} \asymp \|(a_k)_{k=1}^{\infty}\|_{\ell_2} + \sup_{m\ge 1} \Big(\frac{2^{-m}}{w(2^{-m})}\Big)^{1/q} \Big|\sum_{k=1}^{m} a_k\Big|. \tag{15.7}$$

Proof Let $S_n := \sum_{k=1}^{n} a_k r_k$, $n = 1, 2, \ldots$. From definition of the Rademacher functions it follows that for $m = 1, 2, \ldots, n$

$$\int_0^{2^{-m}} |S_n(t)|^q\, dt = 2^{-n} \sum_{\varepsilon_k=\pm 1} |a_1 + \ldots + a_m + \varepsilon_{m+1}a_{m+1} + \ldots + \varepsilon_n a_n|^q\,, \tag{15.8}$$

where the sum is taken over all choices of signs $\varepsilon_{m+1} = \pm 1, \ldots, \varepsilon_n = \pm 1$. Therefore, by (15.1) and Hölder's inequality, for every $m = 1, 2, \ldots, n$

$$\begin{aligned}
\|S_n\|_{K_{w,q}} &\ge \Big(\frac{1}{w(2^{-m})} \int_0^{2^{-m}} |S_n(t)|^q\, dt\Big)^{1/q} \ge \frac{2^{m(1-1/q)}}{w(2^{-m})^{1/q}} \int_0^{2^{-m}} |S_n(t)|\, dt \\
&= \frac{2^{m(1-1/q)-n}}{w(2^{-m})^{1/q}} \sum_{\varepsilon_k=\pm 1} |a_1 + \ldots + a_m + \varepsilon_{m+1}a_{m+1} + \ldots + \varepsilon_n a_n| \\
&\ge \frac{2^{m(1-1/q)-n}}{w(2^{-m})^{1/q}} \Big| \sum_{\varepsilon_k=\pm 1} (a_1 + \ldots + a_m + \varepsilon_{m+1}a_{m+1} + \ldots + \varepsilon_n a_n)\Big| \\
&= \frac{2^{m(1-1/q)-n}}{w(2^{-m})^{1/q}}\, 2^{n-m} \Big|\sum_{k=1}^{m} a_k\Big| = \Big(\frac{2^{-m}}{w(2^{-m})}\Big)^{1/q} \Big|\sum_{k=1}^{m} a_k\Big|.
\end{aligned}$$

Furthermore, since $0 \le w(t) \le 1$ (recall that w is increasing and $w(1) = 1$), then from (15.1) and Theorem 5.4 it follows

$$\|S_n\|_{K_{w,q}} \ge \sup_{0<x\le 1} \Big(\int_0^x |S_n(t)|^q\, dt\Big)^{1/q} = \|S_n\|_q \ge \frac{1}{\sqrt{2}} \|(a_k)_{k=1}^{n}\|_{\ell_2}.$$

As a result, the last two inequalities yield

$$\|S_n\|_{K_{w,q}} \ge \frac{1}{2\sqrt{2}} \Big(\|(a_k)_{k=1}^{n}\|_{\ell_2} + \max_{1\le m\le n} \Big(\frac{2^{-m}}{w(2^{-m})}\Big)^{1/q} \Big|\sum_{k=1}^{n} a_k\Big|\Big).$$

Let us prove the opposite estimate. Noting that the sum $S_n(t)$ is constant on the interval $[0, 2^{-n}]$, we have for all $0 < x \le 2^{-n}$

$$\frac{1}{x}\int_0^x |S_n(t)|^q\,dt = 2^n \int_0^{2^{-n}} |S_n(t)|^q\,dt.$$

Moreover, because w is quasiconcave, we have $x/w(x) \le 2^{-n}/w(2^{-n})$ if $0 < x \le 2^{-n}$, and $w(2^{-j}) \le 2w(x)$ if $x \ge 2^{-j-1}$, $j = 1, 2, \ldots$. Hence,

$$\sup_{x\in(0,1]} \frac{1}{w(x)}\int_0^x |S_n(t)|^q\,dt$$

$$= \max\left[\sup_{x\in(0,2^{-n}]} \frac{1}{w(x)}\int_0^x |S_n(t)|^q\,dt,\ \max_{1\le k\le n} \sup_{x\in[2^{-n+k-1},2^{-n+k}]} \frac{1}{w(x)}\int_0^x |S_n(t)|^q\,dt\right]$$

$$\le \max\left[\sup_{x\in(0,2^{-n}]} \frac{x}{w(x)}\cdot\frac{1}{x}\int_0^x |S_n(t)|^q\,dt,\ \max_{1\le k\le n} \frac{1}{w(2^{-n+k-1})}\int_0^{2^{-n+k}} |S_n(t)|^q\,dt\right]$$

$$\le \max\left[\frac{1}{w(2^{-n})}\int_0^{2^{-n}} |S_n(t)|^q\,dt,\ \max_{1\le k\le n} \frac{2}{w(2^{-n+k})}\int_0^{2^{-n+k}} |S_n(t)|^q\,dt\right]$$

$$\le 2 \max_{0\le k\le n} \frac{1}{w(2^{-n+k})}\int_0^{2^{-n+k}} |S_n(t)|^q\,dt,$$

whence it follows

$$\|S_n\|_{K_{w,q}} \le 2^{1/q} \max_{0\le m\le n} \left(\frac{1}{w(2^{-m})}\int_0^{2^{-m}} |S_n(t)|^q\,dt\right)^{1/q}.$$

Let us estimate the right-hand side of this inequality from above. Using (15.8) together with Minkowski's inequality and Khintchine's L_q-inequality, we obtain for each $m = 0, 1, \ldots, n$

$$\left(\int_0^{2^{-m}} |S_n(t)|^q\,dt\right)^{1/q} \le \left(2^{-n} \sum_{\varepsilon_i=\pm1, i=m+1,\ldots,n} |a_1 + \ldots + a_m|^q\right)^{1/q}$$

$$+ \left(2^{-n} \sum_{\varepsilon_i=\pm1} |\varepsilon_{m+1}a_{m+1} + \ldots + \varepsilon_n a_n|^q\right)^{1/q}$$

$$= 2^{-m/q}\left(\left|\sum_{i=1}^{m} a_i\right| + \left\| \sum_{j=m+1}^{n} a_j r_j \right\|_q\right)$$

$$\le 2^{-m/q}\left(\left|\sum_{i=1}^{m} a_i\right| + \sqrt{q}\,\|(a_j)_{j=m+1}^n\|_{\ell_2}\right).$$

Since the function $t/w(t)$ increases, this and the preceding inequalities imply that

$$\|S_n\|_{K_{w,q}} \le 2^{1/q}\sqrt{q}\left(\|\{a_k\}_{k=1}^n\|_{\ell_2} + \max_{1\le m\le n}\left(\frac{2^{-m}}{w(2^{-m})}\right)^{1/q}\left|\sum_{k=1}^{m} a_k\right|\right).$$

Thus,

$$\|S_n\|_{K_{w,q}} \asymp \|(a_k)_{k=1}^n\|_{\ell_2} + \max_{1\le m\le n}\left(\frac{2^{-m}}{w(2^{-m})}\right)^{1/q}\left|\sum_{k=1}^{m} a_k\right|$$

with a constant independent of $n \in \mathbb{N}$ and $a_k \in \mathbb{R}$, $k = 1, 2, \ldots, n$. Appealing to the Fatou property, from the last equivalence we derive (15.7). □

Next, we clarify, under which conditions on a weight w the Rademacher functions span ℓ_2 in the space $K_{w,q}$.

Theorem 15.3 *The Rademacher system is equivalent in the space $K_{w,q}$ to the unit vector basis in ℓ_2 if and only if there exists a constant $c > 0$ such that*

$$w(t) \ge c\,t\,\log_2^{q/2}(2/t),\quad 0 < t \le 1. \tag{15.9}$$

Proof Assume first that inequality (15.9) holds. Then for all $1 \le m \le n$ we have

$$\left(\frac{2^{-m}}{w(2^{-m})}\right)^{1/q} m^{1/2} \le c^{-1/q}.$$

On the other hand, by the Cauchy-Schwarz-Bunyakovskii inequality,

$$\left|\sum_{k=1}^{m} a_k\right| \le m^{1/2}\|(a_k)_{k=1}^m\|_{\ell_2}.$$

Combining these estimates with (15.7), we conclude that the sequence $\{r_n\}$ is equivalent in $K_{w,q}$ to the unit vector basis in ℓ_2.

To prove the converse, we suppose that inequality (15.9) fails. Then, using the quasiconcavity of w, we can find a sequence of positive integers $m_k \to \infty$ such that

$$\left(\frac{2^{-m_k}}{w(2^{-m_k})}\right)^{1/q} \cdot m_k^{1/2} \to \infty \text{ as } k \to \infty.$$

Consider the sums $S_k := m_k^{-1/2} \sum_{i=1}^{m_k} r_i$, $k = 1, 2, \ldots$. From the last relation and equivalence (15.7) it follows that $\|S_k\|_{K_{w,q}} \to \infty$ as $k \to \infty$. Hence, $\{r_n\}$ is not equivalent in $K_{w,q}$ to the unit vector basis in ℓ_2. □

Remark 15.1 Let $1 \leq q < \infty$ and let w be a quasiconcave function on $[0, 1]$. Consider the s.s. $M(w, q)$ equipped with the norm

$$\|f\|_{M(w,q)} := \sup_{0<x\leq 1} \left(\frac{1}{w(x)} \int_0^x f^*(t)^q \, dt\right)^{1/q}.$$

This space can be treated as a natural "symmetrization" of $K_{w,q}$. In particular, in the case $q = 1$ it is the usual Marcinkiewicz space $M(w)$. If G denotes, as above, the separable part of the Orlicz space L_{N_2}, then one can readily check that the embedding $G \subset M(w, q)$ holds if and only if w satisfies condition (15.9). Hence, according to Theorems 15.3 and 2.3, the Rademacher functions span ℓ_2 in $M(w, q)$ if and only if they span ℓ_2 in $K_{w,q}$.

A completely different result appears when comparing the behaviour of Rademacher sums in the space $K_{w,1}$ with $w(t) = t \log_2^{1/r}(2/t)$, where $r > 2$, and in its "symmetrization", the Marcinkiewicz space $M(w)$. Indeed, since (15.9) is no longer true, from (15.7) it follows that

$$\left\|\sum_{k=1}^{\infty} a_k r_k\right\|_{K_{w,1}} \asymp \|(a_k)\|_{\ell_2} + \sup_{m\geq 1} m^{-1/r} \left|\sum_{k=1}^{m} a_k\right|.$$

In contrast to that, the Rademacher sequence is equivalent in $M(w)$ to the unit vector basis in the Marcinkiewicz sequence space $l_{r,\infty}$ with the norm

$$\|(a_k)\|_{l_{r,\infty}} = \sup_{m=1,2,\ldots} m^{-1/r} \sum_{k=1}^{m} a_k^*,$$

where (a_k^*) is the decreasing permutation of the sequence $(|a_k|)$ (see Example 3.2).

In the case $w(x) = x$ from Theorems 15.2 and 15.3 we immediately get the following statements.

Corollary 15.1 *For every* $1 \le q < \infty$ *we have*

$$\Big\| \sum_{k=1}^{\infty} a_k r_k \Big\|_{K_q} \asymp \|(a_k)_{k=1}^{\infty}\|_{\ell_2} + \sup_{m \ge 1} \Big| \sum_{k=1}^{m} a_k \Big|. \tag{15.10}$$

Corollary 15.2 *Let* $1 \le q < \infty$. *A series* $\sum_{k=1}^{\infty} a_k r_k$ *converges in the space* K_q *if and only if both series* $\sum_{k=1}^{\infty} a_k^2$ *and* $\sum_{k=1}^{\infty} a_k$ *converge.*

Corollary 15.3 $\{r_n\}_{n=1}^{\infty}$ *is a basic sequence in* K_q, $1 \le q < \infty$, *equivalent to each of its subsequences.*

Corollary 15.4 $\{r_n\}_{n=1}^{\infty}$ *does not contain unconditional basic subsequences in* K_q, $1 \le q < \infty$.

Remark 15.2 Comparing equivalence (15.10) with Theorem 14.2 from the preceding chapter, we can conclude that there is a complete similarity in the behaviour of Rademacher sums in Cesàro spaces K_q and in BMO. In particular, one can readily prove that the statement similar to Proposition 14.3 for block basic sequences of $\{r_n\}$ holds also for K_q, $1 \le q < \infty$. At the same time, it is not hard to check that these spaces are not comparable, that is, no one of them is embedded into the other. Moreover, while K_q is a Banach function lattice, BMO is not. As we shall see later, some geometrical properties of their Rademacher subspaces also substantially differ.

15.3 Complementability of Rademacher Subspaces in the Cesàro Type Spaces

The key role in studying the complementability problem in Cesàro type spaces will be played by the fact that these spaces contain "natural" subspaces isomorphic to the space $L_1[0, 1]$. This will allow us to apply in the proofs two well-known results from the theory of L_1-spaces both connected with the research by Dunford and Pettis. The first one is the Dunford–Pettis criterion of weak compactness, which reads that a subset of $L_1[0, 1]$ is relatively weakly compact if and only if it is equi-integrable [1, Theorem 5.2.9]. Recall that a set $A \subset L_1[0, 1]$ is said to be *equi-integrable* whenever

$$\lim_{m(E) \to 0} \sup_{f \in A} \int_E |f(t)|\, dt = 0.$$

The second result asserts that L_1-spaces have the so-called Dunford–Pettis property. This fact implies that a weakly compact linear operator $T : L_1[a, b] \to X$, where X is a Banach space, is weak-to-norm sequentially continuous, i.e., if a sequence $\{f_n\}_{n=1}^{\infty} \subset L_1[a, b]$ converges to f weakly, then $\{Tf_n\}_{n=1}^{\infty}$ converges to Tf in the X-norm [1, Theorem 5.4.5].

Theorem 15.4 *For every quasiconcave function* w *the Rademacher subspace* $\mathcal{R}K_{w,1} := [r_n]_{K_{w,1}}$ *is not complemented in* $K_{w,1}$.

Proof On the contrary, assume that $\mathcal{R}K_{w,1}$ is a complemented subspace in $K_{w,1}$, and let Q be a bounded linear projection in this space whose image is $\mathcal{R}K_{w,1}$. Since the Rademacher functions form, by Theorem 15.2, a basic sequence in $K_{w,1}$, then there are functionals $\varphi_n \in K_{w,1}{}^*$, $n = 1, 2, \ldots$, such that

$$Qf(x) = \sum_{n=1}^{\infty} \varphi_n(f) r_n(x), \quad f \in K_{w,1}. \tag{15.11}$$

Clearly, $K_{w,1} \subset L_1$. Therefore, the Köthe dual space $K_{w,1}{}'$ contains L_∞, and hence $K_{w,1}{}'$ is a total set of functionals on $K_{w,1}$ (i.e., from $\int_0^1 f(s)g(s)\,ds = 0$ for all $g \in K'_{w,1}$ it follows that $f = 0$). Consequently, according to the known representation theorem (see e.g. [151, Theorem X.3.6], we have

$$K_{w,1}{}^* = (K_{w,1}{}^*)^c \oplus (K_{w,1}{}^*)^s, \tag{15.12}$$

where $(K_{w,1}{}^*)^c$ (resp. $(K_{w,1}{}^*)^s$) denotes the set of all order continuous functionals, generated by the space $K_{w,1}{}'$ (resp. the set of all singular bounded linear functionals). Note that every functional from $(K_{w,1}{}^*)^s$ vanishes on the separable part $K_{w,1}{}^\circ$ of the space $K_{w,1}$, consisting of all $f \in K_{w,1}$ such that $\lim_{m(E)\to 0} \|f\chi_E\|_{K_{w,1}} = 0$.

From representation formula (15.12) it follows that $\varphi_n = \psi_n + \theta_n$, where $\psi_n \in (K_{w,1}{}^*)^c$ and $\theta_n \in (K_{w,1}{}^*)^s$, $n = 1, 2, \ldots$ Moreover, since Q is a projection onto the subspace $\mathcal{R}K_{w,1}$, then we have

$$\psi_n(r_n) + \theta_n(r_n) = 1 \ \text{ and } \ \psi_n(r_i) + \theta_n(r_i) = 0 \ \text{ if } \ i \neq n. \tag{15.13}$$

Clearly, $r_i - \chi_{[0,1]} \in K_{w,1}{}^\circ$ for each $i = 1, 2, \ldots$. Therefore, $\theta_n(r_i) = \theta_n(\chi_{[0,1]}) =: c_n$ for all positive integers n and i. Combining this with (15.13), we obtain that $\psi_n(r_i) = -c_n$ if $i \neq n$. On the other hand, by the definition of $(K_{w,1}{}^*)^c$,

$$\psi_n(f) = \int_0^1 g_n(t) f(t)\,dt, \tag{15.14}$$

where $g_n \in K_{w,1}{}'$, $n = 1, 2, \ldots$. Since $K_{w,1} \supset L_\infty$, then $K_{w,1}{}' \subset L_1$. Consequently, taking into account that $\{r_i\}$ is an uniformly bounded orthonormal system, we have $\psi_n(r_i) \to 0$ as $i \to \infty$ for every $n = 1, 2, \ldots$. Thus, $c_n = 0$, $n = 1, 2, \ldots$, and from (15.13) and (15.14) it follows that

$$\int_0^1 g_n(t) r_n(t)\,dt = 1 \ \text{ and } \ \int_0^1 g_n(t) r_i(t)\,dt = 0 \ \text{ if } \ i \neq n. \tag{15.15}$$

In addition, if $f \in K_{w,1}{}^{\circ}$, then $\theta_n(f) = 0$, $n = 1, 2, \dots$, whence

$$Qf(x) = \sum_{n=1}^{\infty} \int_0^1 g_n(t) f(t)\, dt \cdot r_n(x).$$

Next, we prove that there exist $h \in (0, 1)$ and a positive integer n_0 such that for all $n \geq n_0$ it holds

$$\left| \int_h^1 g_n(t) r_n(t)\, dt \right| \geq \frac{1}{2}. \tag{15.16}$$

Indeed, if it is not the case, in view of the first equation in (15.15), there is a subsequence $\{g_{n_i}\} \subset \{g_n\}$ satisfying the condition

$$\int_0^{1/i} |g_{n_i}(t)|\, dt \geq \left| \int_0^{1/i} g_{n_i}(t) r_{n_i}(t)\, dt \right| \geq \frac{1}{2}, \quad i = 1, 2, \dots \tag{15.17}$$

At the same time, for every $f \in K_{w,1}{}^{\circ}$, by Theorem 15.2, we have

$$\sum_{n=1}^{\infty} \left(\int_0^1 g_n(t) f(t)\, dt \right)^2 \leq C \|Q\| \|f\|_{K_{w,1}}. \tag{15.18}$$

Since for any $f \in L_\infty$ and every $\varepsilon > 0$ the function $f \chi_{(\varepsilon,1]}$ belongs to $K_{w,1}{}^{\circ}$ and $\|f \chi_{(\varepsilon,1]}\|_{K_{w,1}} \leq \|f\|_{K_{w,1}}$, the latter estimate implies that

$$\sum_{n=1}^{\infty} \left(\int_\varepsilon^1 g_n(t) f(t)\, dt \right)^2 \leq C \|Q\| \|f\|_{K_{w,1}}$$

for all $f \in L_\infty$ and $\varepsilon > 0$. Because $g_n f \in L_1[0, 1], n \in \mathbb{N}$, we may take here $\varepsilon \to 0$ to obtain inequality (15.18) for all $f \in L_\infty$. This yields $g_n \xrightarrow{w} 0$ in $L_1[0, 1]$. Therefore, by the above-mentioned Dunford–Pettis criterion [1, Theorem 5.2.9], $\{g_n\}_{n=1}^{\infty}$ is an equi-integrable set in $L_1[0, 1]$, which contradicts (15.17). Thus, inequality (15.16) holds for some $h \in (0, 1)$ and all sufficiently large $n \in \mathbb{N}$.

Further, since $f \chi_{[h,1]} \in K_{w,1}^{\circ}$ for any function $f \in K_{w,1}$, we can define the operator Q_h by $Q_h f := Q(f \chi_{[h,1]})$, $f \in K_{w,1}$. From the estimate

$$\|Q_h f\|_{K_{w,1}} = \|Q(f \chi_{[h,1]})\|_{K_{w,1}} \leq \|Q\| \|f\|_{K_{w,1}}$$

it follows that Q_h is bounded on the space $K_{w,1}$. Moreover, if f is measurable on $[0, 1]$ and $\operatorname{supp} f$ is contained in the interval $[h, 1]$, we have

$$\|f\|_{L_1[h,1]} \leq \|f\|_{K_{w,1}} = \sup_{h \leq x \leq 1} \frac{1}{w(x)} \int_h^x |f(t)|\, dt \leq \frac{1}{w(h)} \|f\|_{L_1[h,1]}. \tag{15.19}$$

Combining this together with the embedding $K_{w,1} \subset L_1[0,1]$, we conclude that Q_h is bounded in $L_1[0,1]$ as well. By Theorem 1.4, the Rademacher system is equivalent in $L_1[0,1]$ to the unit vector basis in ℓ_2. Hence, the image of the operator Q_h is a Hilbert subspace in $L_1[0,1]$, and so Q_h is weakly compact. As was said above, the space $L_1[0,1]$ possesses the Dunford–Pettis property, which implies that Q_h is weak-to-norm sequentially continuous. Therefore, since $r_n\chi_{[h,1]} \xrightarrow{w} 0$ in $L_1[0,1]$, we have

$$\|Q_h(r_n\chi_{[h,1]})\|_{L_1} \to 0 \text{ as } n \to \infty.$$

On the other hand, from Khintchine's L_1-inequality and estimate (15.16) it follows that

$$\|Q_h(r_n\chi_{[h,1]})\|_{L_1} \asymp \Big(\sum_{i=1}^{\infty}\Big(\int_h^1 g_i(t)r_n(t)\,dt\Big)^2\Big)^{1/2} \ge \Big|\int_h^1 g_n(t)r_n(t)\,dt\Big| \ge \frac{1}{2}$$

for all $n \ge n_0$. This contradiction concludes the proof. □

Remark 15.3 Theorem 15.4 reveals a substantial difference in properties of the Rademacher subspaces in the Cesàro space $K_{w,1}$ and in its "symmetrization", the Marcinkiewicz space $M(w)$. Indeed, while the subspace $\mathcal{R}K_{w,1}$ fails to be complemented in $K_{w,1}$ for every weighted function w, from Theorem 2.5 it follows that the Rademacher subspace $[r_n]_{M(w)}$ is complemented in $M(w)$ provided if $G \subset M(w) \subset G'$.

The situation is completely different in the case when $q > 1$.

Theorem 15.5 *Let $1 < q < \infty$. If a quasiconcave function w satisfies condition (15.9), then the subspace $\mathcal{R}K_{w,q} := [r_n]_{K_{w,q}}$ is complemented in $K_{w,q}$.*

Proof According to the proof of Theorem 2.5, the orthogonal projection

$$Pf(x) := \sum_{n=1}^{\infty}\int_0^1 r_n(t)f(t)\,dt \cdot r_n(x)$$

is bounded in the space $L_q[0,1]$, $1 < q < \infty$. Therefore, by Theorems 15.3 and 1.4, we have

$$\|Pf\|_{K_{w,q}} \le C\|Pf\|_q \le C\|P\|_{L_q\to L_q}\|f\|_q \le C\|P\|_{L_q\to L_q}\|f\|_{K_{w,q}}.$$

Hence, it follows that P is bounded in the space $K_{w,q}$ as well. Since the image of P coincides with the subspace $\mathcal{R}K_{w,q}$, we obtain the desired result. □

The complementability problem for the subspace $[r_n]_{Ces_p}(= \mathcal{R})$ (see Theorem 15.1) in the Cesàro space Ces_p, $1 \le p < \infty$, can be treated much in the same way as in Theorem 15.4, and even somewhat simpler. The reason is that this space is separable, and so the dual space $Ces_p{}^*$ coincides with the Köthe dual $Ces_p{}'$. As above (see (15.19)), one can show that for each $h \in (0, 1)$ there exists a constant $C_h > 0$ such that inequality $\|f\|_{C(p)} \le C_h\|f\|_{L_1}$ holds whenever $\operatorname{supp} f \subset [h, 1]$. Hence, applying Theorem 15.1, as in the proof of Theorem 15.4, we obtain the following result.

Theorem 15.6 *For every $1 \le p < \infty$ the Rademacher subspace $\mathcal{R}$ is not complemented in the space Ces_p.*

15.4 The Structure of Rademacher Subspaces in the Spaces K_q

Geometrical properties of the Rademacher space $\mathcal{R}K_q := [r_n]_{K_q}$ can be investigated in more detail. As the following theorem shows, they are substantially different in the cases $q = 1$ and $1 < q < \infty$.

Theorem 15.7

1. *Every infinite-dimensional subspace of the space $\mathcal{R}K_1$ is either isomorphic to ℓ_2 and is not complemented in $K_1 = Ces_\infty[0, 1]$ or contains a further subspace that is isomorphic to c_0 and complemented in $\mathcal{R}K_1$.*
2. *Let $1 < q < \infty$. Every infinite-dimensional subspace of the space $\mathcal{R}K_q$ is either isomorphic to ℓ_2 and is complemented in K_q or contains a further subspace that is isomorphic to c_0 and complemented in $\mathcal{R}K_q$.*

Comparing the second part of Theorem 15.7 with Theorem 14.5, we see how like the structure of Rademacher subspaces in the spaces K_q, $1 < q < \infty$, and BMO. And this is not surprising if we take into account the analogy in the behaviour of Rademacher sums in these spaces, which was mentioned in Remark 15.2. At the same time, the result of the first part of Theorem 15.7 is in marked contrast to what we had in BMO. This may be clarified as follows. While the continuous embedding $BMO \subset L_p$ is valid for each $p \in [1, \infty)$ (which allows us to use complementability of the Rademacher subspace $\mathcal{R}$ in L_p, $1 < p < \infty$), the embedding $K_1 \subset L_p$ holds only in the case when $p = 1$ (but $\mathcal{R}$ is not complemented in L_1). Taking into account this observation, we prove here only that part of Theorem 15.7, which differs the Cesàro spaces from BMO. The rest can be obtained in the same way as in the case of BMO (see Theorem 14.5 and Corollary 14.6). So, we intend to prove the following statement.

Theorem 15.8 *Every subspace X of the Rademacher space $\mathcal{R}K_1$ that is isomorphic to ℓ_2 is not complemented in K_1.*

Proof To get a contradiction, assume that a subspace X, satisfying the conditions of the theorem, is complemented in K_1. Let $\{x_j\}_{j=1}^\infty \subset X$ be a sequence, equivalent in K_1 to the unit vector basis $\{e_j\}_{j=1}^\infty$ in the space ℓ_2. Clearly, the relation $e_j \xrightarrow{w} 0$

in ℓ_2 implies that $x_j \xrightarrow{w} 0$ in K_1. Since the Rademacher sequence is a basis in $\mathcal{R}K_1$, $x_j \in \mathcal{R}K_1$, and $\|x_j\|_{K_1} \asymp 1$, $j = 1, 2, \ldots$, then applying the Bessaga–Pełczyński selection principle (see Theorem B.3), we can find a subsequence $\{x_{j_n}\} \subset \{x_j\}$, equivalent to a block basic sequence $\{u_n\}$ of the Rademacher system, such that $\|x_{j_n} - u_n\|_{K_1} \to 0$ as $n \to \infty$. Let

$$u_n = \sum_{k=m_n+1}^{m_{n+1}} a_k r_k, \quad n \in \mathbb{N}, \quad \text{where } 0 \le m_1 < m_2 < \cdots < m_n < \ldots$$

Then the sequence $\{u_n\}$ is equivalent in K_1 to the unit vector basis in ℓ_2 as well, i.e.,

$$\Big\| \sum_{n=1}^{\infty} b_n u_n \Big\|_{K_1} \asymp \Big(\sum_{n=1}^{\infty} b_n^2 \Big)^{1/2}, \quad b_n \in \mathbb{R}. \tag{15.20}$$

Further, if $\varphi \in K_1{}^*$, then

$$|\varphi(u_n)| \le |\varphi(u_n - x_{j_n})| + |\varphi(x_{j_n})| \le \|\varphi\| \, \|u_n - x_{j_n}\| + |\varphi(x_{j_n})|.$$

Since $x_j \xrightarrow{w} 0$ in K_1 and $\|u_n - x_{j_n}\|_{K_1} \to 0$, from this inequality it follows that $u_n \xrightarrow{w} 0$ in K_1. Consequently, by Remark 15.2, in the same way as in the concluding part of the proof of Theorem 14.5, we obtain that

$$\sum_{k=m_n+1}^{m_{n+1}} a_k \to 0 \text{ as } n \to \infty. \tag{15.21}$$

Therefore, assuming that

$$\liminf_{n\to\infty} \sum_{k=m_n+1}^{m_{n+1}} a_k^2 = 0, \tag{15.22}$$

one can select from the sequence $\{u_n\}_{n=1}^{\infty}$ a subsequence $\{u_{n_i}\}_{i=1}^{\infty}$, satisfying conditions (a), (b), and (c) of Proposition 14.3. Using this proposition (more precisely, its version for the space K_1; see Remark 15.2), we infer that the subspace $[u_{n_i}]_{K_1} \subset [u_n]_{K_1}$ is isomorphic to c_0. Since this contradicts equivalence (15.20), equation (15.22) does not hold, and so, by Theorem 15.2, we have

$$\|u_n\|_{K_1} \asymp \Big(\sum_{k=m_n+1}^{m_{n+1}} a_k^2 \Big)^{1/2} \asymp 1, \; n = 1, 2, \ldots \tag{15.23}$$

Moreover, thanks to the principle of small perturbations (see Theorem B.2), we may assume that the subspace $[u_n]$ (together with $[x_{j_n}]$) is complemented in K_1.

Let Q be a bounded linear projection in K_1 whose image is the subspace $[u_n]$. Since $\{u_n\}$ is a basic sequence in K_1, there are functionals $\varphi_n \in K_1{}^*$, $n = 1, 2, \dots,$ such that

$$Qf(x) = \sum_{n=1}^{\infty} \varphi_n(f)\, u_n(x), \;\; f \in K_1.$$

Precisely as in the proof of Theorem 15.4 we have

$$\varphi_n = \psi_n + \theta_n, \;\text{ with } \psi_n \in (K_1^*)^c \text{ and } \theta_n \in (K_1^*)^s, n = 1, 2, \dots,$$

where $(K_1^*)^c$ (resp. $(K_1^*)^s$) is the set of all order continuous functionals, generated by the Köthe dual space $K_1{}'$ (resp. singular bounded linear functionals). Additionally, since Q is a projection onto $[u_n]$, we have

$$\psi_n(u_n)+\theta_n(u_n) = 1, \; n = 1, 2, \dots, \;\text{ and } \psi_n(u_i)+\theta_n(u_i) = 0 \text{ if } i \neq n. \tag{15.24}$$

From (15.20) it follows that

$$\|Qf\|_{K_1} \asymp \Big(\sum_{n=1}^{\infty} \varphi_n(f)^2\Big)^{1/2} < \infty \text{ for every } \; f \in K_1.$$

Therefore, $\varphi_n \overset{w^*}{\to} 0$ in $K_1{}^*$, whence

$$\|\varphi_n\|_{K_1{}^*} \le A \;\text{ for some }\; A > 0 \;\text{ and all }\; n = 1, 2, \dots \tag{15.25}$$

On the other hand, since singular functionals vanish on the set $K_1{}^\circ$, then for all $f \in K_1{}^\circ$

$$Qf(x) = \sum_{n=1}^{\infty} \psi_n(f)\, u_n(x). \tag{15.26}$$

Let us show that the formal extension of Q to the whole space K_1, i.e., the operator

$$\bar{Q}f(x) := \sum_{n=1}^{\infty} \psi_n(f)\, u_n(x), \;\;\; f \in K_1,$$

is bounded on this space.

First of all, since $\psi_n \in (K_1^*)^c$, we have

$$\psi_n(f) = \int_0^1 g_n(t) f(t)\, dt, \;\text{ where }\; g_n \in K_1{}'. \tag{15.27}$$

Furthermore, for every $f \in K_1$ we have $|f|\chi_{[1/m,1]}\cdot\text{sign}\, g_n \in K_1{}^\circ, m, n = 1, 2, \ldots$, and so, by (15.25),

$$\int_0^1 |g_n(t)f(t)\chi_{[1/m,1]}(t)|\,dt = \psi_n(|f|\chi_{[1/m,1]}\cdot \text{sign} g_n)$$
$$= \varphi_n\big(|f|\chi_{[1/m,1]}\,\text{sign} g_n\big)$$
$$\leq A\,\|f\|_{K_1},\ m, n = 1, 2, \ldots$$

Taking here $m \to \infty$ and using the Fatou lemma, we get

$$\int_0^1 |g_n(t)f(t)|\,dt \leq A\,\|f\|_{K_1} \text{ for all } f \in K_1 \text{ and } n = 1, 2, \ldots,$$

whence $\|\psi_n\|_{K_1{}^*} \leq A$. From this inequality and (15.25) it follows that

$$\|\theta_n\|_{K_1{}^*} \leq 2A,\ n = 1, 2, \ldots \tag{15.28}$$

Moreover, by (15.20), (15.26), and (15.27),

$$\sum_{n=1}^{\infty}\Big(\int_0^1 g_n(t)f(t)\chi_{[1/m,1]}(t)\,dt\Big)^2 \asymp \|Q(f\chi_{[1/m,1]})\|_{K_1}^2 \leq \|Q\|^2\|f\|_{K_1}^2.$$

Observe that $g_n f \in L_1$ for all $n = 1, 2, \ldots$. In consequence, as in the proof of Theorem 15.4, passing in this inequality to the limit as $m \to \infty$ and using definition of the operator $\bar{Q}$ together with equivalence (15.20), we get

$$\|\bar{Q}f\|_{K_1}^2 \asymp \sum_{n=1}^{\infty}\Big(\int_0^1 g_n(t)f(t)\,dt\Big)^2 \leq C\|f\|_{K_1}^2. \tag{15.29}$$

Thus, the above claim is proved. In particular, from (15.29) it follows that $g_n \overset{w^*}{\to} 0$ in $K_1{}^*$.

Further, since $r_i - \chi_{[0,1]} \in K_1{}^\circ, i = 1, 2, \ldots$, then $\theta_n(r_i) = \theta_n(\chi_{[0,1]}) =: c_n$ for all $n, i = 1, 2, \ldots$, which implies that

$$\theta_n(u_i) = c_n \cdot \sum_{k=m_i+1}^{m_{i+1}} a_k,\ n, i = 1, 2, \ldots$$

Also, according to (15.28), for all $n = 1, 2, \ldots$ we have

$$|c_n| = |\theta_n(\chi_{[0,1]})| \leq \|\theta_n\|_{K_1{}^*}\|\chi_{[0,1]}\|_{K_1} \leq 2A.$$

Therefore, from (15.21) it follows

$$\lim_{i\to\infty}\sup_{n\in\mathbb{N}}|\theta_n(u_i)|=0.$$

On the other hand, in view of (15.24),

$$\psi_n(u_i)=-\theta_n(u_i),\ i\neq n,\ \text{and}\ \psi_n(u_n)=1-\theta_n(u_n),\ n=1,2,\ldots,$$

and so

$$\lim_{n\to\infty}\psi_n(u_n)=1\ \text{ and }\ \lim_{i\to\infty}\sup_{n\neq i}|\psi_n(u_i)|=0.$$

Thus, passing to suitable subsequences of sequences $\{u_n\}$ and $\{\psi_n\}$ (and preserving the notation), we may assume that

$$\psi_n(u_n)\geq 1-2^{-n},\ n=1,2,\ldots,\ \text{ and } |\psi_n(u_i)|\leq 2^{-i},\ n\neq i. \tag{15.30}$$

It is clear that, by equivalence (15.20), the corresponding operator, which will be denoted still by $\bar{Q}$, is bounded on K_1.

Since $K_1\subset L_1$, the operator $\bar{Q}$ is simultaneously bounded from K_1 in L_1. Show that $\bar{Q}: K_1\to L_1$ is weakly compact. To this end, in view of the Dunford–Pettis criterion (see the very beginning of Sect. 15.3), it suffices to check that the set of functions $\{\bar{Q}f:\|f\|_{K_1}\leq 1\}$ is equi-integrable.

Indeed, using the Cauchy-Schwarz-Bunyakovskii inequality, the fact that $\{r_k\}$ is an orthonormal sequence, and relations (15.23), (15.29), we get the following estimate for every function $f\in K_1$, $\|f\|_{K_1}\leq 1$, and any set $E\subset[0,1]$:

$$\begin{aligned}\|\bar{Q}f\cdot\chi_E\|_{L_1}&\leq m(E)^{1/2}\|\bar{Q}f\|_{L_2}=m(E)^{1/2}\Big(\sum_{n=1}^{\infty}\psi_n(f)^2\cdot\sum_{k=m_n+1}^{m_{n+1}}a_k^2\Big)^{1/2}\\&\asymp m(E)^{1/2}\Big(\sum_{n=1}^{\infty}\psi_n(f)^2\Big)^{1/2}\leq C\|\bar{Q}\|\,m(E)^{1/2}.\end{aligned}$$

Hence,

$$\lim_{m(E)\to 0}\sup\big\{\|\bar{Q}f\cdot\chi_E\|_{L_1}:\|f\|_{K_1}\leq 1\big\}=0,$$

which implies that $\bar{Q}: K_1\to L_1$ is a weakly compact operator.

Next, we consider two cases separately. Assume first that there are $\delta\in(0,1)$ and a subsequence $\{u_{n_k}\}\subset\{u_n\}$ such that

$$|\psi_{n_k}(u_{n_k}\cdot\chi_{[\delta,1]})|=\Big|\int_\delta^1 g_{n_k}(t)u_{n_k}(t)\,dt\Big|\geq\frac{1}{2},\ k=1,2,\ldots \tag{15.31}$$

Then, precisely as in the proof of Theorem 15.4, for any f with supp $f \subset [\delta, 1]$ we get

$$\|f\|_{L_1[\delta,1]} \le \|f\|_{K_1} = \sup_{\delta \le x \le 1} \frac{1}{x} \int_\delta^1 |f(t)|\, dt \le \frac{1}{\delta} \|f\|_{L_1[\delta,1]}.$$

Therefore, $\bar{Q}_\delta f := \bar{Q}(f\chi_{[\delta,1]})$ is a weakly compact operator in $L_1[0, 1]$. Since $L_1[0, 1]$ has the Dunford–Pettis property (see again the beginning of Sect. 15.3 or [1, Theorem 5.4.5]), then $\bar{Q}_\delta$ is weak-to-norm sequentially continuous. Moreover, one can readily see that from $u_n \xrightarrow{w} 0$ in K_1 it follows $u_n\chi_{[\delta,1]} \xrightarrow{w} 0$ in $L_1[0, 1]$. Combining these facts, we conclude that

$$\|\bar{Q}(u_n\chi_{[\delta,1]})\|_{L_1} \to 0 \text{ as } n \to \infty.$$

On the other hand, applying Khintchine's L_1-inequality and taking into account equivalence (15.23) together with estimate (15.31), we have

$$\begin{aligned} \|\bar{Q}(u_{n_k} \cdot \chi_{[\delta,1]})\|_{L_1} &\asymp \Big\| \sum_{i=1}^\infty \psi_i(u_{n_k} \cdot \chi_{[\delta,1]}) \sum_{j=m_i+1}^{m_{i+1}} a_j r_j \Big\|_{L_1} \\ &\asymp \Big(\sum_{i=1}^\infty \psi_i(u_{n_k} \cdot \chi_{[\delta,1]})^2 \sum_{j=m_i+1}^{m_{i+1}} a_j^2 \Big)^{1/2} \\ &\asymp \Big(\sum_{i=1}^\infty \psi_i(u_{n_k} \cdot \chi_{[\delta,1]})^2 \Big)^{1/2} \ge |\psi_{n_k}(u_{n_k} \cdot \chi_{[\delta,1]})| \ge \frac{1}{2} \end{aligned}$$

for all $k = 1, 2, \ldots$. Clearly, the last relations contradict each other; hence, if (15.31) is fulfilled, the proof of the theorem is complete.

Otherwise, assume that (15.31) does not hold. Then, by (15.30) and (15.27), for any $\delta \in (0, 1)$ and all sufficiently large $n \in \mathbb{N}$ we have

$$\Big| \int_0^\delta g_n(t) u_n(t)\, dt \Big| > \frac{1}{4}. \tag{15.32}$$

Setting $\delta_0 = 1/2$, we find $n_1 \in \mathbb{N}$ and $\delta_1 \in (0, \delta_0)$ such that

$$\Big| \int_{\delta_1}^{\delta_0} g_{n_1}(t) u_{n_1}(t)\, dt \Big| > \frac{1}{4}.$$

Let $v_{n_1} := u_{n_1}\chi_{(\delta_1,\delta_0)}$. Since $g_n \overset{w^*}{\to} 0$ in $K_1{}^*$ and $u_n \overset{w}{\to} 0$ in K_1, then according to (15.32) there exists $n_2 > n_1$ with

$$\left|\int_0^{\delta_1} g_{n_1}(t)u_{n_2}(t)\,dt\right| < \frac{1}{2^3},$$

$$\left|\int_0^1 g_{n_2}(t)v_{n_1}(t)\,dt\right| < \frac{1}{2^3},$$

and

$$\left|\int_0^{\delta_1} g_{n_2}(t)u_{n_2}(t)\,dt\right| > \frac{1}{4}.$$

Similarly, there is $\delta_2 \in (0, \delta_1)$ such that for the functions g_{n_i} and v_{n_i}, $i = 1, 2$, where $v_{n_2} := u_{n_2}\chi_{[\delta_2,\delta_1]}$, we have

$$\left|\int_0^1 g_{n_j}(t)v_{n_i}(t)\,dt\right| < \frac{1}{2^3} \text{ if } 1 \le i \ne j \le 2,$$

and

$$\left|\int_0^1 g_{n_2}(t)v_{n_2}(t)\,dt\right| > \frac{1}{4}.$$

Suppose that positive integers $n_1 < n_2 < \ldots < n_k$ and real numbers $1/2 = \delta_0 > \delta_1 > \ldots > \delta_k > 0$ are already chosen so that the functions g_{n_i} and $v_{n_i} := u_{n_i}\chi_{[\delta_i,\delta_{i-1}]}$, $i = 1, 2, \ldots, k$, satisfy the conditions:

$$\left|\int_0^1 g_{n_j}(t)v_{n_i}(t)\,dt\right| < \frac{1}{2^{i+j}},\quad 1 \le i \ne j \le k, \tag{15.33}$$

and

$$\left|\int_0^1 g_{n_i}(t)v_{n_i}(t)\,dt\right| > \frac{1}{4},\quad i = 1, 2, \ldots, k. \tag{15.34}$$

Using now the fact that $u_n \overset{w}{\to} 0$ in K_1, $g_n \overset{w^*}{\to} 0$ in $K_1{}^*$, and again inequality (15.32), we can find a positive integer $n_{k+1} > n_k$ such that

$$\left|\int_0^{\delta_k} g_{n_i}(t)u_{n_{k+1}}(t)\,dt\right| < \frac{1}{2^{i+k+1}},\quad i = 1, 2, \ldots, k,$$

$$\left|\int_0^1 g_{n_{k+1}}(t)v_{n_i}(t)\,dt\right| < \frac{1}{2^{i+k+1}},\quad i = 1, 2, \ldots, k,$$

and

$$\left|\int_0^{\delta_k} g_{n_{k+1}}(t)u_{n_{k+1}}(t)\,dt\right| > \frac{1}{4}.$$

Obviously, there exists $\delta_{k+1} \in (0, \delta_k)$ such that the functions g_{n_i} and v_{n_i}, where $v_{n_{k+1}} := u_{n_{k+1}}\chi_{[\delta_{k+1},\delta_k]}$, satisfy inequalities (15.33) for $1 \le i \ne j \le k+1$ and (15.34) for $i = 1, 2, \ldots, k+1$.

As a result, we obtain sequences of positive integers $n_1 < n_2 < \ldots < n_i < \ldots$ and real numbers $1/2 = \delta_0 > \delta_1 > \ldots \delta_i > \ldots > 0$ such that the functions g_{n_i} and $v_{n_i} := u_{n_i}\chi_{[\delta_i,\delta_{i-1}]}$, $i = 1, 2, \ldots$, satisfy the conditions:

$$\left|\int_0^1 g_{n_j}(t)v_{n_i}(t)\,dt\right| \le \frac{1}{2^{i+j}},\quad 1 \le i \ne j < \infty, \tag{15.35}$$

and

$$\left|\int_0^1 g_{n_i}(t)v_{n_i}(t)\,dt\right| > \frac{1}{4},\quad i = 1, 2, \ldots. \tag{15.36}$$

According to Proposition 15.1, we may assume that

$$\sup_{m=1,2,\ldots}\left\|\sum_{i=1}^m v_{n_i}\right\|_{K_1} \le 4\sup_{i=1,2,\ldots}\|v_{n_i}\|_{K_1} \le 4\sup_{n=1,2,\ldots}\|u_n\|_{K_1}.$$

Moreover, since $\bar{Q}$ is bounded in K_1, so is the operator Q' defined by

$$Q'f(x) := \sum_{i=1}^\infty \int_0^1 g_{n_i}(t)f(t)\,dt\cdot u_{n_i}(x).$$

Hence, on the one hand, we have

$$\left\|Q'\Big(\sum_{i=1}^m v_{n_i}\Big)\right\|_{K_1} \le \|Q'\|\left\|\sum_{i=1}^m v_{n_i}\right\|_{K_1} \le 4\|Q'\|\sup_{n=1,2,\ldots}\|u_n\|_{K_1}\quad \text{for all } m \in \mathbb{N}.$$

On the other hand, by (15.20), (15.35), and (15.36), for every $m = 1, 2, \ldots$

$$\begin{aligned}\left\|Q'\Big(\sum_{i=1}^m v_{n_i}\Big)\right\|_{K_1} &= \left\|\sum_{j=1}^\infty\sum_{i=1}^m \int_0^1 g_{n_j}(s)v_{n_i}(s)\,ds\cdot u_{n_j}\right\|_{K_1}\\ &\ge \left\|\sum_{i=1}^m \int_0^1 g_{n_i}(s)v_{n_i}(s)\,ds\cdot u_{n_i}\right\|_{K_1}\end{aligned}$$

$$- \sum_{i,j=1,i\neq j}^{\infty} \left| \int_0^1 g_{n_j}(s) v_{n_i}(s)\, ds \right|$$

$$\geq \left\| \sum_{i=1}^{m} \int_0^1 g_{n_i}(s) v_{n_i}(s)\, ds \cdot u_{n_i} \right\|_{K_1} - 1$$

$$\geq c\Big(\sum_{i=1}^{m} \left| \int_0^1 g_{n_i}(s) v_{n_i}(s)\, ds \right|^2 \Big)^{1/2} - 1 \geq \frac{cm^{1/2}}{4} - 1,$$

where $c > 0$. Since the last relations contradict each other, the theorem is proved. □

Comments and References The space $Ces_\infty[0,1](= K_1)$, known as the Korenblyum–Krein–Levin space, arose first as early as in 1948 in connection with investigations of some problems from the theory of singular integrals (see [164] and [287]). Later on, the spaces $Ces_p[0,\infty)$, $1 \leq p \leq \infty$, have been studied in a few papers, among them [133, 266, 272]. However, for a long time Cesàro function spaces have not attracted a lot of attention contrary to their sequence counterparts.

The Cesàro sequence spaces ces_p, $1 \leq p \leq \infty$, appeared explicitly in 1968 when the Dutch Mathematical Society posted the problem to find a description of their duals. In 1974, this problem was resolved by Jagers [138]. Recall that the space ces_p consists of all sequences $x = (x_k)_{k=1}^\infty$ of real numbers such that

$$\|x\|_{c(p)} := \Big[\sum_{n=1}^{\infty} \Big(\frac{1}{n} \sum_{k=1}^{n} |x_k| \Big)^p \Big]^{1/p} < \infty$$

(with a natural modification in the case $p = \infty$).

Interest in exploring the structure of Cesàro function spaces has grown recently; quite rich literature concerning to various geometrical properties of these spaces can be found in the survey [49].

Proposition 15.1 was proved in [29], where this assertion was applied for identification of the set of all $1 \leq q < \infty$ such that the space $Ces_p[0,1]$ contains a complemented subspace isomorphic to ℓ_q. Almost all results of Sects. 15.2 and 15.3 are taken from the paper [48], the authors of which have initiated studying the behaviour of the Rademacher sequence in function Cesàro type spaces. Theorems 15.7 and 15.8, constituting the contents of Sect. 15.4, were proved in [50]. The first of them can be treated as a version of the well-known Kadec-Pełczyński alternative [147] for the Rademacher space $\mathcal{R}K_q$ (cf. also Theorem 14.5).

There are many other Banach spaces, the norms of which are generated by positive sublinear operators and the L_p-norms. Examples are the real interpolation and extrapolation spaces, the Besov spaces $B^s_{p,q}$, the Triebel spaces $F^s_{p,q}$, the "tent" spaces (for general results related to their properties, see [195] and [30]).

Chapter 16
Rademacher Functions in Morrey Spaces

Another class of function spaces, whose norms are determined by the L_p-norms, is constituted by the Morrey spaces that have become the object of intensive research in recent decades.

In this chapter, we examine the behaviour of the Rademacher functions and the subspaces spanned by them in the Morrey spaces. It will be instructive to compare the results obtained here with similar ones for symmetric and Cesàro type spaces.

16.1 Some Preliminaries Related to Morrey Spaces

Let $1 \le p < \infty$, and let w be a nonnegative function on $[0, 1]$. The *Morrey space* $M_{w,p} = M_{w,p}[0, 1]$ consists of all Lebesgue measurable real-valued functions f on $[0, 1]$ such that

$$\|f\|_{M_{w,p}} := \sup_I w(m(I)) \left(\frac{1}{m(I)} \int_I |f(t)|^p \, dt \right)^{1/p} < \infty, \tag{16.1}$$

where the supremum is taken over all intervals I from $[0, 1]$. In view of Hölder's inequality, the family of these spaces is monotone with respect to p; more precisely, the following embeddings hold:

$$M_{w,p_1} \overset{1}{\subset} M_{w,p_0} \text{ if } 1 \le p_0 \le p_1 < \infty. \tag{16.2}$$

Recall that the norm of the weighted Korenblyum–Krein–Levin space $K_{v,p} = K_{v,p}[0, 1]$, $1 \le p < \infty$, is defined as follows

$$\|f\|_{K_{v,p}} := \sup_{0<x\le 1} \left(\frac{1}{v(x)} \int_0^x |f(t)|^p \, dt \right)^{1/p}.$$

S. V. Astashkin, *The Rademacher System in Function Spaces*,
https://doi.org/10.1007/978-3-030-47890-2_16

Let us slightly modify this definition setting $\widetilde{K}_{w,p} = K_{v,p}$ if $v(x) = xw(x)^{-p}$. Then, we have

$$\|f\|_{\widetilde{K}_{w,p}} = \sup_{0<x\leq 1} w(x) \left(\frac{1}{x} \int_0^x |f(t)|^p \, dt \right)^{1/p}. \tag{16.3}$$

The only difference of this formula from (16.1) is that the supremum in (16.3) is taken over a smaller class of intervals of the form $(0, x)$, $0 < x \leq 1$. In the case when the supremum in (16.3) is taken over all subsets of $[0, 1]$ of measure x we immediately obtain, by using the properties of the decreasing rearrangement f^*, the norm of the space $\widetilde{M(w, p)}$ (a result of "symmetrization" of $M_{w,p}$ and $\widetilde{K}_{w,p}$):

$$\|f\|_{\widetilde{M(w,p)}} = \sup_{0<x\leq 1} w(x) \left(\frac{1}{x} \int_0^x f^*(t)^p \, dt \right)^{1/p}. \tag{16.4}$$

In what follows, we assume that w is a quasiconcave function on $[0, 1]$ (i.e., $w(0) = 0$, w increases, and $w(t)/t$ decreases on $(0, 1]$). Moreover, let $w(1) = 1$. Then, we have

$$L_\infty \overset{1}{\subset} \widetilde{M(w, p)} \overset{1}{\subset} M_{w,p} \overset{1}{\subset} \widetilde{K}_{w,p} \overset{1}{\subset} L_p, \tag{16.5}$$

because the corresponding suprema in (16.3), (16.1), and (16.4) are taken over increasing classes of subsets of $[0, 1]$.

Observe that in the case when $\lim_{t\to 0^+} w(t) > 0$, by Lebesgue's differentiation theorem (see e.g. [215, Theorem IX.4.2]), it follows that $M_{w,p} = \widetilde{M(w, p)} = L_\infty$. On the other hand, if $\sup_{0<t\leq 1} w(t)\, t^{-1/p} < \infty$ we have $M_{w,p} = \widetilde{M(w, p)} = L_p$ with equivalent norms. To avoid these trivial cases, throughout this chapter we shall assume that

$$\lim_{t\to 0^+} w(t) = \liminf_{t\to 0^+} \frac{t^{1/p}}{w(t)} = 0. \tag{16.6}$$

Let us show that, under some conditions imposed on a weight w, the second and third inclusions in (16.5) are proper.

Proposition 16.1 *For every $1 \leq p < \infty$ the following assertions hold:*

(i) there exists a function $f \in \widetilde{K}_{w,p} \setminus M_{w,p}$;
(ii) if the function $w(t)\, t^{-1/p}$ is decreasing on $(0, 1]$, then there is a function $g \in M_{w,p} \setminus \widetilde{M(w, p)}$.

Proof

(*i*) By (16.6), we can find a sequence $\{t_k\}_{k=1}^\infty \subset (0, 1]$ such that $1/2 \geq t_1 > t_2 > \cdots > 0$, $\lim_{k\to\infty} t_k = 0$, $w(t_1)t_1^{-1/p} < w(t_2)t_2^{-1/p} < \ldots$, and

$\lim_{k\to\infty} w(t_k)t_k^{-1/p} = \infty$. Let us denote $w_1(t) := w(t)\, t^{-1/p}$ and

$$f_1(s) := \sum_{k=1}^{\infty} \left(w_1(t_k)^{-p/2} - w_1(t_{k+1})^{-p/2} \right)^{1/p} (t_k - t_{k+1})^{-1/p} \chi_{(t_{k+1},t_k]}(s).$$

Clearly, $f_1 \geq 0$ and $\operatorname{supp} f_1 \subset [0, 1/2]$. Moreover, for every $k \in \mathbb{N}$ we have

$$\begin{aligned}\int_0^{t_k} f_1(s)^p\, ds &= \sum_{i=k}^{\infty} \int_{t_{i+1}}^{t_i} f_1(s)^p\, ds \\ &= \sum_{i=k}^{\infty} \frac{w_1(t_i)^{-p/2} - w_1(t_{i+1})^{-p/2}}{t_i - t_{i+1}} (t_i - t_{i+1}) = w_1(t_k)^{-p/2}. \quad (16.7)\end{aligned}$$

Hence, in particular, it follows that $f_1 \in L_p$. Setting $f(t) := f_1(t + 1/2)$, $0 \leq t \leq 1$, we have $\|f\|_p = \|f_1\|_p$, and so $f \in L_p$. Observe that $\operatorname{supp} f \subset [1/2, 1]$, and therefore $f \in \widetilde{K}_{w,p}$. Indeed,

$$\begin{aligned}\|f\|_{\widetilde{K}_{w,p}} &= \sup_{0<x\leq 1} w(x) \left(\frac{1}{x} \int_0^x f(t)^p\, dt \right)^{1/p} = \sup_{1/2\leq x\leq 1} \frac{w(x)}{x^{1/p}} \left(\int_{1/2}^x f(t)^p\, dt \right)^{1/p} \\ &\asymp \sup_{1/2\leq x\leq 1} \left(\int_{1/2}^x f(t)^p\, dt \right)^{1/p} = \|f\|_p < \infty.\end{aligned}$$

On the other hand, if $I_k := [1/2, t_k + 1/2]$, $k = 1, 2, \ldots$, then by (16.7),

$$w(m(I_k)) \left(\frac{1}{m(I_k)} \int_{I_k} f(t)^p\, dt \right)^{1/p} = w_1(t_k) \left(\int_0^{t_k} f_1(s)^p\, ds \right)^{1/p} = w_1(t_k)^{1/2}.$$

Since $\lim_{k\to\infty} w_1(t_k) = \infty$, this implies that $f \notin M_{w,p}$.

(ii) By (i), there is a function $f \in \widetilde{K}_{w,p} \setminus M_{w,p}$. Then, appealing to embeddings (16.5), we obtain that $f \in L_p$ and $f \notin \widetilde{M(w,p)}$. At the same time, as it is shown in [2], the assumptions of the proposition ensure that there exist a function $g \in M_{w,p}$ and constants $c_0 > 0$, $\lambda_0 > 0$ such that

$$m\{t \in [0,1] : |g(t)| > \lambda\} \geq c_0\, m\{t \in [0,1] : |f(t)| > \lambda\}, \quad \lambda \geq \lambda_0.$$

Since $f \notin \widetilde{M(w,p)}$, from the last inequality it follows immediately that $g \notin \widetilde{M(w,p)}$ as well. As a result, we get $g \in M_{w,p} \setminus \widetilde{M(w,p)}$. □

Remark 16.1 The above proof shows that, under the assumptions of Proposition 16.1(ii), the Morrey space $M_{w,p}$ is not symmetric for each $1 \leq p < \infty$.

16.2 Rademacher Sums in Morrey Spaces

We start with studying the behaviour of Rademacher sums in the Morrey spaces $M_{w,p}$. As usual, $\Delta_n^k = ((k-1)\,2^{-n}, k\,2^{-n})$, $n = 0, 1, 2, \dots$ and $k = 1, 2, \dots, 2^n$.

Theorem 16.1 *Let $1 \le p < \infty$. With a constant independent of a weight w we have*

$$\Big\| \sum_{k=1}^{\infty} a_k\, r_k \Big\|_{M_{w,p}} \asymp \|(a_k)_{k=1}^{\infty}\|_{\ell_2} + \sup_{l\in\mathbb{N}} \Big(w(2^{-l}) \sum_{k=1}^{l} |a_k| \Big).$$

Proof Let Δ be an arbitrary dyadic interval of rank l, that is, $\Delta = \Delta_l^i$ with some $i = 1, 2, \dots, 2^l$. Then, for every $f = \sum_{k=1}^{\infty} a_k\, r_k$ we have

$$\Big(\int_\Delta |f(t)|^p\, dt\Big)^{1/p} = \Big(\int_\Delta \Big|\sum_{k=1}^{l} a_k\varepsilon_k + \sum_{k=l+1}^{\infty} a_k r_k(t)\Big|^p dt\Big)^{1/p},$$

where $\varepsilon_k = \operatorname{sign} r_k{}_{|\Delta}$, $k = 1, 2, \dots, l$. Since the functions

$$\sum_{k=1}^{l} a_k\varepsilon_k + \sum_{k=l+1}^{\infty} a_k r_k(t) \quad\text{and}\quad \sum_{k=1}^{l} a_k\varepsilon_k - \sum_{k=l+1}^{\infty} a_k r_k(t)$$

are identically distributed when they are restricted to the interval Δ, it follows that

$$\Big(\int_\Delta |f(t)|^p\, dt\Big)^{1/p} = \frac{1}{2}\Big(\int_\Delta \Big|\sum_{k=1}^{l} a_k\varepsilon_k + \sum_{k=l+1}^{\infty} a_k r_k(t)\Big|^p dt\Big)^{1/p}$$
$$+ \frac{1}{2}\Big(\int_\Delta \Big|\sum_{k=1}^{l} a_k\varepsilon_k - \sum_{k=l+1}^{\infty} a_k r_k(t)\Big|^p dt\Big)^{1/p}.$$

Therefore, by Minkowski's inequality, we infer

$$\Big(\int_\Delta |f(t)|^p\, dt\Big)^{1/p} \ge \Big(\int_\Delta \Big|\sum_{k=1}^{l} a_k\varepsilon_k\Big|^p dt\Big)^{1/p} 2^{-l/p} \Big|\sum_{k=1}^{l} a_k\varepsilon_k\Big|.$$

Clearly, one may find $i_0 \in \{1, 2, \dots, 2^l\}$ such that $r_k{}_{|\Delta_l^{i_0}} = \operatorname{sign} a_k$ for all $k = 1, 2, \dots, l$. Consequently, rewriting the last inequality for the interval $\Delta_l^{i_0}$, we have

$$\Big(\frac{1}{m(\Delta_l^{i_0})}\int_{\Delta_l^{i_0}} |f(t)|^p\, dt\Big)^{1/p} \ge \sum_{k=1}^{l} |a_k|.$$

Since $l \in \mathbb{N}$ is arbitrary, then combining this together with (16.1), we get

$$\|f\|_{M_{w,p}} \geq \sup_{l\in\mathbb{N}} w(2^{-l}) \sum_{k=1}^{l} |a_k|. \tag{16.8}$$

On the other hand, from embeddings (16.5) and Khintchine's inequality (see Theorem 5.4) it follows that

$$\|f\|_{M_{w,p}} \geq \|f\|_p \geq \frac{1}{\sqrt{2}} \|(a_k)_{k=1}^{\infty}\|_{\ell_2}.$$

As a result, combining the last estimates, we obtain

$$\|f\|_{M_{w,p}} \geq \frac{1}{2\sqrt{2}} \Big(\|(a_k)_{k=1}^{\infty}\|_{\ell_2} + \sup_{l\in\mathbb{N}} w(2^{-l}) \sum_{k=1}^{l} |a_k| \Big).$$

Let us prove the reverse estimate. For a given interval $I \subset [0, 1]$ we can find two adjacent dyadic intervals Δ_1 and Δ_2 of the same length such that

$$I \subset \Delta_1 \cup \Delta_2 \text{ and } \frac{1}{2} m(\Delta_1) \leq m(I) \leq 2\, m(\Delta_1). \tag{16.9}$$

If $m(\Delta_1) = m(\Delta_2) = 2^{-l}$, then by the triangle inequality for the L_p-norm and Khintchine's inequality (see Theorem 1.4), we have

$$\begin{aligned}
\Big(\int_{\Delta_1} |f(t)|^p\, dt\Big)^{1/p} &= \Big(\int_{\Delta_1} \Big|\sum_{k=1}^{l} a_k \varepsilon_k + \sum_{k=l+1}^{\infty} a_k r_k(t)\Big|^p dt\Big)^{1/p} \\
&\leq \Big(\int_{\Delta_1} \Big|\sum_{k=1}^{l} a_k \varepsilon_k\Big|^p dt\Big)^{1/p} + \Big(\int_{\Delta_1} \Big|\sum_{k=l+1}^{\infty} a_k r_k(t)\Big|^p dt\Big)^{1/p} \\
&\leq 2^{-l/p} \sum_{k=1}^{l} |a_k| + 2^{-l/p} \Big(\int_0^1 \Big|\sum_{k=l+1}^{\infty} a_k r_{k-l}(t)\Big|^p dt\Big)^{1/p} \\
&\leq 2^{-l/p} \Big(\sum_{k=1}^{l} |a_k| + \sqrt{p}\, \|(a_k)_{k=1}^{\infty}\|_{\ell_2}\Big).
\end{aligned}$$

A similar estimate holds also for the integral $\left(\int_{\Delta_2}|f(t)|^p\,dt\right)^{1/p}$. Therefore, (16.9) implies that

$$\left(\frac{1}{m(I)}\int_I|f(t)|^p\,dt\right)^{1/p}\le 2^{1/p}\left(\frac{1}{m(\Delta_1)}\int_{\Delta_1}|f(t)|^p\,dt+\frac{1}{m(\Delta_2)}\int_{\Delta_2}|f(t)|^p\,dt\right)^{1/p}$$
$$\le 4^{1/p}\sqrt{p}\left(\sum_{k=1}^{l}|a_k|+\|(a_k)_{k=1}^\infty\|_{\ell_2}\right),$$

and so, in view of the quasiconcavity of w,

$$w(m(I))\left(\frac{1}{m(I)}\int_I|f(t)|^p\,dt\right)^{1/p}\le w(2\cdot 2^{-l})\,4^{1/p}\sqrt{p}\left(\sum_{k=1}^{l}|a_k|+\|(a_k)_{k=1}^\infty\|_{\ell_2}\right)$$
$$\le 2\cdot 4^{1/p}\sqrt{p}\,w(2^{-l})\left(\sum_{k=1}^{l}|a_k|+\|(a_k)_{k=1}^\infty\|_{\ell_2}\right).$$

Thus, from definition of the $M_{w,p}$-norm it follows that

$$\|f\|_{M_{w,p}}\le 2\cdot 4^{1/p}\sqrt{p}\left(\sup_{l\in\mathbb{N}}w(2^{-l})\sum_{k=1}^{l}|a_k|+\|(a_k)_{k=1}^\infty\|_{\ell_2}\right).$$

□

The following result should be compared with Theorem 15.3.

Corollary 16.1 *Let $1\le p<\infty$. The Rademacher sequence is equivalent in $M_{w,p}$ to the unit vector basis in ℓ_2 if and only if*

$$\sup_{0<t\le 1} w(t)\log_2^{1/2}(2/t)<\infty. \tag{16.10}$$

Proof Assuming first that (16.10) is fulfilled, we have

$$\|\log_2^{1/2}(2/\cdot)\|_{\widetilde{M(w,p)}}\asymp\sup_{0<x\le 1}w(x)\left(\frac{1}{x}\int_0^x\log_2^{p/2}(2/t)\,dt\right)^{1/p}$$
$$\asymp\sup_{0<x\le 1}w(x)\log_2^{1/2}(2/x)<\infty.$$

Therefore, since the function $\log_2^{1/2}(2/t)$ has the greatest (up to a constant) decreasing rearrangement in the Orlicz space L_{N_2} with $N_2(u) = \exp(u^2) - 1$ (see Proposition 2.4), the embedding $L_{N_2} \subset \widetilde{M(w,p)}$ holds. Hence, by Theorem 2.2 (see also Remark 2.1), the Rademacher functions span ℓ_2 in the space $\widetilde{M(w,p)}$. Observe that, according to Khintchine's L_p-inequality (see Theorem 1.4), these functions span the same subspace in L_p. Then, from embeddings (16.5) it follows that the sequence $\{r_n\}$ is equivalent in $M_{w,p}$ to the unit vector basis in ℓ_2.

Conversely, suppose that condition (16.10) does not hold. Then, by the quasi-concavity of w, there exists a sequence of positive integers $m_k \to \infty$ such that

$$w(2^{-m_k})\, m_k^{1/2} \to \infty \text{ as } k \to \infty. \tag{16.11}$$

Consider the sums $S_k(t) = m_k^{-1/2} \sum_{i=1}^{m_k} r_i(t)$, $k = 1, 2, \ldots$. Then, from (16.11) and Theorem 16.1 it follows that $\|S_k\|_{M_{w,p}} \to \infty$ as $k \to \infty$, though the corresponding sequences of coefficients are normalized in ℓ_2. □

Remark 16.2 According to the proof of Corollary 16.1, Theorem 15.3, and embeddings (16.5), relation (16.10) is a necessary and sufficient condition, under which the sequence $\{r_n\}$ is equivalent to the unit vector basis in ℓ_2 in each of the spaces $\widetilde{M(w,p)}$, $M_{w,p}$, and $\widetilde{K}_{w,p}$, $1 \le p < \infty$.

It is instructive also to compare the behaviour of Rademacher sums in the spaces $\widetilde{M(w,1)}$, $M_{w,1}$, and $\widetilde{K}_{w,1}$ in the case when $w(t) = \ln^{-1/q}(2/t)$, where $q > 2$. Then (16.10) does not hold, and we have respectively

$$\Big\| \sum_{k=1}^{\infty} a_k r_k \Big\|_{\widetilde{M(w,1)}} \asymp \sup_{m \in \mathbb{N}} m^{-1/q} \sum_{k=1}^{m} a_k^*,$$

where (a_k^*) is the decreasing permutation of the sequence $(|a_k|)_{k=1}^{\infty}$ (Example 3.2),

$$\Big\| \sum_{k=1}^{\infty} a_k r_k \Big\|_{M_{w,1}} \asymp \|(a_k)_{k=1}^{\infty}\|_{\ell_2} + \sup_{m \in \mathbb{N}} m^{-1/q} \sum_{k=1}^{m} |a_k|$$

(Theorem 16.1), and

$$\Big\| \sum_{k=1}^{\infty} a_k r_k \Big\|_{\widetilde{K}_{w,1}} \asymp \|(a_k)_{k=1}^{\infty}\|_{l_2} + \sup_{m \in \mathbb{N}} m^{-1/q} \Big| \sum_{k=1}^{m} a_k \Big|$$

(Theorem 15.2).

Next, we proceed with a detailed investigation of the complementability problem of the closed linear span of the Rademacher functions in the Morrey space $M_{w,p}$. Since the results (and their proofs) turn out to be different for $p > 1$ and $p = 1$, we consider these cases separately.

16.3 Complementability of Rademacher Subspaces in the Spaces $M_{w,p}$, $p > 1$

Theorem 16.2 *Let $1 < p < \infty$. The subspace $\mathcal{R}M_{w,p} := [r_n]_{M_{w,p}}$ is complemented in the Morrey space $M_{w,p}$ if and only if condition (16.10) holds.*

To prove this result, we need the following auxiliary assertion.

Proposition 16.2 *If condition (16.10) does not hold, then $\mathcal{R}M_{w,p}$ contains a complemented subspace isomorphic to c_0.*

Proof Since the function w is quasiconcave, by the assumption, we have

$$\limsup_{n\to\infty} w(2^{-n})\sqrt{n} = \infty. \tag{16.12}$$

We shall select an increasing sequence of positive integers as follows. For n_1 we take the least positive integer satisfying the inequality $w(2^{-n_1})\sqrt{n_1} \geq 2$. Then, as it is easy to see, $w(2^{-n_1})\sqrt{n_1} < 2^2$. By induction, assume that the numbers $n_1 < n_2 < \ldots < n_{k-1}$ are already chosen. In view of (16.12), there exists the least positive integer n_k such that

$$w(2^{-n_k})\sqrt{n_k - n_{k-1}} \geq 2^k. \tag{16.13}$$

Then, obviously,

$$w(2^{-n_k})\sqrt{n_k - n_{k-1}} < 2^{k+1}. \tag{16.14}$$

Thus, we obtain a sequence $0 = n_0 < n_1 < \ldots$, satisfying inequalities (16.13) and (16.14) for all $k \in \mathbb{N}$. Consider the block basic sequence $\{v_k\}_{k=1}^{\infty}$ of the Rademacher system defined by

$$v_k := \sum_{i=n_{k-1}+1}^{n_k} a_i\, r_i, \quad \text{where } a_i = \frac{1}{(n_k - n_{k-1})\, w(2^{-n_k})} \quad \text{for } n_{k-1} < i \leq n_k.$$

Let us estimate the norms $\|v_k\|_{M_{w,p}}$, $k = 1, 2, \ldots$.

Recall that, by Theorem 16.1,

$$\left\| \sum_{k=1}^{\infty} b_k\, r_k \right\|_{M_{w,p}} \asymp \|(b_k)\|_{\ell_2} + \|(b_k)\|_w, \tag{16.15}$$

where

$$\|(b_k)\|_w := \sup_{q\in\mathbb{N}} \left(w(2^{-q}) \sum_{k=1}^{q} |b_k| \right).$$

First, from (16.13) it follows that

$$\Big(\sum_{i=n_{k-1}+1}^{n_k} a_i^2\Big)^{1/2} = \frac{1}{\sqrt{n_k - n_{k-1}}\, w(2^{-n_k})} \le 2^{-k},\ k = 1, 2, \ldots \tag{16.16}$$

Moreover, taking into account (16.13), (16.14) and the choice of n_k, for every $k \in \mathbb{N}$ and all $n_{k-1} < i \le n_k$ we have

$$w(2^{-i}) \sum_{j=n_{k-1}+1}^{i} a_j = \frac{w(2^{-i})(i - n_{k-1})}{(n_k - n_{k-1})w(2^{-n_k})} \le \frac{2^{k+1}\sqrt{i - n_{k-1}}}{2^k\sqrt{n_k - n_{k-1}}} \le 2.$$

As a result, from the last estimates, inequality (16.8), and Theorem 16.1 it follows

$$1 = w(2^{-n_k}) \sum_{i=n_{k-1}+1}^{n_k} a_i \le \|v_k\|_{M_{w,p}} \le C \tag{16.17}$$

for some constant $C > 0$ and any $k \in \mathbb{N}$.

Further, let us select a subsequence $\{m_i\} \subset \{n_k\}$ such that

$$w(2^{-m_{i+1}}) \le \frac{1}{2} w(2^{-m_i}),\ \ i = 1, 2, \ldots \tag{16.18}$$

(see (16.6)) and denote by $\{u_i\}_{i=1}^\infty$ the corresponding subsequence of the block basic sequence $\{v_k\}_{k=1}^\infty$. Then, the functions u_i can be represented in the form:

$$u_i = \sum_{k=l_i}^{m_i} a_k r_k, \ \text{ where } l_i = n_{j_i - 1} + 1,\ m_i = n_{j_i},\ j_1 < j_2 < \ldots.$$

Show that the sequence $\{u_i\}_{i=1}^\infty$ is equivalent in $M_{w,p}$ to the unit vector basis in c_0.

Let $f = \sum_{i=1}^\infty \beta_i u_i$, $\beta_i \in \mathbb{R}$. Then, we have

$$f = \sum_{i=1}^\infty \beta_i \sum_{k=l_i}^{m_i} a_k r_k = \sum_{k=1}^\infty b_k r_k,$$

where $b_k = \beta_i a_k$ if $l_i \le k \le m_i$ for some $i = 1, 2, \ldots$, and $b_k = 0$, otherwise.

To estimate $\|(b_k)\|_w$ (see (16.15)), we assume first that $m_s \le q < l_{s+1}$ for some $s \in \mathbb{N}$. Then,

$$\sum_{k=1}^q |b_k| = \sum_{i=1}^s |\beta_i| \sum_{k=l_i}^{m_i} a_k = \sum_{i=1}^s |\beta_i| \frac{1}{w(2^{-m_i})} \le \|(\beta_i)\|_{c_0} \sum_{i=1}^s \frac{1}{w(2^{-m_i})}.$$

Hence, taking into account that w increases, from inequality (16.18) we get

$$w(2^{-q}) \sum_{k=1}^{q} |b_k| \le \|(\beta_i)\|_{c_0} \sum_{i=1}^{s} \frac{w(2^{-m_s})}{w(2^{-m_i})} \le \|(\beta_i)\|_{c_0} \sum_{i=0}^{\infty} 2^{-i} = 2\,\|(\beta_i)\|_{c_0}.$$

In the case when $l_s \le q < m_s$ for some $s \in \mathbb{N}$, similarly,

$$\begin{aligned}\sum_{k=1}^{q} |b_k| &\le \left(\sum_{i=1}^{s-1} \frac{1}{w(2^{-m_i})} + \sum_{k=l_s}^{q} a_k \right) \|(\beta_i)\|_{c_0} \\ &= \left(\sum_{i=1}^{s-1} \frac{1}{w(2^{-m_i})} + \frac{q - l_s + 1}{(m_s - l_s + 1)\, w(2^{-m_s})} \right) \|(\beta_i)\|_{c_0}.\end{aligned}$$

Since $l_s = n_{j_s - 1} + 1$ and $m_s = n_{j_s}$ for some $j_s \in \mathbb{N}$, in view of (16.13), (16.18), and the choice of n_{j_s}, we obtain

$$\begin{aligned}w(2^{-q}) \sum_{k=1}^{q} |b_k| &\le \left(\sum_{i=1}^{s-1} \frac{w(2^{-m_{s-1}})}{w(2^{-m_i})} + \frac{w(2^{-q})(q - l_s + 1)}{(m_s - l_s + 1)\, w(2^{-m_s})} \right) \|(\beta_i)\|_{c_0} \\ &\le \left(\sum_{i=0}^{\infty} 2^{-i} + \frac{2^{j_s+1} \sqrt{q - l_s + 1}}{2^{j_s} \sqrt{m_s - l_s + 1}} \right) \|(\beta_i)\|_{c_0} \le 4\,\|(\beta_i)\|_{c_0}.\end{aligned}$$

Summarizing all, we see that this estimate holds for any $q \in \mathbb{N}$ and hence $\|(b_k)\|_w \le 4\,\|(\beta_i)\|_{c_0}$. Furthermore, from (16.16) it follows that

$$\|(b_k)\|_{\ell_2}^2 \le \sum_{i=1}^{\infty} \beta_i^2 \sum_{k=l_i}^{m_i} a_k^2 \le \|(\beta_i)\|_{c_0}^2.$$

Thus, by (16.15),

$$\|f\|_{M_{w,p}} \le C \left(\|(b_k)\|_{\ell_2} + \|(b_k)\|_w \right) \le 5\,C\,\|(\beta_i)\|_{c_0}.$$

In the opposite direction, since $\{u_i\}$ is an unconditional basic sequence in $M_{w,p}$ and by (16.17) $\|u_i\|_{M_{w,p}} \ge 1$, $i \in \mathbb{N}$, we get

$$\|f\|_{M_{w,p}} \ge c \sup_{i \in \mathbb{N}} |\beta_i|\, \|u_i\|_{M_{w,p}} \ge c\,\|(\beta_i)\|_{c_0}$$

with some $c > 0$. So, the subspace $[u_i]_{M_{w,p}}$ is isomorphic to c_0. Moreover, because $\mathcal{R}M_{w,p}$ is separable, the Sobczyk theorem (see, for example, [1, Corollary 2.5.9]) implies that $[u_i]_{M_{w,p}}$ is a complemented subspace in $\mathcal{R}M_{w,p}$. This completes the proof of the proposition. □

Proof of Theorem 16.2 Let us assume first that condition (16.10) holds. Then, by Corollary 16.1, the sequence $\{r_n\}$ is equivalent in $M_{w,p}$ to the unit vector basis in ℓ_2. As we know (see the proof of Theorem 2.5), the orthogonal projection P, generated by the Rademacher system, is bounded in L_p if $1 < p < \infty$. Therefore, since $M_{w,p} \overset{1}{\subset} L_p$, then from Khintchine's L_p-inequality it follows that

$$\|Pf\|_{M_{w,p}} \asymp \|Pf\|_p \le \|P\|_{L_p \to L_p} \|f\|_p \le \|P\|_{L_p \to L_p} \|f\|_{M_{w,p}}.$$

Thus, $P : M_{w,p} \to M_{w,p}$ is bounded, and so the subspace $\mathcal{R}M_{w,p}$ is complemented in $M_{w,p}$.

Conversely, we argue in a similar way as in the proof of Theorem 14.4. Suppose that the subspace $\mathcal{R}M_{w,p}$ is complemented in $M_{w,p}$, and let $P_1 : M_{w,p} \to M_{w,p}$ be a bounded linear projection whose image is $\mathcal{R}M_{w,p}$. If we assume that condition (16.10) is not fulfilled, then by Proposition 16.2, there is a subspace E that is complemented in $\mathcal{R}M_{w,p}$ and is isomorphic to c_0. Denote by P_2 a bounded linear projection in $\mathcal{R}M_{w,p}$ such that $P_2(\mathcal{R}M_{w,p}) = E$. Then $P := P_2 \cdot P_1$ is a linear projection bounded in $M_{w,p}$, whose image coincides with E. Hence, $M_{w,p}$ contains a complemented subspace E isomorphic to c_0.

On the other hand, $M_{w,p}$ is a dual space; namely, $M_{w,p} = (H^{q,u})^*$, where $H^{q,u}$, $1/p + 1/q = 1$, is the so-called "block space" (see, for example, [289, Proposition 5] or [75]). Therefore, the existence of a complemented subspace in $M_{w,p}$ that is isomorphic to c_0 contradicts the well-known Bessaga–Pełczyński theorem [73, Corollary 4] (see also Theorem 4 and its proof in [72]). This finishes the proof. □

16.4 Complementability of Rademacher Subspaces in the Spaces $M_{w,1}$

In the case when $p = 1$ the answer to the question whether the Rademacher subspace is complemented or not in the Morrey space does not depend on properties of a weighted function w.

Theorem 16.3 *For every quasiconcave function w the Rademacher subspace $\mathcal{R}M_{w,1} := [r_n]_{M_{w,1}}$ is not complemented in the Morrey space $M_{w,1}$.*

Here, we start also with proving some auxiliary results.

Denote by $M_{w,p}^d$, $1 \le p < \infty$, a dyadic version of the space $M_{w,p}$. Specifically, $M_{w,p}^d$ consists of all measurable functions $f : [0,1] \to \mathbb{R}$ such that

$$\|f\|_{M_{w,p}^d} = \sup_{\Delta \in \mathfrak{D}} w(m(\Delta)) \left(\frac{1}{m(\Delta)} \int_\Delta |f(t)|^p \, dt \right)^{1/p} < \infty,$$

where $\mathfrak{D}$ is the set of all dyadic intervals from $[0, 1]$. It turns out that, in contrast to the BMO-spaces, this dyadic norm is equivalent to the $M_{w,p}$-norm itself.

Lemma 16.1 *For every* $1 \le p < \infty$ *we have* $M_{w,p} = M_{w,p}^d$ *and*

$$\|f\|_{M_{w,p}^d} \le \|f\|_{M_{w,p}} \le 4\,\|f\|_{M_{w,p}^d}. \tag{16.19}$$

Proof The left-hand side inequality in (16.19) is obvious. To prove the right-hand side one, for any interval $I \subset [0, 1]$ we find, precisely as in the proof of Theorem 16.1, two adjacent dyadic intervals Δ_1 and Δ_2 of the same length such that $I \subset \Delta_1 \cup \Delta_2$ and $\frac{1}{2}m(\Delta_1) \le m(I) \le 2m(\Delta_1)$. Then, by the quasiconcavity of w, we have

$$\begin{aligned}
w(m(I))\left(\frac{1}{m(I)}\int_I |f(t)|^p\,dt\right)^{1/p} &= \frac{w(m(I))}{m(I)}\left(m(I)^{p-1}\int_I |f(t)|^p\,dt\right)^{1/p} \\
&\le \frac{2w(\frac{1}{2}m(\Delta_1))}{m(\Delta_1)}\left[2^{p-1}m(\Delta_1)^{p-1}\left(\int_{\Delta_1}|f(t)|^p\,dt + \int_{\Delta_2}|f(t)|^p\,dt\right)\right]^{1/p} \\
&\le 2^{2-1/p}\,w(m(\Delta_1))\left(\frac{1}{m(\Delta_1)}\int_{\Delta_1}|f(t)|^p\,dt + \frac{1}{m(\Delta_2)}\int_{\Delta_2}|f(t)|^p\,dt\right)^{1/p} \\
&\le 4\sup_{\Delta\in\mathfrak{D}} w(m(\Delta))\left(\frac{1}{m(\Delta)}\int_\Delta |f(t)|^p\,dt\right)^{1/p} = 4\,\|f\|_{M_{w,p}^d}.
\end{aligned}$$

Taking the supremum over all intervals $I \subset [0, 1]$, we obtain the right-hand side inequality in (16.19) and thus complete the proof. □

As above, let P be the orthogonal projection generated by the Rademacher system, i.e.,

$$Pf(t) := \sum_{k=1}^{\infty}\int_0^1 f(s)r_k(s)\,ds \cdot r_k(t).$$

Observe that, by Lemma 16.1, we have $[r_n]_{M_{w,p}^d} = \mathcal{R}M_{w,p}$.

Proposition 16.3 *Let* $1 \le p < \infty$. *If* $\mathcal{R}M_{w,p}$ *is a complemented subspace in* $M_{w,p}^d$, *then the projection* P *is bounded in* $M_{w,p}^d$.

Proof Let $t = \sum_{i=1}^{\infty}\alpha_i 2^{-i}$ and $u = \sum_{i=1}^{\infty}\beta_i 2^{-i}$ $(\alpha_i, \beta_i = 0, 1)$ be the binary expansions of numbers $t, u \in [0, 1]$. As in the proof of a similar result for s.s.'s (see Theorem 2.5), we define the operation:

$$t \oplus u = \sum_{i=1}^{\infty} 2^{-i}[(\alpha_i + \beta_i) \bmod 2],$$

which turns the interval $[0, 1]$ to a compact Abelian group. For each $u \in [0, 1]$ the transformation $w_u(s) = s \oplus u$ preserves the Lebesgue measure on $[0, 1]$, i.e., for any measurable $E \subset [0, 1]$ its inverse image $w_u^{-1}(E)$ is measurable and $m(w_u^{-1}(E)) = m(E)$. Moreover, w_u maps every dyadic interval onto some dyadic interval as well. Hence, the operators $T_u f := f \circ w_u$, $0 \le u \le 1$, act isometrically in $M_{w,p}^d$. Furthermore, from definition of the Rademacher functions it follows that the subspace $\mathcal{R}M_{w,p}$ is invariant with respect to the action of these operators. Thus, by the Rudin theorem (see e.g. [255, Theorem 1.5.18]), there exists a bounded linear projection Q in $M_{w,p}^d$, commuting with all operators T_u, $0 \le u \le 1$, such that the image of Q coincides with $\mathcal{R}M_{w,p}$. Further, completely in the same way as in the proof of Theorem 2.5, we can show that $Q = P$. So, the proposition is proved. □

Lemma 16.2 *Suppose that the Rademacher sequence is equivalent in a Banach function lattice X on $[0, 1]$ to the unit vector basis in ℓ_2, i.e., for some $C > 0$ and all $a = (a_k)_{k=1}^{\infty} \in \ell_2$*

$$C^{-1} \|a\|_{\ell_2} \le \Big\| \sum_{k=1}^{\infty} a_k r_k \Big\|_X \le C \|a\|_{\ell_2}. \tag{16.20}$$

Moreover, assume that $\{r_k\} \subset X'$, where X' is the Köthe dual space for X.

Then, the orthogonal projection P is bounded in X if and only if there exists a constant $C_1 > 0$ such that for every sequence $a = (a_k)_{k=1}^{\infty} \in \ell_2$ we have

$$\Big\| \sum_{k=1}^{\infty} a_k r_k \Big\|_{X'} \le C_1 \|a\|_{\ell_2}. \tag{16.21}$$

Proof First, let us suppose that condition (16.21) is fulfilled. For arbitrary function $f \in X$ we denote by $c_k(f)$ its Fourier-Rademacher coefficients defined by

$$c_k(f) := \int_0^1 f(s) r_k(s)\, ds, \ k = 1, 2, \dots$$

Then, from (16.21) for every $n \in \mathbb{N}$ it follows that

$$\sum_{k=1}^{n} c_k(f)^2 = \int_0^1 f(s) \sum_{k=1}^{n} c_k(f) r_k(s)\, ds \le \|f\|_X \Big\| \sum_{k=1}^{n} c_k(f) r_k \Big\|_{X'}$$
$$\le C_1 \|f\|_X \Big(\sum_{k=1}^{n} c_k(f)^2 \Big)^{1/2},$$

and therefore, taking into account (16.20), we obtain

$$\|Pf\|_X \le C\Big(\sum_{k=1}^{\infty} c_k(f)^2\Big)^{1/2} \le C\cdot C_1\|f\|_X.$$

Thus, the projection P is bounded in X.

Conversely, let P be bounded in X. Then, from (16.20) it follows that

$$\begin{aligned}\int_0^1 f(t)\sum_{k=1}^{n} a_k r_k(t)\,dt &= \sum_{k=1}^{n} a_k\cdot c_k(f) \le \|a\|_{\ell_2}\Big(\sum_{k=1}^{n} c_k(f)^2\Big)^{1/2}\\ &\le C\,\|a\|_{\ell_2}\,\|Pf\|_X \le C\,\|P\|_{X\to X}\,\|a\|_{\ell_2}\,\|f\|_X\end{aligned}$$

for any $n\in\mathbb{N}$, $a=(a_k)_{k=1}^{\infty}\in\ell_2$, and $f\in X$. Hence,

$$\Big\|\sum_{k=1}^{\infty} a_k\,r_k\Big\|_{X'} \le C\,\|P\|_{X\to X}\,\|a\|_{\ell_2},$$

and inequality (16.21) is proved. □

Proof of Theorem 16.3 Suppose, towards a contradiction, that $\mathcal{R}M_{w,1}$ is complemented in $M_{w,1}$.

Consider first the simpler case when condition (16.10) does not hold. If Q is a bounded linear projection in $M_{w,1}$ whose image is $\mathcal{R}M_{w,1}$, then by Theorem 16.1 and embeddings (16.2), for every $1<p<\infty$ and all $f\in M_{w,p}$, we have

$$\|Qf\|_{M_{w,p}} \asymp \|Qf\|_{M_{w,1}} \le \|Q\|\,\|f\|_{M_{w,1}} \le \|Q\|\,\|f\|_{M_{w,p}}.$$

Hence, Q is a bounded linear projection in $M_{w,p}$ as well, which clearly contradicts Theorem 16.2.

Let us pass to the more complicated case when condition (16.10) is satisfied by w. According to Corollary 16.1, the sequence $\{r_n\}$ is equivalent in $M_{w,1}$ to the unit vector basis in ℓ_2. Then, by Lemma 16.1, the Rademacher functions span ℓ_2 in the space $M_{w,1}^d$ as well. Moreover, by the same reason, the assumption on complementability of the subspace $\mathcal{R}M_{w,1}$ in $M_{w,1}$ implies that $\mathcal{R}M_{w,1}$ is complemented in $M_{w,1}^d$ as well. Thus, taking into account Proposition 16.2 and Lemma 16.2, to get a contradiction, it suffices to show that

$$\limsup_{n\to\infty}\frac{1}{\sqrt{n}}\Big\|\sum_{k=1}^{n} r_k\Big\|_{(M_{w,1}^d)'} = \infty. \tag{16.22}$$

For every $l \in \mathbb{N}$ such that $\sqrt{l/2} \in \mathbb{N}$ we consider the set

$$E_l := \{t \in [0,1] : 0 \le \sum_{k=1}^{2l} r_k(t) \le \sqrt{l/2}\}.$$

Clearly, $E_l = \bigcup_{k \in S_l} \Delta_{2l}^k$, where S_l is a subset of $\{1, 2, \ldots, 2^{2l}\}$. Also, it is easy to see that $m(E_l) \to 0$ as $l \to \infty$. Denoting

$$f_l := \frac{\chi_{E_l}}{w(m(E_l))}, \quad l \in \mathbb{N},$$

we shall prove that

$$\|f_l\|_{M_{w,1}^d} \le 1 \quad \text{for all} \quad l \in \mathbb{N}. \tag{16.23}$$

Indeed, let Δ be a dyadic interval from $[0, 1]$. Without loss of generality, we may assume that $\Delta \cap E_l \ne \varnothing$. Then, by using the quasiconcavity of w, we have

$$\frac{w(m(\Delta))}{m(\Delta)} \int_\Delta |f_l(t)|\, dt = \frac{w(m(\Delta))}{m(\Delta)} \cdot \frac{m(\Delta \cap E_l)}{w(m(E_l))} \le \frac{w(m(\Delta))}{m(\Delta)} \cdot \frac{m(\Delta \cap E_l)}{w(m(\Delta \cap E_l))} \le 1,$$

and so (16.23) is proved.

From (16.23) it follows that

$$\begin{aligned} \Big\|\sum_{k=1}^{2l} r_k\Big\|_{(M_{w,1}^d)'} &\ge \int_0^1 \Big|\sum_{k=1}^{2l} r_k(t)\Big| \cdot f_l(t)\, dt = \frac{1}{w(m(E_l))} \int_{E_l} \Big|\sum_{k=1}^{2l} r_k(t)\Big|\, dt \\ &= \frac{1}{w(m(E_l))} 2^{-2l} \sum_{i \in S_l} \Big|\sum_{k=1}^{2l} \varepsilon_k^i\Big|, \end{aligned}$$

where $\varepsilon_k^i = \operatorname{sign} r_k|_{\Delta_{2l}^i}$, $k = 1, 2, \ldots, 2l$, $i \in S_l$. Putting $\sigma_l := \sum_{i \in S_l} |\sum_{k=1}^{2l} \varepsilon_k^i|$, by definition of the set E_l, we obtain that

$$\sigma_l = 2 \cdot \sum_{l - \sqrt{l/2} \le k \le l} C_{2l}^k (l - k) = 2 \cdot \sum_{k=1}^{\sqrt{l/2}} C_{2l}^{l-k} \cdot k \tag{16.24}$$

(here, as usual, $C_n^i := \frac{n!}{i!(n-i)!}$, $n \in \mathbb{N}$, $i = 0, 1, \ldots, n$).

Let us estimate the ratio $\dfrac{C_{2l}^{l-k}}{C_{2l}^{l}}$ from below for $1 \le k \le \sqrt{l/2}$. We observe first that

$$\ln\left(\frac{1-t}{1+t}\right) + 2t + 2t^3 \ge 0 \ \text{ for all } \ 0 \le t \le \frac{1}{2}. \tag{16.25}$$

In fact, setting

$$\varphi(t) := \ln\left(\frac{1-t}{1+t}\right) + 2t + 2t^3,$$

we see that $\varphi(0) = 0$ and

$$\varphi'(t) = -\frac{2}{1-t^2} + 2 + 6t^2 = \frac{2t^2(2-3t^2)}{1-t^2} \ge 0$$

for all $t \in [0, 1/2]$. Hence, $\varphi(t)$ increases on the interval $[0, 1/2]$, which implies (16.25). Further, since

$$\begin{aligned}
\frac{C_{2l}^{l-k}}{C_{2l}^{l}} &= \frac{(l!)^2}{(l-k)!(l+k)!} = \frac{(l-k+1)\cdot \ldots \cdot (l-1)\cdot l}{(l+1)\cdot \ldots \cdot (l+k-1)\cdot (l+k)} \\
&= \frac{l}{l+k}\cdot \frac{(l-k+1)\cdot \ldots \cdot (l-1)}{(l+1)\cdot \ldots \cdot (l+k-1)} = \frac{l}{l+k}\cdot \prod_{j=1}^{k-1} \frac{1-\frac{j}{l}}{1+\frac{j}{l}} \\
&= \frac{l}{l+k}\cdot \exp\left(\sum_{j=1}^{k-1} \ln \frac{1-\frac{j}{l}}{1+\frac{j}{l}}\right),
\end{aligned}$$

then from inequality (16.25) and the condition $1 \le k \le \sqrt{l/2}$ it follows that

$$\begin{aligned}
\frac{C_{2l}^{l-k}}{C_{2l}^{l}} &\ge \frac{l}{l+k} \exp\Big(-\frac{2}{l}\sum_{j=1}^{k-1} j - \frac{2}{l^3}\sum_{j=1}^{k-1} j^3\Big) \\
&= \frac{l}{l+k} \exp\left(\frac{-k(k-1)}{l}\right) \exp\left(\frac{-(k-1)^2k^2}{2l^3}\right) \\
&\ge \frac{1}{2} \exp\Big(-\frac{k^2}{l} - \frac{1}{l}\Big).
\end{aligned}$$

Combining this estimate with equation (16.24), we infer

$$\sigma_l = 2 \cdot \sum_{k=1}^{\sqrt{l/2}} \frac{C_{2l}^{l-k}}{C_{2l}^{l}} \cdot k \cdot C_{2l}^{l} \geq C_{2l}^{l} \cdot e^{-1/l} \cdot \sum_{k=1}^{\sqrt{l/2}} e^{-k^2/l} \cdot k. \tag{16.26}$$

Note that $\psi(u) = e^{-\frac{u^2}{l}} \cdot u$ is an increasing function on the interval $[0, \sqrt{l/2}]$ because of

$$\psi'(u) = e^{-u^2/l} + ue^{-u^2/l}(-2u/l) = e^{-u^2/l}\left(1 - 2u^2/l\right) \geq 0, \quad 0 \leq u \leq \sqrt{l/2}.$$

Consequently,

$$\sum_{k=1}^{\sqrt{l/2}} e^{-k^2/l} \cdot k > \sum_{k=1}^{\sqrt{l/2}} \int_{k-1}^{k} e^{-u^2/l} \cdot u\, du = \frac{l}{2}\left(1 - \frac{1}{\sqrt{e}}\right) \geq \frac{l}{3}.$$

Moreover, an easy calculation, by the Stirling formula (1.13), shows that

$$\lim_{l \to \infty} C_{2l}^{l} 4^{-l} \sqrt{\pi l} = 1.$$

Thus, from the last estimate and (16.26) it follows that

$$\begin{aligned}
\left\| \sum_{k=1}^{2l} r_k \right\|_{(M_{w,1}^d)'} &\geq \frac{1}{w(m(E_l))} 2^{-2l} \sum_{i \in S_l} \left| \sum_{k=1}^{2l} \varepsilon_k^i \right| = \frac{1}{w(m(E_l))} 2^{-2l} \sigma_l \\
&\geq \frac{1}{w(m(E_l))} 2^{-2l} \cdot C_{2l}^{l} \cdot e^{-1/l} \cdot \sum_{k=1}^{\sqrt{l/2}} e^{-k^2/l} \cdot k \\
&\geq \frac{1}{w(m(E_l))} 4^{-l} \cdot C_{2l}^{l} \cdot e^{-1/l} \cdot \frac{l}{3} \asymp \frac{\sqrt{l}}{w(m(E_l))}
\end{aligned}$$

for all $l \in \mathbb{N}$ such that $\sqrt{l/2} \in \mathbb{N}$. Since $m(E_l) \to 0$, then, by (16.10), we have $w(m(E_l)) \to 0$ as $l \to \infty$. Hence, the last inequality implies (16.22), as we wished.

□

16.5 The Structure of Rademacher Subspaces of Morrey Spaces

Theorem 16.1 allows us also to obtain the following alternative related to geometrical properties of the Rademacher subspaces of the spaces $M_{w,p}$.

Theorem 16.4 *Let $1 \le p < \infty$. Then, every infinite-dimensional subspace of the space $\mathcal{R}M_{w,p} := [r_n]_{M_{w,p}}$ is either isomorphic to ℓ_2 or contains a further subspace that is isomorphic to c_0 and is complemented in $\mathcal{R}M_{w,p}$.*

The next two propositions will be the main tools in the proof of this theorem. As above, we assume that a weighted function w satisfies condition (16.6).

Proposition 16.4 *Let $1 \le p < \infty$. Then, the Rademacher functions form a shrinking basis in the space $\mathcal{R}M_{w,p}$.*

Proof Since $\{r_n\}_{n=1}^{\infty}$ is a basis in the space $\mathcal{R}M_{w,p}$, it suffices only to prove the shrinking property of this sequence, i.e., that for every $\varphi \in M_{w,p}^*$

$$\|\varphi|_{[r_n]_{n=m}^{\infty}}\|_{M_{w,p}^*} \to 0 \text{ as } m \to \infty. \tag{16.27}$$

Assume that (16.27) does not hold. Then, there exist $\varepsilon \in (0,1), \varphi \in M_{w,p}^*$, $\|\varphi\|_{M_{w,p}^*} = 1$, and a sequence of functions

$$f_n = \sum_{k=m_n}^{\infty} a_k^{m_n} r_k, \text{ with } m_1 < m_2 < \dots,$$

such that $\|f_n\|_{M_{w,p}} = 1, n = 1, 2, \dots$, and

$$\varphi(f_n) \ge \varepsilon \text{ for all } n = 1, 2, \dots \tag{16.28}$$

Let us construct two sequences of positive integers $\{q_i\}_{i=1}^{\infty}$ and $\{p_i\}_{i=1}^{\infty}$, $1 \le q_1 < p_1 < q_2 < p_2 < \dots$ as follows (cf. the proof of Proposition 14.4). Setting $q_1 = m_1$, we may find $p_1 > q_1$ so that

$$\Big\| \sum_{n=p_1+1}^{\infty} a_k^{q_1} r_k \Big\|_{M_{w,p}} \le \frac{\varepsilon}{2}.$$

By induction, suppose that the numbers $1 \le q_1 < p_1 < q_2 < p_2 < \dots q_{i-1} < p_{i-1}$, for some $i \ge 2$, are already chosen. Then, for q_i we take the least of numbers $m_n > p_{i-1}$ such that

$$w(2^{-q_i}) \le \frac{1}{2} w(2^{-q_{i-1}}). \tag{16.29}$$

(it is possible because of $\lim_{t \to 0^+} w(t) = 0$). Also, let $p_i > q_i$ be satisfied the following inequality:

$$\Big\| \sum_{n=p_i+1}^{\infty} a_k^{q_i} r_k \Big\|_{M_{w,p}} \le \varepsilon/2. \tag{16.30}$$

We set $\alpha_k^i := a_k^{q_i}$ if $q_i \le k \le p_i$, and $\alpha_k^i := 0$ if $p_i < k < q_{i+1}$, $i = 1, 2, \ldots$. Then, the functions

$$u_i := \sum_{k=q_i}^{q_{i+1}-1} \alpha_k^i r_k, \quad i = 1, 2, \ldots$$

constitute a block basic sequence of the Rademacher sequence. Moreover, by the definition of u_i, we have

$$\sup_{i=1,2,\ldots} \|u_i\|_{M_{w,p}} \le 2, \tag{16.31}$$

and from the choice of the functional φ and inequality (16.30) it follows that

$$\begin{aligned}\varphi(u_i) &= \varphi\Big(\sum_{k=q_i}^{p_i} a_k^{q_i} r_k\Big) = \varphi(f_i) - \varphi\Big(\sum_{k=p_i+1}^{\infty} a_k^{q_i} r_k\Big) \\ &\ge \varphi(f_i) - \Big\|\sum_{k=p_i+1}^{\infty} a_k^{q_i} r_k\Big\|_{M_{w,p}} \ge \frac{\varepsilon}{2}. \end{aligned} \tag{16.32}$$

Let $\{\gamma_n\}_{n=1}^{\infty}$ be an arbitrary sequence of positive numbers such that

$$\sum_{n=1}^{\infty} \gamma_n^2 < \infty \text{ and } \sum_{n=1}^{\infty} \gamma_n = \infty. \tag{16.33}$$

Show that the series $\sum_{n=1}^{\infty} \gamma_n u_n$ converges in the space $M_{w,p}$. To this end, we put $b_k := \alpha_k^i \cdot \gamma_i$ if $q_i \le k < q_{i+1}$. By Theorem 16.1, for every $m \in \mathbb{N}$ we have

$$\Big\|\sum_{n=m}^{\infty} \gamma_n u_n\Big\|_{M_{w,p}} = \Big\|\sum_{k=q_m}^{\infty} b_k r_k\Big\|_{M_{w,p}} \asymp \Big(\sum_{k=q_m}^{\infty} b_k^2\Big)^{1/2} + \sup_{l \ge q_m} w(2^{-l}) \cdot \sum_{k=q_m}^{l} |b_k|. \tag{16.34}$$

We estimate now both terms in the right-hand side of (16.34) from above. First, from (16.31) and Theorem 16.1 it follows that

$$\sum_{k=q_m}^{\infty} b_k^2 = \sum_{i=m}^{\infty} \gamma_i^2 \sum_{k=q_i}^{q_{i+1}-1} (\alpha_k^i)^2 \le C_1 \sum_{i=m}^{\infty} \gamma_i^2. \tag{16.35}$$

Similarly, if $q_m < \ldots < q_{m+r} \le l < q_{m+r+1}$ for some $r = 1, 2, \ldots$, then

$$\sum_{k=q_m}^{l} |b_k| = \sum_{i=m}^{m+r-1} \gamma_i \sum_{k=q_i}^{q_{i+1}-1} |\alpha_k^i| + \gamma_{m+r} \sum_{k=q_{m+r}}^{l} |\alpha_k^{m+r}|$$

$$\le C_2 \Big(\sum_{i=m}^{m+r-1} \frac{\gamma_i}{w(2^{-q_{i+1}})} + \frac{\gamma_{m+r}}{w(2^{-l})} \Big).$$

Combining this inequality together with (16.29), we obtain

$$w(2^{-l}) \sum_{k=q_m}^{l} |b_k| \le C_2 \Big(\sum_{i=m}^{m+r-1} \gamma_i \frac{w(2^{-q_{m+r}})}{w(2^{-q_{i+1}})} + \gamma_{m+r} \Big)$$

$$\le C_2 \Big(\sum_{i=m}^{m+r-1} \gamma_i \, 2^{-m-r+i+1} + \gamma_{m+r} \Big)$$

$$\le C_2 \max_{i \ge m} \gamma_i \Big(\sum_{j=0}^{r-1} 2^{1+j-r} + 1 \Big) < 3\, C_2 \max_{i \ge m} \gamma_i .$$

Clearly, the last estimate holds also in the simpler case when $q_m \le l < q_{m+1}$. Thus, for every $m \in \mathbb{N}$

$$\sup_{l \ge q_m} w(2^{-l}) \sum_{k=q_m}^{l} |b_k| \le 3\, C_2 \max_{i \ge m} \gamma_i . \tag{16.36}$$

From (16.33)–(16.36) it follows that the series $\sum_{n=1}^{\infty} \gamma_n \, u_n$ converges in the Morrey space $M_{w,p}$. At the same time, since $\varphi \in M_{w,p}^*$, according to (16.32) and (16.33), we have

$$\varphi\Big(\sum_{n=1}^{\infty} \gamma_n \, u_n \Big) = \sum_{n=1}^{\infty} \gamma_n \, \varphi(u_n) \ge \frac{\varepsilon}{2} \sum_{n=1}^{\infty} \gamma_n = \infty.$$

This contradiction indicates that (16.27) is valid, and so the proposition is proved. □

Corollary 16.2 *For every* $1 \le p < \infty$

(i) $r_k \xrightarrow{w} 0$ *in* $M_{w,p}$ *as* $k \to \infty$;
(ii) the Rademacher functions form a basis in the dual space $\mathcal{R}M_{w,p}^*$.

Proof Assertion (i) is an immediate consequence of Proposition 16.4. Since $\{r_k\}_{k=1}^{\infty}$ is an orthogonal system, then assertion (ii) follows from Proposition 16.4 and [181, Proposition 1.b.1]. □

Proposition 16.5 *Let $1 \le p < \infty$. Suppose that*

$$u_n = \sum_{k=m_n+1}^{m_{n+1}} a_k r_k, \quad 0 \le m_1 < m_2 < \ldots,$$

is a block basic sequence of $\{r_k\}$ such that $\|u_n\|_{M_{w,p}} = 1$, $n \in \mathbb{N}$, and $\sum_{k=m_n+1}^{m_{n+1}} a_k^2 \to 0$ as $n \to \infty$. Moreover, let

$$w(2^{-m_{n+1}}) \le \frac{1}{2} w(2^{-m_n}), \ n = 1, 2, \ldots \tag{16.37}$$

Then, $\{u_n\}_{n=1}^{\infty}$ contains a subsequence equivalent in the space $M_{w,p}$ to the unit vector basis in c_0.

Proof Passing to a subsequence if it is necessary (and preserving the notation), without loss of generality, we may assume that

$$\sum_{k=m_n+1}^{m_{n+1}} a_k^2 \le 2^{-n}, \quad n = 1, 2, \ldots \tag{16.38}$$

Let $f \in \mathcal{R}M_{w,p}$ belong to the closed linear span of $\{u_n\}_{n=1}^{\infty}$. Clearly, then $f = \sum_{n=1}^{\infty} \beta_n u_n$. Setting $b_k = a_k \beta_i$ if $m_i < k \le m_{i+1}$, $i = 1, 2, \ldots$, by (16.38), we obtain

$$\sum_{k=1}^{\infty} b_k^2 = \sum_{i=1}^{\infty} \beta_i^2 \sum_{k=m_i+1}^{m_{i+1}} a_k^2 \le \sup_{i=1,2,\ldots} \beta_i^2 \cdot \sum_{i=1}^{\infty} 2^{-i} \le \|(\beta_i)\|_{c_0}^2.$$

Moreover, precisely in the same way as in the proof of Proposition 16.4, from (16.37) and the equality $\|u_i\|_{M_{w,p}} = 1$, $i = 1, 2, \ldots$, it follows that for some constant $C' > 0$

$$\sup_{l=1,2,\ldots} w(2^{-l}) \sum_{k=1}^{l} |b_k| \le C' \|(\beta_i)\|_{c_0}.$$

Combining the last two inequalities together with equivalence (16.15), we conclude that $\|f\|_{M_{w,p}} \le C \|(\beta_i)\|_{c_0}$ for some $C > 0$.

Conversely, since $\{u_n\}$ is an unconditional basic sequence in $M_{w,p}$, then

$$\|f\|_{M_{w,p}} \geq c \sup_{i\in\mathbb{N}} |\beta_i| \|u_i\|_{M_{w,p}} = c\,\|(\beta_i)\|_{c_0}$$

for some $c > 0$, which completes the proof. □

Proof of Theorem 16.4 Let X be an infinite-dimensional subspace of the space $\mathcal{R}M_{w,p}$. Suppose first that for every function $f = \sum_{k=1}^{\infty} b_k r_k \in X$ we have

$$\|f\|_{M_{w,p}} \asymp \Big(\sum_{k=1}^{\infty} b_k^2\Big)^{1/2}$$

with a constant independent of coefficients $b_k \in \mathbb{R}$, $k = 1, 2, \ldots$. Then, clearly, X is isomorphic to an infinite-dimensional subspace of ℓ_2, and hence to ℓ_2 itself. Therefore, the assumption that X is not isomorphic to ℓ_2 implies the existence of a sequence $\{f_n\}_{n=1}^{\infty} \subset X$, $f_n = \sum_{k=1}^{\infty} b_{n,k} r_k$, such that $\|f_n\|_{M_{w,p}} = 1$, $n = 1, 2, \ldots$, and

$$\sum_{k=1}^{\infty} b_{n,k}^2 \to 0 \text{ as } n \to \infty. \tag{16.39}$$

Observe that no subsequence of $\{f_n\}_{n=1}^{\infty}$ converges in the $M_{w,p}$-norm. Indeed, on the contrary, if $\|f_{n_k} - f\|_{M_{w,p}} \to 0$ for some subsequence $\{f_{n_k}\} \subset \{f_n\}$ and $f \in X$, then from Theorem 16.1 and (16.39) it follows that $f = 0$. On the other hand, obviously, $\|f\|_{M_{w,p}} = 1$, and we come to a contradiction. Thus, passing to a subsequence if it is necessary (and preserving the notation), we may assume that

$$\|f_n - f_m\|_{M_{w,p}} \geq c \text{ if } n \neq m, \tag{16.40}$$

for some $c > 0$. Recall that, by Corollary 16.2, the sequence $\{r_k\}_{k=1}^{\infty}$ is a basis in the dual space $\mathcal{R}M_{w,p}^*$. Applying the standard diagonal procedure, we can find a sequence of positive integers $\{n_k\}_{k=1}^{\infty}$, $n_1 < n_2 < \ldots$, such that for every $i = 1, 2, \ldots$ there exists the limit $\lim_{k\to\infty} \int_0^1 r_i(s) f_{n_k}(s)\, ds$. Then, we have

$$\lim_{k\to\infty} \int_0^1 r_i(s)\big(f_{n_{2k+1}}(s) - f_{n_{2k}}(s)\big)\, ds = 0 \text{ for all } i = 1, 2, \ldots$$

Since the sequence $\{f_{n_{2k+1}} - f_{n_{2k}}\}_{k=1}^{\infty}$ is bounded in $M_{w,p}$, this implies that $f_{n_{2k+1}} - f_{n_{2k}} \overset{w}{\to} 0$. Consequently, taking into account (16.40) and using the Bessaga–Pełczyński selection principle (see Theorem B.3), we can construct a subsequence of $\{f_{n_{2k+1}} - f_{n_{2k}}\}_{k=1}^{\infty}$ (again preserving the notation) and a block basic sequence

$$u_k = \sum_{j=m_k+1}^{m_{k+1}} a_j r_j,\ 0 < m_1 < m_2 < \ldots,$$

such that

$$\|u_k - (f_{n_{2k+1}} - f_{n_{2k}})\|_{M_{w,p}} \le B^{-1} \cdot 2^{-k-1}, \quad k = 1, 2, \ldots, \tag{16.41}$$

where B is the basis constant of the sequence $\{r_k\}$ in $\mathcal{R}M_{w,p}$, and

$$w(2^{-m_{k+1}}) \le \frac{1}{2} \cdot w(2^{-m_k}), \quad k = 1, 2, \ldots.$$

From inequality (16.41) and the principle of small perturbations (see Theorem B.2) it follows that the sequences $\{u_k\}_{k=1}^{\infty}$ and $\{f_{n_{2k+1}} - f_{n_{2k}}\}_{k=1}^{\infty}$ are equivalent in the space $M_{w,p}$. Moreover, by Theorem 16.1 and (16.39), we have

$$\sum_{j=m_k+1}^{m_{k+1}} a_j^2 \to 0 \text{ as } k \to \infty.$$

Therefore, applying Proposition 16.5, we conclude that the sequence $\{u_k\}_{k=1}^{\infty}$ (and so also $\{f_{n_{2k+1}} - f_{n_{2k}}\}_{k=1}^{\infty}$) contains a subsequence equivalent to the unit vector basis in c_0. Since $\{f_{n_{2k+1}} - f_{n_{2k}}\}_{k=1}^{\infty} \subset X$, then hence X contains a subspace isomorphic to c_0. The fact that this subspace is complemented in $\mathcal{R}M_{w,p}$ is an immediate consequence of the Sobczyk theorem [1, Corollary 2.5.9]. Thus, the proof is completed. □

Comments and References The Morrey spaces, introduced by Morrey [211] in 1938 in connection with the study of partial differential equations, were intensively investigated during the past decades. In particular, a lot of papers were devoted to exploring properties of various operators of harmonic analysis (maximal, singular, potential and others) both in the classical Morrey spaces and in their numerous generalized versions. In the theory of partial differential equations, along with the weighted Lebesgue spaces, the Morrey type spaces play an especially important role. They appeared to be quite useful in the study of the local behaviour of the solutions of partial differential equations, a priori estimates and other topics.

In applications, usually, one considers the Morrey type spaces of functions defined on a domain Ω in $\mathbb{R}^n$. Let $1 \le p < \infty$, and let v be a nonnegative increasing function on $[0, \infty)$. The Morrey space $M_{p,v}(\Omega)$ is the class of all Lebesgue measurable real-valued functions f on Ω such that

$$\|f\|_{M_{p,v}} := \sup_{0<r<\mathrm{diam}(\Omega),\, x_0\in\Omega} v(r) \left(\frac{1}{m(B_r(x_0))} \int_{B_r(x_0)\cap\Omega} |f(t)|^p \, dt \right)^{1/p} < \infty,$$

where $B_r(x_0)$ is the Euclidean ball with the center at x_0 and radius r. In particular, if $v(r) = r^{1/q}$, where $0 < p \le q < \infty$, we get the classical Morrey space $M_{p,q}(\Omega)$ (see [169, Part 4.3], [179, 229], and [289]).

All the main results of this chapter have been obtained in the paper [51]. Their comparison with results on the behaviour of Rademacher functions in other function spaces shows that the Morrey spaces $M_{w,p}$ occupy, in this relation, in a sense an intermediate position between their "symmetrizations" $\widetilde{M(w, p)}$ and the corresponding weighted spaces of the Korenblyum–Krein–Levin type (see the beginning of the chapter).

Appendix A
Some Basics on Probability Theory

Definition A.1 Random variables (r.v.'s) ξ_k, $k = 1, 2, \ldots, n$, defined on a probability space $(\Omega, \Sigma, \mathbb{P})$ are called *independent* if for arbitrary intervals $I_k \subset \mathbb{R}$, $k = 1, 2, \ldots, n$, the following equation holds:

$$\mathbb{P}\{\omega \in \Omega : \xi_k(\omega) \in I_k,\ k = 1, 2, \ldots, n\} = \prod_{k=1}^{n} \mathbb{P}\{\omega \in \Omega : \xi_k(\omega) \in I_k\}.$$

We shall say that $\{\xi_k\}_{k=1}^{\infty}$ is a *sequence of independent r.v.'s* provided that for every $n \in \mathbb{N}$ the r.v.'s ξ_k, $k = 1, 2, \ldots, n$, are independent.

For any independent r.v.'s $\xi_k \in L_1(\Omega, \Sigma, \mathbb{P})$, $k = 1, 2, \ldots, n$, we have

$$\int_{\Omega} \prod_{k=1}^{n} \xi_k(\omega)\, d\mathbb{P} = \prod_{k=1}^{n} \int_{\Omega} \xi_k(\omega)\, d\mathbb{P} \tag{A.1}$$

[153, Theorem 2.2.4].

Throughout by $\mathbb{E}\xi$ we denote the expectation of an r.v. ξ, that is, $\mathbb{E}\xi := \int_{\Omega} \xi(\omega)\, d\mathbb{P}(\omega)$. The *distribution function* F_ξ and the *characteristic function* θ_ξ of an r.v. ξ are defined by

$$F_\xi(x) := \mathbb{P}\{\omega \in \Omega : \xi(\omega) < x\}$$

and

$$\theta_\xi(t) := \mathbb{E}\exp(it\xi) = \int_{-\infty}^{\infty} \exp(itx)\, dF_\xi(x),$$

S. V. Astashkin, *The Rademacher System in Function Spaces*,
https://doi.org/10.1007/978-3-030-47890-2

respectively. If r.v.'s ξ_k, $k = 1, 2, \dots, n$, are independent and $S := \sum_{k=1}^{n} \xi_k$, then from (A.1) it follows easily that

$$\theta_S(t) = \prod_{k=1}^{n} \theta_{\xi_k}(t), \quad t \in \mathbb{R}.$$

R.v.'s ξ and η are *identically distributed* if $F_\xi(x) = F_\eta(x)$ for all $x \in \mathbb{R}$. Following to [148], we shall say that two collections of r.v.'s $\{\xi_k\}_{k=1}^{n}$ and $\{\eta_k\}_{k=1}^{n}$ defined on probability spaces $(\Omega, \Sigma, \mathbb{P})$ and $(\Omega', \Sigma', \mathbb{P}')$, respectively, are *similar* whenever for arbitrary intervals $I_k \subset \mathbb{R}$, $k = 1, 2, \dots, n$, we have

$$\mathbb{P}\{\omega \in \Omega : \xi_k(\omega) \in I_k,\ k = 1, 2, \dots, n\} = \mathbb{P}'\{\omega' \in \Omega' : \eta_k(\omega') \in I_k,\ k = 1, 2, \dots, n\}.$$

Sequences $\{\xi_k\}_{k=1}^{\infty}$ and $\{\eta_k\}_{k=1}^{\infty}$ are *similar* if the collections $\{\xi_k\}_{k=1}^{n}$ and $\{\eta_k\}_{k=1}^{n}$ are similar for every $n \in \mathbb{N}$. In particular, if $\{\xi_k\}_{k=1}^{\infty}$ and $\{\eta_k\}_{k=1}^{\infty}$ are sequences of independent r.v.'s such that ξ_k and η_k are identically distributed for every $k = 1, 2, \dots$, these sequences are similar.

According to the well-known continuity theorem [81, Theorem 6.4.3], the convergence of distribution functions $F_{\xi_n}(x)$ to a distribution function $F_\xi(x)$ at all continuity points of $F_\xi(x)$ (in particular, the convergence of ξ_n to ξ in probability) implies the pointwise convergence on $\mathbb{R}$ of the characteristic functions $\theta_{\xi_n}(t)$ to the characteristic function $\theta_\xi(t)$. Moreover, by the uniqueness theorem (see e.g. [81, Corollary of Theorem 6.3.2, p. 141]), every characteristic function uniquely determines the distribution function of a given r.v., i.e., from the equation $\theta_\xi(t) = \theta_\eta(t)$, $t \in \mathbb{R}$, for r.v.'s ξ and η it follows that ξ and η are identically distributed.

Let $\mathfrak{g}$ be a standard Gaussian r.v., that is, the distribution function of $\mathfrak{g}$ is defined by

$$\Phi(x) = \frac{1}{\sqrt{2\pi}} \int_{-\infty}^{x} e^{-t^2/2}\, dt, \quad x \in \mathbb{R}. \tag{A.2}$$

Then, an easy calculation shows that $\mathbb{E}|\mathfrak{g}| = (2/\pi)^{1/2}$, and the even moments of $\mathfrak{g}$ are expressed via the Γ-function $\Gamma(x) = \int_0^\infty t^{x-1} e^{-x}\, dt$, $x > 0$, as follows:

$$\mathbb{E}\mathfrak{g}^{2m} = \frac{1}{\sqrt{2\pi}} \Gamma\left(\frac{m+1}{2}\right).$$

Hence, by the formula

$$\Gamma\left(\frac{m+1}{2}\right) = \frac{(2m)!}{2^m m!}\sqrt{2\pi}, \quad m \in \mathbb{N},$$

we have

$$\mathbb{E}\mathfrak{g}^{2m} = \frac{(2m)!}{2^m m!}, \quad m \in \mathbb{N}.$$

Recall the central limit theorem, where by the variance of an r.v. ξ is understood, as usual, the expectation $\mathbb{E}(\xi - \mathbb{E}\xi)^2$ (see e.g. [81, Theorem 7.2]).

Theorem A.1 *Let* $\{\xi_k\}_{k=1}^{\infty}$ *be a sequence of independent identically distributed r.v.'s, defined on a probability space* $(\Omega, \Sigma, \mathbb{P})$, *with expectation a and variance* σ^2. *If* $S_n := \xi_1 + \xi_2 + \ldots + \xi_n$ *and*

$$U_n := \frac{S_n - an}{\sigma\sqrt{n}},$$

then for any $x \in \mathbb{R}$ *we have*

$$\lim_{n\to\infty} \mathbb{P}\{\omega \in \Omega : U_n(\omega) < x\} = \Phi(x),$$

where $\Phi(x)$ *is the distribution function of a standard Gaussian r.v.* $\mathfrak{g}$ *(see* (A.2)*).*

Let $(\Omega, \Sigma, \mathbb{P})$ be a probability space, and let F be a Banach space. Denote by $\mathcal{B}$ the Borel σ-algebra of F. Then, every measurable mapping $\Xi : \Omega \to F$ (with respect to the measurable space $(F, \mathcal{B})$) is said to be an *random vector*. Precisely, as in the case of r.v.'s, we may define *independent* and *identically distributed* random vectors as well as *similar* sequences of random vectors. A random vector $\Xi : \Omega \to F$ is called *symmetrically distributed* if the vectors Ξ and $-\Xi$ are identically distributed, i.e. $\mathbb{P}\{\omega : \Xi(\omega) \in B\} = \mathbb{P}\{\omega : -\Xi(\omega) \in B\}$ for each $B \in \mathcal{B}$.

Let $\{\Xi_k\}_{k=1}^{\infty}$ be a sequence of independent symmetrically distributed random vectors with values in a Banach space F. We set

$$\Theta_m := \sum_{k=1}^{m} \Xi_k, \quad m = 1, 2, \ldots, \quad \text{and} \quad M(\omega) := \sup_{m=1,2,\ldots} \|\Theta_m(\omega)\|_F.$$

The following classical *maximal inequality* is due to P. Levy (see e.g. [148, Lemma II.1]).

Theorem A.2 *Let a series* $S = \sum_{k=1}^{\infty} \Xi_k$ *converge a.e. Then, for every* $u > 0$ *it holds*

$$\mathbb{P}\{\omega : M(\omega) > u\} \le 2\mathbb{P}\{\omega : \|S(\omega)\|_F > u\}.$$

Suppose that ξ is an integrable r.v. defined on a probability space $(\Omega, \Sigma, \mathbb{P})$. Then, if $\mathcal{A}$ is a σ-subalgebra of the σ-algebra Σ, then there exists a unique (up to zero measure) $\mathcal{A}$-measurable integrable r.v. $\mathbb{E}[\xi \mid \mathcal{A}]$ such that for all $B \in \mathcal{A}$

$$\int_B \xi(\omega)\, d\mathbb{P}(\omega) = \int_B \mathbb{E}[\xi \mid \mathcal{A}](\omega)\, d\mathbb{P}(\omega)$$

(see e.g. [97, Theorem 1.1]). The r.v. $\mathbb{E}[\xi \mid \mathcal{A}]$ is called the *conditional expectation* of ξ with respect to $\mathcal{A}$. If $\mathcal{A}$ is the σ-subalgebra of Σ, generated by r.v.'s $\eta_1, \eta_2, \ldots, \eta_n$, then the conditional expectation with respect to $\mathcal{A}$ will be denoted by $\mathbb{E}[\xi \mid \eta_1, \eta_2, \ldots, \eta_n]$ or $\mathbb{E}_{\eta_1,\eta_2,\ldots,\eta_n}\xi$.

Let $U : \mathbb{R} \to \mathbb{R}$ be a convex mapping, and let ξ be an r.v. such that $\xi \in L_1(\Omega, \Sigma, \mathbb{P})$ and $U(\xi) \in L_1(\Omega, \Sigma, \mathbb{P})$. Then, the conditional expectations of the r.v.'s ξ and $U(\xi)$ satisfy the following Jensen inequality:

$$U(\mathbb{E}[\xi \mid \mathcal{A}]) \le \mathbb{E}[U(\xi) \mid \mathcal{A}] \quad \text{a.e.}$$

Hence, in particular, it follows that the operator $\xi \mapsto \mathbb{E}[\xi \mid \mathcal{A}]$ is bounded with norm 1 in the space $L_p(\Omega, \Sigma, \mathbb{P})$ for every $1 \le p \le \infty$.

In conclusion, we recall the definition of a martingale with respect to an increasing sequence of σ-algebras.

Definition A.2 Let $\{\mathcal{A}_i\}_{i \in I}$ be a finite or infinite sequence of σ-subalgebras of a σ-algebra Σ such that $\mathcal{A}_i \subset \mathcal{A}_j$ if $i, j \in I$ and $i \le j$. A sequence of r.v.'s $\{\xi_i\}_{i \in I}$ is said to be a *martingale* with respect to the family $\{\mathcal{A}_i\}_{i \in I}$ whenever we have

(i) ξ_i is $\mathcal{A}_i$-measurable and $\xi_i \in L_1(\Omega, \Sigma, \mathbb{P})$ for each $i \in I$;
(ii) $\mathbb{E}[\xi_j \mid \mathcal{A}_i] = \xi_i$ for any $i, j \in I$ such that $i \le j$.

A sequence of r.v.'s (in particular, a martingale) $\{\xi_i\}_{i=1}^{\infty}$ is called *equi-integrable* provided that

$$\lim_{\mathbb{P}(E) \to 0} \sup_{i=1,2,\ldots} \int_E |\xi_i(\omega)|\, d\mathbb{P} = 0.$$

For a detailed account of the above definitions and facts, see the books on probability theory (for example, [81, 97, 118, 265]).

Appendix B
Basic Sequences and Lacunary Systems

Definition B.1 A sequence $\{x_n\}_{n=1}^{\infty}$ from a real Banach space X is said to be *basic* if every x from the closed linear span $[x_n]_X$ of $\{x_n\}_{n=1}^{\infty}$ is uniquely represented in the form

$$x = \sum_{i=1}^{\infty} a_i x_i, \quad \text{where } a_i \in \mathbb{R}.$$

It is well known (see e.g. [1, Proposition 1.1.9] or [181, Proposition 1.a.3]) that $\{x_n\}_{n=1}^{\infty} \subset X$ is a basic sequence in X if and only if the following conditions hold:

(1) $x_n \neq 0$ for all $n \in \mathbb{N}$;
(2) there exists a constant $C > 0$ such that for all positive integers m and n such that $m < n$ and any $a_i \in \mathbb{R}$ we have

$$\Big\| \sum_{i=1}^{m} a_i x_i \Big\|_X \leq C \Big\| \sum_{i=1}^{n} a_i x_i \Big\|_X.$$

If $\{x_n\}_{n=1}^{\infty}$ is a basic sequence in a Banach space X, then the least constant C such that condition (2) holds is called the *basis constant* of $\{x_n\}_{n=1}^{\infty}$. In the case when the basis constant of $\{x_n\}_{n=1}^{\infty}$ is equal to 1 one says that $\{x_n\}_{n=1}^{\infty}$ is a *monotone* basic sequence.

If a basic sequence $\{x_n\}_{n=1}^{\infty}$ is *complete* in X, i.e., the closed linear span $[x_n]_X$ coincides with the whole space X, then $\{x_n\}_{n=1}^{\infty}$ is called a *(Schauder) basis* of X. In this case there exists a unique sequence of functionals $\{x_n^*\}_{n=1}^{\infty}$ from the dual space X^* such that

(i) $x_k^*(x_k) = 1$, and $x_k^*(x_j) = 0$ if $k \neq j$;
(ii) $x = \sum_{n=1}^{\infty} x_n^*(x) x_n$ for each $x \in X$.

S. V. Astashkin, *The Rademacher System in Function Spaces*,
https://doi.org/10.1007/978-3-030-47890-2

The functionals x_n^*, $n = 1, 2, \ldots$, satisfying the condition (i), are called the *biorthogonal functionals* associated with the system $\{x_n\}_{n=1}^{\infty}$.

As usual, ℓ_p, $1 \le p < \infty$ (resp. c_0) is the Banach space of all sequences $a = (a_n)_{n=1}^{\infty}$ of real numbers such that

$$\|a\|_{\ell_p} := \Big(\sum_{n=1}^{\infty} |a_n|^p\Big)^{1/p} < \infty$$

(resp. $\lim_{n\to 0} a_n = 0$ with the norm

$$\|a\|_{c_0} := \max_{n=1,2,\ldots} |a_n|).$$

Clearly, the canonical unit vectors $e_n = (e_n(k))_{k=1}^{\infty}$, $n = 1, 2, \ldots$, where $e_n(n) = 1$, and $e_n(k) = 0$ if $k \ne n$, form a monotone basis in each of these spaces.

A basis $\{x_n\}_{n=1}^{\infty}$ in a Banach space X is said to be *shrinking* whenever for every $x^* \in X^*$ we have

$$\lim_{N\to\infty} \big\|x^*|_{[x_n]_{n>N}}\big\| = 0, \quad \text{where } \big\|x^*|_{[x_n]_{n>N}}\big\| := \sup\big\{|x^*(y)| : y \in [x_n]_{n>N}, \|y\|_X \le 1\big\}.$$

Theorem B.1 ([1, Proposition 3.2.6] or [181, Proposition 1.b.1])
A basis $\{x_n\}_{n=1}^{\infty}$ in a Banach space X is shrinking if and only if the sequence of biorthogonal functionals $\{x_n^*\}_{n=1}^{\infty}$ is a basis in the dual space X^*.

Definition B.2 We say that a basic sequence $\{x_n\}_{n=1}^{\infty}$ in a Banach space X is *equivalent* to a basic sequence $\{y_n\}_{n=1}^{\infty}$ in a Banach space Y if for each sequence $(a_n)_{n=1}^{\infty}$ of real numbers the series $\sum_{n=1}^{\infty} a_n x_n$ converges in X if and only if the series $\sum_{n=1}^{\infty} a_n y_n$ converges in Y. In other words, there is a constant $C > 0$ such that for every $m \in \mathbb{N}$ and all $a_n \in \mathbb{R}$, $n = 1, 2, \ldots, m$, we have

$$C^{-1}\Big\|\sum_{n=1}^{m} a_n y_n\Big\|_Y \le \Big\|\sum_{n=1}^{m} a_n x_n\Big\|_X \le C\Big\|\sum_{n=1}^{m} a_n y_n\Big\|_Y.$$

In particular, a basic sequence $\{x_n\}_{n=1}^{\infty} \subset X$ is equivalent to the unit vector basis $\{e_n\}_{n=1}^{\infty}$ in ℓ_p, $1 \le p < \infty$ (resp. in c_0) whenever for some $C > 0$ and all $m \in \mathbb{N}$, $a_n \in \mathbb{R}$, $n = 1, 2, \ldots, m$, it holds

$$C^{-1}\Big(\sum_{n=1}^{m} |a_n|^p\Big)^{1/p} \le \Big\|\sum_{n=1}^{m} a_n x_n\Big\|_X \le C\Big(\sum_{n=1}^{m} |a_n|^p\Big)^{1/p}$$

$$\Big(\text{resp. } C^{-1} \max_{n=1,2,\ldots,m} |a_n| \le \Big\|\sum_{n=1}^{m} a_n x_n\Big\|_X \le C \max_{n=1,2,\ldots,m} |a_n|\Big).$$

In what follows, by a subspace of a Banach space X we understand a closed linear subspace of X. Recall that a subspace F of a Banach space X is said to be *complemented* in X if there exists a bounded linear projection in X, whose image is F.

The following important result [1, Theorem 1.3.9] (see also [181, Proposition 1.a.9]) reveals a certain stability of bases and their complementability.

Theorem B.2 (The Principle of Small Perturbations) *Let $\{x_n\}_{n=1}^{\infty}$ be a basic sequence in a Banach space X with basis constant K. Then, if $\{y_n\}_{n=1}^{\infty}$ is a sequence in X such that*

$$\sum_{n=1}^{\infty} \frac{\|x_n - y_n\|}{\|x_n\|} < \frac{1}{2K},$$

then the sequences $\{x_n\}_{n=1}^{\infty}$ and $\{y_n\}_{n=1}^{\infty}$ are equivalent in X. In particular, $\{y_n\}_{n=1}^{\infty}$ is a basic sequence in X. If $\{x_n\}_{n=1}^{\infty}$ is a basis, so is $\{y_n\}_{n=1}^{\infty}$. And finally, if the subspace $[x_n]$ is complemented in X, then $[y_n]$ is complemented in X as well.

Definition B.3 Let $\{x_n\}_{n=1}^{\infty}$ be a basic sequence in a Banach space X. Suppose that $(m_n)_{n=0}^{\infty}$ is a strictly increasing sequence of integers with $m_0 = 0$ and that $(a_n)_{n=1}^{\infty}$ is an arbitrary sequence of real numbers. Then a sequence of nonzero vectors $\{u_n\}_{n=1}^{\infty}$ in X of the form

$$u_n := \sum_{j=m_{n-1}+1}^{m_n} a_j x_j$$

is called a *block basic sequence* of $\{x_n\}_{n=1}^{\infty}$.

It is easy to check (see [1, Lemma 1.3.5]) that each block basic sequence of a basic sequence with basis constant K is also a basic sequence with basis constant less than or equal to K.

When dealing with basic sequences, the following result known as the *Bessaga–Pełczyński selection principle* is very useful (see [1, Proposition 1.3.10] or [181, Proposition 1.a.12]).

Theorem B.3 *Let $\{x_n\}_{n=1}^{\infty}$ be a basis in a Banach space X with basis constant K and the sequence of biorthogonal functionals $\{x_n^*\}_{n=1}^{\infty}$. Suppose that a sequence $\{y_n\}_{n=1}^{\infty}$ in X such that*

(i) $\inf_{n=1,2,\dots} \|y_n\|_X > 0$, *but*
(ii) $\lim_{n\to\infty} x_k^*(y_n) = 0$ *for all* $k \in \mathbb{N}$.

Then $\{y_n\}_{n=1}^{\infty}$ contains a subsequence $\{y_{n_k}\}_{k=1}^{\infty}$ that is equivalent in X to some block basic sequence $\{u_k\}_{k=1}^{\infty}$ of $\{x_n\}_{n=1}^{\infty}$. Moreover, for every $\varepsilon > 0$ it is possible to choose a strictly increasing sequence of positive integers $(n_k)_{k=1}^{\infty}$ so that $\{y_{n_k}\}_{k=1}^{\infty}$ has basis constant at most $K + \varepsilon$. In particular, the same result holds if $\{y_n\}_{n=1}^{\infty}$ converges to 0 weakly but not in the norm of X.

As usual, a *permutation* π of the set of positive integers is a one-to-one mapping $\pi : \mathbb{N} \to \mathbb{N}$.

Definition B.4 A basic sequence $\{x_n\}_{n=1}^\infty$ of a Banach space X is called *unconditional* if for any permutation $\pi : \mathbb{N} \to \mathbb{N}$ the sequence $\{x_{\pi(n)}\}_{n=1}^\infty$ is basic in X as well.

It is known (see [153, Theorem 1.1.1] or [181, Proposition 1.c.1]) that a basic sequence $\{x_n\}_{n=1}^\infty$ is unconditional in X if and only if from the convergence of a series $\sum_{i=1}^\infty a_i x_i$, $a_i \in \mathbb{R}$, it follows the convergence of the series $\sum_{i=1}^\infty \theta_i a_i x_i$ for arbitrary sequence $(\theta_i)_{i=1}^\infty$, $\theta_i = \pm 1$. Therefore, according to the Closed Graph theorem and the Uniform Boundedness Principle, it follows that a basic sequence $\{x_n\}_{n=1}^\infty$ is unconditional in X if and only if the family of operators

$$T_\theta\Big(\sum_{i=1}^\infty a_i x_i\Big) = \sum_{i=1}^\infty \theta_i a_i x_i, \quad \theta = (\theta_i), \theta_i = \pm 1,$$

is uniformly bounded in the closed linear span $[x_n]_X$. This means that there is a constant $C > 0$ such that for arbitrary $n \in \mathbb{N}$ and all $\theta_i = \pm 1$, $a_i \in \mathbb{R}$, $i = 1, 2, \dots, n$, we have

$$\Big\| \sum_{i=1}^n \theta_i a_i x_i \Big\|_X \le C \Big\| \sum_{i=1}^n a_i x_i \Big\|_X.$$

The least constant C so that the last inequality holds is called the *unconditional basis constant* of $\{x_n\}_{n=1}^\infty$.

The following definition may be treated as a natural generalization of the notion of unconditional basic sequence.

Definition B.5 Let X be a Banach space and let X^* be its dual. Suppose that a sequence $\{x_n\}_{n=1}^\infty \subset X$ has a sequence of biorthogonal functionals $\{x_n^*\}_{n=1}^\infty \subset X^*$. Then, we say that $(x_n, x_n^*)_{n=1}^\infty$ is a *system of random unconditional convergence* (briefly an *RUC-system*) if there exists a constant $C > 0$ such that for all $m \in \mathbb{N}$ and arbitrary $x \in [x_n]_X$ we have

$$\int_0^1 \Big\| \sum_{i=1}^m r_n(s) x_n^*(x) x_n \Big\|_X ds \le C \|x\|_X,$$

where $r_n(s)$ are the Rademacher functions, $n = 1, 2, \dots$

Sometimes the biorthogonal functionals x_n^* can be identified with the elements x_n themselves (for example, if $\{x_n\}_{n=1}^\infty$ is an orthonormal sequence of functions defined on a measure space). Then, if the condition of the last definition is satisfied we say that $\{x_n\}_{n=1}^\infty$ is an RUC-system.

Definition B.6 An unconditional basic sequence $\{x_n\}_{n=1}^{\infty}$ in a Banach space X is called *symmetric* if it is equivalent in X to arbitrary sequence of the form $\{x_{\pi(n)}\}_{n=1}^{\infty}$, where π is a permutation of $\mathbb{N}$.

It is not difficult to show (see also [1, Lemma 9.2.1]) that a basic sequence $\{x_n\}_{n=1}^{\infty}$ is symmetric in X if and only if there is a constant $C > 0$ such that for all $x = \sum_{n=1}^{\infty} a_n x_n \in X$ the inequality

$$\Big\| \sum_{n=1}^{\infty} \theta_n a_n x_{\pi(n)} \Big\|_X \le C \Big\| \sum_{n=1}^{\infty} a_n x_n \Big\|_X .$$

holds for all choices of $\theta_n = \pm 1$ and all permutations π of $\mathbb{N}$. The least constant C with the above property is called the *symmetric constant* of $\{x_n\}_{n=1}^{\infty}$.

Clearly, if $\{x_n\}_{n=1}^{\infty}$ has symmetric constant 1, then its unconditional basis constant is also 1 (see [181, the beginning of Section 3a, p. 113]).

For more details on bases, basic sequences and their properties, see the monographs [1, 153, 181, 182], and the survey paper [206].

We turn now to the so-called *lacunary* systems of functions (or r.v.'s). The term "lacunarity" means that some properties of such a system resemble those of sequences of independent r.v.'s. As a rule, a lacunary system is being a quite sparse subsystem of some more "massive" sequence that can be complete in one or other function space.

A typical example of lacunary systems is the main object of this book—the sequence of Rademacher functions defined on the interval [0, 1] by

$$r_n(t) = \operatorname{sign} \sin 2^n \pi t, \quad n = 1, 2, \ldots$$

$\{r_n\}_{n=1}^{\infty}$ is a subsystem of the Walsh system, which is complete in $L_p[0, 1]$ for each $1 \le p < \infty$. The Rademacher sequence is orthonormal and, what is more, is strongly multiplicative in the sense of the next definition.

Definition B.7 A finite or countable system $\{f_n\}$ of r.v.'s defined on a probability space $(\Omega, \Sigma, \mathbb{P})$ is said to be *multiplicative* provided that for any pairwise distinct positive integers $n_1, n_2, \ldots, n_k$, $k \in \mathbb{N}$, we have

$$\mathbb{E}(f_{n_1} \cdot f_{n_2} \ldots f_{n_k}) = 0.$$

If additionally

$$\mathbb{E}(f_{n_1} \cdot f_{n_2} \dots f_{n_k} \cdot f_n^2) = 0$$

for any pairwise distinct positive integers $n_1, n_2, \dots, n_k$ and $n \neq n_i$, $i = 1, 2, \dots, k$, then $\{f_n\}$ is called a *strongly multiplicative* system.

Let us proceed with some important examples of lacunary systems.

Example B.1 (Lacunary Trigonometric System) Let $f_n(x) = \sin(2\pi k_n x)$, $x \in [0, 1]$, where $(k_n)_{n=1}^{\infty}$ is a strictly increasing sequence of positive integers. One can readily check that this system is multiplicative if $k_{n+1}/k_n \geq 2$ and it is strongly multiplicative if $k_{n+1}/k_n \geq 3$. Note that the difference between these conditions is substantial. For example, the system $\{\sin(2^n \pi x)\}_{n=1}^{\infty}$ is clearly multiplicative, but not strongly multiplicative. Indeed, an easy calculation shows that

$$\begin{aligned}\int_0^1 \sin^2(2\pi x)\sin(4\pi x)\sin(8\pi x)\,dx &= \frac{1}{2}\Big(\int_0^1 \sin(4\pi x)\sin(8\pi x)\,dx \\ &\quad - \int_0^1 \cos(4\pi x)\sin(4\pi x)\sin(8\pi x)\,dx\Big) \\ &= -\frac{1}{4}\int_0^1 \sin^2(8\pi x)\,dx < 0.\end{aligned}$$

Example B.2 (Independent Mean Zero r.v.'s) A finite or countable sequence $\{f_n\}$ of independent r.v.'s such that $f_n \in L_2(\Omega, \Sigma, \mathbb{P})$ and $\mathbb{E} f_n = 0$, $n = 1, 2, \dots$, is a strongly multiplicative system. This is an immediate consequence of equation (A.1).

Example B.3 (Martingale Differences) Let $\{\xi_i\}$ be a finite or countable martingale with respect to an increasing sequence of σ-algebras $\{\mathcal{A}_i\}$. Then, the *martingale differences* $f_i := \xi_i - \xi_{i-1}$, $i = 1, 2, \dots$, where $\xi_0 := \mathbb{E}\xi_1$, form a multiplicative system. In fact, for all $i_1 < i_2 < \cdots < i_k$, due to the well-known properties of the conditional expectation [97, Lemma 1.13], we have

$$\int_\Omega f_{i_1} \dots f_{i_k}\,d\mathbb{P} = \int_\Omega \mathbb{E}\left[f_{i_1} \dots f_{i_k} | \mathfrak{F}_{i_k-1}\right] d\mathbb{P} = \int_\Omega f_{i_1} \dots f_{i_{k-1}} \cdot \mathbb{E}\left[f_{i_k} | \mathfrak{F}_{i_k-1}\right] d\mathbb{P} = 0,$$

because

$$\mathbb{E}\left[f_{i_k} | \mathcal{A}_{i_k-1}\right] = \mathbb{E}\left[\xi_{i_k} | \mathcal{A}_{i_k-1}\right] - \xi_{i_k-1} = \xi_{i_k-1} - \xi_{i_k-1} = 0,$$

by the definition of martingales.

As was already mentioned, lacunary systems share certain properties of systems of independent r.v.'s. Here are the definitions related to the *absolute convergence* of lacunary series. An initial motivation for their introduction was the classical Sidon theorem on the absolute convergence of lacunary (in the Hadamard sense) trigonometric Fourier series of bounded functions [267] (see also Comments and References to Chap. 8).

Definition B.8 Let $(\Omega, \Sigma, \mathbb{P})$ be a probability space, and let $\{f_n\}_{n=1}^{\infty}$ be a basic sequence in $L_2(\Omega, \Sigma, \mathbb{P})$. It is called a *Sidon system* provided that from the condition $f = \sum_{n=1}^{\infty} a_n f_n \in L_\infty(\Omega, \Sigma, \mathbb{P})$ (convergence in $L_2(\Omega, \Sigma, \mathbb{P})$) it follows that $(a_n)_{n=1}^{\infty} \in \ell_1$.

Definition B.9 A basic sequence $\{f_n\}_{n=1}^{\infty}$ in $L_2[0, 1]$ is said to be a *Sidon–Zygmund system* if there exists a set $E \subset [0, 1]$ of positive Lebesgue measure such that for each interval $I \subset [0, 1]$, with $m(I \cap E) > 0$, from the condition $\sum_{n=1}^{\infty} a_n f_n \in L_\infty(I)$ (convergence in $L_2[0, 1]$) it follows that $(a_n)_{n=1}^{\infty} \in \ell_1$. If we may take here as E the whole interval $[0, 1]$, then we say that $\{f_n\}_{n=1}^{\infty}$ is a *strong Sidon–Zygmund system*.

Clearly, any strong Sidon–Zygmund system is being a Sidon–Zygmund system, and any Sidon–Zygmund system, in turn, is a Sidon system on $[0, 1]$ with the Lebesgue measure.

Another property of lacunary systems is connected with the *integrability* of sums of series; it owes its appearance, first of all, to Khintchine's inequality (see Chap. 1) and also to a similar result for lacunary trigonometric series due to Zygmund [290].

Definition B.10 A sequence of r.v.'s $\{f_n\}_{n=1}^{\infty}$, $f_n \in L_p(\Omega, \Sigma, \mathbb{P})$, $p > 2$, is a S_p*-system* if

$$\Big\| \sum_{n=1}^{m} a_n f_n \Big\|_p \le K_p \Big\| \sum_{n=1}^{m} a_n f_n \Big\|_2,$$

where a constant $K_p > 0$ does not depend on $m \in \mathbb{N}$ and $a_n \in \mathbb{R}$, $n = 1, 2, \ldots, m$.

If a sequence $\{f_n\}$ is a S_p-system for every $p > 2$, then it is said to be a S_∞-*system*.

For more detailed information on lacunary systems, see the survey paper [122] as well as the monographs [146] and [153].

Appendix C
Banach Function Lattices and Symmetric Spaces

Let (T, Σ, μ) be a σ-finite measure space, that is, μ is a σ-finite measure on a σ-algebra Σ of subsets of T. The set of all a.e. finite real-valued functions (of equivalence classes) defined on T with natural algebraic operations and the topology of convergence in measure μ on sets of finite measure is a linear topological space. We denote it by $S(T, \Sigma, \mu)$. The notation $x \le y$, where $x, y \in S(T, \Sigma, \mu)$, means that there is a set $A \subset T$ such that $\mu(T \setminus A) = 0$ and $x(t) \le y(t)$ for $t \in A$ (we say that $x \le y$ a.e. on T). In particular, the functions, which coincide a.e., are identified.

As a rule, (T, Σ, μ) will be one of the following measure spaces:

(a) $I := [0, 1]$ with the Lebesgue measure;
(b) $I \times I$ with the Lebesgue measure;
(c) $[0, \infty)$ with the Lebesgue measure;
(d) $\mathbb{Z}$ (resp. $\mathbb{N}$) with the counting measure μ defined by $\mu(\{k\}) = 1, \; k \in \mathbb{Z}$ (resp. $k \in \mathbb{N}$).

In what follows, we denote by $m(A)$ the Lebesgue measure of a measurable set $A \subset \mathbb{R}^n$, $n \in \mathbb{N}$. Moreover, $\chi_B(t)$ denotes the *characteristic* function (or the indicator) of a set $B \subset T$, that is, $\chi_B(t) = 1$ if $t \in B$, and $\chi_B(t) = 0$ if $t \notin B$.

A linear subspace $E \subset S(T, \Sigma, \mu)$ is said to be a *function lattice* (or *ideal space*) if $\{x \in S(T, \Sigma, \mu) : |x| \le |y|\} \subset E$ for every $y \in E$. Any function lattice E, equipped with a *monotone norm* (i.e., $\|x\| \le \|y\|$ for any $x, y \in E$ such that $|x| \le |y|$), is called a *normed function lattice* (or *normed ideal space*). If E is complete as a normed linear space, we shall say that E is a *Banach function lattice* (or *Banach ideal space*).

In the case when a measure space (T, Σ, μ) is of type (d) we say, respectively, about *sequence lattices*, *normed* and *Banach sequence lattices*.

Every normed lattice E over a σ-finite measure space (T, Σ, μ) is linearly and continuously embedded into the linear topological space $S(T, \Sigma, \mu)$ (see e.g. [151, Theorem 4.3.1]). This means that from the convergence in norm it follows the convergence in measure on all subsets of T of finite measure.

S. V. Astashkin, *The Rademacher System in Function Spaces*,
https://doi.org/10.1007/978-3-030-47890-2

If E is a Banach function lattice, then the *Köthe dual* (or *associated*) function lattice E' consists of all $y \in S(T, \Sigma, \mu)$ such that

$$\|y\|_{E'} := \sup\left\{\int_T x(t)y(t)\,d\mu : \|x\|_E \leq 1\right\} < \infty$$

(in the case when E is a Banach sequence lattice the integral should be replaced with the sum). Any Köthe dual function lattice E' is complete with respect to the norm $y \mapsto \|y\|_{E'}$ and is embedded isometrically into (Banach) dual space E^*. Every Banach function lattice E is continuously embedded into its second Köthe dual E''; moreover, $\|x\|_{E''} \leq \|x\|_E$ for $x \in E$. A Banach function lattice E *has the Fatou property* (or *maximal*) if from $x_n \in E$, $n = 1, 2, \ldots$, $\sup_{n=1,2,\ldots} \|x_n\|_E < \infty$, $x \in S(T, \Sigma, \mu)$ and $x_n \to x$ a.e. on T it follows that $x \in E$ and $||x||_E \leq \liminf_{n\to\infty} ||x_n||_E$. Observe that such a lattice E has the Fatou property whenever from $x_n \in E$, $x_n \geq 0$, $n = 1, 2, \ldots$, $x \in S(T, \Sigma, \mu)$, and $x_n \uparrow x$ a.e. on T it follows that $x \in E$ and $\lim_{n\to\infty} \|x_n\|_E = \|x\|_E$. Moreover, a Banach function lattice E has the Fatou property if and only if the natural inclusion of E into E'' is a surjective isometry [151, Theorem 6.1.7].

Let E be a Banach function lattice. We say that a function $x \in E$ has *absolute continuous norm* in E whenever for arbitrary decreasing sequence of measurable sets $B_n \subset T$, $n = 1, 2, \ldots$, such that $\bigcap_{n=1}^{\infty} B_n = \varnothing$ we have $\|x \cdot \chi_{B_n}\|_E \to 0$. If E is a Banach function lattice over one of the measure spaces listed in (a)–(d), then it is separable if and only if any function $x \in E$ has in E absolute continuous norm. Each of these conditions is equivalent as well to the (isometric) equality $E' = E^*$.

The norm in a Banach function lattice E is called *order semi-continuous* if the conditions $x_n \in E$, $n = 1, 2, \ldots$, $x \in E$, and $x_n \to x$ a.e. on T imply that $||x||_E \leq \liminf_{n\to\infty} ||x_n||_E$. This is equivalent to the fact that E is embedded into E'' isometrically, that is,

$$||x||_E = \sup\left\{\int_T x(t)y(t)\,d\mu : ||y||_{E'} \leq 1\right\}$$

[151, Theorem 6.1.6]. In particular, if a Banach function lattice is separable or has the Fatou property, then its norm is order semi-continuous.

For a detailed exposition of the theory of Banach function and sequence lattices see, for example, the monographs [70, 151, 182].

An important class of Banach lattices is formed by the so-called symmetric spaces. The theory of symmetric spaces is widely represented in monographs [70, 168, 182, 253]; we recall here only some definitions and results that relate mainly to symmetric spaces of functions defined on [0, 1].

For any $x = x(t) \in S[0, 1] := S([0, 1], m)$, $x \geq 0$, we introduce the *distribution function* $n_x(\tau)$ by

$$n_x(\tau) := m\{t \in [0, 1] : x(t) > \tau\}, \quad \tau > 0$$

(it is worth to note that this function somewhat differs from its "probabilistic" version from Appendix A). The function $n_x(\tau)$ is nonnegative, decreasing, and right-continuous. Two functions $x, y \in S[0,1]$ are called *equimeasurable* if $n_{|x|}(\tau) = n_{|y|}(\tau)$ for all $\tau > 0$. In particular, if functions $x, y \in S[0,1]$ are *identically distributed*, i.e.,

$$m\{t \in [0,1] : x(t) > \tau\} = m\{t \in [0,1] : y(t) > \tau\}, \quad \tau > 0,$$

then they are equimeasurable. Also, any function $x \in S[0,1]$ is equimeasurable with its decreasing, left-continuous rearrangement $x^*(t)$ defined by

$$x^*(t) := \inf\{\tau \geq 0 : n_{|x|}(\tau) < t\}, \quad 0 < t \leq 1.$$

Definition C.1 A Banach function lattice $X \subset S[0,1]$ is said to be a *symmetric space*[1] (briefly s.s.) whenever from the fact that $x \in X$ and $y \in S[0,1]$ are equimeasurable it follows $y \in X$ and $\|y\|_X = \|x\|_X$.

We shall call also every such a norm *symmetric*.

Let us give the most important examples of s.s.'s. First, these are the spaces L_p, $1 \leq p \leq \infty$, with the usual norms

$$\|x\|_p = \begin{cases} \left(\int_0^1 |x(t)|^p\,dt\right)^{1/p}, & 1 \leq p < \infty \\ \operatorname{ess\,sup}_{t\in[0,1]} |x(t)|, & p = \infty \end{cases},$$

and their generalization, the family of spaces $L_{p,q}$, $1 < p < \infty$, $1 \leq q \leq \infty$, equipped with the norms

$$\|x\|_{p,q} = \begin{cases} \left[\frac{q}{p}\int_0^1 \left(t^{1/p}x^*(t)\right)^q \frac{dt}{t}\right]^{1/q}, & 1 \leq q < \infty, \\ \operatorname{ess\,sup}_{t\in[0,1]} \left(t^{1/p}x^*(t)\right), & q = \infty. \end{cases}$$

Though the functional $x \mapsto \|x\|_{p,q}$ does not satisfy the triangle inequality for $p < q \leq \infty$, it is equivalent to a symmetric norm. Namely, for all $1 < p < \infty$, $1 \leq q \leq \infty$, and every $x \in L_{p,q}$ we have

$$c_{p,q}\|x^{**}\|_{p,q} \leq \|x\|_{p,q} \leq \|x^{**}\|_{p,q}, \quad \text{where } x^{**}(t) := \frac{1}{t}\int_0^t x^*(s)\,ds$$

[1] It is widely known also the more restrictive notion of *rearrangement invariant* space. Such a space is meant either a symmetric space that is separable or has the Fatou property [182], or a symmetric space with the Fatou property [70].

(the constant $c_{p,q}$ depends only on p and q). One can easily check [168, Lemma II.6.5] that $L_{p,q_1} \subset L_{p,q_2}$ if $1 \le q_1 \le q_2 \le \infty$, and $L_{p,p} = L_p$ isometrically.

Another natural generalization of L_p-spaces are Orlicz spaces (for a detailed account of their theory see the monographs [167, 191]). Let M be an *Orlicz function,* that is, a continuous, strictly increasing and convex function on $[0, \infty)$, $M(0) = 0$. Then, the *Orlicz space* L_M consists of all $x \in S[0, 1]$ such that there exists $\lambda > 0$ satisfying the inequality

$$\int_0^1 M\left(\frac{|x(t)|}{\lambda}\right) dt \le 1.$$

This space is equipped with the *Luxemburg norm* $\|x\|_{L_M} = \inf \lambda$, where the infimum is taken over all λ such that the last inequality holds. The Köthe dual space L'_M can be identified with the Orlicz space L_{M^*}, where $M^*(u)$ is the complementary (in the Young sense) Orlicz function to the function $M(u)$, i.e.,

$$M^*(u) := \sup_{v \ge 0}(uv - M(v))$$

(see, for example, [167, § 14]).

In the theory of s.s.'s an important role is played by the following class of functions. A function $\varphi(t)$ defined on $[0, 1]$ is called *quasiconcave* if $\varphi(0) = 0$, $\varphi(t)$ increases, and $\varphi(t)/t$ decreases. Any increasing concave function on $[0, 1]$ that vanishes at zero is quasiconcave [168, § II.1, p. 67]. On the other hand, any quasiconcave function $\varphi(t)$ is equivalent to its least concave majorant $\bar{\varphi}(t)$; more precisely, we have

$$\frac{1}{2}\bar{\varphi}(t) \le \varphi(t) \le \bar{\varphi}(t), \quad 0 < t \le 1$$

[168, Theorem II.1.1].

If φ is a quasiconcave function and $p \in [1, \infty)$, then the *Lorentz space* $\Lambda_p(\varphi)$ consists of all $x \in S[0, 1]$ such that

$$\|x\|_{\varphi,p} := \left(\int_0^1 (x^*(t))^p \, d\bar{\varphi}(t)\right)^{1/p} < \infty.$$

In particular, the space $\Lambda_1(\varphi)$ will be denoted by $\Lambda(\varphi)$. The *Marcinkiewicz space* $M(\varphi)$ is the set of all $x \in S[0, 1]$ such that

$$\|x\|_{M(\varphi)} := \sup_{0<t\le 1} \frac{1}{\varphi(t)} \int_0^t x^*(s)\, ds < \infty.$$

The spaces $\Lambda(\varphi)$ and $M(\varphi)$ are Köthe dual to each other, i.e., $\Lambda(\varphi)' = M(\varphi)$ and $M(\varphi)' = \Lambda(\varphi)$ [168, Theorems II.5.2 and II.5.4].

We say that a nonnegative functional V defined on a s.s. X is *convex* if $V(x+y) \leq V(x) + V(y)$, $x, y \in X$, and $V(\alpha x) = |\alpha| V(x)$, $x \in X$, $\alpha \in \mathbb{R}$.

The following property characterizes the Lorentz spaces in the class of all s.s.'s [168, Lemma II.5.2].

Theorem C.1 *Let φ be an increasing concave function on $[0, 1]$, $\lim_{t\to 0} \varphi(t) = 0$. If a convex functional V defined on the Lorentz space $\Lambda(\varphi)$ is bounded on the set of characteristic functions, i.e., for some $C > 0$ and all $A \subset [0, 1]$ we have $V(\chi_A) \leq C\varphi(m(A))$, then V is bounded on the whole space $\Lambda(\varphi)$.*

The spaces L_1 and L_∞, respectively, are the largest and the smallest in the class of s.s.'s on $[0, 1]$, that is, for every s.s. X the continuous embeddings $L_\infty \subset X \subset L_1$ hold [168, Theorem II.4.1].

By X° we shall denote the *separable part* of a s.s. X, i.e., the closure of L_∞ in X. Observe that X° is a s.s., which is separable whenever $X \neq L_\infty$ [168, Lemma II.4.4]. Then, X° consists of all functions of X that have absolute continuous norm in the latter space. In particular, the space $M(\varphi)^\circ$ is the set of all functions $x(t)$ satisfying

$$\lim_{t\to 0+} \frac{1}{\varphi(t)} \int_0^t x^*(s)\, ds = 0.$$

Moreover, we have $(M(\varphi)^\circ)^* = (M(\varphi)^\circ)' = \Lambda(\varphi)$ [168, Theorem II.5.4].

An important characteristic of a s.s. X is the *fundamental function* $\phi_X(s) := \|\chi_{(0,s)}\|_X$, $0 \leq s \leq 1$. One can readily check that

$$\phi_{L_{p,q}}(s) = s^{1/p} \text{ for all } 1 < p < \infty,\ 1 \leq q \leq \infty,$$

$$\phi_{\Lambda(\varphi)}(s) = \bar{\varphi}(s) \text{ and } \phi_{M(\varphi)}(s) = \tilde{\varphi}(s), \quad \text{with } \tilde{\varphi}(s) = s/\varphi(s).$$

Also,

$$\phi_{L_M}(s) = \frac{1}{M^{-1}(1/s)},$$

where M^{-1} is the inverse function for M.

For an arbitrary s.s. X the fundamental function ϕ_X is quasiconcave on $[0, 1]$ [168, Theorem II.4.7]. Furthermore, the Lorentz and Marcinkiewicz spaces are extremal in the class of s.s.'s with the same fundamental function. Specifically, if φ is the fundamental function of a s.s. X, then we have

$$\Lambda(\varphi) \subset X \subset M(\tilde{\varphi})$$

(the embeddings are continuous) [168, Theorems II.5.5 and II.5.7].

The *dilation function* $\mathcal{M}_f$ of a nonnegative function f on the interval $(0, l)$, where $l = 1$ or $l = \infty$, is defined by

$$\mathcal{M}_f(t) := \sup\left\{\frac{f(st)}{f(s)} : s \in (0, l \cdot \min(1, 1/t))\right\}, \quad t > 0.$$

Since $\mathcal{M}_f$ is submultiplicative on $(0, \infty)$ (i.e., $\mathcal{M}_f(t_1 t_2) \leq \mathcal{M}_f(t_1) \cdot \mathcal{M}_f(t_2)$ for $t_1, t_2 > 0$), then there exist the numbers

$$\gamma_f := \lim_{t \to 0+} \frac{\ln \mathcal{M}_f(t)}{\ln t} = \sup_{0<t<1} \frac{\ln \mathcal{M}_f(t)}{\ln t}$$

and

$$\delta_f := \lim_{t \to +\infty} \frac{\ln \mathcal{M}_f(t)}{\ln t} = \inf_{1<t<\infty} \frac{\ln \mathcal{M}_f(t)}{\ln t},$$

called the *lower* and *upper dilation exponents* of f, respectively [168, § II.1, p. 75–76]. In the case when φ is a quasiconcave function, we have $0 \leq \gamma_\varphi \leq \delta_\varphi \leq 1$, $\gamma_{\tilde{\varphi}} = 1 - \delta_\varphi$, and $\delta_{\tilde{\varphi}} = 1 - \gamma_\varphi$.

The following result often simplifies the calculation (up to equivalence) of the norm in a Marcinkiewicz space (see [168, Theorem II.5.3] and [168, Lemma II.1.4]).

Theorem C.2 *Let φ be a quasiconcave function on* $[0, 1]$. *The following conditions are equivalent:*

(a) we have

$$\|x\|_{M(\varphi)} \asymp \sup_{0<t\leq 1} \frac{t}{\varphi(t)} x^*(t)$$

with a constant independent of $x \in M(\varphi)$;

(b) $\gamma_\varphi > 0$.

Moreover, if one of these conditions holds, then $\mathcal{M}_\varphi(1/2) < 1$ and

$$\|x\|_{M(\varphi)} \leq \frac{2 \ln 2}{1 - \mathcal{M}_\varphi(1/2)} \sup_{0<t\leq 1} \frac{t}{\varphi(t)} x^*(t).$$

The *dilation operator*

$$\sigma_\tau x(t) := x(t/\tau) \cdot \chi_{[0, \min(1,\tau)]}(t), \quad \tau > 0,$$

is bounded in every s.s. X and $\|\sigma_\tau\|_{X \to X} \leq \max(1, \tau)$ [168, Theorem II.4.5]. The numbers α_X and β_X given by

$$\alpha_X := \lim_{\tau \to 0+} \frac{\ln \|\sigma_\tau\|_{X \to X}}{\ln \tau} = \sup_{0<\tau<1} \frac{\ln \|\sigma_\tau\|_{X \to X}}{\ln \tau} \tag{C.1}$$

and

$$\beta_X := \lim_{\tau\to+\infty} \frac{\ln \|\sigma_\tau\|_{X\to X}}{\ln \tau} = \inf_{\tau>1} \frac{\ln \|\sigma_\tau\|_{X\to X}}{\ln \tau}. \tag{C.2}$$

are called the *Boyd indices* of X. For each s.s. X we have

$$0 \le \alpha_X \le \gamma_{\phi_X} \le \delta_{\phi_X} \le \beta_X \le 1,$$

where γ_{ϕ_X} and δ_{ϕ_X} are the lower and upper dilation exponents of the fundamental function ϕ_X [168, § II.1, p. 138]. Also, one can check that

$$\alpha_{L_{p,q}} = \beta_{L_{p,q}} = 1/p\ , \ \alpha_{\Lambda(\varphi)} = \gamma_\varphi\ , \ \beta_{\Lambda(\varphi)} = \delta_\varphi\ , \ \alpha_{M(\varphi)} = \gamma_{\tilde\varphi}\ , \ \beta_{M(\varphi)} = \delta_{\tilde\varphi}$$

(here, as above, $\tilde\varphi(t) = t/\varphi(t)$). Moreover, if a s.s. X is isometrically embedded into its second Köthe dual X'', then $\alpha_X + \beta_{X'} = \alpha_{X'} + \beta_X = 1$, $\alpha_{X''} = \alpha_X$, and $\beta_{X''} = \beta_X$ [168, Theorem II.4.11].

The next useful result is an immediate consequence of the definition of s.s.'s and the boundedness of the dilation operator in such a space (see [168, Corollary II.4.2, p. 133]).

Theorem C.3 *Let X be a s.s., $x \in S[0,1]$ and $y \in X$ be such that $n_{|x|}(u) \le C n_{|y|}(u)$, for some constant $C > 0$ and all $u > 0$. Then, $x \in X$ and $\|x\|_X \le \max(1, C)\|y\|_X$.*

We shall need also the notion of a symmetric space of r.v.'s defined on a probability space $(\Omega, \Sigma, \mathbb{P})$. Specifically, if X is a s.s. on $[0,1]$, then the corresponding space $X(\Omega) := X(\Omega, \Sigma, \mathbb{P})$ consists of all r.v.'s $f : \Omega \to \mathbb{R}$ such that $f^* \in X$. Here, f^* is the decreasing, left-continuous rearrangement of an r.v. f, i.e.,

$$f^*(t) := \inf\{\tau \ge 0 : \mathbb{P}\{\omega \in \Omega : |f(\omega)| > \tau\} < t\}, \quad 0 < t \le 1.$$

Moreover, we set $\|f\|_{X(\Omega)} := \|f^*\|_X$. This definition allows us to transfer, in a natural way, all basic concepts of the theory of s.s.'s on $[0,1]$ to the case of s.s's on a probability space $(\Omega, \Sigma, \mathbb{P})$.

Symmetric sequence spaces may be defined in a quite similar way. Namely, a sequence Banach lattice E is said to be a *symmetric space* (briefly, a *sequence s.s.*) whenever for any $a = (a_k)_{k=1}^\infty \in E$ and arbitrary permutation π of $\mathbb{N}$ the sequence $a_\pi = (a_{\pi(k)})_{k=1}^\infty$ belongs to E and $\|a_\pi\|_E = \|a\|_E$. The most important examples of sequence s.s.'s are the spaces ℓ_p, $1 \le p \le \infty$, with the usual norms

$$\|a\|_{\ell_p} := \begin{cases} \left(\sum_{k=1}^\infty |a_k|^p\right)^{1/p}, & 1 \le p < \infty \\ \sup\limits_{k=1,2,\dots} |a_k|\ , & p = \infty \end{cases}.$$

Their generalization, the $\ell_{p,q}$-spaces, $1 < p < \infty$, $1 \le q \le \infty$, are equipped with the norms

$$\|a\|_{\ell_{p,q}} := \Big\{ \sum_{k=1}^{\infty} (a_k^*)^q k^{q/p-1} \Big\}^{1/q} \quad \text{if } 1 \le q < \infty,$$

and

$$\|a\|_{\ell_{p,\infty}} := \sup_{k=1,2,\dots} a_k^* k^{1/p},$$

where $(a_k^*)_{k=1}^{\infty}$ is the decreasing permutation of the sequence $(|a_k|)_{k=1}^{\infty}$ (as in the case of function spaces, the functional $a \mapsto \|a\|_{\ell_{p,q}}$ does not satisfy the triangle inequality for $p < q \le \infty$, but it is equivalent to a symmetric norm).

One more example is the sequence Marcinkiewicz space $m(\varphi)$ (φ is a quasiconcave function on $[0, \infty)$), which consists of all bounded sequences of real numbers $a = (a_k)_{k=1}^{\infty}$ such that

$$\|a\|_{m(\varphi)} := \sup_{n=1,2,\dots} \frac{1}{\varphi(n)} \sum\nolimits_{k=1}^{n} a_k^* < \infty.$$

For each $m \in \mathbb{N}$ by $\tilde{\sigma}_m$ and $\tilde{\sigma}_{1/m}$ we define the *dilation operators* as follows: if $a = (a_n)_{n=1}^{\infty}$, then

$$\tilde{\sigma}_m a = ((\tilde{\sigma}_m a)_n)_{n=1}^{\infty} = \left(a_{[\frac{m-1+n}{m}]}\right)_{n=1}^{\infty} = \Big(\overbrace{a_1, a_1, \dots, a_1}^{m}, \overbrace{a_2, a_2, \dots, a_2}^{m}, \dots\Big)$$

and

$$\tilde{\sigma}_{1/m} a = \left((\tilde{\sigma}_{1/m} a)_n\right)_{n=1}^{\infty} = \Big(\frac{1}{m} \sum_{k=(n-1)m+1}^{nm} a_k\Big)_{n=1}^{\infty}$$

$$= \left(\frac{1}{m} \sum_{k=1}^{m} a_k, \frac{1}{m} \sum_{k=m+1}^{2m} a_k, \dots, \frac{1}{m} \sum_{k=(n-1)m+1}^{nm} a_k, \dots \right).$$

(see, for example, [168, p. 223]). Similarly as in the case of function spaces, these operators are bounded in every sequence s.s. E. In the same way as in (C.1) and (C.2), we may define the *lower* and *upper Boyd indices* $\tilde{\alpha}_E$ and $\tilde{\beta}_E$ of E, only instead of $\|\sigma_\tau\|_{X \to X}$ we use the operator norms $\|\tilde{\sigma}_{1/m}\|_{E \to E}$ and $\|\tilde{\sigma}_m\|_{E \to E}$, respectively. It is easy to show that $0 \le \tilde{\alpha}_E \le \tilde{\beta}_E \le 1$ for every sequence s.s. E. Moreover, $\tilde{\alpha}_{\ell_p} = \tilde{\beta}_{\ell_p} = 1/p$ for all $1 \le p \le \infty$ (here, $1/\infty = 0$).

In many issues related to the theory of function and sequence s.s.'s an important role is played by the Hardy–Littlewood operators

$$\mathcal{H}f(s) = \frac{1}{s}\int_0^s f(u)\,du \ , \quad \tilde{\mathcal{H}}(a_j) = \Big(\frac{1}{k}\sum_{j=1}^{k} a_j\Big)_{k=1}^{\infty},$$

and by the corresponding conjugate operators

$$\mathcal{H}'f(s) = \int_s^1 \frac{f(u)}{u}\,du \ , \quad \tilde{\mathcal{H}}'(a_j) = \Big(\sum_{j=k}^{\infty} \frac{a_j}{j}\Big)_{k=1}^{\infty}.$$

Theorem C.4 ([168, § II.8])

(1) The operator $\mathcal{H}$ (resp. $\tilde{\mathcal{H}}$) is bounded in a function s.s. X (resp. sequence s.s. E) if and only if $\beta_X < 1$ (resp. $\tilde{\beta}_E < 1$).
(2) The operator $\mathcal{H}'$ (resp. $\tilde{\mathcal{H}}'$) is bounded in a function s.s. X (resp. sequence s.s. E) if and only if $\alpha_X > 0$ (resp. $\tilde{\alpha}_E > 0$).

In particular, the following classical Hardy inequality holds:

$$\Big\| \frac{1}{t}\int_0^t f(s)\,ds \Big\|_p \le p'\|f\|_p,$$

where $1 < p \le \infty$ and $1/p + 1/p' = 1$. This inequality, proved by Hardy [132] as early as in 1925, has became later on a starting point for further investigations of more general integral inequalities (see e.g. [170]).

Appendix D
Interpolation of Operators and Spaces of the Real Method of Interpolation

A pair $\vec{X} = (X_0, X_1)$ of Banach spaces is called a *Banach couple* if X_0 and X_1 are both linearly and continuously embedded in some Hausdorff linear topological space. In particular, arbitrary Banach lattices E_0 and E_1 over the same underlying σ-finite measure space (T, Σ, μ) form a Banach couple, because every such a Banach lattice is continuously embedded into the space $S(T, \Sigma, \mu)$ (see Appendix C).

For a Banach couple (X_0, X_1) we define the *intersection* $X_0 \cap X_1$ and the *sum* $X_0 + X_1$ as the Banach spaces equipped with the norms

$$\|x\|_{X_0\cap X_1} := \max\left\{\|x\|_{X_0},\ \|x\|_{X_1}\right\}$$

and

$$\|x\|_{X_0+X_1} := \inf\left\{\|x_0\|_{X_0} + \|x_1\|_{X_1} : x = x_0 + x_1, x_i \in X_i, i = 0, 1\right\},$$

respectively. A Banach space X is called *intermediate* for a Banach couple $\vec{X} = (X_0, X_1)$ if $X_0 \cap X_1 \subset X \subset X_0 + X_1$.

For any Banach couples $\vec{X} = (X_0, X_1)$ and $\vec{Y} = (Y_0, Y_1)$ denote by $L(\vec{X}, \vec{Y})$ the Banach space of all linear operators $T : X_0 + X_1 \to Y_0 + Y_1$ that are bounded from X_i in Y_i, $i = 0, 1$, with the norm

$$\|T\|_{L(\vec{X},\vec{Y})} := \max_{i=0,1} \|T\|_{X_i\to Y_i}.$$

Let $\vec{X} = (X_0, X_1)$ and $\vec{Y} = (Y_0, Y_1)$ be two Banach couples, and let X and Y be intermediate Banach spaces for $\vec{X}$ and $\vec{Y}$, respectively. We say that (X_0, X_1, X) is an *interpolation triple with respect to the triple* (Y_0, Y_1, Y) if every operator $T \in L(\vec{X}, \vec{Y})$ is bounded from X in Y. Then, due to the Closed Graph theorem, there exists $C > 0$ such that for any $T \in L(\vec{X}, \vec{Y})$ we have $\|T\|_{X\to Y} \le C\|T\|_{L(\vec{X},\vec{Y})}$. If we can take here $C = 1$, then (X_0, X_1, X) is called an *exact interpolation triple with*

S. V. Astashkin, *The Rademacher System in Function Spaces*,
https://doi.org/10.1007/978-3-030-47890-2

respect to the triple (Y_0, Y_1, Y). In particular, in the case when $X_0 = Y_0$, $X_1 = Y_1$, and $X = Y$ the space X is said to be an *(exact) interpolation space with respect to the couple* (X_0, X_1) (or *between the spaces* X_0 and X_1). In this case, we write: $X \in \mathcal{I}(X_0, X_1)$. The least constant C, satisfying the inequality

$$\|T\|_{X \to X} \le C\|T\|_{L(\vec{X},\vec{X})},$$

is called the *interpolation constant* of X.

Historically, the first interpolation result was the following classical Riesz–Thorin theorem (see e.g. [71, Theorem 1.1.1]).

Theorem D.1 *Suppose that* (T, Σ, μ) *and* (T', Σ', μ') *are two* σ*-finite measure spaces,* $1 \le p_0, p_1, q_0, q_1 \le \infty$, $p_0 \ne p_1$, $q_0 \ne q_1$. *Then, if all the spaces are considered over the complex field and for some* $\theta \in [0, 1]$ *we have*

$$\frac{1}{p} = \frac{1-\theta}{p_0} + \frac{\theta}{p_1}, \quad \frac{1}{q} = \frac{1-\theta}{q_0} + \frac{\theta}{q_1},$$

then $(L_{p_0}(T, \mu), L_{p_1}(T, \mu), L_p(T, \mu))$ *is an exact interpolation triple with respect to the triple* $(L_{q_0}(T', \mu'), L_{q_1}(T', \mu'), L_q(T', \mu'))$.

Remark D.1 In the real case we can guarantee only that the following weaker inequality holds:

$$\|T\|_{L_p(T,\mu)\to L_q(T',\mu')} \le 2 \cdot \|T\|_{L(\vec{X},\vec{Y})},$$

where $\vec{X} = (L_{p_0}(T, \mu), L_{p_1}(T, \mu))$, $\vec{Y} = (L_{q_0}(T', \mu'), L_{q_1}(T', \mu'))$ [166, § 1.2, p. 38].

Theorem D.1 is valid for the vector-valued L_p-spaces as well [71, Theorem 5.1.2]. Let us recall one of its numerous applications.

Let $\mathcal{S}$ be a compact Abelian group with a probabilistic Haar measure μ, and let ν be a σ-finite complex-valued function, defined on the Borel σ-algebra of $\mathcal{S}$, with the finite variation $\|\nu\|$. If $f * \nu$ is the convolution of ν and a vector-valued function $f : \mathcal{S} \to F$ (F is a Banach space), i.e.,

$$(f * \nu)(s) := \int_{\mathcal{S}} f(s-\sigma)\, d\nu(\sigma), \quad s \in \mathcal{S},$$

then it is not hard to check that $\|f * \nu\|_{L_1(\mathcal{S})} \le \|\nu\|\|f\|_{L_1(\mathcal{S})}$ and $\|f * \nu\|_{L_\infty(\mathcal{S})} \le \|\nu\|\|f\|_{L_\infty(\mathcal{S})}$. Here and in the subsequent Theorem D.2, we set

$$\|g\|_{L_p(\mathcal{S})} := \left(\int_{\mathcal{S}} \|g(s)\|_F^p\, d\mu\right)^{1/p}, \quad 1 \le p \le \infty$$

(with a natural modification for $p = \infty$). Applying Theorem D.1 in the vector-valued setting for $p_0 = q_0 = 1$ and $p_1 = q_1 = \infty$, we come to the following result.

Theorem D.2 (Vector-Valued Young Inequality) *For every* $1 \le p \le \infty$ *we have*

$$\|f * \nu\|_{L_p(\mathcal{S})} \le \|\nu\| \|f\|_{L_p(\mathcal{S})}.$$

Another significant result, which deals with interpolation properties of operators bounded in s.s.'s on [0, 1], is the following Boyd theorem (see e.g. [182, Theorem 2.b.11]).

Theorem D.3 *Suppose that a linear operator* T *is bounded in the space* $L_p = L_p[0, 1]$ *for each* $1 < p < \infty$. *If the Boyd indices of a s.s.* X *satisfy the condition* $0 < \alpha_X \le \beta_X < 1$, *then* $X \subset \bigcup_{1<p<\infty} L_p$, *and* T *is bounded in* X.

Sometimes, it is possible to find a description of all interpolation spaces of a certain class with respect to some Banach couple. So, it is known the following useful identification of interpolation Marcinkiewicz spaces with respect to an arbitrary couple of Marcinkiewicz spaces (see Theorem 4 and Lemma 2 from [7]).

Theorem D.4 *Let* ψ_0 *and* ψ_1 *be quasiconcave functions on* $[0, 1]$ *such that the function* ψ_1/ψ_0 *is increasing on* $[0, 1]$ *and* $\lim_{t\to 0} \psi_1(t)/\psi_0(t) = 0$. *Then, a Marcinkiewicz space* $M(\psi)$ *is an interpolation space between the spaces* $M(\psi_0)$ *and* $M(\psi_1)$ *if and only if for some* $C > 0$ *and all* $0 < s, t \le 1$ *we have*

$$\frac{\psi(s)}{\psi(t)} \le C \max\left\{\frac{\psi_0(s)}{\psi_0(t)}, \frac{\psi_1(s)}{\psi_1(t)}\right\}.$$

One of the most important ways of constructing interpolation spaces is based on use of the so-called *Peetre $\mathcal{K}$-functional* of a Banach couple (X_0, X_1), which is defined as follows:

$$\mathcal{K}(t, x; X_0, X_1) := \inf\{||x_0||_{X_0} + t||x_1||_{X_1} : x = x_0 + x_1, x_i \in X_i\},$$

where $x \in X_0 + X_1$ and $t > 0$. One can easily show that $\mathcal{K}(t, x; X_0, X_1)$, with a fixed $x \in X_0 + X_1$, is a continuous increasing concave nonnegative function for $t > 0$ [71, Lemma 3.1.1]. Conversely, fixing $t > 0$, we obtain a norm on the sum $X_0 + X_1$ that is equivalent to the $X_0 + X_1$-norm. Moreover, if $X_0 \overset{1}{\subset} X_1$ (resp. $X_1 \overset{1}{\subset} X_0$), then $\mathcal{K}(t, x; X_0, X_1) = t\|x\|_{X_1}$, $0 \le t \le 1$ (resp. $\mathcal{K}(t, x; X_0, X_1) = \|x\|_{X_0}$, $t \ge 1$).

Sometimes the $\mathcal{K}$-functional admits a quite simple equivalent expression. In particular, for each σ-finite measure space and arbitrary $p \ge 1$ we have

$$\mathcal{K}(t, x; L_p, L_\infty) \asymp \left(\int_0^{t^p} x^*(s)^p\, ds\right)^{1/p}, \quad t > 0, \tag{D.1}$$

with the equivalence constant independent of $x \in L_p + L_\infty$ and $t > 0$ [71, Theorem 5.2.1]. Note that equivalence (D.1) becomes an equality for $p = 1$.

Next, the following result for a couple of Lorentz spaces holds: if φ_0 and φ_1 are two increasing concave functions on [0, 1] such that the function $\varphi_0(s)/\varphi_1(s)$ is decreasing, then

$$\mathcal{K}(t, x; \Lambda(\varphi_0), \Lambda(\varphi_1)) = \int_0^1 x^*(s)\, d(\min\{\varphi_0(s), t\varphi_1(s)\}), \quad 0 < t \le 1 \qquad \text{(D.2)}$$

[168, Theorem II.5.9].

Under the above conditions on φ_0 and φ_1, we can identify (up to equivalence) also the $\mathcal{K}$-functional for a couple of Marcinkiewicz spaces. Indeed, from the equation $(E_0 \cap E_1)' = E_0' + E_1'$, which holds for any Banach lattices E_0 and E_1, a formula for the norm of the intersection of Lorentz spaces [168, Theorem II.5.10], and the duality relations for Lorentz and Marcinkiewicz spaces from Appendix C it follows that

$$\mathcal{K}(t, x; M(\varphi_0), M(\varphi_1)) \asymp \sup_{0<u\le 1} \frac{\int_0^u x^*(s)\, ds}{\max(\varphi_0(u), \varphi_1(u)/t)}, \quad t > 0 \qquad \text{(D.3)}$$

(see also [103]).

Let $\{u_k\}_{k=-\infty}^{\infty}$ be an arbitrary sequence of positive numbers, and let F be a Banach lattice of two-sided sequences. Denote by $F(u_k)$ the *weighted space* consisting of all sequences $a = (a_k)_{k=-\infty}^{\infty}$ such that $(a_k u_k)_{k=-\infty}^{\infty} \in E$ with the following natural norm:

$$\|a\|_{F(u_k)} := \|(a_k u_k)\|_F$$

(a *weighted function space* is defined similarly). We say that a Banach lattice F of two-sided sequences, with $F \supset l_\infty \cap l_\infty(2^{-k})$, is a *parameter of the* $\mathcal{K}$*-method of interpolation*. If F is a parameter of the $\mathcal{K}$-method of interpolation and (X_0, X_1) is a Banach couple, then the *space of the real* $\mathcal{K}$*-method* $(X_0, X_1)_F^{\mathcal{K}}$ consists of all $x \in X_0 + X_1$ such that

$$(\mathcal{K}(2^k, x; X_0, X_1))_{k=-\infty}^{\infty} \in F,$$

and it is equipped with the norm

$$\|x\| := \|(\mathcal{K}(2^k, x; X_0, X_1))_k\|_F.$$

One can readily check that $\left(X_0, X_1, (X_0, X_1)_F^{\mathcal{K}}\right)$ is an exact interpolation triple with respect to the triple $\left(Y_0, Y_1, (Y_0, Y_1)_F^{\mathcal{K}}\right)$ for arbitrary Banach couples (X_0, X_1) and (Y_0, Y_1) [88, Proposition 1.4.2]. Moreover, each sublinear operator bounded from a

Banach couple (X_0, X_1) in a Banach couple (Y_0, Y_1) is bounded from $(X_0, X_1)_F^{\mathcal{K}}$ in $(Y_0, Y_1)_F^{\mathcal{K}}$ [88, § 11].

In the special case when $F = \ell_p(2^{-k\theta})$, where $0 < \theta < 1$ and $1 \le p \le \infty$, we obtain the scale of the classical *Lions–Peetre interpolation spaces* $(X_0, X_1)_{\theta,p}$ with the norms

$$\|x\|_{(X_0,X_1)_{\theta,p}} := \Big(\sum_{k=-\infty}^{\infty} \big(\mathcal{K}(2^k, x; X_0, X_1)2^{-k\theta}\big)^p\Big)^{1/p}$$

(with a natural modification for $p = \infty$). For a detailed information related to their properties, we refer to the monograph [71].

As parameters of the $\mathcal{K}$-method of interpolation, we can use equivalently also function Banach lattices on the semi-axis $(0, \infty)$. In particular, it is not hard to check (see [71, Lemma 3.1.3]) that for arbitrary $0 < \theta < 1$ and $1 \le p \le \infty$ we have

$$2^{-\theta}(\ln 2)^{1/p}\|x\|_{(X_0,X_1)_{\theta,p}} \le \Big(\int_0^{\infty} \big(t^{-\theta}\mathcal{K}(t, x; X_0, X_1)\big)^p \frac{dt}{t}\Big)^{1/p} \le 2^{1+\theta}\|x\|_{(X_0,X_1)_{\theta,p}}$$

(again with a natural modification for $p = \infty$). Hence, the middle integral expression in the last inequality, being the norm of the $\mathcal{K}$-functional $\mathcal{K}(t, x; X_0, X_1)$ in the certain weighted function L_p-space, is equivalent to the norm of $(X_0, X_1)_{\theta,p}$.

In some cases, it can be found a rather satisfactory description of the spaces $(X_0, X_1)_{\theta,p}$. So, if (X_0, X_1) is a couple the L_p-spaces, these are the $L_{p,q}$-spaces (see Appendix C). More precisely, the following result holds [71, Theorem 5.2.1].

Theorem D.5 *Let* $L_{p_i} := L_{p_i}(T, \Sigma, \mu)$, $i = 0, 1$, *where* (T, Σ, μ) *is an arbitrary* σ*-finite measure space,* $1 \le p_0 < p_1 \le \infty$, $1 \le q \le \infty$, $1/p = (1-\theta)/p_0+\theta/p_1$. *Then,* $(L_{p_0}, L_{p_1})_{\theta,q} = L_{p,q}$ *(with equivalence of norms).*

In particular, we have $(\ell_{p_0}, \ell_{p_1})_{\theta,q} = \ell_{p,q}$.

It is worth to note that for a quite wide class of Banach couples we can obtain, by using the real $\mathcal{K}$-method, all interpolation spaces with respect to them.

Definition D.1 A Banach couple $\vec{X} = (X_0, X_1)$ is called a *$\mathcal{K}$-monotone* (or *Calderón–Mityagin*) couple if from $x \in X_0 + X_1$, $y \in X_0 + X_1$, and the inequality

$$\mathcal{K}(t, y; X_0, X_1) \le \mathcal{K}(t, x; X_0, X_1), \quad t > 0,$$

it follows the existence of an operator $U \in L(\vec{X}, \vec{X})$ such that $y = Ux$.

The second name of $\mathcal{K}$-monotone couples is justified by the fact that historically the first result in this direction was the theorem on a description of interpolation spaces with respect to the Banach couple (L_1, L_∞), proved independently by Calderón [94] and Mityagin [208]. Let us present one of equivalent formulations of this statement [168, Theorem II.4.3], using the above expression (D.1) for the $\mathcal{K}$-functional for the couple (L_1, L_∞).

Theorem D.6 *A s.s. X on $[0, 1]$ is an interpolation space with respect to the couple (L_1, L_∞) if and only if the following condition holds: if $x \in X$, $y \in L_1$, and*

$$\int_0^t y^*(s)\,ds \le \int_0^t x^*(s)\,ds, \quad 0 \le t \le 1, \tag{D.4}$$

then $y \in X$ and $\|y\|_X \le C\|x\|_X$, where $C > 0$ is the interpolation constant of the space X.

In particular, it follows from Theorem D.6 that the class of interpolation spaces with respect to the couple (L_1, L_∞) includes all s.s.'s, which are separable or have the Fatou property [168, Theorems II.4.9 and II.4.10].

Inequality (D.4) (in brief, $y \prec x$) generates on the space $L_1[0, 1]$ a (partial) ordering, which is often called the *Hardy–Littlewood–Pólya submajorization*.

One of the central results in the interpolation theory of operators is the following Brudnyi–Kruglyak theorem [88, Theorem 15.1], which shows that all interpolation spaces with respect to a $\mathcal{K}$-monotone couple are generated by the real $\mathcal{K}$-method.

Theorem D.7 *If a Banach couple (X_0, X_1) is $\mathcal{K}$-monotone, then every interpolation space X with respect to this couple is representable in the form*

$$X = (X_0, X_1)_F^{\mathcal{K}}$$

for some parameter F of the real $\mathcal{K}$-method.

According to the well-known Sparr theorem [268], an arbitrary couple of weighted L_p-spaces $(L_{p_0}(u_0), L_{p_1}(u_1))$ is $\mathcal{K}$-monotone. Therefore, the statement of Theorem D.7, in particular, is valid for such a couple.

A detailed exposition of the above-mentioned and many other results and facts of the interpolation theory of operators can be found in the monographs [70, 71, 88, 89, 166, 168, 191].

References

1. F. Albiac, N.J. Kalton, in *Topics in Banach Space Theory*. Graduate Texts in Mathematics, vol. 233 (Springer, New York, 2006)
2. J. Alvarez Alonso, The distribution function in the Morrey space. Proc. Am. Math. Soc. **83**, 693–699 (1981)
3. D.S. Anisimov, S.V. Kislyakov, Double singular integrals: interpolation and correction. St. Petersburg Math. J. **16**, 749–772 (2005)
4. N. Annoni, L. Grafakos, P. Honzik, On an inequality of Sagher and Zhou concerning Stein's lemma. Collect. Math. **60**, 297–306 (2009)
5. N. Annoni, L. Grafakos, P. Honzik, An inequality concerning the Rademacher functions. Preprint (Institute Math., AS CR, Prague, 2007-12-10), 8pp.
6. J. Arias-de-Reyna, in *Pointwise Convergence of Fourier Series*. Lecture Notes in Math. vol. 1785 (Springer, 2002)
7. I.U. Asekritova, in *On $\mathcal{K}$-functional of the couple* $(\mathcal{K}_{\Phi_0}(\vec{X}), \mathcal{K}_{\Phi_1}(\vec{X}))$. Studies in the Theory of Functions of Several Variables (Yaroslav. Gos. Univ., Yaroslavl', 1980), pp. 3–32 (in Russian)
8. N.H. Asmar, S. Montgomery-Smith, On the distribution of Sidon series. Arkiv för Mat. **31**, 13–26 (1993)
9. S.V. Astashkin, in *On Some Bilinear Multiplicative Functional*. Studies in the Theory of Functions of Several Variables (Yaroslav. Gos. Univ., Yaroslavl', 1982), pp. 3–15 (in Russian)
10. S.V. Astashkin, Interpolation of operators in quasinormed groups of measurable functions. Sib. Math. J. **35**, 1075–1082 (1994)
11. S.V. Astashkin, On multiplier of a rearrangement invariant space with respect to the tensor product. Funct. Anal. Appl. **30**, 267–269 (1996)
12. S.V. Astashkin, On interpolation of subspaces of symmetric spaces generated by the Rademacher system. Izv. RAEN ser. MMMIU **1**(1), 18–35 (1997)
13. S.V. Astashkin, Operators of weak type and stability of the real interpolation method in symmetric spaces. Izv. RAEN, ser. MMMIU **1**(4), 3–25 (1997)
14. S.V. Astashkin, Tensor product in symmetric function spaces, in *Proc. of Fourth Intern. Conf. Funct. Spaces* (Zielona Gora, 1995). Collect. Math. **48**, 375–391 (1997)
15. S.V. Astashkin, Rademacher chaos in symmetric spaces. East J. Approx. **4**, 311–336 (1998)
16. S.V. Astashkin, On series with respect to the Rademacher system in symmetric spaces "close" to L_∞. Funct. Anal. Appl. **32**, 192–194 (1998)
17. S.V. Astashkin, Multiple Rademacher series in rearrangement invariant spaces. Funct. Anal. Appl. **33**, 141–143 (1999)

S. V. Astashkin, *The Rademacher System in Function Spaces*,
https://doi.org/10.1007/978-3-030-47890-2

18. S.V. Astashkin, Extraction of subsystems "majorized" by the Rademacher system. Math. Notes **65**, 407–417 (1999)
19. S.V. Astashkin, On the selection of subsystems equivalent in distribution to the Rademacher system. Math. Notes **66**, 384–388 (1999)
20. S.V. Astashkin, Rademacher chaos in symmetric spaces, 2. East J. Approx. **6**, 71–86 (2000)
21. S.V. Astashkin, Systems of random variables equivalent in distribution to the Rademacher system and $\mathcal{K}$-closed representability of Banach couples. Sbornik Math. **191**, 779–807 (2000)
22. S.V. Astashkin, Fourier-Rademacher coefficients of functions in symmetric spaces. Sib. Math. J. **41**, 601–610 (2000)
23. S.V. Astashkin, About interpolation of subspaces of rearrangement invariant spaces generated by Rademacher system. Int. J. Math. Math. Sci. **25**, 451–465 (2001)
24. S.V. Astashkin, Extrapolation properties of the scale of L_p-spaces. Sbornik Math. **194**, 23–42 (2003)
25. S.V. Astashkin, On the multiplier space generated by the Rademacher system. Math. Notes **75**, 158–165 (2004)
26. S.V. Astashkin, Independent functions in symmetric spaces and Kruglov property. Sbornik Math. **199**, 945–963 (2008)
27. S.V. Astashkin, Rademacher functions in symmetric spaces. Contemporary mathematics. Fundamental Directions **32**, 3–161 (2009); English transl. in: J. Math. Sci. (N.Y.) **169**, 725–886 (2010)
28. S.V. Astashkin, On the comparison of distribution functions of random variables. Math. Notes **87**, 15–22 (2010)
29. S.V. Astashkin, On the geometric properties of Cesàro spaces. Sbornik Math. **203**, 514–533 (2012)
30. S.V. Astashkin, Geometrical properties of Banach spaces generated by sublinear operators. Positivity **17**, 223–234 (2012)
31. S.V. Astashkin, On subspaces generated by independent functions in symmetric spaces with Kruglov property. St. Petersburg Math. J. **25**, 513–527 (2014)
32. S.V. Astashkin, Khintchine inequality on sets of small measure. Funct. Anal. Appl. **48**, 235–241 (2014)
33. S.V. Astashkin, Martingale transforms of Rademacher sequence in symmetric spaces. St. Petersburg Math. J. **27**, 191–206 (2016)
34. S.V. Astashkin, Rademacher series in weighted symmetric spaces. Isr. J. Math. **218**, 371–390 (2017)
35. S.V. Astashkin, On comparing systems of random variables with the Rademacher sequence. Izvestiya Math. **81**, 1063–1079 (2017)
36. S.V. Astashkin, M.Sh. Braverman, *A Subspace of a Symmetric Space Generated by the Rademacher System with Vector Coefficients*, in *Operator Equations in Function Spaces* (Voron. Gos. Univ., Voronezh, 1986), pp. 3–10 (in Russian)
37. S.V. Astashkin, G.P. Curbera, Symmetric kernel of Rademacher multiplicator spaces. J. Funct. Anal. **226**, 173–192 (2005)
38. S.V. Astashkin, G.P. Curbera, Rademacher multiplicator spaces equal to L^∞. Proc. Am. Math. Soc. **136**, 3493–3501 (2008)
39. S.V. Astashkin, G.P. Curbera, Rearrangement invariance of Rademacher multiplicator spaces. J. Funct. Anal. **256**, 4071–4094 (2009)
40. S.V. Astashkin, G.P. Curbera, A weighted Khintchine inequality. Rev. Mat. Iberoam. **30**, 237–246 (2014)
41. S.V. Astashkin, G.P. Curbera, Local Khintchine inequality in rearrangement invariant spaces. Ann. Matem. Pura Appl. **194**, 619–643 (2015)
42. S.V. Astashkin, G.P. Curbera, K.E. Tikhomirov, On the existence of RUC systems in rearrangement invariant spaces. Math. Nachr. **289**, 175–186 (2016)
43. S.V. Astashkin, F.L. Hernandez, E.M. Semenov, Strictly singular inclusions of rearrangement invariant spaces and Rademacher spaces. Studia Math. **193**, 269–283 (2009)

44. S.V. Astashkin, M. Leibov, L. Maligranda, Rademacher functions in *BMO*. Studia Math. **205**, 83–100 (2011)
45. S.V. Astashkin, K.V. Lykov, Strong extrapolation spaces and interpolation. Sib. Math. J. **50**, 199–213 (2009)
46. S.V. Astashkin, K.V. Lykov, *Extrapolation Description of Rearrangement Invariant Spaces and Related Problems*, in *Proc. Intern. Symp. on Banach and Function Spaces III* (Kitakyushu, Japan, 2009) (Yokohama Publishers, 2011), pp. 1–52
47. S.V. Astashkin, K.V. Lykov, Sparse Rademacher chaos in symmetric spaces. St. Petersburg Math. J. **28**, 1–20 (2017)
48. S.V. Astashkin, L. Maligranda, Rademacher functions in Cesáro type spaces. Studia Math. **198**, 235–247 (2010)
49. S.V. Astashkin, L. Maligranda, Structure of Cesáro function spaces: a survey, in *Proc. Conf. "Function Spaces X"*, vol. 102 (Banach Center Publ., Warszawa, 2014), pp. 13–40
50. S.V. Astashkin, L. Maligranda, Structure of Rademacher subspaces in Cesáro type spaces. Studia Math. **226**, 259–279 (2015)
51. S.V. Astashkin, L. Maligranda, Rademacher functions in Morrey spaces. J. Math. Anal. Appl. **441**, 1133–1154 (2016)
52. S.V. Astashkin, L. Maligranda, E.M. Semenov, On complementability of subspaces generated by contractions and shifts of functions. Doklady Math. **387**, 583–585 (2002)
53. S.V. Astashkin, L. Maligranda, E.M. Semenov, Multiplicator space and complemented subspaces of rearrangement invariant space. J. Funct. Anal. **202**, 247–276 (2003)
54. S.V. Astashkin, L. Maligranda, R.S. Sukhanov, Sequences of independent Walsh functions in *BMO*. Sib. Math. J. **54**, 205–211 (2013)
55. S.V. Astashkin, E.M. Semenov, Fourier-Rademacher coefficients. Doklady Math. **375**, 7–9 (2000)
56. S.V. Astashkin, E.M. Semenov, Spaces defined by Paley function. Sbornik Math. **204**, 937–957 (2013)
57. S.V. Astashkin, E.M. Semenov, Paley spaces. Sib. Math. J. **56**, 21–27 (2015)
58. S.V. Astashkin, R.S. Sukhanov, On some properties of Rademacher chaos. Math. Notes **91**, 613–624 (2012)
59. S.V. Astashkin, F.A. Sukochev, Comparison of sums of independent and disjoint functions in symmetric spaces. Math. Notes **76**, 449–454 (2004)
60. S.V. Astashkin, F.A. Sukochev, Series of independent random variables in rearrangement invariant spaces: an operator approach. Isr. J. Math. **145**, 125–156 (2005)
61. S.V. Astashkin, F.A. Sukochev, Series of independent mean zero random variables in rearrangement invariant spaces with the Kruglov property. J. Math. Sci. **148**, 795–809 (2008)
62. S.V. Astashkin, F.A. Sukochev, Independent functions and the geometry of Banach spaces. Russ. Math. Surv. **65**(396), 1003–1081 (2011)
63. S.V. Astashkin, E.M. Semenov, F.A. Sukochev, D.V. Zanin, Kruglov operator and operators defined by random permutations. Funct. Anal. Appl. **43**, 83–95 (2009)
64. S.V. Astashkin, F.A. Sukochev, D.V. Zanin, Disjointification inequalities in symmetric quasi-Banach spaces and their applications. Pac. J. Math. **270**, 257–285 (2014)
65. S. Banach, Sur les séries lacunaires. Bull. Int. Acad. Polon. Sci. A. 4–8, 149–154 (1933)
66. W. Beckner, Inequalities in Fourier analysis. Ann. Math. **102**, 159–182 (1975)
67. W. Bednorz, R. Latała, On the suprema of Bernoulli processes. C. R. Math. Acad. Sci. Paris **351**, 131–134 (2013)
68. W. Bednorz, R. Latała, On the boundedness of Bernoulli processes. Ann. Math. **140**, 1167–1203 (2014)
69. A.S. Belov, V.A. Rodin, Norms of lacunary polynomials in functional spaces. Math. Notes **51**, 318–320 (1992)
70. C. Bennett, R. Sharpley, *Interpolation of Operators* (Academic Press, London, 1988)
71. J. Bergh, J. Löfström, *Interpolation Spaces. An Introduction* (Springer, Berlin, 1976)
72. C. Bessaga, A. Pełczyński, On bases and unconditional convergence of series in Banach spaces. Studia Math. **17**, 151–164 (1958)

73. C. Bessaga, A. Pełczyński, Some remarks on conjugate spaces containing subspaces isomorphic to the space c_0. Bull. Acad. Polon. Sci. Sér. Sci. Math. Astr. Phys. **6**, 249–250 (1958)
74. P. Billard, S. Kwapień, A. Pełczyński, Ch. Samuel, Biorthogonal systems of random unconditional convergence in Banach spaces. Longhorn Notes. Texas Func. Anal. Seminar., 1985–1986, pp.'13–35
75. O. Blasco, A. Ruiz, L. Vega, Non-interpolation in Morrey-Campanato and block spaces. Ann. Scuola Norm. Sup. Pisa Cl. Sci. (4) **28**, 31–40 (1999)
76. R. Blei, *Analysis in Integer and Fractional Dimensions* (Cambridge University Press, Cambridge, UK, 2001)
77. R. Blei, S. Janson, Rademacher chaos: tail estimates versus limit theorems. Ark. Mat. **42**, 13–29 (2004)
78. A. Boccuto, A.V. Bukhvalov, A.R. Sambucini, Some inequalities in classical spaces with mixed norms. Positivity **6**, 393–411 (2002)
79. V.I. Bogachev, O.G. Smolyanov, *Real and Functional Analysis* (Springer, 2020)
80. C. Borell, A note on the integrability of Walsh polynomials, 1979 (unpublished)
81. A.A. Borovkov, *Probability Theory* (Springer, London, 2013)
82. J. Bourgain, Embedding L^1 in L_1/H^1. Trans. Am. Math. Soc. **278**, 689–702 (1983)
83. J. Bourgain, M. Lewko, Sidonicity and variants of Kaczmarz's problem. Ann. Inst. Fourier (Grenoble) **67**, 1321–1352 (2017)
84. J. Bourgain, L. Tzafriri, Invertibility of "large" submatrices with applications to the geometry of Banach spaces and harmonic analysis. Isr. J. Math. **57**, 137–224 (1987)
85. M.Sh. Braverman, Complementability of subspaces generated by independent functions in a symmetric space. Funct. Anal. Appl. **16**, 129–130 (1982)
86. M.Sh. Braverman, I.Ya. Novikov, Subspaces of symmetric spaces, generated by independent random variables. Sib. Math. J. **25**, 361–370 (1984)
87. M.Sh. Braverman, *Independent Random Variables and Rearrangement Invariant Spaces* (Cambridge University Press, 1994)
88. Yu.A. Brudnyi, N.Ya. Kruglyak, *Real Interpolation Functors* (Yaroslavl', 1981) (Deposited in VINITI 13.05.1981, Dep. N 2620-81) 211 pp. (in Russian)
89. Yu.A. Brudnyi, N.Ya. Kruglyak, *Interpolation Functors and Interpolation Spaces 1* (North-Holland, 1991)
90. A.V. Bukhvalov, On connection of symmetric spaces on the square with mixed norm spaces. Proc. Conf. Tartus. Gos. Univ. (Tartu, 1980), 100–102 (in Russian)
91. D.L. Burkholder, Martingale transforms. Ann. Math. Stat. **37**, 1494–1504 (1966)
92. D.L. Burkholder, Independent sequences with the Stein property. Ann. Math. Stat. **39**, 1282–1288 (1968)
93. D.L. Burkholder, R.F. Gandy, Extrapolation and interpolation of quasi-linear operators on martingales. Acta Math. **124**, 249–304 (1970)
94. A.P. Calderón, Spaces between L^1 and L^∞ and the theorem of Marcinkiewicz. Studia Math. **26**, 273–299 (1966)
95. N.L. Carothers, in *A Short Course on Banach Space Theory*. London Mathematical Society Student Texts, vol. 64 (Cambridge University Press, Cambridge, UK, 2005)
96. J. Carrillo-Alanís, On local Khintchine inequalities for spaces of exponential integrability. Proc. Am. Math. Soc. **139**, 2753–2757 (2011)
97. S.N. Cohen, R.J. Elliott, in *Stochastic Calculus and Applications*, 2nd edn. Probability and Its Applications (Springer, New York, 2015)
98. G.P. Curbera, Banach space properties of L^1 of a vector measure. Proc. Am. Math. Soc. **123**, 3797–3806 (1995)
99. G.P. Curbera, A note on function spaces generated by Rademacher series. Proc. Edinburgh Math. Soc. **40**, 119–126 (1997)
100. G.P. Curbera, V.A. Rodin, Multipliers on the set of Rademacher series in symmetric spaces. Funct. Anal. Appl. **36**, 244–246 (2002)
101. G.P. Curbera, V.A. Rodin, Multiplication operators on the space of Rademacher series in rearrangement invariant spaces. Math. Proc. Camb. Phil. Soc. **134**, 153–162 (2003)

102. M. Cwikel, Monotonicity properties of interpolation spaces 2. Arkiv för Mat. **19**, 123–136 (1981)
103. M. Cwikel, P. Nilsson, Interpolation of Marcinkiewicz spaces. Math. Scand. **56**, 29–42 (1985)
104. V.H. De la Peña, E. Giné, in *Decoupling: From Dependence to Independence*. Prob. Its Appl. (Springer, New York, Berlin, Heidelberg, 1999)
105. V.H. De la Peña, S.J. Montgomery-Smith, Bounds on the tail probability of U-statistics and quadratic forms. Bull. Am. Math. Soc. **31**, 223–227 (1994)
106. V.H. De la Peña, S.J. Montgomery-Smith, Decoupling inequalities for the tail probabilities of multivariate U-statistics. Ann. Probab. **23**, 806–816 (1995)
107. V.H. De la Peña, S.J. Montgomery-Smith, J. Szulga, Contraction and decoupling inequalities for multilinear forms and U-statistics. Ann. Probab. **22**, 1745–1765 (1994)
108. S.J. Dilworth, S. Montgomery-Smith, The distribution of vector-valued Rademacher series. Ann. Probab. **21**, 2046–2052 (1993)
109. J. Diestel, in *Sequences and Series in Banach Spaces*. Graduate Texts in Mathematics, vol. 92 (Springer, New York, 1984)
110. J. Diestel, H. Jarchow, A. Tonge, *Absolutely Summing Operators* (Cambridge University Press, Cambridge, 1995)
111. P.G. Dodds, E.M. Semenov, F.A. Sukochev, RUC-systems in rearrangement invariant spaces. Studia Math. **151**, 161–173 (2002)
112. P.G. Dodds, E.M. Semenov, F.A. Sukochev, The Banach-Saks property in rearrangement invariant spaces. Studia Math. **162**, 263–294 (2004)
113. N. Dunford, J. Schwartz, *Linear Operators Part II: Spectral Theory* (Interscience, New York, 1963)
114. M.L. Eaton, A note on symmetric Bernoulli random variables. Ann. Math. Stat. **41**, 1223–1226 (1970)
115. R.E. Edwards, in *Fourier Series. A Modern Introduction, Vol. 2*, 2nd edn. Graduate Texts in Mathematics, vol. 85 (Springer, New York, 1982)
116. R.E. Edwards, K.A. Ross, p-Sidon sets. J. Funct. Anal. **15**, 404–427 (1974)
117. P. Fatou, Séries trigonométrique et séries de Taylor. Acta Math. **30**, 335–400 (1906)
118. W. Feller, *An Introduction to Probability Theory and Its Applications, Vol. 1*, 3rd edn. (Wiley, New York, 1968)
119. T. Figiel, P. Hitczenko, W.B. Johnson, G. Schechtman, J. Zinn, Extremal properties of Rademacher functions with applications to the Khintchine and Rosenthal inequalities. Trans. Am. Math. Soc. **349**, 997–1027 (1997)
120. T. Figiel, T. Iwaniec, A. Pełczyński, Computing norms and critical exponents of some operators in L^p-spaces. Studia Math. **79**, 227–274 (1984)
121. C. Franchetti, E.M. Semenov, A Hilbert space characterization among function spaces. Anal. Math. **21**, 85–93 (1995)
122. V.F. Gaposhkin, Lacunary series and independent functions. Russ. Math. Surv. **21**, 3–82 (1966)
123. V.F. Gaposhkin, Convergence and limit theorems for sequences of random variables. Probab. Appl. **17**, 379–400 (1972)
124. J.B. Garnett, in *Bounded Analytic Functions*. Graduate Texts in Mathematics, vol. 236 (Springer, New York, 2007)
125. J.B. Garnett, P.W. Jones, The distance in BMO to L^∞. Ann. Math. **108**, 373–393 (1978)
126. B. Golubov, A. Efimov, V. Skvortsov, *Walsh Series and Transforms: Theory and Applications* (Springer, 1991). [Mathematics and Its Applications, Soviet series, vol. 64]
127. U. Haagerup, The best constants in the Khintchine inequality. Studia Math. **70**, 231–283 (1982)
128. U. Haagerup, M. Musat, On the best constants in noncommutative Khintchine-type inequalities. J. Funct. Anal. **250**, 588–624 (2007)
129. U. Haagerup, M. Musat, Factorization and dilation problems for completely positive maps on von Neumann algebras. Comm. Math. Phys. **303**, 555–594 (2011)
130. R.R. Hall, On a conjecture of Littlewood. Math. Proc. Camb. Phil. Soc. **78**, 443–445 (1975)

131. K.E. Hare, R. (Xu) Yang, Sidon sets are proportionally Sidon with small Sidon constants. Can. Math. Bull. **62**(4), 798–809 (2019)
132. G.H. Hardy, Notes on some points in the integral calculus, LX, An inequality between integrals. Messenger Math. **54**, 150–156 (1925)
133. B.D. Hassard, D.A. Hussein, On Cesàro function spaces. Tamkang J. Math. **4**, 19–25 (1973)
134. F.L. Hernandez, S.Ja. Novikov, E.M. Semenov, Strictly singular embeddings between rearrangement invariant spaces. Positivity **7**, 119–124 (2003)
135. P. Hitczenko, Domination inequality for martingale transforms of a Rademacher sequence. Isr. J. Math. **84**, 161–178 (1993)
136. T. Holmstedt, Interpolation of quasi-normed space. Math. Scand. **26**, 177–199 (1970)
137. T.P. Hÿtonen, J.M.A.M. van Neerven, M.C. Veraar, L.W. Weis, in *Analysis in Banach Spaces, Vol. II: Probabilistic Methods and Operator Theory*. A Series of Modern Surveys in Mathematics, vol. 67 (Springer, 2017)
138. A.A. Jagers, A note on Cesàro sequence spaces. Nieuw Arch. Wiskund (3) **22**, 113–124 (1974)
139. J. Jakubowski, S. Kwapień, On multiplicative systems of functions. Bull. Acad. Pol. Sci. Ser. Sci. Math. **27**, 689–694 (1979)
140. S. Janson, Interpolation of subcouples and quotient couples. Arkiv Math. **31**, 307–338 (1993)
141. F. John, L. Nirenberg, On functions of bounded mean oscillation. Comm. Pure Appl. Math. **14**, 415–426 (1961)
142. W.B. Johnson, B. Maurey, G. Schechtman, L. Tzafriri, Symmetric structures in Banach spaces. Mem. Am. Math. Soc. **19**, 298 pp. (1979)
143. W.B. Johnson, G. Schechtman, Martingale inequalities in rearrangement invariant function spaces. Isr. J. Math. **64**, 267–275 (1988)
144. M. Junge, Embedding of the operator space OH and the logarithmic "little Grothendieck inequality". Invent. Math. **161**, 389–406 (2005)
145. S. Kaczmarz, G. Steinhaus, Le systéme orthogonal de M. Rademacher. Studia Math. **2**, 231–247 (1930) (in French)
146. S. Kaczmarz, G. Steinhaus, *Theorie der Orthogonalreihen* (Warszawa-Lwów, 1935) (in German)
147. M.I. Kadec, A. Pełczyński, Bases, lacunary sequences and complemented subspaces in the spaces L_p. Studia Math. **21**, 161–176 (1961/1962)
148. J.-P. Kahane, *Some Random Series of Functions*, 2nd edn. (Cambridge University Press, 1985)
149. N. Kalton, Rademacher series and decoupling. New York J. Math. **11**, 563–595 (2005)
150. A. Kamińska, L. Maligranda, L.E. Persson, Type, cotype and convexity properties of Orlicz spaces. Publ. del Depart. de Anal. Matem. Fac. de Matem. Univ. Compl. de Madrid **42**, 113–126 (1996–1997)
151. L.V. Kantorovich, G.P. Akilov, *Functional Analysis*, 2nd edn. (Pergamon Press, Oxford-Elmsford, New York, 1982)
152. B.S. Kashin, On some properties of the space of trigonometric polynomials with a uniform norm. Proc. Steklov Inst. Math. **145**, 121–127 (1981)
153. B.S. Kashin, A.A. Saakyan, *Orthogonal Series* (Amer. Math. Soc., Providence RI, 1989)
154. A. Khintchine, Über dyadische Bruche. Math. Zeit. **18**, 109–116 (1923)
155. A. Khintchine, A.N. Kolmogorov, Über Konvergenz von Reihen, dieren Glieder durch den Zufall bestimmt werden. Mat. Sb. **32**, 668–677 (1925)
156. M. Kikuchi, Characterization of Banach function spaces that preserve the Burkholder square-function inequality. Ill. J. Math. **47**, 867–882 (2003)
157. S.V. Kislyakov, Interpolation of H^p-spaces: some recent developments. Funct. spaces, interpol. spaces, and rel. topics (Haifa, 1995). Isr. Math. Conf. Proc. **113**, 102–140 (1999)
158. S.V. Kislyakov, On BMO-regular lattices of measurable functions. St. Petersburg Math. J. **14**, 273–286 (2003)
159. S.V. Kislyakov, On BMO-regular lattices of measurable functions. II. J. Math. Sci. (N.Y.) **129**, 4018–4021 (2005)

160. S.V. Kislyakov, Q.H. Xu, Real interpolation and singular integrals. St. Petersburg Math. J. **8**, 593–615 (1997)
161. A. Kolmogoroff, Über das Gesetz des iterierten Logarithmus. Math. Ann. **101**, 126–135 (1929)
162. R. Komorowski, On the best possible constants in the Khintchine inequality for $p \geq 2$. Bull. Lond. Math. Soc. **20**, 73–75 (1988)
163. H. König, C. Schütt, N. Tomczak-Jaegermann, Proection constants of symmetric spaces and variants of Khintchine's inequality. J. Reine Angew. Math. **511**, 1–42 (1999)
164. B.I. Korenblyum, S.G. Krein, B.Y. Levin, On certain nonlinear questions of the theory of singular integrals. Doklady Akad. Nauk SSSR (N.S.) **62**, 17–20 (1948) (in Russian)
165. A. Korenovskii, *Mean Oscillations and Equimeasurable Rearrangements of Functions* (Springer, Berlin, Heidelberg, 2007)
166. M.A. Krasnosel'skii, P.P. Zabreiko, E.I. Pustylnik, P.E. Sobolevskii, *Integral Operators in Spaces of Summable Functions* (Noordhoff, Leiden, 1976)
167. M.A. Krasnosel'skii, Ya.B. Rutickii, *Convex Functions and Orlicz Spaces* (Noordhooff, Groningen, 1961)
168. S.G. Krein, Yu.I. Petunin, E.M. Semenov, *Interpolation of Linear Operators* (Amer. Math. Soc., Providence, RI, 1982)
169. A. Kufner, O. John, S. Fučik, *Function Spaces* (Noordhoff, Leyden, 1977)
170. A. Kufner, L. Maligranda, L.E. Persson, *The Hardy Inequality — About Its History and Some Related Results* (Pilsen, 2007)
171. S. Kwapień, A theorem on the Rademacher series with vector valued coefficients, in *Proc. Confer. "Probab. Theory on Banach Spaces"*, Oberwolfach, 1975, pp. 157–158
172. S. Kwapień, Decoupling inequalities for polynomial chaos. Ann. Probab. **15**, 1062–1071 (1987)
173. S. Kwapień, R. Latała, K. Oleszkiewicz, Comparison of moments of sums of independent random variables and differential inequalities. J. Funct. Anal. **136**, 258–268 (1996)
174. S. Kwapień, W.A. Woyczyński, *Random Series and Stochastic Integrals. Single and Multiple* (Birkhäuser, Boston, 1992)
175. R. Latała, On weak tail domination of random vectors. Bull. Polish Acad. Sci. Math. **57**, 75–80 (2009)
176. R. Latała, K. Oleszkiewicz, On the best constant in the Khintchine-Kahane inequality. Studia Math. **109**, 101–104 (1994)
177. M. Ledoux, M. Talagrand, *Probability in Banach Spaces* (Springer, Berlin, 1991)
178. M.V. Leibov, *The Geometry of the Function Space BMO*, Candidate's Dissertation, Rostov-na-Donu, 1985, 133 pp. (in Russian)
179. P.G. Lemarié-Rieusset, Multipliers and Morrey spaces. Potential Anal. **38**, 741–752 (2013)
180. O.V. Lelond, E.M. Semenov, S.N. Uksusov, The space of Fourier–Haar multipliers. Sib. Math. J. **46**, 103–110 (2005)
181. J. Lindenstrauss, L. Tzafriri, *Classical Banach Spaces I. Sequence Spaces* (Springer, Berlin, 1977)
182. J. Lindenstrauss, L. Tzafriri, *Classical Banach Spaces II, Function Spaces* (Springer, Berlin, 1979).
183. J.E. Littlewood, On the mean values of power series. Proc. Lond. Math. Soc. **25**, 328–337 (1926)
184. J.E. Littlewood, On bounded bilinear forms in an infinite number of variables. Quart. J. Math. Oxford. **1**, 164–174 (1930)
185. G.G. Lorentz, Relations between function spaces. Proc. Am. Math. Soc. **12**, 127–132 (1961)
186. A.A. Lunin, Operator norms of submatrices. Math. Notes **45**, 248–252 (1989)
187. F. Lust-Piquard, In égalit és de Khintchine dans C_p $(1 < p < \infty)$, (French) [Khinchine inequalities in C_p $(1 < p < \infty)$]. C. R. Acad. Sci. Paris Sér. I Math. **303**(7), 289–292 (1986)
188. F. Lust-Piquard, G. Pisier, Noncommutative Khintchine and Paley inequalities. Ark. Mat. **29**(2), 241–260 (1991)

189. L. Maligranda, in *The K-Functional for Symmetric Spaces*. Interpolation Spaces and Allied Topics in Analysis (Lund, 1983). Lecture Notes in Math., vol. 1070 (Springer, Berlin, 1984), pp. 169–182
190. L. Maligranda, Review of the Russian version on the paper [27]. Math. Rev. **MR2525624**, 9 p.
191. L. Maligranda, in *Orlicz Spaces and Interpolation*. Seminars in Math., vol. 5 (Univ. of Campinas, Campinas SP, Brazil, 1989), 206 pp.
192. L. Maligranda, Józef Marcinkiewicz (1910–1940)-on the centenary of his birth. Banach Center Publ. **95**, 133–234 (2011)
193. J. Marcinkiewicz, A. Zygmund, Sur les fonctions indépendantes. Fund. Math. **29**, 60–90 (1937); reprinted in: "Józef Marcinkiewicz, Collected Papers", PWN, Warszawa, 1964, 233–259
194. M.B. Marcus, G. Pisier, Characterizations of almost surely continuous p-stable random Fourier series and strongly stationary processes. Acta Math. **152**, 245–301 (1984)
195. M. Mastyło, Banach spaces via sublinear operators. Math. Jpn. **36**(1), 85–92 (1991)
196. W. Matuszewska, W. Orlicz, A note on modular spaces. IX. Bull. Acad. Polon. Sci. Sér. Sci. Math. Astronom. Phys. **16**, 801–808 (1968)
197. R.D. Mauldin (ed.), *The Scottish Book. Mathematics from the Scottish Café, with Selected Problems from the New Scottish Book*, 2nd edn. (Springer International Publishing, Switzerland, 2015)
198. B. Maurey, in *Type, Cotype and K-Convexity*. Handbook of the Geometry of Banach Spaces, vol. 2 (North-Holland, Amsterdam, 2003), pp. 1299–1332
199. B. Maurey, G.-Pisier, Séries de variables aléatoires vectorielles indépendantes et propriétés géométriques des espaces de Banach. Studia Math. **58**, 45–90 (1976)
200. T. Mei, BMO is the intersection of two translates of dyadic BMO. C. R. Math. Acad. Sci. Paris **336**, 1003–1006 (2003)
201. Y. Meyer, *Algebraic Numbers and Harmonic Analysis* (North-Holland, 1972)
202. M. Milman, Tensor products of function spaces. Proc. Am. Math. Soc. **82**, 626–628 (1976)
203. M. Milman, Embeddings of $L(p, q)$ spaces and Orlicz spaces with mixed norms. Notas Mat. **13**, 1–7 (1977)
204. M. Milman, Some new function spaces and their tensor products. Notas Mat. **20**, 1–128 (1978)
205. M. Milman, Embeddings of Lorentz-Marcinkiewicz spaces with mixed norms. Anal. Math. **4**, 215–223 (1978)
206. V.D. Milman, Geometric theory of Banach spaces. Part I. The theory of basis and minimal systems. Russ. Math. Surv. **25**, 111–170 (1970)
207. V.D. Milman, H. Wolfson, Minkowski spaces with extremal distance from the Euclidean space. Isr. J. Math. **29**, 113–131 (1978)
208. B.S. Mityagin, An interpolation theorem for modular spaces, in *Proc. Conf. "Interpol. Spaces and Allied Topics in Anal."* (Lund, 1983). Lecture Notes in Math. **1070**, 10–23 (1984) [translated from Mat. Sb. **66** (1965), 473–482 (in Russian)]
209. S.J. Montgomery-Smith, The distribution of Rademacher sums. Proc. Am. Math. Soc. **109**, 517–522 (1990)
210. S.J. Montgomery-Smith, E. Semenov, Embeddings of rearrangement invariant spaces that are not strictly singular. Positivity **4**, 397–404 (2000)
211. C.B. Morrey, On the solutions of quasi-linear elliptic partial differential equations. Trans. Am. Math. Soc. **43**, 126–166 (1938)
212. D.Kh. Mushtari, On an extremal property of the symmetric Bernoulli distribution. Russ. Math. (Iz. VUZ) **39**, 60–62 (1995)
213. P.F. Müller, in *Isomorphisms Between H^1 Spaces*. Math. Monographs (New Series), vol. 66 (Birkhäuser-Verlag, Basel, 2005)
214. P.F. Müller, G. Schechtman, On complemented subspaces of H^1 and VMO. Lect. Notes Math. **1376**, 113–125 (1989)
215. I.P. Natanson, *Theory of Functions of a Real Variable, vol. 1* (Frederic Ungar, New York, 1955)

216. I.P. Natanson, *Theory of Functions of a Real Variable, vol. 2* (Frederic Ungar, New York, 1960)
217. P. Nayar, K. Oleszkiewicz, Khintchine type inequalities with optimal constants via ultra log-concavity. Positivity **109**, 359–371 (2012)
218. F.L. Nazarov, A.N. Podkorytov, Ball, Haagerup, and distribution functions. Compl. Anal. Oper. Rel. Topics. Oper. Theory Adv. Appl. **113**, 247–267 (2000)
219. E.M. Nikishin, On a property of sums of independent variables. Math. Notes **16**, 1015–1017 (1974)
220. I.Ya. Novikov, *Orthogonal Series in Symmetric Spaces*, Candidate's Dissertation, Voronezh, 1984, 93 pp. (in Russian)
221. K. Oleszkiewicz, Precise moment and tail bounds for Rademacher sums in terms of weak parameters. Isr. J. Math. **203**, 429–443 (2014)
222. R. O'Neil, Convolution operators and $L(p, q)$ spaces. Duke Math. J. **30**, 129–142 (1963)
223. W. Orlicz, Über unbedingte Konvergenz in Funktionräumen. Studia Math. **1**, 83–85 (1930)
224. E.V. Orlov, On Fourier coefficients over Sidon sets. Anal. Math. **8**, 277–285 (1982) (in Russian, with English summary)
225. A. Ostrowski, Sur quelques applications des fonctions convexes et concave au sens re I. Schur. J. Math. Pures Appl. **31**, 253–259 (1952) (in French)
226. R.E.A.C. Paley, A remarkable series of orthogonal functions (I). Proc. Lond. Math. Soc. **34**, 241–264 (1932)
227. R.E.A.C. Paley, A. Zygmund, On some series of functions. I, Proc. Camb. Phil. Soc. **26**, 337–357 (1930); reprinted in *Selected Papers of Antoni Zygmund*, vol. 1 (Kluwer, Dordrecht, 1989), pp. 337–357
228. J. Peetre, *A Theory of Interpolation of Normed Spaces*. Lecture Notes, Brasilia: 1963 [Notas Math. **39**, 1–86 (1968)]
229. J. Peetre, On the theory of $\mathcal{L}_{p,\lambda}$ spaces. J. Funct. Anal. **4**, 71–87 (1969)
230. J. Peetre, in *Interpolation Functors and Banach Couples*. Actes Congrès Intern. Math. 1970, vol. 2 (Gauthier-Villars, Paris, 1971), pp. 373–378
231. A. Pełczyński, Commensurate sequences of characters. Proc. Am. Math. Soc. **104**, 525–531 (1988)
232. G. Pěskir, Best constants in Kahane–Khintchine inequalities in Orlicz spaces. J. Multivar. Anal. **45**, 183–216 (1993)
233. G. Pěskir, A.N. Shiryaev, The Khintchine inequalities and martingale expanding sphere of their action. Russ. Math. Surv. **50**, 849–904 (1995)
234. I. Pinelis, An asymptotically Gaussian bound on the Rademacher tails. Electron. J. Probab. **17**(35), 1–22 (2012)
235. I. Pinelis, *On the supremum of the tails of normalized sums of independent Rademacher random variables*, arXiv:1204.1761v1 [math.PR] 8 Apr. 2012. 6 pp.
236. I. Pinelis, *Explicit additive decomposition of norms on* $\mathbb{R}^2$, arXiv:1506.00537v2 [math.FA] 3 Jun. 2015. 6 pp.
237. G. Pisier, *Les inégalités de Kahane–Khintchin d'après C. Borell*, Séminaire sur la géometrie des espaces de Banach, Exposé N 7 (1977–1978), École Polytechnique, Palaiseau (in French)
238. G. Pisier, in *De nouvelles caract é risations des ensembles de Sidon*. Mathematical Anal. and Appl. Part B, vol. 7b. Adv. in Math. Suppl. Studies (Academic Press, New York, 1981), pp. 685–726
239. G. Pisier, Arithmetic characterizations of Sidon sets. Bull. Am. Math. Soc. **8**, 87–89 (1983)
240. G. Pisier, Interpolation between H^p spaces and non-commutative generalizations 1. Pac. J. Math. **155**, 341–368 (1992)
241. G. Pisier, Remarks on the non-commutative Khintchine inequalities for $0 < p < 2$. J. Funct. Anal. **256**(12), 4128–4161 (2009)
242. G. Pisier, Grothendieck's Theorem, past and present. Bull. Am. Math. Soc. **49**, 237–323 (2012)
243. G. Pisier, On uniformly bounded orthonormal Sidon systems. Math. Res. Lett. **24**, 893–932 (2017)

244. G. Pisier, Completely Sidon sets in discrete groups. arXiv:1706.03844v6 [math.OA] 6 Sep. 2018
245. G. Pisier, Q. Xu, in *Non-commutative L_p-Spaces*. Handbook of the Geometry of Banach Spaces, vol. 2 (North-Holland, Amsterdam, 2003), pp. 1459–1517
246. A.M. Plichko, M.M. Popov, in *Symmetric Function Spaces on Atomless Probability Spaces*. Diss. Math. (Rozprawy Mat.). vol. 306 (Polish Acad. Sci., Warsaw, 1990)
247. A.N. Podkorytov, O.I. Reinov, On the Khinchin–Kahane inequality. St. Petersburg Math. J. **10**, 211–215 (1999)
248. H. Rademacher, Einige Sätze über Reihen von allgemeinen Orthogonalfunktionen. Math. Ann. **87**, 111–138 (1922) (in German)
249. Y. Raynaud, Complemented hilbertian subspaces in rearrangement invariant function spaces. Ill. J. Math. **39**, 212–250 (1995)
250. V.A. Rodin, *Fourier Series in a Symmetric Space and Interpolation of Operators*, Candidate's Dissertation, Voronezh, 1973 (in Russian)
251. V.A. Rodin, E.M. Semenov, Rademacher series in symmetric spaces. Anal. Math. **1**, 207–222 (1975)
252. V.A. Rodin, E.M. Semenov, Complementability of the subspace generated by the Rademacher system in a symmetric space. Funct. Anal. Appl. **13**, 150–151 (1979)
253. B.-Z.A. Rubshtein, G.Ya. Grabarnik, M.A. Muratov, Yu.S. Pashkova, in *Foundations of Symmetric Spaces of Measurable Functions. Lorentz, Marcinkiewicz and Orlicz Spaces*. Developments in Mathematics, vol. 45 (Springer, Switzerland, 2016)
254. W. Rudin, Trigonometric series with gaps. J. Math. Mech. **9**, 203–227 (1960)
255. W. Rudin, *Functional Analysis*, 2nd edn. (MacGraw-Hill, New York, 1991)
256. Ya.B. Rutickii, On some classes of measurable functions. Usp. Mat. Nauk **20**, 205–208 (1965)
257. D.V. Rutsky, Weighted Calderón-Zygmund decomposition with some applications to interpolation. J. Math. Sci. (N. Y.) **209**, 783–791 (2015)
258. Y. Sagher, K. Zhou, A local version of a theorem of Khintchine, in *Proc. "Analysis and Partial Differential Equations"*, vol. 122. Lect. Notes in Pure and Appl. Math. (Dekker, New York, 1990), pp. 327–330
259. Y. Sagher, K. Zhou, On a theorem of Burkholder. Ill. J. Math. **37**, 637–642 (1993)
260. Y. Sagher, K. Zhou, Exponential integrability of Rademacher series, in *Convergence in Ergodic Theory and Probability* (de Gruyter, Berlin, 1996), pp. 389–395
261. G. Schechtman, Two remarks on 1-unconditional basic sequences in L_p, $3 \leq p < \infty$, in *Operator Theory: Advances and Applications*, vol. 77 (Birkhäuser-Verlag, Basel/Switzerland, 1995), pp. 251–254
262. I. Schur, Über einer Klasse von Mittelbildungen mit Anwendungen auf die Determinantentheorie. Sitzber. Berl. Math. Ges. **22**, 9–20 (1923)
263. E.M. Semenov, RUC-basic properties of the Haar system. Doklady Math. **348**, 590–591 (1996) (in Russian)
264. I. Şerb, New variants of Khintchine's inequality. Collect. Math. **52**, 193–203 (2001)
265. A.N. Shiryaev, in *Probability-1*, 3rd edn. Graduate Texts in Mathematics, vol. 95 (Springer, New York, 2016)
266. J.S. Shiue, A note on Cesàro function space. Tamkang J. Math. **1**, 91–95 (1970)
267. S. Sidon, Verallgemeinerung eines Satzes über die Absolute Konvergenz von Fourier-Reihen mit Lücken. Math. Ann. **97**, 675–676 (1927)
268. G. Sparr, Interpolation of weighted L_p-spaces. Studia Math. **62**, 229–271 (1978)
269. S.B. Stechkin, On absolute convergence of Fourier series. Izv. Acad. Nauk SSSR. Ser. Matem. **20**, 385–412 (1956)
270. S.B. Stechkin, On best lacunary systems of functions. Izv. Acad. Nauk SSSR. Ser. Matem. **25**, 357–366 (1961)
271. F.A. Sukochev, RUC-bases in Orlicz and Lorentz operator spaces. Positivity **2**, 265–279 (1998)
272. P.W. Sy, W.Y. Zhang, P.Y. Lee, The dual of Cesàro function spaces. Glas. Mat. Ser. III **22**, 103–112 (1987)

273. S.J. Szarek, On the best constant in the Khintchine inequality. Studia Math. **58**, 197–208 (1976)
274. S.J. Szarek, in *General Fourier Sums with Vector Coefficients and Analogs of* $\Lambda(p)$*-Sets*. Probability Theory and Harmonic Analysis (Cleveland, Ohio, 1983). Monogr. Textbooks. Pure Appl. Math., vol. 98 (Dekker, New York, 1986), pp. 195–208
275. M. Talagrand, An isoperimetric theorem on the cube and the Khintchine–Kahane inequalities. Proc. Am. Math. Soc. **104**, 905–909 (1988)
276. M. Talagrand, Regularity of infinitely divisible processes. Ann. Probab. **21**, 362–432 (1993)
277. M. Talagrand, *The Generic Chaining. Upper and Lower Bounds of Stochastic Processes* (Springer, Berlin, 2005)
278. M. Talagrand, *Upper and Lower Bounds for Stochastic Processes. Modern Methods and Classical Problems* (Springer, Berlin, Heidelberg, 2014)
279. E.V. Tokarev, Complemented subspaces of symmetric function spaces. Math. Notes **32**, 726–728 (1982)
280. N.N. Vakhania, V.I. Tarieladze, S.A. Chobanyan, in *Probability Distributions on Banach Spaces*. Math. Its Appl. (Soviet Series) (Kluwer, Holland, 1987)
281. M. Veraar, On Khintchine inequalities with a weight. Proc. Am. Math. Soc. **138**, 4119–4121 (2011)
282. D.W. Walkup, Polya sequences, binomial convolution and the union of random sets. J. Appl. Probab. **13**, 76–85 (1976)
283. J. Weidmann, in *Spectral Theory of Ordinary Differential Operators*. Lecture Notes in Math., vol. 1258 (Springer, New York/Berlin, 1987)
284. P. Whittle, Bounds for the moments of linear and quadratic forms in independent random variables. Theory Probab. Appl. **17**, 302–305 (1960)
285. N. Wiener, The homogeneous chaos. Am. J. Math. **60**, 897–936 (1941)
286. N. Wiener, A. Wintner, The discrete chaos. Am. J. Math. **65**, 279–298 (1943)
287. W. Wnuk, *Banach Lattices with Order Continuous Norms* (Polish Scientific Publ. PWN, Warszawa, 1999)
288. P. Wojtaszczyk, in *Banach Spaces for Analysts*. Cambridge Studies in Advanced Mathematics, vol. 25 (Cambridge University Press, Cambridge, UK, 1991)
289. C.T. Zorko, Morrey spaces. Proc. Am. Math. Soc. **98**, 586–592 (1986)
290. A. Zygmund, Sur les séries trigonométric lacunaires. J. Lond. Math. Soc. **5**, 138–145 (1930)
291. A. Zygmund, On the convergence of lacunary trigonometric series. Fund. Math. **16**, 90–107 (1930)
292. A. Zygmund, On lacunary trigonometric series. Trans. Am. Math. Soc. **34**, 435–446 (1932)
293. A. Zygmund, *Trigonometric Series, Vol. I and II. With a Foreword by Robert Fefferman*, 3rd edn. (Cambridge University Press, Cambridge, 2002)

Index

S. V. Astashkin, *The Rademacher System in Function Spaces*,
https://doi.org/10.1007/978-3-030-47890-2

T

The manufacturer's authorised representative in the EU is Springer Nature Customer Service Centre GmbH, Europaplatz 3, 69115 Heidelberg, Germany. If you have any concerns regarding our products, please contact ProductSafety@springernature.com

Printed and bound by CPI Group (UK) Ltd, Croydon, CR0 4YY
30/09/2026
02225646-0019